Proceedings of the International Symposium on Engineering under Uncertainty: Safety Assessment and Management (ISEUSAM - 2012)

Subrata Chakraborty • Gautam Bhattacharya
Editors

Proceedings of the International Symposium on Engineering under Uncertainty: Safety Assessment and Management (ISEUSAM - 2012)

Editors
Subrata Chakraborty
Civil Engineering
Bengal Engineering and Science University
Howrah, India

Gautam Bhattacharya
Civil Engineering
Bengal Engineering and Science University
Howrah, India

Printed in 2 Volumes
ISBN 978-81-322-0756-6 ISBN 978-81-322-0757-3 (eBook)
DOI 10.1007/978-81-322-0757-3
Springer New Delhi Heidelberg New York Dordrecht London

Library of Congress Control Number: 2012944989

Printed on acid-free paper

Springer is part of Springer Science+Business Media (www.springer.com)

Preface

In engineering applications, it is important to model and treat adequately all the available information during the analysis and design phase. Typically, the information is originated from different sources like field measurements, experts' judgments, objective and subjective considerations. Over these features, the influences originated from the human errors, imperfections in the construction techniques and production process and influence of the boundary and environmental conditions are added. All these aspects can be brought under one common denominator: that is, "presence of uncertainty." Thus, reliability and safety are the core issues which need to be addressed during the analysis, design, construction and operation of engineering systems under such uncertainties. In this backdrop, the aim of ISEUSAM 2012 is to facilitate the discussion for a better understanding and management of uncertainty and risk, encompassing various aspects of safety and reliability of engineering systems. To be specific, the overall theme of the symposium is modelling, analysis and design of engineering systems and decision-making under uncertainties relevant to all engineering disciplines.

The symposium, being the first of its kind organized in India, received overwhelming response from national as well as international scholars, experts and delegates from different parts of the world. Papers were received from authors from several parts of the world including Australia, Canada, China, Germany, Italy, Sharjah, the UK and the USA, besides India. More than 200 authors from India and abroad have shown their interest in the symposium, out of which a total of 90 papers were presented in various technical sessions comprising 4 plenary sessions and 12 parallel technical sessions.

The proceedings began on January 4, 2012, on a grand scale amidst rousing welcome to the delegates and great enthusiasm amongst the organizers and the participants with the opening ceremony hosting, amongst other dignitaries, Dr. Rakesh Kumar Bhandari, Director, Variable Energy Cyclotron Centre, Kolkata, who inaugurated the 3-day international event, Professor Ajoy Kumar Ray, Vice Chancellor, Bengal Engineering and Science University,

Shibpur, who presided over the opening session, and Professor Arun Kumar Majumdar, Deputy Director, Indian Institute of Technology Kharagpur, who graced the occasion as the guest of honour.

The inaugural keynote address, delivered by Achintya Haldar, Emeritus Professor, University of Arizona, USA, on *Past, Present and Future of Engineering under Uncertainty: Safety Assessment and Management,* set the tone for the rest of the proceedings. This was followed by a series of technical sessions including plenary sessions on numerous sub-themes spread over all three days. The other keynote addresses include *Techniques of Analysis of Imprecision in Engineering Systems* by Ajoy K. Ray, Vice Chancellor, BESU, Shibpur; *Uncertainty Quantification in System Risk Assessment and Decision-Making* by Sankaran Mahadevan, John R. Murray Sr. Chair in Engineering, Vanderbilt University, USA; *Advancing Geotechnics in the Twenty-First Century – Dealing with Uncertainty and Other Challenges* by Robin Chowdhury, Emeritus Professor, University of Wollongong, Australia; *State of the Art on Stochastic Control of Structures for Seismic Excitation* by T.K. Datta, Emeritus Professor, IIT Delhi; *Discovering Hidden Structural Degradations* by Abhijit Mukherjee, Director, Thapar University; *SHM of Prestressed Concrete Girder Bridge* by Pradipta Banerji, Director, IIT Roorkee; *Nanotoxicology: A Threat to the Environment and to Human Beings* by D. Dutta Majumder, Professor Emeritus, Indian Statistical Institute, Kolkata; *Uncertainty in Interpreting the Scale Effect of Plate Load Tests in Unsaturated Soils* by Sai K. Vanapalli, Professor and Chair, Civil Engineering Department, University of Ottawa, Canada; and *Response Control of Tall Buildings Using Tuned Liquid Damper* by S. K. Bhattacharyya, Director, CBRI, Roorkee. Two other eminent personalities who had accepted the invitation to deliver keynote lectures, but due to unavoidable circumstances could not be present on the occasion, also sent their contributions for inclusion in the symposium proceedings. These are *Uncertainties in Transportation Infrastructure Development and Management* by Kumares C. Sinha, Olson Distinguished Professor of Civil Engineering, Purdue University, USA, and *Physical Perspective Towards Stochastic Optimal Controls of Engineering Structures* by Jie Li, State Key Laboratory of Disaster Reduction in Civil Engineering, Tongji University, Shanghai, China.

In order to accommodate a wide spectrum of highly relevant sub-themes across the major engineering disciplines, presentations of the invited and the contributory papers were held in two parallel technical sessions. Amongst the presenters were senior professors and chairs from reputed universities from India and abroad, most of the IITs and the IISc Bangalore on the one hand, and experts from the R&D organizations such as BARC, AERB, DRDO, CBRI, CRRI, SERC and leading industry houses such as UltraTech Cement, M.N. Dastur & Company (P) Ltd., Petrofac International Ltd., UAE, and L&T on the other. All the technical sessions invariably concluded with a highly animated discussion session which enthralled the participants and brought their applause in appreciation.

The closing ceremony marked a fitting finale to the 3-day event. Professor Achintya Haldar summed up the proceedings over the last 3 days. He also spelt out the future direction of the symposium by mooting a proposal of organizing it on a regular basis. Dr. Milan Kumar Sanyal, Director, Saha Institute of Nuclear Physics, Kolkata, in his role as the chief guest, enlightened the audience about the uncertainties involved in a nuclear project. Dr. T. K. Datta, Professor Emeritus, IIT Delhi, was the Guest of Honour on the occasion. "Death is certain, yet, when, is uncertain. . ." echoed Dr. Datta on a philosophical note and went on to applaud the Department of Civil Engineering, BESU, for putting in such a wonderful effort in organizing this symposium on uncertainty, the first of its kind in India. Representatives from the delegates, including the well-known academician Professor G. Venkatachalam, expressed that they had found the sessions truly engrossing and also that they were highly satisfied with the arrangements. It thus appears that the symposium could at least partly fulfil the objective with which it was organized.

The organizers sincerely regret that this volume could not be made ready well in advance, and, therefore, at the time of the symposium, the delegates and the participants could be handed over only a CD version of their contributions. But, better late than never, that the proceedings could eventually be published, is a matter of some satisfaction.

Professor of Civil Engineering Subrata Chakraborty
Bengal Engineering and Science University, Shibpur
Organizing Secretary, ISEUSAM-2012

Professor of Civil Engineering Gautam Bhattacharya
Bengal Engineering and Science University, Shibpur
Joint Organizing Secretary, ISEUSAM-2012

About the Editors

Dr. Subrata Chakraborty is currently a Professor at the Bengal Engineering and Science University, Shibpur. He is a fellow of the Indian National Academy Engineering and the Institution of Engineers (India). He has obtained his Ph.D. in structural engineering from IIT Kharagpur in 1995. He was postdoctoral researcher at University of Cambridge, UK and University of Arizona, USA and Technical University of Aachen, Germany. In general, Prof. Chakraborty's research interest lies in the field of computational mechanics under uncertainty, structural health monitoring, vibration control, composite mechanics etc. While his inspiring teaching coupled with innate urge for intensive research has already established him as a distinguished academician at the national level, several awards and laurels have come his way. The Humboldt Fellowship for Experienced Researchers, the INAE Young Engineer Award, the BOYSCAST Fellowship, and the Young Faculty Research Award deserve special mention.

Dr. Gautam Bhattacharya, an experienced academic and researcher, is one of the senior Professors of Civil Engineering at the Bengal Engineering and Science University, Shibpur, and currently the Vice-Chairman, Kolkata chapter of the Indian Geotechnical Society. Having obtained his B.E. and M.E. in Civil Engineering from the erstwhile B.E. College, Shibpur, he went to IIT Kanpur to pursue his Doctoral study during which (1985–1990) he developed great interest in the subject of slope stability, and worked on the application of advanced numerical methods in slope analysis for his PhD thesis. He has since been engaged in teaching soil mechanics and foundation engineering and in pursuing research on both deterministic and probabilistic approaches of analysis of unreinforced and reinforced slopes, retaining structures and foundations under static and seismic conditions. He has published several scientific articles in peer reviewed journals and also co-authored a book with the CRC Press/Balkema. He has teaching, research and consultancy experience in the field of geotechnical engineering for about three decades.

Contents of Volume I

Contents of Volume II

Past, Present, and Future of Engineering under Uncertainty: Safety Assessment and Management

Achintya Haldar

Abstract The author's perspective of engineering under uncertainty is presented. In the first part of this chapter, past, present, and future trends of reliability assessment methods, applicable to many branches of engineering, are presented. The discussions cover the cases for both explicit and implicit limit state functions. Finite element-based reliability evaluation methods for large structures satisfying underlying physics are emphasized. The necessity of estimating risks for both strength and serviceability limit states is documented. Concept of several energy dissipation mechanisms recently introduced to improve performance and reduce risk during seismic excitations can be explicitly incorporated in the formulation. Reliability evaluation of very large structures requiring over several hours of continuous running of a computer for one deterministic evaluation is briefly presented. Since major sources of uncertainty cannot be completely eliminated from the analysis and design of an engineering system, the risk needs to be managed appropriately. Risk management in the context of decision analysis framework is also briefly presented. In discussing future directions, the use of artificial neural networks and soft computing, incorporation of cognitive sources of uncertainty, developing necessary computer programs, and education-related issues are discussed.

Keywords Reliability analysis • Seismic analysis • Nonlinear response • Partially restrained connections • Shear walls • Post-Northridge connections • Computer programs • Education • Uncertainty management

A. Haldar (✉)
Department of Civil Engineering and Engineering Mechanics,
University of Arizona, Tucson, AZ, USA
e-mail: Haldar@u.arizona.edu

S. Chakraborty and G. Bhattacharya (eds.), *Proceedings of the International Symposium on Engineering under Uncertainty: Safety Assessment and Management (ISEUSAM - 2012)*,
DOI 10.1007/978-81-322-0757-3_1, © Springer India 2013

1 Introduction

Uncertainty must have been present from the beginning of time. Our forefathers must have experienced it through observations and experiences. In engineering practices, the probability concept is essentially an attempt to incorporate uncertainty in the formulation. The probability concept can be defined in two ways: (1) an expression of relative frequency and (2) degree of belief. The underlying mathematics of probability are based on three axioms, well developed and accepted by experts; however, sometimes it is used in a philosophical sense. Since the relative frequency concept is almost never used [22], a measure of confidence in expressing uncertain events leads to the degree of belief statements. Laplace (1749–1827), a famous mathematician in "A Philosophical Essay on Probabilities," wrote "It is seen in this essay that the theory of probabilities is at bottom only common sense reduced to calculus; it makes us appreciate with exactitude that which exact minds feel by a sort of instinct without being able ofttimes to give a reason for it. It leaves no arbitrariness in the choice of opinions and sides to be taken; and by its use can always be determined the most advantageous choice. Thereby, it supplements most happily the ignorance and weakness of the human mind" (translation, [25]).

These timeless remarks sum up the importance of probability, reliability, and uncertainty concepts in human endeavor. In my way of thinking, I will try to give my understanding or assessment of "Engineering under Uncertainty – Past, Present, and Future." Obviously, the time lines when past meets present and present becomes future are very difficult to establish. According to Albert Einstein, "People like us, who believe in physics, know that the distinction between past, present, and future is only a stubbornly persistent illusion."

Although the area of probability has a glorious past, I will try to emphasize the present and future in reliability assessment and management in this chapter. I believe that structural engineering provided leadership in developing these areas, and I will emphasize it in my presentation.

2 Reliability Assessment: Past

Many brilliant scholars such as Einstein did not believe in probability. His famous comment that "I am convinced that He (God) does not play dice" is known to most present scholars. On the other hand, the Rev. Thomas Bayes (1702–1761), a Presbyterian Minister at Tunbridge Wells, wrote the Bayes' theorem in 1763 in "An Essay towards Solving a Problem in the Doctrine of Chances" [14]. It has become one of the most important branches of statistics in the modern time. Please note that the paper was published 2 years after his death. One of Bayes' friends, Richard Price, sent the paper for publication by adding an introduction, examples, and figures. Most likely, Bayes was not confident about the paper. At present, the

Bayesian approach provides a mechanism to incorporate uncertainty information in cases of inadequate reliable data by combining experience and judgment.

It is reasonable to state that Freudenthal [15] formally introduced the structural reliability discipline. Obviously, it has gone through monumental developments since then under the leadership of many scholars ([2, 3, 6, 13, 18–20, 33, 34, 42], and many others). A more complete list can be found in Haldar and Mahadevan [18].

2.1 *Available Reliability Evaluation Methods*

It will be informative to discuss very briefly how the reliability evaluation methods evolved in the past few decades. After four decades of extensive work in different engineering disciplines, several reliability evaluation procedures of various degrees of complexity and sophistication are now available. First-generation structural design guidelines and codes are being developed and promoted worldwide using some of these procedures. Obviously, depending upon the procedure being used, the estimated risk could be different. The engineering profession has not yet officially accepted a particular risk evaluation method. Thus, design of structures satisfying an underlying risk can be debated; even the basic concept of acceptable risk is often openly debated in the profession. Moreover, uncertainty introduced due to human error is not yet understood and thus cannot be explicitly introduced in the available reliability assessment methods. Also, risk or reliability estimated using these methods may not match actual observations. Sometimes, to be more accurate, scholars denote the estimated risk as the notional or relative risk. The main idea is if risk can be estimated using a reasonably acceptable procedure, the design alternatives will indicate different levels of relative risk. When the information is used consistently, it may produce a risk-aversive appropriate design.

Before introducing the reliability-based design concept, it may be informative to study different deterministic structural design concepts used in the recent past and their relationship with the reliability-based concept. The fundamental concept behind any design is that the resistance or capacity or supply should at least satisfy demand with some conservatism or safety factor built in it. The level of conservatism is introduced in the design in several ways depending on the basic design concept being used. In structural design using the allowable stress design (ASD) approach, the basic concept is that the allowable stresses should be greater than the unfactored nominal loads or load combinations expected during the lifetime of a structure. The allowable stresses are calculated using safety factors. In other words, the nominal resistance R_n is divided by a safety factor to compute the allowable resistance R_a, and safe design requires that the nominal load effect S_n is less than R_a. In the ultimate strength design (USD) method, the loads are multiplied by load factors to determine the ultimate load effects, and the members are required to resist the ultimate load. In this case, the safety factors are used in the loads and load combinations. Since the predictabilities of different types of load are expected to be different, the USD significantly improved the ASD concept. In the current risk-based design concept, widely known as the load and resistance factor design (LRFD), safety factors are introduced to both load and resistance under the constraint of an underlying risk, producing improved designs. It should be pointed out that LRFD-based designs are calibrated with the old time-tested ASD designs. Thus, the final design may be very similar but LRFD design is expected to be more risk consistent.

2.2 Fundamental Concept of Reliability-Based Design

Without losing any generality, suppose R and S represent the resistance or capacity and demand or load effect, respectively, and both are random variables since they are functions of many other random variables. The uncertainty in R and S can be completely defined by their corresponding probability density functions (PDFs) denoted as $f_R(r)$ and $f_S(s)$, respectively. Then the probability of failure of the structural element can be defined as the probability of the resistance being less than the load effect or simply $P(R < S)$. Mathematically, it can be expressed as [18]

$$P(\text{failure}) = P(R \langle S) = \int_0^\infty \left[\int_0^s f_R(r)\mathrm{d}r\right] f_S(s)\mathrm{d}s = \int_0^\infty F_R(s) f_S(s)\mathrm{d}s \tag{1}$$

where $F_R(s)$ is the cumulative distribution function (CDF) of R evaluated at s. Conceptually, Eq. (1) states that for a particular value of the random variable $S = s$, $F_R(s)$ is the probability of failure. However, since S is also a random variable, the integration needs to be carried out for all possible values of S, with their respective likelihood represented by the corresponding PDF. Equation (1) can be considered as the fundamental equation of the reliability-based design concept. It is shown in Fig. 1. In Fig. 1, the nominal values of resistance and load effect are denoted as R_n and S_n, respectively, and the corresponding PDFs of R and S are shown. The overlapped (dashed) area between the two PDFs provides a *qualitative measure* of the probability of failure. Controlling the size of the overlapped area is essentially the idea behind reliability-based design. Haldar and Mahadevan [18] pointed out that the area could be controlled by changing the relative locations of the two PDFs by separating the mean values of R and S (μ_R and μ_S), the uncertainty expressed in terms of their standard deviations (σ_R and σ_S), and the shape of the PDFs ($f_R(r)$ and $f_S(s)$).

In general, the CDF and the PDF of S may not be available in explicit forms, and the integration of Eq. (1) may not be practical. However, Eq. (1) can be evaluated, without performing the integration if R and S are both statistically independent normal [11] or lognormal [41] random variables. Considering the practical aspects of a design, since R and S can be linear and nonlinear functions of many other

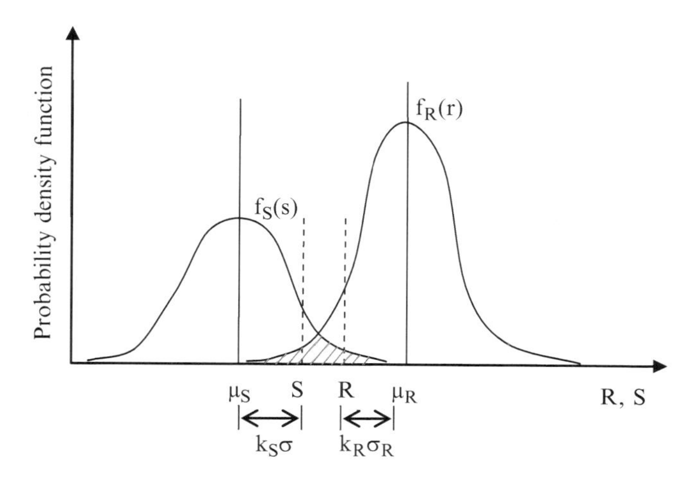

Fig. 1 Reliability-based design concept [18]

random variables, their normality or log normality assumptions can rarely be satisfied.

If the risk cannot be evaluated in closed form, it needs to be evaluated in approximate ways. This led to the development of several reliability analysis techniques. Initially, in the late 1960s, the first-order second-moment (FOSM) method, also known as the mean value first-order second-moment (MVFOSM), was proposed neglecting the distributional information on the random variables present in the problem. This important deficiency was overcome by the advanced first-order second-moment (AFOSM) method applicable when all the variables are assumed to be normal and independent [20]. A more general formulation applicable to different types of distribution was proposed by Rackwitz and Fiessler [34]. In the context of AFOSM, the probability of failure can be estimated using two types of approximations to the limit state at the design point: first order (leading to the name FORM) and second order (leading to the name SORM). Since FORM is a major reliability evaluation technique commonly used in the profession, it is discussed in more detail below.

The basic idea behind reliability-based design is to design satisfying several performance criteria and considering the uncertainties in the relevant load- and resistance-related random variables, called the basic variables X_i. Since the R and S random variables in Eq. (1) are functions of many other load- and resistance-related random variables, they are generally treated as basic random variables. The relationship between the basic random variables and the performance criterion, known as the performance or limit state function, can be mathematically represented as

$$Z = g(X_1, X_2, \ldots, X_n) \tag{2}$$

The failure surface or the limit state of interest can then be defined as $Z = 0$. The limit state equation plays an important role in evaluating reliability using

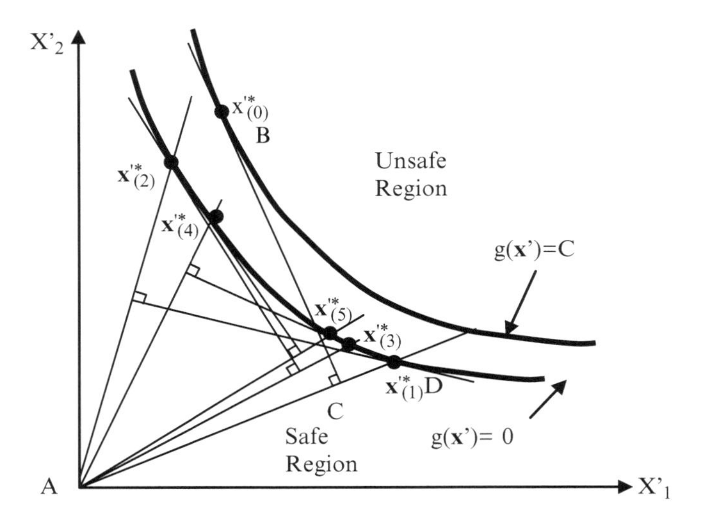

Fig. 2 Limit state concept [18, 19]

FORM/SORM. It represents the boundary between the safe and unsafe regions and a state beyond which a structure can no longer fulfill the function for which it was designed. Assuming *R* and *S* are the two basic random variables, the limit state equation and the safe and unsafe regions are shown in Fig. 2.

A limit state equation can be an explicit or implicit function of the basic random variables and can be linear or nonlinear. For nonlinear limit state functions, an iterative strategy is required to estimate the probability of failure as discussed by Haldar and Mahadevan [18], elsewhere. Two types of performance functions are generally used in engineering: strength and serviceability. Strength performance functions relate to the safety of the structures, and serviceability performance functions are related to the serviceability (deflection, vibration, etc.) of the structure. The reliabilities underlying the strength and serviceability performance functions are expected to be different.

2.3 *Reliability Assessment: Implicit Limit State Functions*

Reliability evaluation using FORM is relatively simple if the limit state function is explicit in terms of design variables. In this case, the derivatives of the limit state functions with respect to design variables are readily available; the iterative process necessary for the reliability evaluation becomes very straightforward. However, in many cases of practical importance, particularly for complicated large systems, the explicit expressions for the limit state functions may not be available. For such systems, the required functions need to be generated numerically such as the finite element analysis. In such cases, the derivatives are not readily available. Their numerical evaluation could be time-consuming. Some alternatives are necessary.

Several computational approaches can be pursued for reliability analysis of systems with implicit performance functions. They can be broadly divided in to three categories, based on their essential philosophy, as (1) Monte Carlo simulation (MCS), (2) response surface method (RSM), and (3) sensitivity-based analysis. Monte Carlo simulation uses randomly generated samples of input variables for each deterministic analysis. Its efficiency can be increased using intelligent schemes, as will be discussed later. It can be used for both explicit and implicit limit state functions. The RSM approximately constructs a polynomial (mainly first or second order) using a few selected deterministic analyses and in some cases regression analysis of these results. The approximate closed-form expression thus obtained is then used to estimate reliability using FORM/SORM. In the sensitivity-based approach, the sensitivity of the structural response to the input variables is computed, and it can be integrated with the FORM approach to extract the information on the underlying reliability. The value of the performance function is evaluated using deterministic structural analysis. The gradient is computed using sensitivity analysis. When the limit state function is implicit, the sensitivities can be computed in three different ways: (1) through a finite difference approach, (2) through classical perturbation methods that apply the chain rule of differentiation to finite element analysis, and (3) through iterative perturbation analysis techniques [19]. The sensitivity-based reliability analysis approach is more elegant and in general more efficient than the simulation and response surface methods. Haldar and Mahadevan [19] suggested the use of the iterative perturbation technique in the context of the basic nonlinear stochastic finite element method (SFEM)-based algorithm.

2.4 Unified Stochastic Finite Element Method

Without losing any generality, the limit state function can be expressed in terms of the set of basic random variables $\mathbf{x}$ (e.g., loads, material properties, and structural geometry), the set of displacements $\mathbf{u}$, and the set of load effects $\mathbf{s}$ (except the displacements) such as internal forces and stresses. The displacement $\mathbf{u} = \mathbf{QD}$, where $\mathbf{D}$ is the global displacement vector and $\mathbf{Q}$ is a transformation matrix. The limit state function can be expressed as $g(\mathbf{x}, \mathbf{u}, \mathbf{s}) = 0$. For reliability computation, it is convenient to transform $\mathbf{x}$ into the standard normal space $\mathbf{y} = \mathrm{y}(\mathbf{x})$ such that the elements of $\mathbf{y}$ are statistically independent and have a standard normal distribution. An iteration algorithm can be used to locate the design point (the most likely failure point) on the limit state function using the first-order approximation. During each iteration, the structural response and the response gradient vectors are calculated using finite element models. The following iteration scheme can be used for finding the coordinates of the design point:

$$\mathbf{y}_{i+1} = \left[\mathbf{y}_i^t \alpha_i + \frac{g(\mathbf{y}_i)}{|\nabla g(\mathbf{y}_i)|}\right] \alpha_i \tag{3}$$

where

$$\nabla g(\mathbf{y}) = \left[\frac{\partial g(\mathbf{y})}{\partial y_1}, \ldots, \frac{\partial g(\mathbf{y})}{\partial y_n}\right]^t \quad \text{and} \quad \alpha_i = -\frac{\nabla g(\mathbf{y}_i)}{|\nabla g(\mathbf{y}_i)|} \tag{4}$$

To implement the algorithm, the gradient $\nabla g(\mathbf{y})$of the limit state function in the standard normal space can be derived as [19]:

$$\nabla g(\mathbf{y}) = \left[\frac{\partial g(\mathbf{y})}{\partial \mathbf{s}} \mathbf{J}_{s,x} + \left(\mathbf{Q}\frac{\partial g(\mathbf{y})}{\partial \mathbf{u}} + \frac{\partial g(\mathbf{y})}{\partial \mathbf{s}} \mathbf{J}_{s,D}\right) \mathbf{J}_{D,x} + \frac{\partial g(\mathbf{y})}{\partial \mathbf{x}}\right] \mathbf{J}_{y,x}^{-1} \tag{5}$$

where $\mathbf{J}_{i,j}$'s are the Jacobians of transformation (e.g., $\mathbf{J}_{s,x} = \partial \mathbf{s}/\partial \mathbf{x}$) and y_i's are statistically independent random variables in the standard normal space. The evaluation of the quantities in Eq. (5) will depend on the problem under consideration (linear or nonlinear, two- or three-dimensional, etc.) and the performance functions used. The essential numerical aspect of SFEM is the evaluation of three partial derivatives, $\partial g/\partial \mathbf{s}$, $\partial g/\partial \mathbf{u}$, and $\partial g/\partial \mathbf{x}$, and four Jacobians, $\mathbf{J}_{s,x}$, $\mathbf{J}_{s,D}$, $\mathbf{J}_{D,x}$, and $\mathbf{J}_{y,x}$. They can be evaluated by procedures suggested by Haldar and Mahadevan [19] for linear and nonlinear, two- or three-dimensional structures. Once the coordinates of the design point $\mathbf{y}^*$ are evaluated with a preselected convergence criterion, the reliability index β can be evaluated as

$$\beta = \sqrt{(\mathbf{y}^*)^t(\mathbf{y}^*)} \tag{6}$$

The evaluation of Eq. (5) will depend on the problem under consideration and the limit state functions used. The probability of failure, P_f, can be calculated as

$$P_\mathrm{f} = \Phi(-\beta) = 1.0 - \Phi(\beta) \tag{7}$$

where Φ is the standard normal cumulative distribution function. Equation (7) can be considered as a notational failure probability. When the reliability index is larger, the probability of failure will be smaller. The author and his team published numerous papers to validate the above procedure.

3 Reliability Assessment: Present

3.1 Available Risk Evaluation Methods for Large Structures

As mentioned earlier, one of the alternatives for reliability analysis of large structures with implicit limit state functions is the use of the RSM [7]. The primary purpose of applying RSM in reliability analysis is to approximate the original complex and implicit limit state function using a simple and explicit polynomial [8, 24, 46]. Three basic weaknesses of RSM that limits its application potential are (1) it cannot incorporate distribution information of random variables; (2) if the response surface (RS) is not generated in the failure region, it may not be directly applicable; and (3) for large systems, it may not give the optimal sampling points. Thus, a basic RSM-based reliability method may not be applicable for large structures.

Before suggesting strategies on how to remove deficiencies in RSM, it is necessary to briefly discuss other available methods to generate RS. In recent past, several methods with the general objective of approximately developing multivariate expressions for RS for mechanical engineering applications were proposed. One such method is high-dimensional model representation (HDMR) [9, 37, 45]. It is also referred to as "decomposition method," "univariate approximation," "bivariate approximation," "S – variate approximation," etc. HDMR captures the high-dimensional relationships between sets of input and output model variables in such a way that the component functions of the approximation are ordered starting from a constant and adding terms such as first order, second order, and so on. The concept appears to be reasonable if higher-order variable correlations are weak, allowing the physical model to be captured by the first few lower-order terms.

Another major work is known as the explicit design space decomposition (EDSD). It can be used when responses can be classified into two classes, e.g., safe and unsafe. The classification is performed using explicitly defined boundaries in space. A machine learning technique known as support vector machines (SVM) [5] is used to construct the boundaries separating distinct classes. The failure regions corresponding to different modes of failure are represented with a single SVM boundary, which is refined through adaptive sampling.

3.2 Improvement of RSM

To bring distributional information of random variables and to efficiently locate the failure region for large complicated systems, the author proposed to integrate RSM and FORM. The integration can be carried out with the help of following tasks.

3.2.1 Degree of Polynomial

The degree of polynomial used to generate a response surface (RS) should be kept to a minimum to increase efficiency. At present, second-order polynomial without and with cross terms are generally used to generate response surfaces. Recently, Li et al. [27] proposed high-order response surface method (HORSM). The method employs Hermite polynomials and the one-dimensional Gaussian points as sampling points to determine the highest power of each variables. Considering the fact that higher-order polynomial may result in ill-conditional system of equations for unknown coefficients and exhibit irregular behavior outside of the domain of samples, for complicated large systems, second-order polynomial, without and with cross terms, can be used. They can be represented as

$$\hat{g}(\mathbf{X}) = b_0 + \sum_{i=1}^{k} b_i X_i + \sum_{i=1}^{k} b_{ii} X_i^2 \tag{8}$$

$$\hat{g}(\mathbf{X}) = b_0 + \sum_{i=1}^{k} b_i X_i + \sum_{i=1}^{k} b_{ii} X_i^2 + \sum_{i=1}^{k-1} \sum_{j>1}^{k} b_{ij} X_i X_j \tag{9}$$

where X_i ($i = 1, 2, \ldots, k$) is the ith random variable and b_0, b_i, b_{ii}, and b_{ij} are unknown coefficients to be determined; they need to be estimated using the response information at the sampling points by conducting several deterministic FE analyses, and k represents the total number of sensitive random variables after making less sensitive variables constants at their mean values, by conducting sensitivity analyses [18]. The numbers of coefficients necessary to define Eqs. (8 and 9) are $p = 2k + 1$ and $= (k + 1)(k + 2)/2$, respectively. The coefficients can be fully defined either by solving a set of linear equations or from regression analysis using responses at specific data points called experimental sampling points around a center point.

3.2.2 Detection of Failure Region

In the context of iterative scheme of FORM, to locate the coordinates of the most probable failure point and the corresponding reliability index, the initial center point $\mathbf{x}_{C_1}$ can be selected to be the mean values of the random variable X_i's. The response surface $\hat{g}(\mathbf{X})$ can be generated explicitly in terms of the random variables X_i's by conducting deterministic FE analyses at all the experimental sampling points, as will be discussed next. Once an explicit expression of the limit state function $\hat{g}(\mathbf{X})$ is obtained, the coordinates of the checking point $\mathbf{x}_{D_1}$ (iterative process to identify the coordinates of the most probable failure point) can be estimated using FORM, using all the statistical information on X_i's. The actual response can be evaluated again at the checking point $\mathbf{x}_{D_1}$, i.e., $g(\mathbf{x}_{D_1})$ and a new center point $\mathbf{x}_{C_2}$ can be selected using a linear interpolation [8, 36] as

$$\mathbf{x}_{C_2} = \mathbf{x}_{C_1} + (\mathbf{x}_{D_1} - \mathbf{x}_{C_1}) \frac{g(\mathbf{x}_{C_1})}{g(\mathbf{x}_{C_1}) - g(\mathbf{x}_{D_1})} \quad \text{if } g(\mathbf{x}_{D_1}) \geq g(\mathbf{x}_{C_1}) \tag{10}$$

$$\mathbf{x}_{C_2} = \mathbf{x}_{D_1} + (\mathbf{x}_{C_1} - \mathbf{x}_{D_1}) \frac{g(\mathbf{x}_{D_1})}{g(\mathbf{x}_{D_1}) - g(\mathbf{x}_{C_1})} \quad \text{if } g(\mathbf{x}_{D_1}) < g(\mathbf{x}_{C_1}) \tag{11}$$

A new center point $\mathbf{x}_{C_2}$ then can be used to develop an explicit performance function for the next iteration. This iterative scheme can be repeated until a preselected convergence criterion of $(\mathbf{x}_{C_{i+1}} - \mathbf{x}_{C_i})/\mathbf{x}_{C_i} \leq \varepsilon$ is satisfied. ε can be considered to be |0.05|. The second deficiency of RSM will be removed by locating the failure region using the above scheme.

3.2.3 Selection of Sampling Points

Saturated design (SD) and central composite design (CCD) are the two most promising schemes that can be used to generate experimental sampling points around the center point. SD is less accurate but more efficient since it requires only as many sampling points as the total number of unknown coefficients to define the response surface. CCD is more accurate but less efficient since a regression analysis needs to be carried out to evaluate the unknown coefficients. The details of experimental design procedures can be found in [7, 24]. In any case, the use of SD or CCD will remove the third deficiency of RSM.

Since the proposed algorithm is iterative and the basic SD and CCD require different amounts of computational effort, considering efficiency without compromising accuracy, several schemes can be followed. Among numerous schemes, one basic and two promising schemes are:

Scheme 0 – Use SD with second-order polynomial without the cross terms throughout all the iterations.
Scheme 1 – Use Eq. (8) and SD for the intermediate iterations and Eq. (9) and full SD for the final iteration.
Scheme 2 – Use Eq. (8) and SD for the intermediate iterations and Eq. (9) and CCD for the final iteration.

To illustrate the computational effort required for the reliability evaluation of large structural system, suppose the total number of sensitive random variables present in the formulation is, $k = 40$. The total number of coefficients necessary to define Eq. (8) will be $2 \times 40 + 1 = 81$ and to define Eq. (9) will be $(40 + 1)(40 + 2)/2 = 861$. It can also be shown that if Eq. (8) and SD scheme are used to generate the response surface, the total number of sampling points, essentially the total number of deterministic FE-based response analyses, will be 81. However, if Eq. (9) and full SD scheme are used, the corresponding deterministic analyses will be 861. If Eq. (9) and CCD scheme are used, the corresponding deterministic analyses will be $2^{40} + 2 \times 40 + 1 = 1{,}099{,}511{,}160{,}081$.

3.2.4 Mathematical Representation of Large Systems for Reliability Evaluation

The phrase "probability of failure" implies that the risk needs to be evaluated just before failure in the presence of several sources of nonlinearities. Finite element

(FE)-based formulations are generally used to realistically consider different sources of nonlinearity and other performance-enhancing features with improved energy dissipation mechanism now being used after the Northridge earthquake of 1994. Thus, for appropriate reliability evaluation, it is essential that structures are represented realistically by FEs and all major sources of nonlinearity and uncertainty are appropriately incorporated in the formulation.

To study the behavior of frame structures satisfying underlying physics, consideration of appropriate rigidities of connections is essential. In a typical design, all connections are considered to be fully restrained (FR), i.e., the angles between the girders and columns, before and after the application of loads, will remain the same. However, extensive experimental studies indicate that they are essentially partially restrained (PR) connection with different rigidities. In a deterministic analysis, PR connections add a major source of nonlinearity. In a dynamic analysis, it adds a major source of energy dissipation. In reliability analysis, it adds a major source of uncertainty. In general, the relationship between the moment M, transmitted by the connection, and the relative rotation angle θ is used to represent the flexible behavior. Among the many alternatives (Richard model, piecewise linear model, polynomial model, exponential model, B-Spline model, etc.), the Richard four-parameter moment-rotation model is chosen here to represent the flexible behavior of a connection. It is expressed as [39]

$$M = \frac{(k - k_p)\theta}{\left(1 + \left|\frac{(k - k_p)\theta}{M_0}\right|^N\right)^{\frac{1}{N}}} + k_p\theta \tag{12}$$

where M is the connection moment, θ is the relative rotation between the connecting elements, k is the initial stiffness, k_p is the plastic stiffness, M_0 is the reference moment, and N is the curve shape parameter. These parameters are identified in Fig. 3. To incorporate flexibility in the connections, a beam-column element can be introduced to represent each connection. However, its stiffness needs to be updated at each iteration since the stiffness representing the partial rigidity depends on θ. The tangent stiffness of the connection element, $K_C(\theta)$, can be shown to be

$$K_C(\theta) = \frac{dM}{d\theta} = \frac{(k - k_p)}{\left(1 + \left|\frac{(k - k_p)\theta}{M_0}\right|^N\right)^{\frac{N+1}{N}}} + k_p \tag{13}$$

The Richard model discussed above represents only the monotonically increasing loading portion of the M-θ curves. However, the unloading and reloading behavior of the M-θ curves is also essential for any nonlinear seismic analysis [10]. Using the Masing rule and the Richard model, Huh and Haldar [21]

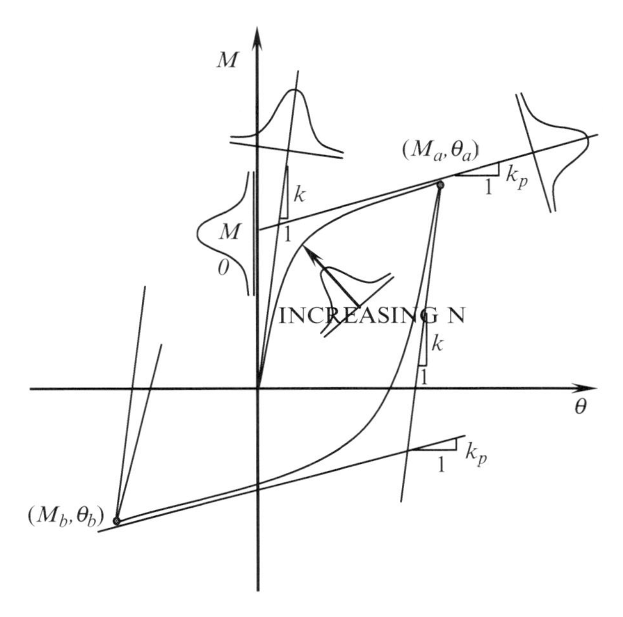

Fig. 3 M-θ curve using the Richard model, Masing rule, and uncertainty

theoretically developed the unloading and reloading parts of the M-θ curves. The tangent stiffness for the unloading and reloading behavior of a PR connection can be represented as

$$K_{\mathrm{C}}(\theta) = \frac{\mathrm{d}M}{\mathrm{d}\theta} = \frac{(k - k_{\mathrm{p}})}{\left(1 + \left|\frac{(k - k_{\mathrm{p}})(\theta_{\mathrm{a}} - \theta)}{2M_0}\right|^N\right)^{\frac{N+1}{N}}} + k_{\mathrm{p}} \tag{14}$$

As shown in Fig. 3, this represents hysteretic behavior at the PR connections. The basic FE formulation of the structure remains unchanged.

3.2.5 Pre- and Post-Northridge PR Connections

During the Northridge earthquake of 1994, several connections in steel frames fractured in a brittle and premature manner. A typical connection, shown in Fig. 4, was fabricated with the beam flanges attached to the column flanges by full penetration welds (field-welded) and with the beam web bolted (field-bolted) to single plate shear tabs [40], denoted hereafter as the pre-NC.

In the post-Northridge design practices, the thrusts were to make the connections more flexible than the pre-NC and to move the location of formation of any plastic hinge away from the connection and to provide more ductility to increase the energy absorption capacity. Several improved connections can be found in the literature

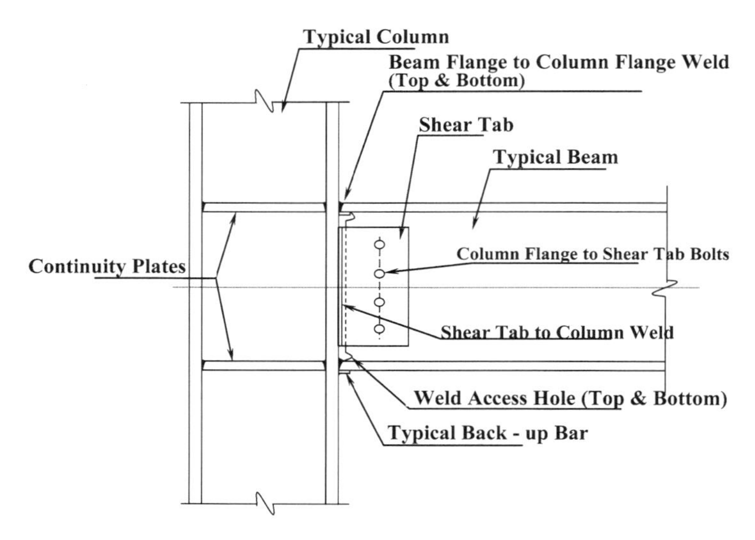

Fig. 4 A typical pre-NC

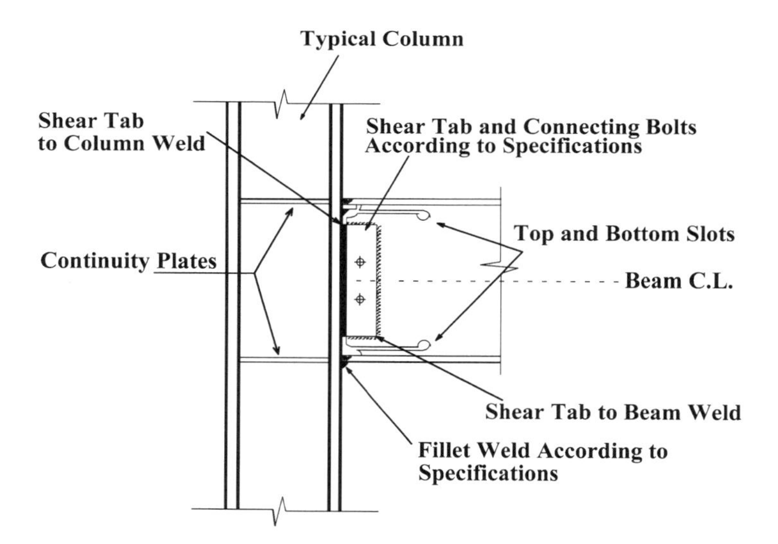

Fig. 5 A typical post-NC

including cover-plated connections, spliced beam connections, side-plated connections, bottom haunch connections, connections with vertical ribs, and connections with a reduced beam sections (RBS) or dog-boned (FEMA 350-3). Seismic Structural Design Associates, Inc. (SSDA) proposed a unique proprietary slotted web (SSDA SlottedWeb^{TM}) moment connection (Richard et al. [40]), as shown in Fig. 5, denoted hereafter as the post-NC. The author was given access to some of the actual SSDA full-scale test results. Using the four-parameter Richard model, the research team first proposed a mathematical model to represent moment-relative rotation (M-θ) curves for this type of connections [30].

4 Examples

A three-story three-bay steel frame, as shown in Fig. 6, is considered. Section sizes of beams and columns, using A36 steel, are also shown in the figure. It was excited by a seismic time history shown in Fig. 7 [17, 21].

The four parameters of the Richard model are calculated by PRCONN [38], a commercially available computer program for both pre-NC and post-NC connections. For the example under consideration, considering the sizes of columns and beams, three types of connection are necessary. They are denoted as types A, B, and C, hereafter. Four Richard parameters for both pre-NC and post-NC connections are summarized in Table 1.

4.1 Limit States or Performance Functions

In structural engineering, both strength and serviceability limit states are used for reliability estimation. The strength limit states mainly depend on the failure modes. Most of the elements in the structural system considered are beam columns. The interaction equations suggested by the American Institute of Steel Construction's

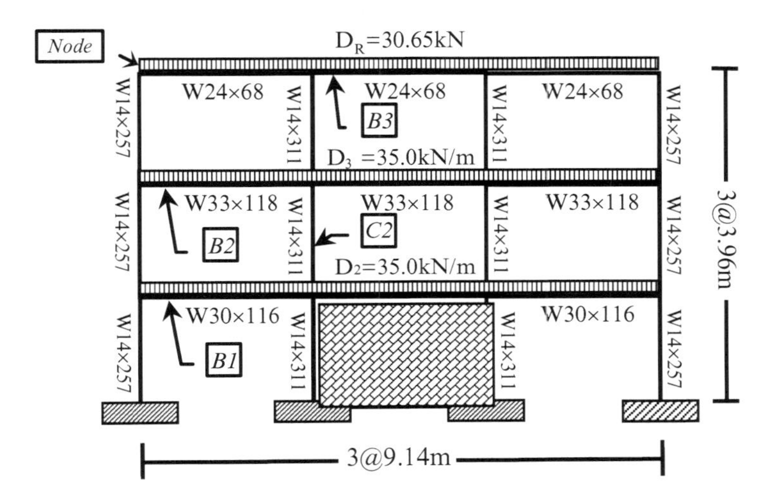

Fig. 6 A 3-story 3-bay SMRF structure

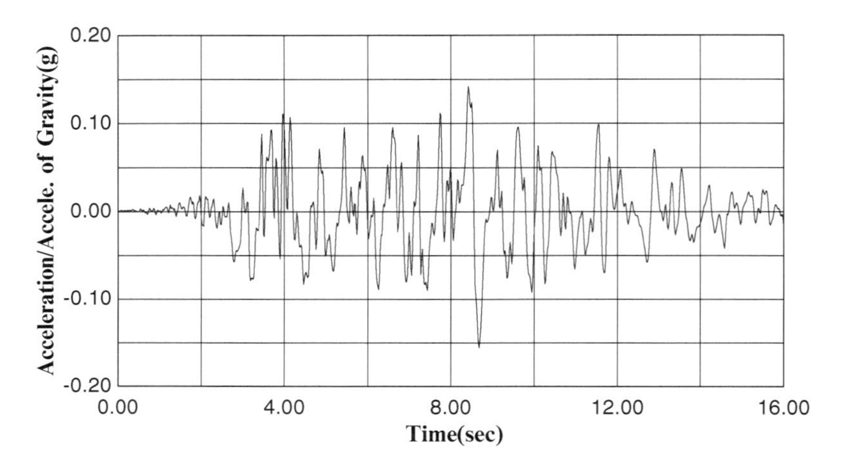

Fig. 7 Earthquake time history

Table 1 Parameters of Richard equation for M-θ curves

Connection assembly type				Connection parameters			
ID.		Beam	Column	k^a	$k_p{}^a$	$M_0{}^b$	N
Pre-NC	A	W24 × 68	W14 × 257	2.51E7	5.56E5	4.16E4	1.1
		W24 × 68	W14 × 311				
	B	W33 × 118	W14 × 257	5.08E7	1.14E5	6.79E4	1.1
		W33 × 118	W14 × 311				
	C	W30 × 116	W14 × 257	3.95E7	9.19E5	5.65E4	1.1
		W30 × 116	W14 × 311				
Post-NC	A	W24 × 68	W14 × 257	1.00E9	4.52E5	9.64E4	1.0
		W24 × 68	W14 × 311				
	B	W33 × 118	W14 × 257	2.34E9	4.52E5	2.44E5	1.0
		W33 × 118	W14 × 311				
	C	W30 × 116	W14 × 257	2.14E9	4.52E5	2.21E5	1.0
		W30 × 116	W14 × 311				

[a]kN cm/rad
[b]kN cm

Load and Resistance Factor Design [1] manual for two-dimensional structures are used in this study. The serviceability limit states can be represented as

$$g(\mathbf{X}) = \delta_{\text{allow}} - y_{\max}(\mathbf{X}) = \delta_{\text{allow}} - \hat{g}(\mathbf{X}) \tag{15}$$

where δ_{allow} is the allowable interstory drift or overall lateral displacement specified in codes and $y_{\max}(\mathbf{X})$ is the corresponding the maximum interstory drift or overall lateral displacement estimated.

Table 2 Statistical information on the design variables

	Item	Random variable	Mean value	COV	Dist.
Member	All	E(kN/m^2)	2.0E+8	0.06	Ln
		F_y(kN/m^2)	2.48E+5	0.10	Ln
	Column	I_x^{C1}(m^4)	1.42E-3	0.05	Ln
	W14 × 257	Z_x^{C1}(m^3)	7.98E-3	0.05	Ln
	Column	A^{C2}(m^2)	5.90E-2	0.05	Ln
	W14 × 311	I_x^{C2}(m^4)	1.80E-3	0.05	Ln
	Beam	I_x^{B2}(m^4)	2.46E-3	0.05	Ln
	W33 × 118				
	Beam	I_x^{B3}(m^4)	2.05E-3	0.05	Ln
	W30 × 116	Z_x^{B3}(m^3)	6.19E-3	0.05	Ln
Seismic load		ξ	0.05	0.15	Type I
		g_e	1.0	0.20	Type I
Connection	Richard model parameter	K^a	Refer to values in Table 3	0.15	N
		$k_p{}^a$		0.15	N
		$M_0{}^b$		0.15	N
		N		0.05	N
Shear wall[c]		E_C(kN/m^2)	2.14E+7	0.18	Ln
		v	0.17	0.10	Ln

Ln lognormal distribution
[a]kN cm/rad
[b]kN cm
[c]$f_C' = 2.068 \times 10^4$ (kN/m^2)

4.2 *Reliability Evaluations of Frame with Different Connection Conditions, Without RC Shear Wall*

The statistical characteristics of all the design variables used in the formulation are summarized in Table 2. The probabilities of failure of the frame for the lateral deflection at the top of the frame for serviceability limit state and the strength limit state of the weakest members are estimated assuming all the connections are of FR type. The results are summarized in Table 3. To verify the results, 20,000 MCS for the serviceability and 30,000 MCS for the strength limit states were carried out. The results clearly indicate that the bare steel frame will not satisfy the serviceability requirement. Then, the reliabilities of the frame are estimated assuming all the connections are post-NC and pre-NC types, and the results are summarized in Table 4. The behavior of the frame in the presence of FR and post-NC for both serviceability and strength limit states are very similar. This was also observed during the full-scale experimental investigations. This observation clearly indicates that the method can predict realistic behavior of complex structural systems. In any case, the lateral stiffness of the frame needs to be increased.

Table 3 Reliability evaluation of FR frame

Limit state		Service-ability	Strength limit state	
		Node at 1	Beam ($B1$)	Column ($C1$)
MCS	P_f	0.08740	N/A[a]	N/A[a]
	$\beta \approx \Phi^{-1}(1 - P_f)$	1.357	N/A	N/A
	NOS[b]	20,000	30,000	30,000
Proposed algorithm	No. of RV	8	6	6
	Scheme	1	2	2
	β	1.330	4.724	5.402
	Error w.r.t β	1.99%	N/A	N/A
	TNSP[c]	79	103	103

[a]Not a single failure observed for 30,000cycles of simulation since large reliability indexes are expected in the strength limit state
[b]Number of simulation for deterministic FEM analyses
[c]Total number of sampling points (total number of deterministic FEM analyses)

Table 4 Reliability evaluations of frame without and with shear wall

Steel frame without shear wall					Steel frame with shear wall			
		Connection type				Connection type		
		FR	Post-NC	Pre-NC		FR	Post-NC	Pre-NC
Serviceability limit state (node 1)								
	β	1.330	1.329	0.463	β	3.667	3.534	1.685
	$P_f \approx \Phi(-\beta)$	0.092	0.092	0.322	$P_f \approx \Phi(-\beta)$	1.2E-4	2.0E-4	4.6E-2
	No. of R.V.	8	20	20	No. of R.V.	10	22	22
	TNSP	79	313	313	TNSP	108	366	366
Strength limit state								
	β	4.724	4.756	3.681	β	6.879	6.714	4.467
Beam	$P_f \approx \Phi(-\beta)$	1.16E-6	9.87E-7	1.16E-4	$P_f \approx \Phi(-\beta)$	3.01E-12	9.477E-12	3.97E-6
	No. of R.V.	6	18	18	No. of R.V.	8	20	20
	TNSP	103	264	264	TNSP	79	313	313
Column	β	5.402	5.376	4.154	β	6.879	6.714	4.467
	$P_f \approx \Phi(-\beta)$	3.30E-8	3.81E-8	1.63E-5	$P_f \approx \Phi(-\beta)$	3.01E-12	9.47E-12	3.97E-6
	No. of R.V.	6	18	18	No. of R.V.	8	20	20
	TNSP	103	264	264	TNSP	79	313	313

4.3 Reliability Evaluations of Frames with Different Connection Conditions with RC Shear Wall

To increase the lateral stiffness, the steel frame is strengthened with a reinforced concrete (RC) shear wall at the first floor level, as shown in Fig. 6. For the steel and concrete dual system, all the steel elements in the frame are modeled as beam-column elements. A four-node plane stress element is introduced for the shear wall in the frame. To consider the presence of RC shear wall, the modulus of elasticity, E_C, and the Poisson ratio of concrete,ν, are necessary in the deterministic formulation. Cracking may develop at a very early stage of loading. It was observed that the degradation of the stiffness of the shear walls occurs after cracking and can be considered effectively by reducing the modulus of elasticity of the shear walls [26]. The rupture strength of concrete, f_r, is assumed to be $f_r = 7.5 \times \sqrt{f'_c}$, where f_c' is the compressive strength of concrete. After the tensile stress of each shear wall exceeds the prescribed tensile stress of concrete, the degradation of the shear wall stiffness is assumed to be reduced to 40% of the original stiffness [26]. The uncertainty in all the variables considered for the bare steel frame will remain the same. However, two additional sources of uncertainty, namely, in E_C and v, need to be considered, as given in Table 2.

The frame is again excited by the same earthquake time history as shown in Fig. 7. The probabilities of failure for the combined dual system in presence of FR, post-NC, and pre-NC connections are calculated using the proposed algorithm for the strength and serviceability limit states. The results are summarized in Table 4. The results indicate that the presence of shear wall at the first floor level significantly improves both the serviceability and strength behavior of the steel frame. If the probabilities of failure need to be reduced further, RC shear walls can be added in the second and/or third floor. Again, this improved behavior can be observed and quantified by carrying out about hundred deterministic evaluations instead of thousands of MCS. The improved behavior of the frame in the presence of RC shear wall is expected; however, the proposed algorithm can quantify the amount of improvement in terms of probability of failure for different design alternatives.

5 Reliability Assessment: Future

5.1 Reliability Evaluation of Large Structural Systems

In some studies considered by the author, one deterministic nonlinear dynamic analysis of large structures may take over 10 h of computer time. If one has to use very small, say only 100 simulations, it may take 1,000 h or over 41 days of uninterrupted running of a computer. The author proposed to estimate reliability of such systems using only tens instead of hundreds or thousands of deterministic evaluations at intelligently selected points to extract the reliability information. The procedure is still under development. The concept is briefly discussed below.

Scheme M1: To improve the efficiency of Scheme 1 discussed earlier, the cross terms (edge points), $k\ (k - 1)$, are suggested to be added only for the most important variables in the last iteration. Since the proposed algorithm is an integral part of FORM, all the random variables in the formulation can be arranged in descending order of their sensitivity indexes $\alpha(X_i)$, i.e., $\alpha(X_1) > \alpha(X_2) > \alpha(X_3)$.........$>\alpha(X_k)$. The sensitivity of a variable X, $\alpha(X)$, is the directional cosines of the unit normal vector at the design point. In the last iteration, the cross terms are added only for the sensitive random variables, m, and the corresponding reliability index is calculated. The total number of FEM analyses required for Scheme 1 and *M1* is $(k + 1)(k + 2)/2$ and $2k + 1 + m(2k - m - 1)/2$, respectively. For an example, suppose for a large structural system, $k = 40$ and $m = 3$. The total number of required FEM analyses will be 861 and 195, respectively, for the two schemes.

Scheme M2: Instead of using full factorial plan in CCD, Myers et al. [31] recently proposed quarter factorial plan. This improved and efficient version of Scheme 2 will be denoted hereafter as Scheme M2. In Scheme M2, it is proposed that only quarter of the factorial points corresponding to the most sensitive random variables are to be considered. In other words, in the last iteration, the variables are to be arranged in descending order according to their sensitivity indexes $\alpha(X_i)$, i.e., $\alpha(X_1) > \alpha(X_2) > \alpha(X_3)\ldots\ldots\ldots > \alpha(X_k)$. Then, the factorial sampling points are determined by using the values +1 and −1, in the coded space, for $X_1, X_2, X_3, \ldots$, until the number of quarter of factorial sampling $= 0.25 \times 2^k = 2^{k-2}$.

6 Risk Management

Since risk cannot be completely eliminated in engineering analysis and design, it needs to be managed appropriately. It is undesirable and uneconomical, if not impossible, to design a risk-free structure. According to Raiffa [35], the decision or management can be subdivided in to risk or data analysis, inference, and decision. Risk analysis is discussed in the first part of this chapter. Inference attempts to incorporate additional scientific knowledge in the formulation that may have been ignored in the previous data analysis. The decision is the final outcome of any risk-based engineering design. For a given structure, the risks for different design alternatives can be estimated. The information on risk and the corresponding consequences of failure, including the replacement cost, can be combined using a decision analysis framework to obtain the best alternative. Thus, the probability concept provides a unified framework for quantitative analysis of risk as well as the formulation of trade-off studies for decision-making, planning, and design considering economic aspects of the problem.

The relevant components of a decision analysis are generally described in the form of a decision tree [16]. The decision tree helps to organize all necessary information in a systematic manner suitable for analytical evaluation of the optimal alternative. A decision-making process starts by choosing an action, a, from among the available alternatives actions ($a_1, a_2, \ldots, a_n$), called decision variables. They branch out from a square node or a decision node. Once a decision has been made, a natural outcome θ from among all possible states, $\theta_1, \theta_2, \ldots, \theta_n$, would materialize. All possible states are beyond the control of a decision-maker; they are shown to originate from a circular node, called the chance node. Each natural outcome based on the action taken is expected to have a different risk of success or failure and expressed in terms of probability $P(\theta i|ai)$. As a result of having taken action a and having found true state θ, a decision-maker will obtain a utility value $u(a, \theta)$, a numerical measure of the consequences of this action-state pair. A general decision analysis framework may contain the following necessary components: (1) decision variables, (2) consequence, (3) risk associated with each alternative, and (4) identification of the decision-maker and the decision criteria.

7 Future Directions

7.1 *Artificial Neural Networks and Soft Computing*

The application of artificial neural networks (ANN), the more generic term used by the research community as soft computing, in civil engineering has been significant in the very recent past [23]. In addition to ANN, the other soft computing techniques include genetic algorithm, evolutionary computation, machine learning, organic computing, probabilistic reasoning, etc. The applicability of these techniques could be problem specific, some of them can be combined, or one technique can be used when another failed to meet the objectives of the study. Soft computing differs from conventional hard computing. Unlike hard computing, soft computing is tolerant of imprecision, uncertainty, partial truth, and approximation. To some extent, it essentially plays a role similar to human mind.

7.2 *Incorporation of Cognitive Sources of Uncertainty*

Being a reliability person, I feel I should address this subject very briefly. Most of the works on reliability-based structural engineering incorporate noncognitive (quantitative) sources of uncertainty using crisp set theory. Cognitive or qualitative sources of uncertainty are also important. They come from the vagueness of the problem arising from the intellectual abstractions of reality. To incorporate cognitive sources of uncertainty, fuzzy set theory [4] is generally used.

7.3 *Education*

Lack of education could be a major reason for avoidance of using the reliability-based design concept by the profession. If closed-form reliability analysis is not possible and the design codes do not cover the design of a particular structure, at the minimum, simulation can be used to satisfy the intent of the codes. In Europe, highway and railway companies are using simulation for assessment purposes. In the USA, the general feeling is that we are safe if we design according to the design code. Designers should use all available means to satisfy performance requirements, according to a judge. The automotive industry satisfied the code requirements in one case. However, according to a judge, they should have used simulation to address the problem more comprehensively.

Some of the developments in the risk-based design using simulation are very encouraging. Simulation could be used in design in some countries, but it is also necessary to look at its legal ramification. Unlike in Europe, in the USA, a code is not a government document. It is developed by the profession and its acceptance is voted by the users and developers. It was pointed out that in some countries, code guidelines must be followed to the letters, and other countries permit alternative methods if they are better. In Europe, two tendencies currently exist: Anglo-Saxon – more or less free to do anything, and middle-European – fixed or obligatory requirements. Current Eurocode is obligatory. We need to change the mentality and laws to implement simulation or reliability-based design concept in addressing real problems.

In the context of education of future structural engineers, the presence of uncertainty must be identified in design courses. Reliability assessment methods can contribute to the transition from deterministic to probabilistic way of thinking of students as well as designers. In the USA, the Accreditation Board of Engineering and Technology (ABET) now requires that all civil engineering undergraduate students demonstrate knowledge of the application of probability and statistics to engineering problems, indicating its importance in civil engineering education. Most of the risk-based design codes are the by-product of education and research at the graduate level. In summary, the profession is moving gradually in accepting the reliability-based design concept.

7.4 Computer Programs

The state of the art in reliability estimation is quite advanced; however, it is not popular with the practicing engineers. One issue could be the lack of availability of the user-friendly software. Two types of issues need to be addressed at this stage. Reliability-based computer software should be developed for direct applications, or the reliability-based design feature should be added to the commercially available deterministic software. Some of the commercially available reliability-based computer software is briefly discussed next.

NESSUS (Numerical Evaluation of Stochastic Structures Under Stress) was developed by the Southwest Research Institute [43] under the sponsorship of NASA Lewis Research Center. It combines probabilistic analysis with a general-purpose finite element/boundary element code. The probabilistic analysis features an advanced mean value (AMV) technique. The program also includes techniques such as fast convolution and curvature-based adaptive importance sampling.

PROBAN (PROBability ANalysis) was developed at Det Norske Veritas, Norway, through A.S. Veritas Research [44]. PROBAN was designed to be a general-purpose probabilistic analysis tool. It is capable of estimating the probability of failure using FORM and SORM for a single event, unions, intersections, and unions of intersections. The approximate FORM/SORM results can be updated through importance sampling simulation scheme. The probability of general events can be computed by Monte Carlo simulation and directional sampling.

CALREL (CAL-RELiability) is a general-purpose structural reliability analysis program designed to compute probability integrals [28]. It incorporates four general techniques for computing the probability of failure: FORM, SORM, directional simulation with exact or approximate surfaces, and Monte Carlo simulation. It has a library of probability distributions of independent as well as dependent random variables.

Under the sponsorship of the Pacific Earthquake Engineering Research (PEER), McKenna et al. [29] have developed a finite element reliability code within the framework of OpenSees.

Structural engineers without formal education in risk-based design may not be able to use the computer programs mentioned above. They need to be retrained with very little effort. They may be very knowledgeable using exiting deterministic analysis software including commercially available finite element packages. This expertise needs to be integrated with risk-based design concept. Thus, probabilistic features may need to be added to the deterministic finite element packages. Proppe et al. [32] discussed the subject in great detail. For proper interface of deterministic software, they advocated for graphical user interface, communication interface which must be flexible enough to cope with different application programming interfaces and data format, and the reduction of the problem sizes before undertaking reliability analysis. COSSAN [12] software attempted to implement the concept.

The list of computer programs given here may not be exhaustive. However, they are being developed and are expected to play a major role in implementing reliability-based engineering analysis and design in the near future.

8 Conclusions

Engineering under uncertainty has evolved in the past several decades. It has attracted multidisciplinary research interest. A brief overview of the past, present, and future in the author's assessment is given here. Albert Einstein stated that "The important thing is not to stop questioning. Curiosity has its own reason for existing." The profession is very curious on the topic, and there is no doubt that future analysis and design of engineering structures will be entirely conducted using probability concept.

Acknowledgments I would like to thank all my teachers for teaching me subjects that helped to develop my career and understanding of my life in a broader sense. I also would like to thank all my former and current students who taught me subjects for which I did not have any formal education. They helped me explore some of the unchartered areas. I also appreciate financial support I received from many funding agencies to explore several challenging and risky research areas. I would also like to thank the organizing committee of the International Symposium on Engineering Under Uncertainty: Safety Assessment and Management (ISEUSAM-2012) for inviting me to give the inaugural keynote speech.

References

1. American Institute of Steel Construction (2005) Manual of steel construction: load and resistance factor design. Illinois, Chicago
2. Ang AH-S, Cornell CA (1974) Reliability bases of structural safety and design. J Struct Eng ASCE 100(ST9):1755–1769

3. Ang AH-S, Tang WH (1975) Probability concepts in engineering design, vol. I: Basic principles. Wiley, New York
4. Ayyub BM, Klir GJ (2006) Uncertainty modeling and analysis in engineering and the sciences. Chapman and Hall/CRC, Boca Raton
5. Basudhar A, Missoum S, Harrison Sanchez A (2008) Limit state function identification using support vector machines for discontinuous responses and disjoint failure domains. Probab Eng Mech 23(1):1–11
6. Benjamin JR, Cornell CA (1970) Probability, statistics, and decision for civil engineers. McGraw-Hill, New York
7. Box GP, William GH, Hunter JS (1978) Statistics for experimenters: an introduction to design, data analysis and modeling building. Wiley, New York
8. Bucher CG, Bourgund U (1990) A fast and efficient response surface approach for structural reliability problems. Struct Saf 7:57–66
9. Chowdhury R, Rao BN, Prasad AM (2008) High dimensional model representation for piece wise continuous function approximation. Commun Numer Methods Eng 24(12):1587–1609
10. Colson A (1991) Theoretical modeling of semi-rigid connection behavior. J Construct Steel Res 19:213–224
11. Cornell CA (1969) A probability-based structural code. J Am Concr Inst 66(12):974–985
12. COSSAN (Computational Stochastic Structural Analysis) – Stand – Alone Toolbox (1996) User's manual. IfM – Nr: A, Institute of Engineering Mechanics, Leopold – Franzens University, Innsbruck
13. Ellingwood BR, Galambos TV, MacGregor JG, Cornell CA (1980) Development of probability based load criterion for American National Standard A58. NBS special publication 577. U.S. Department of Commerce, Washington, DC
14. Fisher RA (1959) Statistical methods and scientific inference. Hafner, New York
15. Freudenthal AM (1947) Safety of structures. Trans ASCE 112:125–180
16. Haldar A (1980) Liquefaction study – a decision analysis framework. J Geotech Eng Div ASCE 106(GT12): 1297–1312
17. Haldar A, Farag R, Huh J (2010) Reliability evaluation of large structural systems. Keynote lecture. International symposium on reliability engineering and risk management (ISRERM2010), Shanghai, China, pp. 131–142
18. Haldar A, Mahadevan S (2000) Probability, reliability and statistical methods in engineering design. Wiley, New York
19. Haldar A, Mahadevan S (2000) Reliability assessment using stochastic finite element analysis. Wiley, New York
20. Hasofar AM, Lind NC (1974) Exact and invariant second moment code format. J Eng Mech ASCE 100(EM1):111–121
21. Huh J, Haldar A (2011) A novel risk assessment method for complex structural systems. IEEE Trans Reliab 60(1):210–218
22. Jeffreys H (1961) Theory of probability. Oxford University Press, New York
23. Kartam N, Flood I, Garrett JH (1997) Artificial neural networks for civil engineers. American Society of Civil Engineers, New York
24. Khuri AI, Cornell CA (1996) Response surfaces designs and analyses. Marcel Dekker, New York
25. Laplace PSM (1951) A philosophical essay on probabilities (translated from the sixth French edition by Truscott FW and Emory FL). Dover Publications, New York
26. Lefas D, Kotsovos D, Ambraseys N (1990) Behavior of reinforced concrete structural walls: strength, deformation characteristics, and failure mechanism. ACI Struct J 87(1):23–31
27. Li H, Lu Z, Qiao H (2008) A new high-order response surface method for structural reliability analysis, 2008, personal communication
28. Liu P-L, Lin H-Z, Der Kiureghian A (1989) CALREL. University of California, Berkeley
29. McKenna F, Fenves GL, Scott MH (2002) Open system for earthquake engineering simulation. Pacific Earthquake Engineering Research Center, Berkeley. http://opensees.berkeley.edu/

30. Mehrabian A, Haldar A, Reyes AS (2005) Seismic response analysis of steel frames with post-Northridge connection. Steel Comp Struct 5(4):271–287
31. Myers RH, Montgomery DC, Anderson-Cook CM (2009) Response surface methodology: process and product optimization using designed experiments. Wiley, New York
32. Proppe C, Pradlwarter HJ, Schueller GI (2001) Software for stochastic structural analysis – needs and requirements. In: Corotis RB, Schueller GI, Shinizuka M (eds) Proceedings of the 4th international conference on structural safety and reliability, AA Balkema Publishers, The Netherlands
33. Rackwitz R (1976) Practical probabilistic approach to design. Bulletin No. 112, Comite European du Beton, Paris
34. Rackwitz R, Fiessler B (1978) Structural reliability under combined random load sequences. Comput Struct 9(5):484–494
35. Raiffa H (1968) Decision analysis. Addison-Wesley, Reading
36. Rajashekhar MR, Ellingwood BR (1993) A new look at the response surface approach for reliability analysis. Struct Saf 12:205–220
37. Rao BN, Chowdhury R (2009) Enhanced high dimensional model representation for reliability analysis. Int J Numer Methods Eng 77(5):719–750
38. Richard RM (1993) PRCONN manual. RMR Design Group, Tucson
39. Richard RM, Abbott BJ (1975) Versatile elastic-plastic stress-strain formula. J Eng Mech ASCE 101(EM4):511–515
40. Richard RM, Allen CJ, Partridge JE (1997)Proprietary slotted beam connection designs. Mod Steel Constr 37(3):28–35
41. Richard RM, Radau RE (1998) Force, stress and strain distribution in FR bolted welded connections. In: Proceedings of structural engineering worldwide, San Francisco, CA
42. Rosenbleuth E, Esteva L (1972) Reliability bases for some Mexican codes. ACI Publ SP-31:1–41
43. Shinozuka M (1983) Basic analysis of structural safety. J Struct Eng ASCE 109(3):721–740
44. Southwest Research Institute (1991) NEUSS, San Antonio, Texas
45. Veritas Sesam Systems (1991) *PROBAN*, Houston, Texas
46. Xu H, Rahman S (2005) Decomposition methods for structural reliability analysis. Probab Eng Mech 20:239–250
47. Yao TH-J, Wen YK (1996) Response surface method for time-variant reliability analysis. J Struct Eng ASCE 122(2):193–201

Geotechnics in the Twenty-First Century, Uncertainties and Other Challenges: With Particular Reference to Landslide Hazard and Risk Assessment

Robin Chowdhury, Phil Flentje, and Gautam Bhattacharya

Abstract This chapter addresses emerging challenges in geotechnics in the context of the significant challenges posed by hazards, both natural and human-induced. The tremendous importance of dealing with uncertainties in an organized and systematic way is highlighted. The chapter includes reflections on responding to the need for multidisciplinary approaches. While the concepts and ideas are pertinent to diverse applications of geotechnics or to the whole of geotechnical engineering, illustrative examples will be limited to research trends in slope stability and landslide management.

From time to time, researchers, academics, and practicing engineers refer to the need for interdisciplinary approaches in geotechnical engineering. However, surveys of the relevant literature reveal few examples of documented research studies based within an interdisciplinary framework. Meanwhile there is a broad acceptance of the significant role of uncertainties in geotechnics.

This chapter includes reflections on what steps might be taken to develop better approaches for analysis and improved strategies for managing emerging challenges in geotechnical engineering. For example, one might start with the need to highlight different types of uncertainties such as geotechnical, geological, and hydrological. Very often, geotechnical engineers focus on variability of soil properties such as shear strength parameters and on systematic uncertainties. Yet there may be more important factors in the state of nature which are ignored because of the lack of a multidisciplinary focus. For example, the understanding of the potential for

Robin Chowdhury, Invited Keynote Speaker, ISEUSAM-2012 Conference, January 2012, BESU, Shibpur, India

R. Chowdhury (✉) • P. Flentje
University of Wollongong, New South Wales, Australia
e-mail: robin@uow.edu.au; pflentje@uow.edu.au

G. Bhattacharya
Bengal Engineering & Science University, Shibpur, Howrah, West Bengal 711103, India
e-mail: gautam@civil.becs.ac.in

S. Chakraborty and G. Bhattacharya (eds.), *Proceedings of the International Symposium on Engineering under Uncertainty: Safety Assessment and Management (ISEUSAM - 2012)*, DOI 10.1007/978-81-322-0757-3_2, © Springer India 2013

progressive failure within a soil mass or a slope may require careful consideration of the geological context and of the history of stress and strain. The latter may be a consequence of previous seismic activity and fluctuations in rainfall and groundwater flow.

The frequency and consequences of geotechnical failures involving soil and rock continue to increase globally. The most significant failures and disasters are often associated with major natural events but not exclusively so.

It is expected that climate change will lead to even more unfavorable conditions for geotechnical projects and thus to increasing susceptibility and hazard of landsliding. This is primarily because of the expected increase in the variability of rainfall and the expected increase in sea levels. Responding to the effects of climate change will thus require more flexible and robust strategies for assessment of landslide susceptibility and to innovative engineering solutions.

Keywords Geotechnics • Uncertainty • Hazards • Risk • Landslide

1 Introduction

There are significant challenges for the future development and application of geotechnical engineering. Developments in research, analysis, and practice have taken place to advance knowledge and practice. While the scope of the profession and its discipline areas is already vast, significant extension is required in the areas of hazard and risk assessment and management. In particular, the field of natural disaster reduction requires the development of innovative approaches within a multidisciplinary framework. Very useful and up-to-date information on the occurrence frequency and impact of different natural disasters is being assessed and analyzed by a number of organizations around the world. However, geotechnical engineers have not played a prominent part in such activities so far. Reference may be made to the research and educational materials developed on a regular basis by the Global Alliance for Disaster Reduction (GADR) with the aim of information dissemination and training for disaster reduction. Some selected illustrations from GADR are presented in an Appendix to this chapter. The role of geotechnical engineers in implementing such goals is obvious from these illustrations.

The variability of soil and rock masses and other uncertainties have always posed unique challenges to geotechnical engineers. In the last few decades, the need to identify and quantify uncertainties on a systematic basis has been widely accepted. Methods for inclusion of such data in formal ways include reliability analysis within a probabilistic framework. Considerable progress has been made in complementing traditional deterministic methods with probabilistic studies. Nevertheless, the rate of consequent change to geotechnical practice has been relatively slow and sometimes halfhearted. Reviewing all the developments in geotechnical engineering which have taken place over the last 30 years or more would require painstaking and critical reviews from a team of experts over a considerable period of time and the subsequent reporting of the findings in a series of books. In comparison, the scope of this keynote chapter is humble. Experienced academics who have been engaged in serious scholarship, research, and consulting over several decades should be able to reflect on recent and continuing trends as well as warning signs of complacency or lack of vision. In this spirit, an attempt is made to highlight some pertinent issues and challenges for the assessment and management of geotechnical risk with particular reference to slope stability and landslides.

The writers of the present chapter present some highlights of their own research through case study examples. These relate to aspects of regional slope stability and hazard assessment such as a landslide inventory map, elements of a relational database, rainfall intensity duration for triggering landslides, continuous monitoring of landslide sites in near-real time, landslide susceptibility, and hazard maps. The chapter concludes with reflections on continuing and emerging challenges. For further details, the reader may refer to Chowdhury and Flentje [5] and Flentje et al. [11, 12], and a comprehensive book [8].

In order to get a sense of global trends in geotechnical analysis and the assessment and management of risk, reference may be made to the work of experts and professionals in different countries as reported in recent publications. The applications include the safety of foundations, dams, and slopes against triggering events such as rainstorms, floods earthquakes, and tsunamis.

The following is a sample of five papers from a 2011 conference related to geotechnical risk assessment and management, GeoRisk 2011. Despite covering a wide range of topics and techniques, it is interesting that GIS-based regional analysis for susceptibility and hazard zoning is not included among these publications. Such gaps are often noted and reveal that far greater effort is required to establish multidisciplinary focus in geotechnical research. This is clearly a continuing challenge for geotechnics in the twenty-first century.

- A comprehensive paper on geohazard assessment and management involving the need for integration of hazard, vulnerability, and consequences and the consideration of acceptable and tolerable risk levels [16].
- Risk assessment of Success Dam, California, is discussed by Bowles et al. [2] with particular reference to the evaluation of operating restrictions as an interim measure to mitigate earthquake risk. The potential modes of failure related to earthquake events and flood events are discussed in two companion papers.
- The practical application of risk assessment in dam safety (the practice in USA) is discussed in a paper by Scott [23].
- Unresolved issues in Geotechnical Risk and Reliability [9].
- Development of a risk-based landslide warning system [24].

The first paper [16] has a wide scope of topics and discusses the following six case studies:

Hazard assessment and early warning for a rock slope over a fjord arm on the west coast of Norway – the slope is subject to frequent rockslides usually with volumes in the range 0.5–5 million cubic meters.

Vulnerability assessment – Norwegian clay slopes in an urban area on the south coast of Norway
Risk assessment – 2004 tsunami in the Indian Ocean
Risk mitigation – quick clay in the city of Drammen along the Drammensfjord and the Drammen River.
Risk mitigation – Early Warning System for landslide dams, Lake Sarez in the Pamir Mountain Range in eastern Tajikistan
Risk of tailings dam break – probability of nonperformance of a tailings management facility at Rosia Montana in Romania

2 Uncertainties Affecting Geotechnics

The major challenges in geotechnical engineering arise from uncertainties and the need to incorporate them in analysis, design, and practice. The geotechnical performance of a specific site, facility, system, or regional geotechnical project may be affected by different types of uncertainty such as the following (with examples in brackets):

- Geological uncertainty (geological details)
- Geotechnical parameter uncertainty (variability of shear strength parameters and of pore water pressure)
- Hydrological uncertainty (aspects of groundwater flow)
- Uncertainty related to historical data (frequency of slides, falls, or flows)
- Uncertainty related to natural or external events (magnitude, location and timing of rainstorm, flood, earthquake, and tsunami)
- Project uncertainty (construction quality, construction delays)
- Uncertainty due to unknown factors (effects of climate change)

On some projects, depending on the aims, geotechnical engineers may be justified in restricting their attention to uncertainties arising from geological, geotechnical, and hydrological factors. For example, the limited aim may be to complement deterministic methods of analysis with probabilistic studies to account for imperfect knowledge of geological details and limited data concerning measured soil properties and pore water pressures. It is necessary to recognize that often pore pressures change over time, and therefore, pore pressure uncertainty has both spatial and temporal aspects which can be critically important.

During the early development of probabilistic analysis methods, researchers often focused on the variability of soil properties in order to develop the tools for probabilistic analysis. It was soon realized that natural variability of geotechnical parameters such as shear strength must be separated from systematic uncertainties such as measurement error and limited number of samples. Another advance in understanding has been that the variability of a parameter, measured by its standard deviation, is a function of the spatial dimension over which the variability is considered. In some problems, consideration of spatial variability on a formal basis is important and leads to significant insights.

An important issue relates to the choice of geotechnical parameters and their number for inclusion in an uncertainty analysis. The selection is often based on experience and can be justified by performing sensitivity studies. A more difficult issue is the consideration of "new" geotechnical parameters not used in traditional deterministic or even in probabilistic studies. Thus, one must think "outside the box" for "new" parameters which might have significant influence on geotechnical reliability. Otherwise, the utility and benefits of reliability analyses may not be fully realized. As an example, the "residual factor" (defined as proportion of a slip surface over which shear strength has decreased to a residual value) is rarely used as a variable in geotechnical slope analysis. Recently, interesting results have been revealed from a consideration of "residual factor" in slope stability as a random variable [1, 4]. Ignoring the residual factor can lead to overestimate of reliability and thus lead to unsafe or unconservative practice.

For regional studies such as zoning for landslide susceptibility and hazard assessment, historical data about previous events are very important. Therefore, uncertainties with respect to historical data must be considered and analyzed carefully. Such regional studies are different in concept and implementation from traditional site-specific deterministic and probabilistic studies and often make use of different datasets. A successful knowledge-based approach for assessment of landslide susceptibility and hazard has been described by Flentje [10].

If the aim of a geotechnical project is to evaluate geotechnical risk, it is necessary to consider the uncertainty related to the occurrence of an external event or events that may affect the site or the project over an appropriate period of time such as the life of the project.

Consideration of project uncertainty would require consideration of economic, financial, and administrative factors in addition to the relevant technical factors considered above. In this regard, the reader may refer to a recent paper on georisks in the business environment by Brumund [3]; the paper also makes reference to unknown risk factors.

For projects which are very important because of their size, location, economic significance, or environmental impact, efforts must be made to consider uncertainty due to unknown factors. Suitable experts may be co-opted by the project team for such an exercise.

3 Slope Analysis Methods

3.1 Limit Equilibrium and Stress-Deformation Approaches

Deterministic methods can be categorized as limit equilibrium methods and stress-deformation methods. Starting from simple and approximate limit equilibrium methods based on simplifying assumptions, several advanced and relatively rigorous methods have been developed.

The use of advanced numerical methods for stress-deformation analysis is essential when the estimation of strains and deformations within a slope is required. In most cases, two-dimensional (2D) stress-deformation analyses would suffice. However, there are significant problems which need to be modeled and analyzed in three dimensions. Methods appropriate for 3D stress-deformation analysis have been developed and used successfully. Advanced stress-deformation approaches include the finite difference method, the finite element method, the boundary element method, the distinct element method, and the discontinuous deformation analysis method.

3.2 Progressive Failure

Progressive failure of natural slopes, embankment dams, and excavated slopes is a consequence of nonuniform stress and strain distribution and the strain-softening behavior of earth masses. Thus, shear strength of a soil element, or the shear resistance along a discontinuity within a soil or rock mass, may decrease from a peak to a residual value with increasing strain or increasing deformation. Analysis and simulation of progressive failure require that strain-softening behavior be taken into consideration within the context of changing stress or strain fields. This may be done by using advanced methods such as an initial stress approach or a sophisticated stress-deformation approach. Of the many historical landslides in which progressive failure is known to have played an important part, perhaps the most widely studied is the catastrophic Vaiont slide which occurred in Italy in 1964. The causes and mechanisms have not been fully explained by any one study, and there are still uncertainties concerning both the statics and dynamics of the slide. For further details and a list of some relevant references, the reader may refer to Chowdhury et al. [8].

3.3 Probabilistic Approaches and Simulation of Progressive Failure

A probabilistic approach should not be seen simply as the replacement of a calculated "factor of safety" as a performance index by a calculated "probability of failure." It is important to consider the broader perspective and greater insight offered by adopting a probabilistic framework. It enables a better analysis of observational data and enables the modeling of the reliability of a system. Updating of reliability on the basis of observation becomes feasible, and innovative approaches can be used for the modeling of progressive failure probability and for back-analysis of failed slopes. Other innovative applications of a probabilistic approach with pertinent details and references are discussed by Chowdhury et al. [8].

An interesting approach for probabilistic seismic landslide analysis which incorporates the traditional infinite slope limit equilibrium model as well as the rigid block displacement model has been demonstrated by Jibson et al. [15].

A probabilistic approach also facilitates the communication of uncertainties concerning hazard assessment and slope performance to a wide range of end-users including planners, owners, clients, and the general public.

3.4 Geotechnical Slope Analysis in a Regional Context

Understanding geology, geomorphology, and groundwater flow is of key importance. Therefore, judicious use must be made of advanced methods of modeling in order to gain the best possible understanding of the geological framework and to minimize the role of uncertainties on the outcome of analyses [17, 21].

Variability of ground conditions, spatial and temporal, is important in both regional and site-specific analysis. Consequently, probability concepts are very useful in both cases although they may be applied in quite different ways.

Spatial and temporal variability of triggering factors such as rainfall have a marked influence on the occurrence and distribution of landslides in a region [8, 18]

This context is important for understanding the uncertainties in the development of critical pore water pressures. Consequently, it helps in the estimation of rainfall threshold for onset of landsliding. Regional and local factors both would have a strong influence on the combinations of rainfall magnitude and duration leading to critical conditions.

Since earthquakes trigger many landslides which can have a devastating impact, it is important to understand the causative and influencing factors. The occurrence, reach, volume, and distribution of earthquake-induced landslides are related to earthquake magnitude and other regional factors. For further details and a list of some relevant references, the reader may refer to Chowdhury et al. [8].

4 Regional Slope Stability Assessments

4.1 Basic Requirements

Regional slope stability studies are often carried out within the framework of a Geographical Information System (GIS) and are facilitated by the preparation of relevant datasets relating to the main influencing factors such as geology, topography, and drainage characteristics and by developing a comprehensive inventory of existing landslides. The development of a Digital Elevation Model (DEM) facilitates GIS-based modeling of landslide susceptibility, hazard, and risk within a GIS framework. Regional slope stability and hazard studies facilitate the development of effective landslide risk management strategies in an urban area. The next section of this chapter is devoted to a brief discussion of GIS as a versatile and powerful system for spatial and even temporal analysis. This is followed by a section providing a brief overview of sources and methods for obtaining accurate spatial data. The data may relate to areas ranging from relatively limited zones to very large regions. Some of these resources and methods have a global reach and applicability. Such data are very valuable for developing Digital Elevation Models (DEMs) of increasing accuracy. For regional analysis, a DEM is, of course, a very important and powerful tool.

4.2 Landslide Inventory

The development of comprehensive databases including a landslide inventory is most desirable if not essential especially for the assessment of slope stability in a regional context. It is important to study the occurrence and spatial distribution of first-time slope failures as well as reactivated landslides.

Identifying the location of existing landslides is just the beginning of a systematic and sustained process with the aim of developing a comprehensive landslide inventory. Among other features, it should include the nature, size, mechanism, triggering factors, and date of occurrence of existing landslides. While some old landslide areas may be dormant, others may be reactivated by one or more regional triggering factors such as heavy rainfall and earthquakes.

One comprehensive study of this type has been discussed in some detail in Chapter 11 of Chowdhury et al. [8]. This study was made for the Greater Wollongong region, New South Wales, Australia, by the University of Wollongong (UOW) Landslide Research Team (LRT). In this chapter, this study is also referred to as the Wollongong Regional Study.

A small segment for the Wollongong Landslide Inventory for the Wollongong Regional Study is shown in Fig. 1. The elements of a Landslide Relational Database are shown in Fig. 2. Some details of the same are shown in Figs. 3 and 4. A successful knowledge-based approach for assessment of landslide susceptibility and hazard has been described by Flentje [10] and is covered in some detail in a separate section of this chapter.

4.3 Role of Geographical Information Systems (GIS)

GIS enables the collection, organization, processing, managing, and updating of spatial and temporal information concerning geological, geotechnical, topographical, and other key parameters. The information can be accessed and applied by a

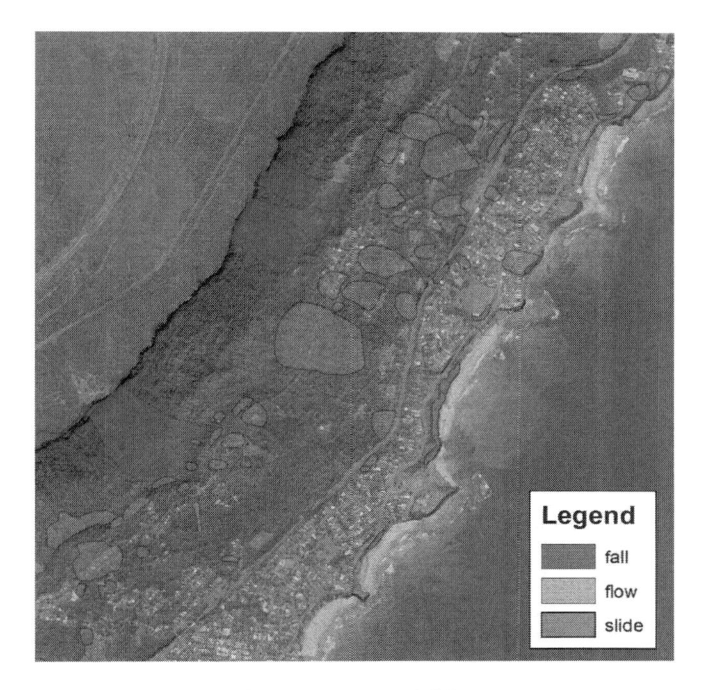

Fig. 1 Segment of the University of Wollongong Landslide Inventory

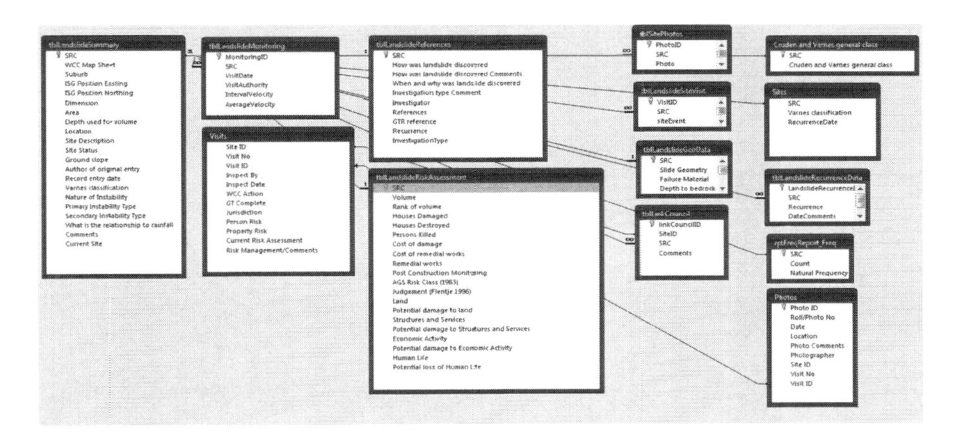

Fig. 2 Elements of a Landslide Relational Database

range of professionals such as geotechnical engineers, engineering geologists, civil engineers, and planners for assessing hazard of landsliding as well as for risk management. Traditional slope analysis must, therefore, be used within the context of a modern framework which includes GIS. Among the other advantages of GIS

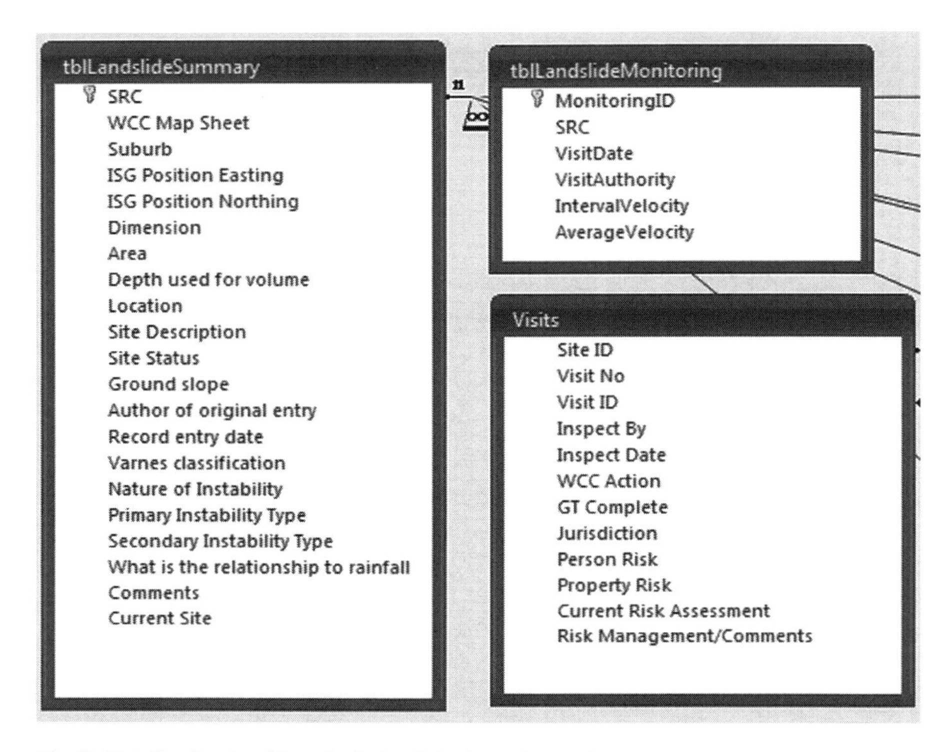

Fig. 3 Details of main tables of relational database shown above

are the ability to deal with multiple hazards, the joining of disparate data, and the ability to include decision support and warning systems [13].

Papers concerning the application of basic, widely available, GIS systems as well as about the development of advanced GIS systems continue to be published. For instance, Reeves and West [22], covering a conference session on "Geodata for the urban environment," found that 11 out of 30 papers were about the "Development of Geographic Information Systems" while Gibson and Chowdhury [13] pointed out that the input of engineering geologists (and, by implication, geotechnical engineers) to urban geohazards management is increasingly through the medium of GIS.

Consequently, 3D geological models have been discussed by a number of authors such as Rees et al. [21] who envisage that such models should be the basis for 4D process modeling in which temporal changes and factors can be taken into consideration. They refer, in particular, to time-series data concerning precipitation, groundwater, sea level, and temperature. Such data, if and when available, can be integrated with 3D spatial modeling.

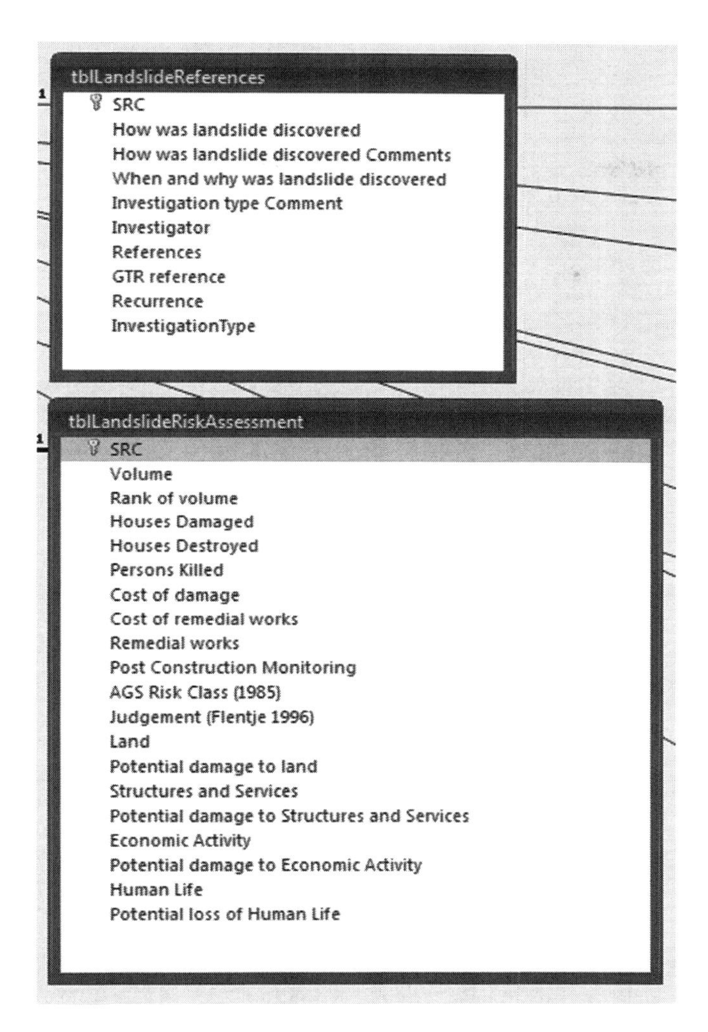

Fig. 4 Details of selected tables of relational database shown above

4.4 Sources of Accurate Spatial Data Relevant to the Development of Digital Elevation Models

Over the last decade, Airborne Laser Scan (ALS) or Light Detection and Ranging (LiDAR) techniques are increasingly being applied across Australia to collect high-resolution terrain point datasets. When processed and used to develop Geographic Information System (GIS) Digital Elevation Models (DEMs), the data provides high-resolution contemporary terrain models that form fundamental GIS datasets. Prior to the advent of this technology, DEMs were typically derived from

10- to 50-year-old photogrammetric contour datasets. When processed, ALS datasets can comprise point clouds of many millions of ground-reflected points covering large areas up to hundreds of square kilometers, with average point densities exceeding one point per square meter. Collection, processing, and delivery of these data types are being enhanced and formalized over time. Increasingly, this data is also being collected in tandem with high-resolution geo-referenced imagery.

Airborne and Satellite-Derived Synthetic Aperture Radar (SAR) techniques are also being increasingly developed and applied internationally to develop terrain models and specifically differential models between return visits over the same area in order to highlight the changes in ground surfaces with time. This is being used to monitor landslide movement, ground subsidence, and other environmental change.

NASA and the Japan Aerospace Exploration Society have just recently (mid-October 2011) and freely released via the Internet the Advanced Spaceborne Thermal Emission and Reflection Radiometer (ASTER) Global Digital Elevation Model (GDEM) – ASTER GDEM v2 global 30 m Digital Elevation Model as an update to the year 2000 vintage NASA SRTM Global DEM at 90 and 30 m pixel resolutions. This global data release means moderately high-resolution global Digital Elevation Model data are available to all.

The development of ALS terrain models and the free release of the global ASTER GDEM v2 have important implications for the development of high-resolution landslide inventories and zoning maps worldwide. These datasets mean one of the main barriers in the development of this work has been eliminated.

4.5 Observational Approach: Monitoring and Alert Systems

Geotechnical analysis should not be considered in isolation since a good understanding of site conditions and field performance is essential. This is particularly important for site-specific as well as regional studies of slopes and landslides. Observation and monitoring of slopes are very important for understanding all aspects of performance: from increases in pore water pressures to the evidence of excessive stress and strain, from the development of tension cracks and small shear movements to initiation of progressive failure, and from the development of a complete landslide to the postfailure displacement of the landslide mass.

Observation and monitoring also facilitate an understanding of the occurrence of multiple slope failures or widespread landsliding within a region after a significant triggering event such as rainfall of high magnitude and intensity [10, 12]. Observational approaches facilitate accurate back-analyses of slope failures and landslides. Moreover, geotechnical analysis and the assessment of hazard and risk can be updated with the availability of additional observational data on different parameters such as pore water pressure and shear strength. The availability of continuous monitoring data obtained in near-real time will also contribute to more accurate assessments and back-analyses. Consequently, such continuous monitoring will lead to further advancement in the understanding of slope behavior.

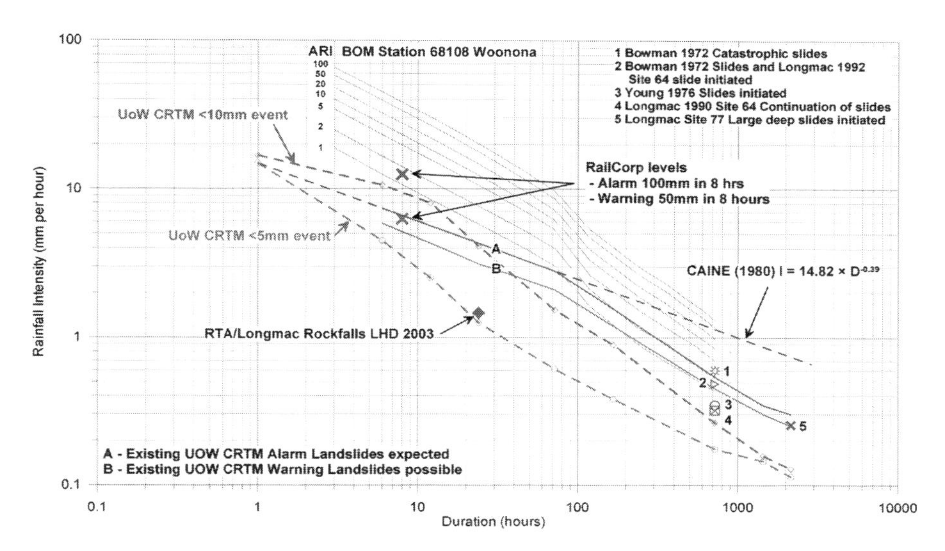

Fig. 5 Interpreted threshold curves for landsliding in Wollongong, superimposed on annual recurrence interval curves for a selected rainfall station

One part of the Wollongong Regional Study is the development of rainfall-intensity-duration curves for the triggering of landslides overlaid with historical rainfall average recurrence interval (ARI) curves as shown in Fig. 5. From the very beginning of this research, the potential use of such curves for alert and warning systems was recognized. In fact, this research facilitated risk management in the Wollongong study area during intense rainfalls of August 1998 when widespread landsliding occurred.

More recent improvement and extension of this work involves the use of data from our growing network of continuous real-time monitoring stations where we are also introducing the magnitude of displacement as an additional parameter. Aspects of this research are shown in Fig. 5, and as more data become available from continuous monitoring, additional displacement (magnitude)-based curves can be added to such a plot.

Two examples of continuous landslide performance monitoring are shown in Figs. 6 and 7. Figure 6 relates to a coastal urban landslide site (43,000 m^3) with limited trench drains installed. The relationship between rainfall, pore water pressure rise, and displacement is clearly evident at two different time intervals in this figure. Figure 7 shows data from a complex translational landslide system (720,000 m^3) which is located on a major highway in NSW Australia. In the 1970s, landsliding severed this artery in several locations resulting in road closures and significant losses arising from damage to infrastructure and from traffic disruptions.

After comprehensive investigations, remedial measures were installed. At this site, a dewatering pump system was installed, which continues to operate to this day. However, this drainage system has been reviewed and upgraded from time to

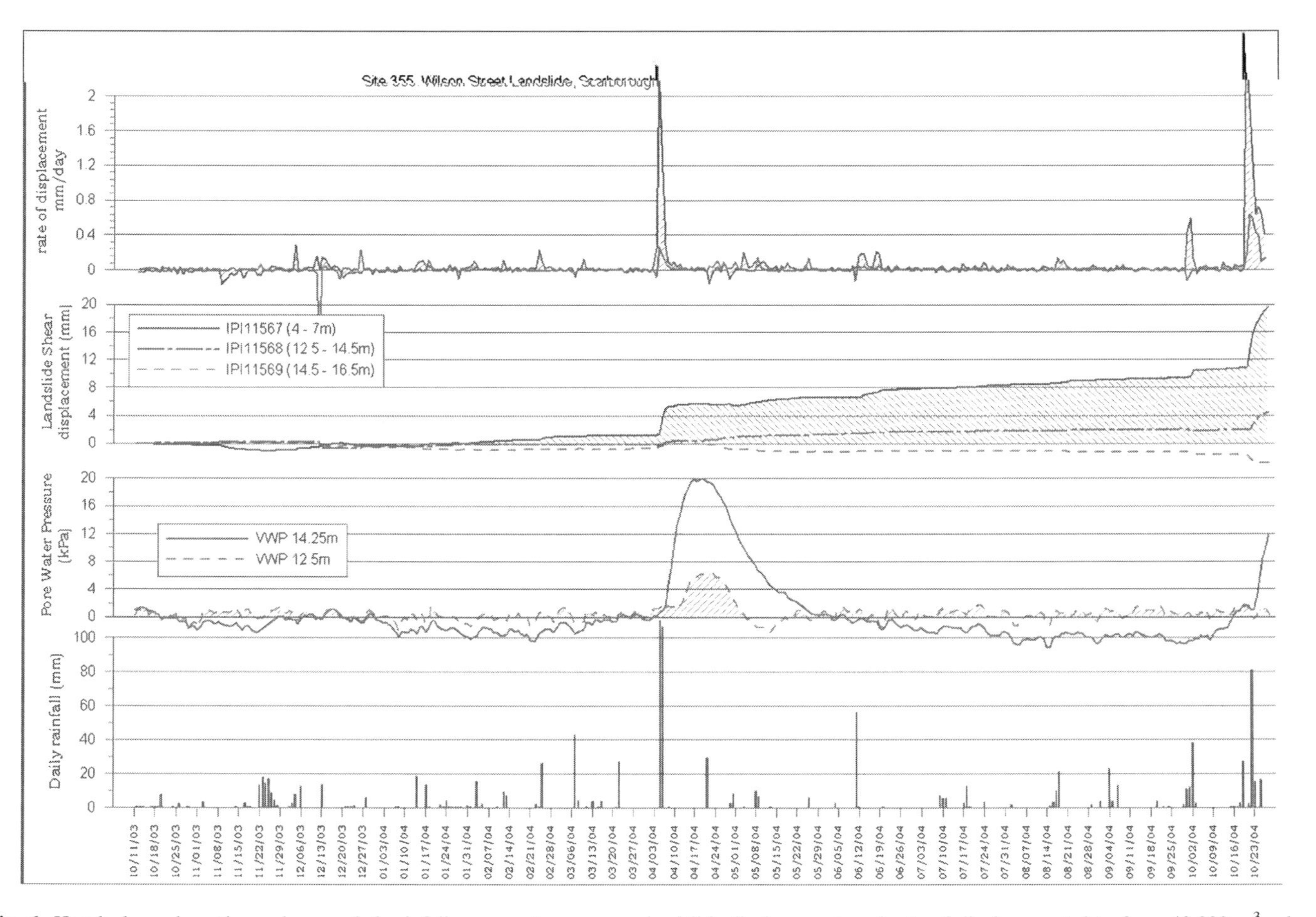

Fig. 6 Hourly logged continuously recorded rainfall, pore water pressure, landslide displacement, and rate of displacement data for a 43,000 m^3 urban landslide site in Wollongong

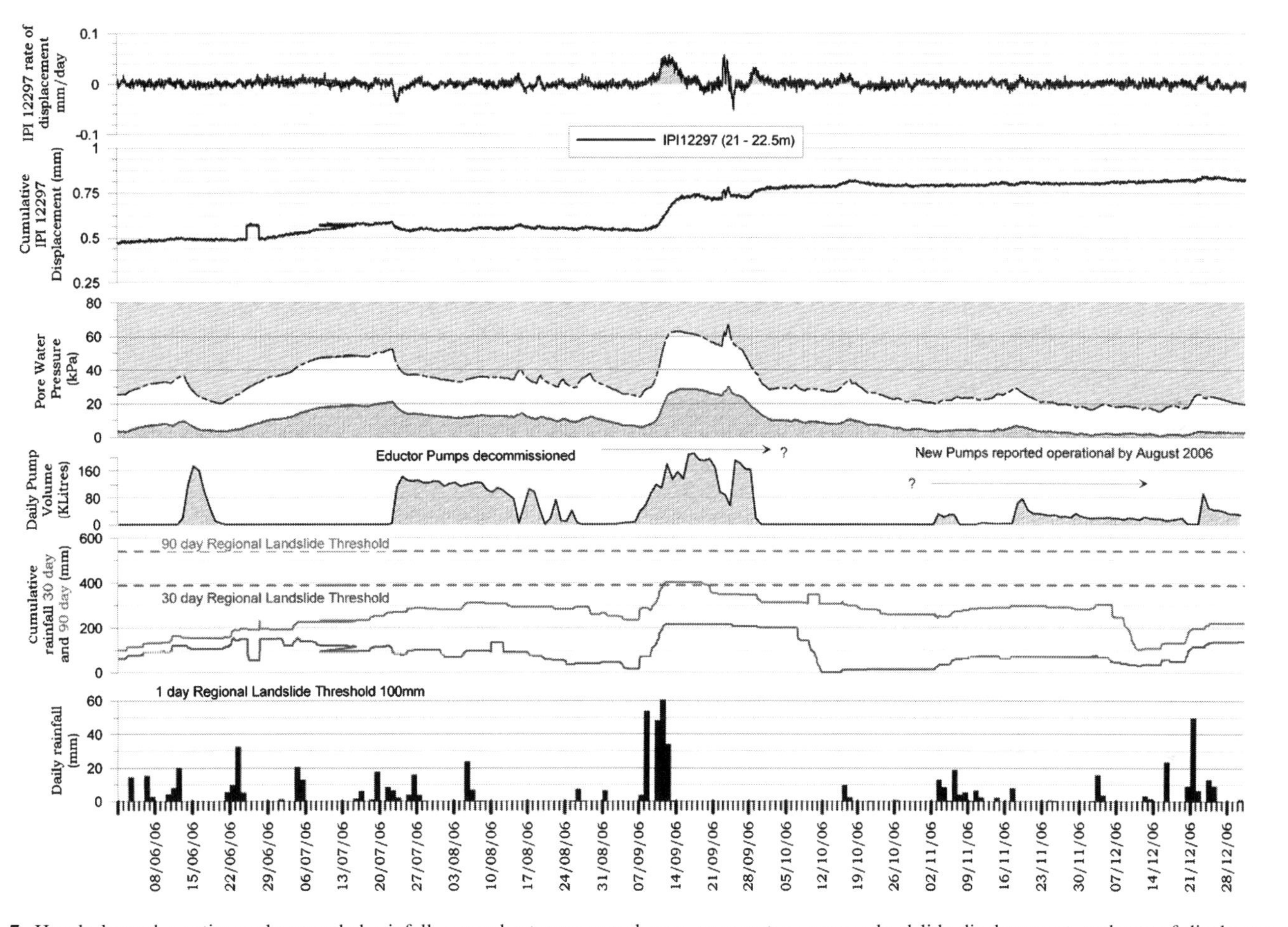

Fig. 7 Hourly logged, continuously recorded rainfall, groundwater pump volumes, pore water pressure, landslide displacement, and rate of displacement data for a 720,000 m^3 landslide affecting a major transport artery in Wollongong

time. Since 2004, this site has been connected to the Continuous Monitoring Network of the University of Wollongong Landslide Research Team. Interpretation of the monitoring data shows that movement has been limited to less than 10 mm since the continuous monitoring commenced as shown in Fig. 7 [11]. However, the occurrence of events of this small movement was considered unacceptable by the authorities. Hence, pump and monitoring system upgrades commenced in 2006 and have been completed in 2011.

4.6 Susceptibility and Hazard Assessment (Wollongong Regional Study)

4.6.1 The Susceptibility Model Area and the Datasets

The area chosen within the Wollongong Region for modeling landslide susceptibility (Susceptibility Model Area) is 188 km^2 in extent and contains 426 slide category landslides.

The datasets used for this study include:

- Geology (mapped geological formations, 21 variables)
- Vegetation (mapped vegetation categories, 15 variables)
- Slope inclination (continuous floating point distribution)
- Slope aspect (continuous floating point distribution)
- Terrain units (buffered water courses, spur lines, and other intermediate slopes)
- Curvature (continuous floating point distribution)
- Profile curvature (continuous floating point distribution)
- Plan curvature (continuous floating point distribution)
- Flow accumulation (continuous integer)
- Wetness index (continuous floating point distribution)

4.6.2 Landslide Inventory

The landslide inventory for this study has been developed over a 15-year period and comprises a relational MS Access and ESRI ArcGIS Geodatabase with 75 available fields of information for each landslide site. It contains information on a total of 614 landslides (falls, flows, and slides) including 480 slides. Among the 426 landslides within the Susceptibility Model Area, landslide volumes have been calculated for 378 of these sites. The average volume is 21,800 m^3 and the maximum volume is 720,000 m^3.

4.6.3 Knowledge-Based Approach Based on Data Mining Model

The specific knowledge-based approach used for analysis and synthesis of the datasets for this study is the data mining (DM) process or model. The DM learning process is facilitated by the software "See5" which is a fully developed application of "C4.5" [20]. The DM learning process helps extract patterns from the databases related to the study. Known landslide areas are used for one half of the model training, the other half comprising randomly selected points from within the model area but outside the known landslide boundaries. Several rules are generated during the process of modeling. Rules which indicate potential landsliding are assigned positive confidence values and those which indicate potential stability (no-landsliding) are assigned negative confidence values. The rule set is then reapplied within the GIS software using the ESRI ModelBuilder extension to produce the susceptibility grid. The complete process of susceptibility and hazard zoning is described in Flentje [10] and in Chapter 11 of Chowdhury et al. [8].

4.6.4 Susceptibility and Hazard Zones

On the basis of the analysis and synthesis using the knowledge-based approach, it has been possible to demarcate zones of susceptibility and hazard into four categories:

1. Very low susceptibility (or hazard) of landsliding (VL)
2. Low susceptibility (or hazard) of landsliding (L)
3. Moderate susceptibility (or hazard) of landsliding (M)
4. High susceptibility (or hazard) of landsliding (H)

A segment of the landslide susceptibility map is shown in Fig. 8 below. A segment of the landslide hazard map, an enlarged portion from the bottom left of Fig. 8, is reproduced as Fig. 9. Relative likelihoods of failure in different zones, estimated from the proportion of total landslides which occurred in each zone over a period of 126 years, are presented in columns 1 and 2 of Table 1 below. This information is only a part of the full table presented as Table 11.3 in Chowdhury et al. [8].

5 Estimated Reliability Indices and Factors of Safety

An innovative concept has been proposed by Chowdhury and Flentje [7] for quantifying failure susceptibility from zoning maps developed on the basis of detailed knowledge-based methods and techniques within a GIS framework. The procedure was illustrated with reference to the results of the Wollongong Regional Study and the relevant tables are reproduced here. Assuming that the factor of safety has a normal distribution, the reliability index was calculated for each zone

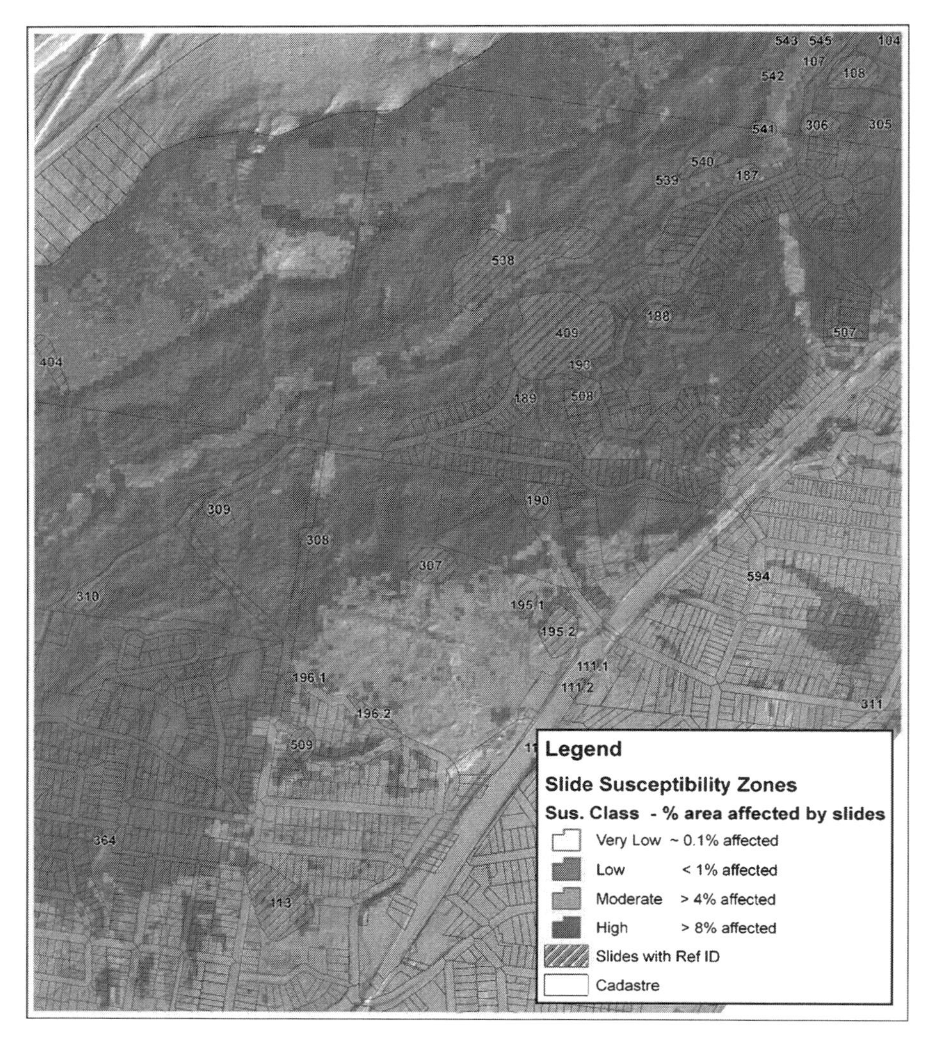

Fig. 8 Segment of landslide inventory and susceptibility zoning map, Wollongong Local Government Area, New South Wales, Australia

based on the associated failure likelihood which is assumed to represent the probability of failure. These results are presented in the third or last column of Table 1.

Assuming that the coefficient of variation of the factor of safety is 10%, the typical values of mean factor of safety for each zone are shown in Table 2. The results were also obtained for other values of the coefficient of variation of the factor of safety (5, 10, 15, and 20%). These results are shown in Table 3.

Most of the landslides have occurred during very high rainfall events. It is assumed here, in the first instance, that most failures are associated with a pore

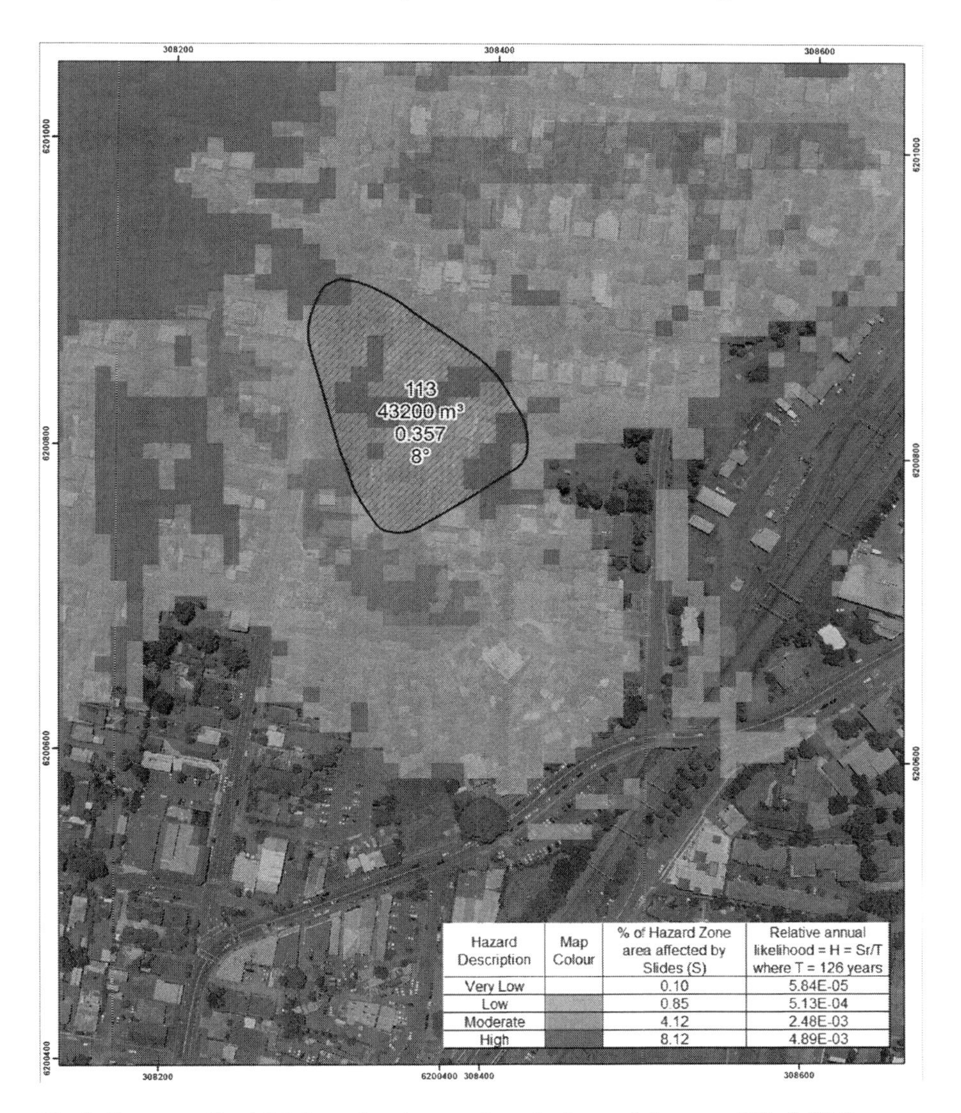

Fig. 9 Segment of landslide hazard zoning map from the *bottom left* corner of Fig.8, Wollongong Local Government Area, New South Wales, Australia. Landslide label shows four important particulars of each landslide stacked vertically. These are (1) site reference code, (2) landslide volume, (3) annual frequency of reactivation derived from inventory, and (4) landslide profile angle. Hazard zoning in legend shows relative annual likelihood as explained in the text

Table 1 Failure likelihood and reliability index for each hazard zone (After Chowdhury et al. [8])

Hazard zone description	Failure likelihood	Reliability index
Very low	7.36×10^{-3}	2.44
Low	6.46×10^{-2}	1.51
Moderate	3.12×10^{-1}	0.49
High	6.16×10^{-1}	−0.30

Table 2 Typical mean value of factor of safety (F) for each hazard zone considering coefficient of variation to be 10% (After Chowdhury and Flentje [7])

Hazard zone description	Reliability index	Mean of factor of safety, F ($V_F = 10\%$)
Very low	2.44	1.32
Low	1.51	1.18
Moderate	0.49	1.05
High	−0.3	0.97

Table 3 Typical mean values of factor of safety for different values of coefficient of variation (After Chowdhury and Flentje [7])

	Mean of F for different hazard zones			
V_F%	Very low	Low	Moderate	High
5	1.14	1.08	1.02	0.98
10	1.32	1.18	1.05	0.97
15	1.57	1.29	1.08	0.96
20	1.95	1.43	1.11	0.94

Table 4 Typical mean factor of safety with different values of pore pressure ratio (slope inclination $i = 12°$, $V_F = 10\%$) (After Chowdhury and Flentje [7])

	Mean of F for different hazard zones			
Pore water pressure ratio	Very low	Low	Moderate	High
0.5	1.32	1.18	1.05	0.97
0.4	1.61	1.44	1.28	1.18
0.3	1.90	1.70	1.51	1.40
0.2	2.19	1.95	1.74	1.61

water pressure ratio of about 0.5 (full seepage condition in a natural slope). Furthermore, assuming that the "infinite slope" model applies to most natural slopes and that cohesion intercept is close to zero, the values of factor of safety can be calculated for other values of the pore pressure ratio (0.2, 0.3, and 0.4) for any assumed value of the slope inclination. The results shown below in Table 4 are for a slope with an inclination of 12 degrees for pore pressure ratios in the range 0.2–0.5.

5.1 Discussion on the Proposed Concept and Procedure

The above results were obtained as a typical F value or a set of F values referring to each hazard zone. However, taking into consideration the spatial variation of slope angle, shear strength, and other factors, this approach may facilitate the calculation F at individual locations. Well-documented case studies of site-specific analysis would be required for such an extension of the procedure. Other possibilities include estimation of the variation of local probability of failure. The approach may also be used for scenario modeling relating to the effects of climate change. If reliable data concerning pore pressure changes become available, failure susceptibility under those conditions can be modeled, and the likelihood and impact of potential catastrophic slope failures can be investigated.

6 Discussion, Specific Lessons or Challenges

The focus of this chapter has been on hazard and risk assessment in geotechnical engineering. Advancing geotechnical engineering requires the development and use of knowledge which facilitates increasingly reliable assessments even when the budgets are relatively limited. Because of a variety of uncertainties, progress requires an astute combination of site-specific and regional assessments. For some projects, qualitative assessments within the framework of a regional study may be sufficient. In other projects, quantitative assessments, deterministic and probabilistic, may be essential.

In this chapter, different cases have been discussed in relation to the Wollongong Regional Study. Firstly, reference was made to the basis of an alert and warning system for rainfall-induced landsliding based on rainfall-intensity-duration plots supplemented by continuous monitoring. The challenges here are obvious. How do we use the continuous pore pressure data from monitoring to greater advantage? How do we integrate all the continuous monitoring data to provide better alert and warning systems? This research has applications in geotechnical projects generally well beyond slopes and landslides.

The examples concerning continuous monitoring of two case studies discussed in this chapter illustrate the potential of such research for assessing remedial and preventive measures. The lesson from the case studies is that, depending on the importance of a project, even very low hazard levels may be unacceptable. As emphasized earlier, the decision to upgrade subsurface drainage at the cost of hundreds of thousands of dollars over several years was taken and implemented despite the shear movements being far below disruptive magnitudes as revealed by continuous monitoring. The challenge in such problems is to consolidate this experience for future applications so that costs and benefits can be rationalized further.

The last example from the Wollongong Regional Study concerned the preparation of zoning maps for landslide susceptibility and hazard. Reference was made to an innovative approach for quantitative interpretation of such maps in terms of well-known performance indicators such as "factor of safety" under a variety of pore pressure conditions. The challenge here is to develop this methodology further to take into consideration the spatial and temporal variability within the study region.

7 Challenges Due to External Factors

Beyond the scope of this chapter, what are the broad challenges in geotechnical hazard and risk assessment? How do we deal with the increasing numbers of geotechnical failures occurring globally including many disasters and how do we mitigate the increasingly adverse consequences of such events? What strategies, preventive, remedial, and others, are necessary?

Often catastrophic landslides are caused by high-magnitude natural events such as rainstorms and earthquakes. It is also important to consider the contribution of human activities such as indiscriminate deforestation and rapid urbanization to landslide hazard. There is an increasing realization that poor planning of land and infrastructure development has increased the potential for slope instability in many regions of the world.

Issues concerned with increasing hazard and vulnerability are very complex and cannot be tackled by geotechnical engineers alone. Therefore, the importance of working in interdisciplinary teams must again be emphasized. Reference has already been made to the use of geological modeling (2D, 3D, and potentially 4D) and powerful tools such as GIS which can be used in combination with geotechnical and geological models.

At the level of analysis methods and techniques, one of the important challenges for the future is to use slope deformation (or slip movement) as a performance indicator rather than the conventional factor of safety. Also, at the level of analysis, attention needs to be given to better description of uncertainties related to construction of slopes including the quality of supervision.

Research into the effects of climate change and, in particular, its implications for geotechnical engineering is urgently needed [19, 21]. The variability of influencing factors such as rainfall and pore water pressure can be expected to increase. However, there will be significant uncertainties associated with estimates of variability in geotechnical parameters and other temporal and spatial factors. Consequently, geotechnical engineers need to be equipped with better tools for dealing with variability and uncertainty. There may also be other changes in the rate at which natural processes like weathering and erosion occur. Sea level rise is another important projected consequence of global warming and climate change, and it would have adverse effects on the stability of coastal slopes.

8 Concluding Remarks

A wide range of methods, from the simplest to the most sophisticated, are available for the geotechnical analysis of slopes. This includes both static and dynamic conditions and a variety of conditions relating to the infiltration, seepage, and drainage of water. Considering regional slope stability, comprehensive databases and powerful geological models can be combined within a GIS framework to assess and use information and data relevant to the analysis of slopes and the assessment of the hazard of landsliding. The use of knowledge-based systems for assessment of failure susceptibility, hazard, or performance can be facilitated by these powerful tools. However, this must all be based on a thorough fieldwork ethic.

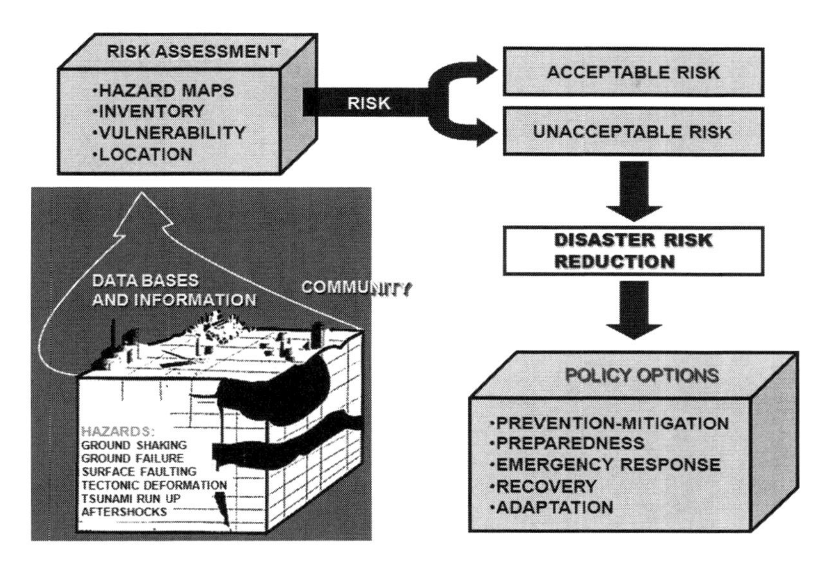

Fig. A.1 Elements of risk assessment and management for natural disasters (Courtesy of Walter Hays [14])

Fig. A.2 Components of risk (Courtesy of Walter Hays [14])

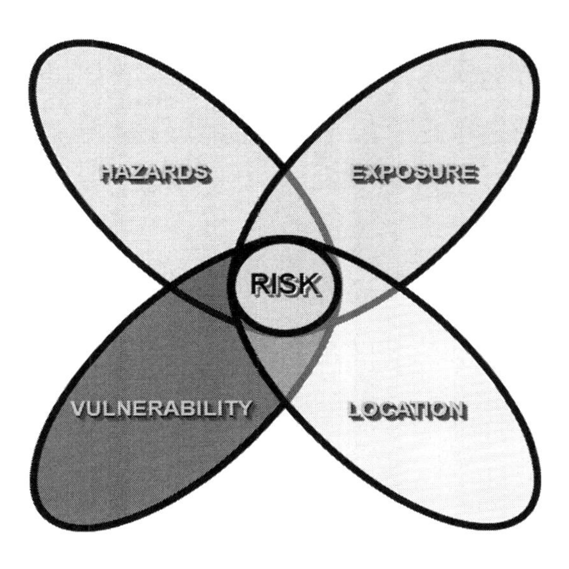

It is important to understand the changes in geohazards with time. In particular, geotechnical engineers and engineering geologists will face long-term challenges due to climate change. Research is required to learn about the effects of climate change in greater detail so that methods of analysis and interpretation can be improved and extended. Exploration of such issues will be facilitated by a proper understanding of the basic concepts of geotechnical slope analysis and the fundamental principles on which the available methods of analysis are based.

Appendix A: Selected Figures from PowerPoint Slide Set Entitled "Understanding Risk and Risk Reduction" [14]

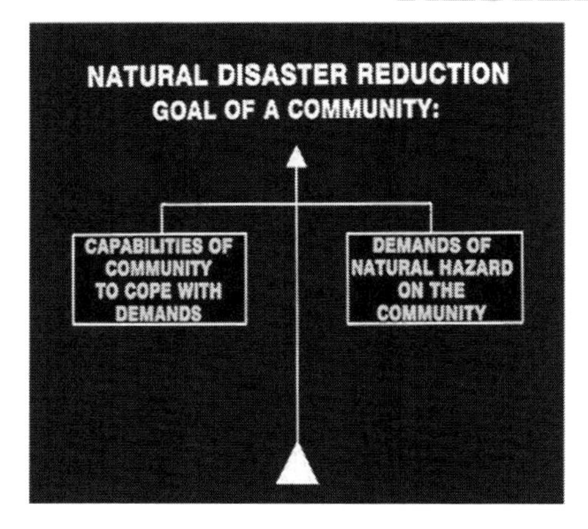

Fig. A.3 Common agenda for natural disaster resilience (Courtesy of Walter Hays [14])

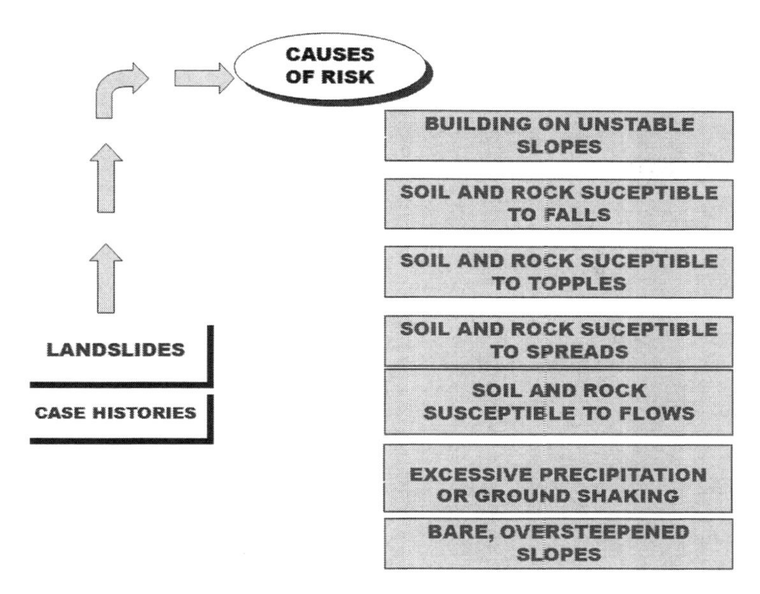

Fig. A.5 Some causes of risk for landslides (Courtesy of Walter Hays [14])

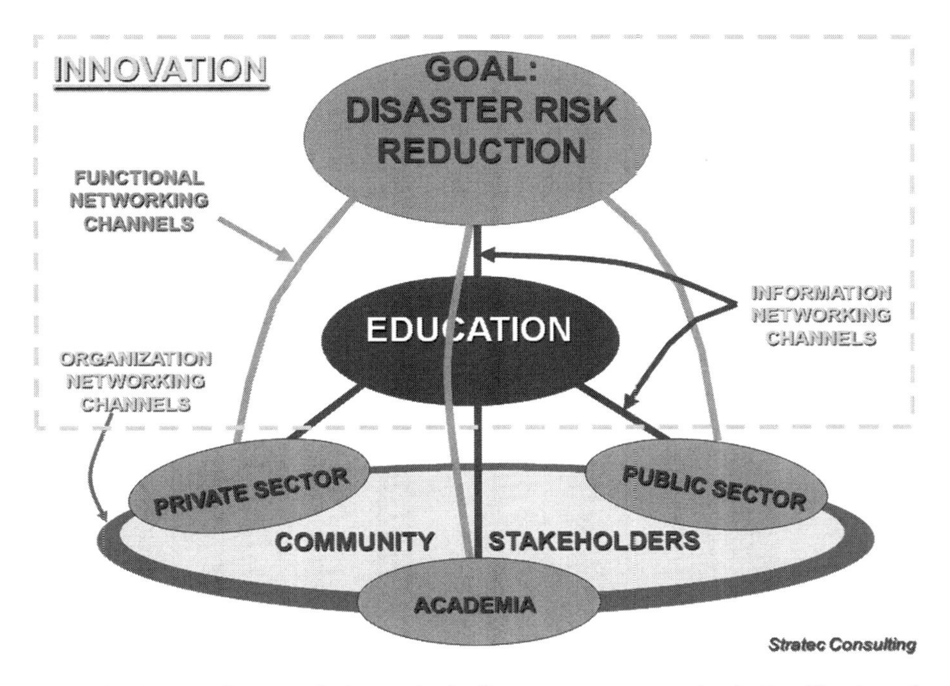

Fig. A.4 The overall context for innovation in disaster management and reduction (Courtesy of Walter Hays [14])

References

1. Bhattacharya G, Chowdhury R (2011) Continuing research concerning the residual factor as a random variable. Progress report, September 2011
2. Bowles D, Rutherford M, Anderson L (2011) Risk assessment of success Dam, California: evaluating of operating restrictions as an interim measure to mitigate earthquake risk. In: Juang CH, Phoon KK, Puppala AJ, Green RA, Fenton GA (eds) Proceedings of GeoRisk 2011. Geotechnical risk assessment and management, Geo-Institute, A.S.C.E
3. Brumund F (2011) Geo – risks in the business environment. In: Juang CH, Phoon KK, Puppala AJ, Green RA, Fenton GA (eds) Proceedings of GeoRisk 2011. Geotechnical risk assessment and management, Geo-Institute, A.S.C.E
4. Chowdhury R, Bhattacharya G (2011) Reliability analysis of strain-softening slopes. In: Proceedings of the 13th international conference of IACMAG, Melbourne, Australia, May, 2011, vol II, pp. 1169–1174
5. Chowdhury R, Flentje P (2008) Strategic approaches for the management of risk in geomechanics. Theme paper. In: Proceedings of the 12 IACMAG conference, Goa, India, CD-ROM, pp 3031–3042
6. Chowdhury R, Flentje P (2010) Geotechnical analysis of slopes and landslides: achievements and challenges (Paper number 10054). In: Proceedings of the 11th IAEG congress of the international association of engineering geology and the environment, Auckland, New Zealand, 6 pp
7. Chowdhury R, Flentje P (2011) Practical reliability approach to urban slope stability. Proceedings of the ICASP11, the 11th international conference on application of statistics and probability in civil engineering, 1–4 August. ETH, Zurich, 5 pp
8. Chowdhury R, Flentje P, Bhattacharya G (2010) Geotechnical slope analysis, CRC Press, Balkema, Taylor and Francis Group, Boca Raton, 746 pp
9. Christian JT, Baecher GB (2011) Unresolved problems in geotechnical risk and reliability. In: Juang CH, Phoon KK, Puppala AJ, Green RA, Fenton GA (eds) Proceedings of GeoRisk 2011. Geotechnical risk assessment and management, Geo-Institute, A.S.C.E
10. Flentje P (2009) Landslide inventory development and landslide susceptibility zoning in the Wollongong City Council Local Government Area. Unpublished report to Industry Partners-Wollongong City Council, RailCorp and the Roads and Traffic Authority, University of Wollongong, Australia, 73 pp
11. Flentje P, Chowdhury R, Miner AS, Mazengarb C (2010) Periodic and continuous monitoring to assess landslide frequency-selected Australian examples. In: Proceedings of the 11th IAEG congress of the international association of engineering geology and the environment, Auckland, New Zealand, 6 pp.
12. Flentje P, Stirling D, Chowdhury R (2007) Landslide susceptibility and hazard derived from a Landslide Inventory using data mining – an Australian case study. In: Proceedings of the first North American landslide conference, landslides and society: integrated science, engineering, management, and mitigation, Vail, Colorado, 3–8 June 2007, 10 pp. CD, Paper number 17823–024,
13. Gibson AD, Chowdhury R (2009) Planning and geohazards. In: Culshaw MG, Reeves HJ, Jefferson I, Spink TW (eds) Engineering geology for tomorrow's cities, vol 22. Engineering Geology Special Publication, London, pp 113–123
14. Hays W (2011) Understanding risk and risk reduction- a set of power point slides. Global Alliance for Disaster Reduction (GADR)
15. Jibson RW, Harp EL, Michael JA (2000) A method for producing digital probabilistic seismic landslide hazard maps. Eng Geol 58(3–4):271–289
16. Lacasse S, Nadim F (2011) Learning to live with geohazards: from research to practice. In: Juang CH, Phoon KK, Puppala AJ, Green RA, Fenton GA (eds) Proceedings of GeoRisk 2011. Geotechnical risk assessment and management, Geo-Institute, A.S.C.E

17. Marker BR (2009) Geology of mega-cities and urban areas. In: Culshaw MG, Reeves HJ, Jefferson I, Spink TW (eds) Engineering geology for tomorrow's cities, vol 22, Engineering Geology Special Publication. Geological Society, London, pp 33–48
18. Murray E (2001) Rainfall thresholds for landslide initiation in the Wollongong Region. Internal report to Australian Geological Survey Organisation and SPIRT Project Team at the University of Wollongong
19. Nathanail J, Banks V (2009) Climate change: implications for engineering geology practice 2009. In: Culshaw MG, Reeves HJ, Jefferson I, Spink TW (eds) Engineering geology for tomorrow's cities, vol 22, Engineering Geology Special Publication. Geological Society, London, pp 65–82
20. Quilan R (1993) C 4.5: programs for machine learning. Morgan, San Mateo
21. Rees JG, Gibson AD, Harrison M, Hughes A, Walsby JC (2009) Regional modeling of geohazards change. In: Culshaw MG, Reeves HJ, Jefferson I, Spink TW (eds) Engineering geology for tomorrow's cities, vol 22, Engineering Geology Special Publication. Geological Society, London, pp 49–64
22. Reeves HJ, West TR (2009) Geodata for the urban environment. In: Culshaw MG, Reeves HJ, Jefferson I, Spink TW (eds) Engineering geology for tomorrow's cities, vol 22, Engineering Geology Special Publication. Geological Society, London, pp 209–213
23. Scott GA (2011) The practical application of risk assessment to Dam safety. In: Juang CH, Phoon KK, Puppala AJ, Green RA, Fenton GA (eds) Proceedings of GeoRisk 2011. Geotechnical risk assessment and management, Geo-Institute, A.S.C.E
24. Tang WH, Zhang LM (2011) Development of a risk-based landslide warning system. In: Juang CH, Phoon KK, Puppala AJ, Green RA, Fenton GA (eds) Proceedings of GeoRisk 2011. Geotechnical risk assessment and management, Geo-Institute, A.S.C.E

Uncertainties in Transportation Infrastructure Development and Management

Kumares C. Sinha, Samuel Labi, and Qiang Bai

Abstract The development and management of transportation infrastructure is a continuous process that includes the phases of planning and design, construction, operations, maintenance, preservation, and reconstruction. Uncertainties at each phase include variability in demand estimation, reliability of planning and design parameters, construction cost overruns and time delay, unexpected outcomes of operational policies and maintenance and preservation strategies, and risks of unintended disruption due to incidents or sudden extreme events. These variabilities, which are due to inexact levels of natural and anthropogenic factors in the system environment, are manifest ultimately in the form of variable outcomes of specific performance measures established for that phase. Transportation infrastructure managers seek to adequately identify and describe these uncertainties through a quantitative assessment of the likelihood and consequence of each of possible level of the performance outcome and to incorporate these uncertainties into the decision-making process. This chapter identifies major sources of uncertainties at different phases of transportation infrastructure development and management and examines the methods of their measurements. Finally, this chapter presents several approaches to incorporate uncertainties in transportation infrastructure decision-making and provides future directions for research.

Keywords Transportation infrastructure • Uncertainty • Extreme events • Decision-making

---

K.C. Sinha (✉) • S. Labi • Q. Bai
School of Civil Engineering, Purdue University, 550 Stadium Mall Drive, West Lafayette, IN 47907, USA
e-mail: ksinha@purdue.edu; labi@purdue.edu; qbai@purdue.edu

S. Chakraborty and G. Bhattacharya (eds.), *Proceedings of the International Symposium on Engineering under Uncertainty: Safety Assessment and Management (ISEUSAM - 2012)*,
DOI 10.1007/978-81-322-0757-3_3, © Springer India 2013

1 Introduction

> The future is uncertain...but this uncertainty is at the very heart of human creativity.
> Ilya Prigogine (1917–2003)

The *Oxford English Dictionary* [25] defines uncertainty as "the quality of being uncertain in respect of duration, continuance, occurrence, etc.; the state of not being definitely known or perfectly clear; or the amount of variation in a numerical result that is consistent with observation." As it is in everyday life, the uncertainty is an inevitable aspect of transportation infrastructure development and management and is receiving much attention because of several recent catastrophic events. For instance, the collapses of the Autoroute 19 bridge in Laval of Quebec, Canada, in 2006 and the Minnesota I-35W Mississippi River bridge in 2007 caused many deaths and injuries and created severe travel disruptions. Also, in the earthquake/tsunami of the Pacific coast of Tohoku in Japan in 2011, the sole bridge connecting to Miyatojima was destroyed, which isolated the island's residents; many sections of Tōhoku Expressway were damaged; and the Sendai Airport was flooded and partially damaged. In fact, almost all similar extreme events, such as earthquakes and flooding, always cause damages to transportation infrastructures. In addition to catastrophic collapses and damages due to extreme events, there are other inherent uncertainties in transportation infrastructure development, such as variability in demand estimation, reliability in planning, and construction cost and time estimation. These uncertainties create not only tremendous economic and property losses, but also cause loss of human lives, and pose a serious public health and safety problem. Inability to identify the sources of uncertainty and inadequate assessment of its degree of occurrence introduces significant unreliability in infrastructure decisions. The process of infrastructure development and management could be greatly enhanced if potential uncertainties could be identified and explicitly incorporated in decision-making, in order to minimize potential risks.

2 Transportation Infrastructure Development and Management Process

2.1 *Phases of Infrastructure Development and Management*

Transportation infrastructure development and management is a multiphase process, which can be divided into four key phases: (1) planning and design, (2) construction, (3) operations, maintenance, and preservation, and (4) reconstruction caused by obsolescence/disruption. At the planning and design phase, the infrastructure need is assessed through demand estimation and the anticipated cost and performance impacts associated with alternative locations and designs are evaluated. The construction phase carries the infrastructure to its physical realization. The longest phase – operations, maintenance, and preservation – simply involves the use of the system and is carried out continually. Reconstruction is done in response to the physical failure of the system either due to deterioration or due to natural or man-made disasters. At each phase, uncertainties arise from variabilities in natural factors in system environment and anthropogenic factors such as inputs of the infrastructure decision-makers and are manifest ultimately in the form of variable levels of relevant performance measures established for that phase.

2.2 *Sources of Uncertainties at Each Phase*

2.2.1 The Planning and Design Phase

Transportation infrastructure planning and design is an inherently complex process that is inextricably tied to social, economic, environmental, and political concerns, each of which is associated with significant uncertainty. At the planning stage, possible sources of uncertainty are associated mainly with travel demand and land-use and environmental impacts.

Transportation Demand Uncertainty

Transportation demand is a basic input for establishing the capacity or sizing of an infrastructure. For instance, future traffic volume is an important factor for determining highway geometrics, design speed, and capacity; passenger volume is a prerequisite to design airport terminals; and expected ridership is the key variable for the design of the routing stock and other features of urban transit. Also, transportation demand is the key determinant of the financial viability of projects [26, 27]. In fact, the level of anticipated travel demand serves as the basis for impact analysis of transportation infrastructure investments from social, environmental, and economic perspectives. In practice, however, demand estimation is plagued with a very significant degree of uncertainty in spite of multiple, sustained efforts to enhance demand estimation ([17]). Flyvbjerg et al. [13] conducted a survey on the differences between forecast and actual travel demand of the first year of the operation by examining 210 transportation projects (27 rail projects and 183 road projects) around the world. It was found that in 72% of rail projects, passenger forecasts were overestimated by more than 67%; 50% of road projects had a difference between actual and forecasted traffic of more than ±20% and 25% of projects had the difference more than ±40%. Major sources of these variabilities are shown in Fig. 1.

It is seen that the trip distribution part of travel demand modeling was one of the main sources for the discrepancy in both rail and highway projects, while land-use development had more effect on highway projects than on rail projects. It is also seen that the "deliberately slanted forecast" was a major cause of rail travel demand inaccuracy, which indicates that political influences can introduce significant uncertainty in some project types.

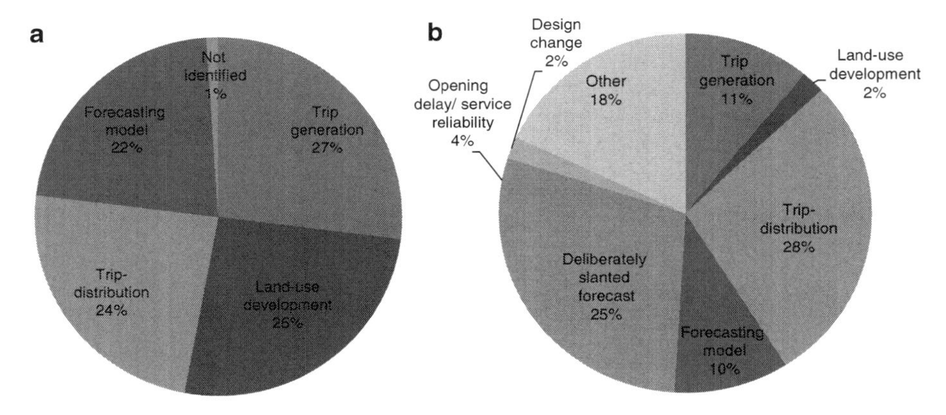

Fig. 1 Causes of uncertainties in traffic forecasts in transportation projects. (**a**) Highway projects and (**b**) Rail projects (Derived from Flyvbjerg et al. [13])

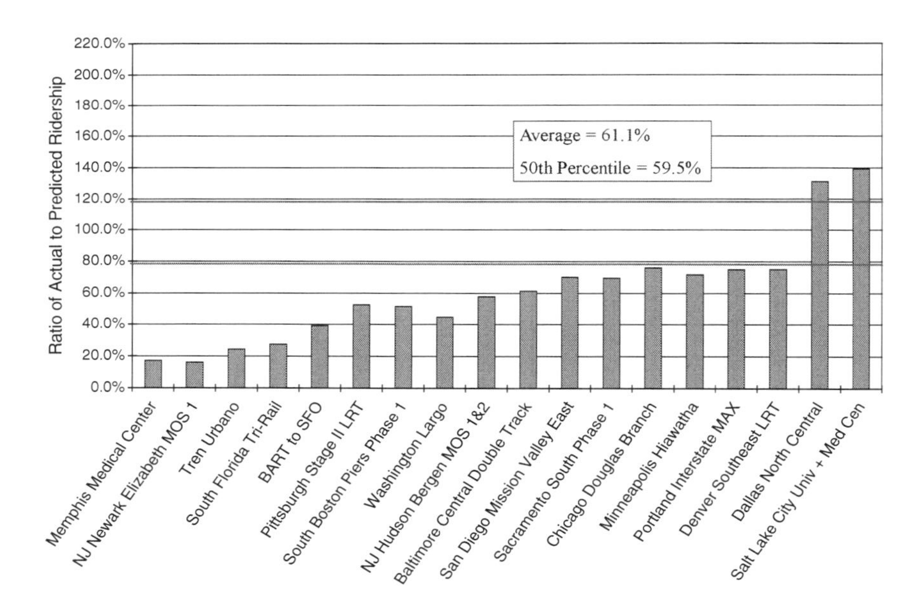

Fig. 2 Transit actual ridership versus predicted ridership (Source: Workman [35])

Figure 2 presents the ratios of the actual ridership to the predicted ridership for 17 urban rail projects in USA. In most cases, actual values were much lower than predicted values. Consequences of travel demand uncertainty can be significant. Overestimation of demand would lead to excess supply and its underutilization, resulting lower revenue and financial problem to the operating agencies.

Underestimation of demand, on the other hand, would cause congestion and excessive user delay and might require expensive upgrading. For instance, the Shenyang-Dalian Freeway, China's first freeway, was completed in 1990 with four lanes, two in each direction. According to the initial planning, the travel demand was not expected to reach the design traffic volume before 2010, the design year of the freeway. However, due to the high traffic volume, the freeway had to be upgraded to eight lanes in 2002, 8 years before the design year. There are many similar cases in China, including the Guangzhou-Qinagyuan Freeway (constructed with four lanes in 1999, expanded to eight lanes in 2009), Fuzhou-Quanzhou Freeway (constructed with four lanes in 1999, expanded to eight lanes in 2010), and Xi'an-Baoji Freeway (constructed with four lanes in 1995, expanded to eight lanes in 2011). If the forecasted traffic volume has less uncertainty, the construction and the upgrading could be done in a more cost-efficient and timely way. For public-private partnership projects, the consequences of demand uncertainty can be severe; if the actual travel demand is far less than as predicted, the investor fails to breakeven, and the investment may run into a loss.

Environmental Impact Uncertainty

Uncertainty associated with environmental impacts of transportation projects is another source of great concern to transportation decision-makers and the general public. To comply with the National Environmental Policy Act (NEPA) of 1969, transportation agencies in the USA evaluate the potential impacts of projects on social and natural environments, and projects are planned and designed accordingly including the provision of appropriate mitigation measures. However, because of the largely unpredictable nature of the climate, weather, traffic characteristics, and emission volumes, the estimation of expected environmental impacts with high degree of uncertainty becomes a challenge.

2.2.2 The Construction Phase

At the construction phase, the uncertainty is mainly associated with construction duration and cost. Table 1 presents results from a study of highway projects in several states in the USA [4], indicating that high percentages of projects had cost and time overruns.

The *Engineering News Record* [11] also reported that for most large transportation infrastructure projects, the final costs significantly exceeded the estimated costs. For example, the San Francisco Bay Bridge replacement project had a 30% cost overrun [10], the Tokyo Oedo Subway in Japan 105% [35], and the Springfield Interchange project in Northern Virginia 180% [11].

Based on data from 708 Florida highway projects during 1999–2001, Vidalis and Najafi [33] found that the construction time overruns could be attributed to unexpected site conditions such as poor geotechnical conditions, utility relocations, and

Table 1 Cost and time overruns at selected states (Source: Bordat et al. [4])

State	Period	Percentage of projects with cost overruns (%)	Percentage of projects with time overruns (%)
Idaho	1997–2001	55–67	–
Indiana	1996–2002	55	12
Missouri	1999–2002	60–64	–
New Mexico	2002	62	10
Ohio	1994–2001	80–92	44–56
Oregon	1998–2002	18–33	15–65
Tennessee	1998–2002	61	14
Texas	1998–2002	66–75	52–55

other environmental problems; deviations between design drawings and actual site conditions; and inclement weather, while cost overruns were due to errors and omissions in plans and modifications and subsequent change orders. Majid and McCaffer [21] found that the top five causes of construction time overrun were (1) late delivery or slow mobilization, (2) damaged materials, (3) poor planning, (4) equipment breakdown, and (5) improper equipment. In Indiana, Bhargava et al. [2] observed that the factors that caused cost overrun mostly included changes in project scope and site conditions. Consequences of construction cost and time overruns can lead to delayed use of the facility, defer the economic development, and cause financial problems for transportation agencies. In most cases, cost and time overruns worsen each other interactively.

2.2.3 The Operations, Maintenance, and Preservation Phase

This phase involves the estimation of annual maintenance and operating costs and timing and cost of periodic preservation. The factors that affect these items are the prevailing infrastructure physical conditions/performances and traffic characteristics. Operational strategies, including traffic management decisions, are typically based on average traffic and the typical pattern of traffic changes during a certain period. However, both the traffic and the change pattern are not deterministic and often exhibit marked variations. These variations may render the operational strategies ineffective. Also, the uncertainty associated with user behavior may cause uncertainty in infrastructure operations, safety, and security. For tolled facilities, one of the serious concerns has been the uncertainty in demand which leads to uncertainty in toll revenue.

With regard to infrastructure maintenance and preservation, a key context of decision-making is to identify specific maintenance and rehabilitation (M&R) treatments at a given time or specific M&R activity profile or schedule that optimizes the cost efficiency within performance constraints. In these contexts, the optimal solution is heavily influenced by prevailing infrastructure attributes such as physical condition, rate of deterioration, M&R cost and effectiveness, amount of traffic, climatic severity, and prices of raw materials, labor, and equipment use. Variabilities in condition inspection outcomes, the stochastic nature of the infrastructure deterioration, the uncertainty in M&R cost and effectiveness, and variation of traffic levels and distribution all introduce a great deal of uncertainty in the treatment/schedule selection processes. Uncertainties associated with infrastructure deterioration rates and processes have been very extensively studied. In maintenance management, infrastructure performance models are usually applied to simulate the infrastructure deterioration process and to predict the future performance for optimal maintenance decision-making. Often Markov chain process is used for this purpose as it can incorporate the stochastic property [16, 23, 20, 37, 15].

2.2.4 Obsolescence/Disruption and Reconstruction Phase

There can be four types of processes which make an infrastructure facility obsolete: (1) natural deterioration, (2) the changed demand, (3) catastrophic physical failure, and (4) extreme events such as natural and man-made disasters. A transportation facility usually deteriorates gradually over time. When the physical condition reaches a certain level that does not allow cost-effective maintenance and preservation, it requires reconstruction. The uncertainty associated with this type of obsolescence arises from the lack of precise knowledge of the deterioration process of the facility. The causes of the deterioration may include weather, traffic load, and the infrastructure design itself. Often a facility requires reconstruction because the change in demand characteristics, such as vehicle size and weight, makes the geometrics and structural aspects of the facility obsolete. Even though most infrastructure facilities are reconstructed before their failure, there are some cases where infrastructures experience sudden catastrophic failures. This type of disruption is of high uncertainty and is very hard to predict. While better inspection and monitoring of facility conditions might minimize such failures, there will still be a certain degree of uncertainty because of the lack of precise knowledge of the deterioration process.

Extreme events constitute another important source of uncertainty. Extreme events can be defined as occurrences that, relative to some class of related occurrences, are either notable, rare, unique, profound, or otherwise significant in terms of its impacts, effects, or outcomes [29]. Common extreme events that can cause catastrophic disruptions can be categorized into natural disasters, such as the earthquake, tsunami, flooding, landslide, hurricane; and man-made events, such as terrorism attacks and collisions. Figure 3 presents examples of extreme events. The degree of uncertainty associated with infrastructure disruption/destruction due to extreme events depends on both the uncertainty of extreme events themselves as well as the uncertainty of infrastructure resilience/vulnerability. While it is difficult to prevent the occurrence of extreme events, the resilience of the transportation infrastructure can be strengthened to decrease the level of damages and network disruptions under such events.

Fig. 3 Example consequences of extreme events. (**a**) Chehalis River flooding (Source: Blogspot.com [3]), (**b**) Japanese earthquake (Source: Buzzfeed.com [6]), (**c**) California highway landslide (Source: Cbslocal.com [7]), and (**d**) Jintang Bridge collision in China (Source: Xinhuanet.com [36])

3 Measurement of Uncertainty

In order to incorporate the effect of uncertainty in transportation decision-making, it is important to establish a yardstick that could serve as a basis for quantifying the level of uncertainty from each source. As indicated in the definition of uncertainty, the possible outcome of the consequence is unknown in an uncertainty situation. In practice, uncertainty can be further categorized into two cases: (1) risk case where the probability distribution of the outcome is known, and (2) total uncertainty case where the distribution or even the range of the outcome is not known [19]. In past research, several objective approaches, such as the use of expected value, probability distribution, likelihood value, and confidence interval, were developed to quantify uncertainties in the risk situation. In a total uncertainty situation, approaches based on uncertainty or vulnerability ratings are often used. The choice of a specific approach or technique depends on the availability of data and the context of decision-making, which, in turn, is influenced by the phase of transportation infrastructure development in question. There is no universal method to quantify all types of uncertainties. In this section, a number of common methods are presented.

3.1 *Probability Distributions*

For some parameters associated with infrastructure development and management, such as the pavement condition rating, bridge remaining service life, and construction cost, probability distributions, rather than fixed values (as it is implicitly assumed in deterministic decision-making), are widely used to describe or measure the degree of uncertainty. From the probability distribution of the parameter of interest, a number of statistical measures can help quantify the degree of uncertainty:

(a) The statistical range of values (minimum and maximum values of the parameter).
(b) The standard deviation or variance.
(c) The coefficient of variation (ratio of the mean to the standard deviation).
(d) A visual examination of the shape of the distribution. This can provide clues regarding the degree of uncertainty of the parameter of interest. For example, there is greater certainty when the distribution is compact compared to a diffused distribution.
(e) Confidence interval. In practice, the decision-maker may be relatively unconcerned about the value of even the exact distribution of the investment outcome and may be more concerned about what the range of the outcome will be under a certain confidence level or conversely, at what confidence level the outcome can be expected to fall within a certain range.
(f) The probability that the outcome is more/less than a certain specified value. The greater the probability, the lower the uncertainty associated with that parameter; the smaller the probability, the greater the uncertainty associated with that parameter. Thus, such probabilities can serve as a measure of uncertainty and they can be determined from the cumulative probability function of the outcome.

In practice, the probability distribution of a parameter can be obtained through several ways. If available, the probability distribution can be developed from the historical data. In some cases, when historical data is not available, expert opinions can be used to assign a distribution to the parameter of interest. Also, the distribution of the consequence can be generated using such techniques as Monte Carlo simulation.

The probability distribution approach has been widely used to measure uncertainty of some parameters in transportation infrastructure development and management. For instance, Li and Sinha [19] applied binomial distribution to quantify the uncertainty of bridge condition ratings and beta distributions to measure the uncertainty of construction expenditure, delay time, and crash rate, in the development of a comprehensive highway infrastructure management system. Ford et al. [14] applied probability distribution to measure the uncertainty in bridge replacement needs assessment.

3.2 *Uncertainty Rating*

In some situations, there may exist certain parameters which are qualitative in nature, or there are inappropriate, inadequate, or unreliable data for probability distributions to be developed. For instance, it is hard to evaluate the probability that an in-service bridge will be destroyed in an earthquake since we cannot do real experiment to obtain the data. In such a case, ratings based on the structure and current condition of the bridge may be useful. In some other situations, particularly where human perspectives are involved, objective attempts to quantify uncertainties on a numerical scale may not be possible. In these cases, expert opinions could be used to develop a representative description of the degree of uncertainty. An example of this approach is Shackle's surprise function [30] to evaluate the subjective degree of uncertainty as perceived by decision-makers. NYSDOT (24) applied infrastructure vulnerability rating to capture the uncertainty associated with infrastructure resilience to sudden disruption through disaster, threat or likelihood of such disaster events, exposure to disaster (or consequence), or any two or all three of these attributes.

4 Incorporating Uncertainty into Decision-Making

With the realization that the parameters involved in transportation infrastructure development process are highly variable and subject to significant uncertainty, the importance of incorporating such uncertainty in decision-making cannot be overemphasized. A number of researchers have attempted, to varying degrees of success, to develop procedures that duly account for uncertainty, as discussed below.

4.1 *Probability Models*

Of the probability models, Markov chain models probably are the most widely used, particularly for modeling the time-related performance of infrastructure. A Markov chain, which describes the transition from one state to another in a chain-like manner stochastically, is a random and memoryless process, in which the next state depends only on the current state and not on the entire past. A transportation facility at a certain current condition can transform subsequently, due to deterioration, to any one of several possible condition states each with a specific probability. Figure 4 presents a simple example of transition probability matrix for superstructure rating of steel bridges in Indiana [32]. From the figure, it is seen that the probability of a new bridge with superstructure rating of 9 remaining in the same condition in the next period is 0.976 while the probability of transferring to the next lower condition state of 8 is 0.024.

Condition States

Condition States	9	8	7	6	5	4	3	2
9	0.976	0.024						
8		0.936	0.064					
7			0.89	0.11				
6				0.885	0.115			
5					0.92	0.08		
4						0.93	0.07	
3							0.92	0.08
2								1

Fig. 4 Example of Markov transition probability matrix for bridge superstructure rating (Source: Sinha et al. [32])

Weaknesses of this method are that the subsequent condition states actually are influenced by past states and that there is an assumption of time homogeneity. Nevertheless, Markov chain models have been extensively used, with appropriate modifications, in transportation infrastructure performance modeling for pavements [5], bridges [16], harbors and coastal structures [38], and railroads [1, 28].

4.2 Monte Carlo Simulation

Figure 5 presents the process of Monte Carlo simulation. Distributions of the parameters are the basic inputs; the output is the distribution of the outcome for evaluation or decision-making. A randomly generated number from each parameter's distribution is used to yield the final evaluation outcome in each iteration. Thousands of iterations are typically conducted and thus thousands of potential final outcomes are generated and their distribution is determined. Monte Carlo simulation has been widely used in transportation area, including life-cycle cost analysis in pavement design [34], the variability of construction cost escalation pattern [2], and others [9, 12, 22].

4.3 Stochastic Dominance

Using the distribution of the evaluation outcome, the decision-maker can calculate the mean value of the outcome and then make a decision based on the mean value. For example, if the outcome mean value of alternative A is superior to that of alternative B, then alternative A is preferred. However, if the two outcomes are described by a probability distribution, then the mean alone cannot guarantee that A is always superior to B. For instance, in the case 1 of Fig. 6, distribution A has a higher mean value (30) than that of B (25). However, from the cumulative probability function, when the value of the input parameter is less than 22, the probability

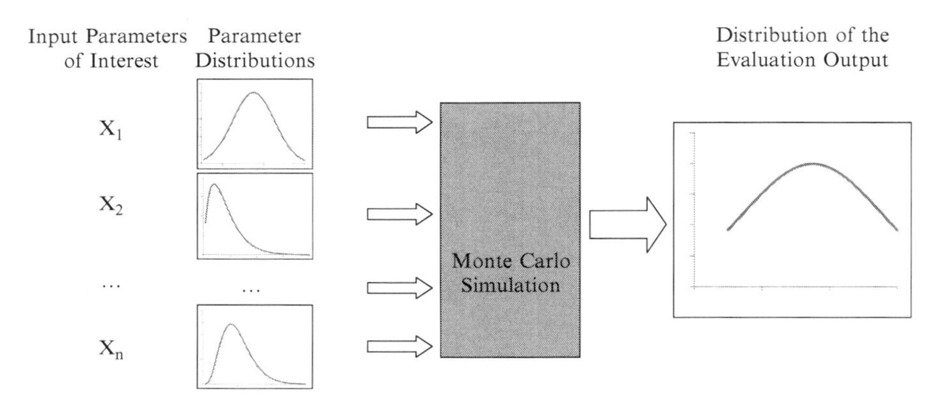

Fig. 5 Illustration of Monte Carlo simulation process

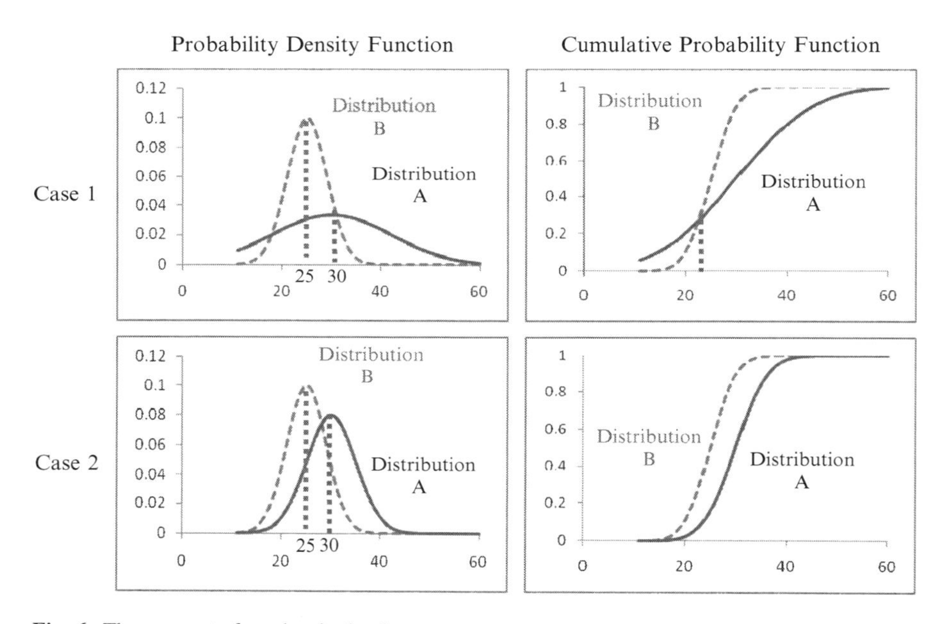

Fig. 6 The concept of stochastic dominance

that *A*'s outcome is less than a certain value is greater than the probability that *B*'s outcome is less than that value; this means that when the outcome is less than 22, alternative *B* is superior to alternative *A* under conditions of uncertainty. As this example demonstrates, using the expected value of probability distributions corresponding to each alternative is not reliable. Stochastic dominance is a method to compare two distributions [8]. In the case 2 of Fig. 6, *A* and *B* have different distributions, not all the possible values of *A* are superior than the value of *B*, but from their cumulative functions, it is seen that for any given outcome level, the probability that A is smaller than the given level is equal or less than the probability that B is smaller than the given level. Obviously, A is superior to B. In other words, A stochastically dominates B. Thus, it is seen that if the expected value of A is greater than B, that situation does not guarantee that A stochastically dominates B, but if A stochastically dominates B, then the expected value of A is greater than B for sure.

The concept of stochastic dominance avoids the possible bias of just using the expected value in comparing two alternatives. It can yield a more robust evaluation result. Cope et al. [9] applied stochastic dominance to evaluate the effectiveness of using stainless steel as the bridge deck reinforcement. It can also be used in multicriteria decision-making [39].

4.4 Expected Utility Theory

In order to incorporate both parameter uncertainty and decision-maker preferences, Keeney and Raiffa [18] developed the expected utility theory, where a utility function $u(x)$ is first developed as a function of the outcome (x) in its original unit, to represent the degree of preference on different values of the outcome. Then the probability density function of x, i.e., $f(x)$, is used to calculate the expected utility, i.e., $\mathrm{EU} = \int_{x_{\min}}^{x_{\max}} u(x) * f(x)\mathrm{d}x$. The alternative with the highest expected utility becomes the preferred choice. Expected utility theory has been used in numerous studies in transportation infrastructure development. For example, Li and Sinha [19] applied expected utility theory to deal with the uncertainty in transportation asset management for Indiana Department of Transportation.

4.5 Shackle's Model

As seen in previous sections, if the distribution of a parameter is known, it is relatively easy to use the expected value, expected utility value, or stochastic dominance concepts to make a decision. But in practice, it may not be possible to get the distribution. In this case, Shackle's model [30] can be applied to incorporate uncertainties in the decision-making process [19]. There are three main steps in this approach.

1. Establish the degree of surprise function. Degree of surprise is used to measure the decision-maker's degree of uncertainty with gains (positive returns) and losses (negative returns) from the expectation. Usually, the values of degree of surprise range from 0 (no surprise) to 10 (extremely surprised).
2. Develop priority function and focus values. Priority function is developed to evaluate the weighting index of each pair outcome and its degree of surprise.

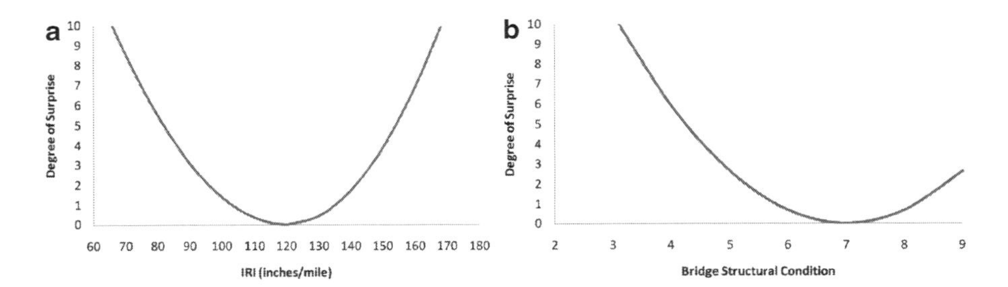

Fig. 7 Examples of surprise function. (**a**) IRI (expectation: 120 in./mile). (**b**) Bridge Structural condition (expectation: 7)

The priority function is usually from 0 (lowest priority) to 10 (highest priority). Based on the priority function, the focus gain (*G*), the gain with the maximum priority value, and the focus loss (*L*), the loss with the maximum priority value can be found.

3. Calculate standardized focus gain-over-loss ratio. The degrees of surprise of the focus gain and focus loss are usually nonzero. Then, there is a need to find out the standardized gain and loss values, which are the values of the outcomes that on the same priority indifference curves that have zero degree of surprise. Next, the standardized focus gain-over-loss ratio can be calculated. Usually, the project with the higher standardized focus gain-over-loss ratio is more desirable to decision-makers.

Surprise and priority functions are developed from surveys of potential decision-makers. As an example, Fig. 7 presents the surprise functions for the pavement International Roughness Index (IRI) and the structural condition rating for bridges developed by Li and Sinha [19] for their study to incorporate uncertainty in highway asset management. It is seen that the further the parameter is from its expectation, the larger is the level of surprise the decision-maker experiences.

In addition to the methods discussed above, there are a host of other methods that can be used to incorporate uncertainty into decision-making process, such as sensitivity analysis, fuzzy set theory, mean-variance utility theory, and Bayes' method. Sensitivity analysis is the most widely used approach; it examines the degree of changes in the outcome if there are some changes in the input/dependent variables. Also, fuzzy set theory has been extensively used to deal with uncertainty in transportation infrastructure management. For example, Shoukry et al. [31] developed a universal measure capable of formally assessing the condition of a pavement section based on fuzzy set theory.

The choice of an approach to deal with uncertainty in transportation infrastructure development and management will depend on the specific context of the decision being made. As mentioned earlier, there is no single method that can be applied universally to all problem types and contexts. Also, not all types of uncertainties can be successfully incorporated into the decision-making process.

5 Summary and Conclusions

The development and management of transportation infrastructure is characterized by significant uncertainty at each of its phase. To make optimal and robust decisions at each phase, it is critical not only to be aware of the possible sources of uncertainty, but also to quantify these uncertainties and more importantly, to incorporate them in decision-making processes. While much information is already in existence, further work is necessary to apply the information in infrastructure development and management process so that improved decisions can be made. Some possible research directions include:

1. Extreme events related research. There is an urgent need to address the uncertainty associated with infrastructure disruptions due to sudden catastrophic failures and natural or man-made disasters in order to provide appropriate resilience and sustainability in future infrastructure facilities.
2. Vulnerability assessment related to climate changes, network resilience, and sustainability evaluation and vulnerability management, including how to minimize the effect of network disruption.
3. The integration of optimization and Monte Carlo simulation to incorporate uncertainty in infrastructure investment decision-making. Optimization and Monte Carlo simulation are both computationally time-consuming processes and novel heuristic algorithms are necessary to solve the integration problem.
4. Future research is needed on the development of methods for quantifying uncertainty associated with human input. Decision-makers' preference structure and their perception of uncertainty are important in an effective decision-making process.
5. Trade-off between uncertainty/risk and benefit/return. In most situations, high benefit/return is associated with high risk. There is a need to develop a robust trade-off methodology to reach a balance between uncertainty/risk and benefit/return.

References

1. Bell MGH, Schmoecker JD, Iida Y, Lam WHK (2002) Transit network reliability: an application of absorbing Markov chains. Transportation and traffic theory in the 21st century. In: Proceedings of the 15th international symposium on transportation and traffic theory. University of South Australia, Adelaide, Australia. Accessed 10 Oct 2011
2. Bhargava A, Labi S, Sinha KC (2010) Development of a framework for Ex Post Facto evaluation of highway project costs in Indiana. Publication# FHWA/IN/JTRP-2009/33. Joint Transportation Research Program, Indiana Department of Transportation and Purdue University, West Lafayette, IN
3. Blogspot.com (2007) Light rail will save us. http://victoriataftkpam.blogspot.com/2007/12/light-rail-will-save-us.html. Accessed October 10 2011

4. Bordat C, McCullouch BG, Sinha KC, Labi S (2004) An Analysis of Cost Overruns and Time Delays of INDOT Projects. Joint Transportation Research Program, Indiana Department of Transportation and Purdue University, West Lafayette, Indiana
5. Butt AA, Shahin MY, Feighan KJ, Carpenter SH (1987) Pavement performance prediction model using the Markov process. Transportation research record no. 1123, pp 12–19
6. Buzzfeed.com (2011) Road split in two by Japanese earthquake. http://www.buzzfeed.com/burnred/road-split-in-two-by-japanese-earthquake-281t. Accessed 10 Oct 2011
7. Cbslocal.com (2011) Third mudslide closes Highway 1 on Big Sur coast. http://sanfrancisco.cbslocal.com/2011/03/28/third-mudslide-closes-highway-1-on-big-sur-coast/. Accessed 10 Oct 2011
8. Clemen RT (1996) Making hard decisions. Duxbury Press, Pacific Grove
9. Cope A, Bai Q, Samdariya A, Labi S (2011) Assessing the efficacy of stainless steel for bridge deck reinforcement under uncertainty using Monte Carlo simulation. Struct Infrastruct Eng. doi:10.1080/15732479.2011.602418
10. ENR (2001) Bay bridge replacement comes in above estimate. ENR, p. 5, Dec. 31, 2001
11. ENR (2002) Virginia's big 'mixing bowl' is 180% over budget. ENR, p. 7, Dec. 2, 2002
12. Ersahin T, McCabe B, Doyle M (2003) Monte Carlo simulation analysis at Lester B Pearson International Airport development project. Construction Research Congress. Winds of change: integration and innovation in construction. Proceedings of the Congress, Honolulu, Hawaii, United States
13. Flyvbjerg B, Holm MKS, Buhl SL (2006) Inaccuracy in traffic forecasts. Transp Rev 26(1):1–24
14. Ford K, Arman M, Labi S, Sinha KC, Shirole A, Thompson P, Li Z (2011) Methodology for estimating life expectancies of highway assets (draft). School of Civil Engineering, Purdue University, West Lafayette
15. Fu G, Devaraj D (2008) Methodology of Homogeneous and Non-Homogeneous Markov Chains for Modeling Bridge Element Deterioration. Wayne State University, Detroit, MI
16. Jiang Y, Sinha KC (1989) Bridge service life prediction model using the Markov chain. Transportation research record no. 1223, pp 24–30
17. Kanafani A (1981) Transportation Demand Analysis. John Wiley and Sons, New York, NY
18. Keeney RL, Raiffa H (1993) Decisions with multiple objectives: Preferences and value tradeoffs. Cambridge University Press, New York
19. Li Z, Sinha KC (2004) Methodology for the development of a highway asset management system for Indiana. Purdue University, West Lafayette
20. Li N, Xie WC, Haas R (1996) Reliability-based processing of Markov chains for modeling pavement network deterioration. Transportation research record no. 1524, pp 203–213
21. Majid MZA, McCaffer R (1998) Factors of Non-Excusable Delays that Influence Contractors' Performance. J Manag Eng 14(3):42–48
22. Nagai K, Tomita Y, Fujimoto Y (1985) A fatigue crack initiation model and the life estimation under random loading by Monte Carlo method. J Soc Naval Archit Jpn 158(60):552–564
23. Nesbitt DM, Sparks GA, Neudorf RD (1993) A semi-Markov formulation of the pavement maintenance optimization problem. Can J Civil Eng XX(III):436–447
24. NYSDOT (2002) Vulnerability manuals. Bridge Safety Program, New York State DOT
25. Oxford University Press (2011) Uncertainty. In: Oxford English Dictionary Online. http://dictionary.oed.com. Accessed October 10 2011
26. Pickrell DH (1990) Urban rail transit projects: forecast versus actual ridership and cost. US Department of Transportation, Washington, DC
27. Richmond JED (1998) New rail transit investments: a review. John F. Kennedy School of Government, Harvard University, Cambridge, MA
28. Riddell WT, Lynch J (2005) A Markov chain model for fatigue crack growth, inspection and repair: the relationship between probability of detection, reliability and number of repairs in fleets of railroad tank cars. In: Proceedings of the ASME pressure vessels and piping conference 2005 – operations, applications and components, Denver, Colorado, United States

29. Sarewitz D, Pielke RA (2001) Extreme events: a research and policy framework for disasters in context, Int Geol Rev, 43(5):406–418
30. Shackle GLS (1949) Expectation in economics, 2nd edn. Cambridge University Press, Cambridge
31. Shoukry SN, Martinelli DR, Reigle JA (1997) Universal pavement distress evaluator based on fuzzy sets. Transportation research record, no. 1592, pp 180–186
32. Sinha KC, Labi S, McCullouch B, Bhargava A, Bai Q (2009) Updating and enhancing the Indiana Bridge Management System (IBMS). Joint Transportation Research Program, Purdue University, West Lafayette
33. Vidalis SM, Najafi FT (2002) Cost and time overruns in highway construction. In: 4th transportation specialty conference of the Canadian Society for Civil Engineering, Montréal, QC, Canada, 5–8 June 2002
34. Walls J, Smith MR (1998) Life-cycle cost analysis in pavement design – interim technical bulletin. Federal Highway Administration, SW Washington, DC
35. Workman SL (2008) Predicted vs. actual costs and ridership of new starts projects. In: 88th annual meeting of Transportation Research Board, Washington, DC
36. Xinhuanet.com (2008) Four missing after cargo vessel hits bridge in China. http://news.xinhuanet.com/english/2008-03/27/content_7868845.htm. Accessed 10 Oct 2011
37. Yang JD, Lu JJ, Gunaratne M, Dietrich B (2006) Modeling crack deterioration of flexible pavements: comparison of recurrent Markov chains and artificial neural networks. Transportation research record no. 1974, pp 18–25
38. Yokota H, Komure K (2004) Estimation of structural deterioration process by Markov-chain and costs for rehabilitation. Life-cycle performance of deteriorating structures. Third IABMAS workshop on life-cycle cost analysis and design of civil infrastructure systems and the JCSS workshop on probabilistic modeling of deterioration processes in concrete structures. Lausanne, Switzerland, pp 424–431
39. Zhang Y, Fan ZP, Liu Y (2010) A method based on stochastic dominance degrees for stochastic multiple criteria decision making. Comput Ind Eng 58:544–552

Physical Perspective Toward Stochastic Optimal Controls of Engineering Structures

Jie Li and Yong-Bo Peng

Abstract In the past few years, starting with the thought of physical stochastic systems and the principle of preservation of probability, a family of probability density evolution methods (PDEM) has been developed. It provides a new perspective toward the accurate design and optimization of structural performance under random engineering excitations such as earthquake ground motions and strong winds. On this basis, a physical approach to structural stochastic optimal control is proposed in the present chapter. A family of probabilistic criteria, including the criterion based on mean and standard deviation of responses, the criterion based on Exceedance probability, and the criterion based on global reliability of systems, is elaborated. The stochastic optimal control of a randomly base-excited single-degree-of-freedom system with active tendon is investigated for illustrative purposes. The results indicate that the control effect relies upon control criteria of which the control criterion in global reliability operates efficiently and gains the desirable structural performance. The results obtained by the proposed method are also compared against those by the LQG control, revealing that the PDEM-based stochastic optimal control exhibits significant benefits over the classical LQG control. Besides, the stochastic optimal control, using the global reliability

J. Li (✉)
State Key Laboratory of Disaster Reduction in Civil Engineering, Tongji University, Shanghai 200092, China

School of Civil Engineering, Tongji University, Shanghai 200092, China
e-mail: lijie@tongji.edu.cn

Y.-B. Peng
State Key Laboratory of Disaster Reduction in Civil Engineering, Tongji University, Shanghai 200092, China

Shanghai Institute of Disaster Prevention and Relief, Tongji University, Shanghai 200092, China
e-mail: pengyongbo@tongji.edu.cn

S. Chakraborty and G. Bhattacharya (eds.), *Proceedings of the International Symposium on Engineering under Uncertainty: Safety Assessment and Management (ISEUSAM - 2012)*,
DOI 10.1007/978-81-322-0757-3_4,

criterion, of an eight-story shear frame structure is carried out. The numerical example elucidates the validity and applicability of the developed physical stochastic optimal control methodology.

Keywords Probability density evolution method • Stochastic optimal control • Control criteria • Global reliability • LQG control

1 Introduction

Stochastic dynamics has gained increasing interests and has been extensively studied. However, although the original thought may date back to Einstein [4] and Langevin [11] and then studied in rigorous formulations by mathematicians [8, 10, 37], the random vibration theory, a component of stochastic dynamics, was only regarded as a branch of engineering science until the early of 1960s (Crandall [2]; Lin [28]). Till early 1990s, the theory and pragmatic approaches for random vibration of linear structures were well developed. Meanwhile, researchers were challenged by nonlinear random vibration, despite great efforts devoted coming up with a variety of methods, including the stochastic linearization, equivalent nonlinearization, stochastic averaging, path-integration method, FPK equation, and the Monte Carlo simulation (see, e.g., [29, 30, 41]). The challenge still existed. On the other hand, investigations on stochastic structural analysis (or referred to stochastic finite element method by some researchers), as a critical component of stochastic dynamics, in which the randomness of structural parameters is dealt with, started a little later from the late 1960s. Till middle 1990s, a series of approaches were presented, among which three were dominant: the Monte Carlo simulation [32, 33], the random perturbation technique [6, 9], and the orthogonal polynomial expansion [5, 12]. Likewise with the random vibration, here the analysis of nonlinear stochastic structures encountered huge challenges as well [31].

In the past 10 years, starting with the thought of physical stochastic systems [13] and the principle of preservation of probability [19], a family of probability density evolution methods (PDEM) has been developed, in which a generalized density evolution equation was established. The generalized density evolution equation profoundly reveals the essential relationship between the stochastic and deterministic systems. It is successfully employed in stochastic dynamic response analysis of multi-degree-of-freedom systems [20] and therefore provides a new perspective toward serious problems such as the dynamic reliability of structures, the stochastic stability of dynamical systems, and the stochastic optimal control of engineering structures.

In this chapter, the application of PDEM on the stochastic optimal control of structures will be summarized. Therefore, the fundamental theory of the generalized density evolution equation is firstly revisited. A physical approach to stochastic optimal control of structures is then presented. The optimal control criteria, including those based on mean and standard deviation of responses and those based on exceedance probability and global reliability of systems, are elaborated. The stochastic optimal control of a randomly base-excited single-degree-of-freedom system with active tendon is investigated for illustrative purposes. Comparative studies of these probabilistic criteria and the developed control methodology against the classical LQG control are carried out. The optimal control strategy is then further employed in the investigation of the stochastic optimal control of an eight-story shear frame. Some concluding remarks are included.

2 Principle Equation

2.1 Principle of Preservation of Probability Revisited

It is noted that the probability evolution in a stochastic dynamical system admits the principle of preservation of probability, which can be stated as the following: if the random factors involved in a stochastic system are retained, the probability will be preserved in the evolution process of the system. Although this principle may be faintly cognized quite long ago (see, e.g., [36]), the physical meaning has been only clarified in the past few years from the state description and random event description, respectively [16–19]. The fundamental logic position of the principle of preservation of probability was then solidly established with the development of a new family of generalized density evolution equations that integrates the ever-proposed probability density evolution equations, including the classic Liouville equation, Dostupov-Pugachev equation, and the FPK equation [20].

To revisit the principle of preservation of probability, consider an n-dimensional stochastic dynamical system governed by the following state equation:

$$\mathbf{Y} = \mathbf{A}(\mathbf{Y}, t), \mathbf{Y}(t_0) = \mathbf{Y}_0 \tag{1}$$

where $\mathbf{Y} = (Y_1, Y_2, \cdots, Y_n)^{\mathrm{T}}$ denotes the n-dimensional state vector, $\mathbf{Y}_0 = (Y_{0,1}, Y_{0,2}, \cdots, Y_{0,n})^{\mathrm{T}}$ denotes the corresponding initial vector, and $\mathbf{A}(\cdot)$ is a deterministic operator vector. Evidently, in the case that $\mathbf{Y}_0$ is a random vector, $\mathbf{Y}(t)$ will be a stochastic process vector.

The state equation (1) essentially establishes a mapping from $\mathbf{Y}_0$ to $\mathbf{Y}(t)$, which can be expressed as

$$\mathbf{Y}(t) = g(\mathbf{Y}_0, t) = \mathrm{G}_t(\mathbf{Y}_0) \tag{2}$$

where $g(\cdot), \mathrm{G}_t(\cdot)$ are both mapping operators from $\mathbf{Y}_0$ to $\mathbf{Y}(t)$.

Since $\mathbf{Y}_0$ denotes a random vector, $\{\mathbf{Y}_0 \in \Omega_{t_0}\}$ is a random event. Here Ω_{t_0} is any arbitrary domain in the distribution range of $\mathbf{Y}_0$. According to the stochastic state equation (1), $\mathbf{Y}_0$ will be changed to $\mathbf{Y}(t)$ at time t. The domain Ω_{t_0} to which $\mathbf{Y}_0$

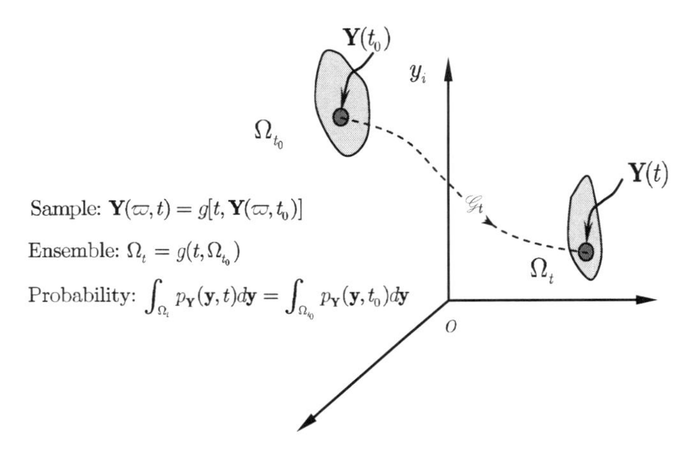

Fig. 1 Dynamical system, mapping, and probability evolution

belongs at time t_0 is accordingly changed to $\boldsymbol{\Omega}_t$ to which $\mathbf{Y}(t)$ belongs at time t; see Fig. 1.

$$\boldsymbol{\Omega}_t = g(\boldsymbol{\Omega}_{t_0}, t) = \mathrm{G}_t(\boldsymbol{\Omega}_{t_0}) \tag{3}$$

Since the probability is preserved in the mapping of any arbitrary element events, we have

$$\int_{\Omega_{t_0}} p_{\mathbf{Y}_0}(\mathbf{y}, t_0)d\mathbf{y} = \int_{\Omega_t} p_{\mathbf{Y}}(\mathbf{y}, t)d\mathbf{y} \tag{4}$$

It is understood that Eq. (4) also holds at $t + \Delta t$, which will then result in

$$\frac{\mathrm{D}}{\mathrm{D}t}\int_{\Omega_t} p_{\mathbf{Y}}(\mathbf{y}, t)d\mathbf{y} = 0 \tag{5}$$

where $\mathrm{D}(\cdot)/\mathrm{D}t$ operates its arguments with denotation of total derivative.

Equation (5) is clearly the mathematical formulation of the principle of preservation of probability in a stochastic dynamical system. Since the fact of probability invariability of a random event is recognized here, we refer to Eq. (5) as the random event description of the principle of preservation of probability. The meaning of the principle of preservation of probability can also be clarified from the state-space description. These two descriptions are somehow analogous to the Lagrangian and Eulerian descriptions in the continuum mechanics, although there also some distinctive properties particularly in whether overlapping is allowed. For details, refer to Li and Chen [17, 19].

2.2 Generalized Density Evolution Equation (GDEE)

Without loss of generality, consider the equation of motion of a multi-degree-of-freedom (MDOF) system as follows:

$$\mathbf{M}(\eta)\ddot{\mathrm{X}} + \mathbf{C}(\eta)\dot{\mathrm{X}} + \mathbf{f}(\eta, \mathbf{X}) = \Gamma\xi(t) \tag{6}$$

where $\eta = (\eta_1, \eta_2, \cdots, \eta_{s_1})$ are the random parameters involved in the physical properties of the system. If the excitation is a stochastic ground accelerogram $\xi(t) = \ddot{X}_{\mathrm{g}}(t)$, for example, then $\Gamma = -\mathbf{M1}$, $\mathbf{1} = (1, 1, \cdots, 1)^{\mathrm{T}}$. Here $\ddot{\mathrm{X}}, \dot{\mathrm{X}}, \mathbf{X}$ are the accelerations, velocities, and displacements of the structure relative to ground. $\mathbf{M}(\cdot), \mathbf{C}(\cdot), \mathbf{f}(\cdot)$ denote the mass, damping, and stiffness matrices of the structural system, respectively.

In the modeling of stochastic dynamic excitations such as earthquake ground motions, strong winds, and sea waves, the thought of physical stochastic process can be employed [14, 15, 27]. For general stochastic processes or random fields, the double-stage orthogonal decomposition can be adopted such that the excitation could be represented by a random function [22]

$$\ddot{X}_{\mathrm{g}}(t) = \ddot{X}_{\mathrm{g}}(\zeta, t) \tag{7}$$

where $\zeta = (\zeta_1, \zeta_2, \cdots, \zeta_{s_2})$.

For notational consistency, denote

$$\Theta = (\eta, \zeta) = (\eta_1, \eta_2, \cdots, \eta_{s_1}, \zeta_1, \zeta_2, \cdots, \zeta_{s_2}) = (\Theta_1, \Theta_2, \cdots, \Theta_s) \tag{8}$$

in which $s = s_1 + s_2$ is the total number of the basic random variables involved in the system. Equation (6) can thus be rewritten into

$$\mathbf{M}(\Theta)\ddot{\mathrm{X}} + \mathbf{C}(\Theta)\dot{\mathrm{X}} + \mathbf{f}(\Theta, \mathbf{X}) = \mathbf{F}(\Theta, t) \tag{9}$$

where $\mathbf{F}(\Theta, t) = \Gamma\ddot{\mathrm{X}}_{\mathrm{g}}(\zeta, t)$.

This is the equation to be resolved in which all the randomness from the initial conditions, excitations, and system parameters is involved and exposed in a unified manner. Such a stochastic equation of motion can be further rewritten into a stochastic state equation which was firstly formulated by Dostupov and Pugachev [3].

If, besides the displacements and velocities, we are also interested in other physical quantities $\mathbf{Z} = (Z_1, Z_2, \cdots, Z_m)^{\mathrm{T}}$ in the system (e.g., the stress, internal forces), then the augmented system $(\mathbf{Z}, \Theta)$ is probability preserved because all the random factors are involved; thus, according to Eq. (5), we have [19]

$$\frac{\mathrm{D}}{\mathrm{D}t}\int_{\Omega_t \times \Omega_\theta} p_{\mathbf{Z}\Theta}(\mathbf{z}, \theta, t) d\mathbf{z} d\theta = 0 \tag{10}$$

where $\Omega_t \times \Omega_\theta$ is any arbitrary domain in the augmented state space $\Omega \times \Omega_\Theta$, Ω_Θ is the distribution range of the random vector Θ, and $p_{\mathbf{Z}\Theta}(\mathbf{z}, \theta, t)$ is the joint probability density function (PDF) of $(\mathbf{Z}(t), \Theta)$.

After a series of mathematical manipulations, including the use of Reynolds' transfer theorem, we have

$$\int_{\Omega_{t_0} \times \Omega_\theta} \left(\frac{\partial p_{\mathbf{Z}\Theta}(\mathbf{z}, \theta, t)}{\partial t} + \sum_{j=1}^{m} \dot{Z}_j(\theta, t) \frac{\partial p_{\mathbf{Z}\Theta}(\mathbf{z}, \theta, t)}{\partial z_j} \right) d\mathbf{z} d\theta = 0 \tag{11}$$

which holds for any arbitrary $\Omega_{t_0} \times \Omega_\theta \in \Omega \times \Omega_\Theta$. Thus, we have for any arbitrary $\Omega_\theta \in \Omega_\Theta$

$$\int_{\Omega_\theta} \left(\frac{\partial p_{\mathbf{Z}\Theta}(\mathbf{z}, \theta, t)}{\partial t} + \sum_{j=1}^{m} \dot{Z}_j(\theta, t) \frac{\partial p_{\mathbf{Z}\Theta}(\mathbf{z}, \theta, t)}{\partial z_j} \right) d\theta = 0 \tag{12}$$

and also the following partial differential equation:

$$\frac{\partial p_{\mathbf{Z}\Theta}(\mathbf{z}, \theta, t)}{\partial t} + \sum_{j=1}^{m} \dot{Z}_j(\theta, t) \frac{\partial p_{\mathbf{Z}\Theta}(\mathbf{z}, \theta, t)}{\partial z_j} = 0 \tag{13}$$

Specifically, as $m = 1$, Eqs. (12) and (13) become, respectively,

$$\int_{\Omega_\theta} \left(\frac{\partial p_{Z\Theta}(z, \theta, t)}{\partial t} + \dot{Z}(\theta, t) \frac{\partial p_{Z\Theta}(z, \theta, t)}{\partial z} \right) d\theta = 0 \tag{14}$$

and

$$\frac{\partial p_{Z\Theta}(z, \theta, t)}{\partial t} + \dot{Z}(\theta, t) \frac{\partial p_{Z\Theta}(z, \theta, t)}{\partial z} = 0 \tag{15}$$

which is a one-dimensional partial differential equation.

Equations (13) and (15) are referred to as generalized density evolution equations (GDEEs). They reveal the intrinsic connections between a stochastic dynamical system and its deterministic counterpart. It is remarkable that the dimension of a GDEE is not relevant to the dimension (or degree-of-freedom) of the original system; see Eq. (9). This distinguishes GDEEs from the traditional probability density evolution equations (e.g., Liouville, Dostupov-Pugachev, and FPK equations), of which the dimension must be identical to the dimension of the original state equation (twice the degree-of-freedom).

Clearly, Eq. (14) is mathematically equivalent to Eq. (15). But it will be seen later that Eq. (14) itself may provide additional insight into the problem. Particularly, if the physical quantity Z of interest is the displacement X of the system, Eq. (15) becomes

$$\frac{\partial p_{X\Theta}(x,\theta,t)}{\partial t} = -\dot{X}(\theta,t)\frac{\partial p_{X\Theta}(x,\theta,t)}{\partial x} \tag{16}$$

Here we can see the rule clearly revealed by the GDEE: in the evolution of a general dynamical system, the time variant rate of the joint PDF of displacement and source random parameters is proportional to the space variant rate with the coefficient being instantaneous velocity. In other words, the flow of probability is determined by the change of physical states. This demonstrates strongly that the evolution of probability density is not disordered but admits a restrictive physical law. Clearly, this holds for the general physical system with underlying randomness. This rule could not be exposed in such an explicit way in the traditional probability density evolution equations.

Although in principle the GDEE holds for any arbitrary dimension, in most cases, one- or two-dimensional GDEEs are adequate. For simplicity and clarity, in the following sections, we will be focused on the one-dimensional GDEE. Generally, the boundary condition for Eq. (15) is

$$p_{Z\Theta}(z,\theta,t)|_{z\to\pm\infty} = 0 \quad \text{or} \quad p_{Z\Theta}(z,\theta,t) = 0, z \in \Omega_{\mathrm{f}} \tag{17}$$

the latter of which is usually adopted in first-passage reliability evaluation where Ω_{f} is the failure domain, while the initial condition is usually

$$p_{Z\Theta}(z,\theta,t)|_{t=t_0} = \delta(z-z_0)p_{\Theta}(\theta) \tag{18}$$

where z_0 is the deterministic initial value.

Solving Eq. (15), the instantaneous PDF of $Z(t)$ can be obtained by

$$p_Z(z,t) = \int_{\Omega_\Theta} p_{Z\Theta}(z,\theta,t)d\theta \tag{19}$$

The GDEE was firstly obtained as the uncoupled version of the parametric Liouville equation for linear systems [16]. Then for nonlinear systems, the GDEE was reached when the formal solution was employed [18]. It is from the above derivation that the meanings of the GDEE were thoroughly clarified and a solid physical foundation was laid [19].

2.3 Point Evolution and Ensemble Evolution

Since Eq. (14) holds for any arbitrary $\Omega_\theta \in \Omega_\Theta$, then for any arbitrary partition of probability-assigned space [1], of which the sub-domains are Ω_q's, $q = 1, 2, \cdots, n_{\mathrm{pt}}$

satisfying $\boldsymbol{\Omega}_i \cap \boldsymbol{\Omega}_j = \varnothing, \forall i \neq j$ and $\bigcup_{q=1}^{n_{\rm pt}} \boldsymbol{\Omega}_q = \boldsymbol{\Omega}_\Theta$, Eq. (14) constructed in the sub-domain then becomes

$$\int_{\Omega_q} \left(\frac{\partial p_{Z\Theta}(z,\theta,t)}{\partial t} + \dot{Z}(\theta,t) \frac{\partial p_{Z\Theta}(z,\theta,t)}{\partial z} \right) d\theta = 0, q = 1, 2, \cdots, n_{\rm pt} \tag{20}$$

It is noted that

$$P_q = \int_{\Omega_q} p_\Theta(\theta) d\theta, \quad q = 1, 2, \cdots, n_{\rm pt} \tag{21}$$

is the assigned probability over $\boldsymbol{\Omega}_q$ [1], and

$$p_q(z,t) = \int_{\Omega_q} p_{Z\Theta}(z,\theta,t) d\theta, \quad q = 1, 2, \cdots, n_{\rm pt} \tag{22}$$

then Eq. (20) becomes

$$\frac{\partial p_q(z,t)}{\partial t} + \int_{\Omega_q} \left[\dot{Z}(\theta,t) \frac{\partial p_{Z\Theta}(z,\theta,t)}{\partial z} \right] d\theta = 0, q = 1, 2, \cdots, n_{\rm pt} \tag{23}$$

According to Eq. (19), it follows that

$$p_Z(z,t) = \sum_{q=1}^{n_{\rm pt}} p_q(z,t) \tag{24}$$

There are two important properties that can be observed here:

1. Partition of probability-assigned space and the property of independent evolution
   The functions $p_q(z,t)$ defined in Eq. (22) themselves are not probability density functions because $\int_{-\infty}^{\infty} p_q(z,t) dz = P_q \neq 1$, that is, the consistency condition is not satisfied. However, except for this violation, they are very similar to probability density functions in many aspects. Actually, a normalized function $\tilde{p}_q(z,t) = p_q(z,t)/P_q$ meets all the conditions of a probability density function, which might be called the partial-probability density function over $\boldsymbol{\Omega}_q$. Equation (24) can then be rewritten into

$$p_Z(z,t) = \sum_{q=1}^{n_{\rm pt}} P_q \cdot \tilde{p}_q(z,t) \tag{25}$$

It is noted that P_q's are specified by the partition and are time invariant. Thus, the probability density function of $Z(t)$ could be regarded as the weighted sum of a set of partial-probability density functions. What is interesting regarding the partial-probability density functions is that they are in a sense mutually independent, that is, once a partition of probability-assigned space is determined (consequently Ω_q's are specified), then a partial-probability density function $\tilde{p}_q(z,t)$ is completely governed by Eq. (23) (it is of course true if the function $p_q(z,t)$ is substituted by $\tilde{p}_q(z,t)$); the evolution of other partial-probability density functions, $\tilde{p}_r(z,t), r \neq q$, has no effects on the evolution of $\tilde{p}_q(z,t)$. This property of independent evolution of partial-probability density function means that the original problem can be partitioned into a series of independent subproblems, which are usually easier than the original problem. Thus, the possibility of new approaches is implied but still to be explored. It is also stressed that such a property of independent evolution is not conditioned on any assumption of mutual independence of basic random variables.

2. Relationship between point evolution and ensemble evolution
The second term in Eq. (23) usually cannot be integrated explicitly. It is seen from this term that to capture the partial-probability density function $\tilde{p}_q(z,t)$ over Ω_q, the exact information of the velocity dependency on $\theta \in \Omega_q$ is required. This means that the evolution of $\tilde{p}_q(z,t)$ depends on all the exact information in Ω_q; in other words, the evolution of $\tilde{p}_q(z,t)$ is determined by the evolution of information of the ensemble over Ω_q. This manner could be called ensemble evolution.

To uncouple the second term in Eq. (23), we can assume

$$\dot{X}(\theta,t) \doteq \dot{X}(\theta_q,t), \quad \text{for } \theta \in \Omega_q \tag{26}$$

where $\theta_q \in \Omega_q$ is a representative point of Ω_q. For instance, θ_q could be determined by the Voronoi cell [1], by the average $\theta_q = \frac{1}{P_q}\int_{\Omega_q} \theta p_\Theta(\theta)d\theta$, or in some other appropriate manners. By doing this, Eq. (23) becomes

$$\frac{\partial p_q(z,t)}{\partial t} + \dot{Z}(\theta_q,t)\frac{\partial p_q(z,t)}{\partial z} = 0, \quad q = 1,2,\cdots,n_{\mathrm{pt}} \tag{27}$$

The meaning of Eq. (26) is clear that the ensemble evolution in Eq. (23) is represented by the information of a representative point in the sub-domain, that is, the ensemble evolution in a sub-domain is represented by a point evolution. Another possible manner of uncoupling the second term in Eq. (23) implies a small variation of $p_{Z\Theta}(z,\theta,t)$ over the sub-domain Ω_q. In this case, it follows that

$$\frac{\partial p_q(z,t)}{\partial t} + E_q[\dot{Z}(\theta,t)]\frac{\partial p_q(z,t)}{\partial z} = 0, \ q = 1,2,\cdots,n_{\mathrm{pt}} \tag{28}$$

where $E_q[\dot{Z}(\theta,t)] = \frac{1}{P_q}\int_{\Omega_q}\dot{Z}(\theta,t)p_\Theta(\theta)d\theta$ is the average of $\dot{Z}(\theta,t)$ over Ω_q. In some cases, $E_q[\dot{Z}(\theta,t)]$ might be close to $\dot{Z}(\theta_q,t)$, and thus, Eqs. (27) and (28) coincide.

2.4 Numerical Procedure for the GDEE

In the probability density evolution method, Eq. (9) is the physical equation, while Eq. (15) is the GDEE with initial and boundary conditions specified by Eqs. (17) and (18). Hence, solving the problem needs to incorporate physical equations and the GDEE. For some very simple cases, a closed-form solution might be obtained, say, by the method of characteristics [18]. While for most practical engineering problems, numerical method is needed. To this end, we start with Eq. (14) instead of Eq. (15) because from the standpoint of numerical solution, usually an equation in the form of an integral may have some advantages over an equation in the form of a differential.

According to the discussions in the preceding section, Eqs. (23), (27), or (28) could be adopted as the governing equation for numerical solution. Equation (23) is an exact equation equivalent to the original Eqs. (14) and (15). In the present stage, numerical algorithms for Eq. (27) were extensively studied and will be outlined here.

It is seen that Eq. (27) is a linear partial differential equation. To obtain the solution, the coefficients should be determined first, while these coefficients are time rates of the physical quantity of interest as $\{\Theta = \theta\}$ and thus can be obtained through solving Eq. (9). Therefore, the GDEE can be solved in the following steps:

Step 1: Select representative points (RPs for short) in the probability-assigned space and determine their assigned probability. Select a set of representative points in the distribution domain Ω_Θ. Denote them by $\theta_q = (\theta_{q,1}, \theta_{q,2}, \cdots, \theta_{q,s})$; $q = 1, 2, \cdots, n_{\text{pt}}$, where n_{pt} is the number of the selected points. Simultaneously, determine the assigned probability of each point according to Eq. (22) using the Voronoi cells [1].

Step 2: Solve deterministic dynamical systems. For the specified $\Theta = \theta_q, q = 1, 2, \cdots, n_{\text{pt}}$, solve the physical equation (Eq. 9) to obtain time rate (velocity) of the physical quantities $\dot{Z}(\theta_q, t)$. Through steps 1 and 2, the ensemble evolution is replaced by point evolution as representatives.

Step 3: Solve the GDEE (Eq. 27) under the initial condition, as a discretized version of Eq. (18),

$$p_q(z,t)\big|_{t=t_0} = \delta(z - z_0)P_q \tag{29}$$

by the finite difference method with TVD scheme to acquire the numerical solution of $p_q(z,t)$.

Step 4: Sum up all the results to obtain the probability density function of $Z(t)$ via the Eq. (24).

It is seen clearly that the solving process of the GDEE is to incorporate a series of deterministic analysis (point evolution) and numerical solving of partial differential equations, which is just the essential of the basic thought that the physical mechanism of probability density evolution is the evolution of the physical system.

3 Performance Evolution of Controlled Systems

Extensive studies have been done on the structural optimal control, which serves as one of the most effective measures to mitigate damage and loss of structures induced by disastrous actions such as earthquake ground motions and strong winds [7]. However, the randomness inherent in the dynamics of the system or its operational environment and coupled with the nonlinearity of structural behaviors should be taken into account so as to gain a precise control of structures. The reliability of structures, otherwise, associated with structural performance still cannot be guaranteed even if the responses are greatly reduced compared to the uncontrolled counterparts. Thus, the methods of stochastic optimal control have usually been relied upon to provide a rational mathematical context for analyzing and describing the problem.

Actually, pioneering investigations of stochastic optimal control by mathematician were dated back to semi-century ago and resulted in fruitful theorems and approaches [39]. These advances mainly hinge on the models of Itô stochastic differential equations (e.g., LQG control). They limit themselves in application to white noise or filtered white noise that is quite different from practical engineering excitations. The seismic ground motion, for example, exhibits strongly nonstationary and non-Gaussian properties. In addition, stochastic optimal control of multi-dimensional nonlinear systems is still a challenging problem in open. It is clear that the above two challenges both stem from the classical framework of stochastic dynamics. Therefore, a revolutionary scheme through physical control methodology based on PDEM is developed in the last few years [23–26].

Consider the multi-degree-of-freedom (MDOF) system represented by Eq. (9) is exerted a control action, of which the equation of motion is given by

$$\mathbf{M}(\Theta)\ddot{\mathrm{X}} + \mathbf{C}(\Theta)\dot{\mathrm{X}} + \mathbf{f}(\Theta, \mathbf{X}) = \mathbf{B}_s\mathbf{U}(\Theta, t) + \mathbf{D}_s\mathbf{F}(\Theta, t) \tag{30}$$

where $\mathbf{U}(\Theta, t)$ is the control gain vector provided by the control action, $\mathbf{B}_s$ is a matrix denoting the location of controllers, and $\mathbf{D}_s$ is a matrix denoting the location of excitations.

In the state space, Eq. (30) becomes

$$\dot{Z}(t) = \mathbf{AZ}(t) + \mathbf{BU}(t) + \mathbf{DF}(\Theta, t) \tag{31}$$

where **A** is a system matrix, **B** is a controller location matrix, and **D** is a excitation location vector.

In most cases, Eq. (30) is a well-posed equation, and relationship between the state vector $\mathbf{Z}(t)$ and control gain $\mathbf{U}(t)$ can be determined uniquely. Clearly, it is a function of $\boldsymbol{\Theta}$ and might be assumed to take the form

$$\mathbf{Z}(t) = \mathbf{H}_{\mathbf{Z}}(\boldsymbol{\Theta}, t) \tag{32}$$

$$\mathbf{U}(t) = \mathbf{H}_{\mathbf{U}}(\boldsymbol{\Theta}, t) \tag{33}$$

It is seen that all the randomness involved in this system comes from $\boldsymbol{\Theta}$; thus, the augmented systems of components of state and control force vectors $(Z(t), \boldsymbol{\Theta})$, $(U(t), \boldsymbol{\Theta})$ are both probability preserved and satisfy the GDEEs, respectively, as follows [25]:

$$\frac{\partial p_{Z\Theta}(z, \theta, t)}{\partial t} + \dot{Z}(\theta, t)\frac{\partial p_{Z\Theta}(z, \theta, t)}{\partial z} = 0 \tag{34}$$

$$\frac{\partial p_{U\Theta}(u, \theta, t)}{\partial t} + \dot{U}(\theta, t)\frac{\partial p_{U\Theta}(u, \theta, t)}{\partial u} = 0 \tag{35}$$

The corresponding instantaneous PDFs of $Z(t)$ and $U(t)$ can be obtained by solving the above partial differential equations with given initial conditions

$$p_Z(z, t) = \int_{\Omega_\Theta} p_{Z\Theta}(z, \theta, t)d\theta \tag{36}$$

$$p_U(u, t) = \int_{\Omega_\Theta} p_{U\Theta}(u, \theta, t)d\theta \tag{37}$$

where Ω_Θ is the distribution domain of $\boldsymbol{\Theta}$ and the joint PDFs $p_{Z\Theta}(z, \theta, t)$ and $p_{U\Theta}$ (u, θ, t) are the solutions of Eqs. (34) and (35), respectively.

As mentioned in the previous sections, the GDEEs reveal the intrinsic relationship between stochastic systems and deterministic systems via the random event description of the principle of preservation of probability. It is thus indicated, according to the relationship between point evolution and ensemble evolution, that the structural stochastic optimal control can be implemented through a collection of representative deterministic optimal controls and their synthesis on evolution of probability densities. Distinguished from the classical stochastic optimal control scheme, the control methodology based on the PDEM is termed as the physical scheme of structural stochastic optimal control.

Figure 2 shows the discrepancy among the deterministic control (DC), the LQG control, and the physical stochastic optimal control (PSC) tracing the performance evolution of optimal control systems. One might realize that the performance trajectory of the deterministic control is point to point, and obviously, it lacks the ability of governing the system performance due to the randomness of external

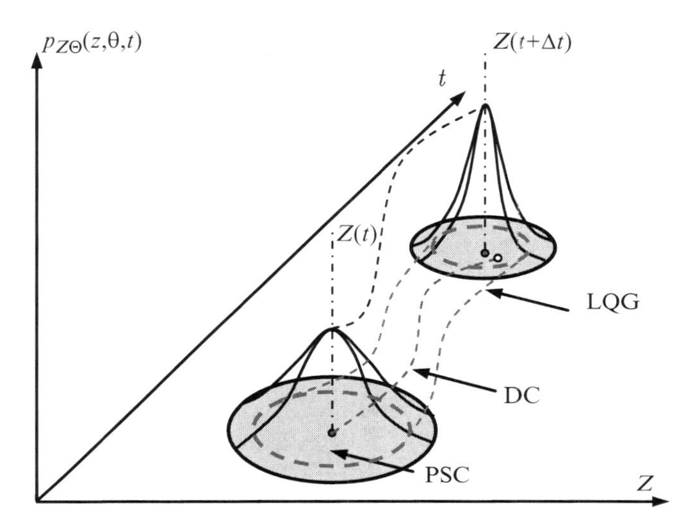

Fig. 2 Performance evolution of optimal control systems: comparison of determinate control (*DC*), LQG control, and physical stochastic optimal control (*PSC*)

excitations. The performance trajectory of the LQG control, meanwhile, is circle to circle. It is remarked here that the classical stochastic optimal control is essentially to govern the system statistics to the general stochastic dynamical systems since there still lacks of efficient methods to solve the response process of the stochastic systems with strong nonlinearities in the context of classical random mechanics. The LQG control, therefore, just holds the system performance in mean-square sense and cannot reach its high-order statistics. The performance trajectory of the PSC control, however, is domain to domain, which can achieve the accurate control of the system performance since the system quantities of interest all admit the GDEEs, Eqs. (34) and (35).

4 Probabilistic Criteria of Structural Stochastic Optimal Control

The structural stochastic optimal control involves maximizing or minimizing the specified cost function, whose generalized form is typically the quadratic combination of displacement, velocity, acceleration and control force. A standard quadratic cost function is given by the following expression [34]:

$$J_1(\mathbf{Z}, \mathbf{U}, \Theta, t) = \frac{1}{2}\mathbf{Z}^T(t_f)\mathbf{P}(t_f)\mathbf{Z}(t_f) + \frac{1}{2}\int_{t_0}^{t_f} \left[\mathbf{Z}^T(t)\mathbf{Q}\mathbf{Z}(t) + \mathbf{U}^T(t)\mathbf{R}\mathbf{U}(t)\right]dt \quad (38)$$

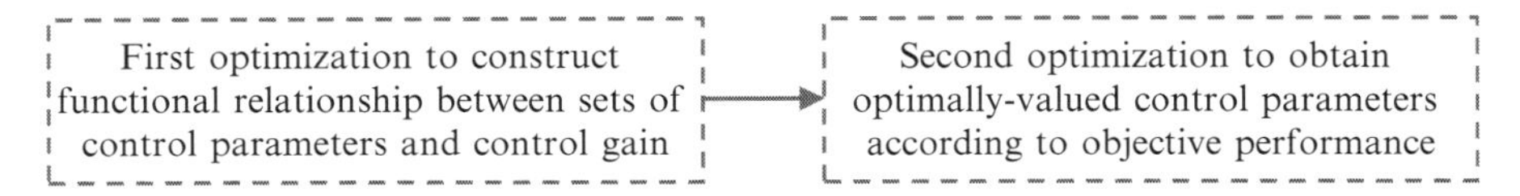

Fig. 3 Two step optimizations included in the physical stochastic optimal control

where **Q** is a positive semi-definite matrix, **R** is a positive definite matrix, and t_f is the terminal time, usually longer than that of the excitation. As should be noted, the cost function of the classical LQG control is defined as the ensemble-expected formula of Eq. (38) that is a deterministic function in dependence upon the time argument. Its minimization is to obtain the minimum second-order statistics of the state as the given parameters of control policy and construct the corresponding control gain under Gaussian process assumptions. In many cases of practical interests, the probability distribution function of the state related to structural performance is unknown, and the control gain essentially relies on second-order statistics, while the cost function represented by Eq. (38) is a stochastic process, of which minimization is to make the representative solution of the system state globally optimized in case of the given parameters of control policy. This treatment would result in a minimum second-order statistics or the optimum shape of the PDF of system quantities of interests. It is thus practicable to construct a control gain relevant to a predetermined performance of engineering structures since the procedure developed in this chapter adapts to the optimal control of general stochastic systems. In brief, the procedure involves two step optimizations; see Fig. 3. In the first step, for each realization θ_q of the stochastic parameter Θ, the minimization of the cost function Eq. (38) is carried out to build up a functional mapping from the set of parameters of control policy to the set of control gains. In the second step, the specified parameters of control policy to be used are obtained by optimizing the control gain according to the objective structural performance.

Therefore, viewed from representative realizations, the minimum of J_1 results in a solution of the conditional extreme value of cost function. The functional mapping, for a closed-loop control system, from the set of control parameters to the set of control gains is yield by [25]

$$\mathbf{U}(\Theta, t) = -\mathbf{R}^{-1}\mathbf{B}^T\mathbf{P}\mathbf{Z}(\Theta, t) \tag{39}$$

where P is the Riccati matrix function.

As indicated previously, the control effectiveness of stochastic optimal control relies on the specified control policy related to the objective performance of the structure. The critical procedure of designing control system actually is the determination of parameters of control policy, that is, weighting matrices **Q** and **R** in Eq. (38). There were a couple of strategies regarding to the weighting matrix choice in the context of classical LQG control such as system statistics assessment based on the mathematical expectation of the quantity of interest [40], system robustness analysis in probabilistic optimal sense [35], and comparison of weighting matrices in the context of Hamilton theoretical framework [42]. We are attempting to, nevertheless, develop a family of probabilistic criteria of weight matrices optimization in the context of the physical stochastic optimal control of structures.

4.1 System Second-Order Statistics Assessment (SSSA)

A probabilistic criterion of weight matrices optimization based on the system second-order statistics assessment, including constraint quantities and assessment quantities, is proposed as follows:

$$\min(J_2) = \underset{Q,R}{\arg\min}\,\{E[\tilde{Y}]\ or\ \sigma[\tilde{Y}]|F[\tilde{X}] \leq \tilde{X}_{\text{con}}\} \tag{40}$$

Where J_2 denotes a performance function, $\tilde{Y} = \max_t [\max_i \ Y_i(\Theta, t)|]$ is the equivalent extreme-value vector of the quantities to be assessed, $\tilde{X} = \max_t [\max_i |X_i(\Theta, t)|]$ is the equivalent extreme-value vector of the quantities to be used as the constraint, $\tilde{X}_{\text{con}}$ is the threshold of the constraint, the hat "~" on symbols indicates the equivalent extreme-value vector or equivalent extreme-value process [21], and $F[\cdot]$ is the characteristic value function indicating confidence level. The employment of the control criterion of Eq. (40) is to seek the optimal weighting matrices such that the mean or standard deviation of the assessment quantity $\tilde{Y}$ is minimized when the characteristic value of constraint quantity $\tilde{X}$ less than its threshold $\tilde{X}_{\text{con}}$.

4.2 Minimum of Exceedance Probability of Single System Quantity (MESS)

An exceedance probability criterion in the context of first-passage failure of single system quantity can be specified as follows:

$$\min(J_2) = \underset{\text{Q,R}}{\arg\min}\,\{\Pr(\tilde{Y} - \tilde{Y}_{\text{thd}} > 0) + (H(\tilde{X}_{\max} - \tilde{X}_{\text{con}}))\} \tag{41}$$

where $\Pr(\cdot)$ operates its arguments with denotation of exceedance probability, equivalent extreme-value vector $\tilde{Y}$ is the objective system quantity, and $H(\cdot)$ is the Heaviside step function. The physical meaning of this criterion is that the exceedance probability of the system quantity is minimized [26].

4.3 Minimum of Exceedance Probability of Multiple System Quantities (MEMS)

An exceedance probability criterion in the context of global failure of multiple system quantities is defined as follows:

$$\min(J_2) = \arg\min_{\mathrm{Q,R}} \left\{ \begin{array}{l} \frac{1}{2}[\Pr_{\tilde{Z}}^{T}(\tilde{Z} - \tilde{Z}_{\mathrm{thd}} > \mathbf{0})\Pr_{\tilde{Z}}(\tilde{Z} - \tilde{Z}_{\mathrm{thd}} > 0) + \Pr_{\tilde{U}}^{T}(\tilde{U} - \tilde{U}_{\mathrm{thd}} > \mathbf{0}) \\ \Pr_{\tilde{U}}(\tilde{U} - \tilde{U}_{\mathrm{thd}} > \mathbf{0})] + \left(H(\tilde{X}_{\max} - \tilde{X}_{\mathrm{con}})\right) \end{array} \right\} \tag{42}$$

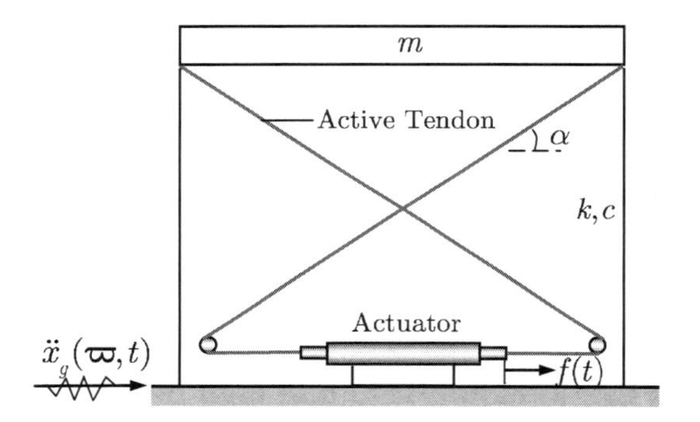

Fig. 4 Base-excited single-story structure with active tendon control system

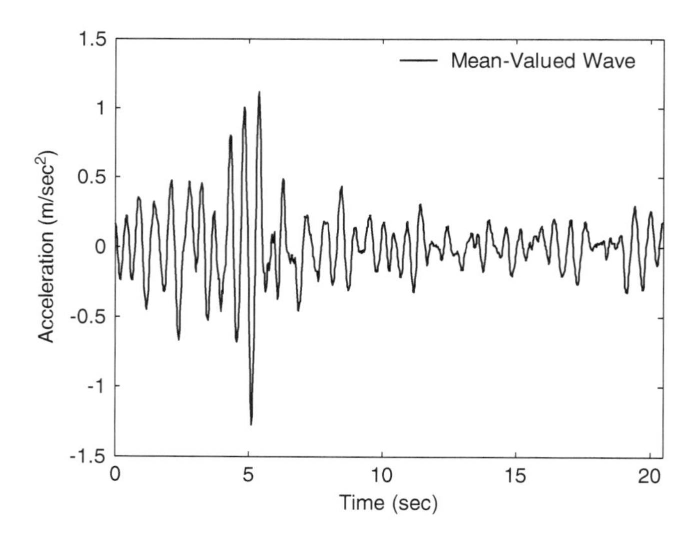

Fig. 5 Mean-valued time history of ground motion

where equivalent extreme-value vectors of state and control force $\tilde{Z}, \tilde{U}$ are the objective system quantities. It is indicated that this control criterion characterizes system safety (indicated in the controlled inter-story drift), system serviceability (indicated in the controlled inter-story velocity), system comfortability (indicated in the constrained story acceleration), controller workability (indicated in the limit control force), and their trade-off.

5 Comparative Studies

A base-excited single-story structure with an active tendon control system (see Fig. 4) is considered as a case for comparative studies of the control policies deduced from the above probabilistic criteria and the developed control methodology against the classical LQG. The properties of the system are as follows: the mass of the story is $m = 1 \times 10^5$ kg; the natural circular frequency of the uncontrolled structural system is $\omega_0 = 11.22$ rad/s; the control force of the actuator is denoted by $f(t)$, α representing the inclination angle of the tendon with respect to the base, and the acting force $u(t)$ on the structure is simulated; and the damping ratio is assumed to be 0.05. A stochastic earthquake ground motion model is used in this case [15], and the mean-valued time history of ground acceleration with peak 0.11 g is shown in Fig. 5.

The objective of stochastic optimal control is to limit the inter-story drift such that the system locates the reliability state, to limit the inter-story velocity such that the system provides the desired serviceability, to limit the story acceleration such that the system provides the desired comfortability, and to limit the control force such that the controller sustains its workability. The thresholds/constraint values of the inter-story drift, of the inter-story velocity, of the story acceleration, and of the control force are 10 mm, 100 mm/s, 3,000 mm/s, and 200 kN, respectively.

5.1 Advantages in Global Reliability-Based Probabilistic Criterion

For the control criterion of system second-order statistics assessment (SSSA), the inter-story drift is set as the constraint, and the assessment quantities include the inter-story drift, the story acceleration, and the control force. The characteristic value function is defined as mean plus three times of standard deviation of equivalent extreme-value variables. For the control criterion of minimum of exceedance probability of single system quantity (MESS), the inter-story drift is set as the objective system quantity, and the constraint quantities include the story acceleration and the control force. For the control criterion of minimum of exceedance probability of multiple system quantities (MEMS), the inter-story drift, inter-story

Table 1 Comparison of control policies

Ext. values	SSSA: Q = diag {80,80}, $R = 10^{-12}$	MESS: Q = diag {101.0,195.4}, $R = 10^{-10}$	MEMS: Q = diag {1073.6,505.0}, $R = 10^{-10}$
Dis-Mn (mm)	28.47[a]	28.47	28.47
	1.16[b]	6.23	4.15
	95.93%[c]	78.12	85.42%
Dis-Std (mm)	13.78	13.78	13.78
	0.20	1.41	0.81
	98.55%	89.77%	94.12%
Acc-Mn (mm/s)	3602.66	3602.66	3602.66
	1069.79	1235.60	1141.00
	70.31%	65.70%	68.33%
Acc-Std (mm/s)	1745.59	1745.59	1745.59
	360.81	348.92	331.78
	79.33%	80.01%	80.99%
CF-Mn (kN)	105.56	86.55	94.93
CF-Std (kN)	35.59	30.71	32.46

[a]Indicates the uncontrolled system quantities
[b]Indicates the controlled system quantities
[c]Indicates the control efficiency defined as $(a - b)/a$

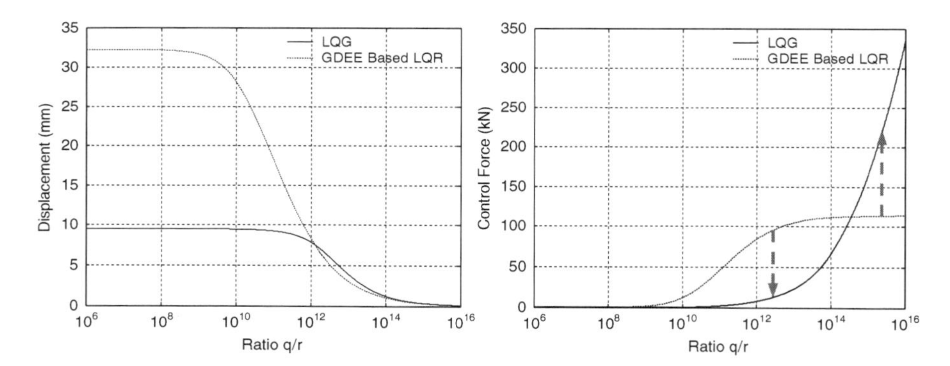

Fig. 6 Comparison of root-mean-square equivalent-extreme quantity vs. weight ratio between PSC and LQG (**a**) Equivalent-extreme relative displacement (**b**) Equivalent-extreme control force

velocity, and control force are set as the objective system quantities, while the constraint quantity is the story acceleration.

The comparison between the three control policies is investigated. The numerical results are listed in Table 1. It is seen that the effectiveness of response control hinges on the physical meanings of the optimal control criteria. As indicated in this case, the control criterion SSSA exhibits the larger control force due to the inter-story drift being only considered as the constraint quantity, which thus has lower inter-story drift. The control criterion MESS, however, exhibits the smaller control force due to the story acceleration and control force being simultaneously considered as the constraint quantities that result in a less reduction on the inter-story drift.

The control criterion MEMS, as seen from Table 1, achieves the best trade-off between control effectiveness and economy in that the objective system quantities include the inter-story drift, together with inter-story velocity and control force. It thus has reason to believe that the multi-objective criterion in the global reliability sense is the primary criterion of structural performance controls.

5.2 *Control Gains Against the Classical LQG*

It is noted that the classical stochastic optimal control strategies also could be applied to a class of stochastic dynamical systems and synthesize the moments or the PDFs of the controlled quantities. The class of systems is typically driven by independent additive Gaussian white noise and usually modeled as the Itô stochastic differential equations. The response processes, meanwhile, exhibit Markov property, of which the transition probabilities are governed by the Fokker-Planck-Kolmogorov equation (FPK equation). It remains an open challenge in the civil engineering system driven by non-Gaussian noise. The proposed physical stochastic optimal control methodology, however, occupies the validity and applicability to the civil engineering system. As a comparative study, Fig. 6 shows the discrepancy of root-mean-square quantity vs. weight ratio, using the control criterion of SSSA, between the advocated method and the LQG control.

One could see that the LQG control would underestimate the desired control force when the coefficient ratio of weighting matrices locates at the lower value, and it would overestimate the desired control force when the coefficient ratio of weighting matrices locates at the higher value. It is thus remarked that the LQG control using the nominal Gaussian white noise as the input cannot design the rational control system for civil engineering structures.

6 Numerical Example

An eight-story single-span shear frame fully controlled by active tendons is taken as a numerical example, of which the properties of the uncontrolled structure are identified according to Yang et al. [38]. The floor mass of each story unit is $m = 3.456 \times 10^5$ kg; the elastic stiffness of each story is $k = 3.404 \times 10^2$ kN/mm; and the internal damping coefficient of each story unit $c = 2.937$ kN × sec/mm,

Table 2 Optimization results of example

Parameters	Q_d	Q_v	R_u
Initial value	100	100	10^{-12}
Optimal value	102.8	163.7	10^{-12}
Objective value	11.22×10^{-6} ($P_{f,d} = 0.0023$, $P_{f,v} = 0.0035$, $P_{f,u} = 0.0022$)		

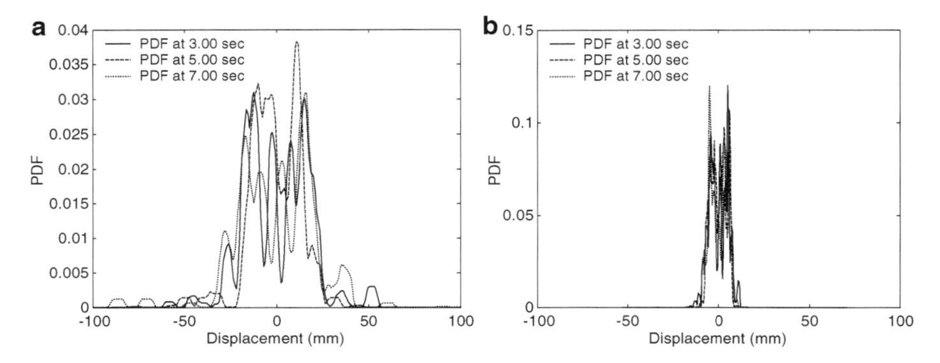

Fig. 7 Typical PDFs of inter-0-1 drift at typical instants of time (**a**) Without control (**b**) With control

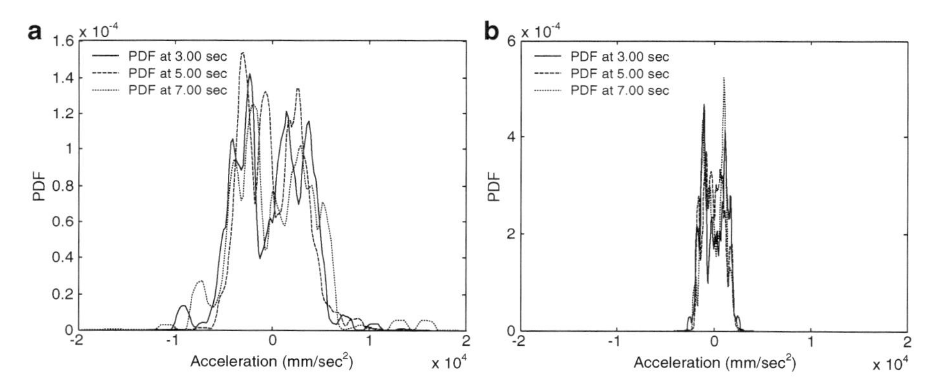

Fig. 8 Typical PDFs of the 8th story acceleration at typical instants of time (**a**) Without control (**b**) With control

which corresponds to a 2% damping ratio for the first vibrational mode of the entire building. The external damping is assumed to be zero. The computed natural frequencies are 5.79, 17.18, 27.98, 37.82, 46.38, 53.36, 58.53, and 61.69 rad/s, respectively. The earthquake ground motion model is the same as that of the preceding SDOF system, and the peak acceleration is 0.30 g. The control criterion MEMS is employed, and the thresholds/constraint values of the structural inter-story drifts, inter-story velocities, story acceleration, and the control forces are 15 mm, 150 mm/s, 2,000 kN, and 8,000 mm/s, respectively. For simplicity, the form of the weighting matrices in this case takes

$$\mathbf{Q} = \mathrm{diag}\{Q_d, ..., Q_d; Q_v, ..., Q_v\}, \quad R = \mathrm{diag}\{R_u, ..., R_u\} \tag{43}$$

The optimization results of the numerical example are shown in Table 2. It is seen that the exceedance probability of system quantities, rather than the ratio of reduction of responses, is provided when the objective value of performance function reaches to the minimum, indicating an accurate control of structural performance implemented. The optimization results also show that the stochastic optimal control achieves a best trade-off between effectiveness and economy.

Figure 7 shows typical PDFs of the inter-0-1-story drift of the controlled/uncontrolled structures at typical instants of time. One can see that the variation of the inter-story drift is obviously reduced. Likewise, the PDFs of the eight-story acceleration at typical instants show a reduction of system response since that distribution of the story acceleration has been narrowed (see Fig. 8). It is indicated that the seismic performance of the structure is improved significantly in case that the stochastic optimal control employing the exceedance probability criterion is applied.

7 Concluding Remarks

In this chapter, the fundamental theory of the generalized density evolution equation is firstly revisited. Then, a physical scheme of structural stochastic optimal control based on the probability density evolution method is presented for the stochastic optimal controls of engineering structures excited by general nonstationary and non-Gaussian processes. It extends the classical stochastic optimal control approaches, such as the LQG control, of which the random dynamic excitations are exclusively assumed as independent white noises or filter white noises. A family of optimal control criteria for designing the controller parameter, including the criterion based on mean and standard deviation of responses, the criterion based on Exceedance probability, and the criterion based on global reliability of systems, is elaborated by investigating the stochastic optimal control of a base-excited single-story structure with an active tendon control system. It is indicated that the control effect relies upon the probabilistic criteria of which the control criterion in global reliability operates efficiently and gains the desirable structural performance. The proposed stochastic optimal control scheme, meanwhile, of structures exhibits significant benefits over the classical LQG control. An eight-story shear frame controlled by active tendons is further investigated, employing the control criterion in global reliability of the system quantities. It is revealed in the numerical example that the seismic performance of the structure is improved significantly, indicating the validity and applicability of the developed PDEM-based stochastic optimal control methodology for the accurate control of structural performance.

Acknowledgements The supports of the National Natural Science Foundation of China (Grant Nos. 50621062, 51108344) and the Exploratory Program of State Key Laboratory of Disaster Reduction in Civil Engineering at Tongji University (Grant No. SLDRCE11-B-04) are highly appreciated.

References

1. Chen JB, Ghanem R, Li J (2009) Partition of the probability-assigned space in probability density evolution analysis of nonlinear stochastic structures. Probab Eng Mech 24(1):27–42
2. Crandall SH (1958) Random Vibration. Technology Press of MIT; John Wiley and Sons, New York
3. Dostupov BG, Pugachev VS (1957) The equation for the integral of a system of ordinary differential equations containing random parameters. Automatikai Telemekhanika 18:620–630
4. Einstein A (1905) Über Die Von Der Molecular-Kinetischen Theorie Der Wärme Geforderte Bewegung Von in Rhuenden Flüssigkeiten Sus-Pendierten Teilchen. Ann Phys (Leipzig) 17:549–560
5. Ghanem RG, Spanos PD (1991) Stochastic finite elements: a spectral approach. Springer, Berlin
6. Halder A, Mahadevar S (2000) Reliability assessment using stochastic finite element analysis. Wiley, New York
7. Housner GW, Bergman LA, Caughey TK et al (1997) Structural control: past, present, and future. J Eng Mech 123(9):897–971
8. Itô K (1942) Differential equations determining a Markoff process. Zenkoku Sizyo Sugaku Danwakasi, 1077
9. Kleiber M, Hien TD (1992) The stochastic finite element method. Wiley, Chichester
10. Kolmogorov A (1931) über die analytischen Methoden in der Wahrscheinlichkeitsrechnung. Math Ann 104(1):415–458
11. Langevin P (1908) Sur La Theorie Du Mouvement Brownien. C. R. Acad. Sci, Paris, pp 530–532
12. Li J (1996) Stochastic structural systems: analysis and modeling. Science Press, Beijing (in Chinese)
13. Li J (2006) A physical approach to stochastic dynamical systems. Sci Paper Online 1(2):93–104 (in Chinese)
14. Li J (2008) Physical stochastic models for the dynamic excitations of engineering structures. In: Advances in theory and applications of random vibration, 119–132, Tongji University Press, Shanghai (in Chinese)
15. Li J, Ai XQ (2006) Study on random model of earthquake ground motion based on physical process. Earthq Eng Eng Vib 26(5):21–26 (in Chinese)
16. Li J, Chen JB (2003) Probability density evolution method for dynamic response analysis of stochastic structures. In: Proceeding of the fifth international conference on stochastic structural dynamics, Hangzhou, China, pp 309–316
17. Li J, Chen JB (2006) Generalized density evolution equations for stochastic dynamical systems. Prog Nat Sci 16(6):712–719
18. Li J, Chen JB (2006) The probability density evolution method for dynamic response analysis of non-linear stochastic structures. Int J Numer Methods Eng 65:882–903
19. Li J, Chen JB (2008) The principle of preservation of probability and the generalized density evolution equation. Struct Saf 30:65–77
20. Li J, Chen JB (2009) Stochastic dynamics of structures. Wiley, Singapore
21. Li J, Chen JB, Fan WL (2007) The equivalent extreme-value event and evaluation of the structural system reliability. Struct Saf 29(2):112–131
22. Li J, Liu ZJ (2006) Expansion method of stochastic processes based on normalized orthogonal bases. J Tongji Univ (Nat Sci) 34(10):1279–1283
23. Li J, Peng YB (2007) Stochastic optimal control of earthquake-excited linear systems. In: Proceedings of 8th Pacific conference on earthquake engineering, Singapore, 5–7 Dec 2007
24. Li J, Peng YB, Chen JB (2008) GDEE-based stochastic control strategy of MR damping systems. In: Proceedings of 10th international symposium on structural engineering for young experts, Changsha, China, pp 1207–1212

25. Li J, Peng YB, Chen JB (2010) A physical approach to structural stochastic optimal controls. Probabilistic Engineering Mechanics 25(1):127–141
26. Li J, Peng YB, Chen JB (2011) Probabilistic criteria of structural stochastic optimal controls. Probab Eng Mech 26(2):240–253
27. Li J, Yan Q, Chen JB (2011) Stochastic modeling of engineering dynamic excitations for stochastic dynamics of structures. Probab Eng Mech 27:19–28
28. Lin YK (1967) Probabilistic Theory of Structural Dynamics. McGraw-Hill, New York
29. Lin YK, Cai GQ (1995) Probabilistic structural dynamics: advanced theory and applications. McGraw-Hill, New York
30. Lutes LD, Sarkani S (2004) Random vibrations: analysis of structural and mechanical systems. Butterworth-Heinemann, Amsterdam
31. Schenk CA, Schuëller GI (2005) Uncertainty assessment of large finite element systems. Springer, Berlin
32. Shinozuka M, Deodatis G (1991) Simulation of stochastic processes by spectral representation. Appl Mech Rev 44(4):191–204
33. Shinozuka M, Jan CM (1972) Digital simulation of random processes and its applications. J Sound Vib 25:111–128
34. Soong TT (1990) Active structural control: theory and practice. Longman Scientific & Technical, New York
35. Stengel RF, Ray LR, Marrison CI (1992) Probabilistic evaluation of control system robustness. IMA workshop on control systems design for advanced engineering systems: complexity, uncertainty, information and organization, Minneapolis, MN, 12–16 Oct 1992
36. Syski R (1967) Stochastic differential equations. Chapter 8. In: Saaty TL (ed) Modern nonlinear equations. McGraw-Hill, New York
37. Wiener N (1923) Differential space. J Math Phys 2(13):131–174
38. Yang JN, Akbarpour A, Ghaemmaghami P (1987) New optimal control algorithms for structural control. J Eng Mech 113(9):1369–1386
39. Yong JM, Zhou XY (1999) Stochastic controls: Hamiltonian systems and HJB equations. Springer, New York
40. Zhang WS, Xu YL (2001) Closed form solution for along-wind response of actively controlled tall buildings with LQG controllers. J Wind Eng Ind Aerodyn 89:785–807
41. Zhu WQ (1992) Random vibration. Science Press, Beijing (in Chinese)
42. Zhu WQ, Ying ZG, Soong TT (2001) An optimal nonlinear feedback control strategy for randomly excited structural systems. Nonlinear Dyn 24:31–51

Uncertainty Quantification for Decision-Making in Engineered Systems

Sankaran Mahadevan

Abstract This chapter discusses current research and opportunities for uncertainty quantification in performance prediction and risk assessment of engineered systems. Model-based simulation becomes attractive for systems that are too large and complex for full-scale testing. However, model-based simulation involves many approximations and assumptions, and thus, confidence in the simulation result is an important consideration in risk-informed decision-making. Sources of uncertainty are both aleatory and epistemic, stemming from natural variability, information uncertainty, and modeling approximations. The chapter draws on illustrative problems in aerospace, mechanical, civil, and environmental engineering disciplines to discuss (1) recent research on quantifying various types of errors and uncertainties, particularly focusing on data uncertainty and model uncertainty (both due to model form assumptions and solution approximations); (2) framework for integrating information from multiple sources (models, tests, experts), multiple model development activities (calibration, verification, validation), and multiple formats; and (3) using uncertainty quantification in risk-informed decision-making throughout the life cycle of engineered systems, such as design, operations, health and risk assessment, and risk management.

Keywords Uncertainty quantification • Model based simulation • Surrogate models

1 Introduction

Uncertainty quantification is important in the assessing and predicting performance of complex engineering systems, especially given limited experimental or real-world data. Simulation of complex physical systems involves multiple levels of

---

S. Mahadevan (✉)
John R. Murray Sr. Professor of Civil and Environmental Engineering, Vanderbilt University, Nashville, TN, USA
e-mail: sankaran.mahadevan@vanderbilt.edu

S. Chakraborty and G. Bhattacharya (eds.), *Proceedings of the International Symposium on Engineering under Uncertainty: Safety Assessment and Management (ISEUSAM - 2012)*, DOI 10.1007/978-81-322-0757-3_5,

modeling, ranging from the material to component to subsystem to system. Interacting models and simulation codes from multiple disciplines (multiple physics) may be required, with iterative analyses between some of the codes. As the models are integrated across multiple disciplines and levels, the problem becomes more complex, and assessing the predictive capability of the overall system model becomes more difficult. Many factors contribute to the uncertainty in the prediction of the system model including inherent variability in model input parameters, sparse data, measurement error, modeling errors, assumptions, and approximations.

The various sources of uncertainty in performance prediction can be grouped into three categories:

- Physical variability
- Data uncertainty
- Model error

1.1 Physical Variability

This type of uncertainty also referred to as aleatory or irreducible uncertainty arises from natural or inherent random variability of physical processes and variables, due to many factors such as environmental and operational variations, construction processes, and quality control. This type of uncertainty is present both in system properties (e.g., material strength, porosity, diffusivity, geometry variations, chemical reaction rates) and external influences and demands on the system (e.g., concentration of chemicals, temperature, humidity, mechanical loads). As a result, in model-based prediction of system behavior, there is uncertainty regarding the precise values for model parameters and model inputs, leading to uncertainty about the precise values of the model output. Such quantities are represented in engineering analysis as random variables, with statistical parameters, such as mean values, standard deviations, and distribution types, estimated from observed data or in some cases assumed. Variations over space or time are modeled as random processes.

1.2 Data Uncertainty

This type of uncertainty falls under the category of epistemic uncertainty (i.e., knowledge or information uncertainty) or reducible uncertainty (i.e., the uncertainty is reduced as more information is obtained). Data uncertainty occurs in different forms. In the case of a quantity treated as a random variable, the accuracy of the statistical distribution parameters depends on the amount of data available. If the data is sparse, the distribution parameters themselves are uncertain and may need to be treated as random variables. Alternatively, information may be imprecise or qualitative, or as a range of values, based on expert opinion. Both probabilistic and non-probabilistic methods have been explored to represent epistemic uncertainty. Measurement error (either in the laboratory or in the field) is another important source of data uncertainty.

1.3 Model Error

This results from approximate mathematical models of the system behavior and from numerical approximations during the computational process, resulting in two types of error in general – solution approximation error and model form error. The performance assessment of a complex system involves the use of several analysis models, each with its own assumptions and approximations. The errors from the various analysis components combine in a complicated manner to produce the overall model error (described by both bias and uncertainty).

The roles of several types of uncertainty in the use of model-based simulation for performance assessment can be easily seen in the case of reliability analysis. Consider the probability of an undesirable event denoted by $g(\mathbf{X}) < k$, which can be computed from Eq. (1):

$$P(g(\mathbf{X})<k) = \int_{g(\mathbf{X})<k} f_{\mathbf{X}}(\mathbf{x})\mathrm{d}\mathbf{x} \tag{1}$$

where $\mathbf{X}$ is the vector of input random variables, $f_{\mathbf{X}}(\mathbf{x})$ is the joint probability density function of $\mathbf{X}$, $g(\mathbf{X})$ is the model output, and k is the regulatory requirement in performance assessment. Every term on the right-hand side of Eq. (1) has uncertainty. There is inherent variability represented by the vector of random variables $\mathbf{X}$, data uncertainty (due to inadequate data) regarding the distribution type and distribution parameters of $f_{\mathbf{X}}(\mathbf{x})$, and model errors in the computation of $g(\mathbf{X})$. Thus, it is necessary to systematically identify the various sources of uncertainty and develop the framework for including them in the overall uncertainty quantification in the performance assessment of engineering systems.

The uncertainty analysis methods covered in this chapter are grouped by sections along the four major groups of analysis activities that are needed for performance assessment under uncertainty:

1. Input uncertainty quantification
2. Uncertainty propagation analysis (includes model error quantification)
3. Model calibration, verification, validation, and extrapolation
4. Probabilistic performance assessment

A brief summary of the analysis methods covered in the four groups is as follows:

Input uncertainty quantification: Physical variability of parameters can be quantified through random variables by statistical analysis. Parameters that vary in time or space are modeled as random processes or random fields with appropriate correlation structure. Data uncertainty that leads to uncertainty in the distribution parameters and distribution types can be addressed using confidence intervals and Bayesian statistics. Recent methods to include several sources of data uncertainty, namely, sparse data, interval data, and measurement error, are discussed in this chapter.

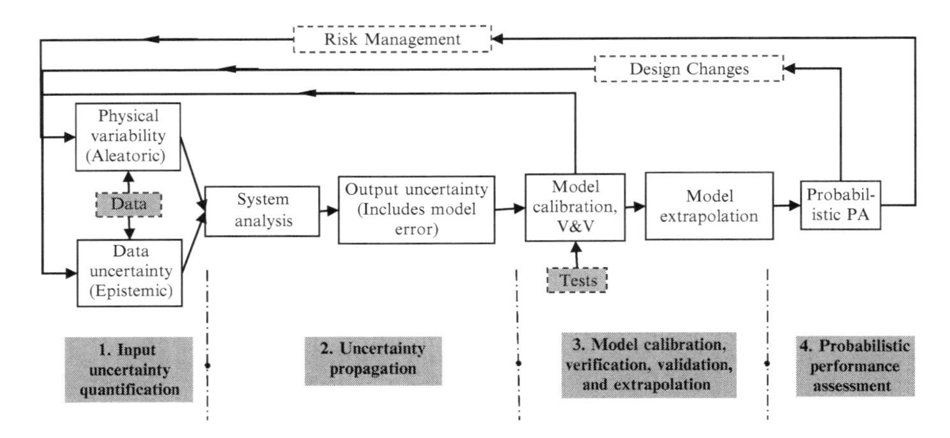

Fig. 1 Uncertainty quantification, propagation, and management

Uncertainty propagation analysis: Both classical and Bayesian probabilistic approaches can be investigated to propagate inherent variability and data uncertainty through individual sub-models and the overall system model. To reduce the computational expense, surrogate models can be constructed using several different techniques. Methods for sensitivity analyses in the presence of uncertainty are discussed. The uncertainty in the overall model output also includes model errors and approximations in each step of the analysis; therefore, approaches to quantify model error are included in the discussion.

Model calibration, verification, validation, and extrapolation: Model calibration is the process of adjusting model parameters to obtain good agreement between model predictions and experimental observations. Both classical and Bayesian statistical methods are discussed for model calibration with available data. One particular concern is how to properly integrate different types of data, available at different levels of the model hierarchy. Assessment of the "correct" implementation of the model is called verification, and assessment of the degree of agreement of the model response with the available physical observation is called validation. Model verification and validation activities help to quantify model error (both model form error and solution approximation error). A Bayesian network framework is discussed for quantifying the confidence in model prediction based on data, models, and activities at various levels of the system hierarchy. Such information is available in heterogeneous formats from multiple sources, and a consistent framework to integrate such disparate information is important.

Performance assessment: Limit state-based reliability analysis methods are available to help quantify the assessment results in a probabilistic manner. Monte Carlo simulation with high-fidelity analyses modules is computationally expensive; hence, surrogate (or abstracted) models are frequently used with Monte Carlo simulation. In that case, the uncertainty or error introduced by the surrogate model also needs to be quantified.

Figure 1 shows the four groups of activities within a conceptual framework for systematic quantification, propagation, and management of various types of uncertainty. The methods discussed in this chapter address all the four steps shown in Fig. 1. The different steps of analysis in Fig. 1 are not strictly sequential. While uncertainty has been dealt with using probabilistic as well as non-probabilistic (e.g., fuzzy sets, possibility theory, evidence theory) formats in the literature, this chapter will only focus on probabilistic analysis.

In Fig. 1, the box "Data" in the input uncertainty quantification step includes laboratory data, historical field data, literature sources, and expert opinion. The box "Design changes" may refer to conceptual, preliminary, or detailed design, depending on the development stage. The boxes "Design changes" and "Risk management" are outside the scope of this chapter, although they are part of the overall uncertainty management framework.

2 Input Uncertainty Quantification

2.1 Physical Variability

Examples of model input variables with physical variability (i.e., inherent, natural variability) include:

(a) Material properties (e.g., mechanical and thermal properties, soil properties, chemical properties)
(b) Geometrical properties (e.g., Structural dimensions, concrete cover depth)
(c) External conditions (e.g., mechanical loading, boundary conditions, physical processes such as freeze-thaw, chemical processes such as carbonation, chloride, or sulfate attack)

Many uncertainty quantification studies have only focused on quantifying and propagating the inherent variability in the input parameters. Well-established statistical (both classical and Bayesian) methods are available for this purpose.

In probabilistic analysis, the sample to sample variations (random variables) in the parameters are addressed by defining them as random variables with probability density functions (PDFs). Some parameters may vary not only from sample to sample (as is the case for random variables) but also in spatial or time domain. Parameter variation over time and space can be modeled as *random processes or random fields*.

Some well-known methods for simulating random processes are spectral representation (SR) [13], Karhunen-Loeve expansion (KLE) ([10, 18, 30]), and polynomial chaos expansion (PCE) ([18, 30, 37]). The PCE method has been used to represent the stochastic model output as a function of stochastic inputs.

Consider an example of representing a random process using KLE, expressed as

$$\varpi(x,\chi) = \varpi(x) + \sum_{i=1}^{\infty} \sqrt{\lambda_i}\xi_i(\chi)f_i(x) \tag{2}$$

where $\varpi(x)$ is the mean of the random process $\varpi(x, \chi)$, λ_i and $f_i(x)$ are eigenvalues and eigenfunctions of $C(x_1, x_2)$, and $\xi_i(\chi)$ is a set of uncorrelated standard normal random variables (x is a space or time coordinate, and χ is an index representing different realizations of the random process). Using Eq. (2), realizations of the random process $\varpi(x, \chi)$ can be easily simulated by generating samples of the random variables $\xi_i(\chi)$, and these realizations of $\varpi(x, \chi)$ can be used in the reliability analysis.

Some boundary conditions (e.g., temperature and moisture content) might exhibit a recurring pattern over shorter periods and also a trend over longer periods. Both can be numerically represented by a seasonal model using an autoregressive integrated moving average (ARIMA) method generally used for linear[1] nonstationary[2] processes [5]. This method can be used to predict the temperature or the rainfall magnitudes in the future so that it can be used in the durability analysis of the structures under future environmental conditions.

It may also be important to quantify the statistical correlations between some of the input random variables. Many previous studies on uncertainty quantification simply assume either zero or full correlation, in the absence of adequate data. A Bayesian approach may be pursued for this purpose, as described in Sect. 2.2.

2.2 *Data Uncertainty*

This section discusses methods to quantify uncertainty due to limited statistical data and measurement errors (ε_{exp}). Data may also be available in interval format (e.g., expert opinion). A Bayesian approach, consistent with the framework proposed in Fig. 1, can be used in the presence of data uncertainty. The prior distributions of different physical variables and their distribution parameters can be based on available data and expert judgment, and these are updated as more data becomes available through experiments, analysis, or real-world experience.

Data qualification is an important step in the consideration of data uncertainty. All data points may not have equal weight; a careful investigation of data quality will help to assign appropriate weights to different data sets.

2.2.1 Sparse Statistical Data

For any random variable that is quantitatively described by a probability density function, there is always uncertainty in the corresponding distribution parameters due to small sample size. As testing and data collection activities are performed, the state of knowledge regarding the uncertainty changes and a Bayesian updating

[1] The current observation can be expressed as a linear function of past observations.

[2] A process is said to be nonstationary if its probability structure varies with the time or space coordinate.

approach can be implemented. The Bayesian approach also applies to joint distributions of multiple random variables, which also helps to include the uncertainty in correlations between the variables. A prior joint distribution is assumed (or individual distributions and correlations are assumed) and then updated as data becomes available.

Instead of assuming a well-known prior distribution form (e.g., uniform, normal) for sparse data sets, either empirical distribution functions or flexible families of distributions based on the data can be constructed. A bootstrapping[3] technique can then be used to quantify the uncertainty in the distribution parameters. The *empirical distribution function* is constructed by ranking the observations from lowest to highest value and assigning a probability value to each observation. Examples of flexible distribution families include the Johnson family, Pearson family, gamma distribution, and stretched exponential distribution (e.g., [48]). Recently, Sankararaman and Mahadevan [43] developed a likelihood-based approach to construct nonparametric probability distributions in the presence of both sparse and interval data.

Transformations have been proposed from a non-probabilistic to probabilistic format, through the maximum likelihood approach [25, 40]. Such transformations have attracted the criticism that information is either added or lost in the process. Two ways to address the criticism are to (1) construct empirical distribution functions based on interval data collected from multiple experts or experiments [9] and (2) construct flexible families of distributions with bounds on distribution parameters based on the interval data, without forcing a distribution assumption (McDonald et al. 2008). These can then be treated as random variables with probability distribution functions and combined with other random variables in a Bayesian framework to quantify the overall system model uncertainty. The use of families of distributions will result in multiple probability distributions for the output, representing the contributions of both physical variability and data uncertainty. The nonparametric approach of Sankararaman and Mahadevan [43] also has the ability to quantify the contributions of aleatory and epistemic uncertainty to the probabilistic representation of an uncertain variable.

2.2.2 Measurement Error

The measurement error in each input variable in many studies (e.g., [1]) is assumed to be independent and identically distributed (IID) normal with zero mean and an assumed variance, i.e., $\varepsilon_{\text{exp}} \sim N\left(0, \sigma_{\text{exp}}^2\right)$. Due to the measurement uncertainty, the distribution parameter σ_{exp} cannot be obtained as a deterministic value. Instead, it is a random variable with a prior density $\tau(\sigma_{\text{exp}})$. Thus, when new data is available

[3] Bootstrapping is a data-based simulation method for statistical inference by resampling from an existing data set [7].

after testing, the distribution of σ_{exp} can be easily updated using the Bayesian theorem. Another way to represent measurement error ε_{exp} is through an interval only, and not as a random variable.

3 Uncertainty Propagation Analysis

In this section, methods to quantify the contributions of different sources of uncertainty and error as they propagate through the system analysis model, including the contribution of model error, are discussed, in order to quantify the overall uncertainty in the system model output.

This section covers two issues: (1) quantification of model output uncertainty, given input uncertainty (both physical variability and data uncertainty), and (2) quantification of model error (due to both model form selection and solution approximations).

Several uncertainty analysis studies, including a study with respect to the proposed Yucca Mountain high-level waste repository, have recognized the distinction between physical variability and data uncertainty [16, 17]. As a result, these methods evaluate the variability in an inner loop calculation and data uncertainty in an outer loop calculation.

3.1 Propagation of Physical Variability

Various probabilistic methods (e.g., Monte Carlo simulation and first-order or second-order analytical approximations) have been studied for the propagation of physical variability in model inputs and model parameters [14] expressed through random variables and random process or fields. Stochastic finite element methods (e.g., [10, 15]) have been developed for single discipline problems, in structural, thermal, and fluid mechanics. An example of such propagation is shown in Fig. 2. Several types of combinations of system analysis model and statistical analysis techniques are available:

- Monte Carlo simulation with the deterministic system analysis as a black-box (e.g., [39]) to estimate model output statistics or probability of regulatory compliance
- Monte Carlo simulation with a surrogate model to replace the deterministic system analysis model (e.g., [10, 18, 19, 47]), to estimate model output statistics or probability of regulatory compliance
- Local sensitivity analysis using finite difference, perturbation, or adjoint analyses, leading to estimates of the first-order or second-order moments of the output (e.g., [3])
- Global sensitivity and effects analysis and analysis of variance in the output (e.g., [4])

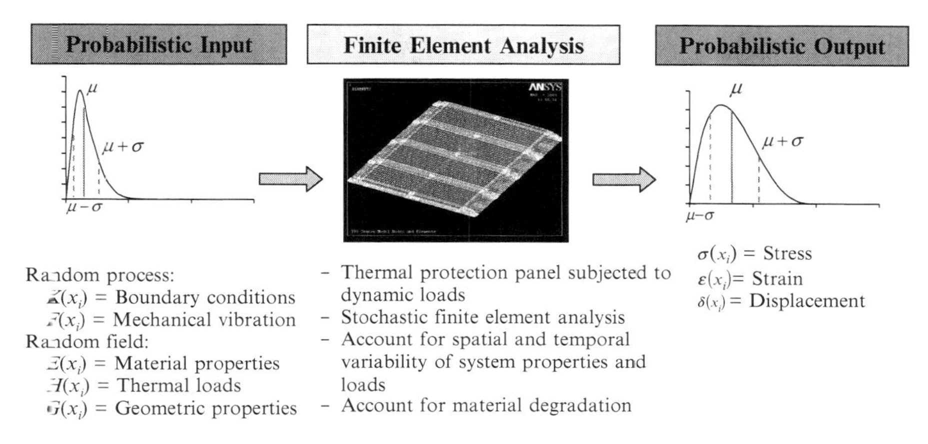

Fig. 2 Example of physical variability propagation

These techniques are generic and can be applied to engineering systems with multiple component modules and multiple physics. However, most applications of these techniques have only considered physical variability. The techniques need to include the contribution of data uncertainty and model error to the overall model prediction uncertainty. Computational effort is a significant issue in practical applications, since these techniques involve a number of repeated runs of the system analysis model. The system analysis may be replaced with an inexpensive surrogate model in order to achieve computational efficiency; this is discussed in Sect. 3.3 of this report. Efficient Monte Carlo techniques have also been pursued to reduce the number of system model runs, including *Latin hypercube sampling* (LHS) [8, 32] and *importance sampling* [28, 50].

When multiple requirements are defined, computation of the overall probability of satisfying multiple performance criteria requires integration over a multidimensional space defined by unions and intersections of individual events (of satisfaction or violation of individual criteria).

3.2 Propagation of Data Uncertainty

Three types of data uncertainty were discussed in Sect. 2. Sparse point data results in uncertainty about the parameters of the probability distributions describing quantities with physical variability. In that case, uncertainty propagation analysis takes a nested implementation. In the outer loop, samples of the distribution parameters are randomly generated, and for each set of sampled distribution parameter values, probabilistic propagation analysis is carried out as in Sect. 3.1. This results in the computation of multiple probability distributions of the output or confidence intervals for the estimates of probability of failure.

In the case of measurement error, choice of the uncertainty propagation technique depends on how the measurement error is represented. If the measurement error is represented as a random variable, it is simply added to the measured quantity, which is also a random variable due to physical variability. Thus, a sum of two random variables may be used to include both physical variability and measurement error in a quantity of interest. If the measurement error is represented as an interval, one way to implement probabilistic analysis is to represent the interval through families of distributions or upper and lower bounds on probability distributions, as discussed in Sect. 2.2.1. In that case, multiple probabilistic analyses, using the same nested approach as in the case of sparse data, can be employed to generate multiple output distributions or confidence intervals for the model output. The same approach is possible for interval variables that are only available as a range of values, as in the case of expert opinion.

Propagation of uncertainty is conceptually very simple but computationally quite expensive to implement, especially when both physical variability and data uncertainty are to be considered. The presence of both types of uncertainty requires a nested implementation of uncertainty propagation analysis (simulation of data uncertainty in the outer loop and simulation of physical variability in the inner loop). If the system model runs are time-consuming, then uncertainty propagation analysis could be prohibitively expensive. One way to overcome the computational hurdle is to use an inexpensive surrogate model to replace the detailed system model, as discussed next.

3.3 Surrogate Models

Surrogate models (also known as response surface models) are frequently used to replace the expensive system model and used for multiple simulations to quantify the uncertainty in the output. Many types of surrogate modeling methods are available, such as linear and nonlinear regression, polynomial chaos expansion, Gaussian process modeling (e.g., Kriging model), splines, moving least squares, support vector regression, relevance vector regression, neural nets, or even simple lookup tables. For example, Goktepe et al. [12] used neural network and polynomial regression models to simulate expansion of concrete specimens under sulfate attack. All surrogate models require training or fitting data, collected by running the full-scale system model repeatedly for different sets of input variable values. Selecting the sets of input values is referred to as statistical design of experiments, and there is extensive literature on this subject. Two types of surrogate modeling methods are discussed below that might achieve computational efficiency while maintaining high accuracy in output uncertainty quantification. The first method expresses the model output in terms of a series expansion of special polynomials such as Hermite polynomials and is referred to as a stochastic response surface method (SRSM). The second method expresses the model output through a Gaussian process and is referred to as Gaussian process modeling.

3.3.1 Stochastic Response Surface Method (SRSM)

The common approach for building a surrogate or response surface model is to use least squares fitting based on polynomials or other mathematical forms based on physical considerations. In SRSM, the response surface is constructed by approximating both the input and output random variables and fields through series expansions of standard random variables (e.g., [18, 19, 47]). This approach has been shown to be efficient, stable, and convergent in several structural, thermal, and fluid flow problems. A general procedure for SRSM is as follows:

(a) Representation of random inputs (either random variables or random processes) in terms of Standard Random Variables (SRVs) by *K-L* expansion, as in Eq. (2).
(b) Expression of model outputs in chaos series expansion. Once the inputs are expressed as functions of the selected SRVs, the output quantities can also be represented as functions of the same set of SRVs. If the SRVs are Gaussian, the output can be expressed a Hermite polynomial chaos series expansion in terms of Gaussian variables. If the SRVs are non-Gaussian, the output can be expressed by a general Askey chaos expansion in terms of non-Gaussian variables [10].
(c) Estimation of the unknown coefficients in the series expansion. The improved probabilistic collocation method [19] is used to minimize the residual in the random dimension by requiring the residual at the collocation points equal to zero. The model outputs are computed at a set of collocation points and used to estimate the coefficients. These collocation points are the roots of the Hermite polynomial of a higher order. This way of selecting collocation points would capture points from regions of high probability [45].
(d) Calculation of the statistics of the output that has been cast as a response surface in terms of a chaos expansion. The statistics of the response can be estimated with the response surface using either Monte Carlo simulation or analytical approximation.

3.3.2 Kriging or Gaussian Process Models

Gaussian process (GP) models have several features that make them attractive for use as surrogate models. The primary feature of interest is the ability of the model to "account for its own uncertainty." That is, each prediction obtained from a Gaussian process model also has an associated variance or uncertainty. This prediction variance primarily depends on the closeness of the prediction location to the training data, but it is also related to the functional form of the response. For example, see Fig. 3, which depicts a one-dimensional Gaussian process model. Note how the uncertainty bounds are related to both the closeness to the training points, as well as the shape of the curve.

The basic idea of the GP model is that the output quantities are modeled as a group of multivariate normal random variables. A parametric covariance function is

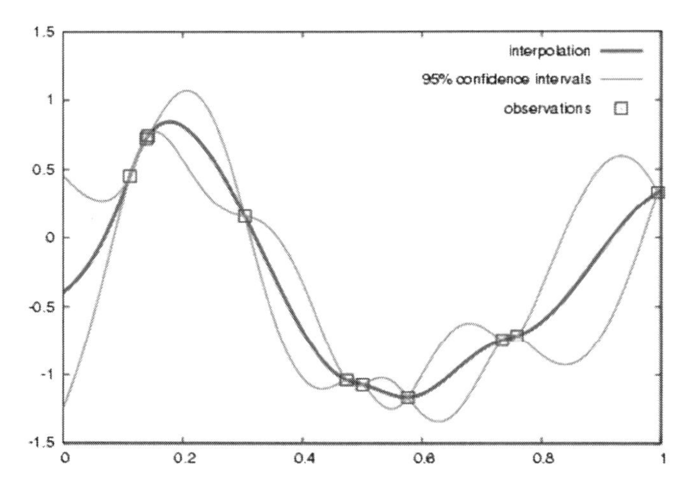

Fig. 3 Gaussian process model with uncertainty bounds

then constructed as a function of the inputs. The covariance function is based on the idea that when the inputs are close together, the correlation between the outputs will be high. As a result, the uncertainty associated with the model prediction is small for input values that are close to the training points and large for input values that are not close to the training points. In addition, the GP model may incorporate a systematic trend function, such as a linear or quadratic regression of the inputs (in the notation of Gaussian process models, this is called the mean function, while in Kriging, it is often called a trend function). The effect of the mean function on predictions which interpolate the training data is small, but when the model is used for extrapolation, the predictions will follow the mean function very closely.

Within the GP modeling technique, it is also possible to adaptively select the design of experiments to achieve very high accuracy. The method begins with an initial GP model built from a very small number of samples, and then one intelligently chooses where to generate subsequent samples to ensure the model is accurate in the vicinity of the region of interest. Since the GP model provides the expected value and variance of the output quantity, the next sample may be chosen in the region of highest variance, if the objective is to minimize the prediction variance. The method has been shown to be both accurate and computationally efficient for arbitrarily shaped functions [2].

3.4 Sensitivity Analysis

Sensitivity analysis serves several important functions: (1) identification of dominant variables or sub-models, thus helping to focus data collection resources efficiently; (2) identification of insignificant variables or sub-models of limited significance, helping to reduce the size of the problem and computational effort; and (3) quantification of the contribution of solution approximation error. Both local and global sensitivity analysis techniques are available to investigate the quantitative effect of different sources of variation (physical parameters, models, and measured data) on the variation of the model output. The primary benefit of sensitivity analysis to uncertainty analysis is to enable the identification of which physical parameters have the greatest influence on the output [6, 42]).

Sensitivity analysis can be local or global. Local sensitivity analysis utilizes first-order derivatives of system output quantities with respect to the parameters. It is usually performed for a nominal set of parameter values. Global sensitivity analysis typically uses statistical sampling methods, such as Latin Hypercube Sampling, to determine the total uncertainty in the system output over the entire range of the input uncertainty and to apportion that uncertainty among the various parameters.

3.5 *Model Error Quantification*

Model errors may relate to governing equations, boundary and initial condition assumptions, loading description, and approximations or errors in solution algorithms (e.g., truncation of higher order terms, finite element discretization, curve-fitting models for material damage such as *S-N* curve). Overall model error may be quantified by comparing model prediction and experimental observation, properly accounting for uncertainties in both. This overall error measure combines both model form and solution approximation errors, so it needs to be considered in two parts. Numerical errors in the model prediction can be quantified first, using sensitivity analysis, uncertainty propagation analysis, discretization error quantification, and truncation (residual) error quantification. The measurement error in the input variables can be propagated to the prediction of the output. The error in the prediction of the output due to the measurement error in the input variables is approximated by using a first-order sensitivity analysis [36]. Then the model form error can be quantified based on all the above errors, following the approach illustrated for a heat transfer problem by Rebba et al. [36].

3.5.1 Solution Approximation Error

Several components of prediction error, such as discretization error (denoted by ε_d) and uncertainty propagation analysis error (ε_s) can be considered. Several methods to quantify the discretization error in finite element analysis are available in the literature. However, most of these methods do not quantify the actual error; instead, they only quantify some indicator measures to facilitate adaptive mesh refinement. The Richardson extrapolation (RE) method comes closest to quantifying the actual discretization error [38]. (In some applications, the model is run with different levels of resolution, until an acceptable level of accuracy is achieved; formal error quantification may not be required).

Errors in uncertainty propagation analysis (ε_s) are method-dependent, i.e., sampling error occurs in Monte Carlo methods and truncation error occurs in response surface methods (either conventional or polynomial chaos-based). For example, sampling error could be assumed to be a Gaussian random variable with zero mean and variance given by σ^2/N where N is the number of Monte Carlo runs and σ^2 is the original variance of the model output [41]. The truncation error is simply the residual error in the response surface.

Rebba et al. [36] and Liang and Mahadevan [26] used the above concept to construct a surrogate model for finite element discretization error in structural analysis, using the stochastic response surface method (SRSM). Gaussian process models may also be employed for this purpose. Both options are helpful in quantifying the solution approximation error.

3.5.2 Model form Error

The overall prediction error is a combination of errors resulting from numerical solution approximations and model form selection. A simple way is to express the total observed error (difference between prediction and observation) as the sum of the following error sources:

$$\varepsilon_{obs} = \varepsilon_{num} + \varepsilon_{model} - \varepsilon_{exp} \tag{3}$$

where ε_{num}, ε_{model}, and ε_{exp} represent numerical solution error, model form error, and output measurement error, respectively. However, solution approximation error results from multiple sources and is probably a nonlinear combination of various errors such as discretization error, round-off and truncation errors, and stochastic analysis errors. One option is to construct a regression model consisting of the individual error components [36]. The residual of such a regression analysis will include the model form error (after subtracting the experimental error effects). By denoting ε_{obs} as the difference between the data and prediction, i.e., $\varepsilon_{obs} = y_{exp} - y_{pred}$, we can construct the following relation by considering a few sources of numerical solution error [36]:

$$\varepsilon_{obs} = f(\varepsilon_h, \varepsilon_d, \varepsilon_s) + \varepsilon_{model} - \varepsilon_{exp} \tag{4}$$

where ε_h, ε_d, and ε_s represent output error due to input parameter measurement error, finite element discretization error, and uncertainty propagation analysis error, respectively, all of which contribute to numerical solution error. Rebba et al. [36] illustrated the estimation of model form error using the above concept for a one-dimensional heat conduction problem, using a polynomial chaos expansion for the input-output model as well as numerical solution error. Kennedy and O'Hagan [24] calibrated Gaussian process surrogate models for both the input-output model and the model form error (which is also referred to as model discrepancy or model inadequacy term). Both approaches incorporate the dependence of model error on input values.

4 Model Calibration, Validation and Extrapolation

After quantifying and propagating the physical variability, data uncertainty, and model error for individual components of the overall system model, the probability of meeting performance requirements (and our confidence in the model prediction) needs to be assessed based on extrapolating the model to field conditions (which are uncertain as well), where sometimes very limited or no experimental data is available. Rigorous verification, validation, and calibration methods are needed to establish credibility in the modeling and simulation. Both classical and Bayesian statistical methodologies have been investigated during recent years. The methods have the capability to consider multiple output quantities or a single model output at different spatial and temporal points.

This section discusses methods for (1) calibration of model parameters, based on observation data; (2) validation assessment of the model, based on observation data; and (3) estimation of confidence in the extrapolation of model prediction from laboratory conditions to field conditions.

4.1 Model Calibration

Two types of statistical techniques may be pursued for model calibration uncertainty, the least squares approach and the Bayesian approach. The least squares approach estimates the values of the calibration parameters that minimize the discrepancy between model prediction and experimental observation. This approach can also be used to calibrate surrogate models or low-fidelity models, based on high-fidelity runs, by treating the high-fidelity results similar to experimental data.

The second approach is Bayesian calibration [24] using Gaussian process surrogate models. This approach is flexible and allows different forms for including the model errors during calibration of model parameters [31]. Recently, Sankararaman and Mahadevan [43] extended least squares, likelihood and Bayesian calibration approaches to include imprecise and unpaired input-output data sets, a commonly occurring situation when using historical data or data from the literature, where all the inputs to the model may not be reported.

Markov Chain Monte Carlo (MCMC) simulation is used for numerical implementation of the Bayesian updating analysis. Several efficient sampling techniques are available for MCMC, such as Gibbs sampling, the Metropolis algorithm, and the Metropolis-Hastings algorithm [11].

4.2 Model Validation

Model validation involves comparing prediction with observation data (either historical or experimental) when both have uncertainty. Since there is uncertainty in both model prediction and experimental observation, it is necessary to pursue rigorous statistical techniques to perform model validation assessment rather than simple graphical comparisons, provided data is even available for such comparisons. Statistical hypothesis testing is one approach to quantitative model validation under uncertainty, and both classic and Bayesian statistics have been explored. Classical hypothesis testing is a well-developed statistical method for accepting or rejecting a model based on an error statistic (see e.g., [46]). Validation metrics have been investigated in recent years based on Bayesian hypothesis testing [29, 33, 34, 49], reliability-based methods [35], and risk-based decision analysis [22, 23]. Ling and Mahadevan [27] provide detailed discussion of the interpretations of various metrics, their mathematical relationships, and implementation issues, with the example of a MEMS device reliability prediction problem and validation data.

In Bayesian hypothesis testing, we assign prior probabilities for the null and alternative hypotheses; let these be denoted as $P(H_0)$ and $P(H_a)$ such that $P(H_0) + P(H_a) = 1$. Here, H_0: model error < allowable limit, and H_a: model error > allowable limit. When data D is obtained, the probabilities are updated as $P(H_0 \mid D)$ and $P(H_a \mid D)$ using the Bayesian theorem. Then, a Bayesian factor [20] B is defined as the ratio of likelihoods of observing D under H_0 and H_a; i.e., the first term in the square brackets on the right-hand side of Eq. (5):

$$\frac{P(H_0|D)}{P(H_a|D)} = \left[\frac{P(D|H_0)}{P(D|H_a)}\right] \frac{P(H_0)}{P(H_a)} \tag{5}$$

If $B > 1$, the data gives more support to H_0 than H_a. Also, the confidence in H_0, based on the data, comes from the posterior null probability $P(H_0 \mid D)$, which can be rearranged from the above equation as $\frac{P(H_0)B}{P(H_0)B+1-P(H_0)}$. Typically, in the absence of prior knowledge, we may assign equal probabilities to each hypothesis, and thus, $P(H_0) = P(H_a) = 0.5$. In that case, the posterior null probability can be further simplified to $B/(B + 1)$. Thus, a B value of 1.0 represents 50 % confidence in the null hypothesis being true.

The Bayesian hypothesis testing is also able to account for uncertainty in the distribution parameters (mentioned in Sect. 2). For such problems, the validation metric (Bayesian factor) itself becomes a random variable. In that case, the probability of the Bayesian factor exceeding a specified value can be used as the decision criterion for model acceptance/rejection.

Notice that model validation only refers to the situation when controlled, target experiments are performed to evaluate model prediction, and both the model runs and experiments are done under the same set of input and boundary conditions. The validation is done only by comparing the outputs of the model and the experiment.

Once the model is calibrated, verified, and validated, it may be investigated for confidence in extrapolating to field conditions different from laboratory conditions. This is discussed in the next section.

4.3 Overall Uncertainty Quantification

While individual methods for calibration, verification, and validation have been developed as mentioned above, it is necessary to integrate the results from these activities for the purpose of overall uncertainty quantification in the model prediction. This is not trivial because of several reasons. First, the solution approximation errors calculated as a result of the verification process need to be accounted for during calibration, validation, and prediction. Second, the result of validation may lead to a binary result, i.e., the model is accepted or rejected; however, even when the model is accepted, it is not completely correct. Hence, it is necessary to account for the degree of correctness of the model in the prediction. Third, calibration and validation are performed using independent data sets, and it is not straightforward to compute their combined effect on the overall uncertainty in the response.

The issue gets further complicated when system-level behavior is predicted based on a hierarchy of models. As the complexity of the system under study increases, there may be several components and subsystems at multiple levels of hierarchy, which integrate to form the overall multilevel system. Each of these components and subsystems is represented using component-level and subsystem-level models which are mathematically connected to represent the overall system model which is used to study the underlying system. In each level, there is a computational model with inputs, parameters, outputs, experimental data (hopefully available for calibration and validation separately), and several sources of uncertainty – physical variability, data uncertainty (sparse or imprecise data, measurement errors, expert opinion), and model uncertainty (parameter uncertainty, solution approximation errors, and model form error).

Recent studies by the author and coworkers have demonstrated that the Bayesian network methodology provides an efficient and powerful tool to integrate multiple levels of models, associated sources of uncertainty and error, and available data at multiple levels and in multiple formats. While the Bayesian approach can be used to perform calibration and validation individually for each model in the multi-level system, it is not straightforward to integrate the information from these activities in order to compute the overall uncertainty in the system-level prediction. Sankararaman and Mahadevan [44] extend the Bayesian approach to integrate and propagate information from verification, calibration, and validation activities in order to quantify the margins and uncertainties in the overall system-level prediction.

Bayesian networks [21] are directed acyclic graphical representations with nodes to represent the random variables and arcs to show the conditional dependencies among the nodes. Data in any one node can be used to update the statistics of all

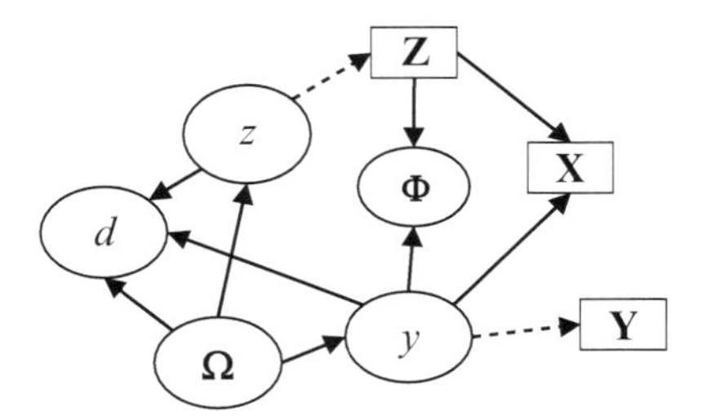

Fig. 4 Bayes network

other nodes. This property makes the Bayesian network a powerful tool to integrate information generated from multiple activities and to quantify the uncertainty in prediction under actual usage conditions [29].

Figure 4 shows an illustrative Bayesian network for confidence extrapolation. An ellipse represents a random variable, and a rectangle represents observed data. A solid line arrow represents a conditional probability link, and a dashed line arrow represents the link of a variable to its observed data if available. The probability densities of the variables $\boldsymbol{\Omega}$, z, and y are updated using the validated data $\mathbf{Y}$. The updated statistics of $\boldsymbol{\Omega}$, z, and y are then used to estimate the updated statistics of the decision variable d (i.e., assessment metric). In addition, both model prediction and predictive experiments are related to input variables $\mathbf{X}$ via physical parameters Φ. Note that there is no observed data available for d; yet we are able to calculate the confidence in the prediction of d, by making use of observed data in several other nodes and propagation of posterior statistics through the Bayesian network.

The Bayesian network thus links the various simulation codes and corresponding experimental observations to facilitate two objectives: (1) uncertainty quantification and propagation and (2) confidence assessment in system behavior prediction in the application domain, based on data from the laboratory domain, expert opinion, and various computational models at different levels of the system hierarchy.

5 Conclusion

Uncertainty quantification in performance assessment involves consideration of three sources of uncertainty – inherent variability, information uncertainty, and model errors. This chapter surveyed probabilistic methods to quantify the uncertainty in model-based prediction due to each of these sources and addressed them in four stages – input characterization based on data; propagation of uncertainties and errors through the system model; model calibration, validation, and extrapolation; and performance assessment. Flexible distribution families as well as a nonparametric Bayesian approach were discussed to handle sparse data and interval data. Methods to quantify model errors resulting from both model form selection and solution approximation were discussed. Bayesian methods were discussed for model calibration, validation, and extrapolation. An important issue is computational expense, when iterative analysis between multiple codes is necessary. Uncertainty quantification multiplies the computational effort of deterministic analysis by an order of magnitude. Therefore, the use of surrogate models, sensitivity and screening analyses, and first-order approximations of overall output uncertainty are available to reduce the computational expense.

Many of the methods described in the chapter have been applied to mechanical systems that are small in size or time-independent, and the uncertainties considered were not very large. None of these simplifications are available in the case of long-term performance assessment of civil infrastructure systems, and real-world data to validate long-term model predictions are not available. Thus, the extrapolations are based on laboratory data or limited term observations and come with large uncertainty. The application of the methods described in this chapter to such complex systems needs to be investigated. However, it should be recognized that the benefit of uncertainty quantification is not so much in predicting the actual failure probability or similar measures but in facilitating engineering decision-making, such as comparing different design and analysis options, performing sensitivity analyses, and allocating resources for uncertainty reduction through further data collection and/or model refinement.

Acknowledgement The research described in this chapter by the author and his students/colleagues has been funded by many sources during the past decade. A partial listing of the recent sources includes the following: (1) National Science Foundation (IGERT project on Reliability and Risk Assessment and Management at Vanderbilt University), (2) Sandia National Laboratories (Bayesian framework for model validation, calibration, and error estimation), (3) US Department of Energy (uncertainty quantification in micro-electro-mechanical systems (MEMS) reliability prediction, long-term durability of cementitious barriers), (4) National Aeronautics and Space Administration (space vehicle performance uncertainty quantification, uncertainty quantification in diagnosis and prognosis, Bayesian network development for testing resource allocation), (5) US Air Force Office of Scientific Research (multidisciplinary uncertainty analysis of aircraft components), and (6) Federal Aviation Administration (uncertainty quantification in fracture mechanics simulation of rotorcraft components). The support is gratefully acknowledged.

References

1. Barford NC (1985) Experimental measurements: precision, error, and truth. Wiley, New York
2. Bichon, BJ, Eldred, MS, Swiler, LP, Mahadevan, S, McFarland, JM (2007) Multimodal reliability assessment for complex engineering applications using efficient global optimization. In: Proceedings of 9th AIAA non-deterministic approaches conference, Waikiki, HI
3. Blischke WR, Murthy DNP (2000) Reliability: modeling, prediction, and optimization. Wiley, New York
4. Box GEP, Hunter WG, Hunter JS (1978) Statistics for experimenters, an introduction to design, data analysis, and model building. Wiley, New York
5. Box GEP, Jenkins GM, Reinsel GC (1994) Time series analysis forecasting and control, 3rd edn. Prentice Hall, Englewood Cliffs

6. Campolongo F, Saltelli A, Sorensen T, Tarantola S (2000) Hitchhiker's guide to sensitivity analysis. In: Saltelli A, Chan K, Scott EM (eds) Sensitivity analysis. Wiley, New York, pp 15–47
7. Efron B, Tibshirani RJ (1994) An introduction to the bootstrap. Chapman & Hall/CRC, New York/Boca Raton
8. Farrar CR, Sohn H, Hemez FM, Anderson MC, Bement MT, Cornwell PJ, Doebling SW, Schultze JF, Lieven N, Robertson AN (2003) Damage prognosis: current status and future needs. Technical report LA–14051–MS, Los Alamos National Laboratory, Los Alamos, New Mexico
9. Ferson S, Kreinovich V, Hajagos J, Oberkampf W, Ginzburg L (2007) Experimental uncertainty estimation and statistics for data having interval uncertainty. Sandia National Laboratories technical report, SAND2003-0939, Albuquerque, New Mexico
10. Ghanem R, Spanos P (2003) Stochastic finite elements: a spectral approach. Springer, New York
11. Gilks WR, Richardson S, Spiegelhalter DJ (1996) Markov Chain Monte Carlo in practice, Interdisciplinary statistics series. Chapman and Hall, Boca Raton
12. Goktepe AB, Inan G, Ramyar K, Sezer A (2006) Estimation of sulfate expansion level of pc mortar using statistical and neural approaches. Constr Build Mater 20:441–449
13. Gurley KR (1997) Modeling and simulation of non-Gaussian processes. Ph.D. thesis, University of Notre Dame, April
14. Haldar A, Mahadevan S (2000) Probability, reliability and statistical methods in engineering design. Wiley, New York
15. Haldar A, Mahadevan S (2000) Reliability analysis using the stochastic finite element method. Wiley, New York
16. Helton JC, Sallabery CJ (2009) Conceptual basis for the definition and calculation of expected dose in performance assessments for the proposed high-level radioactive waste repository at Yucca Mountain, Nevada. Reliab Eng Syst Saf 94:677–698
17. Helton JC, Sallabery CJ (2009) Computational implementation of sampling-based approaches to the calculation of expected dose in performance assessments for the proposed high-level radioactive waste repository at Yucca Mountain, Nevada. Reliab Eng Syst Saf 94:699–721
18. Huang S, Mahadevan S, Rebba R (2007) Collocation-based stochastic finite element analysis for random field problems, Probab Eng Mech 22:194–205
19. Isukapalli SS, Roy A, Georgopoulos PG (1998) Stochastic response surface methods (SRSMs) for uncertainty propagation: application to environmental and biological systems. Risk Anal 18(3):351–363
20. Jeffreys H (1961) Theory of probability, 3rd edn. Oxford University Press, London
21. Jensen FV, Jensen FB (2001) Bayesian networks and decision graphs. Springer, New York
22. Jiang X, Mahadevan S (2007) Bayesian risk-based decision method for model validation under uncertainty. Reliab Eng Syst Saf 92(6):707–718
23. Jiang X, Mahadevan S (2008) Bayesian validation assessment of multivariate computational models. J Appl Stat 35(1):49–65
24. Kennedy MC, O'Hagan A (2001) Bayesian calibration of computer models (with discussion). JR Stat Soc Ser B 63(3):425–464
25. Langley RS (2000) A unified approach to the probabilistic and possibilistic analysis of uncertain systems. ASCE J Eng Mech 126:1163–1172
26. Liang B, Mahadevan S (2011) Error and uncertainty quantification and sensitivity analysis of mechanics computational models. Int J Uncertain Quantif 1:147–161
27. Ling Y, Mahadevan S (2012) Intepretations, relationships, and application issues in model validation. In: Proceedings, 53rd AIAA/ASME/ASCE Structures, Dynamics and Materials (SDM) conference, paper no. AIAA-2012-1366, Honolulu, Hawaii, April 2012
28. Mahadevan S, Raghothamachar P (2000) Adaptive simulation for system reliability analysis of large structures. Comput Struct 77(6):725–734

29. Mahadevan S, Rebba R (2005) Validation of reliability computational models using Bayes networks. Reliab Eng Syst Saf 87(2):223–232
30. Mathelin L, Hussaini MY, Zang TA (2005) Stochastic approaches to uncertainty quantification in CFD simulations. Numer Algorithm 38:209–236
31. McFarland JM (2008) Uncertainty analysis for computer simulations through validation and calibration. Ph.D. dissertation, Vanderbilt University, Nashville, TN
32. Mckay MD, Conover WJ, Beckman RJ (1979) A comparison of three methods for selecting values of input variables in the analysis of output from a computer code. Technometrics 21:239–245
33. Rebba R (2005) Model validation and design under uncertainty. Ph.D. dissertation, Vanderbilt University, Nashville, TN, USA
34. Rebba R, Mahadevan S (2006) Model predictive capability assessment under uncertainty. AIAA J 44(10):2376–2384
35. Rebba R, Mahadevan S (2008) Computational methods for model reliability assessment. Reliab Eng Syst Saf 93:1197–1207
36. Rebba R, Mahadevan S, Huang S (2006) Validation and error estimation of computational models. Reliab Eng Syst Saf 91(10–11):1390–1397
37. Red-Horse JR, Benjamin AS (2004) A probabilistic approach to uncertainty quantification with limited information. Reliab Eng Syst Saf 85:183–190
38. Richards SA (1997) Completed Richardson extrapolation in space and time. Commun Numer Methods Eng 13(7):558–573
39. Robert CP, Casella G (2004) Monte Carlo statistical methods, 2nd edn. Springer, New York
40. Ross TJ, Booker JM, Parkinson WJ (2002) Fuzzy logic and probability applications: bridging the gap. SIAM, Philadelphia
41. Rubinstein RY (1981) Simulation and the Monte Carlo method. Wiley, New York
42. Saltelli A, Chan K, Scott EM (2000) Sensitivity analysis. Wiley, West Sussex
43. Sankararaman S, Mahadevan S (2011) Likelihood-based representation of epistemic uncertainty due to sparse point data and interval data. Reliab Eng Syst Saf 96:814–824
44. Sankararaman S, Mahadevan S (2012) Roll-up of calibration and validation results towards system-level QMU. In: Proceedings of 15th AIAA non-deterministic approaches conference, Honolulu, Hawaii
45. Tatang MA, Pan W, Prinn RG, McRae GJ (1997) An efficient method for parametric uncertainty analysis of numerical geophysical models. J Geophys Res 102(D18):21925–21932
46. Trucano TG, Easterling RG, Dowding KJ, Paez TL, Urbina A, Romero VJ, Rutherford BM, Hills RG (2001) Description of the Sandia validation metrics project. Sandia National Laboratories technical report, SAND2001-1339, Albuquerque, New Mexico
47. Xiu D, Karniadakis GE (2003) Modeling uncertainty in flow simulations via generalized polynomial chaos. J Comput Phys 187(1):137–167
48. Zaman K, McDonald M, Mahadevan S (2011) A probabilistic approach for representation of interval uncertainty. Reliab Eng Syst Saf 96(1):117–130
49. Zhang R, Mahadevan S (2003) Bayesian methodology for reliability model acceptance. Reliab Eng Syst Saf 80(1):95–103
50. Zou T, Mahadevan S, Mourelatos Z (2003) Reliability-based evaluation of automotive wind noise quality. Reliab Eng Syst Saf 82(2):217–224

A Brief Review of Stochastic Control of Structures

T.K. Datta

Abstract Stochastic control of structures is a relatively new area of research. Despite enormous amount of work on deterministic structural control and its application on many civil engineering structures, stochastic structural control is evolving as a promising subject of investigation. The main reason for choosing a stochastic control theory is that most of the feedback and excitation measurements are polluted with noises which are inherently random. Further, environmental dynamic excitations are more realistically modeled as stochastic processes. In this presentation, stochastic control of structures is briefly reviewed in order to highlight different types of researches being carried out in this area in recent years. The review includes linear and nonlinear stochastic control of linear systems, nonlinear stochastic control of nonlinear systems, stochastic control of hysteretic systems, and stochastic control of structures using stochastic linearization technique. A number of interesting applications of stochastic control theory to structures such as stochastic control of hysteretic column, semiactive control of building frame using MR-TLCD, and active control of coupled buildings are discussed. Finally, future scopes of research in this area emerging from the literature review are outlined.

Keywords Structures • Linear and nonlinear stochastic control • Linear systems • Nonlinear systems • Hysteretic system

---

T.K. Datta (✉)
Department of Civil Engineering, Indian Institute of Technology Delhi,
Hauz Khas, New Delhi 110 016, India
e-mail: tushar_k_datta@yahoo.com

S. Chakraborty and G. Bhattacharya (eds.), *Proceedings of the International Symposium on Engineering under Uncertainty: Safety Assessment and Management (ISEUSAM - 2012)*,
DOI 10.1007/978-81-322-0757-3_6,

1 Introduction

Stochastic optimal control is a topic of considerable interest in different fields of engineering since uncertainties are inherent in all types of control problems. These uncertainties include uncertainties of modeling, estimation of parameters, measurements, disturbances, etc. In the field of structural engineering, stochastic optimal control of structures becomes necessary when the structures are subjected to random excitations and the system states are estimated from measurements with random noises. Further, it is also required because uncertainties are also present in the stiffness and damping of the system. As a result, the stochastic optimal control should also be robust, i.e., its stability should be guaranteed.

Mathematical theory of stochastic optimal control is quite well developed. However, only linear quadratic Gaussian control strategy has been used so far for structural control problems mainly because of its simplicity. In recent years, nonlinear stochastic control of structures has attracted the attention of researchers mainly because the nonlinearities are present in most practical control problems. Even without considering the nonlinearities of the system or even after linearizing them, nonlinear control algorithm may be applied to control the system. It is found that more effective control is obtained by applying nonlinear control theory. A number of papers dealing with linear and nonlinear stochastic control of structures are available in the literature [1–14]. A survey of these papers reveals the current state of the art on the subject including the application problems that have been solved.

In this chapter, different types of work which have been carried out on the stochastic control of structures are briefly reviewed. The review includes (1) stochastic optimal control for linear systems with or without observer, (2) stochastic nonlinear optimal control for linear systems with or without observer, (3) stochastic optimal control for nonlinear system using linearization technique, (4) nonlinear optimal control for nonlinear systems, (5) stochastic control using stability criterion, (6) H_α stochastic control for linearized hysteretic system, and (7) stochastic control using stochastic averaging technique. A number of applications such as stochastic seismic control of hysteretic columns, coupled buildings, semiactive control of buildings using MR-TLCD, and MR dampers are also discussed. At the end, future scope of the work in this area is outlined.

2 Stochastic Control Problem

The control problem is mathematically defined as

$$\text{System} \qquad \dot{X} = f(x, u, k, t) \tag{1}$$

$$\text{Observation} \qquad Y = h(x, t) + \eta \tag{2}$$

$$\text{Performance Index} \qquad J = E\left[\int_0^{t_f} F(x, u, t)\mathrm{d}t\right] \tag{3}$$

in which f could be a linear or nonlinear function with or without uncertain system dynamics and noise/excitation (k), η is additive linear or nonlinear noisy component associated with the measurements, u is the control force, F is a nonlinear function of finite or infinite duration, and E denotes expectation. Complexity of the problem increases with increasing nonlinearity in the system and with less availability of measured responses. The minimization of the performance function which provides the control force becomes complex for nonlinear systems. Observed states are determined using some filters. The problem can be formulated with continuous variable or variables described in discrete form. Algorithms of the computations can be derived for both.

3 Linear Stochastic Control

The linear stochastic control takes the following forms [9]:

$$\text{System} \qquad \dot{X} = AX + bu \tag{4}$$

$$\text{Observation} \qquad Y = HX + \eta \tag{5}$$

$$\text{Performance Index} \qquad J = E\left[\int_0^{t_f} \left[X^T QX + ru^2\right]\mathrm{d}t\right] \tag{6}$$

Initial state of the system is normally distributed with mean X_0 and covariance P_0. Observation error η is assumed to be Gaussian white noise with mean zero and mean square value $N(t)$, i.e., $E[\eta(t)\eta(\tau)] = N(t)\delta(t-\tau)$. The conditional mean and covariance of the state $X(t)$ are given by

$$\mu(t) = E[X(t)|Y(\tau)] \tag{7}$$

$$P(t) = E\left[(X-\mu)(X-\mu)^T|Y(\tau)\right] \tag{8}$$

These quantities are governed by Kalman filter equation

$$\dot{\mu} = A\mu + bu + PH^T N^{-1}(Y - H\mu) \tag{9}$$

$$\dot{P} = AP + PA^T - PH^T N^{-1} HP \tag{10}$$

With initial conditions as

$$\mu(0) = X_0 \quad P(0) = P_0$$

$u(t)$ is obtained by minimizing J [1]

$$u(t) = -\frac{1}{r}\mu^T Bb \tag{11}$$

where B satisfies the Riccati equation

$$\dot{B} + BA + A^T B - \frac{1}{r}Bbb^T b + Q = 0 \tag{12}$$

with $B(t_f) = 0$

Discrete state-space model of linear stochastic control system is given by [2]

$$X_{K+1} = AX_K + Bu_K + w_K \tag{13}$$

$$Y_K = CX_K + \eta_K \tag{14}$$

$$Min\,J = Min_{u_K \cdots u_N} E\left[X_{N+1}^T Q_0 X_{N+1} + \sum_{S=K}^{N} X_S^T Q_1 X_S + u_S^T Q_2 u_S\right] \tag{15}$$

$$u_K = -L_K \widehat{X}_{K|K}; \quad L_K = \left[Q_2 + B^T S_{K+1} B\right]^{-1} B^T S_{K+1} A \tag{16}$$

$$S_K = Q_1 + A^T S_{K+1}(A - BL_K); \quad S_{N+1} = Q_0 \tag{17}$$

For fully measured state, $\widehat{X}_{K|K} \to X_K$, i.e., observed state, is the same as the measured state. A backward step computational algorithm is developed without the solution of any Riccati equation. Kalman filter equation in discrete form is used for state estimation, i.e.,

$$\widehat{X}_{K+1} = A\widehat{X}_K + Bu_K + R_K\left(Y_K - C\widehat{X}_K\right) \tag{18}$$

$$R_K = A\sum\nolimits_K C^T\left(C\sum\nolimits_K C^T + N\right)^{-1} \tag{19}$$

$$\sum\nolimits_{K+1} = W + (A - R_K C)\sum\nolimits_K A^T \tag{20}$$

$$\sum\nolimits_K = \text{cov}\widehat{X}_K \tag{21}$$

in which N is defined earlier and W is the covariance matrix of the white noise $w(t)$.

4 Nonlinear Optimal Control

For explaining the nonlinear optimal control for linear system, a linear structure under random ground motion is considered [14]:

$$M\ddot{X} + C\dot{X} + KX = -Mr\ddot{x}_g + PU \tag{22}$$

The solution is carried out in modal coordinate with n modal equations as

$$\ddot{q}_i + 2\eta\omega_i\dot{q}_i + \omega_i^2 q_i = -\beta_i\ddot{x}_g + u_i\,(i = 1\cdots\cdots n) \tag{23}$$

in which β_i is the mode participation factor and u_i is the ith modal control force. Assuming $\ddot{x}_g$ to be a uniformly modulated stationary process,

$$S_{\ddot{x}_g}(\tau) = \sigma^2(\tau)S_{\ddot{x}_g} \tag{24}$$

The uniformly modulated stationary process is considered as filtered white noise with following transformation:

$$\ddot{x}_g = \left(\sum_{j=0}^{r-1} d_j \frac{\mathrm{d}^j}{\mathrm{d}t^j}\right)f\left(\sum c_j \frac{\mathrm{d}^j}{\mathrm{d}t^j}\right)f = \sigma(\tau)w(t) \tag{25}$$

in which $w(t)$ is the white noise process.

With this transformation, a combined Ito equation is formulated as

$$dZ = (AZ + Bu)\mathrm{d}t + CdB(t) \tag{26}$$

$$Z = \left[q^T, \dot{q}^T, f^T\right]^T$$

Observation equations are formed with measured quantities expressed as modal responses and noises. They are converted into equation as

$$\mathrm{d}v = (DZ + Gu)\mathrm{d}t + \sigma_1(\tau)dB_1(t) \tag{27}$$

in which $\dfrac{dv}{dt} = \left[a^T, d^T\right]$ and $\sigma_1(\tau)$ is the accuracy matrix. a and d denote vectors of measured accelerations and floor drifts.

Objective function is taken as

$$J = \lim_{T\to\infty} \frac{1}{T_f} \int_0^{t_f} L(q, \dot{q}, u)dt \tag{28}$$

It is assumed that σ and σ_1 are constants, and as $T \to \infty$, the response of the controlled system is stationary and ergodic. Both system and observation equations are linear. If L is nonquadratic, nonlinear feedback control is formed.

Let state vector estimate be denoted by $\widehat{Z}$. Ito equation with estimated $\widehat{Z}$ is governed by

$$\mathrm{d}\widehat{Z} = \left(A\widehat{Z} + Bu\right)\mathrm{d}t + f(t)\sigma_1(\tau)\widehat{B}(t) \tag{29}$$

$$\widehat{Z}(0) = \widehat{Z}_0 \tag{30}$$

$$f(t) = R_C(t)D^T S_a^{-1} \tag{31}$$

$R_C(t)$ is the covariance matrix of error $\bar{Z} = Z - \widehat{Z}$ which satisfies the Riccati type matrix equation. The objective function is written in the form of

$$\widehat{J} = \int \widehat{L}\left(\widehat{q}, \widehat{\dot{q}}, u\right)dt \tag{32}$$

in which $\widehat{L} = \int Lp\left(Z - \widehat{Z}\right)\mathrm{d}z$; p is Gaussian distribution.

Using Hamiltonian as the variable for performance function,

$$\widehat{H}_i = \frac{\widehat{\dot{q}}_i^2 + \omega_i^2 \widehat{q}_i^2}{2} \tag{33}$$

and using Vanderpol transformation and stochastic averaging, the Ito equation is formed as

$$\mathrm{d}\widehat{H}_i = \left[m_i\left(\widehat{H}_i, \tau\right) + \frac{d\widehat{H}_i}{d\widehat{\dot{q}}_i} u_i\right]\mathrm{d}t + \sigma_i\left(\widehat{H}_i, \tau\right)dB_i(t) \tag{34}$$

Dynamic programming is used to minimize $\widehat{J}$, and optimal $\widehat{u}^*$ is obtained in the form

$$\widehat{u}^* = -\frac{1}{2}R^{-1}P^T \sum_{i=1}^{m} \widehat{q}_i \widehat{\dot{q}}_i \frac{\mathrm{d}v}{\mathrm{d}H_i} \tag{35}$$

$\dfrac{\mathrm{d}v}{\mathrm{d}H_i}$ is obtained from a separate equation, i.e., dynamic programming equation for ergodic control problem [4, 13]

$$\lambda = g(\bar{H}) + \sum_{i=1}^{n_1}\left[m_i\left(\widehat{H}_i\right)\frac{\mathrm{d}v}{\mathrm{d}\widehat{H}_i} - \frac{1}{4}\phi_i^T P_u \phi_i H_i\left(\frac{\mathrm{d}v}{\mathrm{d}\widehat{H}_i}\right)^2 + \frac{1}{2}\sigma_i^2\left(\widehat{H}_i\right)\frac{\mathrm{d}^2 v}{\mathrm{d}\widehat{H}_i^2}\right] \tag{36}$$

$$P_u = PR^{-1}P^T \tag{37}$$

Substituting for $\widehat{u}^*$, Ito equation turns out to be

$$\mathrm{d}\widehat{H}_i = \left[m_i\left(\widehat{H}_i\right) + m_i^u(\bar{H}_i)\right]\mathrm{d}t + \sigma_i\left(\widehat{H}_i\right)\mathrm{d}\bar{B}_i(t) \tag{38a}$$

in which $m_i^u(\bar{H}_i)$ is given by

$$m_i^u(\bar{H}_i) = \left\{\begin{array}{cc} -\frac{1}{2}\phi_i^T P_u \phi_i \widehat{H}_i \dfrac{\mathrm{d}v}{d\widehat{H}_i} & i = 1\ldots n_1 \\ 0 & i = n_1 + 1 \ldots n \end{array}\right\} \tag{38b}$$

in which n_1 is the dominant modes to be controlled.

Solving the associated FPK provides

$$p\left(\widehat{H}_i\right) \text{ and } p\left(\widehat{q}, \dot{\widehat{q}}\right), \text{and then, } E\left[\widehat{q}^2\right] \text{ and } E\left[\dot{\widehat{q}}^2\right] \text{ are obtained.} \tag{39}$$

5 Nonlinear Stochastic Control with Sliding Mode Control

System is represented by [6]

$$\mathrm{d}X = [AX + Bu]\mathrm{d}t + f(X, \xi)\mathrm{d}t + G_x(X)\mathrm{d}v(t) \tag{40}$$

Observer equation is given by

$$\mathrm{d}Y = CX\mathrm{d}t + G_y(X)\mathrm{d}w(t) \tag{41}$$

in which $\mathrm{d}w(t)$ and $\mathrm{d}v(t)$ are Weiner process noises and $\xi(t)$ is deterministic process disturbance which cannot be measured; f is nonlinear uncertain dynamics. It is assumed that A and C are observable and an observer gain matrix K_0 exists. Further, $f(X, \xi)$ consists of two parts, i.e.,

$$f(X, \xi) = f_1(X) + f_2(X, t) \text{ [Nonlinear + unmeasured]} \tag{42}$$

The function and matrices satisfy following conditions: $(A - K_0C)$ satisfies Hurwitz condition, $f_1(X)$ satisfies Lipschitz condition, and $f_2(X, t)$ satisfies matching condition given by $P^{-1}C^T\xi(Y(t), t)$ in which P is obtained by the solution of the following equation [6]:

$$PA_0 + A_0^TP = -Q; \quad A_0 = A - K_0C \tag{43}$$

Lipschitz constant α satisfies

$$\alpha < \frac{1}{2}\frac{\lambda(Q)}{\lambda(P)} \tag{44}$$

Observer is designed based on sliding mode control

$$\mathrm{d}\bar{X} = [A\bar{X} + Bu + f_1(\bar{X})]\mathrm{d}t + K_0(Y - C\bar{X})\mathrm{d}t + S_0(\bar{X}, Y, \bar{\eta}_i)\mathrm{d}t \tag{45}$$

S_0 is defined as

$$S_0 = P^{-1}C^T\sum_{i=0}^{N}\bar{\eta}_i l_i(Y)\frac{Y - C\bar{X}}{\|Y - C\bar{X}\|\|-\phi\|} \tag{46}$$

$$\phi = h_1 h_2\left[\sum_{i=0}^{N}\bar{\eta}_i l_i\right] \tag{47}$$

Convergence of observation error is proved using Lyapunov technique. Sliding MDOF controller is designed as $S_c = \sigma + G\bar{X}$ with $\mathrm{d}\sigma = [GBK_c\bar{X} - GA\bar{X}]\mathrm{d}t$. G is selected such that GB is nonsingular and $A - BK_c = A_c$ satisfies Hurwitz condition with K_c as the coefficient matrix and $S_c = 0$ is the sliding surface. The control law is based on

$$u(t) = -\gamma S_c - K_c\bar{X} - \rho(t)\mathrm{sgn}(S_0) \tag{48}$$

in which ρ is switching gain defined by some norm.

Reachability and subsequent stay of the state on sliding surface are proved. The sliding mode dynamics is governed by

$$\mathrm{d}\bar{X} = A_c\bar{X}\mathrm{d}t + \left[I - B(GB)^{-1}G\right][K_c(Y - C\bar{X} + f(\bar{X}, t)) + S_0]\mathrm{d}t \tag{49}$$

Stability of the closed-loop system is proved by Lyapunov technique.

6 Nonlinear Stochastic Control Using State Dependant Riccati Equation

The control equation is given by

$$\mathrm{d}X = F(X,u)\mathrm{d}t + G(X,t)\mathrm{d}Bt \tag{50}$$

In discrete form, the equation can be written as

$$X_{K+1} = f_K(X, w_k) \tag{51}$$

The observer equation is given by

$$Y = h(X,t) + v; \quad Y_K = h_K(X, v_K) \tag{52}$$

For estimation of the state of nonlinear systems, extended Kalman filter and simulation-based methods are generally used. The particle filter is based on the latter scheme and has many variants. Using simulation, filter obtains conditional mean and variance of the *k-th* state estimate. Applying separation principle, and state-dependent Riccati equation, *k-th* optimal control force is u_k and is determined as [8]

$$u_k = -R^{-1}(X_K)B^T(X_K)P(X_K)X_K \tag{53}$$

in which the state equation which is used as constraint for minimizing the cost function is

$$\dot{X} = A(X)X + B(X)u \tag{54}$$

Ito-Taylor expansion is utilized to formulate the problem in discrete form and solves the problem in discrete steps with

$$u_{K+1} = u_K + \dot{u}_K h \tag{55}$$

7 Nonlinear Stochastic Control Using Optimal Controller

The system to be controlled is defined by

$$\dot{X} = f(X, u, k, t) \tag{56}$$

The observer equation with uncertainty is given by

$$Y = h(x, t) + \eta \tag{57}$$

The objective function takes the form

$$J = \underset{t\to\infty}{E}\left[\int_0^t F(x, u, t)\mathrm{d}t\right] \tag{58}$$

For the problem, separation rule is employed and nonlinear Kalman is employed for state estimation [9]

$$\dot{\widehat{X}} = f\left(\widehat{X}, u, k, t\right) + Ph_y^T M(Y - h(x, t)) \tag{59}$$

$$\dot{P} = f_x P + P f_x^T + P^T\left[h_y^T M(Y - h(x, t))\right]P \tag{60}$$

in which $\widehat{X}$ is the least square estimate.

P and $M\left(\approx N^{-1}\right)$ are similar to those used in linear quadratic case with μ defined as minimum variance estimate of $X(t)$

Filtering can be used to find the unknown parameters k; $\bar{\eta} = 0$

The state equation can be written as

$$\dot{Z} = g(x, u, t) \text{ or } \dot{\widehat{Z}} = g\left(\widehat{x}, u, t\right) \tag{61}$$

The controlled design is based on a target response X^d and instantaneous deviation with

$$F\left(\widehat{X}\right) = \left(\widehat{X} - X^d\right)^T Q\left(\widehat{X} - X^d\right) \quad \text{with} \quad u^* \geq u \geq u_* \tag{62}$$

Minimization of the above function provides maximum rate of movement of $\widehat{X}$ to X^d. This leads to a maximum control force

$$u = \begin{cases} u^*{}_i & \beta_i \leq 0 \\ u_{*i} & \beta_i \geq 0 \end{cases} \quad \beta_i = 2S(x, u, t)^T Q\left(x - x^d\right) \tag{63}$$

For minimization, it is assumed that

$$\dot{x} = f(x, u, k, t) = W(x, u, k, t) + S(x, k, t)u \tag{64}$$

8 Nonlinear Stochastic Control Using Partially Observable Quasi-Hamiltonian

System is described as

$$dX = [AX + G(X)]\mathrm{d}t + \bar{U}_2\mathrm{d}t + c_1\mathrm{d}B \tag{65}$$

Observation equation is given by

$$\mathrm{d}Y = [DX + E(X)]\mathrm{d}t + F\bar{U}_2\mathrm{d}t + c_2\mathrm{d}B + c_3\mathrm{d}B_1 \tag{66}$$

Objective function is

$$J = \lim_{T\to\infty}\frac{1}{T}\int_0^T L(X, U_2)\mathrm{d}t \tag{67}$$

$$\text{in which } X = \left[Q^T, P^T\right]^T, \quad Q = \frac{\partial H'}{\partial P}; \quad \text{and} \qquad P' = -\frac{\partial H'}{\partial Q} - {C'}_0\frac{\partial H'}{\partial P} + U + {K'}_0 w(t) \tag{68}$$

H' is the unperturbed Hamiltonian, U is the control force, ${C'}_0$ is the damping coefficient matrix, ${K'}_0$ is the stochastic excitation amplitude matrix, and $w(t)$ is the Gaussian white noise vector with intensity matrix 2D. Using Wong-Zakai correction, a modified Hamiltonian H is obtained which transforms $\frac{\partial H'}{\partial P}$ to $\frac{\partial H}{\partial P}$, $\frac{\partial H'}{\partial Q}$ to $\frac{\partial H}{\partial Q}$, and K' to K_0 through $2{K'}_0 D K_0' T = K_0 K_0^T$ [11].

$\bar{U}$ is split into

$$\bar{U} = [0 \quad U]^T \quad \bar{U} = \bar{U}_1 + \bar{U}_2 \tag{69}$$

$\bar{U}_1$ is combined with uncontrolled system and observation so that original equations

$$\mathrm{d}X = \bar{A}X\mathrm{d}t + \bar{U}\mathrm{d}t + c_1\mathrm{d}B(t) \tag{70}$$

$$\mathrm{d}Y = \bar{D}Xdt + F\bar{U}\mathrm{d}t + c_2\mathrm{d}B(t) + c_3\mathrm{d}B_1(t) \tag{71}$$

are converted to Eqs. (65 and 66). In the above equations,

$$\bar{A} = \begin{bmatrix} \dfrac{\partial H}{\partial P} \\ -\dfrac{\partial H}{\partial Q} - C_0 \dfrac{\partial H}{\partial P} \end{bmatrix}, \quad \bar{U} = \begin{bmatrix} 0 \\ U \end{bmatrix}, \quad C_1 = \begin{bmatrix} 0 \\ K_0 \end{bmatrix}, \tag{72}$$

$$G(X) = \bar{A} + \bar{U}_1 - AX, \quad D = \frac{\mathrm{d}}{\mathrm{d}x}[\bar{D}(0) + F\bar{U}_1(0)], \tag{73}$$

$$E(X) = \bar{D}(X) + F\bar{U}_1 - D(X), \quad \text{and}$$

$$A = \begin{bmatrix} \dfrac{\partial \bar{H}Q}{\partial Q \partial P} & \dfrac{\partial^2 \bar{H}(0)}{\partial P^2} \\ \dfrac{\partial^2 \bar{H}(0)}{\partial Q^2} - \dfrac{\partial}{\partial Q} C_0(0) \dfrac{\partial \bar{H}(0)}{\partial P} & -\dfrac{\partial^2 \bar{H}(0)}{\partial P \partial Q} - \dfrac{\partial}{\partial P} C_0(0) \dfrac{\partial \bar{H}(0)}{\partial P} \end{bmatrix} \tag{74}$$

With these, the problem is finally converted to completely observable linear problem [11]

$$\mathrm{d}\widehat{X} = \left(A\widehat{X} + \bar{U}_2\right)\mathrm{d}t + \left(R_C D^T + C_1 C_2^T\right) C_1^{-1} \mathrm{d}V_1 \tag{75}$$

$$\mathrm{d}V_1 = \mathrm{d}Y - D\widehat{X}\mathrm{d}t \tag{76}$$

$$J_2 = \lim_{T \to \infty} \frac{1}{T} \int_0^T L_2\left(\widehat{X}, U_2\right) \mathrm{d}t \tag{77}$$

R_C satisfies Riccati equation

$$AR_C + R_C A^T - \left(R_C D^T + C_1 C_2^T\right) C^{-1} \left(D R_C + C_2 C_1^T\right) + C_1 C_1^T = 0 \tag{78}$$

$$U_2^* = -\frac{1}{2} R^{-1} \frac{\mathrm{d}\widehat{H}}{\mathrm{d}\widehat{P}} \frac{\mathrm{d}V}{\mathrm{d}\widehat{H}} \tag{79}$$

$$U^* = U_1 + U_2^* \tag{80}$$

$\dfrac{\mathrm{d}V}{\mathrm{d}H}$ is separately obtained from the dynamical programming equation [4, 13]. Using stochastic averaging, corresponding FPK gives the controlled responses.

9 Control of Hysteretic Structures Using H_α

The equation of motion is written in the form

$$M\ddot{x} + C\dot{x} + K_{eL}x + K_{in}v = DU - ME\ddot{x}_g \tag{81}$$

The Bouc-Wen model relates x and v. The state-space formulation gives [7]

$$\dot{X}_1 = g(X_1) + B_1U + G_1\ddot{X}_g; \quad g(X_1) = \left[\{\dot{x}\} - M^{-1}(c\dot{x} + K_{el}x + K_{in}v)\{\dot{V}\}\right]^T;$$

$$X_1 = [x \quad \dot{x} \quad \dot{v}]^T \tag{82}$$

H_α norm is given by

$$||TF|| = Sup\frac{||x_1||_2}{||W||_2} \quad H_\alpha \rightarrow < \gamma \tag{83}$$

The gain matrix K is obtained based on this norm. One of the variants of the K matrix is given by

$$K = K_f\Theta^+ \tag{84}$$

in which K_f is obtained by keeping H_α norm as less than γ and K_f is given by [7]

$$K_f = -R^{-1}S^T + R^{-1/2}\psi(-N_1)^{1/2} \tag{85}$$

Θ^+ is the pseudo-inverse of Θ. ψ is given by

$$\psi = R^{-1/2}S^T\left(I - \Theta^+\Theta\right)(-N_1)^{1/2} \tag{86}$$

$$N_1 = A^TP + PA + \gamma^{-2}PGG^TP + H_1^TH_1 - SR^{-1}S^T + Q < 0 \tag{87}$$

in which P is obtained by solving the Riccati equation $N_1 + Q = 0$

$$R = H_2^TH_2 + \delta I \quad (\delta > 0) \tag{88}$$

$$S = PB + H_1^TH_2 \tag{89}$$

H_1, H_2, etc., are associated with the following linear control problem [7]:

$$\dot{X} = AX + BU + Gw \tag{90}$$

$$Z = H_1X + H_2U \tag{91}$$

$$Y = QX \tag{92}$$

$$U = KY = K\Theta X \tag{93}$$

For stochastic disturbance $\ddot{x}_g$, it is expressed as filtered white noise. Solution of the Lyapunov equation is used to obtained the responses

$$LV^T + VL^T + \bar{F} = 0 \tag{94}$$

$$\bar{F}_{ij} = 0 \qquad \text{except} \quad \bar{F}_{last,last} = 2\pi S_0 \tag{95}$$

When filter variables η are included, the following state-space equation results:

$$\dot{\phi} = L\phi + F \tag{96}$$

$$\phi = \left[X, \dot{X}, u, \eta, \dot{\eta}\right]^T; \quad \eta \text{ are filter variables} \tag{97}$$

V represents covariance matrix of ϕ. Iteration is required as L contains K_e, C_e; the elements K_{ei}, C_{ei} require $E(\dot{x}_i v_i)/\sigma v_i$ to be known. Note that K_e and C_e appear because of the linearization form of Bouc-Wen model as

$$v = -K_e v - C_e \dot{x} \tag{98}$$

10 Semiactive Stochastic Control with MR Damper

For semiactive control of structure with MR damper, the control force is split into [12]

$$\begin{aligned} U_r(Q,P) &= U_r(Q,P) + U_{ra}(Q,P) \\ &= -c_r b_{ir}\dot{Q}_i - F_r \operatorname{sgn}\left(b_{ir}\dot{Q}_i\right) \\ &= -c_r b_{ir}\dot{Q}_i - c_{ra}V_{re}^{\infty} \operatorname{sgn}\left(b_{ir}\dot{Q}_i\right) \end{aligned}$$

$$\text{in which} \quad \dot{Q} = \frac{\mathrm{d}H''}{\mathrm{d}P_i}, \quad \dot{P}_i = -\frac{\mathrm{d}H''}{\mathrm{d}Q_i} - c''_{ij}\frac{\mathrm{d}H''}{\mathrm{d}P_j} + b_{ir}U_{ra} + f_{ik}\xi_k \tag{100}$$

H'' and C'' are the Hamiltonian and damping coefficients modified by the conservative and dissipative part of the passive control of MR dampers $\left(U_{rp}\right)$. If $\xi(t)$ is the

Gaussian white noise with intensities $2D_{ke}$, then Eq. (100) can be converted into a Ito differential equation by adding Wong-Zakai correction $D_{kl}f_{je}\frac{\partial f_{ik}}{\partial P_j}$.

Once modified Hamiltonian with overall conservative force $-\partial H/\partial Q_i$ and dissipative force $-c_{ij}\partial H/\partial P_j$ are obtained, rest of the formulation remains the same as that given before. The state equation takes the form

$$\mathrm{d}Q_i = \frac{\mathrm{d}H}{\mathrm{d}P_i}\mathrm{d}t \tag{101}$$

$$\mathrm{d}P_i = -\left[\frac{\mathrm{d}H}{\mathrm{d}Q_i} + c_{ij}\frac{\mathrm{d}H}{\mathrm{d}P_j} - b_{ir}U_{ra}\right]\mathrm{d}t + \bar{\sigma}_{ik}\mathrm{d}B_k \tag{102}$$

with $\bar{\sigma}\bar{\sigma}^T = 2fDf^T$.

The control force U_{ra} is given by [12]

$$U_{ra}^* = -\frac{1}{2R_{rr}}\frac{\partial V}{\partial H_i}\left|b_{ir}\dot{Q}_i\right|\mathrm{sgn}\left(b_{ir}\dot{Q}_i\right) \tag{103}$$

in which V is the value function of the dynamic programming and R_{rr} is a positive definite diagonal matrix.

11 Application Problems

Three application problems on stochastic active and semiactive control are presented here which use some of the above control methods. These applications include both linear and nonlinear control problems.

11.1 Nonlinear Semiactive Stochastic Control of Hysteretic Column Using MR Damper

The MR damper is modeled by Bouc-Wen model. The same model is used for hysteretic behavior of the column also [12]

$$\dot{Z}_i = A_i\dot{X} - \beta_i\dot{X}|Z_i|^{n_i} - \gamma_i\left|\dot{X}\right|Z_i|Z_i|^{n_i-1} \tag{104}$$

in which Z_i denotes the hysteretic force and $\dot{X}$ denotes the relative velocity between the two ends of the damper.

The damper force is given by

$$F = c_1\dot{x} + \alpha_1 Z_i \tag{105}$$

Both α_1 and c_1 are functions of filtered voltage u, i.e.,

$$c_1 = c_p + c_s u \quad \alpha_1 = \alpha_p + \alpha_s u \tag{106}$$

c_p, c_s, etc., are constants, and filtered voltage u is determined from applied voltage v as [3]

$$\dot{u} = -\eta(u - v) \tag{107}$$

The damper force can be split in two parts

$$F_p = c_p\dot{x} + \alpha_p Z_i \tag{108}$$

$$F_s = (c_s\dot{x} + \alpha_s Z_i)u \tag{109}$$

F_p is the passive part, and F_s is the semiactive part depending on the filtered voltage u. Integration of Eq. (107) gives

$$u = e^{-\eta t}\int_0^t v\mathrm{d}e^{nt}; \quad v \geq 0 \quad \text{denotes clippings} \tag{110}$$

The equation of motion of the column is given by

$$\ddot{X} + 2\xi\dot{X} + [\alpha - k_1 - k_2\eta(t)]X + (1 - \alpha)Z_1 = \xi(t) - F \tag{111}$$

$\xi(t)$ and $\eta(t)$ are horizontal and vertical random ground acceleration acting as external and parametric excitations. They are Gaussian white noise with intensities of $2D_1$ and $2D_2$, respectively. Restoring force of the column is modeled by the same Bouc-Wen model.

To apply the stochastic averaging method of energy envelop, the hysteretic forces are to be separated as nonlinear elastic forces and nonlinear damping forces.

After doing this, equation is replaced by the following non hysteretic system [3]

$$\begin{aligned} &\ddot{X} + \left[C + 2\xi_1(H) + 2\xi_{2p}(H)\right]\dot{X} + \mathrm{d}U(X)/\mathrm{d}X \\ &\quad = \xi(t) + k_2 X\eta(t) - \left[(c_s + 2\xi_{2s}(H))\dot{X} + \mathrm{d}U_{2s}(X)\right]u \end{aligned} \tag{112}$$

$$H = \frac{\dot{X}^2}{2} + U(X) \tag{113}$$

The nonlinear damping coefficients $2\xi_1$ and $2\xi_{2p}$ are related to the areas of hysteretic loop. The Ito equation finally takes the form

$$\mathrm{d}H = \left\{ m(H) - \left\langle \left[(c_s + 2\xi_{2s}(H)) \left(\frac{\mathrm{d}H}{\mathrm{d}\dot{X}} \right)^2 + \frac{\mathrm{d}U_2}{\mathrm{d}X} \right] u \right\rangle \right\} + \sigma(H)\mathrm{d}B(t) \tag{114}$$

The optimal filtered voltage is given by

$$u^* = \frac{1}{2R} [c_s + 2\xi_{2s}(H)] \frac{\mathrm{d}V}{\mathrm{d}H} \tag{115}$$

Optimal semiactive control force

$$F_s^* = \frac{1}{2R} (c_s \dot{X} + \alpha_s Z_i) [c_s + 2\xi_{2s}(H)] \frac{\mathrm{d}V}{\mathrm{d}H} \tag{116}$$

A suitable value of R may be selected along with $g(H)$ so that $\mathrm{d}V/\mathrm{d}H \geq 0$

$\frac{\mathrm{d}V}{\mathrm{d}H}$ is obtained from the dynamic programming equation. Solution of the corresponding FPK equation gives the controlled responses.

11.2 Stochastic Control by MR-TLCD for Controlling Wind-Induced Vibration

The equation of motion of the MR-TLCD can be written as [5]

$$m_D \ddot{y} + u(\dot{y}) + k_D y = -\lambda m_D \ddot{y}_C \tag{117}$$

in which $m_D = \rho A_D L_D, k_D = 2\rho g A_D, \lambda = \frac{B_D}{L_D}, B_D$ is the width of the tube and L_D is the effective fluid length. Magnetic field passes across the bottom part of the tube over length L_p. The damping force can be split into passive and semiactive parts:

$$u(\dot{y}) = u_p(\dot{y}) + u_s(\dot{y}) \tag{118}$$

$$u_p(\dot{y}) = \frac{1}{2} \rho \delta A_D \dot{y}^2 \operatorname{sgn}(\dot{y}) \tag{119}$$

$$u_s(\dot{y}) = \tau_y \left(c A_D \frac{L_p}{h} \right) \operatorname{sgn}(\dot{y}) \tag{120}$$

in which h is the depth of flow between the fixed poles

$$\delta = \frac{48}{R_c \left(1 + \frac{H}{W}\right)^2} \frac{L_D}{H} + \sum \xi_i \tag{121}$$

δ is the overall head loss coefficient; W is the width of the flow; H is the depth of tube; R_c is Reynolds's number; $\sum \xi_i$ is the coefficient of minor head losses.

The equation of motion of the MR-TLCD fitted building in state space is given by

$$\dot{Z} = AZ + F(t) + U \tag{122}$$

A is an appropriate matrix consisting of A_L and A_N. A_N contains a C_{eq} (equivalent linearized damping) term given by [5]

$$C_{eq} = \zeta^{\delta} A_D \sqrt{\frac{2E[\dot{y}]^2}{\pi}} \tag{123}$$

$$U = -Bu_s(\dot{y}) \tag{124}$$

Semiactive part of the control force is obtained through dynamic programming. If LQR is used, the Lagrangian L and value function V are represented such that the control damping face u_s becomes Riccati solution based. It is given as

$$u_s = R^{-1}B^T PZ \tag{125}$$

$$u_s^* = \begin{cases} F^* \mathrm{sgn}(\dot{y}) & F^* \geq 0 \\ 0 & F^* \leq 0 \end{cases} \tag{126}$$

$$\text{with} \quad F^* = R^{-1}B^T PZ\mathrm{sgn}(\dot{y}) \tag{127}$$

The solution of the equation of motion is obtained using spectral analysis. A statistical linearization is required to obtain the solutions as described by Eq. (123).

11.3 Stochastic Control of Coupled Structures Under Random Excitation

For the coupled buildings (shear type) under lateral excitation and the control forces through n number of interconnected control devices, equations of motion are represented as

$$M_1\ddot{x}_1 + C_1\dot{x}_1 + K_1x_1 = -\ddot{x}_g M_1 E_1 + P_1 U \tag{128}$$

$$M_2\ddot{x}_2 + C_2\dot{x}_2 + K_2x_2 = -\ddot{x}_g M_2 E_2 + P_2 U \tag{129}$$

in which $\ddot{x}_g$ is the random ground motion which is modeled as Gaussian process and having Kanai Tajimi PSDF. Other vector and matrices are self-explanatory.

Assuming the system to be fully observed and making use of modal transformation, the modal equations take the form

$$\ddot{q}_{1i} + 2\zeta_{1i}\omega_{1i}\dot{q}_{1i} + \omega_{1i}^2 q_{1i} = -\beta_{1i}\ddot{x}_g(t) + v_{1i} \quad (i = 1 \cdots\cdots m_3) \tag{130}$$

$$\ddot{q}_{2i} + 2\zeta_{2i}\omega_{2i}\dot{q}_{2i} + \omega_{2i}^2 q_{2i} = -\beta_{2i}\ddot{x}_g(t) + v_{2i} \quad (i = 1 \cdots\cdots m_4) \tag{131}$$

The Hamiltonian in modal coordinate system is

$$\bar{H}_j = \sum H_{ji} \tag{132}$$

$$\text{where} \quad H_{ji} = \left(\dot{q}_{ji}^2 + \omega_{ji}^2 q_{ji}^2\right)/2 \tag{133}$$

The Ito stochastic differential equation for modal vibrational energy after stochastic averaging method is given by [10]

$$\mathrm{d}\bar{H} = \left[\bar{m}(\bar{H}) + \left\langle \frac{\mathrm{d}\bar{H}}{\mathrm{d}Q}\bar{U}_i \right\rangle_t \right]\mathrm{d}t + \bar{\sigma}(\bar{H})\mathrm{d}\bar{w}(t) \tag{134}$$

$$\text{in which} \quad \bar{m} = [\bar{m}_1, \bar{m}_2]^T = [m_{11}, m_{12} \cdots\cdots m_{21}, m_{22} \cdots\cdots]^T \tag{135}$$

$$\bar{\sigma} = \mathrm{diag}\{\bar{\sigma}_1, \bar{\sigma}_2\} = \mathrm{diag}\{\sigma_{11}, \sigma_{12} \cdots\cdots \sigma_{21}, \sigma_{22} \cdots\cdots\}^T \tag{136}$$

$$\bar{\omega} = [\omega_{11}, \omega_{12} \cdots\cdots \omega_{21}, \omega_{22} \cdots\cdots]^T \tag{137}$$

$$\bar{m}_{1i}(\bar{H}_{1i}) = -2\zeta_{1i}\omega_{1i}H_{1i} + \frac{1}{2}\beta_{1i}^2 S_g(\omega_{1i}) \tag{138}$$

$$\bar{n}_{2i}(\bar{H}_{2i}) = -2\zeta_{2i}\omega_{2i}H_{2i} + \frac{1}{2}\beta_{2i}^2 S_g(\omega_{2i}) \tag{139}$$

$$\bar{\sigma}_{1i}^2(\bar{H}_{1i}) = \beta_{1i}^2 \bar{H}_{1i} S_g(\omega_{1i}); \quad \bar{\sigma}_{2i}^2(\bar{H}_{2i}) = \beta_{2i}^2 \bar{H}_{2i} S_g(\omega_{2i}) \tag{140}$$

Using dynamic programming technique, the optimal control force vector is given by [10]

$$U^* = -\frac{1}{2}R_p^{-1}\left(P_1^T \bar{\phi}_1 \frac{\partial \bar{H}_1}{\partial \dot{\bar{Q}}_1}\frac{\partial V}{\partial \bar{H}_1} + P_2^T \bar{\phi}_2 \frac{\partial \bar{H}_2}{\partial \dot{\bar{Q}}_2}\frac{\partial V}{\partial \bar{H}_2}\right) \tag{141}$$

The solution of value function provides $\dfrac{\mathrm{d}V}{\mathrm{d}\bar{H}_i}$

Substituting for U^* and performing stochastic averaging leads to

$$d\bar{H} = [m(\bar{H}) + m_u(\bar{H})]dt + \sigma(\bar{H})dw(t) \tag{142}$$

in which

$$m_{u1} = -\frac{1}{2}\left\{\phi_{11}^T P_u \phi_{11} \bar{H}_{11} \frac{\partial V}{\partial \bar{H}_{11}} \cdots\cdots \phi_{1m3}^T P_u \phi_{1m3} \bar{H}_{1m3} \frac{\partial V}{\partial \bar{H}_{1m3}}\right\}^T \tag{143}$$

$$m_{u2} = -\frac{1}{2}\left\{\phi_{21}^T P_u \phi_{21} \bar{H}_{21} \frac{\partial V}{\partial \bar{H}_{21}} \cdots\cdots \phi_{2m4}^T P_w \phi_{2m4} \bar{H}_{2m4} \frac{\partial V}{\partial \bar{H}_{2m4}}\right\}^T \tag{144}$$

$$P_u = P_1 R_p^{-1} P_1^T \quad P_w = P_2 R_p^{-1} P_2^T \tag{145}$$

12 Conclusions

Stochastic control of structures for environmental excitations is a relatively new area of research. It is also the realistic way of tackling the structural control problems as both environmental excitations and uncertainties associated with the problem are best modeled as stochastic processes. The literature search shows that a number of problems in this area have remained unattempted such as:

1. ANN-based stochastic control of structures
2. Application of fuzzy (stochastic) rule base in developing control algorithms
3. Verification of stochastic control algorithms by real time simulation in on-line control
4. More exhaustive studies on the application of stochastic control using MR and VE dampers
5. Use of different control devices for semiactive and hybrid stochastic control like AMD – TMD, spring-connected MR-TLCD, and base isolated – semiactive dampers, friction damper, and adjacent connectors
6. Stochastic control in post yield state of structures
7. Stochastic control in aeroelastic vibrations of structures
8. Application in the vibration control of offshore structures
9. Application in the vibration control of bridges
10. Stability analysis/robustness of stochastic control strategies

References

1. Ariaratnam ST, Loh NK (1968) Optimal control and filtering of linear stochastic systems. Int J Control 7(5):433–445
2. Astrom KJ (1970) Introduction to stochastic control theory. Academic, New York

3. Cheng H, Zhu WQ, Ying ZG (2006) Stochastic optimal semi-active control of hysteretic systems by using a magneto-rheological damper. Smart Mater Struct 15:711–718
4. Fleming WH, Rishel RW (1975) Deterministic and stochastic optimal control. Springer, New York
5. Ni YQ, Ying ZG, Wang JY, Ko JM, Spencer BF Jr (2004) Stochastic optimal control of wind-excited tall buildings using semi-active MR-TLCDs. Probab Eng Mech 19:269–277
6. Qiao F, Zhu MQ, Liu J, Zhang F (2004) Adaptive observer-based non-linear stochastic system control with sliding mode schemes. J Syst Control Eng 222(1):681–690
7. Sadek F, Ftima MB, El-Borgi S, McCormick J, Riley MA (1994) Control of hysteretic structures using H∞ algorithm and stochastic linearization techniques. J Sound Vib 210(5):540–548
8. Sajeeb R, Manohar CS, Roy D (2007) Use of particle filters in active control algorithm for noisy nonlinear structural dynamical systems. J Sound Vib 306:111–135
9. Seinfeld JH (1970) Optimal stochastic control of nonlinear systems. AICHE J 16:1016–1022
10. Ying ZG, Ni YQ, Ko JM (2004) Non-linear stochastic optimal control for coupled-structures system of multi-degree-of-freedom. J Sound Vib 274:843–861
11. Ying ZG, Zhu WQ (2008) A stochastic optimal control strategy for partially observable nonlinear quasi-Hamiltonian systems. J Sound Vib 310:184–196
12. Ying ZG, Zhu WQ (2003) A stochastic optimal semi-active control strategy for ER/MR dampers. J Sound Vib 259(1):45–62
13. Yong JM, Zhou XY (1999) Stochastic control, Hamiltonian systems and HJB equations. Springer, New York
14. Zhu WQ, Ying ZG (2002) Nonlinear stochastic optimal control of partially observable linear structures. Eng Struct 24:333–342

Uncertainties in Interpreting the Scale Effect of Plate Load Tests in Unsaturated Soils

Sai K. Vanapalli and Won Taek Oh

Abstract The applied stress versus surface settlement (*SVS*) behavior from in situ plate load tests (*PLT*s) is valuable information that can be used for the reliable design of shallow foundations (*SF*s). In situ *PLT*s are commonly conducted on the soils that are typically in a state of unsaturated condition. However, in most cases, the influence of matric suction is not taken into account while interpreting the *SVS* behavior of *PLT*s. In addition, the sizes of plates used for load tests are generally smaller in comparison to real sizes of footings used in practice. Therefore, in situ *PLT* results should be interpreted taking account of not only matric suction but also the scale effects. In the present study, discussions associated with the uncertainties in interpreting the *SVS* behavior of *PLT*s taking account of matric suction and scale effects are detailed and discussed.

Keywords Plate load test • Unsaturated soil • Matric suction • Shallow foundation • Bearing capacity • Settlement

1 Introduction

Bearing capacity and settlement are two key parameters required in the design of foundations. There are several techniques available today to determine or estimate both the bearing capacity and settlement behavior of foundations based on experimental methods, in situ tests, and numerical models including finite element analysis. In addition, there are different ground improvement methods to increase the bearing capacity and reduce the settlements. However, in spite of these advancements, various types of damages still can be caused to the superstructures placed on shallow foundations (hereafter referred to as *SF*s) due to the problems

S.K. Vanapalli (✉) • W.T. Oh
Civil Engineering Department, University of Ottawa, Ottawa, ON, Canada
e-mail: vanapall@eng.uottawa.ca; oh.wontaek@gmail.com

S. Chakraborty and G. Bhattacharya (eds.), *Proceedings of the International Symposium on Engineering under Uncertainty: Safety Assessment and Management (ISEUSAM - 2012)*,
DOI 10.1007/978-81-322-0757-3_7,

associated with the settlements leading to cracks, tilts, differential settlements, or displacements. This is particularly true for coarse-grained soils such as sands in which foundation settlements occur quickly after construction. Due to this reason, the settlement behavior is regarded as a governing parameter in the design of *SF*s in coarse-grained soils [25, 26, 34]. Foundation design codes suggest restricting the settlement of *SF*s placed in coarse-grained soils to 25 mm and also limit their differential settlements (e.g., [13]). Such design code guidelines suggest that the rational design of *SF*s can be achieved by estimating the applied stress versus surface settlement (hereafter referred to as *SVS*) behavior of *SF*s reliably instead of estimating the bearing capacity and settlement separately.

The most reliable testing method to estimate the *SVS* behaviors of *SF*s is in situ plate load tests (hereafter referred to as *PLTs)*. In situ *PLT*s are commonly performed on the soils that are typically in a state of unsaturated condition. This is particularly true in arid or semiarid regions where the natural groundwater table is deep. Hence, the stresses associated with the constructed infrastructures such as *SF*s are distributed in the zone above the groundwater table, where the pore water pressures are negative with respect to the atmospheric pressure (i.e., matric suction). Several researchers showed that the *SVS* behaviors from model footings [35, 40, 42, 45] or in situ *PLTs* [16, 39] are significantly influenced by matric suction. However, in most cases, the in situ *PLT* results are interpreted without taking account of the negative pore water pressure above groundwater table. In other words, the influence of capillary stress or matric suction toward the *SVS* behavior is ignored in engineering practice. Moreover, the *PLT*s are generally conducted with small sizes of plates (either steel or concrete) in comparison to the real sizes of foundations. Due to this reason, the scale effect has been a controversial issue in implementing the *PLT* results into the design of *SF*s. These details suggest that the reliability of the design of *SF*s based on the *PLT* results can be improved by taking account of the influence of not only matric suction but also plate size on the *SVS* behaviors.

In this present study, two sets of in situ plate and footing load test results in unsaturated sandy and clayey soils available in the literature are revisited. Based on the results of these studies, an approach is presented such that the uncertainties associated with the scale effects are reduced or eliminated. In addition, discussions on how to interpret the in situ *PLT* results taking account of matric suction are also presented and discussed. Moreover, a methodology to estimate the variation of *SVS* behavior with respect to matric suction using finite element analysis (hereafter referred as *FEA*) is introduced.

2 Plate Load Test

In situ *PLT*s are generally conducted while designing *SF*s [3] or pavement structures [4, 5, 11] to estimate the reliable design parameters (i.e., bearing capacity and displacement) or to confirm the design assumptions. Figure 1 shows typical "applied stress" versus "surface settlement" (*SVS*) behavior from a *PLT*. The peak

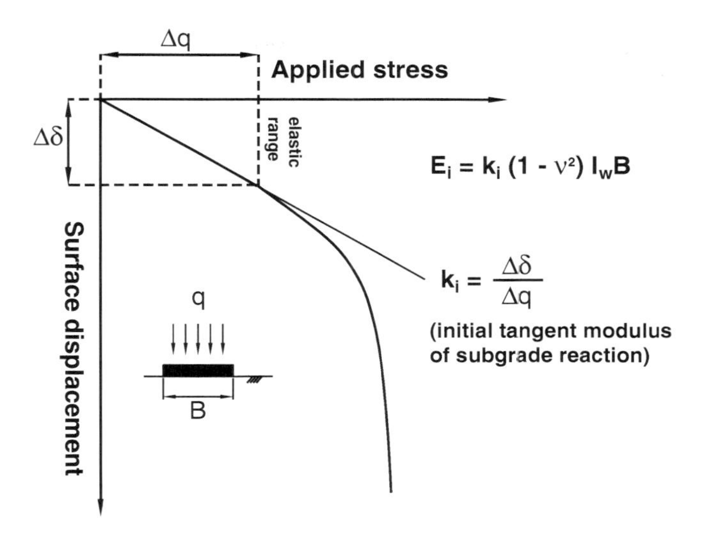

Fig. 1 Typical stress versus displacement behavior from plate load test

stress is defined as ultimate bearing capacity, q_{ult}, for general failure; however, in the case where well-defined failure is not observed (i.e., local or punching failure), the stress corresponding to the 10% of the width of a foundation (ASTM D1194-94) or the stress corresponding to the intersection of elastic and plastic lines of the *SVS* behavior is regarded as q_{ult} [16, 42, 48].

The elastic modulus can be estimated based on the modulus of subgrade reaction, k, that is a slope of *SVS* behavior (i.e., δ versus q) using the theory of elasticity as shown in Eq. (1). The maximum elastic modulus (i.e., initial tangent elastic modulus, E_i) can be computed using the k_i value in the elastic range of *SVS* behavior (initial tangent modulus of subgrade reaction):

$$E = \frac{(1 - \nu^2)}{(\delta/q)} B I_{\mathrm{w}} = k\left(1 - \nu^2\right) I_{\mathrm{w}} B \tag{1}$$

where E = elastic modulus, ν = Poisson's ratio, δ, q = settlement and corresponding stress, B = width (or diameter) of bearing plate, I_{w} = factor involving the influence of shape and flexibility of loaded area, and k = modulus of subgrade reaction.

Ultimate bearing capacity, q_{ult}, and elastic modulus, E, estimated based on the *SVS* behavior from a *PLT* are representative of soils within a depth zone which is approximately 1.5*B*–2.0*B* [38]. Agarwal and Rana [1] performed model footing tests in sands to study the influence of groundwater table on settlement. The results of the study showed that the settlement behavior of relatively dry sand is similar to that of sand with a groundwater table at a depth of 1.5*B* below the model footing. These results indirectly support that the increment of stress due to the load applied on the model footing is predominant in the range of 0–1.5*B* below model footing. These observations are also consistent with the modeling studies results by Oh and Vanapalli [31]. This fact also indicates that the *SVS* behavior from *PLT* is influenced by plate (or footing) size since different plate sizes result in different sizes of stress bulbs and mean stresses in soils. This phenomenon which is conventionally defined as "scale effect" needs to be investigated more rigorously to rationally design the shallow foundations.

3 Scale Effect in Plate Load Test

3.1 Scale Effect and Critical State Line

The Terzaghi's bearing capacity factor, N_{γ}, decreases with an increase in the width of footings [18]. Various attempts have been made by several researchers to understand the causes of scale effects. Three main explanations for the scale effect that generally accepted are as follows:

1. Reduction in the internal friction angle, ϕ', with increasing footing size (i.e., nonlinearity of the Mohr–Coulomb failure envelop) [7, 18, 21]
2. Progressive failure (i.e., different ϕ' along the slip surfaces below a footing) [43, 49]
3. Particle size effect (i.e., the ratio of soil particles to footing size) [41, 43]

According to Hettler and Gudehus [21], there is lack of consistent theory to explain the progressive failure mechanism in the soils below different sizes of footings. In addition, the particle size effect for the in situ plate (or footing) load test (hereafter the term "plate" is used to indicate both steel plate and concrete footing) can be neglected since the ratio of plate size, *B* to d_{50} (i.e., grain size corresponding to 50% finer from the grain size distribution curve), for in situ *PLT*s are mostly greater than 50–100 [24]. The focus of the present study is to better understand the scale effects of *PLT* results and suggest some guidelines of how they can be used in practice.

The reduction in ϕ' with an increasing footing size is attributed to the fact that the larger footing size contributes to a higher mean stress in the soils. In other words, the larger footing induces higher mean stress that contributes to lower ϕ' due to the nonlinearity of the Mohr–Coulomb failure envelop when tested rigorously over a large stress range. This phenomenon can be better explained using the critical state concept ([19]; Fig. 2).

In Fig. 2, the points plotted on the lines *a–b*, *c–d*, and *e–f* simulate the following scenarios:

1. Line *a–b*: Different sizes of footings placed at different depths in sand that have the same initial void ratio value, but the distances to the critical state line are different.

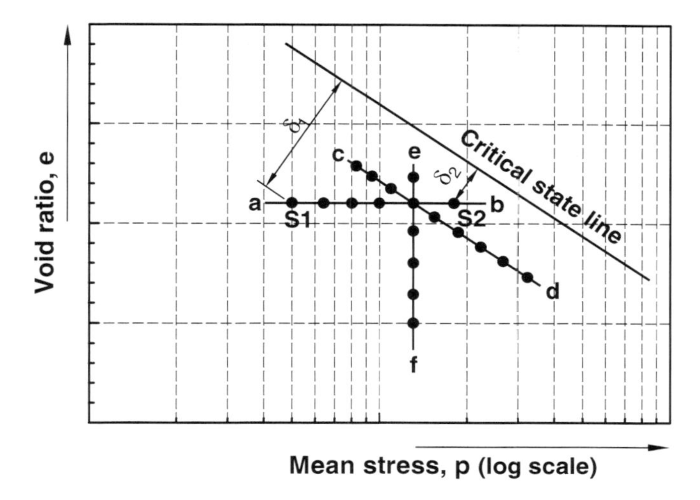

Fig. 2 Relationship between the initial states (i.e., void ratio and mean stress) of soils below footings and critical state line (After Fellenius and Altaee [19])

2. Line *c–d*: Same sizes of footings places at the same depth in sand that have different initial void ratio values, but the distances to the critical state line are the same.
3. Line *e–f*: Different sizes of footings placed at different depths in sands that have different void ratio values, but the distances to the critical state line are the same.

The main concept shown in Fig. 2 is that the behavior of sand below a footing is governed by a distance from the initial state to the critical state line. In other words, the initial states plotted on the line *e–f* will show the same *SVS* behaviors regardless of footing size since the distance to the critical state line for each initial state is the same. On the other hand, the sand below a larger footing (e.g., *S*1 in Fig. 2) will have larger displacement at a certain applied stress in comparison to a smaller footing (e.g., *S*2 in Fig. 2) due to the greater mean stress (i.e., closer to the critical state line) even though the initial void ratio is the same.

3.2 Plate Load Test Results

In this present study, two sets of in situ *PLT*s in sandy and clayey soils available in the literature are revisited to discuss scale effect of *PLT*s.

The Federal Highway Administration (*FHWA*) has encouraged investigators to study the performance of *SF*s by providing research funding. As part of this research project, several series of in situ footing (i.e., 1, 2, 2.5, and 3 m) load tests were conducted on sandy soils. These studies were summarized in a symposium held at the Texas A&M University in 1994 [9] (Fig. 3). Consoli et al. [15]

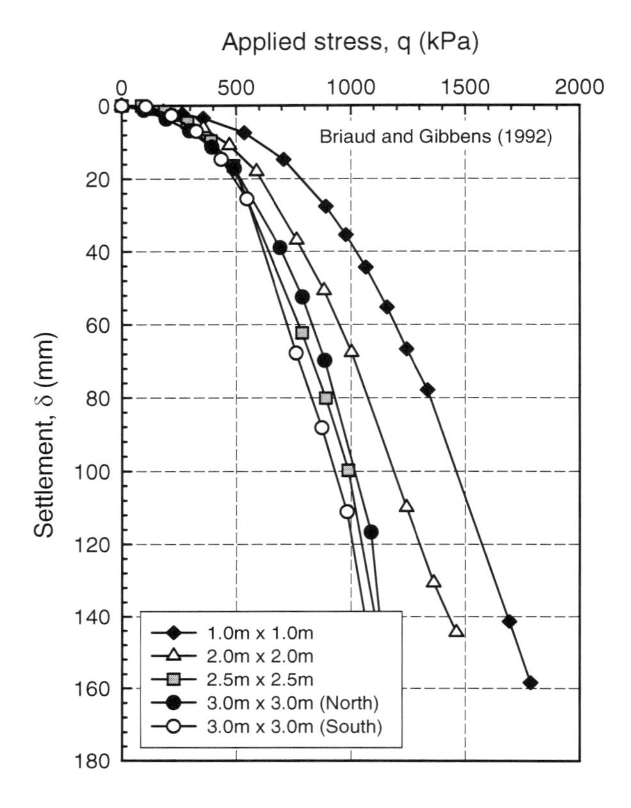

Fig. 3 Stress versus displacement behavior from in situ footing load tests (After Briaud and Gibbens [9])

conducted in situ *PLT*s in unsaturated clayey soils using three steel circular plates (i.e., 0.3, 0.45, and 0.6 m; *PLT*) and three concrete square footings (i.e., 0.4, 0.7, and 1.0 m; *FLT*) (Fig. 4).

As can be seen in Figs. 3 and 4, the bearing capacity increases with decreasing plate size, and different displacement values are observed under different stresses. The *SVS* behaviors clearly show that the *SVS* behavior is dependent of plate size (i.e., scale effect). These observations are consistent with the *SVS* behaviors along the line *a–b* shown in Fig. 2. In other words, the soil below a larger footing induces higher mean stress; therefore, the initial state is closer to the critical state. This phenomenon makes the soil below a larger footing behave as if it is loose soil compared to a smaller footing [12].

3.3 Elimination of Scale Effect of Shallow Foundations

Briaud [8] suggested that scale effect (Fig. 3) can be eliminated by plotting the *SVS* behaviors as "applied stress" versus "settlement/width of footing" (i.e., δ/B) curves

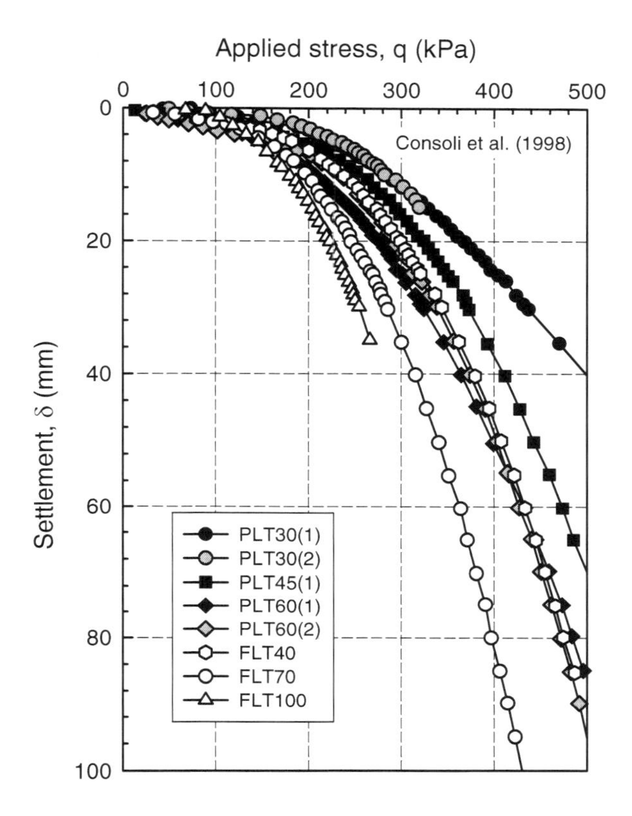

Fig. 4 Stress versus displacement behavior from in situ plate (*PLT*) and footing (*FLT*) load tests (After Consoli et al. [15])

(i.e., normalized settlement; Eq. (2)). Similar trends of results were reported by Osterberg [36] and Palmer [37]:

$$\frac{\delta}{B} = \frac{q(1 - v^2)}{E} I_{\mathrm{w}} \tag{2}$$

According to the report published by *FHWA* [10], this behavior can be explained using triaxial test analogy (Fig. 5). If triaxial tests are conducted for identical sand samples under the same confining pressure where the top platens are different sizes of footings, the stress versus strain behaviors for the samples are unique regardless of the diameter of the samples (i.e., the same stress for the same strain). This concept is similar to relationship between q and δ/B from *PLT*s since the term δ/B can be regarded as strain.

Consoli et al. [15] suggested that the scale effect of *PLT*s (Fig. 4) can be eliminated when the applied stress and displacement are normalized with unconfined compressive strength, q_{u}, and footing width, B, respectively, as shown in Eq. (3). They also analyzed *PLT*s results available in literature [17, 22] and

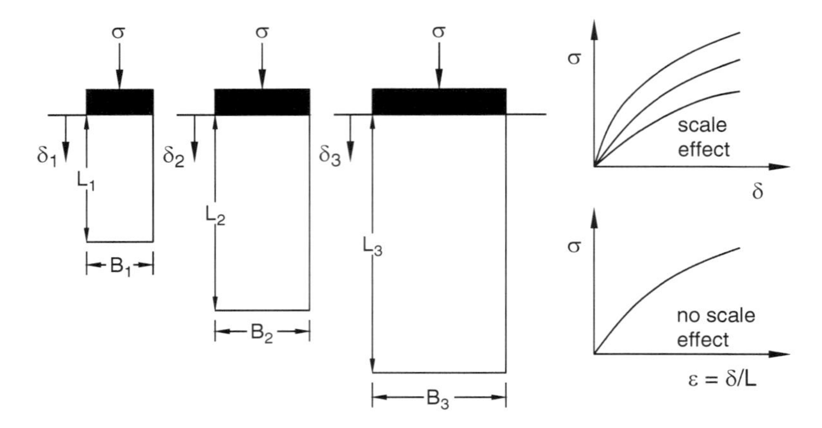

Fig. 5 Triaxial test/shallow foundation analogy [10]

showed that the concept in Eq. (3) can be extended to the *PLT* results in sandy soils as well:

$$\left(\frac{q}{q_u}\right) = \left(\frac{1}{q_u}\right)\left(\frac{E}{1-\nu^2}\right)\left(\frac{1}{C_s}\right)\left(\frac{\delta}{B}\right) = \left\{\frac{C_d}{q_u C_s}\right\}\left(\frac{\delta}{B}\right) \tag{3}$$

where q = applied stress, q_u = unconfined compressive strength at the depth of embedment, δ = surface settlement, B = width of footing, C_s = coefficient involving shape and stiffness of loaded area (I_w in Eq. 1), and C_d = coefficient of deformation ($=E/(1-\nu^2)$).

As can be seen in Figs. 6 and 7, the curves (δ/B versus q) fall in a narrow range. From an engineering practice point of view, these curves can be considered to be unique. Consoli et al. [15] suggested that uniqueness of the normalized curves can be observed at sites where the soils are homogeneous and isotropic in nature.

4 Scale Effect of Plate Size in Unsaturated Soils

The critical state concept discussed above can be effectively used to explain the scale effect of *SF*s in saturated or dry sands. However, this concept may not be applicable to interpret the scale effect of plate size in unsaturated soils since the *SVS* behaviors in unsaturated soils are influenced not only by footing size but also by matric suction value. The influence of matric suction however is typically ignored in conventional engineering practice.

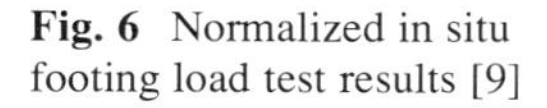

Fig. 6 Normalized in situ footing load test results [9]

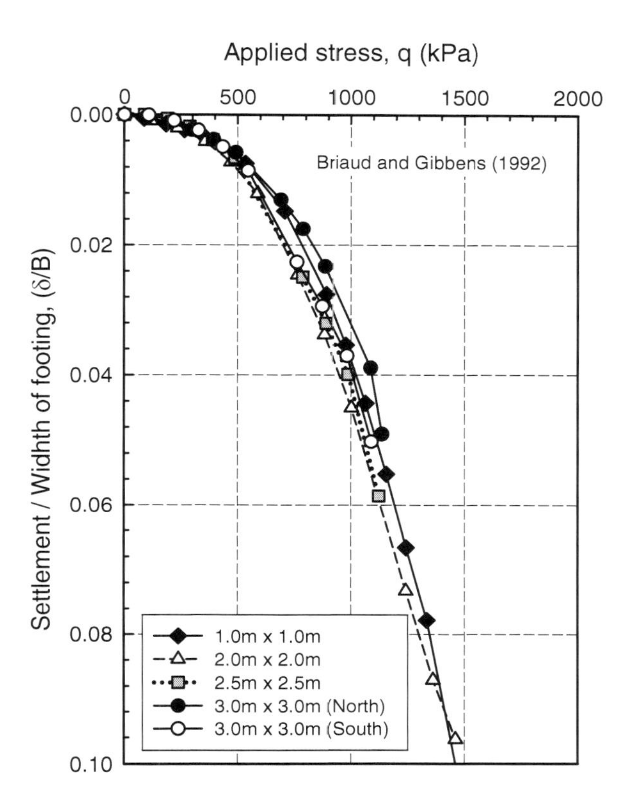

4.1 Average Matric Suction Value

Matric suction distribution profile is mostly not uniform with depth in fields. In this case, the concept of "average matric suction" [45] can be used as a representative matric suction value to interpret mechanical properties of a soil at a certain matric suction distribution profile. The average matric suction value, Ψ, is defined as a matric suction value corresponding to the centroid of the suction distribution diagram from 0 to 1.5B depth (Fig. 8).

As discussed earlier, the stress increment in a soil due to a load or a stress act on a *SF* is predominant in the range of 0–1.5B. Hence, when loads are applied on two different sizes of footings, the sizes of stress bulbs (in the depth zone of 0–1.5B) are different (Fig. 9). In other words, the stress bulb for the smaller footing (i.e., B_1) is shallower in comparison to that of the larger footing (i.e., B_2). These facts indicate that the *SVS* behaviors from *PLT*s are governed by E and ν values within the stress bulb. If a matric suction distribution profile is uniform with depth, the average matric suction value is the same regardless of footing size. However, if the matric suction distribution profile is nonuniform, the average matric suction value is dependent on the footing size. For example, the average matric suction value for

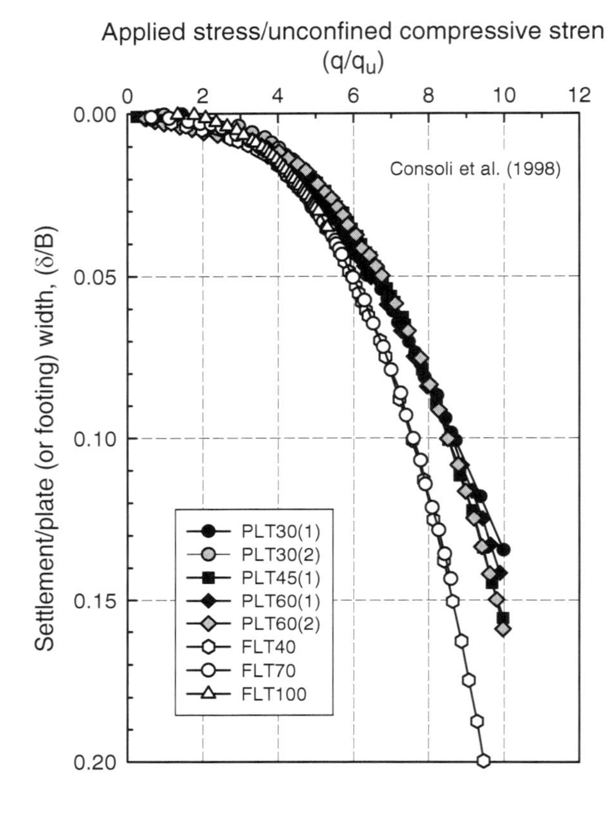

Fig. 7 Normalized in situ plate and footing load test results [15]

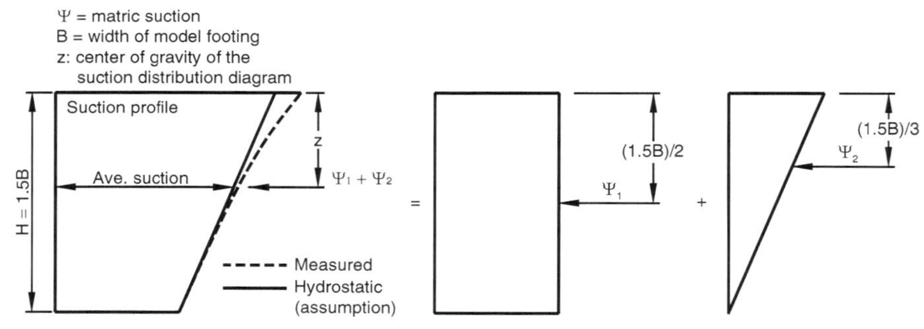

Fig. 8 Estimation of average matric suction value using the centroid of suction distribution diagram

the smaller plate, B_1, (i.e., Ψ_1) is greater than that of larger plate, B_2, (i.e., Ψ_2). In this case, the concept shown in Eqs. (2) and (3) cannot be used to eliminate the scale effect of plate since q_u [29], E_i [34], and v [30, 32, 33] are not constant but vary with respect to matric suction. More discussions are summarized in later sections.

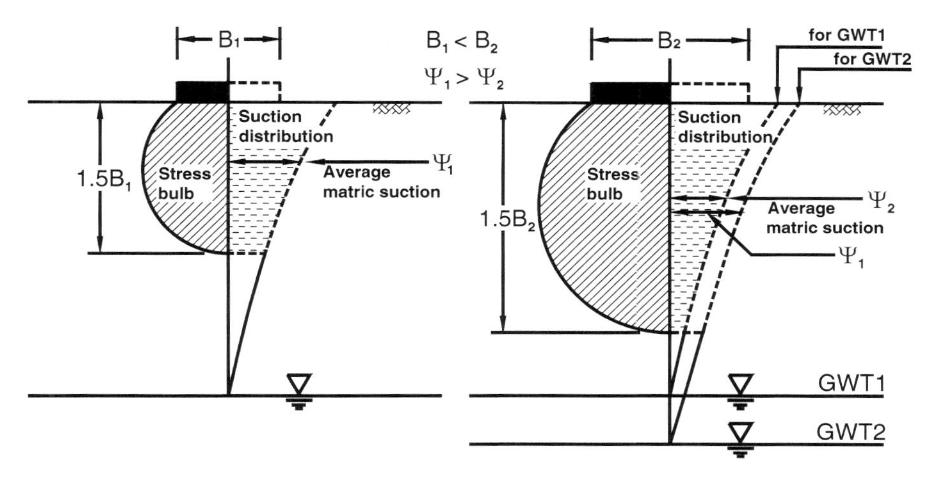

Fig. 9 Average matric suction values for different footing sizes under nonconstant matric suction distribution profile

4.2 *Variation of E_i with Respect to Matric Suction for Coarse-Grained Soils*

Oh et al. [34] analyzed three sets of model footing test results in unsaturated sands [28, 42] and showed that the initial tangent elastic modulus, E_i, is significantly influenced by matric suction. Based on the analyses, they proposed a semiempirical model to estimate the variation of E_i with respect to matric suction using the soil–water characteristic curve (*SWCC*) and the E_i for saturated condition along with two fitting parameters, α and β:

$$E_{i(\text{unsat})} = E_{i(\text{sat})}\left[1 + \alpha\frac{(u_\text{a} - u_\text{w})}{(P_\text{a}/101.3)}(S^\beta)\right] \tag{4}$$

where $E_{i(\text{sat})}$ and $E_{i(\text{unsat})}$ = initial tangent elastic modulus for saturated and unsaturated conditions, respectively, P_a = atmospheric pressure (i.e., 101.3 kPa), and α, β = fitting parameters.

They suggested that the fitting parameter, $\beta = 1$, is required for coarse-grained soils (i.e., $I_\text{p} = 0\%$; *NP*). The fitting parameter, α, is a function of footing size, and the values between 1.5 and 2 were recommended for large sizes of footings in field conditions to reliably estimate E_i (Fig. 10) and elastic settlement (Fig. 11). Vanapalli and Oh [46] analyzed model footing [47], and in situ *PLT* [16, 39] results in unsaturated fine-grained soils and suggested that the fitting parameter, $\beta = 2$, is required for fine-grained soils. The analyses results also showed that the inverse of α (i.e., $1/\alpha$) nonlinearly increases with increasing I_p and the upper and the lower boundary relationship can be used for low and high matric suction values, respectively, at a certain I_p (Fig. 12).

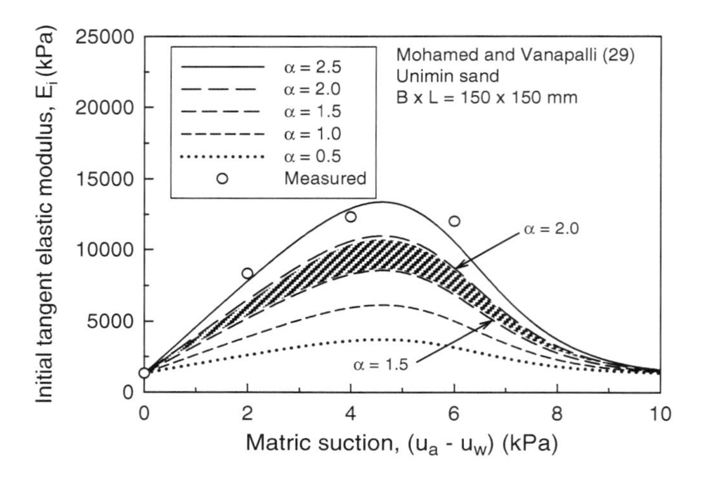

Fig. 10 Variation of modulus of elasticity with the parameter, α, for the 150 mm × 150 mm (Using data from Mohamed and Vanapalli [28])

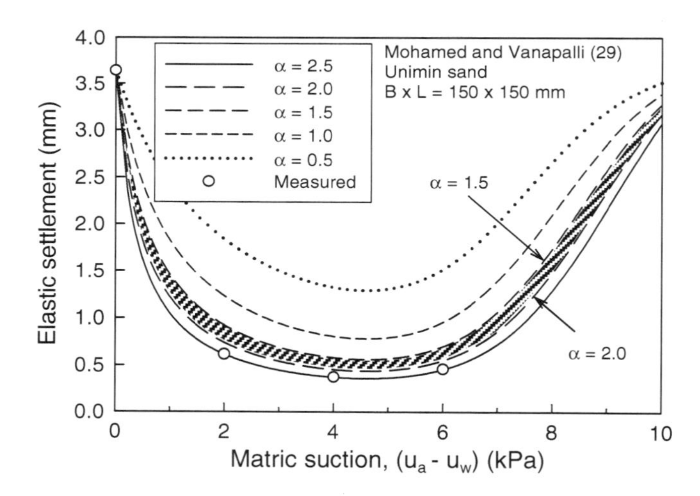

Fig. 11 Variation of elastic settlement with the parameter, α, for the 150 mm × 150 mm footing (Using data from Mohamed and Vanapalli [28])

4.3 Variation of q_u with Respect to Matric Suction for Fine-Grained Soils

Oh and Vanapalli [29] analyzed six sets of unconfined compression test results and showed that the q_u value is a function of matric suction (Figs. 13 and 14). Based on the analyses, they proposed a semiempirical model to estimate the variation of undrained shear strength of unsaturated soils using the *SWCC* and undrained shear

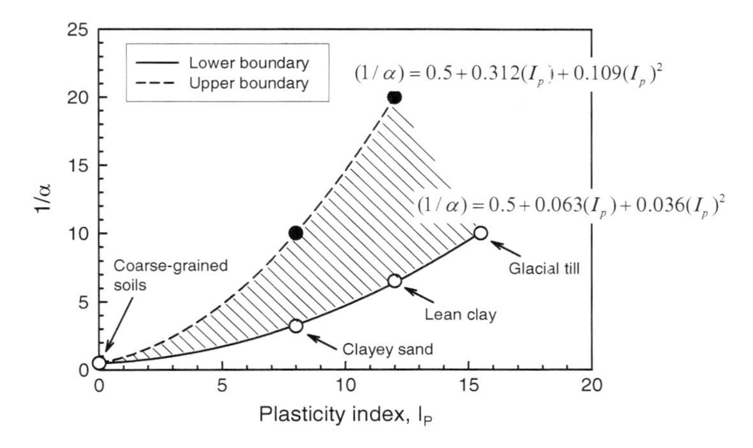

Fig. 12 Relationship between (1/α) and plasticity index, I_p

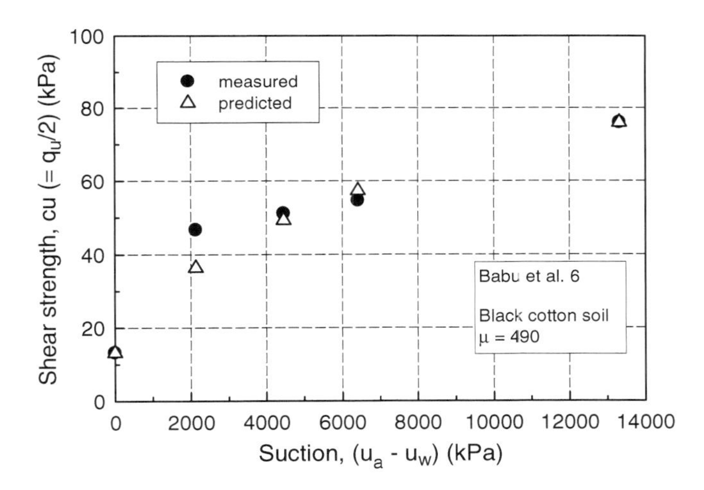

Fig. 13 Comparison between the measured and the estimated shear strength using the data by Babu et al. [6]

strength under saturated condition along with two fitting parameters, v and μ (Eq. 5). Equation (5) is the same in form as Eq. (4):

$$c_{\mathrm{u(unsat)}} = c_{\mathrm{u(sat)}}\left[1 + \frac{(u_{\mathrm{a}} - u_{\mathrm{w}})}{(P_{\mathrm{a}}/101.3)}\frac{(S^{v})}{\mu}\right] \tag{5}$$

where $c_{\mathrm{u(sat)}}$, $c_{\mathrm{u(unsat)}}$ = shear strength under saturated and unsaturated condition, respectively, P_{a} = atmospheric pressure (i.e., 101.3 kPa) and v, μ = fitting parameters.

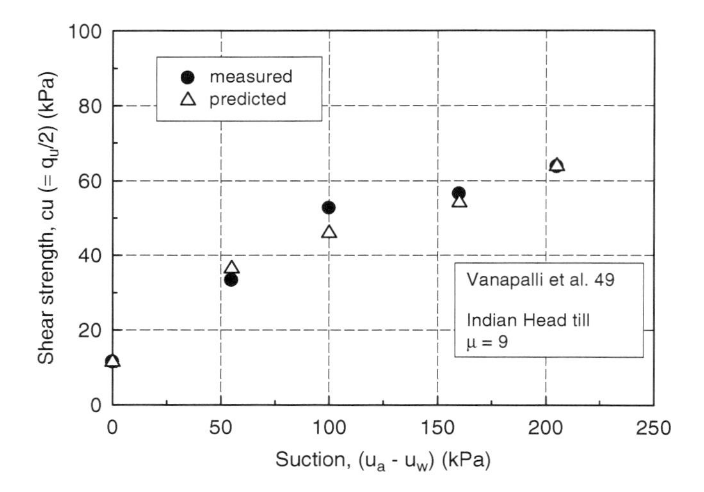

Fig. 14 Comparison between the measured and the estimated shear strength using the data by Vanapalli et al. [47]

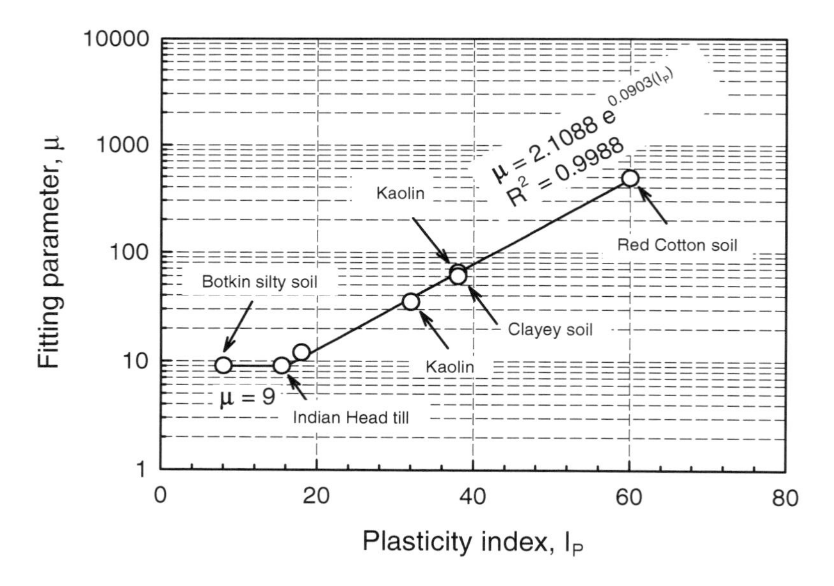

Fig. 15 Relationship between plasticity index, I_p, and the fitting parameter, μ

The fitting parameter, $v = 2$, is required for unsaturated fine-grained soils. Figure 15 shows the relationship between the fitting parameter, μ, and plasticity index, I_p, on semilogarithmic scale for the soils used for the analysis. The fitting parameter, μ, was found to be constant with a value of "9" for the soils that have I_p values in the range of 8 and 15.5%. The value of μ however increases linearly on semilogarithmic scale with increasing I_p.

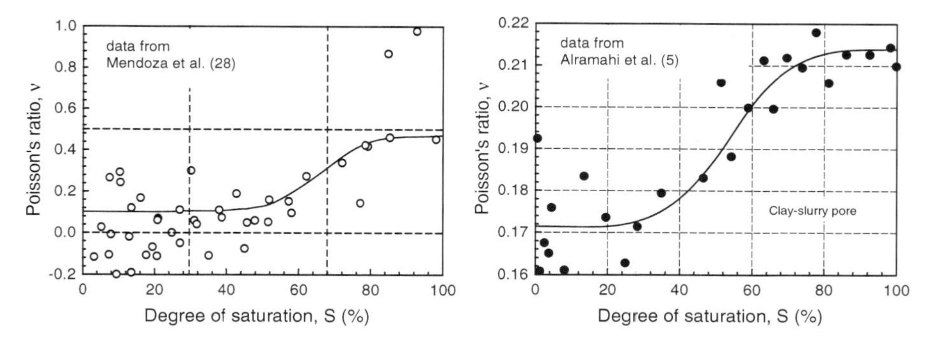

Fig. 16 Variation of Poisson's ratio, ν, with respect to degree saturation

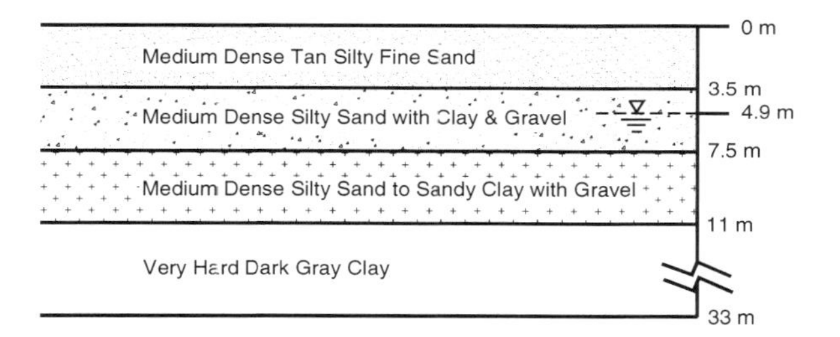

Fig. 17 The average soil profile at the test site (After Briaud and Gibbens [9])

4.4 *Variation of ν with Respect to Matric Suction*

The Poisson's ratio, ν, is typically considered to be a constant value in the elastic settlement analysis of soils. This section briefly highlights how ν varies with matric suction by revisiting published data from the literature. Mendoza et al. [27] and Alramahi et al. [2] conducted bender element tests to investigate the variation of small-strain elastic and shear modulus with respect to degree of saturation for kaolinite and mixture of glass beads and kaolin clay, respectively. Oh and Vanapalli [33] reanalyzed the results and back calculated the Poisson's ratio, ν, with respect to degree of saturation. The analyses of the results show that ν is not constant but varies with the degree of saturation as shown in Fig. 16.

5 Reanalysis of Footing Load Test Results in Briaud and Gibbens [9]

The site selected for the in situ footing load tests was predominantly sand (mostly medium dense silty sand) from 0 to 11 m overlain by hard clay layer (Fig. 17). The groundwater table was observed at a depth of 4.9 m, and the soil above the groundwater table was in a state of unsaturated condition. In this case, different

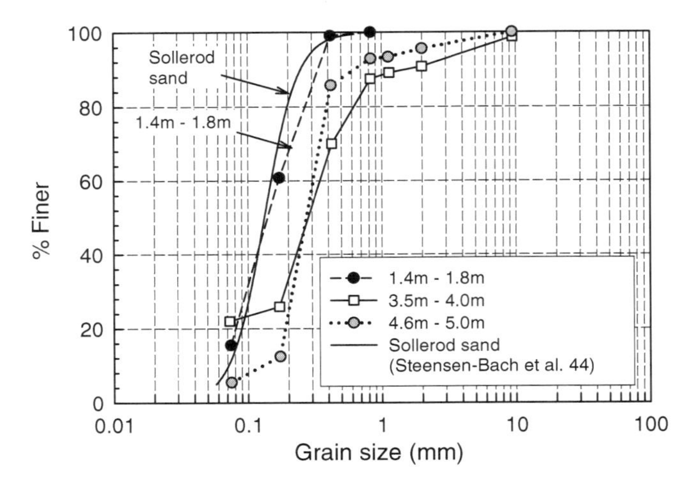

Fig. 18 Grain size distribution curves for the soil samples collected from three different depths [9] and Sollerod sand [42]

Table 1 Summary of the soil properties (From Briaud and Gibbens [9])

Property	Sand (0.6 m)	Sand (3.0 m)
Specific gravity, G_s	2.64	2.66
Water content, w (%)	5.0	5.0
Void ratio, e	0.78	0.75
Effective cohesion, c' (kPa)	0	0
Effective internal friction angle, ϕ' (°)	35.5	34.2

footing sizes may result in different average matric suction values. In other words, scale effect cannot be eliminated with normalized settlement since the soils at the site are not "homogenous and isotropic." Despite this fact, as can be seen in Fig. 6, the *SVS* behaviors from different sizes of footing fall in a narrow range. This behavior can be explained by investigating the variation of matric suction with depth at the site as follows.

Figure 18 shows the grain size distribution curves for the soil samples collected from three different depths (i.e., 1.4–1.8 m, 3.5–4.0 m, and 4.6–5.0 m). The grain size distribution curve the Sollerod sand shown in Fig. 18 is similar to the sand sample collected at the depth of 1.4–1.8 m. The reasons associated with showing the *GSD* curve of Sollerod sand will be discussed later in this chapter. The soil properties used in the analysis are summarized in Table 1.

As shown in Table 1, the water content at the depths of 0.6 and 3.0 m is 5%. This implies that the matric suction value can be assumed to be constant up to the depth of approximately 3.0 m. The field matric suction distribution profile is consistent with the typical matric suction distribution profile above groundwater table for the coarse-grained soils. In other words, matric suction increases gradually (which is close to hydrostatic conditions) up to residual matric suction value and thereafter

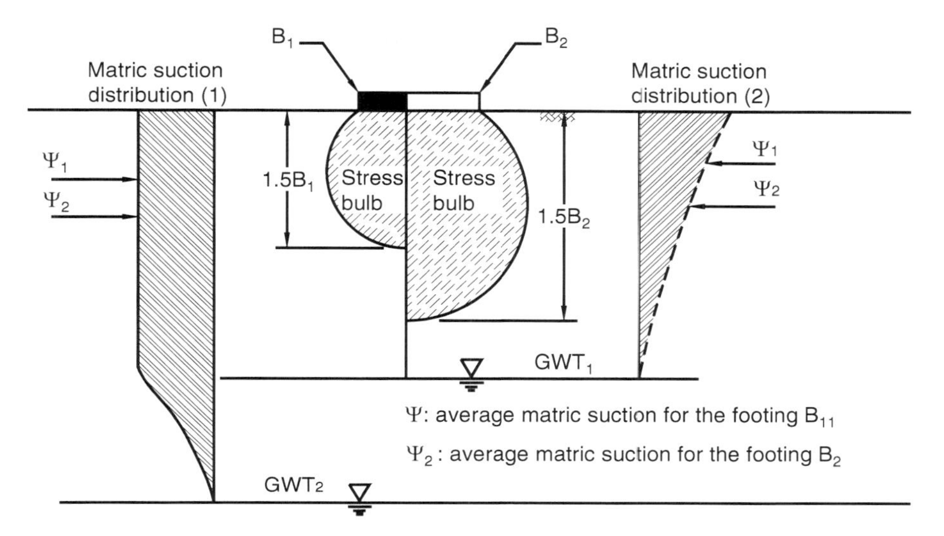

Fig. 19 Average matric suction for different sizes of footing under uniform and nonuniform matric suction distribution

remains close to constant conditions (i.e., matric suction distribution (1) in Fig. 19). This matric suction distribution profile resulted in the same average matric suction value regardless of footing size (for this study). However, it also should be noted that the average matric suction value for each footing can be different if a nonuniform matric suction distribution profile is available below the footings (i.e., matric suction distribution (2) in Fig. 19).

6 Variation of *SVS* Behaviors with Respect to Matric Suction

After construction of *SF*s, the soils below them typically experience wetting–drying cycles due to the reasons mostly associated with the climate (i.e., rain infiltration or evaporation). Hence, it is also important to estimate the variation of *SVS* behaviors with respect to matric suction.

Oh and Vanapalli [32, 33] conducted finite element analysis (*FEA*) using the commercial finite element software SIGMA/W (Geo-Slope 2007; [23]) to simulate *SVS* behavior of in situ footing ($B \times L = 1$ m $\times$ 1 m) load test results ([9]; Fig. 3) on unsaturated sandy soils. The *FEA* was performed using elastic–perfectly plastic model [14] extending the approach proposed by Oh and Vanapalli [30]. The square footing was modeled as a circular footing with an equivalent area (i.e., 1.13 m in diameter, axisymmetric problem).

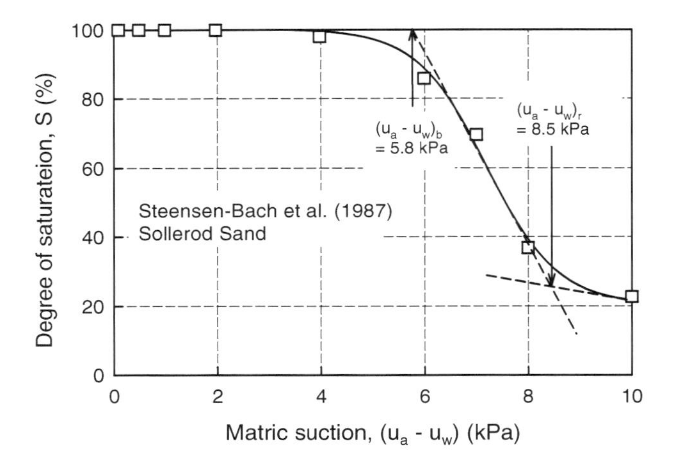

Fig. 20 *SWCC* used for the analysis (Date from Steensen-Bach et al. [42])

The soil–water characteristic curve (*SWCC*) can be used as a tool to estimate the variation of total cohesion, c, (Eq. 6; [44]) and initial tangent elastic modulus, E_i, (Eq. 4) with respect to matric suction:

$$c = c' + (u_a - u_w)(S^{\kappa}) \tan \phi' \tag{6}$$

where c = total cohesion, c' and ϕ' = effective cohesion and internal friction angle for saturated condition, respectively, $(u_a - u_w)$ = matric suction, S = degree of saturation, and κ = fitting parameter ($\kappa = 1$ for sandy soils (i.e. $I_p = 0\%$); [20]).

Information on the *SWCC* was not available in the literature for the site where the in situ footing load test was carried out. Hence, the *SWCC* for the Sollerod sand (Fig. 20) used for the analysis as an alternative based on the following justifications. Among the grain size distribution (hereafter referred as *GSD*) curves shown in Fig. 18, the grain size distribution curve for the range of depth 1.4–1.8 m can be chosen as a representative *GSD* curve since the stress below the footing 1 m × 1 m is predominant in the range of 0–1.5 m (i.e., 1.5*B*) below the footing. This *GSD* curve is similar to that of Sollerod sand (see Fig. 18) used by Steensen-Bach et al. [42] to conduct model footing tests in a sand to understand influence of matric suction on the load carrying capacity. In addition, the shear strength parameters for the Sollerod sand ($c' = 0.8$ kPa and $\phi' = 35.8°$) are also similar to those of the sand where the in situ footing load tests were conducted (see Table 1). The influence of wetting–drying cycles (i.e., hysteresis) and external stresses on the *SWCC* is not taken into account in the analysis due to the limited information.

The variation of *SVS* behavior with respect to matric suction from the *FEA* is shown in Fig. 21. Figure 22a, b shows the variation of settlement under the same stress of 344 kPa and the variation of stress that can cause 25-mm settlement for different matric suction values, respectively. The stress 344 kPa is chosen since the

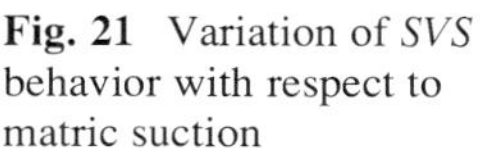

Fig. 21 Variation of *SVS* behavior with respect to matric suction

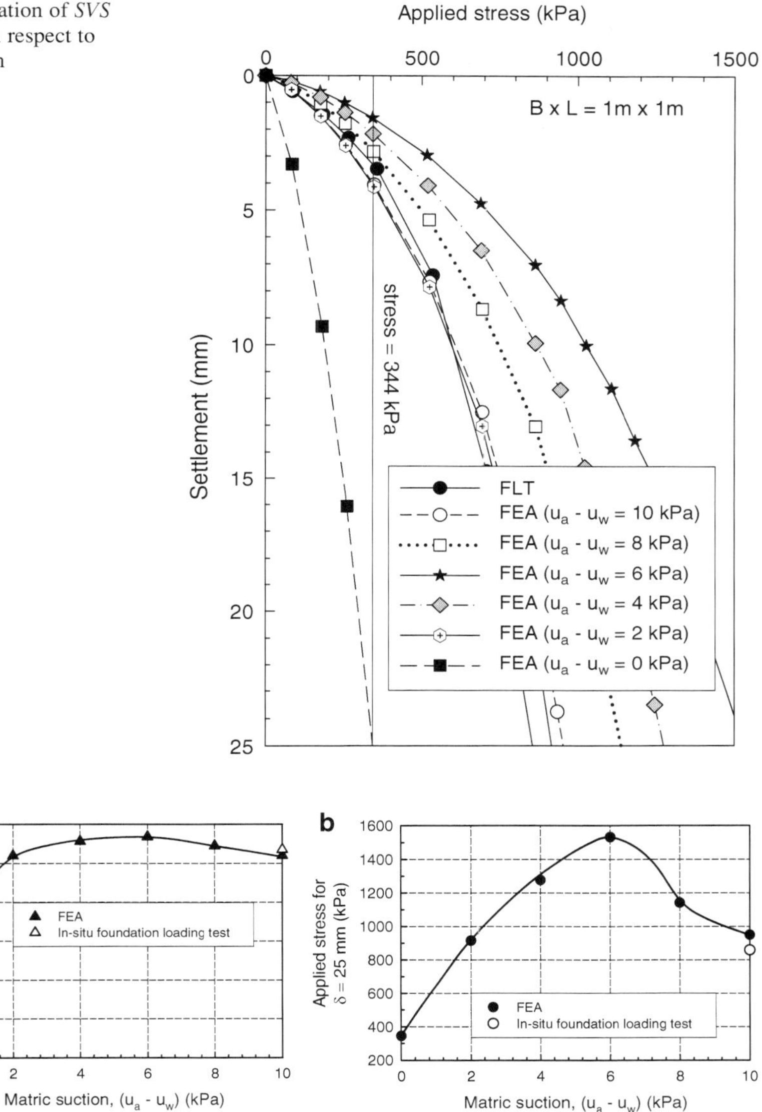

Fig. 22 Variation of (**a**) settlement under the applied stress of 344 kPa and (**b**) stress that can cause 25-mm settlement with respect to matric suction

settlement for saturated condition at this stress is 25 mm. The settlement at the matric suction of 10 kPa (i.e., field condition) is approximately 4 mm and then increases up to 25 mm (i.e., permissible settlement) as the soil approaches saturated conditions under the constant stress (i.e., 344 kPa). The permissible settlement,

25 mm, can be induced at 2.7 times less stress as the soil approaches saturated conditions (i.e., from 10 to 0 kPa). The results imply that settlements can increase due to decrease in matric suction. It is also of interest to note that such a problem can be alleviated if the matric suction of the soil is maintained at 2-kPa value.

7 Summary and Conclusions

Plate load test (*PLT*) is regarded as the most reliable testing method to estimate the applied stress versus surface settlement (*SVS*) behavior of shallow foundations. However, there are uncertainties in interpreting the *PLT* results for soils that are in a state of unsaturated condition. This is mainly attributed to the fact that the *SVS* behavior from the *PLT*s is significantly influenced by both footing size and the capillary stresses (i.e., matric suction). Previous studies showed that the scale effect can be eliminated by normalizing settlement with footing size. This methodology is applicable to the soils that are homogeneous and isotropic with depth in nature such as saturated or dry soils. In case of unsaturated soils, matric suction distribution profile with depth should be taken into account to judge whether or not this methodology is applicable. This is because if the matric suction distribution profile is nonuniform with depth, different plate sizes lead to different average matric suction values. In other words, the soil below the plates cannot be regarded as homogeneous and isotropic since strength, initial tangent elastic modulus, and the Poisson's ratio are function of matric suction. These facts indicate that the reliable deign of shallow foundations based on the *PLT* results can be obtained only when the results are interpreted taking account of the matric suction distribution profile with depth and influence of average matric suction value on the *SVS* behavior.

In case of the shallow foundations resting on unsaturated sandy soils, it is also important to estimate the variation of *SVS* behavior with respect to matric suction. This can be achieved by conducting finite element analysis using the methodology presented in this chapter. According to the finite element analysis for the in situ footing (1 m × 1 m) load test results discussed in this chapter [9], unexpected problems associated with settlement are likely due to decrease in matric suction. Such a problem can be alleviated if the matric suction of the soil is maintained at a low of 2 kPa.

References

1. Agarwal KB, Rana MK (1987) Effect of ground water on settlement of footing in sand. In: Proceedings of the 9th European conference on soil mechanics and foundation engineering, Dublin, pp 751–754.
2. Alramahi B, Alshibli KA, Fratta D (2010) Effect of fine particle migration on the small-strain stiffness of unsaturated soils. J Geotech Geoenviron Eng 136(4):620–628

3. ASTM D1194-94 (2003) Standard test method for bearing capacity of soil for static load and spread footings. American Society for Testing Materials, Philadelphia
4. ASTM D1195-93 (2004) Standard test method for repetitive static plate load tests of soils and flexible pavement components, for use in evaluation and design of airport and highway pavements. American Society for Testing Materials, Philadelphia
5. ASTM D1196 – 93 (2004) Standard test method for nonrepetitive static plate load tests of soils and flexible pavement components, for use in evaluation and design of airport and highway pavements. American Society for Testing Materials, Philadelphia
6. Babu GLS, Rao RS, Peter J (2005) Evaluation of shear strength functions based on soil water characteristic curves. J Test Eval 33(6):461–465
7. Bolton MD, Lau CK (1989) Scale effects in the bearing capacity of granular soils. In: Proceedings of the 12th international conference of soil mechanics and foundation engineering, vol 2. Balkema Publishers, Rotterdam, pp 895–898
8. Briaud J-L (2007) Spread footings in sand: load settlement curve approach. J Geotech Geoenviron Eng 133(8):905–920
9. Briaud J-L, Gibbens R (1994) Predicted and measured behavior of five large spread footings on sand. In: Proceedings of a prediction symposium, ASCE, GSP41
10. Briaud JL, Gibbens RM (1997) Large scale load tests and data base of spread footings on sand. Publication no. FHWA-RD-97-068. Federal Highway Administration, Washington, DC
11. BS 1377–9:1990 Methods for test for soils for civil engineering purposes. In-situ tests, British Standards Institution, 31 Aug 1990, 70 pp
12. Cerato AB, Lutenegger AJ (2007) Scale effects of shallow foundation bearing capacity on granular material. J Geotech Geoenviron Eng 133(3):1192–1202
13. CFEM (2006) Canadian foundation engineering manual, 4th edn. Canadian Geotechnical Society, Toronto
14. Chen WF, Zhang H (1991) Structural plasticity: theory, problems, and CAE software. Springer, New York
15. Consoli NC, Schnaid F, Milititsky J (1998) Interpretation of plate load tests on residual soil site. J Geotech Geoenviron Eng 124(9):857–867
16. Costa YD, Cintra JC, Zornberg JC (2003) Influence of matric suction on the results of plate load tests performed on a lateritic soil deposit. Geotech Test J 26(2):219–226
17. D'Appolonia DJ, D'Appolonia E, Brisette RF (1968) Settlement of spread footings on sand. J Soil Mech Found Div ASCE 3:735–760
18. De Beer EE (1965) The scale effect on the phenomenon of progressive rupture in cohesionless soils. In: Proceedings of the 6th international conference on soil mechanics and foundation engineering, vol 2(3–6), pp 13–17
19. Fellenius BH, Altaee A (1994) Stress and settlement of footings in sand. In: Proceedings of the conference on vertical and horizontal deformations of foundations and embankments, ASCE, GSP40, College Station, vol 2, pp 1760–1773
20. Garven E, Vanapalli SK (2006) Evaluation of empirical procedures for predicting the shear strength of unsaturated soils. In: Proceedings of the 4th international conference on unsaturated soils, ASCE, GSP147, Arizona, vol 2, pp 2570–2581
21. Hettler A, Gudehus G (1988) Influence of the foundation width on the bearing capacity factor. Soils Found 28(4):81–92
22. Ismael NF (1985) Allowable pressure from loading tests on Kuwaiti soils. Can Geotech J 22(2):151–157
23. Krahn J (2007) Stress and deformation modelling with SIGMA/W. Goe-slope International Ltd.
24. Kusakabe O (1995) Foundations. In: Taylor RN (ed) Geotechnical centrifuge technology. Blackie Academic & Professional, London, pp 118–167
25. Lee J, Salgado R (2001) Estimation of footing settlement in sand. Int J Geomech 2(1):1–28

26. Maugeri M, Castelli F, Massimino MR, Verona G (1998) Observed and computed settlements of two shallow foundations on sand. J Geotech Geoenviron Eng 124(7):595–605
27. Mendoza CE, Colmenares JE, Merchan VE (2005) Stiffness of an unsaturated compacted clayey soil at very small strains. In: Proceedings of the international symposium on advanced experimental unsaturated soil mechanics, Trento, Italy, pp 199–204
28. Mohamed FMO, Vanapalli SK (2006) Laboratory investigations for the measurement of the bearing capacity of an unsaturated coarse-grained soil. In: Proceedings of the 59th Canadian geotechnical conference, Vancouver
29. Oh WT, Vanapalli SK (2009) A simple method to estimate the bearing capacity of unsaturated fine-grained soils. In: Proceedings of the 62nd Canadian geotechnical conference, Hailfax, Canada, pp 234–241
30. Oh WT, Vanapalli SK (2010) The relationship between the elastic and shear modulus of unsaturated soils. In: Proceedings of the 5th international conference on unsaturated soils, Barcelona, Spain, pp 341–346.
31. Oh WT, Vanapalli SK (2011) Modelling the applied vertical stress and settlement relationship of shallow foundations in saturated and unsaturated sands. Can Geotech J 48(3): 425–438
32. Oh WT, Vanapalli SK (2011) Modelling the stress versus displacement behavior of shallow foundations in unsaturated coarse-grained soils. In: Proceedings of the 5th international symposium on deformation characteristics of geomaterials, Seoul, Korea, pp 821–828
33. Oh WT, Vanapalli SK (2012) Modelling the settlement behaviour of in-situ shallow foundations in unsaturated sands. In: Proceedings of the geo-congress 2012 (Accepted for publication)
34. Oh WT, Vanapalli SK, Puppala AJ (2009) Semi-empirical model for the prediction of modulus of elasticity for unsaturated soils. Can Geotech J 46(8):903–914
35. Oloo SY (1994) A bearing capacity approach to the design of low-volume traffics roads. Ph.D. thesis, University of Saskatchewan, Saskatoon, Canada
36. Osterberg JS (1947) Discussion in symposium on load tests of bearing capacity of soils. ASTM STP 79. ASTM, Philadelphia, pp 128–139
37. Palmer LA (1947) Field loading tests for the evaluation of the wheel load capacities of airport pavements. ASTM STP 79. ASTM, Philadelphia, pp 9–30
38. Poulos HD, Davis EH (1974) Elastic solutions for soil and rock mechanics. Wiley, New York
39. Rojas JC, Salinas LM, Seja C (2007) Plate-load tests on an unsaturated lean clay. Experimental unsaturated soil mechanics, Springer proceedings in physics, vol 112, pp 445–452
40. Schanz T, Lins Y, Vanapalli SK (2010) Bearing capacity of a strip footing on an unsaturated sand. In: Proceedings of the 5th international conference on unsaturated soils, Barcelona, Spain, pp 1195–1220.
41. Steenfelt JS (1977) Scale effect on bearing capacity factor N_{γ}. In: Proceeding of the 9th international conference of soil mechanics and foundations engineering, vol 1. Balkema Publishers, Rotterdam, pp 749–752
42. Steensen-Bach JO, Foged N, Steenfelt JS (1987) Capillary induced stresses–fact or fiction? In: Proceedings of the 9th European conference on soil mechanics and foundation engineering, Dublin, pp 83–89
43. Tatsuoka F, Okahara M, Tanaka T, Tani K, Morimoto T, Siddiquee MSA (1991) Progressive failure and particle size effect in bearing capacity of a footing on sand, GSP27, vol 2, pp 788–802
44. Vanapalli SK, Fredlund DG, Pufahl DE, Clifton AW (1996) Model for the prediction of shear strength with respect to soil suction. Can Geotech J 33(3):379–392
45. Vanapalli SK, Mohamed FMO (2007) Bearing capacity of model footings in unsaturated soils. In: Proceedings of the experimental unsaturated soil mechanics, Springer proceedings in physics, vol 112, pp 483–493

46. Vanapalli SK, Oh WT (2010) A model for predicting the modulus of elasticity of unsaturated soils using the soil-water characteristic curves. Int J Geotech Eng 4(4):425–433
47. Vanapalli SK, Oh WT, Puppala AJ (2007) Determination of the bearing capacity of unsaturated soils under undrained loading conditions. In: Proceedings of the 60th Canadian geotechnical conference, Ottawa, Canada, pp 1002–1009
48. Xu YF (2004) Fractal approach to unsaturated shear strength. J Geotech Geoenviron Eng 130(3):264–273
49. Yamaguchi H, Kimura T, Fuji N (1976) On the influence of progressive failure on the bearing capacity of shallow foundations in dense sand. Soils Found 16(4):11–22

An Approach for Creating Certainty in Uncertain Environment: A Case Study for Rebuilding a Major Equipment Foundation

Abhijit Dasgupta and Suvendu Dey

Abstract The only thing that makes life possible is permanent, intolerable uncertainty; not knowing what comes next – Ursula Le Guin

To the common man, uncertainty is being in doubt or the state of being unsure about something. In scientific parlance, it is the unpredictable difference between the observed data and the model output. Sources of uncertainty may be many including material, manufacturing, environment, experiments, human factors, assumptions and lack of knowledge. Any one of these or a combination may lead to a significant loss of performance, that is, a large variation in output due to a small variation in the input parameters. The human being craves for certainty because the first priority for every individual on this planet is survival and the process of living contains many risks. This chapter deals with the various uncertainties that confronted a team of engineers during the course of rebuilding and upgrading an existing major equipment in an integrated steel plant. There were multiple challenges and uncertainties involved in every step of the rebuilding process.

Keywords Uncertainty • Equipment • Blasting • Shutdown • Rebuilding • Upgradation

1 Introduction

The potentiality of perfection outweighs actual contradictions. Existence in itself is here to prove that it cannot be an evil – Rabindranath Tagore

The terms risk and uncertainty are intertwined and somewhat complex to analyse and differentiate. Risk can be defined as a state of uncertainty where some of the possibilities involve a loss, catastrophe or other undesirable outcome.

A. Dasgupta • S. Dey (✉)
M.N. Dastur & Company (P) Ltd., Kolkata, West Bengal, India
e-mail: Abhijit.DG@dasturco.com; Suvendu.D@dasturco.com

S. Chakraborty and G. Bhattacharya (eds.), *Proceedings of the International Symposium on Engineering under Uncertainty: Safety Assessment and Management (ISEUSAM - 2012)*, DOI 10.1007/978-81-322-0757-3_8,

Uncertainty, on the other hand, may be defined as the lack of complete certainty, that is, the existence of more than one possibility. The true outcome/state/result/value is not known. Human endeavour has always been to try and minimise risks even if it means working in uncertain conditions. Accordingly, 'one may have uncertainty without risk but not risk without uncertainty'. This chapter presents a case study where uncertainties and contradictions were overcome by the sheer will to achieve perfection and minimise potential risks.

In any integrated steel plant, one of the most important and major equipment is where iron ore, flux and fuel are burned in oxygen-enriched air to produce molten metal and slag. In a premier steel plant of India, such an equipment was first blown in the late 1950s. It was due for relining in the new millennium when the owners decided to upgrade it as well, from the existing 0.64–1.0 MTPA capacity. This was no mean task as every stage of engineering involved uncertainties and risks that had to be mitigated and solutions found.

In this chapter, an effort has been made to identify the uncertainties involved for this rebuilding and upgradation process. This chapter also describes how each uncertainty was analysed and dealt with in a rational manner to reach a level of relative certainty. Some explanatory sketches have also been included for a better understanding of the problem and the solutions.

2 Uncertainties

There were multiple challenges and uncertainties involved in every step of the rebuilding and upgradation process of the equipment. Some of the major uncertainties were:

- Knowledge of the existing foundation system
- Geotechnical data
- Load-carrying capacity
- Time constraint for shutdown of the equipment
- Developing model of foundation and subsequent blasting of part of foundation to simulate results
- Dismantling part of existing foundation by controlled blasting
- Restriction of energy propagation to the base raft of foundation
- Part load transfer through tower and part through existing foundation after partial rebuilding

Each of the above uncertainty has been elaborated in the following sections of this chapter, and steps to overcome them have also been described.

2.1 Existing Foundation System

Since the actual engineering of the equipment was done more than four decades ago, very limited and scanty data could be located from the client's archives. A few

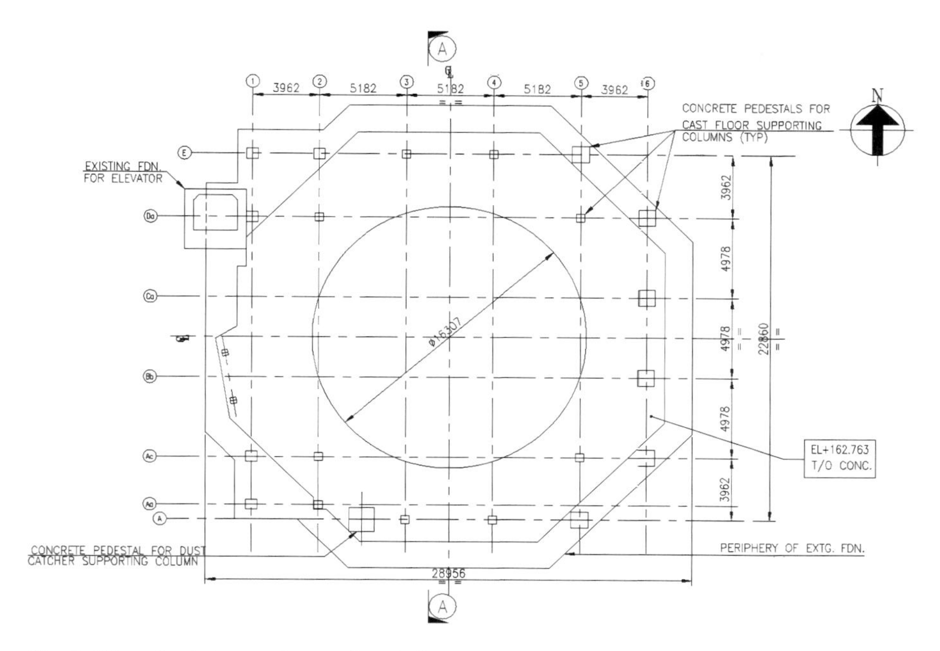

Fig. 1 Plan of existing equipment foundation

old drawings and an article, published in an in-house journal, were all that could be unearthed. Thus, credible information was limited and grossly insufficient. From the very limited data, it was understood that the main equipment shell was supported on a 52-ft.-diameter cylindrical concrete pedestal over a 13-ft.-thick octagonal concrete mat. The mat itself was supported on 28 nos. concrete columns, of size 5 sq. ft, founded on hard mica-schist rock. The column foundations were designed with safe bearing capacity of 4.0–9.0 t/sq. ft. Schematic sketches of the existing equipment foundation are shown in Figs. 1 and 2 below.

The above was the basis of a study undertaken by Dasturco to judge the feasibility of the proposed rebuilding and upgradation of the equipment.

2.2 Geotechnical Data

No soil investigation/geotechnical data could be found catering to the location of the existing equipment. From an old publication, as indicated earlier, it could be inferred that the foundation was designed with a safe bearing capacity between 4.0 and 9.0 t/sq. ft. Due to lack of data in the concerned area, it was decided to use existing geotechnical information from the neighbouring areas. Accordingly, from available soil investigation reports and test data of a nearby mill area, the gross safe bearing capacity of competent rock was estimated to be about 75.0 t/m^2. This corroborated well with the data obtained earlier from the technical article.

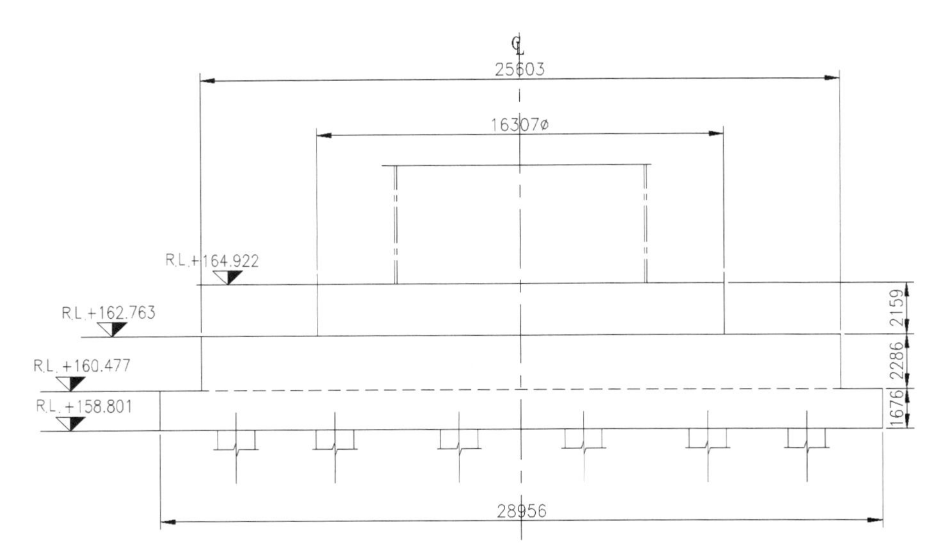

Fig. 2 Sectional view of existing equipment foundation

2.3 *Load-Carrying Capacity*

From the unearthed documents, no detailed load data on foundation could be found except an indicative vertical load of 21,000 Kips (9,258 tons) from the shell and cast house. However, there were necessarily other loads imposed on the foundation proper, and these had to be estimated to study the adequacy of the foundation system. A thorough reassessment of load was done considering additional loading from the elevator and one leg of dust catcher. Moreover, the equipment had undergone intermediate relining and modification works over the years. These would have increased the loading substantially. Considering all these factors, the vertical load on the existing foundation was reassessed and estimated to be of the order of 10,186 tons.

It was decided that the upgraded equipment would be free standing with four (4) tower legs around the shell proper. The design load from the shell and the tower legs considering all possible vertical loads for upgradation of the equipment for capacity enhancement were estimated to be to the tune of 13,000 tons. The existing foundation was not found to be adequate to carry the additional loads. Moreover, some cracks were noted on the foundation shaft and on top of the existing mat indicating signs of distress possibly due to flow of some molten iron on the foundation top. Based on the above, it was decided to get a thorough health study of the existing foundation done. Accordingly, the following studies were carried out:

- Cover metre test
- Carbonation test and pH
- Crack-width measurement and mapping
- Half-cell corrosion potential test
- Schmidt's rebound hammer test
- Ultrasonic pulse velocity test
- Core cutting and crushing tests

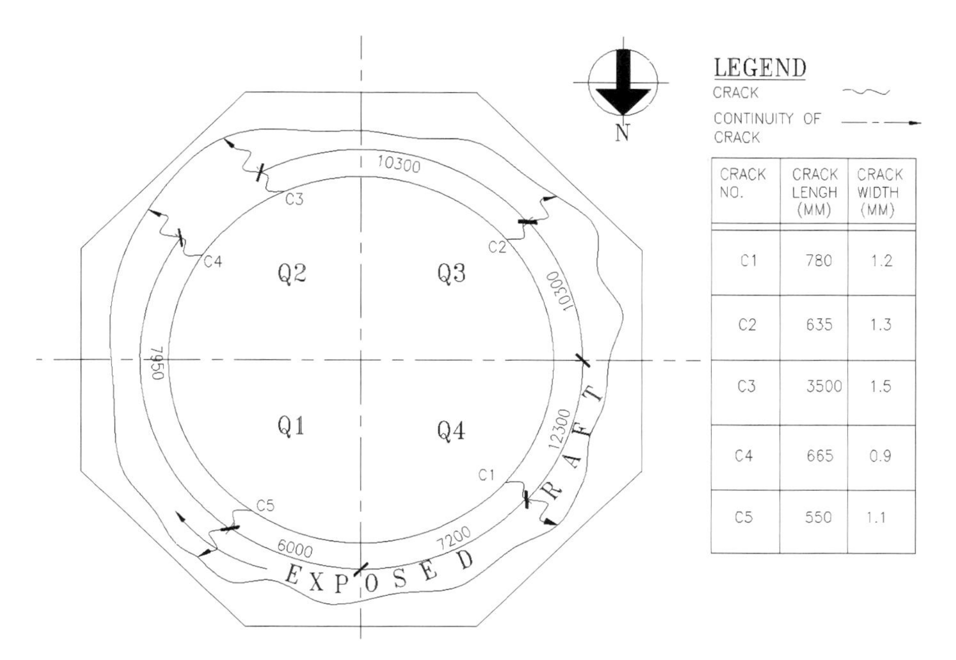

Fig. 3 Crack mapping on existing foundation

Results of the health study indicated that the concrete grade of the existing reinforced concrete foundation was between M15 and M20 (M15 grade shown in drawing/document). Mild steel reinforcement was provided and found to be in excellent condition with no signs of corrosion. Low degree of carbonation and residual alkalinity were also inferred from the tests. Moreover, vertical cracks on outer face of pedestal and horizontal cracks on top surface of mat could be observed. The crack widths were measured and mapped. This is presented in Fig. 3 below.

Thus, it was evident that the structure was under distress for quite some time. Accordingly, Dasturco made the following major recommendations in the feasibility report:

- The proposed four (4) towers, around the shell proper, must be independently supported on 1,000-mm-diameter pile foundations.
- Top pedestal and part of the main foundation raft must be dismantled and rebuilt with heat-resistant concrete (M30 grade) along with new holding down bolts to fix the upgraded equipment base and additional reinforcements, wherever required.
- After modification, the vertical load on the equipment foundation from the shell of the upgraded equipment must not exceed the original design load, that is, 21,000 Kips (9,258 tons).

2.4 Time Constraint for Shutdown of the Equipment

The entire rebuilding and upgradation process had to be done under a very tight time schedule. Since the work involved dismantling and rebuilding of part of the existing foundation, temporary shutdown of the equipment was necessary. However, in a running plant, shutdown of a major equipment is directly related to a loss in production and hence revenue. Thus, it was essential to minimise the shutdown period as far as practicable. A micro schedule was prepared to target a total shutdown period of 100 days, consisting of rebuilding of the furnace shell and other accessories to cater to the capacity enhancement. This period also included erection of steel tower and rebuilding of foundation. Period for dismantling and rebuilding of foundation proper was restricted to 14 days only.

To optimise the shutdown period to a minimum, it was decided to adopt controlled blasting technique for breaking/dismantling part of the foundation, as manual breaking would have taken an enormous amount of time. At the same time, one had to be extremely careful to ensure that the blasting process did not cause any distress or damage to the remaining portion of the structure, proposed to be retained intact. Based on these considerations and to address the uncertainties involved with the after-effects of blasting, it was decided to construct a model of the foundation system and carry out controlled blasting to simulate the actual conditions.

2.5 Model Foundation with Blasting to Simulate Results

As described in the preceding section, it was decided to carry out the trial blasting on a model foundation, similar to the actual equipment foundation, to have hands-on information about the effects of blasting on the portion of the foundation to be retained and reused. Accordingly, the following course of action was decided upon:

- To construct a scaled model foundation based on information gathered from unearthed data about the existing equipment foundation, that is, shape, size, grade of concrete and reinforcements.
- To carry out controlled blasting through 32-mm-diameter vertical holes, at 500-mm centres, circumferentially, in ring formations, with three (3) such rings at radial distance of 500 mm. Sequence of blasting would be from the outer to inner rings with suitable delay per charge.
- To drill 32-mm-diameter horizontal through holes, at 500-mm centres, at 250 mm above the octagonal mat for easy separation and arresting shock wave propagation down below.
- To record shock wave intensities during blasting, near the dismantling level of the model block, engaging suitable sensors.
- To carry out non-destructive/partial destructive tests on balance portion of model foundation block, intended to be retained, before and after blasting, to check for possible health deterioration.

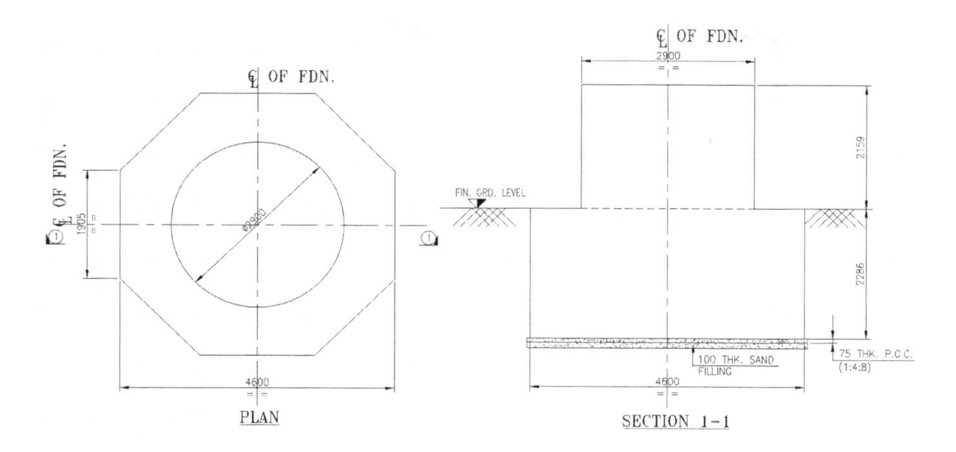

Fig. 4 Details of model foundation

- To simulate all recorded data on model foundation block with the prototype foundation block to ascertain different parameters for controlled blasting of the same and to restrict the disturbance below dismantling level to within acceptable limits.

Details of the model foundation are indicated schematically in Fig. 4 above.

Based on the above guidelines, controlled blasting was successfully carried out, on a model foundation, with extensive recording of data. Analyses of all recorded data led to the following conclusions:

- Results of core samples from octagonal mat before and after blasting did not indicate any deterioration of concrete strength due to the effect of blasting.
- Fragmentation of concrete was less in the area with reduced explosive charge, and the overburden on the pedestal was less than sufficient to contain the fragments from ejecting.
- The mat experienced mainly symmetric compression and a marginal amount of tension due to the effect of blasting. The compressive stresses were well within the allowable limits of concrete, and chances of cracking were remote due to insignificant tensile stresses.
- High vibration levels and high momentary shock-wave velocities were recorded, due to blasting, which were higher than the permissible values. However, these would actually subside to low levels due to attenuation characteristics of the soil surrounding the foundation.

The above observations and analyses of results were found to be quite encouraging thereby emboldening the concerned engineers to finalise dismantling of part of the actual equipment foundation by controlled blasting.

2.6 Dismantling Part of Existing Foundation

Based on the results and simulation studies of blasting of model foundation, actual controlled blasting of the prototype equipment foundation was done. To carry out this work, it was decided to maintain the disposition of vertical and horizontal holes and sequence of blasting in line with the model study. To minimise the critical shutdown period of the equipment, some activities were carried out in the pre-shutdown period, as preparatory work. These mainly included:

- Developing a plain cement-concrete horizontal base on which the reinforcement cage of the foundation pedestal was fabricated with erection framework, bolt sleeves and bolt boxes
- Fabricating lower erection framework followed by fabrication of the upper framework
- Providing permanent steel shuttering with 10-mm-thick steel plates and fixing the same with the reinforcement by welding arrangement
- Providing lifting hooks to each segment of the upper erection frame and strengthening the upper and lower erection frames by bracings to prevent buckling during lifting and transporting

During the initial phase of shutdown, some more preparatory works were done like construction of an RCC overlay on the vertical side of upper octagonal mat with dowel bars and bonding agent, horizontal drilling to create a cut-off plane and drilling about 10% of vertical holes on top of circular pedestal to facilitate blasting.

Dismantling of part of the existing foundation was done during the period allotted for rebuilding of the foundation that was limited to 14 days only to adhere to the schedule for total shutdown period of the furnace. During this period, the balance vertical holes were drilled, in staggered fashion, in a grid of 650 mm square. A temporary safety deck was also erected to act as a barrier for accidental fall of any object during activity above the deck. These were followed by installing the requisite explosive charge in the vertical holes (between 125 and 250 g per hole) and carrying out controlled blasting.

Blasting was carried out in four volleys with the total charge of the explosives being 42.5 kg. Adequate safety precautions were taken by placing sand bags all around the foundation as the operations were done within an existing plant. During blasting, the cylindrical pedestal was loaded with overburden weight of about 100 kg/m^2 to prevent the fragmented pieces from being ejected.

Controlled blasting proved to be a very successful venture. After the operation, it was noted that the total concrete above the octagonal raft could be removed easily leaving a smooth top surface of the octagonal mat. This was because the continuity of reinforcement was limited only along the periphery of the circular shaft and the central portion was actually filled with lean concrete. The resulting debris, post blasting, was cleaned by excavators and disposed to designated dump areas.

A sectional view of the equipment foundation proposed to be partly dismantled and rebuilt is shown in Fig. 5 below. The scheme of blasting along with disposition of blasting holes, sequence of blasting and details of charge placement in each hole is presented in Fig. 6 below.

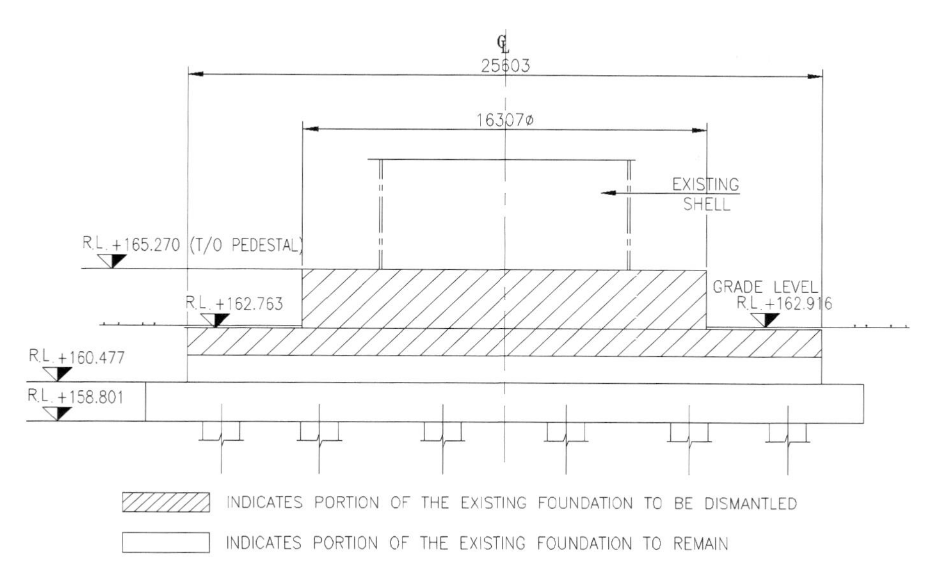

Fig. 5 Sectional view of foundation proposed to be partly dismantled and rebuilt

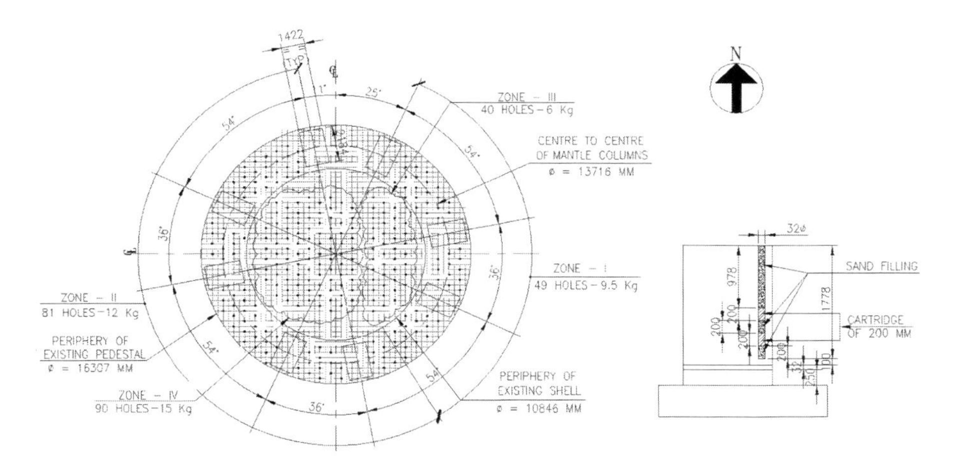

Fig. 6 Disposition of blasting holes, sequence of blasting and details of charge placement

2.7 Restricting Energy Propagation to the Base Raft

A major concern while planning the blasting activities was the propagation of shock waves to the lower portion of the foundation mat, intended to be retained and reused. To restrict wave propagation and avoid possible distress to the lower portion of the foundation mat, some kind of cut-off had to be planned. As indicated earlier, this was planned to be facilitated by drilling horizontal holes, through the cylindrical pedestal, at a suitable height above the top of the octagonal mat. This was implemented during the model study to simulate the actual conditions with extensive measurement of shock-induced strains in the concrete.

The model studies indicated that due to blasting, the vertical reinforcement bars of the cylindrical pedestal had bent outwards from the level of the horizontal holes. The bars held with them chunks of pedestal concrete in the lower portion near the horizontal holes. As a result of this, the level of separation was formed at 200–300 mm above the desired level. As a result of this experience, the vertical reinforcements of the pedestal were cut at the level of the horizontal holes before blasting to facilitate proper separation and fragmentation. Based on the above, a series of 32-mm-diameter horizontal holes were drilled, at 500-mm centres, at 250 mm above the octagonal mat for easy separation and arresting shock wave propagation down below.

After the actual controlled blasting of the prototype equipment foundation took place, the results were there for everyone to see. There were absolutely no signs of distress or crack on the lower portion of the octagonal mat that was planned to be retained. This proved that the series horizontal drill holes, provided at cardinal locations, were indeed effective in creating a cut-off plane for energy dissipation and preventing the shock waves to travel below.

2.8 Load Transfer Through Tower and Existing Foundation

As indicated earlier, an assessment of loading on the existing equipment foundation was done to gauge its present condition. It worked out that the foundation was already overstressed in excess of what it was designed for. Some telltale cracks on the foundation mat also bore testimony to this fact. Thus, there was no way the foundation could be loaded further as per the requirement of upgradation. The engineering solution that was needed to be developed was to design a system that would effectively transfer the enhanced load, from the upgraded equipment, without causing any distress to the foundation. The solution suggested was the following:

- The proposed upgraded equipment should be free-standing type accompanied by four (4) tower legs around the shell proper.
- The entire shell would carry its self-weight including the weight of refractories and weight of offtake, uptake, downcomer and the Compact Bell Less (CBLT) Top charging system.
- The four tower legs would carry, besides their self-weight, the weight of skip bridge, top structure and platforms at different levels.
- To facilitate the above, the main octagonal mat foundation was partly dismantled and rebuilt, as described in the preceding sections. The four (4) towers, on the other hand, were independently supported on 1,000-mm-diameter bored cast-in-situ piles of 300-t capacity each.

Sketches showing the plan and sectional elevations of the rebuilt equipment foundation are shown in Figs. 7, 8, and 9, respectively.

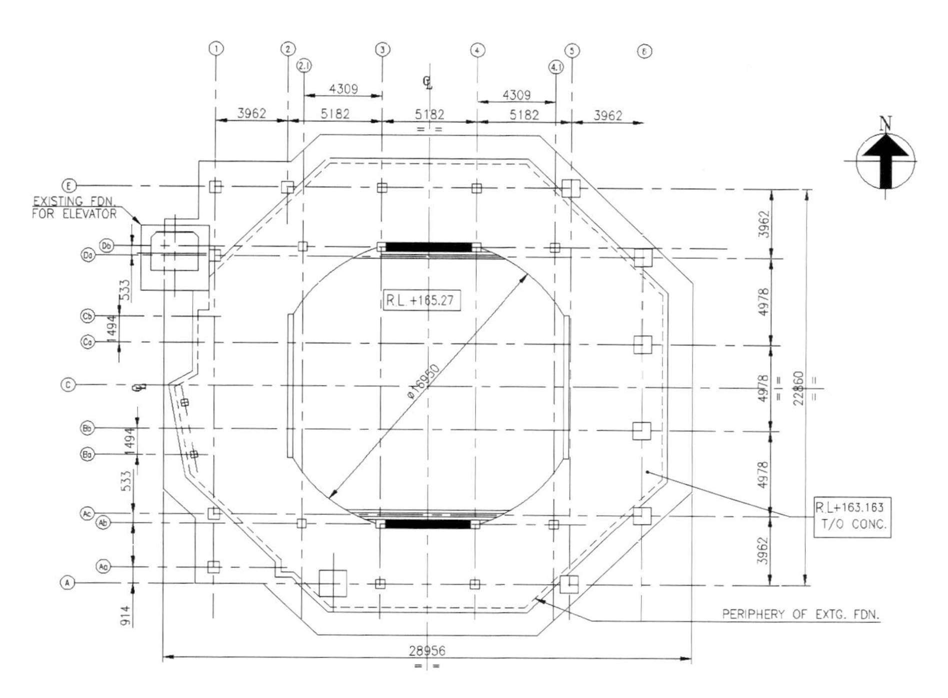

Fig. 7 Plan showing rebuilt equipment foundation

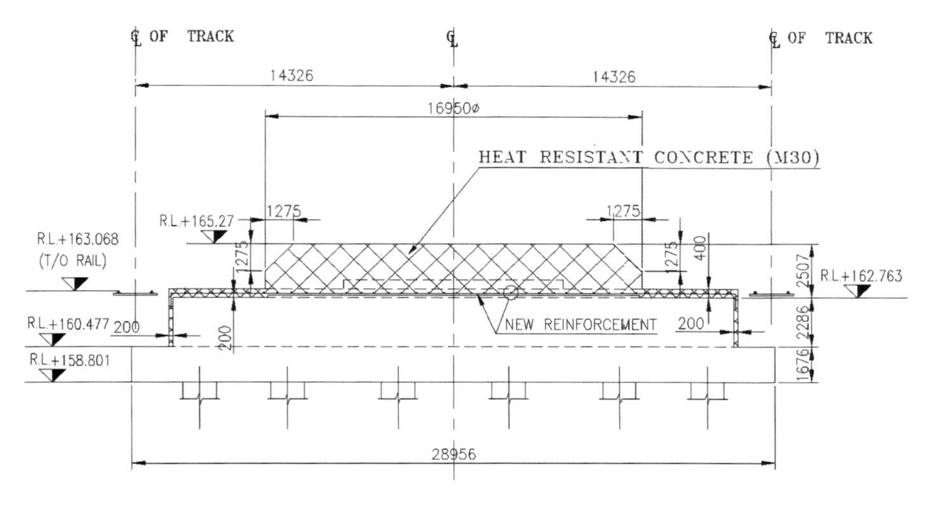

Fig. 8 View from west side showing rebuilt equipment foundation

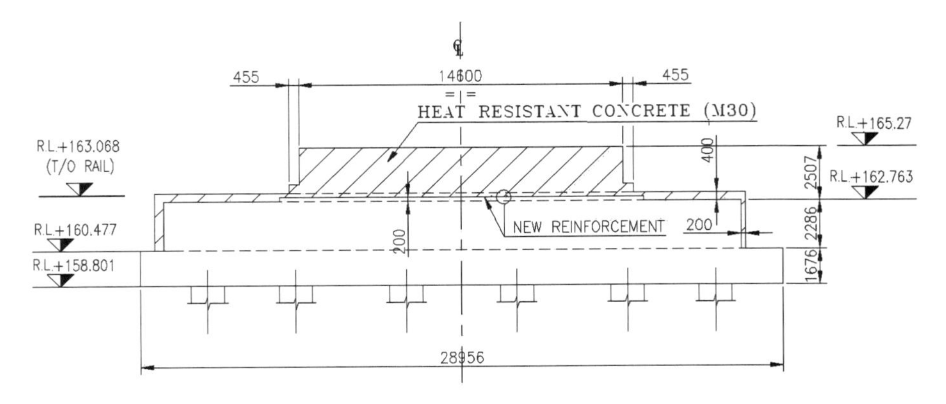

Fig. 9 View from south side showing rebuilt equipment foundation

3 Mitigating Uncertainties in Record Time

There were a number of uncertainties in every step of the rebuilding process that confronted the engineering team. Some of them have been highlighted in the preceding sections of this chapter. However, the biggest and by far the most challenging task was the race against time. As the work was being done in a running plant, any rebuilding and upgradation process necessarily required an optimum period of shutdown of the concerned equipment and some of the associated facilities. It goes without saying that shutting down of a major unit in a running plant hampers production and, consequently, has a direct bearing on revenue. To be fair to the clients, the shutdown period allowed for the work was extremely tight. It seemed impossible and somewhat improbable to complete all the activities within the very stringent time period that was allowed. However, with a dedicated design and construction team working in unison and perfect harmony, the target was achieved a couple of days before the scheduled completion date. The feat was duly recognised and appreciated by the clients in no uncertain terms.

An isometric view of the rebuilt and upgraded equipment foundation is shown in Fig. 10.

4 Conclusion

In a premier integrated steel plant of our country, a major production equipment was due for relining and refurbishment during the early part of the new millennium. However, the clients desired to upgrade the equipment at the same time to enhance their production capacity. This called for a detailed study of the existing foundation

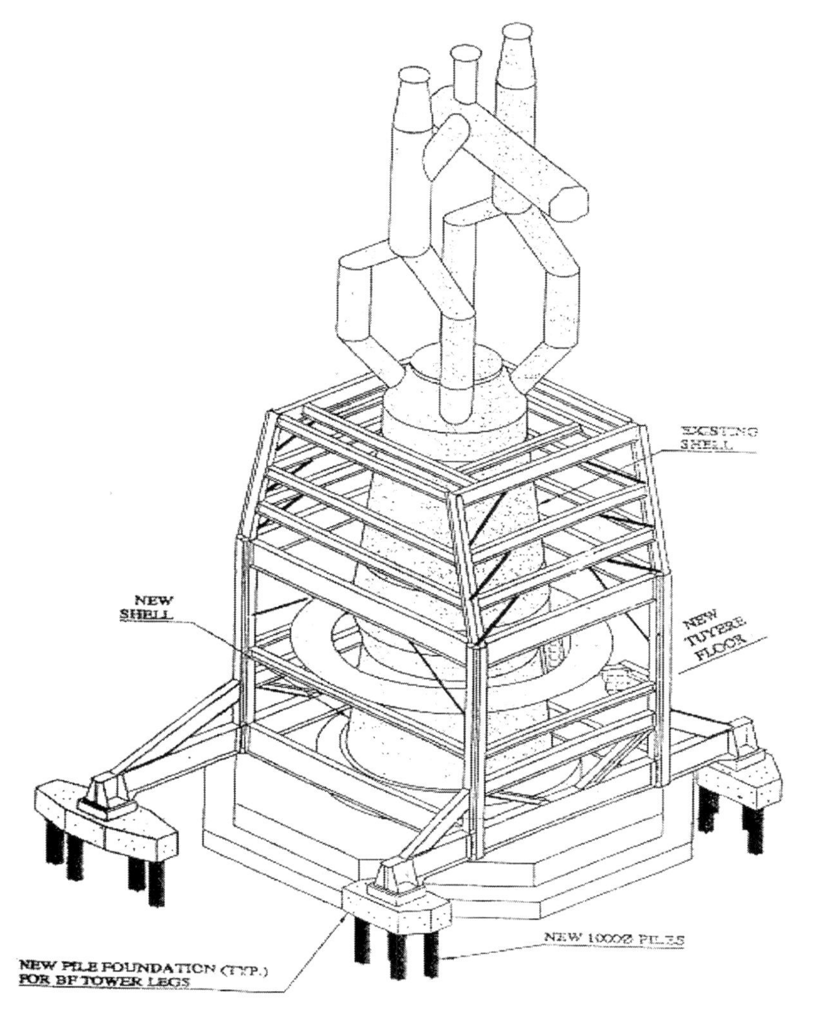

Fig. 10 View of rebuilt and upgraded blast furnace

system of the equipment to assess its feasibility of upgradation. Being constructed and commissioned more than four decades ago, there were very few data available regarding the foundation system of the equipment. A thorough search of the client's archives yielded rather insufficient and scanty data related to the equipment foundation. A team of engineers did a feasibility study based on whatever data could be unearthed and some innovative engineering to develop a workable scheme of rebuilding and upgrading the equipment. There were multiple uncertainties involved in every stage of the work. However, through meticulous planning, brain storming and model studies to simulate results and some innovative engineering, these uncertainties were overcome to reach levels of relative certainty in each and every stage. The equipment was rebuilt and upgraded in record time and handed over to the clients to start production to its planned enhanced capacity.

> So what do we do? Anything. Something. So long as we just don't sit there. If we screw it up, start over. Try something else. If we wait until we've satisfied all the uncertainties, it may be too late. Lee Iacocca

Discovering Hidden Structural Degradations

S. Sharma, S. Sharma, and A. Mukherjee

Abstract Structures degrade due to a variety of reasons such as environmental actions, overloading, and natural calamities. Inspection, diagnosis, and prognosis of damage in installations are imperative for avoiding their catastrophic failures. However, a reliable tool for early detection of damages in large structures has remained elusive. The ability ultrasonic guided waves to travel long distances and pick up the signatures The ability of structural damage makes them most promising among the handful of techniques available. Yet, there are formidable challenges in theoretical understanding and field implementations. In this chapter, we describe some recent developments in utilizing ultrasonic guided waves in discovering hidden damages.

Keywords Ultrasonic • Utilizing • Guided waves • Corrosion • Concrete • Plates

1 Introduction

Many nondestructive evaluation (NDE) techniques such as liquid penetrant dye, radiography, holography, eddy current, magnetic flux, and thermography are available for investigating the insides of a structure, the large size of the installations makes use of these unrealistic techniques. For discovering deep structural degradations, ultrasonic guided waves offer a potentially effective solution since they can travel long distances and pick up signals of early deterioration. It involves introducing a high-frequency stress pulse or "wave packet" into a material and observing the subsequent propagation and reflection of the pulse. The wave characteristics change due to the deterioration in the structure and these are sensitive to the location, extent, and character of damage.

S. Sharma (✉) • S. Sharma • A. Mukherjee
Thapar University, Patiala, Punjab, India
e-mail: shruti.sharma@thapar.edu; sksharma@thapar.edu; abhijit@thapar.edu

S. Chakraborty and G. Bhattacharya (eds.), *Proceedings of the International Symposium on Engineering under Uncertainty: Safety Assessment and Management (ISEUSAM - 2012)*,
DOI 10.1007/978-81-322-0757-3_9,

In this work, suitable ultrasonic guided waves have been used for damage detection of embedded and submerged structures. For embedded systems, suitable guided wave modes have been identified through modeling which are sensitive to different types of corrosion-induced damages encountered in reinforcing bars embedded in concrete. Corrosion of reinforcing steel in concrete is one of the major durability problems faced by civil engineers. Ultrasonic guided waves have been used to develop a non-intrusive corrosion monitoring technique for early detection of corrosion-induced damages in steel embedded in concrete. Corrosion manifests itself in debond and pitting steel bars. But it is imperative to excite the right mode for detection of a particular type of corrosion. A guided wave excited in the reinforcing bar in concrete would be reflected from the defects simulating area reduction due to pitting caused as a result of corrosion, thus facilitating the detection of these defects accurately. Amplitude and time of arrival of these reflections can be used to identify or locate defects. Ultrasonic pulse transmission utilizing two piezoelectric transducers at the two ends of the reinforcing bars is also used in conjunction with pulse echo testing method to quantify the extent of corrosion-induced defects in embedded bars.

On the other hand, many structures like ship hulls, oil storage tanks, and off shore oil platforms are assemblies of large plate like components which are prone to deterioration and damages due to environmental degradation, excessive loads, material fatigue, corrosion, etc. Criticality in such cases is further compounded due to economic constraints like high out of service costs associated with frequent health monitoring checkup schedules. The ultrasonic guided wave methodology is also applied for damage detection in such plate structures seeded with defects like notches especially in submerged states like ship hulls in water. Submerged nature of the specimens also limits the use of available and conventional NDT techniques. Guided waves are a successful means for detection of deterioration of such structures.

2 Guided Waves for Corrosion Monitoring in Reinforcing Bars in Concrete

Reinforced concrete (RC) is the most popular, economical, and versatile construction material as it can be molded into any shape. It has proven to be successful in terms of both structural performance and durability. But in humid conditions, atmospheric pollutants percolate through the concrete cover and cause corrosion of steel reinforcements. The formation of the corrosion products on the surface of the reinforcing bar having higher volume than the corresponding volume of steel results in an outward pressure on concrete. Corrosion, in the form of pitting, may also reduce the ductility of the steel bar by introducing crevices on the surface. But the size and limited accessibility of civil engineering installations prevent adoption of many currently used nondestructive testing methods such as radiography and acoustic emission. Hence, there is a need for nonintrusive, in situ, and real-time corrosion monitoring system for RC structures. In this study, guided waves have been used to develop a non-intrusive corrosion monitoring technique for early detection of corrosion-induced damages in steel embedded in concrete.

Corrosion manifests itself in debond and pitting steel bars. But it is imperative to excite the right mode for detection of a particular type of corrosion. This study investigates the effect of local loss of material and loss of bond on the propagation of ultrasonic waves through the reinforcing bars. Simulated pitting effects were created by notches on the surface of the bar in varying percentages of its cross sectional area. Simulated debond was generated by wrapping a double-sided tape of varying length on the bar embedded in concrete. Conventional techniques of pulse echo and pulse transmission have been used in combination to predict the presence, location, and magnitude of the damages. An experiment is carried out to create accelerated actual corrosion in RC samples. Both the simulated and actual corrosion samples were ultrasonically monitored using guided waves. The results have been compared to estimate the suitability of simulation techniques [1]. The effect of degradation due to corrosion on the ultrasonic signals is reported [2]. Effective combination of suitable guided wave modes could relate to the state of reinforcing bar corrosion [3].

2.1 Description of Experiments

For simulating corrosion in reinforcing bars in concrete, concrete with proportions of cement, sand, and stone aggregates as 1:1.5:2.96 was taken (w/c ratio = 0.45). RC beam specimens of dimensions 150 mm × 150 mm × 700 mm were cast. 12-mm-diameter plain mild steel bars of 1.1 m length were placed at the center of cross section of the beams at the time of casting with a projection of 200 mm on each side of beam. One set of bars with simulated damages in the form of notches (with symmetrical 0, 20, 40, and 60% diameter reduction) were introduced in the middle of the bar before casting them in concrete. Another set of bars were wrapped with a double-sided tape in varying percentages of 0, 12.5, 25, 50,75, and 100% simulating delamination and were then cast in concrete.

The RC beams were ultrasonically monitored using the standard UT setup consisting of a pulser-receiver and transducers. Contact transducers ACCUSCAN "S" series with center frequency of 3.5 MHz is used for the 12-mm bars. The transducers were attached at the two ends of the bars by means of a holder and a coupling gel between the bar and the transducer. Driven by the pulser, the compressional transducer generates a compressive spike pulse that propagates through the embedded bar in the form of longitudinal waves. In a reinforcing bar, different modes can be excited selectively by choosing a frequency bound. To determine the frequency band, standard software, Disperse [4] was used. Only longitudinal modes were considered for excitation since they are least attenuative. They were produced by keeping compressional transducers parallel to the guiding configuration at the

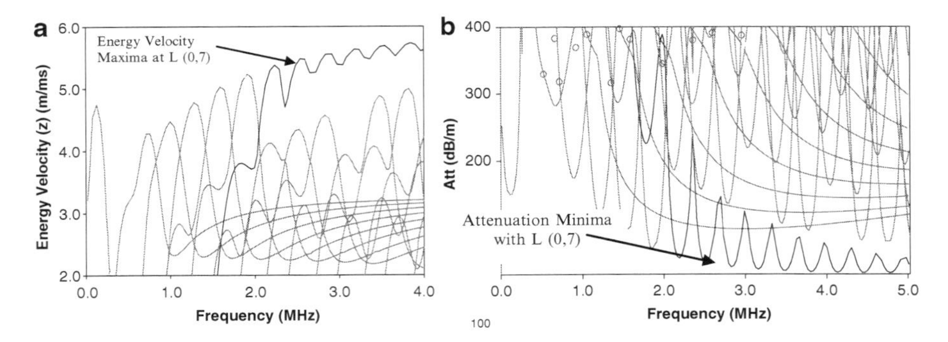

Fig. 1 Dispersion curves for a 12-mm bar in concrete [4]. (**a**) Energy velocity (z) versus frequency. (**b**) Attenuation versus frequency

two ends of the embedded bars by varying the excitation frequencies. The selection of frequencies for testing was done based on the phase velocity dispersion curves and validated by experimentally confirming the signal fidelity. High-frequency low-loss modes as identified by Pavlakovic [5] were chosen. L (0, 7) mode corresponding to maximum energy velocity (Fig. 1a) and minimum attenuation (Fig. 1b) was chosen for study at a frequency of 3.5 MHz. Testing of the embedded bar was done in both pulse echo and transmission modes.

To determine the suitability of the simulated corrosion experiments by ultrasonic guided waves, another set of beam specimens were subjected to actual accelerated corrosion. The projected bar was made an anode by connecting it to a positive terminal of a DC power supply. The middle 300 mm of the beam was selected for exposure to corrosive environment. A thick cotton gauge was placed in this region wrapped with a stainless steel wire mesh around with a dripping mechanism of 5% NaCl fitted on top of it. The negative terminal was connected to the wire mesh and a constant voltage of 30 V was applied between the two terminals. Beams undergoing accelerated chloride corrosion were ultrasonically monitored both in pulse echo and pulse transmission modes. The pulse transmission signatures disappeared in 8 days. Then the corroded beams were taken out and the extracted bars were tested for mass loss and tensile strength. The ultrasonic test results were compared to the actual state of the bar.

3 Results and Discussions

3.1 Simulated Notch Damage Study

Pulse echo records for a 12-mm bar in concrete show the notch echo (NE) as well as the back wall echo (BWE). In a healthy specimen, the peak is the BWE (Fig. 2a).

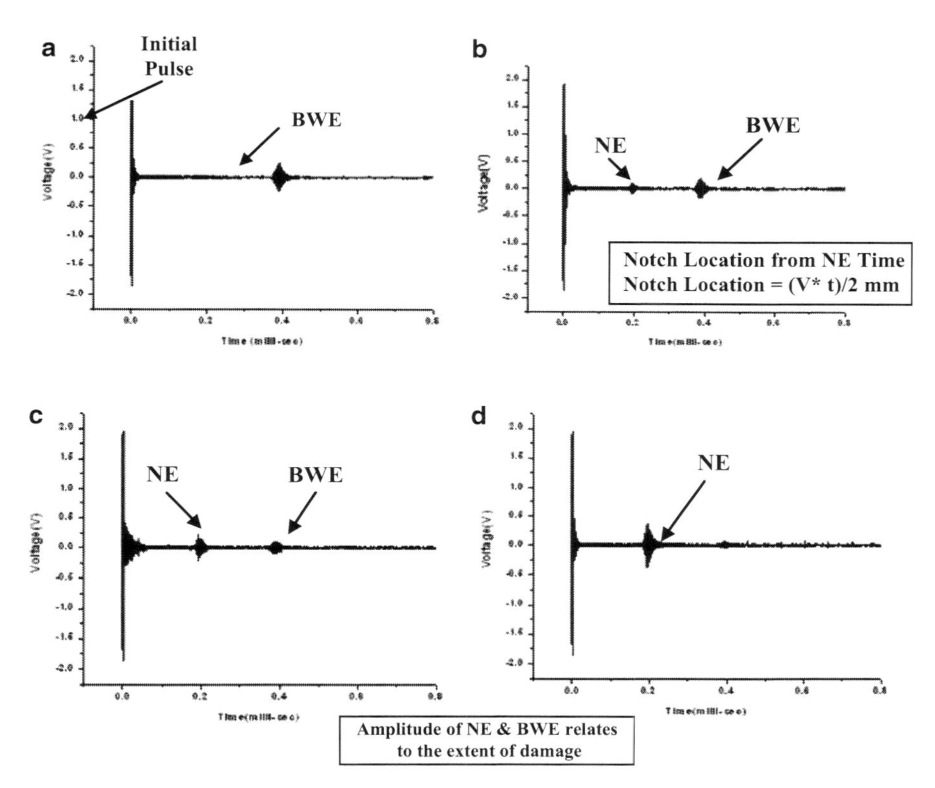

Fig. 2 Pulse echo signatures of a 12-mm, 1.1-m bar embedded in 700 mm of concrete [4]. (**a**) Healthy specimen. (**b**) 20% notch in embedded bar. (**c**) 40% notch in embedded bar. (**d**) 60% notch in embedded bar

In a notched specimen, the first peak is NE and the second peak is BWE (Fig. 2b). Appearance of NE indicates presence of the defect in the embedded bar. By knowing the time of flight of this echo, the location of the damage can be calculated. The magnitude of damage can be directly related to the magnitude of NE as well as BWE. It is observed that the amplitude of NE increased and that of BWE reduced with the increase in the notch dimensions (Fig. 2). As the magnitude of notch increased, more signal energy is reflected back from the notch and less of it is travels to the back wall.

In the pulse transmission signatures, the peaks observed (Fig. 3) are the transmitted peaks obtained after traveling length L of the embedded bar. It may be noted that the arrival time of the pulse is not affected by the presence of the notch. Thus, the notch location is not discernible through pulse transmission. However, studying the relative change in the amplitude of input pulse and the transmitted pulse (P/T-Notch), an assessment of the severity of the damage can be made.

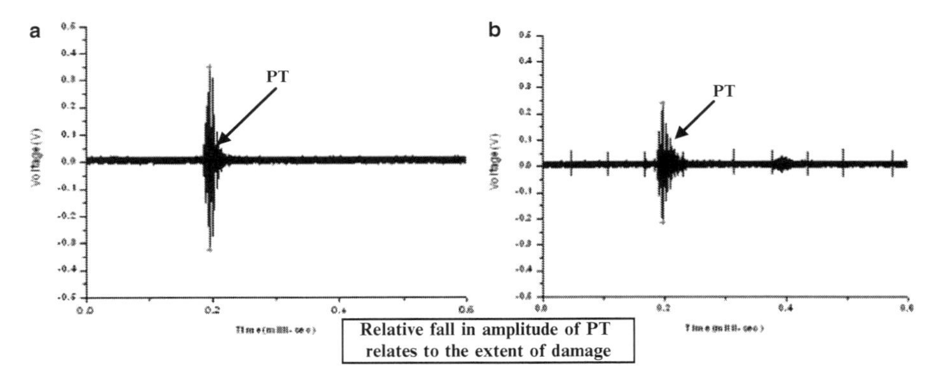

Fig. 3 Pulse transmission signatures of a 12-mm bar embedded in concrete [4]. (**a**) Healthy specimen. (**b**) 60% notch in embedded bar

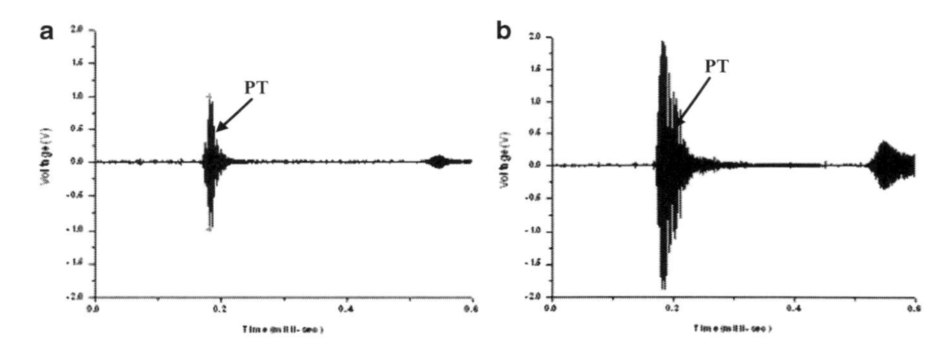

Fig. 4 Pulse transmission signatures in simulated delamination [4]. (**a**) 12.5% delamination. (**b**) 100% delamination

3.2 *Simulated Debond Damage Study*

In simulated debond specimens, as the percentage of delamination increases, the transmitted signal strength (P/T-Debond) keeps on rising (Fig. 4). This is due to decrease in the amount of energy leaking into the surrounding concrete with increase in percentage delamination. Hence, an increase in P/T-Debond can successfully relate to the presence as well as extent of delamination.

3.3 *Comparison of Notch and Debond*

As the percent of damage increased from 0 to 60%, the magnitude of P/T-Notch reduces. This is because as the notch dimensions increased, more energy is reflected

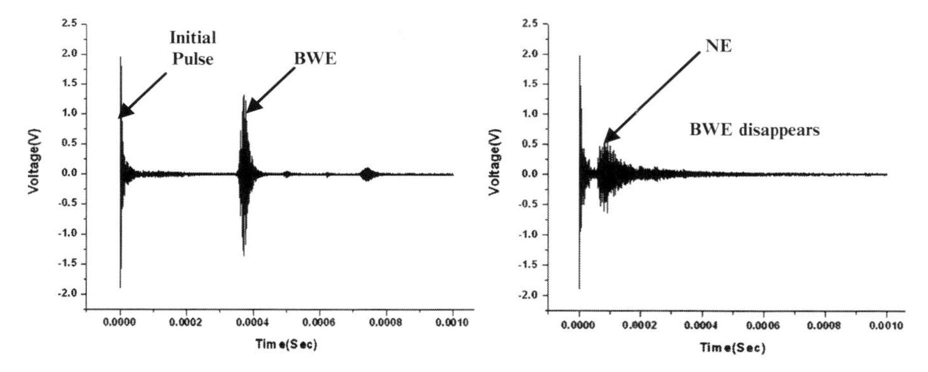

Fig. 5 Pulse echo signatures during accelerated corrosion [5]. (**a**) First day signature. (**b**) Seventh day signature

back and less of it travels through the bar to reach the other end. The counter-balancing effect of the two manifestations of corrosion – debonding and pitting – is clear. Pitting that is similar to a notch reduces the strength of the transmitted pulse (P/T-Notch) while debonding increases the strength of the transmitted pulse (P/T-Debond). Thus, it would be interesting to compare the simulated corrosion with the actual corrosion.

3.4 Actual Corrosion Study

Beams undergoing accelerated chloride corrosion showed reddish brown patches of corrosion products. A longitudinal crack appeared parallel to the bar within 3 days. With the increase in the volume of corrosion products, another crack parallel to the bar appeared on another face of the beam after 6 days. A reddish brown liquid oozed out of the cracks and the ends of the beam. The crack length and width increased with the increase in exposure. At 8 days of exposure, there were two large and wide longitudinal cracks that divided the entire beam into wedges.

Ultrasonic pulse echo signatures were monitored everyday during the exposure to the corrosive environment. In the healthy bar, the signature is characterized by a strong BWE (Fig. 5a). As the exposure proceeds, BWE attenuates rapidly and disappeared completely on fourth day. This is contrary to the expectation if corrosion is manifested through delamination only. Due to the nonuniform loss of material from the bar caused by chlorides, the waveguide is disturbed, thus resulting in scattering of waves. The scattering reduces the strength of the transmitted pulse further.

Another significant observation was the appearance of a peak between the initial pulse and BWE on the second day (Fig. 5b). This indicates that pulses are reflecting from a localized neck formed in the bar due to corrosion. From the time of flight of

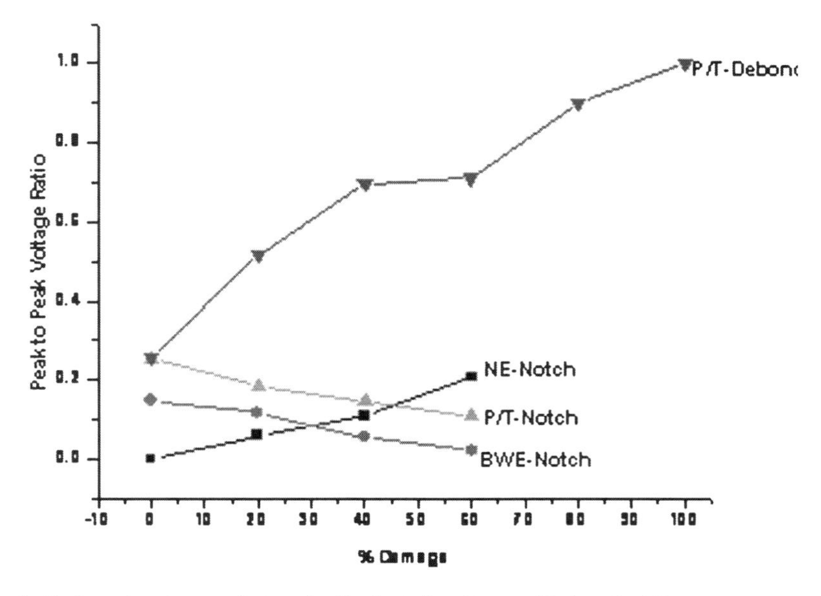

Fig. 6 Peak-peak voltage ratio trends of reflected and transmitted peaks [4]

the peak, the location of the neck was estimated at the bar-beam interface. After completion of the corrosion process, concrete was removed and the extracted bar was observed. A large notch was indeed seen at the estimated location. As corrosion increased, amplitude of this peak increased indicating further loss of area from the interface. Figures 6 and 7 compare the pulse echo results of simulated and actual corrosion. It is observed that corrosion in the presence of chlorides is characterized by localized loss of material similar to notches. In the present sample, one major notch developed due to corrosion resulting in a close match.

The pulse transmission studies where corrosion has been simulated as the loss of bond indicate that the transmitted signal strength goes up. Contrary to this observation, the transmitted pulse steadily lost strength as the corrosion progressed. It disappeared completely on the seventh day (Fig. 8). As discussed, the most likely cause of this phenomenon is the development of large pitting that further restricts the passage of waves. While notches restrict the passage of the waves, smooth delamination would facilitate the passage. Thus, pitting and delamination counteract each other. Figure 9 compares the peak-peak voltage ratios of simulated and actually corroded bars. Clearly, the results of notch specimens are in closer agreement with the actual chloride corrosion. Thus, chloride corrosion is simulated better as notches rather than delamination.

The bars were subjected to a series to destructive tests after the period of exposure was completed. After 8 days, the bar was removed from concrete and its mass loss and tensile strength were calculated. The bar had lost 18.6% of its mass. In tensile test, the bar failed in the region where a huge area loss was observed. The tensile strength reduced to 20% of that of undamaged bar.

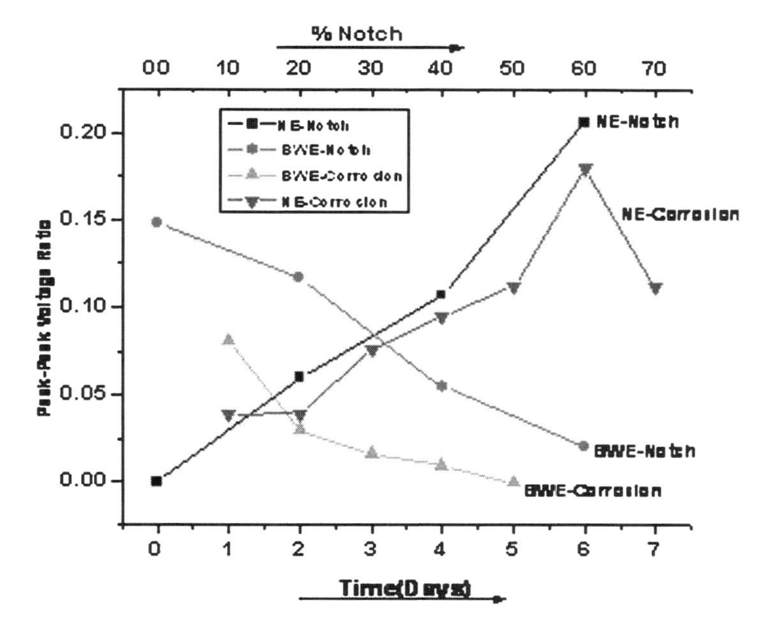

Fig. 7 Peak to peak voltage ratio in pulse echo [5]

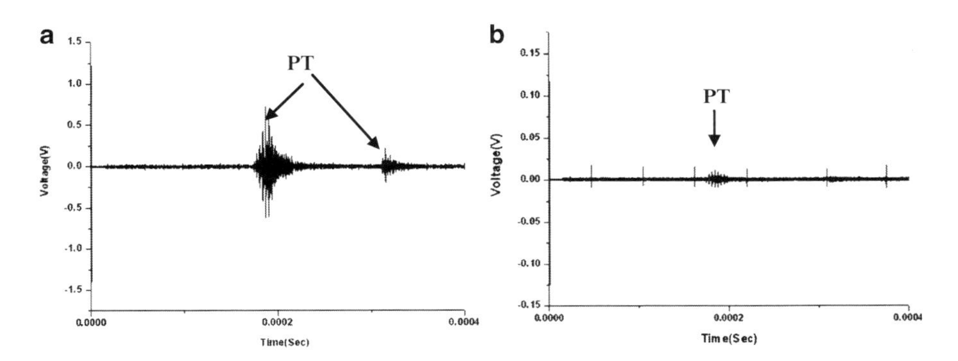

Fig. 8 Pulse transmission signatures during corrosion [5]. (**a**) Second day signature. (**b**) Fifth day signature

These reductions in mass and ultimate tensile strengths correlate well with the ultrasonic monitoring results wherein the signal experiences huge attenuation both in pulse echo and pulse transmission techniques. So the methodology established in the study using ultrasonic pulse echo and pulse transmission can be applied for in situ monitoring of embedded reinforcements undergoing corrosion.

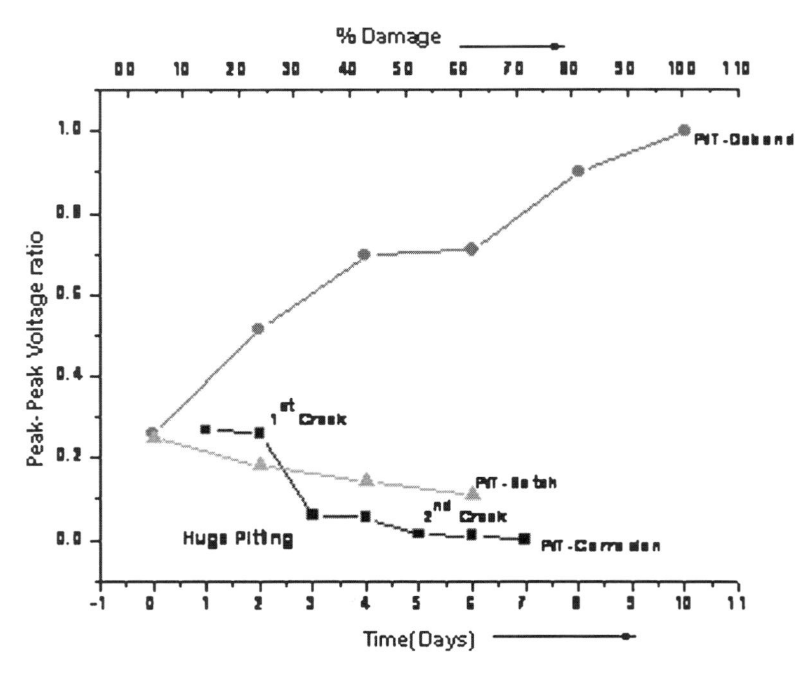

Fig. 9 Peak to peak voltage ratio in pulse transmission [5]

4 Guided Waves for Damage Monitoring in Submerged Structures

Many structures like ship hulls, oil storage tanks, off shore oil platforms are assemblies of large platelike components which are prone to deterioration and damages due to environmental degradation, excessive loads, fatigue loads, corrosion, etc. Criticality in such cases is further compounded due to economic constraints like high out of service costs associated with frequent health monitoring checkup schedules. The ultrasonic guided wave methodology is further applied for damage detection in such plate structures seeded with defects like notches and holes where structure remains in submerged state. Presence of water makes it all the more challenging as water loaded structures exhibit higher attenuative behavior due to comparatively lesser impedance mismatch, but on the other hand, it can be used to an advantage as a natural couplant. This obviates the subjectivity that may creep in while using direct contact techniques. Similar UT setup is used for damage detection in submerged plate specimens with the scanning setup as shown in Fig. 10.

Symmetric Lamb wave modes in plates were preferred because of ease of excitation. Selection of a particular mode is dependent on the type of damage to be monitored. For guided waves in plates, not many studies exist. Basically, two types of Lamb wave modes can exist in isotropic, homogeneous plates – symmetric

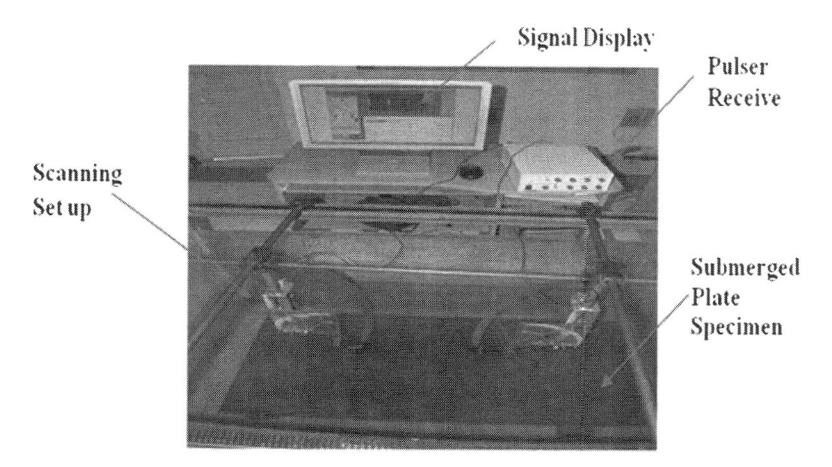

Fig. 10 Experimental setup for submerged specimen

and antisymmetric. Fundamental symmetric and antisymmetric modes describe a longitudinal wave and bending motion of the plate, respectively. As compared to the bulk waves used in UT, these modes are dispersive in nature, i.e., propagation velocity depends on the frequency. Guided waves may be produced by varying the excitation frequency or by changing the angle of incidence. Placing a transducer on the specimen, a guided wave can be excited that interrogates the whole structure.

The ultrasonic guided waves have been used for damage detection in plated steel structures using frequencies of the order of 1–3-MHz ultrasonic techniques of pulse echo and pulse transmission can be effectively used in combination to detect the presence, exact location, and quantification of the damage. Steel plate specimens with simulated defects in the form of localized area reduction were monitored in damaged state and compared with healthy specimens. Guided waves were generated by varying the input frequencies of the transducer. Different modes are excited at different frequencies and the suitable mode was chosen relevant to the defect to be characterized and ones corresponding to highest signal fidelities. Comparison of the wave signatures in healthy and simulated damaged state can easily reveal the existence, location, and extent of damage with reasonable accuracy.

Plate geometry results in the generation of multimode dispersive guided waves which are again obtained using software Disperse [4]. Different modes (symmetric and antisymmetric) of varying orders can be selectively excited by offering subject structure at various critical angles to the incident energy. Different modes have varying levels of sensitivities to different types of defects like notch, dent, and holes. Suitable modes are selected by testing the specimen implanted with a particular defect of varying degree in pulse transmission mode and observing the corresponding response of the received signal for various modes. Then specimen is tested in pulse transmission technique to ascertain the presence of the defect and subsequently pulse echo technique is employed to localize the same. The data obtained from these two techniques can be used to obtain the defect terrain of the specimen in the form of nice tomograms as shown in Fig. 11.

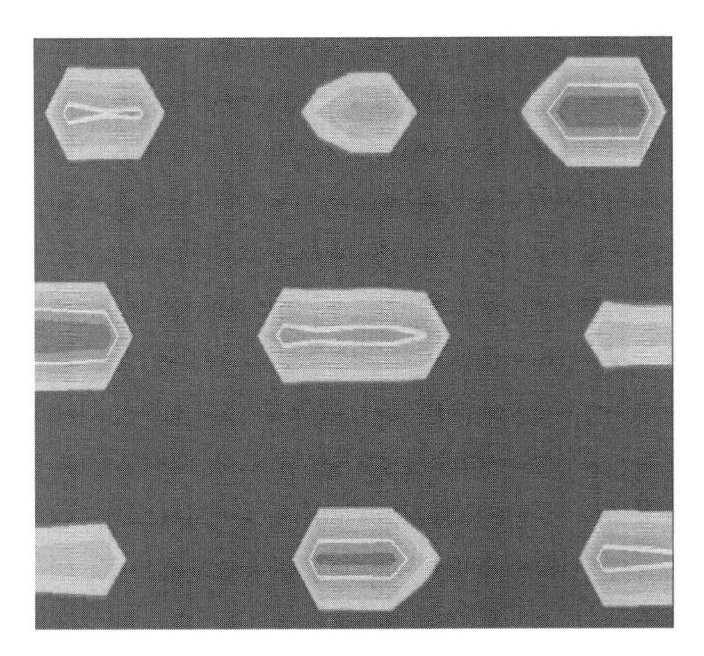

Fig. 11 Defect terrain on a 4-mm-thick submerged steel plate

5 Conclusions

Ultrasonic guided waves can be effectively used for monitoring corrosion damages in reinforcing bars embedded in concrete as well as in submerged steel plates by utilizing its wave guide effects. Corrosion in reinforcing bars has been simulated as loss of bond and loss of area. The results have been compared with that of a bar undergoing accelerated chloride corrosion. The notch specimens had a closer agreement with the corroded bars than the debonded specimens. Chloride corrosion roughens up the surface of the bar and creates large pitting. While it is desirable to perform ultrasonic monitoring when actual corrosion is taking place it is extremely time consuming. This chapter highlights that a judicious contribution of delamination and notch would be essential to closely simulate the corroded bar. It has also been successfully applied to steel plates submerged in water simulating ship hulls and other hidden plate geometries. Guided waves offer a potentially attractive and a practically viable solution to discovering hidden structural degradations in civil and mechanical infrastructural systems.

Acknowledgements The fund received from the Department of Science and Technology, Government of India, Naval Research Board, and All India Council of Technical Education is gratefully acknowledged.

References

1. Sharma S, Mukherjee A (2010) Longitudinal guided waves for monitoring corrosion in reinforcing bars in concrete. Struct Health Monit 13:555–567
2. Sharma S, Mukherjee A (2011) Monitoring corrosion in oxide and chloride environments using ultrasonic guided waves. ASCE J Mater Civil Eng 23(2):207–211
3. Sharma S, Mukherjee A (2012) Non-Destructive Evaluation of Corrosion in varying environments using guided waves. RNDE,. Available Online
4. Pavlakovic BN, Cawley P (2000) DISPERSE user's manual version 2.0.1.1. Imperial College, University of London, London
5. Sharma S, Mukherjee A (2010) Propagating ultrasonic guided waves through concrete reinforcements with simulated and actual corrosion. CINDE J 31(2):6–12

Stochastic Structural Dynamics Using Frequency Adaptive Basis Functions

A. Kundu and S. Adhikari

Abstract A novel Galerkin subspace projection scheme for structural dynamic systems with stochastic parameters is developed in this chapter. The fundamental idea is to solve the discretised stochastic damped dynamical system in the frequency domain by projecting the solution into a reduced subspace of eigenvectors of the underlying deterministic operator. The associated complex random coefficients are obtained as frequency-dependent quantities, termed as spectral functions. Different orders of spectral functions are proposed depending on the order of the terms retained in the expression. Subsequently, Galerkin weighting coefficients are employed to minimise the error induced due to the reduced basis and finite order spectral functions. The complex response quantity is explicitly expressed in terms of the spectral functions, eigenvectors and the Galerkin weighting coefficients. The statistical moments of the solution are evaluated at all frequencies including the resonance and antiresonance frequencies for a fixed value of damping. Two examples involving a beam and a plate with stochastic properties subjected to harmonic excitations are considered. The results are compared to direct Monte Carlo simulation and polynomial chaos expansion for different correlation lengths and variability of randomness.

Keywords Stochastic dynamics • Random field • Spectral decomposition • Karhunen-Loeve Expansion • Stochastic subspace

1 Introduction

The framework of the present work is the parametric uncertainty that is inherent in the mathematical models laid out to describe the governing equations of physical systems. These uncertainties may be intrinsic variability of physical quantities or a

A. Kundu (✉) • S. Adhikari
College of Engineering, Swansea University, Singleton Park SA2 8PP, UK
e-mail: a.kundu.577613@swansea.ac.uk

S. Chakraborty and G. Bhattacharya (eds.), *Proceedings of the International Symposium on Engineering under Uncertainty: Safety Assessment and Management (ISEUSAM - 2012)*,
DOI 10.1007/978-81-322-0757-3_10,

lack of knowledge about the physical behaviours of certain systems. As a result, though the recent advances in computational hardware has enabled the solution of very high resolution problems and even the sophisticated techniques of *a posteriori* error estimation [3], mesh adaptivity or the modelling error analysis has improved the confidence on results, yet these are not enough to determine the credibility of the numerical model.

There has been an increasing amount of research activities over the past three decades to model the governing partial differential equations within the framework of stochastic equations. We refer to few recent review papers [1, 8, 9]. After the discretisation of random fields and displacement fields, the equation of motion can be expressed by [2, 4, 5] a set of stochastic ordinary differential equations

$$\mathbf{M}(\theta)\,\ddot{\mathbf{u}}\,(\theta,\,t) + \mathbf{C}(\theta)\,\dot{\mathbf{u}}\,(\theta,\,t) + \mathbf{K}(\theta)\,\mathbf{u}(\theta,\,t) = \mathbf{f}_0(t) \tag{1}$$

where $\mathbf{M}(\theta) = \mathbf{M}_0 + \sum_{i=1}^{p_1} \mu i(\theta)\mathbf{M}_i \in \mathbb{R}^{n\times n}$ is the random mass matrix and $\mathbf{K}(\theta) = \mathbf{K}_0 + \sum_{i=1}^{p_2} vi(\theta)\mathbf{K}_i \in \mathbb{R}^{n\times n}$ is the random stiffness matrix along with $\mathbf{C}(\theta) \in \mathbb{R}^{n\times n}$ as the random damping matrix. The notation θ is used to denote the random sample space. Here, the mass and stiffness matrices have been expressed in terms of their deterministic components ($\mathbf{M_0}$ and $\mathbf{K_0}$) and the corresponding random contributions ($\mathbf{M_i}$ and $\mathbf{K}_i$) obtained from discretising the stochastic field with a finite number of random variables $(\mu_i(\theta)$ and $v_i(\theta))$ and their corresponding spatial basis functions. In the present work proportional damping is considered for which $\mathbf{C}(\theta) = \varsigma_1\mathbf{M}(\theta) + \varsigma_2\mathbf{K}(\theta)$, where ς_1 and ς_2 are deterministic scalars. For the harmonic analysis of the structural system considered in Eq. (1), it is represented in the frequency domain as

$$\left[-\omega^2\mathbf{M}(\theta) + i\omega\mathbf{C}(\theta) + \mathbf{K}(\theta)\right]\tilde{\mathbf{u}}(\theta,\,\omega) = \tilde{\mathbf{f}}_0(\omega) \tag{2}$$

where $\tilde{\mathbf{u}}(\theta,\,\omega)$ is the complex frequency domain system response amplitude and $\tilde{\mathbf{f}}_0(\omega)$ is the amplitude of the harmonic force.

Now we group the random variables associated with the mass and damping matrices of Eq. (1) as

$$\begin{aligned} \xi_i(\theta) &= \mu_i(\theta) \quad \textit{for} \quad i = 1, 2, \ldots, p_1 \\ \text{and} \quad \xi_{i+p_1}(\theta) &= v_i(\theta) \quad \textit{for} \quad i = 1, 2, \ldots, p_2 \end{aligned}$$

Thus, the total number of random variables used to represent the mass and the stiffness matrices becomes $p = p_1 + p_2$. Following this, the expression for the linear structural system in Eq. (2) can be expressed as

$$\left(\mathbf{A}_0(\omega) + \sum_{i=1}^{p} \xi_i(\theta)\mathbf{A}_i(\omega)\right)\tilde{\mathbf{u}}(\omega, \theta) = \tilde{\mathbf{f}}_0(\omega) \tag{3}$$

where $\mathbf{A_o}$ and $\mathbf{A}_i \in \mathbb{C}^{n\times n}$ represent the complex deterministic and stochastic parts, respectively, of the mass, the stiffness and the damping matrices ensemble. For the case of proportional damping, the matrices $\mathbf{A}_o$ and $\mathbf{A}_i$ can be written as

$$\mathbf{A}_0(\omega) = \left[-\omega^2 + i\omega\varsigma_1\right]\mathbf{M}_0 + [i\omega\varsigma_2 + 1]\mathbf{K}_0 \tag{4}$$

and

$$\begin{aligned} \mathbf{A}_i(\omega) &= \left[-\omega^2 + i\omega\varsigma_1\right]\mathbf{M}_i \quad \text{for} \quad i = 1, 2, ..., p_1 \\ \mathbf{A}_{i+p_1}(\omega) &= [i\omega\varsigma_2 + 1]\mathbf{K}_i \quad \textit{for} \quad i = 1, 2, ..., p_2 \end{aligned} \tag{5}$$

The chapter has been arranged as follows. The projection theory in the vector space is developed in Sect. 2. In Sect. 3 an error minimisation approach in the Hilbert space is proposed. The idea of the reduced orthonormal vector basis is introduced in Sect. 4. Based on the theoretical results, a simple computational approach is shown in Sect. 5 where the proposed method of reduced orthonormal basis is applied to the stochastic mechanics of a Euler-Bernoulli beam. From the theoretical developments and numerical results, some conclusions are drawn in Sect. 6.

2 Spectral Decomposition in the Vector Space

Following the spectral stochastic finite-element method, or otherwise, an approximation to the solution stochastic system can be expressed as a linear combination of functions of random variables and deterministic vectors. Recently Nouy [6, 7] discussed the possibility of an optimal spectral decomposition. The aim is to use small number of terms to reduce the computation without losing the accuracy. We use the eigenvectors $\phi_k \in \mathbb{R}^n$ of the generalised eigenvalue problem

$$\mathbf{K}_0\phi_k = \lambda_k\mathbf{M}_0\phi_k; \quad k = 1, 2, ..n \tag{6}$$

Since the matrices $\mathbf{K_0}$ and $\mathbf{M_0}$ are symmetric and generally non-negative definite, the eigenvectors ϕ_k for $k = 1, 2, ...n$ form an orthonormal basis. Note that in principle, any orthonormal basis can be used. This choice is selected due to the analytical simplicity as will be seen later. For notational convenience, define the matrix of eigenvalues and eigenvectors

$$\lambda_0 = \text{diag}[\lambda_1, \lambda_2, \ldots, \lambda_n] \in \mathbb{R}^{n\times n} \quad \text{and} \quad \Phi = [\phi_1, \phi_2, \ldots, \phi_n] \in \mathbb{R}^{n\times n} \tag{7}$$

Eigenvalues are ordered in the ascending order so that $\lambda_1 < \lambda_2 < \ldots < \lambda_n$. Since Φ is an orthonormal matrix, we have $\Phi^{-1} = \Phi^T$ so that the following identities can easily be established:

$$\begin{aligned} \Phi^T\mathbf{A}_0\Phi &= \Phi^T\left(\left[-\omega^2 + i\omega\varsigma_1\right]\mathbf{M}_0 + [i\omega\varsigma_2 + 1]\mathbf{K}_0\right)\Phi \\ &= \left(-\omega^2 + i\omega\varsigma_1\right)\mathbf{I} + (i\omega\varsigma_2 + 1)\lambda_0 \end{aligned}$$

which gives

$$\Phi^T \mathbf{A}_0 \Phi = \Lambda_0; \quad \mathbf{A}_0 = \Phi^{-T} \Lambda_0 \Phi^{-1} \quad \text{and} \quad \mathbf{A}_0^{-1} = \Phi \Lambda_0^{-1} \Phi^{-T} \tag{8}$$

where $\Lambda_0 = (-\omega^2 + i\omega\varsigma_1)\mathbf{I} + (i\omega\varsigma_2 + 1)\lambda_0$ and $\mathbf{I}$ is the identity matrix. Hence, Λ_0 can also be written as

$$\Lambda_0 = \text{diag}\left[\lambda_{0_1}, \lambda_{0_2}, \ldots, \lambda_{0_n}\right] \in \mathbb{C}^{n\times n} \tag{9}$$

where $\lambda_{0_j} = (-\omega^2 + i\omega\varsigma_1) + (i\omega\varsigma_2 + 1)\lambda_j$ and λ_j is defined in Eq. (7). We also introduce the transformations

$$\tilde{\mathbf{A}}_i = \Phi^T \mathbf{A}_i \Phi \in \mathbb{C}^{n\times n}; i = 0, 1, 2, \ldots, M \tag{10}$$

Note that $\tilde{\mathbf{A}}_i = \Lambda_0$ is a diagonal matrix and

$$\tilde{\mathbf{A}}_i = \Phi^{-T} \tilde{\mathbf{A}}_i \Phi^{-1} \in \mathbb{C}^{n\times n}; i = 0, 1, 2, \ldots, M \tag{11}$$

Suppose the solution of Eq. (3) is given by

$$\hat{\mathbf{u}}(\omega, \theta) = \left[\mathbf{A}_0(\omega) + \sum_{i=1}^{M} \xi_i(\theta)\mathbf{A}_i(\omega)\right]^{-1} \mathbf{f}_0(\omega) \tag{12}$$

Using Eqs. (7), (8), (9), (10), and (11) and the orthonormality of Φ, one has

$$\begin{aligned}\hat{\mathbf{u}}(\omega, \theta) &= \left[\Phi^{-T}\Lambda_0(\omega)\Phi^{-1} + \sum_{i=1}^{M} \xi_i(\theta)\Phi^{-T}\tilde{\mathbf{A}}_i\Phi^{-1}\right]^{-1} \mathbf{f}_0(\omega) \\ &= \Phi\Psi(\omega, \xi(\theta))\Phi^{-T}\mathbf{f}_0(\omega)\end{aligned} \tag{13}$$

where

$$\Psi(\omega, \xi(\theta)) = \left[\Lambda_0(\omega) + \sum_{i=1}^{M} \xi_i(\theta)\tilde{\mathbf{A}}_i(\omega)\right]^{-1} \tag{14}$$

and the M-dimensional random vector

$$\xi(\theta) = \{\xi_1(\theta), \xi_2(\theta), \ldots, \xi_M(\theta)\}^T \tag{15}$$

Now we separate the diagonal and off-diagonal terms of the $\tilde{\mathbf{A}}_i$ matrices as

$$\tilde{\mathbf{A}}_i = \Lambda_i + \Delta_i, \quad i = 1, 2, \ldots, M \tag{16}$$

Here, the diagonal matrix

$$\Lambda_i = \mathrm{diag}\left[\tilde{\mathbf{A}}_i\right] = \mathrm{diag}\left[\lambda_{i_1}, \lambda_{i_2}, \ldots, \lambda_{i_n}\right] \in \mathbb{C}^{n \times n} \tag{17}$$

and the matrix containing only the off-diagonal elements $\Delta_i = \tilde{\mathbf{A}}_i - \Lambda_i$ is such that Trace $(\Delta_i) = 0$. Using these, from Eq. (14) one has

$$\Psi(\omega,\, \xi(\theta)) = \left[\underbrace{\Lambda_0(\omega) + \sum_{i=1}^{M} \xi_i(\theta)\Lambda_i(\omega)}_{\Lambda(\omega,\boldsymbol{\xi}(\theta))} + \underbrace{\sum_{i=1}^{M} \xi_i(\theta)\Delta_i(\omega)}_{\Delta(\omega,\boldsymbol{\xi}(\theta))}\right]^{-1} \tag{18}$$

where $\Lambda(\omega, \boldsymbol{\xi}(\theta)) \in \mathbb{C}^{n \times n}$ is a diagonal matrix and $\Delta(\omega,\, \boldsymbol{\xi}(\theta))$ is an off-diagonal only matrix. In the subsequent expressions, we choose to omit the inclusion of frequency dependence of the individual matrices for the sake of notational simplicity, so that $\Psi(\omega,\, \boldsymbol{\xi}(\theta)) \equiv \Psi(\boldsymbol{\xi}(\theta))$ and so on. Hence, we rewrite Eq. (18) as

$$\Psi(\boldsymbol{\xi}(\theta)) = \left[\Lambda(\boldsymbol{\xi}(\theta))\left[\mathbf{I}_n + \Lambda^{-1}(\boldsymbol{\xi}(\theta))\Delta(\boldsymbol{\xi}(\theta))\right]\right]^{-1} \tag{19}$$

The above expression can be represented using a Neumann type of matrix series [10] as

$$\Psi(\boldsymbol{\xi}(\theta)) = \sum_{s=0}^{\infty} (-1)^s \left[\Lambda^{-1}(\boldsymbol{\xi}(\theta))\Delta(\boldsymbol{\xi}(\theta))\right]^s \Lambda^{-1}(\boldsymbol{\xi}(\theta)) \tag{20}$$

Taking an arbitrary r-th element of $\hat{\mathbf{u}}(\theta)$, Eq. (13) can be rearranged to have

$$\hat{u}_r(\theta) = \sum_{k=1}^{n} \Phi_{rk} \left(\sum_{j=1}^{n} \Psi_{kj}(\boldsymbol{\xi}(\theta)) \left(\phi_j^T \mathbf{f}_0\right) \right) \tag{21}$$

Defining

$$\Gamma_k(\xi(\theta)) = \sum_{j=1}^{n} \Psi_{kj}(\boldsymbol{\xi}(\theta)) \left(\phi_j^T \mathbf{f}_0\right) \tag{22}$$

and collecting all the elements in Eq. (21) for $r = 1,2,\ldots,n$, one has

$$\hat{\mathbf{u}}(\theta) = \sum_{k=1}^{n} \Gamma_k(\boldsymbol{\xi}(\theta))\phi_k \tag{23}$$

This shows that the solution vector $\hat{\mathbf{u}}(\theta)$ can be projected in the space spanned by ϕ_k.

3 Error Minimisation Using the Galerkin Approach

In Sec. 2 we derived the spectral functions such that a projection in an orthonormal basis converges to the exact solution in probability 1. The spectral functions are expressed in terms of a convergent infinite series. First, second- and higher-order spectral functions obtained by truncating the infinite series have been derived. We have also showed that they have the same functional form as the exact solution of Eq. (3). This motivates us to use these functions as 'trial functions' to construct the solution. The idea is to minimise the error arising due to the truncation. A Galerkin approach is proposed where the error is made orthogonal to the spectral functions.

We express the solution vector by the series representation

$$\hat{\mathbf{u}}(\theta) = \sum_{k=1}^{n} c_k \widehat{\Gamma}_k(\boldsymbol{\xi}(\theta))\phi_k \tag{24}$$

Here, the functions $\widehat{\Gamma}_k : \mathbb{C}^M \to \mathbb{C}$ are the spectral functions, and the constants $c_k \in \mathbb{C}$ need to be obtained using the Galerkin approach. The functions $\widehat{\Gamma_k}(\xi(\theta))$ can be the first-order, second-order or any higher-order spectral function (depending on the order of the expansion s in Eq. (20)) and are the complex frequency adaptive weighting coefficient of the eigenvectors introduced earlier in Eq. (6). Substituting the expansion of $\hat{\mathbf{u}}(\theta)$ in the linear system equation (3), the error vector can be obtained as

$$\varepsilon(\theta) = \left(\sum_{i=0}^{M} \mathbf{A}_i \xi_i(\theta) \right) \left(\sum_{k=1}^{n} c_k \widehat{\Gamma_k}(\boldsymbol{\xi}(\theta))\phi_k \right) - \mathbf{f}_0 \in \mathbb{C}^n \tag{25}$$

where $\xi_0 = 1$ is used to simplify the first summation expression. The expression (24) is viewed as a projection where $\left\{ \widehat{\Gamma_k}(\boldsymbol{\xi}(\theta))\phi_k \right\} \in \mathbb{C}^n$ are the basis functions and c_k are the unknown constants to be determined. We wish to obtain the coefficients c_k using the Galerkin approach so that the error is made orthogonal to the basis functions, that is, mathematically

$$\varepsilon(\theta) \perp \left(\widehat{\Gamma}_j(\boldsymbol{\xi}(\theta))\phi_j \right) \quad \text{or} \quad \left\langle \widehat{\Gamma}_j(\boldsymbol{\xi}(\theta))\phi_j, \varepsilon(\theta) \right\rangle = 0 \; \forall \, j = 1, 2, \ldots, n \tag{26}$$

Here, $\langle \mathbf{u}(\theta), \mathbf{v}(\theta) \rangle = \int P(d\theta \mathbf{u}(\theta)\mathbf{v}(\theta))$ defines the inner product norm. Imposing this condition and using the expression of $\varepsilon(\theta)$ from Eq. (25), one has

$$E\left[\left(\widehat{\Gamma}_j(\boldsymbol{\xi}(\theta))\phi_j \right)^T \left(\sum_{i=0}^{M} \mathbf{A}_i \xi_i(\theta) \right) \left(\sum_{k=1}^{n} c_k \widehat{\Gamma_k}(\boldsymbol{\xi}(\theta))\phi_k \right) - \left(\widehat{\Gamma}_j(\boldsymbol{\xi}(\theta))\phi_j \right)^T \mathbf{f}_0 \right] = 0 \tag{27}$$

Interchanging the $E[\bullet]$ and summation operations, this can be simplified to

$$\sum_{k=1}^{n}\left(\sum_{i=0}^{M}\left(\phi_j^T\mathbf{A}_i\phi_k\right)E\left[\xi_i(\theta)\widehat{\Gamma}_j^T(\boldsymbol{\xi}(\theta))\widehat{\Gamma}_k(\boldsymbol{\xi}(\theta))\right]\right)c_k = E\left[\widehat{\Gamma}_j^T(\boldsymbol{\xi}(\theta))\right]\left(\phi_j^T\mathbf{f}_o\right) \quad (28)$$

or

$$\sum_{k=1}^{n}\left(\sum_{i=0}^{M}\widetilde{A}_{i_{jk}}D_{ijk}\right)c_k = b_j \quad (29)$$

Defining the vector $\mathbf{c} = \{c_1, c_2, \ldots, c_n\}^T$, these equations can be expressed in a matrix form as

$$\mathbf{S}\,\mathbf{c} = \mathbf{b} \quad (30)$$

with

$$S_{jk} = \sum_{i=0}^{M}\widetilde{A}_{i_{jk}}D_{ijk}; \quad \forall j, k = 1, 2, \ldots, n \quad (31)$$

where

$$\widetilde{A}_{i_{jk}} = \phi_j^T\mathbf{A}_i\phi_k, \quad (32)$$

$$D_{ijk} = E\left[\xi_i(\theta)\widehat{\Gamma}_j^T(\boldsymbol{\xi}(\theta))\widehat{\Gamma}_k(\boldsymbol{\xi}(\theta))\right] \quad (33)$$

and

$$b_j = E\left[\widehat{\Gamma}_j^T(\boldsymbol{\xi}(\theta))\right]\left(\phi_j^T\mathbf{f}_o\right) \quad (34)$$

Higher-order spectral functions can be used to improve the accuracy and convergence of the series (24). This will be demonstrated in the numerical examples later in the chapter.

4 Model Reduction Using a Reduced Number of Basis

The Galerkin approach proposed in the previous section requires the solution of $n \times n$ algebraic equations. Although in general this is smaller compared to the polynomial chaos approach, the computational cost can still be high for large n as the coefficient matrix is in general a dense matrix. The aim of this section is to reduce it further so that, in addition to large number of random variables, problems with large degrees of freedom can also be solved efficiently.

Suppose the eigenvalues of $\mathbf{A}_0$ are arranged in an increasing order such that

$$\lambda_{01} < \lambda_{02} < \ldots < \lambda_{0n} \tag{35}$$

From the expression of the spectral functions, observe that the eigenvalues appear in the denominator:

$$\widehat{\Gamma}_k^{(1)}(\boldsymbol{\xi}(\theta)) = \frac{\phi_k^T \mathbf{f}_0}{\lambda_{0k} + \sum_{i=1}^{M} \xi_i(\theta)\lambda_{i_k}} \tag{36}$$

The numerator $\left(\phi_k^T \mathbf{f}_0\right)$ is the projection of the force on the deformation mode. Since the eigenvalues are arranged in an increasing order, the denominator of $\left|\Gamma_{k+r}^{(1)}(\boldsymbol{\xi}(\theta))\right|$ is larger than the denominator of $\left|\Gamma_k^{(1)}(\boldsymbol{\xi}(\theta))\right|$ according a suitable measure. The numerator $\left(\phi_k^T \mathbf{f}_0\right)$ depends on the nature of forcing and the eigenvectors. Although this quantity is deterministic, in general an ordering cannot be easily established for different values of k. Because all the eigenvectors are normalised to unity, it is reasonable to consider that $\left({\phi^T}_k f_0\right)$ does not vary significantly for different values of k. Using the ordering of the eigenvalues, one can select a small number ϵ such that $\lambda_1/\lambda_q < \epsilon$ for some value of q, where λ_j is the eigenvalue of the generalised eigenvalue problem defined in Eq. (6). Based on this, we can approximate the solution using a truncated series as

$$\hat{\mathbf{u}}(\theta) \approx \sum_{k=1}^{q} c_k \widehat{\Gamma}_k(\boldsymbol{\xi}(\theta))\phi_k \tag{37}$$

where c_k, $\widehat{\Gamma}_k(\boldsymbol{\xi}(\theta))$ and are obtained following the procedure described in the previous section by letting the indices j, k only up to q in Eqs. (31) and (32). The accuracy of the series (37) can be improved in two ways, namely, (a) by increasing the number of terms q or (b) by increasing the order of the spectral functions $\widehat{\Gamma_k}(\boldsymbol{\xi}(\theta))$. Once the samples of $\mathbf{u} = (\theta)$ are generated, the statistics can be obtained using standard procedures.

5 Illustrative Application: The Stochastic Mechanics of a Euler-Bernoulli Beam

In this section we apply the computational method to a cantilever beam with stochastic bending modulus. We assume that the bending modulus is a homogeneous stationary Gaussian random field of the form

$$EI(x, \theta) = EI_0(1 + a(x, \theta)) \tag{38}$$

where x is the coordinate along the length of the beam, EI_0 is the estimate of the mean bending modulus and $a(x,\theta)$ is a zero mean stationary Gaussian random field. The autocorrelation function of this random field is assumed to be

$$C_a(x_1, x_2) = \sigma_a^2 e^{\frac{-(|x_1 - x_2|)}{\mu_a}} \quad (39)$$

where μ_a is the correlation length and σ_a is the standard deviation. We use the baseline parameters as the length $L = 1$ m, cross section $(b \times h)$ 39×5.93 mm^2 and Young's modulus $E = 2 \times 10^{11}$ Pa.

In study we consider deflection of the tip of the beam under harmonic loads of amplitude $\tilde{f}_0 = 1.0N$. The correlation length considered in this numerical study is $\mu_a = L/2$. The number of terms retained (M) in the Karhunen-Loeve expansion is selected such that $v_M/v_1 = 0.01$ in order to retain 90% of the variability. For this correlation length, the number of terms M comes to 18. For the finite element discretisation, the beam is divided into 40 elements. Standard four degrees of freedom Euler-Bernoulli beam model is used [11]. After applying the fixed boundary condition at one edge, we obtain the number of degrees of freedom of the model to be $n = 80$.

5.1 Results

The proposed method has been compared with a direct Monte Carlo simulation (MCS), where both have been performed with 10,000 samples. For the direct MCS, Eq. (12) is solved for each sample, and the mean and standard deviation is derived by assembling the responses. The calculations have been performed for all the four values of σ_a to simulate increasing uncertainty. This is done to check the accuracy of the proposed method against the direct MCS results for varying degrees of uncertainty.

Figure 1a presents the ratio of the eigenvalues of the generalised eigenvalue problem (6) for which the ratio of the eigenvalues is taken with the first eigenvalue. We choose the reduced basis of the problem based on $\lambda_1/\lambda_q < \in$, where $e = 0.01$, and they are highlighted in the figure. Figure 1b shows the frequency domain response of the deterministic system for both damped and undamped conditions. We have applied a constant modal damping matrix with the damping coefficient $\alpha = 0.02$ (which comes to 1% damping). It is also to be noted that the mass and damping matrices are assumed to be deterministic in nature, while it has to be emphasised that the approach is equally valid for random mass, stiffness and damping matrices. The frequency range of interest for the present study is 0–600 Hz with an interval of 2 Hz. In Fig. 1b, the tip deflection is shown on a log scale for a unit amplitude harmonic force input. The resonance peak amplitudes of the response of the undamped system definitely depend on the frequency resolution of the plot.

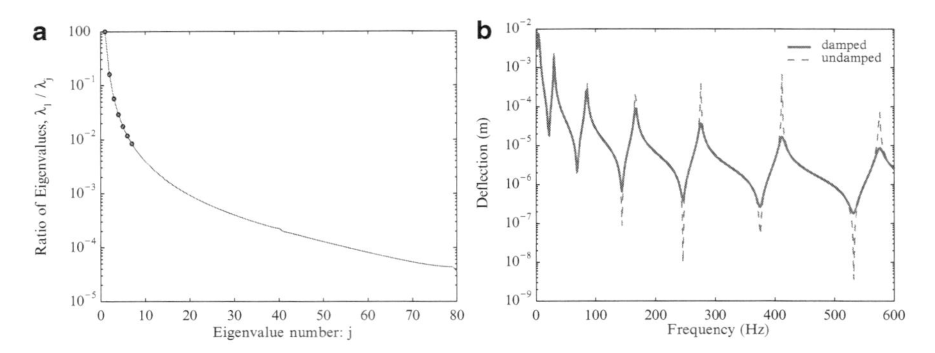

Fig. 1 The eigenvalues of the generalised eigenvalue problem involving the mass and stiffness matrices given in Eq. (6). For $e = 0.01$, the number of reduced eigenvectors $q = 7$ such that $\lambda_1/\lambda_j < \in$. (**a**) Ratio of eigenvalues of the generalised eigenvalue problem. (**b**) Frequency domain response of the tip of the beam under point load for the undamped and damped conditions (constant modal damping)

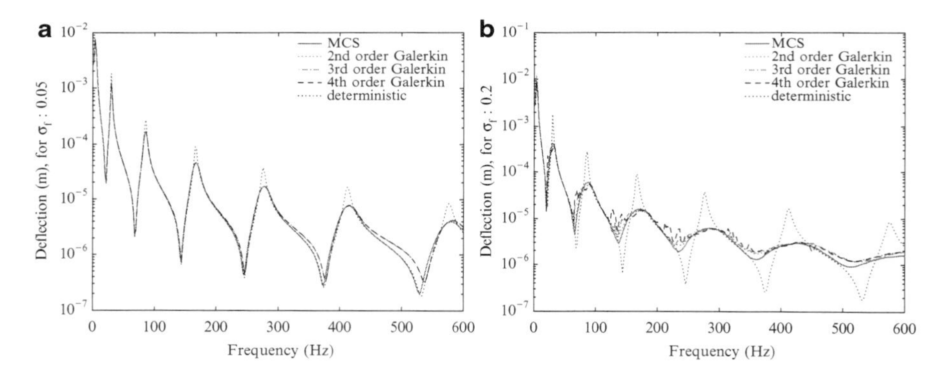

Fig. 2 The frequency domain response of the deflection of the tip of the Euler-Bernoulli beam under unit amplitude harmonic point load at the free end. The response is obtained with 10,000 sample MCS and for $\sigma_a = \{0.05, 0.10, 0.15, 0.20\}$. The proposed Galerkin approach needs solution of a 7×7 linear system of equations only. (**a**) Beam deflection for $\sigma_a = 0.05$. (**b**) Beam deflection for $\sigma_a = 0.2$

The frequency response of the mean deflection of the tip of the beam is shown in Fig. 2 for the cases of $\sigma_a = \{0.05, 0.10, 0.15, 0.20\}$. The figures show a comparison of the direct MCS simulation results with different orders of the solution following Eq. (20), where the orders $s = 2, 3, 4$. A very good agreement between the MCS simulation and the proposed spectral approach can be observed in the figures. All the results have been compared with the response of the deterministic system which shows that the uncertainty has an added damping effect at the resonance peaks. This can be explained by the fact that the parametric variation

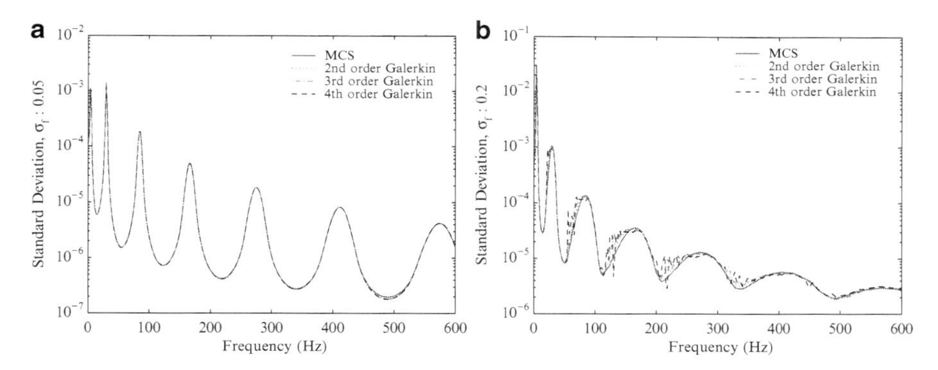

Fig. 3 The standard deviation of the tip deflection of the Euler-Bernoulli beam under unit amplitude harmonic point load at the free end. The response is obtained with 10,000 sample MCS and for $\sigma_a = \{0.05, 0.10, 0.15, 0.20\}$. (**a**) Standard deviation for the reference σ_a=0.05. (**b**) Standard deviation for the reference $\sigma_a = 0.2$

of the beam results in its peak response for the different samples to get distributed around the resonance frequency zones instead of being concentrated at a particular frequency, and when the subsequent averaging is applied, it smoothes out the response peaks to a fair degree. The same explanation holds for the antiresonance frequencies. It can also be observed that increased variability of the parametric uncertainties (as is represented by the increasing value of σ_a) results in an increase of this added damping effect which is consistent with the previous explanation.

The standard deviation of the frequency domain response of the tip deflection for different spectral order of solution of the reduced basis approach is compared with the direct MCS and is shown in Fig. 3, for different values of σ_a. We find that the standard deviation is maximum at the resonance frequencies, which is expected due to the differences in the resonance peak of each sample. It is again observed that the direct MCS solution and the reduced-order approach give almost identical results, which demonstrate the effectiveness of the proposed approach.

The probability density function of the deflection of the tip of the cantilever beam for different degrees of variability of the random field is shown in Fig. 4. The probability density functions have been calculated at the frequency of 412 Hz, which is a resonance frequency of the beam. The results indicate that with the increase in the degree of uncertainty (variance) of the random system, we have long-tailed the density functions which is consistent with the standard deviation curve shown in Fig. 3 and the mean deflection of the stochastic system with the deterministic response in Fig. 2. This shows that the increase in the variability of the stochastic system has a damping effect on the response.

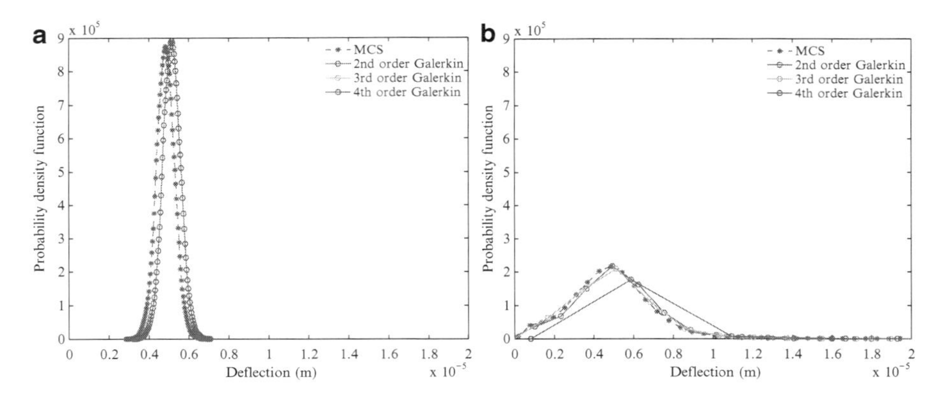

Fig. 4 The probability density function (*PDF*) of the tip deflection of the Euler-Bernoulli beam at 210 Hz under unit amplitude harmonic point load at the free end. The response is obtained with 10,000 samples and for $\sigma_a = \{0.05, 0.10, 0.15, 0.20\}$. (**a**) PDF of the response 210 Hz for $\sigma_a = 0.05$. (**b**) PDF of the response 210 Hz for $\sigma_a = 0.2$

6 Conclusions

Here, we have considered the discretised stochastic partial differential equation for structural systems with generally non-Gaussian random fields. In the classical spectral stochastic finite element approach, the solution is projected into an infinite dimensional orthonormal basis functions, and the associated constant vectors are obtained using the Galerkin type of error minimisation approach. Here an alternative approach is proposed. The solution is projected into a finite dimensional reduced vector basis, and the associated coefficient functions are obtained. The coefficient functions, called as the *spectral functions*, are expressed in terms of the spectral properties of the matrices appearing in the discretised governing equation. It is shown that then the resulting series converge to the exact solution in probability 1. This is a stronger convergence compared to the classical polynomial chaos which converges in the mean-square sense in the Hilbert space. Using an analytical approach, it is shown that the proposed spectral decomposition has the same functional form as the exact solution, which is not a polynomial, but a ratio of polynomials where the denominator has a higher degree than the numerator.

The computational efficiency of the proposed reduced spectral approach has been demonstrated for large linear systems with non-Gaussian random variables. It may be possible to extend the underlying idea to the class of non-linear problems. For example, the proposed spectral approach can be used for every linearisation step or every time step. Further research is necessary in this direction.

Acknowledgement A. Kundu acknowledges the financial support from the Swansea University through the award for Zienkiewicz scholarship. S. Adhikari acknowledges the financial support from the Royal Society of London through the Wolfson Research Merit Award.

References

1. Charmpis DC, Schueeller GI, Pellissetti MF (2007) The need for linking micromechanics of materials with stochastic finite elements: a challenge for materials science. Comput Mater Sci 41(1):27–37
2. Ghanem R, Spanos PD (1991) Stochastic finite elements: a spectral approach. Springer, New York
3. Kelly DW, De SR, Gago JP, Zienkiewicz OC, Babuska I (1983) A posteriori error analysis and adaptive processes in the finite element method: Part I: Error analysis. Int J Numer Methods Eng 19(11):1593–1619
4. Kleiber M, Hien TD (1992) The stochastic finite element method. Wiley, Chichester
5. Matthies HG, Brenner CE, Bucher CG, Soares CG (1997) Uncertainties in probabilistic numerical analysis of structures and solids – stochastic finite elements. Struct Saf 19(3):283–336
6. Nouy A (2007) A generalized spectral decomposition technique to solve a class of linear stochastic partial differential equations. Comput Methods Appl Mech Eng 196(45-48): 4521–4537
7. Nouy A (2008) Generalized spectral decomposition method for solving stochastic finite element equations: invariant subspace problem and dedicated algorithms. Comput Methods Appl Mech Eng 197(51–52):4718–4736
8. Nouy A (2009) Recent developments in spectral stochastic methods for the numerical solution of stochastic partial differential equations. Arch Comput Methods Eng 16:251–285. URL http://dx.doi.org/10.1007/s11831-009-9034-5
9. Stefanou G (2009) The stochastic finite element method: past, present and future. Comput Methods Appl Mech Eng 198(9–12):1031–1051
10. Yamazaki F, Shinozuka M, Dasgupta G (1988) Neumann expansion for stochastic finite element analysis. J Eng Mech ASCE 114(8):1335–1354
11. Zienkiewicz OC, Taylor RL (1991) The finite element method, 4th edn. McGraw-Hill, London

Uncertainties in Transportation Analysis

Partha Chakroborty

Abstract Like in most engineering disciplines, so also in transportation engineering, one often faces situations where the behaviour of the system is stochastic. Given the wide range of situations that comes under transportation engineering, the use of probability theory ranges from simple representation of variability through density functions to complex birth and death processes observed at intersection approaches, parking lots, toll booths, etc.

Unlike in most engineering disciplines, however, one of the basic elements of transportation systems is the human being. Sometimes, they are present as drivers controlling the units that make up a traffic stream, while at other times, they act as users who create demand for a transportation facility (like roads and airports). The uncertainties that human involvement brings to the transport systems often cannot be represented using the theory of probability. One has to look at other means of set descriptions and measures like possibility.

This note outlines the types of uncertainties that arise in the various systems that make up the field of transportation engineering and analysis and suggests appropriate paradigms to deal with them.

Keywords Probability • Fuzzy sets • Possibility • Transportation system

1 Introduction

Transportation is replete with situations which demand analysis through incorporation of uncertainties. However, uncertainties arise in different forms. In order to understand these different forms, the best way is to look at how one can evaluate the truth of the statement "x is A." Here, A is the description of some situation and x is a variable

P. Chakroborty (✉)
Department of Civil Engineering, Indian Institute of Technology Kanpur,
Kanpur 208 016, India
e-mail: partha@iitk.ac.in

S. Chakraborty and G. Bhattacharya (eds.), *Proceedings of the International Symposium on Engineering under Uncertainty: Safety Assessment and Management (ISEUSAM - 2012)*, DOI 10.1007/978-81-322-0757-3_11, © Springer India 2013

which can sometimes be A. For example, (1) x can be queue length, while A could be "greater than 5" or "large," or (2) x can be estimated arrival time at a station, while A could be "before departure of the train" or "desired arrival time."

If x is known deterministically and A can be defined precisely, then the truth of the statement is either 0 or 1. In the simplest and most naïve form, even today, engineering is practiced based on such deterministic evaluations of supposedly precise notions. For example, routinely one would declare that the level of service prevailing on a road section is C even though neither the roads conditions which lead to a particular level of service is deterministic nor the definition of level of service is precise. Engineering analysis can only mature if attempt is made to better understand and model the uncertainties in the processes. Only with a better understanding of these uncertainties and their incorporation in the analysis can tolerances improve and reliance on large "factors of safety" reduce.

In transportation, different situations lead to different types of x and A and consequently to different types of uncertainties. In this short note, a brief description of different types of uncertainty handling paradigms which find a place in transportation is described. This note is divided into five brief sections of which this is the first. The next section describes some situations in transportation engineering which require probability theory and probability theory-based modelling techniques to handle the inherent uncertainties. The third section describes scenarios where the assumption that a situation (like A, above) can be precisely described is no longer tenable. This section looks at how the concept of fuzzy sets can be used in such scenarios. The fourth section dwells briefly on the applicability of possibility theory in transportation problems. The last section summarizes this chapter.

2 Probability Theory and Transportation

In transportation engineering, probability theory and its models are used whenever the chances of occurrences of precisely defined events or sets are sought to be determined. In the following, few situations are described which require use of probabilistic analysis:

- Variation in speeds of vehicles in a traffic stream needs to be represented using probability density functions.
- Birth processes (or inter-birth times), like arrival of vehicles at a section on a road or to a parking lot, must be represented through appropriate means.
- Death processes (or inter-death times), like departure of vehicles from a toll booth or an unsignalized intersection approach, must be represented through appropriate means.
- Combined birth and death processes (i.e. the queueing processes), like the queueing at an intersection, or at a toll plaza or at a parking lot, need to be modelled as proper queues. Note that some queueing systems could be simple single-server queues with Poisson arrival and Poisson departures, while others could be more complicated in terms of the departure process (like at unsignalized intersections), yet others are as complicated as coupled multiple queues (like at toll plazas).

- Choice processes like drivers choosing particular destinations or modes over others or drivers choosing particular toll booths while approaching a plaza over others need to be properly understood and modelled. Often these are modelled using discrete choice analysis which is based on the principles of random utility. Of course, given that this is a human choice process, the use of random utility principles can be questioned.

There are many more examples, but the purpose here is not to list them but to give the reader an idea about the variety of situations that arise in transportation which require probabilistic modelling. A quick survey of the literature will yield many references where probability models have been used to analyse transportation problems.

Yet, at the same time, all the cases that arise in transportation, on a closer scrutiny indicate that they are not fit cases for probabilistic analysis. The next section gives some examples where uncertainty arises due to lack of clarity in defining concepts and discusses how this type of uncertainty needs to be handled.

3 Fuzzy Sets and Transportation

It often happens that a concept is defined using terms which are inherently vague. For example, the level of service (an important design parameter for traffic facilities) is defined using expressions like "slightly and quickly." The highway capacity manual [2–4], from which this concept has originated, undoubtedly considers level of service as a qualitative measure of the driving conditions. Yet, the same manual and various other manuals around the world continue to evaluate the level of service through deterministic means which yield precise levels of service for given stream parameters. Kikuchi and Chakroborty [6] point this out and suggest how this can be corrected.

The problem in the example given here is that of how one defines a set for a concept like level of service A or level of service B. Even if one accepts the proposition that somehow density of the traffic stream is a good indicator of the level of service, the definition of level of service implies that one cannot draw precise or crisp boundaries in the universe of density to indicate different sets representing different levels of service. A more appropriate description would be where the boundaries are not precise and allow elements of the universe to be partial members of a set or that of a set and its complement. Such sets are called fuzzy sets and are defined through membership functions which map the elements of a universe to a unit real scale [0,1] indicating membership to that set.

There are other examples in transportation where the concepts one deals with, ideally, should not be described though traditional sets or crisp sets. This is so because human inference often proceeds with concepts which are perceived (rather than measured) and classified linguistically; typically, such classification does not lend itself to precise boundaries. For example, the concept of "tall" cannot be represented as a set where only people whose heights are greater than, say, 6 ft are members and others are not. This is because a person who is 5 ft 11 in. is also, in the mind of humans, not not-tall.

In transportation, human inference plays a big role. For example, the process of driving is a system where the human driver (or controller) perceives the relevant parameters from the surroundings (like speed, distance headway and relative speed) and infers what action should be taken in terms of acceleration. Theories which describe this inference mechanism through differential equations make the sweeping and reasonably untenable assumption that drivers perceive and infer precisely. Although such analysis has a place in developing the theory of traffic flow, it is important to realize that a fuzzy inference (or control) system will be a truer model of the real-world system. In such models, among other things, concepts are represented as fuzzy sets. Kikuchi and Chakroborty [5] and Chakroborty and Kikuchi [1] highlight these points in terms of the theory of car-following—an essential element of the theory of traffic flow.

Interested readers may refer to Klir and Folger [9], Zimmermann [12] and various other texts for a better understanding of fuzzy sets and other theories built on fuzzy sets.

4 Possibility Theory and Transportation

The first section introduced the concept of looking at the truth of the statement "x is A" as a way to understanding the different types of uncertainty that may exist in a system. The second section described situations in transportation where x is random (or stochastic) while the concept A, that one wish to evaluate, can be precisely defined. For example, in this situation, meaningful questions could be "queue length is greater than 7" or "number of vehicles arriving (in a 5-min interval) is more than 10 but less than 15." As can be seen, what gives rise to the uncertainty in evaluating the truth of the statement "x is A" is the stochastic nature of queue length or number of vehicles arriving (i.e. the x) and not "greater than 7," or "more than 10 but less than 15" (i.e. the A).

In the third section, situations were presented where the source of uncertainty was the way A was defined. For example, concepts like "large delay" or "closing-in quickly" could only be represented, respectively, in the universe of time or relative speed through fuzzy sets which do not have crisp or precise boundaries. Hence, even if x is known precisely, there would be uncertainty associated with the truth of the statement x is A because of how A is defined.

In this section, those cases are considered where the information which helps in determining the nature of x (i.e. whether it is known precisely or it can be best described through a probability density function) is such that a probability density function cannot be used to describe x.

Both probability and possibility are measures which assign to a set a number in the range [0,1] that indicates the strength of the belief that a given element is in the set. This, of course, is a very coarse and naïve definition of measures. The interested reader may refer to Klir and Folger [9] for a simple description of measures. Suffice it to say that probability is but one measure which tries to capture this strength of belief. In fact, it is understood that if the evidence is arranged in a particular way (referred to as *conflicting*) can the associated measure be referred to as probability. While if the evidence is arranged in a different way (referred to as *consonant*), then the associated measure is referred to as a possibility measure. In transportation, often situations arise where the evidence is arranged in a manner which is consonant. Interestingly, Zadeh [11] drew analogies between the theories of fuzzy sets and possibility.

Kikuchi and Chakroborty [7] list many situations where possibility distribution and possibilistic analysis are more relevant; some of them are:

- Representing desire, for example, desired departure time and desired arrival time.
- Representing notion of satisfaction and acceptability, for example, satisfactory cost, acceptable toll and acceptable delay.
- Representing perceived quantities or quantities based on experience, for example, estimated travel time, expected delay and value of time.
- Situations where quantities are represented using possibility measure require analysis using the theory of possibility; a detailed description of this can be found in Kikuchi and Chakroborty [7].

The interested reader may refer to Klir [8] and Klir and Wierman [10] among many others for a comprehensive understanding of how to analyse uncertainty.

5 Conclusion

In this short note, an attempt is made to highlight the different types of uncertainties that exist in transportation and how these types can be handled by using appropriate paradigms. Specifically, the use of theories of probability, possibility and fuzzy sets is described here.

References

1. Chakroborty P, Kikuchi S (1999) Evaluation of general motors' based car-following models and a proposed fuzzy inference model. Transp Res Part C 7:209–235
2. Highway Capacity Manual (1965) HRB. Special report 87. National Research Council, Washington, DC
3. Highway Capacity Manual (1985) TRB. Special report 209, 3rd edn. National Research Council, Washington, DC
4. Highway Capacity Manual (2000) TRB. National Research Council, Washington, DC

5. Kikuchi S, Chakroborty P (1992) Car-following model based on fuzzy inference system. Transp Res Rec 1365:82–91
6. Kikuchi S, Chakroborty P (2006) Frameworks to represent uncertainty when level of service is determined. Transp Res Rec 1968:53–62
7. Kikuchi S, Chakroborty P (2006) Place of possibility theory in transportation analysis. Transp Res Part B 40:595–615
8. Klir GJ (1999) On fuzzy set interpretation of possibility theory. Fuzzy Sets Syst 108:263–273
9. Klir GJ, Folger TA (1988) Fuzzy sets, uncertainty, and information. Prentice-Hall, Englewood Cliffs
10. Klir G, Wierman M (2000) Uncertainty-based Information: elements of generalized information theory. Physica-Verlag, Heidelberg/New York
11. Zadeh LA (1978) Fuzzy sets as basis for a theory of possibility. Fuzzy Sets Syst 1:3–28
12. Zimmermann H-J (1991) Fuzzy set theory and its applications, 2nd edn. Kluwer Academic Publishers, Boston

Reliability Analysis of Municipal Solid Waste Settlement

Sandeep K. Chouksey and G.L. Sivakumar Babu

Abstract It is well-known that the disposal of municipal solid waste (MSW) has become one of the challenges in landfill engineering. It is very important to consider mechanical processes that occur in settlement response of MSW with time. In the recent years, most of the researchers carried out different tests to understand the complex behavior of municipal solid waste and based on the observations and proposed different models for the analysis of stress-strain, time-dependent settlement response of MSW. However, in most of the cases, the variability of MSW is not considered. For the analysis of MSW settlement, it is very important to account for the variability of different parameters representing primary compression, mechanical creep, and effect of biodegradation. In this chapter, an approach is used to represent the complex behavior of municipal solid waste using response surface method constructed based on a newly developed constitutive model for MSW. The variability associated with parameters relating to primary compression, mechanical creep, and biodegradation are used to analyze MSW settlement using reliability analysis framework.

Keywords Municipal solid waste • Mechanical creep • Biodegradation • Response surface method • Reliability analysis

1 Introduction

Landfilling is still the most common treatment and disposal technique for Municipal Solid Waste (MSW) worldwide. In every country, millions of tons of wastes are produced annually and it became one of the mammoth tasks to overcome it. Recently, MSW landfilling has significantly improved and has achieved a stage

S.K. Chouksey • G.L.S. Babu (✉)
Department of Civil Engineering, Indian Institute of Science, Bangalore 560012, India
e-mail: chouksey sandeep@gamil.com; gls@civil.iisc.ernet.in

S. Chakraborty and G. Bhattacharya (eds.), *Proceedings of the International Symposium on Engineering under Uncertainty: Safety Assessment and Management (ISEUSAM - 2012)*, DOI 10.1007/978-81-322-0757-3_12, © Springer India 2013

of highly engineered sanitary landfills in the most developed and developing countries. Evaluation of settlement is one of the critical components in landfill design. The contribution of engineered landfilling requires extensive knowledge of the different processes which occur simultaneously in MSW during settlement. The settlement in MSW is mainly attributed to (1) physical and mechanical processes that include the reorientation of particles, movement of the fine materials into larger voids, and collapse of void spaces; (2) chemical processes that include corrosion, combustion, and oxidation; (3) dissolution processes that consist of dissolving soluble substances by percolating liquids and then forming leachate; and (4) biological decomposition of organics with time depending on humidity and the amount of organics present in the waste.

Due to heterogeneity in the material of MSW, the analysis becomes more complicated because degradation process on MSW is time-dependent phenomena and continuously undergoes degradation with time. In the degradation process, two major mechanisms of biodegradation may occur: aerobic (in the presence of oxygen) and anaerobic (in the absence of oxygen) processes. The production of landfill biogas is a consequence of organic MSW biodegradation. This process is caused by the action of bacteria and other microorganisms that degrade the organic fraction of MSW in wet conditions. To capture this phenomenon in the prediction of settlement and stress-strain response of MSW, several researchers have proposed different models based on the different assumptions [1–3, 7, 8, 10, 12].

Marques et al. [10] presented a model to obtain the compression of MSW in terms of primary compression in response to applied load, secondary mechanical creep, and time-dependent biological decomposition. The model performance was assessed using data from the Bandeirantes landfill, which is a well-documented landfill located in Sao Paulo, Brazil, in which an instrumented test fill was constructed. Machado et al. [8] presented a constitutive model for MSW based on elastoplasticity considering that the MSW contains two component groups: the paste and the fibers. The effect of biodegradation is included in the model using a first-order decay model to simulate gas generation process through a mass-balance approach while the degradation of fibers is related to the decrease of fiber properties with time. The predictions of stress-strain response from the model and observations from the experiments were compared, and guidelines for the use of the model are suggested. Babu et al. [1, 3] proposed constitutive model based on the critical state soil mechanics concept. The model gives the prediction of stress-strain and pore water pressure response and the predicted results were compared with the experimental results. In addition, the model was used to calculate the time-settlement response of simple landfill case. The predicted settlements are compared with the results obtained from the model of Marques et al. [9, 10].

1.1 Settlement Predictive Model

Babu et al. [1] proposed a constitutive model which can be used to determine settlement of MSW landfills based on constitutive modeling approach. In this model, the elastic and plastic behavior as well as mechanical creep and biological decomposition is used to calculate the total volumetric strain of the MSW under loading as follows:

$$\mathrm{d}\varepsilon_v = \mathrm{d}\varepsilon_v^e + \mathrm{d}\varepsilon_v^p + \mathrm{d}\varepsilon_v^c + \mathrm{d}\varepsilon_v^b \tag{1}$$

where $\mathrm{d}\varepsilon_v^e$, $\mathrm{d}\varepsilon_v^p$, $\mathrm{d}\varepsilon_v^c$, and $\mathrm{d}\varepsilon_v^b$ are the increments of volumetric strain due elastic, plastic, time-dependent mechanical creep, and biodegradation effects, respectively. The increment in elastic volumetric strain $\mathrm{d}\varepsilon_v^e$ can be written as:

$$\mathrm{d}\varepsilon_v^e = -\frac{\mathrm{d}e^e}{1+e} = \frac{\kappa}{1+e}\frac{\mathrm{d}p'}{p'} \tag{2}$$

And increment in plastic volumetric strain can be written as

$$\mathrm{d}\varepsilon_v^p = \left(\frac{\lambda-\kappa}{1+e}\right)\left[\frac{\mathrm{d}p'}{p'} + \frac{2\eta \mathrm{d}\eta}{M^2+\eta^2}\right] \tag{3}$$

The above formulations for increments in volumetric strain due to elastic and plastic are well established in critical state soil mechanics literature.

The mechanical creep is a time-dependent phenomenon proposed by Gibson and Lo's [6] model, in exponential function, is given by

$$\varepsilon_v^c = b\Delta p'\left(1 - e^{-ct'}\right) \tag{4}$$

where b is the coefficient of mechanical creep, $\Delta p'$ is the change in mean effective stress, c is the rate constant for mechanical creep, and t' is the time since application of the stress increment. The biological degradation is a function of time and is related to the total amount of strain that can occur due to biological decomposition and the rate of degradation. The time-dependent biodegradation proposed by Park and Lee [12] is given by

$$\varepsilon_v^b = E_{dg}\left(1 - e^{-\mathrm{d}t''}\right) \tag{5}$$

where E_{dg} is the total amount of strain that can occur due to biological decomposition, d is the rate constant for biological decomposition, and t'' is the time since placement of the waste in the landfill.

From Eq. (4), increment in volumetric strain due to creep is written as

$$\mathrm{d}\varepsilon_v^c = cb\Delta p' e^{-ct'}\mathrm{d}t' \tag{6}$$

From Eq. (5), increment in volumetric strain due to biodegradation effect is written as

$$\mathrm{d}\varepsilon_v^b = E_{dg} e^{-\mathrm{d}t''}\mathrm{d}t'' \tag{7}$$

In the present case, t' time since application of the stress increment and t'' time since placement of the waste in the landfill are considered equal to "t."

Using Eqs. (2), (3), (6), and (7) and substituting in Eq. (1), total increment in strain is given by

$$d\varepsilon_v = \frac{\kappa}{1+e}\frac{dp'}{p'} + \left(\frac{\lambda-\kappa}{1+e}\right)\left[\frac{dp'}{p'} + \frac{2\eta d\eta}{M^2+\eta^2}\right] + cb\Delta\sigma' e^{-ct}dt + E_{dg}e^{-dt}dt \quad (8)$$

Calculation procedure of settlement response of MSW using above equations is given in Babu et al. [2].

1.2 Variability of MSW Parameters

Settlement models of Marques et al. [10] and Babu et al. [1, 3] have parameters such as compressibility index, coefficient of mechanical creep (b), creep constant (c), biodegradation constant (E_{dg}) and rate of biodegradation (d). All these parameters are highly variable in nature due to heterogeneity of MSW. For any engineering design of landfill, these parameters are design parameters, and their variability plays vital role in design. Literature review indicates that the influence of all these parameters and their variations have significant effects on prediction of MSW settlement. Based on experimental and field observations, various researchers reported different range of values and percentage of coefficient of variations (COV). For example, Sowers [13] reported that the compression index (c_c) is related to the initial void ratio (e_0) and can vary between 0.15 e_0 and 0.55 e_0 and the value of secondary compression index (c_a) varied between 0.03 e_0 and 0.09 e_0. The upper limit corresponds to MSW containing large quantities of food waste and high decomposable materials. Results of Gabr and Valero [5] indicated c_c values varying from 0.4 to 0.9, and c_a values varying from 0.03 to 0.009 for the initial void ratios (e_0) in the range of approximately 1.0–3.0. Machado et al. [7] obtained the values of primary compression index which varied between 0.52 and 0.92. Marques et al. [10] reported c_c values varying from 0.073 to 1.32 with a coefficient of variation (COV) of 12.6%. The coefficient of mechanical creep (b) was reported in the range of 0.000292–0.000726 and COV of 17.7% and creep constant varying from 0.000969 to 0.00257 with COV of 26.9%. The time-dependent strain due to biodegradation is expressed by equation which uses E_{dg}, the parameter related to total amount of strain that can occur due to biodegradation, and d is the rate constant for biological decomposition. Biodegradation constant depends upon the organic content present in MSW. Marques et al. [10] gave typical range of E_{dg} varying from 0.131 to 0.214 and COV of 12.7% and biodegradation rate constant d varying from 0.000677 to 0.00257 and COV of 42.3%. Foye and Zhao [4] used random field model to analyze differential settlement of existing landfills. They used c_c values 0.22 and 0.29 with COV of 36% and E_{dg} time-dependent strain due to biodegradation equal to 0.03724 and rate constant due to biodegradation (d) equal to

0.00007516. These variables are not certain; their values depend upon the variation conditions like site conditions, initial moisture content, and quantity of biodegradable material present in the existing MSW. Therefore, it is very important to perform settlement analysis of MSW with consideration of variability and their influence on reliability index or probability of failure. In order to simplify the settlement calculations, the above settlement evaluation procedure is used with reference to a typical landfill condition using response surface method (RSM).

1.3 Response Surface Method

RSM is a collection of statistical and mathematical techniques useful for developing, improving, and optimizing process. In the practical application of response surface methodology (RSM), it is necessary to develop an approximating model for the true response surface. A first-order (multilinear) response surface model is given by

$$y_i = \beta_0 + \beta_1 x_1 + \beta_2 x_2 + \cdots\cdots\cdots + \beta_n x_n + \varepsilon \tag{9}$$

Here, y_i is the observed settlement of MSW; the term "linear" is used because Eq. (9) is a linear function of the unknown parameters $\beta_1, \beta_2, \beta_3, \beta_4$, and β_5 that are called the regression coefficients, and $x_1, x_2, x_3, \cdots\cdots\cdots x_n$ are coded variables which are usually defined to be dimensionless with mean zero and same standard deviation. In the multiple linear regression model, natural variables $(b, c, d, E_{dg}, \lambda)$ are converted into coded variable by relationship

$$x_{i1} = \frac{\xi_{i1} - [\max(\xi_{i1}) + \min(\xi_{i1})]/2}{[\max(\xi_{i1}) - \min(\xi_{i1})]/2}$$

In the present study, RSM analysis is performed using single replicate 2^n factorial design to fit first-order linear regression model, where n is the total number of input variables involved in the analysis and corresponding to these variables the number of sample point required is 2^n. For example, in the present case five variables are considered; here, n is equal to 5 and number of sample points required is 32. These 32 sample points are generated using "+"and "−" notation to represent the high and low levels of each factor, the 32 runs in the 2^5 design in the tabular format shown in Table 1.

For the analysis, the maximum and minimum values are assigned based on the one-sigma, two-sigma, and three-sigma rule, i.e., $\sigma \pm \mu, \sigma \pm 2\mu$, and $\sigma \pm 3\mu$, where μ is the mean value and σ is the standard deviation of variables given in Table 2. In order to account the variability of different parameters in the settlement analysis of MSW, five major parameters are used under the loading conditions (from bottom to top) as shown in Fig. 1; the following variable parameters are used for the study.

Table 1 Generation +'s and −'s for generation of response surface equation

S. No.	b (m^2/kN)	c (day^{-1})	d (day^{-1})	E_{dg}	λ
1.	+	+	+	+	+
2.	+	+	+	+	−
3.	+	+	+	−	+
4.	+	+	+	−	−
5.	+	+	−	+	+
6.	+	+	−	+	−
7.	+	+	−	−	+
8.	+	+	−	−	−
9.	+	−	+	+	+
10.	+	−	+	+	−
11.	+	−	+	−	+
12.	+	−	+	−	−
13.	+	−	−	+	+
14.	+	−	−	+	−
15.	+	−	−	−	+
16.	+	−	−	−	−
17.	−	+	+	+	+
18.	−	+	+	+	−
19.	−	+	+	−	+
20.	−	+	+	−	−
21.	−	+	−	+	+
22.	−	+	−	+	−
23.	−	+	−	−	+
24.	−	+	−	−	−
25.	−	−	+	+	+
26.	−	−	+	+	−
27.	−	−	+	−	+
28.	−	−	+	−	−
29.	−	−	−	+	+
30.	−	−	−	+	−
31.	−	−	−	−	+
32.	−	−	−	−	−

Table 2 Parameters used for the regression analysis [10]

	λ	b (m^2/kN)	c (day^{-1})	d (day^{-1})	E_{dg}
Average	0.046	5.27E-04	1.79E-03	1.14E-03	0.15
Standard deviation	0.0059	9.5793E-05	0.000492	0.00049	0.020
COV	12.93	1.82E+01	2.75E+01	4.35E+01	12.99

They are coefficient of time-dependent mechanical creep (b), time-dependent mechanical creep rate constant (c), rate of biodegradation (d), biodegradation constant (E_{dg}), and the slope of normally consolidate line (λ). In the literature, it was reported that they are highly variable in nature, and it is very important to account these variation during the prediction of MSW settlement.

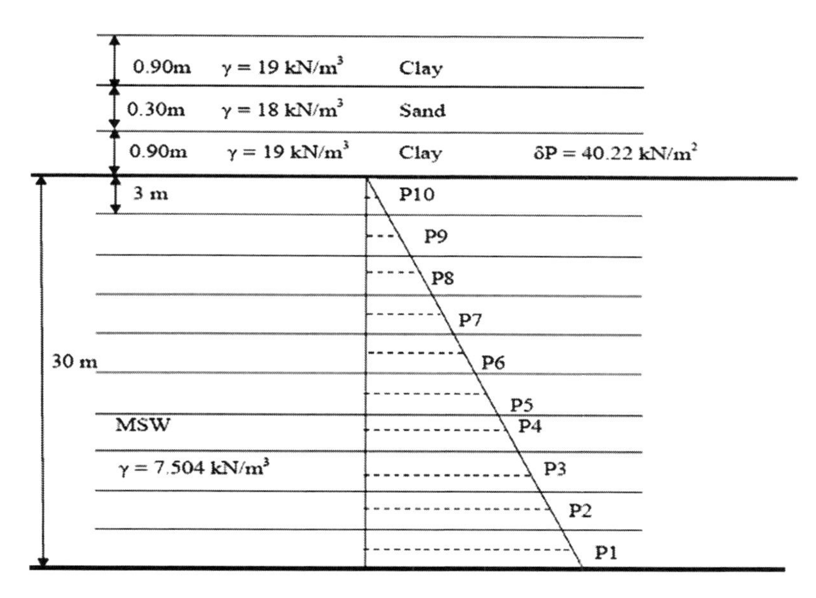

Fig. 1 MSW landfill scenario for estimation of settlement versus time

One-sigma (standard deviation) rule considers about 68%, and two sigma (standard deviations) consider almost 95% of sample variation assuming normal distribution, whereas three sigma (standard deviations) account for 99.7% of the sample variations, assuming the normal distribution. Using one-, two-, and three-sigma rules, the sample points are generated, and corresponding settlements are calculated using proposed model by Babu et al. [1].

The method of least squares is typically used to estimate the regression coefficients in a multiple linear regression model for the simple case of landfill as shown in Fig. 1. Myers and Montgomery [11] gave the multilinear regression in the form of matrix.

$$y = X\beta + \varepsilon \tag{10a}$$

where,

$$\beta = (X'X)^{-1}X'y \tag{10b}$$

Using above method, regression coefficients are calculated, and least square fit with the regression coefficients in terms of natural variables corresponding to the different COVs is presented in Table 3. Settlements are obtained from first-order regression model and from Babu et al. [1]. It is always necessary to examine the fitted model to ensure that it provides an adequate approximation to the true system and verify that none of the least square regression assumptions are violated. To ensure the adequacy of the regression equations, coefficient of regression (R^2 and R^2_{adj}) is calculated. Table 4 presents regression coefficients for the different standard deviations and at different COV.

Table 3 Regression equation in natural variables

(a) One-sigma deviation	
COV 10%	$1.90b + 12.29c + 303.96d + 8.13d + 203.26\lambda - 1.70$
COV 14%	$-11.38b + 16.76c + 426.07d + 11.38d + 284.67\lambda - 6.10$
COV 18%	$-30.35b + 21.78c + 549.94d + 14.62d + 36.20\lambda - 10.50$
COV 20%	$-41.73b + 24.57c + 612.32d + 16.24d + 407\lambda - 12.70$
(b) Two-sigma deviations	
COV 10%	$-41.73b + 25.13c + 613.24d + 16.24d + 407\lambda - 12.70$
COV 14%	$-108.11b + 35.20c + 866.20d + 22.67d + 570.6\lambda - 21.51$
COV 18%	$-201.10b + 47.47c + 1132.40d + 28.98d + 734.94\lambda - 30.33$
COV 20%	$-257.94b + 53.62c + 1273.83d + 32.10d + 817.41\lambda - 34.74$
(c) Three-sigma deviations	
COV 10%	$-127b + 38.54c + 932.10d + 24.26d + 611.62\lambda - 23.71$
COV 14%	$-290.18b + 56.41c + 1346.75d + 33.61d + 858.71\lambda - 36.95$
COV 18%	$-515.88b + 77.63c + 1847.50d + 42.26d + 1107.40\lambda - 50.22$
COV 20%	$-652.44b + 88.80c + 2167.27d + 46.03d + 1231.83\lambda - 56.84$

Table 4 Coefficients of regression (R^2) and R^2_{adj} for 1σ, 2σ, and 3σ for different COVs

COV %	R^2 (1σ)	$R^2_{adj}(1\sigma)$	R^2 (2σ)	R^2_{adj} (2σ)	$R^2(3\sigma)$	$R^2_{adj}(3\sigma)$
10	0.999	0.998	0.999	0.999	0.998	0.998
14	0.998	0.998	0.999	0.998	0.996	0.996
18	0.998	0.997	0.998	0.997	0.994	0.993
20	0.997	0.996	0.997	0.996	0.993	0.992

1.4 Reliability Index Formulation

The reliability index β for the independent variables in n-dimensional space is given as:

$$g(X) = c_0 + \sum_{i=1}^{n} c_i x_i \tag{11}$$

$$\mu_g = c_0 + c_1\mu_{x_1} + c_2\mu_{x_2} + \cdots\cdots + c_n\mu_{x_n} \tag{12}$$

$$\sigma_g = \sqrt{\sum_{i=1}^{n} c_i^2 \sigma_{x_i}^2} \tag{13}$$

$$\beta = \frac{\mu_{\tilde{g}}}{\sigma_{\tilde{g}}} \tag{14}$$

In the present study, using deterministic analysis, with mean values given in Table 2, an ultimate settlement of 9.3 m for 30 years is obtained using the proposed

Table 5 Variation of reliability index with COVs at one-, two-, and three-sigma deviations

COV %	One sigma (β)	Two sigma (β)	Three sigma (β)
10	9.71	2.43	0.81
14	4.96	1.24	0.56
18	2.99	0.75	0.33
20	2.42	0.61	0.28

model as well as the response surface equations. To ascertain the probability of ultimate settlement reaching this value, the limit state function is defined as

$$g(x_i) = 9.3 - y_i \tag{15}$$

2 Methodology

In order to evaluate the reliability index or probability of failure considering variability of different parameters in the calculation of MSW settlement, the settlement given by the model is converted into Eq. (9) to using the procedure described earlier. In this chapter, settlement is evaluated for the one, two, and three standard deviations. Using the mean and standard deviation given in Table 2 and with "+" and "−" or maximum or minimum, values are calculated for all the variables. Table 1 shows the generation of sample points for the one standard deviation considering the normal distribution. These values are used to evaluate performance function based on the multilinear regression analysis as discussed previously for one, two, and three standard deviations at all the percentages of COV. Table 3 present the performance functions for the one, two, and three standard deviation at different percentages (10, 14, 18, and 20%) of COV. The performance functions are in the form of multilinear equations that include all the contributing variables used for the prediction of MSW settlement in the form of natural variables $\left(b, c, d, E_{dg}, \lambda\right)$. Equations (12) and (13) are used to calculate mean and standard deviation of the approximated limit state function. Knowing approximated mean (μ_g) and standard deviation (σ_g), reliability index is calculated using relation given in Eq. (14). Table 5 presents the summary of variation of reliability index values for the one, two, and three standard deviations at different COV. It is observed from Table 5 that the reliability index is inversely proportional to the COV of design variables decreases with increase in percentage of COVs of the design variables and also with increase in standard deviations.

3 Results and Discussion

Using RSM, multilinear equations are developed and these equations are considered as performance functions for the calculation of reliability index using Eqs. (12), (13), and (14) for MSW settlement. It is noted that the reliability index of MSW settlement is inversely related to the coefficient of variation of design variable parameters. The results clearly depict that with increase in percentages of COV, reliability index decreases and probability of failure increases. For example, in case of 10% COV reliability index is calculated 9.71, whereas for the 20% COV this reliability index reduced to 2.42, which is reduction in 75% for one standard deviation. Similar results are observed for other cases. This indicates that probability of failure or reliability index is highly dependent upon COV of variable parameters. It is well-known that the MSW composition is highly heterogenous in nature which leads to higher percentages of COV, and hence, the chance of failure is very high. On the other hand, reliability index of MSW is highly dependent on sampling variations. It is observed that reliability index for 10% COV is 9.71 for one standard deviation, 2.42 for two standard deviations and 0.8078 for the three standard deviations. From one standard deviation to two standard deviation 75% reduction and for three standard deviations approximately 91% reduction in reliability index was observed. These results clearly point out the significance of sampling in the landfilling design for longer time periods.

4 Conclusions

The objective of this chapter is to demonstrate the influence of variability in the estimation of MSW settlement. Settlement is calculated based on the response surface method for the five design variables. Sample points are generated using "+" and "−" method for the one, two, and three standard deviations. Using these sample points, multilinear equations are developed for the estimation of MSW settlement. Based on these equations, reliability index is calculated for all percentages of COV. The results indicate that reliability index is highly dependent upon variability of the design variables and sample variations.

References

1. Babu Sivakumar GL, Reddy KR, Chouksey SK (2010) Constitutive model for municipal solid waste incorporating mechanical creep and biodegradation-induced compression. Waste Manage J 30(1):11–22
2. Babu Sivakumar GL, Reddy KR, Chouksey SK (2011) Parametric study of parametric study of MSW landfill settlement model. Waste Manage J 31(1):1222–1231
3. Babu Sivakumar GL, Reddy KR, Chouksey SK, Kulkarni H (2010) Prediction of long-term municipal solid waste landfill settlement using constitutive model. Pract Period Hazard Toxic Radioact Waste Manage ASCE 14(2):139–150
4. Foye KC, Zhao X (2011) Design criteria for the differential settlement of landfill foundations modeled using random fields. GeoRisk ASCE, Atlanta, GA, 26–28 June
5. Gabr MA, Valero SN (1995) Geotechnical properties of solid waste. ASTM Geotech Test J 18(2):241–251
6. Gibson RE, Lo KY (1961) A theory of soils exhibiting secondary compression. Acta Polytech Scand C-10:1–15

7. Machado SL, Carvalho MF, Vilar OM (2002) Constitutive model for municipal solid waste. J Geotech Geoenviron Eng ASCE 128(11):942–951
8. Machado SL, Vilar OM, Carvalho MF (2008) Constitutive model for long – term municipal solid waste mechanical behavior. Comput Geotech 35:775–790
9. Marques ACM (2001) Compaction and compressibility of municipal solid waste. Ph.D. thesis, Sao Paulo University, Sao Carlos, Brazil
10. Marques ACM, Filz GM, Vilar OM (2003) Composite compressibility model for municipal solid waste. J Geotech Geoenviron Eng 129(4):372–378
11. Myers RH, Montgomery DC (2002) Response surface methodology, process and product optimization using designed experiments, 2nd edn. Wiley, New York
12. Park HI, Lee SR (1997) Long-term settlement behavior of landfills with refuse decomposition. J Resour Manage Technol 24(4):159–165
13. Sowers F (1973) Settlement of waste disposal fills. In: Proceedings of 3rd international conference on soil mechanics and foundation engineering, vol 2, Moscow, pp 207–210

Uncertainty Evaluation in Integrated Risk-Based Engineering

P.V. Varde

Abstract Uncertainty evaluation is one of the important areas that needs to be strengthened toward effective implementation risk-based approach. At the outset, this chapter introduces the broad concepts in respect of integrated risk-based engineering and examines the capability of the current approaches for uncertainty modeling as applicable to integrated risk-based engineering. A brief overview of state of the art in uncertainty analysis for nuclear plants and the limitation of the current approaches in quantitative characterization of uncertainties have been discussed. Role of qualitative or cognitive-based approaches has also been discussed to address the scenario where quantitative approach is not adequate.

Keywords Uncertainty characterization • Risk-based engineering • Probabilistic safety assessment • Nuclear plants • Safety assessment

1 Introduction

Existing literature in safety assessment for nuclear plants deals with two terms, viz., "risk-based decisions" and "risk-informed decisions," for dealing with regulatory cases. In the context of nuclear plant safety evaluation, risk-based engineering deals with the evaluation of safety cases using probabilistic safety assessment (PSA) methods alone, while risk-informed approach decisions are based on, primarily, deterministic methods including design and operational insights, and PSA results either complement or supplement the deterministic findings [1]. These approaches intuitively consider that probabilistic and deterministic methods are two explicit domains. However, ideally speaking, any problem or modeling

P.V. Varde (✉)
Life Cycle Reliability Engineering Lab, Safety Evaluation and Manpower Training & Development Section, Research Reactor Services Division, Bhabha Atomic Research Centre, Mumbai 400 085, India
e-mail: varde@barc.gov.in

S. Chakraborty and G. Bhattacharya (eds.), *Proceedings of the International Symposium on Engineering under Uncertainty: Safety Assessment and Management (ISEUSAM - 2012)*,
DOI 10.1007/978-81-322-0757-3_13,

requires considerations of deterministic as well as probabilistic methods together. Otherwise, the salutation is not adequate and complete. Even though the risk-informed approach that requires considerations of deterministic as primary approach and probabilistic as supplementary/complimentary approach deals with the issues in explicit manner. The fact is that even deterministic variables, like design parameters, process, and nuclear parameters, are often random in nature and require probabilistic treatment. Defense-in-depth along with other principles, viz., redundancy, diversity, and fail-safe design, forms the basic framework of deterministic approach. It will help to characterize the reliability of various barriers of protection – the basic instrument of defense-in-depth. Similarly, probabilistic methods cannot work in isolation and require deterministic input in terms of plant configurations, failure criteria, design inputs, etc. Hence, it can be argued that a holistic approach is required where deterministic and probabilistic methods have to work in an integrated manner in support of decisions related to design, operation, and regulatory review of nuclear plants. The objective should be to remove overconservatism and prescriptive nature of current approach and bring in rationales and make the overall process of safety evaluation scientific, systematic, effective, and integrated in nature.

Integrated risk-based engineering is a new paradigm that is being introduced through this chapter. In this approach, the deterministic as well as probabilistic methods are integrated to form a holistic framework to address the safety issues. However, the key issues that need to be considered for applications are characterization of uncertainty, assessment of safety margins, and requirements of dynamic models for assessment of accident sequence evolution in time domain.

Another significant feature of this chapter is that it is perhaps for the first time that the term "risk-based engineering" has been used and not the traditional "risk-based decisions." The reason is that traditionally the terms "risk-based" and "risk-informed" have been associated with regulatory decisions. However, keeping in view the knowledge base that is available and the tools and methods that have been developed along these years makes the case for risk-based approach to qualify as a discipline as "risk-based engineering." Hence, it is proposed that the term "integrated risk-based engineering" has relevance to any area of engineering, be it design, operations including regulatory reviews.

Keeping in view the theme of this conference, the aspects related to uncertainty have been discussed. Included here is a brief overview of uncertainty evaluation methods in risk-based applications and requirements related to epistemic and aleatory uncertainty. Further, the aspects related to qualitative or cognitive aspects of uncertainty have also been discussed. This chapter treats the subject in a philosophical manner, and there is conscious decision not to cover the specifics that can be found in the referred literature.

2 Integrated Risk-Based Engineering: A Historical Perspective

It is generally felt that the traditional approach to safety assessment is purely deterministic in nature. It is true that most of the cases evaluated as part of safety assessment employ deterministic models and methods. However, if we look at the assumptions, boundary conditions, factors of safety, data, and model, it can be argued that there is a good deal of probabilistic element even as part of traditional safety analysis. These elements or variables had qualitative notions for bounding situations and often provided comparative or relative aspects of two or more prepositions. To understand this point further, let us review the traditional safety analysis report(s) and have a fresh look at the broader aspect of this methodology. The major feature of the traditional safety analysis was based on the maximum credible accident, and for nuclear plants, it was mainly loss-of-coolant accident, loss-of-regulation accident scenario, etc. It was assumed that plant design should consider LOCA and other scenario like station blackout scenario to demonstrate that plant is safe enough. A reference to a safety report will make clear that there is an element of probability in a qualitative manner. Like (a) the possibility of two-out-of-three train failure is very low, (b) possibility of a particular scenario involving multiple failure is very unlikely or low, and (c) series of assumptions that will form bounding conditions. These aspects provided definite observation that probabilistic aspects were part of deterministic methods. These aspects were qualitative in nature. Keeping in view the above and considerations of factor of safety in the design as part of deterministic methods bring out the fact that safety analysis approach was integrated right since inception. This background along with current safety requirements, like process of safety evaluation, should (a) be more rational based and not prescriptive in nature, (b) remove overconservative, (c) be holistic in nature, (d) provide improved framework for addressing uncertainty, (e) allow realistic safety margins, and (f) provide framework for dynamic aspect of the accident scenario evaluation.

The integrated risk-based approach as mentioned above is expected to provide an improved framework for safety engineering. Here, the deterministic and probabilistic approaches treat the issues in an implicit manner unlike risk-informed approach where these two approaches have been employed in explicit manner. In this approach, issues are addressed in an integrated manner employing deterministic and probabilistic approaches. From the point of uncertainty characterization, in this approach, the random phenomenon is addressed as aleatory uncertainty while the model- and data-related uncertainty as epistemic uncertainty. There are host of issues in safety analysis where handling of uncertain issue is more important than quantification. These rather qualitative aspects of uncertainty or cognitive uncertainty need to be addressed by having safety provisions in the plant. The integrated risk-based framework proposes to address these issues.

3 Major Issues for Implementation of Integrated Risk-Based Engineering Approach

One of the major issues that forms the bottleneck to realize application of risk-based engineering is characterization of uncertainty associated with data and model. Even though for internal events there exist reasonable data, characterization of external events poses major challenges. Apart from this availability of probabilistic criteria and safety margins as nation policy for regulation, issues related to new and advanced features of the plants like passive system modeling, digital system reliability in general, and software system reliability pose special challenges. Relatively large uncertainties associated with common cause failures of hardware systems and human action considerations particularly with accident scenarios are one of the major issues.

It can be argued that reduction of uncertainty associated with data and model in and characterization of uncertainties particularly for rare events where data and model are either not available or inadequate is one of the major challenges in implementation of integrated risk-based applications.

4 Uncertainty Analysis in Support of Integrated Risk-Based Decisions: A Brief Overview

Even though there are many definitions of uncertainty given in literature, the one which suits the risk assessment or rather the risk-informed/risk-based decisions has been given by ASME as "representation of the confidence in the state of knowledge about the parameter values and models used in constructing the PSA" [2]. Uncertainty characterization in the form of qualitative assessment and assumptions has been inherent part of risk assessment. However, as the data, tools, and statistical methods developed over the years, the quantitative methods for risk assessment came into being. The uncertainty in estimates has been recognized as inherent part of any analysis results. The actual need of uncertainty characterization was felt while addressing many real-time decisions related to assessment of realistic safety margin.

The major development has been in respect of classification/categorizing uncertainty based on the nature of uncertainty, viz., aleatory uncertainty and epistemic uncertainty. Uncertainty associated due to randomness (chance phenomenon) in the system/process is referred as aleatory uncertainty. This type of uncertainty arises due to inherent characteristic of the system/process and data. Aleatory uncertainty cannot be reduced as it is inherent part of the system. This is the reason aleatory uncertainty is also called irreducible uncertainty [3]. The nature of this uncertainty can be explained further by some examples like results of flipping a coin – head or tail it is matter of chance. Chances of diesel set to start on demand it could be success or failure, etc. On the other hand, the uncertainty associated due to lack of knowledge is referred as epistemic uncertainty. This uncertainty can be reduced either by performing additional number of experiments, more data, and information about the system. This uncertainty is more of subjective in nature.

If we look at the modeling and analysis methods in statistical distributions, we will note that probability distributions provide one of the important and fundamental mechanisms to characterize uncertainty in data by estimating upper bound and lower bounds of the data [4, 5]. Hence, various probability distributions are central to characterization of uncertainty. Apart from this, fuzzy logic approach also provides an important tool to handle uncertainty where the information is imprecise and where probability approach is not adequate to address the issues [6]. Like in many situations, the performance data on system and components is not adequate, and the only input that is available is opinion of the domain experts. Apart from this, there are many situations where it is required to use linguistic variables as an input. Fuzzy approach suits these requirements.

There are many approaches for characterization/modeling of uncertainty. Keeping in the nature of problem is being solved; a judicious selection of applicable method has to be made [7]. Even though the list of approaches listed here is not exhaustive, commonly, the following methods can be reviewed as possible candidate for uncertainty modeling:

1. Probabilistic approach
   - Frequentist approach
   - Bayesian approach
2. Evidence theory – imprecise probability approach
   - Dempster-Shafer theory
   - Possibility theory – fuzzy approach
3. Structural reliability approach (application oriented)
   - First-order reliability method
   - Stochastic response surface method
4. Other nonparametric approaches (application specific)
   - Wilk's method
   - Bootstrap method

Each of the above methods has some merits and limitations. The available literature shows that general practice for uncertainty modeling in PSA is through the probabilistic approach [8–11]. Application of probabilistic distributions to address aleatory as well as epistemic uncertainty forms the fundamentals of this approach. There are two major basic models, classical model which is also referred as frequentist models and subjective model. Frequentist model tends to characterize uncertainty using probability bounds at component level. This approach has some limitations like no information on characterization of distribution and nonavailability of data and information for tail ends. The most popular approach is subjective approach implemented through Bayes theorem called Bayesian approach which allows subjective opinion of the analysts as "prior" knowledge to be integrated with the data or evidence that is available to provide with the estimate of the event called "posteriori" estimate [12]. Even though this approach provides an improved framework for uncertainty characterization in PSA modeling compared to frequentist approach, there are arguments against this approach. The subjectivity that this approach carries with it has become the topic of debate in respect of regulatory decisions. Hence, there are arguments in favor of application of methods that use evidence theory which works on to address "imprecise probabilities" to characterize uncertainty [13–18]. Among the existing approach for imprecise probability, the one involving "coherent imprecise probability" which provides upper and lower bound reliability estimates has been favored by many researchers[19].

Among other methods listed above, each one has its merit for specific applications like response surface method, and FORMs (first-order reliability methods) are used generally for structural reliability modeling [20]. There are some application-specific requirements, like problems involving nonparametric tests where it is not possible to assume any particular distribution (as is the case with probabilistic methods); in such cases, bootstrap nonparametric approach is employed [21]. Even though this method has certain advantages, like it can draw inference even from small samples, estimation of standard error, it is computationally intensive and may become prohibitive for complex problems that are encountered in risk-based applications. Other nonparametric methods that find only limited application in risk-based engineering do not form the scope of this chapter.

From the above, it could be concluded that the probabilistic methods that include classical statistical methods and Bayesian approach form the major approaches for uncertainty characterization in risk-based approach. At times, fuzzy-based approach is used as part where the data deals with imprecise input in the form of linguistic variables. However, fuzzy logic applications need to be scrutinized for methodology that is used to design the membership functions as membership functions have found to introduce subjectivity to final estimates.

5 Major Features of RB Approach Relevant Uncertainty Characterization

Keeping in view the subject of this chapter, i.e., uncertainty characterization for risk-based approach, it is required to understand the nature of major issues that need to be addressed in risk-based characterization and accordingly look for the appropriate approach. At the outset, there appears general consensus that on a case to case basis most of the above listed approaches may provide efficient solution for the specific domain. However, here the aim is to focus on the most appropriate approach that suits the risk-based applications. The PSA in general and Level 1 PSA in particular, as part of risk-based approaches, have following major features [22–25]:

(a) The probabilistic models basically characterize randomness in data and model, and hence, the model at integrated level requires aleatory uncertainty characterization.
(b) The probabilistic models are basically complex and relatively large in size compared to the models developed for other engineering applications.

(c) The uncertainty characterization for PSA models requires an efficient simulation tool/method.
(d) The approach should allow characterization epistemic component of data as well as model.
(e) Confidence intervals for the component, human errors, etc., estimated using statistical analysis form the input for the probabilistic models.
(f) Major part of modeling is performed using fault tree and event tree approaches; hence, the uncertainty modeling approach should be effective for these models.
(g) There should be provision to integrate the prior knowledge about the event for getting the posteriori estimates, i.e., the approach should be able to handle subjective probabilities.
(h) Often, instead of quantitative estimates, the analysts come across situations where it becomes necessary to derive quantitative estimates through "linguistic" inputs. Hence, the framework should enable estimation of variables based on qualitative inputs.
(i) Evaluation of deterministic variable forms part of risk assessment. Hence, provision should exist to characterize uncertainty for structural, thermal hydraulic, and neutronics assessment.
(j) Sensitivity analysis for verifying impact of assumptions, data, etc., forms the fundamental requirements.
(k) The PSA offers improved framework for assessment of safety margin – a basic requirement for risk-based applications.
(l) Even though PSA provides an improved methodology for assessment of common cause failure and human factor data, keeping in view the requirements of risk-based applications further consolidation of data and model is required.

Apart from this, there are specific requirements, like modeling for chemical, environmental, geological, and radiological dose evaluation, which also need to be modeled. As can be seen above, the uncertainty characterization for risk assessment is a complex issue.

5.1 Uncertainty Propagation

The other issue in characterizing uncertainty is consideration of effective methodology for propagation of uncertainty. Here, the literature shows that Monte Carlo simulation and Latin hypercube approach form the most appropriate approach for uncertainty propagation [26]. Even though these approaches are primarily been used for probabilistic methods, there are applications where simulations have been performed in evidence theory or application where the priori has been presented as interval estimates [27]. The risk-based models are generally very complex in terms of (a) size of the model, (b) interconnections of nodes and links, (c) interpretation of results, etc. The available literature shows that the Monte Carlo simulation approach is extensively being used in many applications; it also labeled this method as computationally intensive and approximate in natures. Even with these complexities, the risk-based applications, both Latin hypercube and Monte Carlo, have been working well. Even though it is always expected that higher efficiency in uncertainty modeling is required for selected cases, for overall risk-based models, these approaches can be termed as adequate. In fact, we have developed a risk-based configuration system in which the uncertainty characterization for core damage frequency has been performed using Monte Carlo simulation [28].

6 Uncertainty Characterization: Risk-Informed/Risk-Based Requirements

The scope and objective of risk-informed/risk-based applications determine the major element of Level 1 PSA. However, for the purpose of this chapter, let us consider that development of base Level 1 PSA for regulatory review as the all-encompassing study. The scope of this study includes full-scope PSA which means considerations of (a) internal event (including loss of off-site power and interfacing loss of coolant accident, internal floods, and internal fire); (b) external event, like seismic events, external impacts, and flood; (c) full-power and shut-down PSA; and (d) reactor core as the source of radioactivity (fuel storage pool not included) [25].

The point to be remembered here is that uncertainty characterization should be performed keeping in view the nature of applications [2, 29]. For example, if the application deals with the estimation of surveillance test interval, then the focus will start right from uncertainty in initiating event that demands automatic action of a particular safety system, unavailability for safety significant component, human actions, deterministic parameters that determine failure/success criteria, assumptions which determine the boundary condition for the analysis, etc.

An important reference that deals with uncertainty modeling is USNRC (United State Nuclear Regulatory Commission) document NUREG-1856 (USNRC, 2009) which provides guidance on the treatment of uncertainties in PSA as part of risk-informed decisions [29]. Though the scope of this document is limited to light water reactors, the guidelines with little modification can be adopted for uncertainty modeling in either CANDU (CA-Nadian Deuterium Uranium reactor)/PHWR (pressurized heavy water reactor) or any other Indian nuclear plants. In fact, even though this document provides guidelines on risk-informed decisions, requirements related to risk-based applications can be easily be modeled giving due considerations to the emphasis being placed on the risk metrics used in PSA. Significant contribution of ASME/ANS framework includes incorporation of "state-of-knowledge correlation" [2] which means the correlation that arises between sample values when performing uncertainty analysis for cut sets consisting of basic events using a sampling approach such as the Monte Carlo method; when taken into account, this results, for each sample, in the same value being used for all basic event probabilities to which the same data applies.

As for the standardization of risk-assessment procedure and dealing with uncertainty issues concerned, the PSA community finds itself in relatively comfortable position. The reason is that there is a consensus at international level as to which uncertainty aspects need to be addressed to realize certain quality criteria in PSA applications. The three major references that take care of this aspect are (a) ASME (American Society of Mechanical Engineers)/ANS (American Nuclear Society) Standard on PSA Applications [2], (b) IAEA-TECDOC-1120 (International Atomic Energy Agency-Technical Document) on Quality Attribute of PSA applications [30], and (c) various NEA (Nuclear Energy Agency) documents on PSA [24]. Any PSA applications to qualify as "Quality PSA" need to conform to these quality attributes as laid out for various elements of PSA. For example, the ASME/ANS code provides a very structured framework, wherein there are higher level attributes for an element of PSA, then there are specific attributes that support the higher level attributes, etc. These attributes enable formulating a program in the form of checklists that need to be fulfilled in terms of required attributes to achieve conformance quality level for PSAs. The examples of quality attributes that are required to assure uncertainty analysis requirements following are some examples from ASME/ANS in respect of the PSA element – Initiating event (IE) Modeling.

Examples of some lower level specific attributes from ASME/ANS include:

ASME/ANS attribute IE-C4: "When combining evidence from generic and plant-specific data, USE a Bayesian update process or equivalent statistical process. JUSTIFY the selection of any informative prior distribution used on the basis of industry experience."

Similarly,

ASME/ANS attribute IE-C3: *CALCULATE the initiating event frequency accounting for relevant generic and plant-specific data unless it is justified that there are adequate plant-specific data to characterize the parameter value and its uncertainty.*

Also, the lower support requirement IE-D3 documents the sources of model uncertainty and related assumptions.

The USNRC guide as mentioned above summarizes in details the uncertainty related to supporting requirements of ASME/ANS documents systematically. For details, these documents may be referred. Availability of this ASME/ANS standard, NEA documents, and IAEA-TECDOC is one of the important milestones for risk-based/risk-informed applications as these documents provide an important tool toward standardization of and harmonization of risk-assessment process in general and capturing of important uncertainty assessment aspects that impact the results and insights of risk assessment.

7 Decisions Under Uncertainty

At this point, it is important to understand that the uncertainty in engineering systems creeps basically from two sources, viz., noncognitive generally referred as quantitative uncertainty and cognitive referred as qualitative uncertainty [31].

The major part of this chapter has so far dealt with the noncognitive part of the uncertainty, i.e., uncertainty due to inherent randomness (aleatory) and uncertainty due to lack of knowledge (epistemic). We had enough discussions on this type of uncertainty. However, unless we address the sources of uncertainty due to cognitive aspects, the topic of uncertainty has not been fully addressed. The cognitive uncertainty caused due to inadequate definition of parameters, such as structural performance, safety culture, deterioration/degradation in system functions, level of skill/knowledge base, and experience staff (design, construction, operation, and regulation) [32]. The fact is that dealing with uncertainty using statistical modeling or any other evidence-based approach including approaches that deal with precise or imprecise probabilities has their limitations and cannot address issues involving vagueness of the problem arising from missing information and lack of intellectual abstraction of real-time scenario, be it regulatory decisions, design-related issues, or operational issues. The reason for this is that traditional probabilistic and evidence-based methods for most of the time deal with subjectivities, perceptions, and assumptions that may not form part of the real-time scenarios that require to address cognitive part of the uncertainty. Following subsections bring out the various aspects of cognitive/qualitative part of the uncertainty and methods to address these issues.

7.1 Engineering Design and Analysis

The issues related to "uncertainty" have been part of engineering design and analysis. The traditional working stress design (WSD)/allowable stress design (ASD) in civil engineering deal with uncertainty by defining "suitable factor." The same factor of safety is used for mechanical design to estimate the allowable stress (AS = Yield Stress/FS). This FS accounts for variation in material properties, quality-related issues, degradation during the design life, modeling issues and variation in life cycle loads, and lack of knowledge about the system being designed. The safety factor is essentially based on past experience but does not guarantee safety. Another issue is this approach is highly conservative in nature.

It is expected that an effective design approach should facilitate trade-off between maximizing safety and minimizing cost. Probabilistic- or reliability-based design allows this optimization in an efficient manner. The design problems require treatment of both cognitive and noncognitive sources of uncertainty. It should be recognized that the designer's personal preferences or subjective choices can be source of uncertainties which bring in cognitive aspect of uncertainty. Statistical aspects like variability in assessment of loads, variation in material properties, and extreme loading cycles are the source of noncognitive uncertainties.

In probabilistic-based design approach, considerations of uncertainty when modeled as stress-strength relation for reliability-based design form an integral part of design methodology. The Load and Resistance Factor Design (LRFD), first-order reliability methods (FORM), and second-order reliability methods (SORM) are some of the application of probabilistic approach structural design and analysis. Many of civil engineering codes are based on probabilistic considerations. The available literature shows that design and analysis using probabilistic-based structural reliability approach have matured into an "engineering discipline" [20], and new advances and research have further strengthened this area [32].

The Level 1 PSA models are often utilized in support of design evaluation. During design stage, often complete information and data are not available. This leads to higher level of uncertainty in estimates. On the other hand, the traditional approach using deterministic design methodology involves use of relatively higher safety factors to compensate for the lack of knowledge. The strength of PSA framework is that it provides a systematic framework that allows capturing of uncertainties in data, model, and uncertainty due to missing or fuzzy inputs. Be it probabilistic or evidence-based tools and methods, it provides an improved framework for treatment of uncertainty. Another advantage of PSA framework is that it allows propagation of uncertainty from component level to system level and further up to plant level in terms of confidence bounds in for system unavailability/ initiating event frequency and core damage frequency, respectively.

7.2 *Management of Operational Emergencies*

If we take lessons from the history of nuclear accidents in general and the three major accidents, viz., TMI (Three Mile Island) in 1979, Chernobyl in 1986, and the recent one Fukushima in 2011, it is clear that real-time scenario always require some emergency aids that respond to the actual plant parameters in a given "time window." Even though probabilistic risk analysis framework may address these scenarios, it can only addresses the modeling part of the safety analysis. It is also required to consider the qualitative or cognitive uncertainty aspects and its characteristics for operational emergency scenario.

The major characteristics of the operational emergencies can include:

(a) Deviation of plant condition from normal operations that require safety actions, it could be plant shutdown, actuation of shutdown cooling, etc.
(b) Flooding of plant parameters which include process parameters crossing its preset bounds, parameter trends and indications
(c) Available "time window" for taking a grasp of the situation and action by the operator toward correcting the situation
(d) Feedback in terms of plant parameters regarding the improved/deteriorated situations
(e) Decisions regarding restorations of systems and equipments status if the situation is moving toward normalcy
(f) Decision regarding declaration of emergency which requires a good understanding whether the situation requires declaration of plant emergency, site emergency, or off-site emergency
(g) Interpretation of available safety margins in terms of time window that can be used for designing the emergency operator aids

As the literature shows, that responding to accident/off-normal situations as characterized above calls for modeling that should have following attributes:

(a) Modeling of the anticipated transients and accident conditions in advance such that knowledge-based part is captured in terms of rules/heuristics as far as possible.
(b) Adequate provision to detect and alert plant staff for threat to safety functions in advance.
(c) Unambiguous and automatic plant symptoms based on well-defined criterion like plant safety limits and emergency procedures that guide the operators to take the needed action to arrest further degradation in plant condition.
(d) Considering the plant limits of plant parameters assessment of actual time window that is available for applicable scenarios.
(e) The system for dealing with emergency should take into plant-specific attributes, distribution of manpower, laid down line of communications, other than the standard provisions, the tools, methods, and procedures that can be applied for planned and long-term or extreme situations.
(f) Heuristics on system failure criteria using available safety margins.

Obviously, ball is out of "uncertainty modeling" domain and requires to address the scenarios from other side, i.e., taking decisions such that action part in real-time scenario compensates for the missing knowledge base and brings plant to safe state.

The answer to the above situation is development of knowledge-based systems that not only capture the available knowledge base but also provide advice to maintain plant safety under uncertain situation by maintaining plant safety functions. It may please be noted that here we are not envisaging any role for "risk-monitor" type of systems. We are visualizing an operator support system which can fulfill the following requirements (the list is not exhaustive and only presents few major requirements):

1. Detection of plant deviation based on plant symptoms.
2. The system should exhibit intelligent behavior, like reasoning, learning from the new patterns, conflict resolution capability, pattern recognition capability, and parallel processing of input and information.
3. The system should be able to predict the time window that is available for the safety actions.
4. Takes into account operator training and optimizes the graphic user interface (GUI).
5. The system should be effective in assessment of plant transients – it calls for parallel processing of plant symptoms to present the correct plant deviation.
6. The system should have adequate provision to deal with uncertain and incomplete data.
7. The presentation of results of the reasoning with confidence limits.
8. It should have an efficient diagnostics capability for capturing the basic cause (s) of the failures/transients.
9. The advice should be presented with adequate line of explanations.
10. The system should be interactive and use graphics to present the results.

11. Provisions for presentation of results at various levels, like abstract level advice (like open MV-3001 and Start P-2) to advise with reasonable details (like Open ECCS Valve MV-3001 located in reactor basement area and Start Injection Pump P-2, it can be started from control room L panel).

Even though there are many examples of R&D efforts on development of intelligent operator advisory systems for plant emergencies, readers may refer to the paper by Varde et al. for further details [33]. Here, the probabilistic safety assessment framework is used for knowledge representation. The fault tree models of PSA are used for generating the diagnostics, while the event tree models are used to generate procedure synthesis for evolving emergencies. The intelligent tools like artificial neural network approach are used for identification of transients, while the knowledge-based approach is used for performing diagnostics.

As can be seen above, the uncertain scenarios can be modeled by capturing either from the lessons learned from the past records for anticipated events. Even for the rare events where uncertainty could be of higher levels, the symptom-based models which focus on maintaining the plant safety functions can be used as model plant knowledge base.

8 Regulatory Reviews

In fact, the available literature on decisions under uncertainty has often focused on the regulatory aspects [29]. One of the major differences between operational scenarios and regulatory reviews or risk-informed decisions is that there generally is no preset/specified time window for decisions that directly affect plant safety. The second difference is that in regulatory or risk-informed decisions requires collective decisions and basically a deliberative process unlike operational emergencies where the decisions are taken often by individuals or between a limited set of plant management staff where the available time window and some time resources are often the constraints. Expert elicitation and treatment of the same often form part of the risk-informed decisions. Here, the major question is "what is the upper limit of spread of confidence bounds" that can be tolerated in the decision process. In short, "how much uncertainty in the estimates" can be absorbed in the decision process? It may be noted that the decisions problem should be evaluated using an integrated approach where apart from probabilistic variables even deterministic variables should be subjected to uncertainty analysis. One major aspect of risk assessment from the uncertainty point of view is updating the plant-specific estimates with generic prior data available either in literature or from other plants. This updating brings in subjectivity to the posteriori estimates. Therefore, it is required to justify and document the prior inputs. Bayesian method coupled with Monte Carlo simulation is the conventional approach for uncertainty analysis. The regulatory reviews often deal with inputs in the form of linguistic variables or "perceptions" which require perception-based theory of probabilistic reasoning with imprecise probabilities [13]. In such scenarios, the classical probabilistic approach alone does not work. The literature shows that application of fuzzy logic offers an effective tool to address qualitative and imprecise inputs [34].

The assumptions often form part of any risk-assessment models. These assumptions should be validated by performing the sensitivity analysis. Here, apart from independent parameter assessment, sensitivity analysis should also be carried out for a set of variables. The formations of set of variables require a systematic study of the case under considerations.

The USNRC document NUREG-1855 on "Guidance on the treatment of Uncertainties Associated with PSAs in Risk-informed Decision Making" deals with the subject in details, and readers are recommended to refer to this document for details [29].

9 Conclusions

The available literature shows that there is an increasing trend toward the use of risk assessment or PSA insights in support of decisions. This chapter proposes a new approach called integrated risk-based engineering for dealing with safety of nuclear plants in an integrated and systematic manner. It is explained that this approach is a modification of the existing risk-informed/risk-based approach. Apart from application of PSA models, probabilistic treatment to traditional deterministic variables, success, and failure criteria, assessment of safety margins in general and treatment uncertainties in particular, forms part of the integrated risk-based approach.

There is general consensus that strengthening of uncertainty evaluation is a must for realizing risk-based application. It is expected that integrated risk-based approach will provide the required framework to implement the decisions. This chapter also argues that apart from probabilistic methods, evidence-based approaches need to be used to deal with "imprecise probabilities" which often form important input for the risk-based decisions.

Further other issue that this chapter discusses is that various methods, be it probabilistic or evidence based, cannot provide complete solution for issues related to uncertainty. There are qualitative or cognitive issues that need to be addressed by incorporating management tools for handling real-time situations. This is true for operational applications.

Finally, this chapter drives the point that both the quantitative and quantitative aspects need to be addressed to get toward more holistic solutions. Further research is needed to deal with imprecise probability, while cognitive aspects form the cornerstone of uncertainty evaluation.

Acknowledgments I sincerely thank Shri R. C. Sharma, Head, RRSD, BARC, for his constant support, guidance, and help without which Reliability and PSA activities in Life Cycle Reliability Engineering Lab would not have been possible. I also thank my SE&MTD Section colleagues Mr. Preeti Pal, Shri N. S. Joshi, Shri A. K. Kundu, Shri D. Mathur, and Shri Rampratap for their cooperation and help. I also thank Shri Meshram, Ms Nutan Bhosale, and Shri M. Das for their assistance in various R&D activities in the LCRE lab.

References

1. Chapman JR et al (1999) Challenges in using a probabilistic safety assessment in risk-informed process. Reliab Eng Syst Saf 63:251–255
2. American Society of Mechanical Engineers/American Nuclear Society (2009) Standards for level 1 large early release frequency in probabilistic risk assessment for nuclear power plant applications. ASME/ANS RA-Sa-2009, March 2009
3. Parry GW (1996) The characterization of uncertainty in probabilistic risk assessments of complex systems. Reliab Eng Syst Saf 54:119–126
4. Modarres M (2006) Risk analysis in engineering – techniques, tools and trends. CRC-Taylor & Francis Publication, Boca Raton
5. Modarres M, Mark K, Vasiliy K (2010) Reliability engineering and risk analysis – a practical guide. CRC Press Taylor & Francis Group, Boca Raton
6. Mishra KB, Weber GG (1990) Use of fuzzy set theory for level 1 studies in probabilistic risk assessment. Fuzzy Sets Syst 37(2):139–160
7. Kushwaha HS (ed) (2009) Uncertainty modeling and analysis. A Bhabha Atomic Research Centre Publication, Mumbai
8. Weisman J (1972) Uncertainty and risk in nuclear power plant. Nucl Eng Des 21 (1972):396–405. North–Holland Publishing Company
9. Gábor Lajtha (VEIKI Institute for Electric Power Research, Budapest, Hungary), Attila Bareith, Előd Holló, Zoltán Karsa, Péter Siklóssy, Zsolt Téchy (VEIKI Institute for Electric Power Research, Budapest, Hungary) "Uncertainty of the Level 2 PSA for NPP Paks"
10. Pate-Cornell ME (1986) Probability and uncertainty in nuclear safety decisions. Nucl Eng Des 93:319–327
11. Nilsen T, Aven T (2003) Models and model uncertainty in the context of risk analysis. Reliab Eng Syst Saf 79:309–317
12. Siu NO, Kelly DL (1998) Bayesian parameter estimation in probabilistic risk assessment. Reliab Eng Syst Saf 62:89–116
13. Zadeh LA (2002) Towards a perception-based theory of probabilistic reasoning with imprecise probabilities. J Stat Plan Inference 105:233–264
14. Peter W, Lyle G, Paul B (1996) Analysis of clinical data using imprecise prior probabilities. The Statistician 45(4):457–485
15. Peter W (2000) Towards a unified theory of imprecise probability. Int J Approx Reason 24:125–148
16. Troffaes MCM (2007) Decision making under uncertainty using imprecise probabilities. Int J Approx Reason 45:17–29
17. George KJ (1999) Uncertainty and information measures for imprecise probabilities: an overview. First international symposium on imprecise probabilities and their applications, Belgium, 29 June–2 July 1999
18. Caselton FW, Wuben L (1992) Decision making with imprecise probabilities: Dempster-Shafer theory and application. AGU: Water Resour Res 28(12):3071–3083
19. Kozine IO, Filimonov YV (2000) Imprecise reliabilities: experience and advances. Reliab Eng Syst Saf 67(1):75–83
20. Ranganathan R (1999) Structural reliability analysis and design. Jaico Publishing House, Mumbai
21. Davison AC et al (1997) Bootstrap methods and their applications. Cambridge University Press, Cambridge
22. Winkler RL (1996) Uncertainty in probabilistic risk assessment. Reliab Eng Syst Saf 54:127–132
23. Daneshkhah AR. Uncertainty in probabilistic risk assessment: a Review
24. Nuclear Energy Agency (2007) Use and development of probabilistic safety assessment. Committee on Safety of Nuclear Installations, NEA/CSNI/R(2007)12, November 2007

25. IAEA (2010) Development and application of level 1 probabilistic safety assessment for nuclear power plants. IAEA safety standards – specific safety guide no. SSG-3, Vienna
26. Atomic Energy Regulatory Board (2005) Probabilistic safety assessment for nuclear power plants and research reactors. AERB draft manual. AERB/NF/SM/O-1(R-1)
27. Weichselberger K (2000) The theory of interval-probability as a unifying concept for uncertainty. Int J Approx Reason 24:149–170
28. Agarwal M, Varde PV (2011) Risk-informed asset management approach for nuclear plants. 21st international conference on structural mechanics in reactor technology (SMiRT-21), India, 6–11 Nov 2011
29. Drouin M, Parry G, Lehner J, Martinez-GuridiG, LaChance J, Wheeler T (2009) Guidance on the treatment of uncertainties associated with PSAs in risk-informed decision making (Main report). NUREG-1855 (Vol.1). Office of Nuclear Regulatory Research Office of Nuclear Reactor Regulation, USNRC, USA.
30. International Atomic Energy Agency. Quality attributes for PSA applications. IAEA-TECDOC-1120. IAEA, Vienna
31. Assakkof I. Modeling for uncertainty: ENCE-627 decision analysis for engineering. Making hard decisions. Dept. of Civil Engineering, University of Maryland, College Park
32. Haldar A, Mahadevan S (2000) Probability reliability and statistical methods i±n engineering design. Wiley, New York
33. Varde PV et al (1996) An integrated approach for development of operator support system for research reactor operations and fault diagnosis. Reliab Eng Syst Saf 56
34. Iman K, Hullermeier E (2007) Risk assessment system of natural hazards: a new approach based on fuzzy probability. Fuzzy Set Syst 158:987–999

Past, Resent, and Future of Structural Health Assessment

Achintya Haldar and Ajoy Kumar Das

Abstract Past, present, and future of structural health assessment (SHA) concepts and related areas, as envisioned by the authors, are briefly reviewed in this chapter. The growth in the related areas has been exponential covering several engineering disciplines. After presenting the basic concept, the authors discussed its growth from infancy, that is, hitting something with a hammer and listening to sound, to the use of most recent development of wireless sensors and the associated advanced signal processing algorithms. Available SHA methods are summarized in the first part of this chapter. The works conducted by the research team of the authors are emphasized. Later, some of the future challenges in SHA areas are identified. Since it is a relatively new multidisciplinary area, the education component is also highlighted at the end.

Keywords Structural health assessment • Kalman filter • Substructure • System identification • Uncertainty analysis • Sensors

1 Introduction

The nature and quality of infrastructure have always been one of the indicators of sophistication of a civilization. There is no doubt that we are now at a historical peak. However, keeping the infrastructure at its present level has been a major challenge due to recent financial strain suffered by the global community. We do not have adequate resources to build new infrastructure or replace the aged ones that are over their design lives. The most economical alternative is found to be extending the life of existing infrastructure without compromising our way of living

A. Haldar (✉) • A.K. Das
Department of Civil Engineering and Engineering Mechanics,
University of Arizona, Tucson, AZ, USA
e-mail: haldar@u.arizona.edu; akdas@email.arizona.edu

S. Chakraborty and G. Bhattacharya (eds.), *Proceedings of the International Symposium on Engineering under Uncertainty: Safety Assessment and Management (ISEUSAM - 2012)*,
DOI 10.1007/978-81-322-0757-3_14,

and without exposing public to increased risk. This has been one of the major challenges to the engineering profession and attracted multidisciplinary research interests. The main thrust has been to locate defects in structures at the local element level and then repair them or replace the defective elements, instead of replacing the whole structure. Several advanced theoretical concepts required to detect defects have been proposed. At the same time, improved and smart sensing technologies, high-resolution data acquisition systems, digital communications, and high-performance computational technologies have been developed for implementing these concepts. The general area is now commonly known as structural health assessment (SHA) or structural health monitoring (SHM). In spite of these developments in analytical and sensor technologies, the implementations of these concepts in assessing structural health have been limited due to several reasons. An attempt has been made here to identify some of the major works (emphasizing analytical), their merits and demerits, contributions made by the research team of the authors, and future challenges.

2 Concept of Structural Health Assessment

All civil engineering structures, new and old, may not totally satisfy the intents of the designers. Minor temperature cracks in concrete or lack of proper amount of pretension in bolts cannot completely be eliminated. In that sense, all structures can be assessed as defective. Our past experiences indicate that presence of minor defects that do not alter the structural behavior may not be of interest to engineers. Considering only major defects, all of them are not equally important. Their locations, numbers, and severities will affect the structural behavior. Thus, the concept behind SHA can be briefly summarized as locating major defects, their numbers, and severities in a structure at the local element level. For the sake of completeness of this discussion, available SHA procedures are classified into four levels as suggested by Rytter [39]. They are as follows: level 1 – determination if damage is present in a structure, level 2 – determination of geometric location of the damage, level 3 – assessment of severity of the damage, and level 4 – prediction of remaining life of the structure.

3 Structural Health Assessment: Past

Structural health assessment has been practiced for centuries. Ever since pottery was invented, cracks and cavities in them were detected by listening to the sound generated when tapped by fingers. A similar sonic technique was used by blacksmiths to establish the soundness of the metals they were shaping. Even today, it is not uncommon to observe that inspectors assess structural health by hitting structures with a hammer and listening to the sounds they produce. These types of inspections, with various levels of sophistication, can be broadly termed as nondestructive evaluation (NDE) of health of a structure.

3.1 Early Developments in SHA

Although the awareness of the scientific concepts of many NDE technologies began during 1920s, they experienced major growth during and after the Second World War. However, there had been always problems in the flow of NDE research to everyday use [4]. Besides the use of visual testing (VT), early developments of instrument-based nondestructive detection of defects include penetrate testing (PT), magnetic particle testing (MPT), radiographic testing (RT), ultrasonic testing (UT), Eddy current testing (ET), thermal infrared testing (TIR), and acoustic emission testing (AE). Many of them required the damage/irregularity to be exposed to the surface or within small depth from the open surface. Some of them required direct contact of sensors with the test surface [22]. They mainly focussed on the "hot spot" areas or objects readily available for testing. For instance, RT has been routinely used for detection of internal physical imperfections such as voids, cracks, flaws, segregations, porosities, and inclusions in material at selective location(s). Most of these methods are non-model based, that is, the structure need not be mathematically modeled to identify location and severity of defects.

3.2 Transition from Past to Present: New Challenges

For most large civil infrastructure, the location(s), numbers, and severity of defect(s) may not be known in advance, although sometimes they can be anticipated using past experiences. Also, sometimes, defects may be hidden behind obstructions, for example, cracks in steel members hidden behind fire-proofing materials. Thus, instrument-based nondestructive testing (NDT) may not be practical if the inspector does not know what to inspect or the location of defect is not known *a priori*. During 1970s, detection of cracks was a major thrust. Subsequently, determination of crack size in order to compare with the critical crack size added another level of challenge to the engineering profession. In any case, inspection of "hot spot" areas limited their application potential. Subsequently, a consensus started developing about the use of measured responses to assess current structural health, as discussed next.

3.3 Model-Based SHA

Some of the deficiencies in non-model-based approaches can be removed by using model-based techniques. The aim of this approach is to predict the parameters of the assumed mathematical model of a physical system; that is, the system is considered to behave in predetermined manner represented in algorithmic form using the governing differential equations, finite element (FE) discretization, etc.

The changes in the parameters should indicate the presence of defects. To implement the concept, responses need to be measured by exciting the structure statically or dynamically.

3.3.1 SHA Using Static Responses

Because of its simplicity, initially SHA using static responses were attempted. Static responses are generally measured in terms of displacements, rotations, or strains, and the damage detection problems are generally formulated in an optimization framework employing minimization of error between the analytical and measured quantities. They mostly use FE model for structural representation. Three classes of error functions are reported in the literature: displacement equation error function, output error function, and strain output error function [41]. Recently, Bernal [3] proposed flexibility-based damage localization method, denoted as the damage locating vector (DLV) method. The basic approach is the determination of a set of vectors (i.e., the DLVs), which when applied as static forces at the sensor locations, no stress will be induced in the damaged elements. The method can be a promising damage detection tool as it allows locating damages using limited number of sensor responses. It was verified for truss elements, where axial force remains constant through its length. However, the verification of the procedure for real structures using noise-contaminated responses has yet to be completed.

There are several advantages of SHA using static responses including that the amount of data needed to be stored is relatively small and simple, and no assumption on the mass or damping characteristics is required. Thus, less errors and uncertainties are introduced into the model. However, there are several disadvantages including that the number of measurement points should be larger than the number of unknown parameters to assure a proper solution. Civil engineering structures are generally large and complex with extremely high overall stiffness. It may require extremely large static load to obtain measurable deflections. Fixed reference locations are required to measure deflections which might be impractical to implement for bridges, offshore platforms, etc. Also, static response-based methods are sensitive to measurement errors [1, 2].

3.3.2 SHA Using Dynamic Responses

Recent developments in SHA are mostly based on dynamic responses. There are several advantages of this approach. It is possible to excite structures by dynamic loadings of small amplitude relative to static loadings. In some cases, ambient responses caused by natural sources, for example, wind, earthquake, and moving vehicle, can be used. If acceleration responses are measured, they eliminate the need for fixed physical reference locations. They perform well in presence of high measurement errors [14].

Earlier works on SHA using dynamic responses are mostly modal information based [5, 15, 16, 38, 42]. Changes in modal properties, that is, natural frequencies, damping, and mode shape vectors, or properties derived from these quantities are used as damage indicators. Doebling et al. [15] presented various methods for damage identification including methods based on changes in frequency, mode shapes, mode shape curvature, and modal strain energy. Sohn et al. [42] updated the above report and discussed procedures based on damping, antiresonance, Ritz vectors, a family of autoregressive moving average (ARMA) models, canonical variate analysis, nonlinear features, time-frequency analysis, empirical mode decomposition, Hilbert transform, singular value decomposition, wave propagation, autocorrelation functions, etc. More complete information on them can be found in the literature cited above.

Natural frequency-based methods use change in the natural frequency as the primary feature for damage identification. They are generally categorized as forward problem or inverse problem. The forward problems deal with determination of changes in frequency based on location and severity of damage, whereas the inverse problems deal with determination of damage location and size based on natural frequency measurement. Among the mode shape-based procedures for damage detection, the mode shape/curvature methods generally use two approaches: traditional analysis of mode shape or curvature and modern signal processing methods using mode shapes or curvature. Modal strain energy-based procedures consider fractal modal energy for damage detection [16]. Methods based on damping have the advantage that a larger change in damping can be observed due to small cracks. Also, it is possible to trace nonlinear, dissipative effects produced by the cracks. However, damping properties have not been studied as extensively as natural frequencies and mode shapes [42]. Methods based on dynamically measured flexibility detect damages by comparing flexibility matrix synthesized using the modes of damaged structure to that of undamaged structure or flexibility matrix from FE analysis. The flexibility matrix is most sensitive to changes in the lower frequencies [15].

Modal-based approaches have many desirable features. Instead of using enormous amount of data, the modal information can be expressed in countable form in terms of frequencies and mode shape vectors. Since structural global properties are evaluated, there may be an averaging effect, reducing the effect of noise in the measurements. However, the general consensus is that modal-based approaches fail to evaluate the health of individual structural elements; they indicate overall effect, that is, whether the structure is defective or not [18, 26, 37]. For complicated structural systems, the higher-order calculated modes are unreliable, and the minimum numbers of required modes to identify the system parameters is problem dependent, limiting their applicability. The mode shape vectors may be more sensitive to defects than the frequencies, but the fact remains that they will be unable to predict which element(s) caused the changes. It was reported that even when a member breaks, the natural frequency may not change more than 5%. This type of change can be caused by the noises in the measured responses. A time-domain approach will be preferable.

3.3.3 Damages Initiated During Observations

A considerable amount of work also reported is on damages initiation time, commonly known as time-frequency methods for damage identification. The time-frequency localization capability has been applied for damage feature extractions from sudden changes, breakdown points, discontinuity in higher derivatives of responses, etc. They circumvent the modeling difficulty as they do not require the system to be identified, and the health assessment strategy often reduces to the evaluation of symptoms reflecting the presence and nature of defect [6]. Extensive study on short-time Fourier transform (STFT), Wigner-Ville distribution (WVD), pseudo Wigner-Ville distribution (PWVD), Choi-Williams distribution (CWD), wavelet transform (WT), Hilbert transform (HT), and Hilbert-Huang transform (HHT) for analyzing any nonstationary events localized in time domain has been reported in the literature. STFT is an extension of the Fourier transform allowing for the analysis of nonstationary signals by dividing it into small-time windows and analyzing each using the fast Fourier transform (FFT). The formulation provides localization in time as well as capturing frequency information simultaneously. WT has greater flexibility than STFT in terms of choosing different basis functions or mother wavelets. The wavelets have finite duration, and their energy is localized around a point in time. The WVD gives the energy distribution of a signal as a function of time and frequency; however, it has major shortcoming for multicomponent signals in terms of cross-terms. The CWD provides filtered/ smoothed version of the WVD by removing the cross-terms [40].

These studies are very interesting, but there is no general consensus about the most suitable technique. Recently, Yadav et al. [48] studied some of the time-frequency procedures for defect characterization in a wave-propagation problem. However, the fundamental limitation of STFT, WVD, PWVD, CWD, and CWT is due to the fact that they are based on Fourier analysis and can accommodate only nonstationary phenomena in the data driven from linear systems; they are not suitable to capture nonlinear distortion. In this context, the HT and HHT are suitable for nonlinear and nonstationary data. Application of Hilbert transform to nonlinear data requires the signal to be decomposed to "mono-component" condition without any smaller, riding waves. The real advantage of HT is implemented in HHT proposed by Huang et al. [25]. The procedure consists of empirical mode decomposition (EMD) and Hilbert spectral analysis (HSA). HHT clearly define nonlinearly deformed waveforms; this definition can be the first indication of the existence of damage [24]. They applied the concept for bridge health monitoring using two criteria: nonlinear characteristics of the intra-wave frequency modulations of the bridge response and frequency downshift as an indication of structural yield. Yang et al. [51] proposed two HHT-based procedures for identifying damage time instances, damage locations, and natural frequencies and damping ratios before and after occurrence of damage.

4 Structural Health Assessment: Present

In an attempt to develop an ideal SHA technique for the rapid assessment of structural health, the research team at the University of Arizona identified several desirable features considering theoretical as well as implementation issues. The team concluded that a system identification (SI)-based approach using measured dynamic response information in time domain will have the most desirable attributes. A basic SI-based approach has three essential components: (a) the excitation force(s); (b) the system to be identified, generally represented by some equations in algorithmic form such as by FEs; and (c) the output response information measured by sensors. Using the excitation and response information, the third component, that is, the system, can be identified. The basic concept is that the dynamic responses will change as the structure degrades. Since the structure is represented by FEs, by tracking the changes in the stiffness parameter of the elements, the location and severity of defects can be established.

For a structure with N dynamic degrees of freedom (DDOF), the dynamic governing equation can be written as

$$\mathbf{M}\ddot{\mathbf{x}}(t) + \mathbf{C}\dot{\mathbf{x}}(t) + \mathbf{K}\mathbf{x}(t) = \mathbf{f}(t) \tag{1}$$

where $\mathbf{K}$, $\mathbf{M}$, and $\mathbf{C}$ are $N \times N$ stiffness, mass, and damping matrix, respectively; $\mathbf{x}(t)$, $\dot{\mathbf{x}}(t)$, $\ddot{\mathbf{x}}(t)$, and $\mathbf{f}(t)$ are $N \times 1$ displacement, velocity, acceleration, and load vector, respectively, at time t. The acceleration time histories at the FE node points are expected to be measured by accelerometers. The velocity and displacement time histories can be generated by successively integrating the acceleration time histories, as suggested by Vo and Haldar [44]. Assuming mass is known, $\mathbf{K}$ matrix at the time of inspection can be evaluated. Using the information on the current elements' stiffness properties and comparing them with the "as built" or expected properties or deviation from the previous values if periodic inspections were conducted, the structural health can be assessed.

4.1 General Challenges in Time-Domain SHA

Referring to the SI concept discussed earlier, structural stiffness parameters will be estimated by using information on excitation and measured responses. It is interesting to point out that according to Maybeck [36], deterministic mathematical model and control theories do not appropriately represent the behavior of a physical system, and thus, the SI-based method may not be appropriate for SHA. He correctly pointed out three basic reasons: (a) a mathematical model is incapable of incorporating various sources of uncertainties and thus does not represent true behavior of a system, (b) dynamic systems are driven not only by controlled inputs but also by disturbances that can neither be controlled nor modeled using deterministic formulations, and (c) sensors used for data measurements cannot be perfectly devised to provide complete and perfect data about a system. These concerns and other implementation issues must be addressed before developing a SI-based SHA procedure.

Outside the controlled laboratory environment, measuring input excitation force(s) can be very expensive and problematic during health assessment of an existing structure. In the context of a SI-based approach, it will be desirable if a system can be identified using only measured response information and completely ignoring the excitation information. This task is expected to be challenging since two of the three basic components of SI process will be unknown. Responses, even measured by smart sensors, are expected to be noise contaminated. Depending on the amount of noise, the SI-based approach may be inapplicable. The basic concept also assumes that responses will be available at all DDOFs. For large structural systems, it may be practically impossible or uneconomical to instrument the whole structure; only a part can be instrumented. Thus, the basic challenge is to identify stiffness parameters of a large structural system using limited noise-contaminated response information measured at a small part of the structure. The research team successfully developed such a method in steps, as discussed next.

4.2 SHA Using Responses at All DDOFs

Using noisy responses measured at all DDOFs, Wang and Haldar [46] proposed a procedure, popularly known as iterative least squares with unknown input (ILS-UI). They used viscous damping and verified it for shear buildings. The efficiency of the numerical algorithm was improved later by introducing Rayleigh-type proportional damping, known as modified ILS-UI or MILS-UI [32]. Later, Katkhuda et al. [29] improved the concept further and called it generalized ILS-UI or GILS-UI. All these are least-squares procedures. They were extensively verified using computer-generated response information for shear buildings, two-dimensional trusses, and frames. They added artificially generated white noises in the computer-generated noise-free responses and showed that the methods could assess health of defect-free and defective structures. Recently, the concept has been verified for three-dimensional (3D) structures, denoted as 3D-GILS-UI [11, 12].

For the sake of completeness, other recently proposed least-squares-based SHA procedures need a brief review. Yang et al. [54] proposed a recursive least-squares estimation procedure with unknown inputs (RLSE-UI) for the identification of stiffness, damping, and other nonlinear parameters, and the unmeasured excitations. They implemented an adaptive technique [52] in RLSE-UI to track the variations of structural parameters due to damages. Then, Yang et al. [50, 53] proposed a new data analysis method, denoted as the sequential nonlinear least-square (SNLSE) approach, for the on-line identification of structural parameters. Later, Yang and Huang [49] extended the procedure for unknown excitations and reduce number of sensors (SNLSE-UI-UO). They verified the procedures for simple linear and non-linear structures. Several other methods based on least squares can be found in Choi et al. [10], Chase et al. [8, 9], and Garrido and Rivero-Angeles [17].

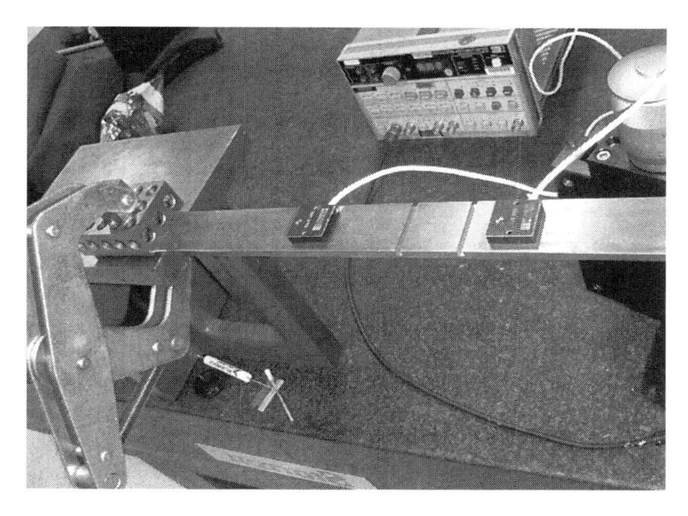

Fig. 1 Laboratory test of defective beams

After analytically establishing the concept that a structure can be identified using only noise-contaminated response information, completely ignoring the excitation information, the research team at the University of Arizona tested a one-dimensional beam [45] and a two-dimensional frame built to one-third scale in the laboratory [34, 35]. The test setups for the two studies are shown in Figs. 1 and 2, respectively. Both studies conclusively confirmed the validity of the basic SI concept without excitation information.

4.3 SHA Using Limited Response Information: Measured at a Small Part of the Structure

It is now established that least-squares concept can be used for SHA without using excitation information, but response information must be available at all DDOFs. This led the team to study cases when response information is available only at a part of the structure. Kalman filter-based algorithm is commonly used when the system is uncertain and the responses are noise-contaminated and not available at all DDOFs.

4.3.1 Kalman Filter

Application of Kalman filter (KF) for assessing health for civil engineering structures is relatively recent. Various forms of Kalman filter can be found in the literature including extended Kalman filter (EKF), unscented Kalman filter (UKF),

Fig. 2 Experimental verification of a scaled 2D frame

particle filter, and ensemble Kalman filter (EnKF). In mathematical sense, the basic KF is a nondeterministic, recursive computational procedure to provide best estimate of the states by minimizing the mean of squared error for a process governed by linear stochastic differential equation expressed as [47]

$$\mathbf{x}(k+1) = \mathbf{F}(k)\mathbf{x}(k) + \mathbf{G}(k)\mathbf{u}(k) + \mathbf{w}(k) \tag{4}$$

with the measurement model of the form:

$$\mathbf{z}(k+1) = \mathbf{H}(k+1)\mathbf{x}(k+1) + \mathbf{v}(k+1) \tag{5}$$

where $\mathbf{x}(k+1)$ and $\mathbf{x}(k)$ are the state vectors at time instant $k + 1$ and k, respectively; vectors $\mathbf{w}(k)$ and $\mathbf{v}(k)$ represent the process and measurement noises, respectively; $\mathbf{F}(k)$ relates the two state vectors in absence of either a driving function or process noise; $\mathbf{G}(k)$ relates the optimal control input $\mathbf{u}(k)$ to the state; and $\mathbf{H}(k + 1)$ in the measurement model relates the state vector to the measurement vector $\mathbf{z}(k + 1)$. $\mathbf{w}(k)$ and $\mathbf{v}(k)$ are considered to be independent, zero-mean, white random vectors with normal probability distributions. Kalman filter is very powerful in several ways. It incorporates the (1) knowledge of the system, (2) statistical description of the system noises, measurement errors and uncertainty in the dynamic models, and (3) any available information on the initial conditions of the variables of interest [36].

The basic KF is essentially applicable for linear structural behavior. For SHA of civil engineering structures, the behavior may not be linear. Moderate to high level of excitation may force the structure to behave nonlinearly. Presence of defects may also cause nonlinearity, even when the excitation is at the low level. This leads to the development of the extended Kalman filter (EKF) concept. For EKF, Eq. (4) is expressed in the following form [47]:

$$\mathbf{x}(k+1) = f[\mathbf{x}(k), \mathbf{u}(k), \mathbf{w}(k)] \tag{6}$$

and the measurement equation, Eq. (5), is modified as

$$\mathbf{z}(k+1) = h[\mathbf{x}(k+1), \mathbf{v}(k+1)] \tag{7}$$

where nonlinear function f relates the states at time k to the current states at time $k + 1$, and it includes the state vector $\mathbf{x}(k)$, driving force $\mathbf{u}(k)$, and process noise $\mathbf{w}(k)$. The nonlinear function h relates the state vector $\mathbf{x}(k + 1)$ to the measurement vector $\mathbf{z}(k + 1)$. Again, $\mathbf{w}(k)$ and $\mathbf{v}(k)$ are the independent, zero-mean, white random vectors with normal distribution, representing the process and measurement noises, respectively. The EKF estimates the state by linearizing the process and measurement equations about the current states and covariances. KF or EKF attempts to predict responses and the model parameters and then updates them at each time point using current measurements. The procedure involving prediction and updating at each time point is generally known as *local iteration*. Completion of local iteration processes covering all time instances in the entire time-history of responses is generally known as first *global iteration*. The global iterations need to be repeated to satisfy the preselected convergence criterion of system parameters. Hoshiya and Saito [23] proposed a weighted global iteration (WGI) procedure with an objective function after the first global iteration to obtain convergence in an efficient way. The entire procedure is denoted as EKF-WGI. Recently, several researchers have improved computational aspects of EKF-WGI [14]. Koh and See [31] proposed an adaptive EKF (AEKF) to estimate both the parameter values and associated uncertainties in the identification. Yang et al. [50, 53] proposed an adaptive tracking technique based on EKF to identify structural parameters and their variation during damage events. Ghosh et al. [19] developed two novel forms of EKF-based parameter identification techniques; these are based on variants of the derivative-free locally transversal linearization (LTL) and multistep transverse linearization (MTrL) procedures. Liu et al. [33] proposed multiple model adaptive estimators (MMAE) that consist of bank of EKF designed in the modal domain (MOKF) and incorporated fuzzy logic-based block in EKF to estimate variance of measurement noise.

When KF or EKF is used to identify a structure using dynamic response information satisfying the governing equation represented by Eq. (1), it requires that the excitation information and the initial values of unknown state vector. As mentioned earlier, to improve implementation potential, the proposed approach needs to identify a structure without using excitation information, and the information on the state vector will be available only at the completion of the identification, not at the beginning. The discussions clearly indicate that the basic KF or EKF cannot be used to identify a structure. To overcome these challenges, the research team proposed a two-stage approach by combining GILS-UI and EKF-WGI procedures. Based on the available measured responses, a substructure can be selected that will satisfy all the requirements to implement the GILS-UI procedure in stage 1. At the completion of stage 1, the information on the unknown excitation and the damping and stiffness parameters of all the elements in the substructure will be available. The identified damping can be assumed to be applicable for the whole structure. Structural members in a structure are expected to have similar cross-sectional properties. Suppose the substructure consists of one beam and one column. The identified stiffness parameters of the beam in the substructure can be assigned to all the beams in the structure. Similarly, the identified stiffness parameters of the column can be assigned to all the columns. This will give information on the initial state vector. With the information on initial state vector and excitation information, the EKF-WGI procedure can be initiated to identify the whole structure in stage 2. This novel concept was developed in stages; they are known as ILS-EKF-UI [47], MILS-EKF-UI [32], and GILS-EKF-UI [28]. These procedures were successfully verified using analytically generated responses, primarily for two-dimensional structures [21, 27]. Martinez-Flores et al. [35] then successfully verified GILS-EKF-UI in the laboratory for a two-dimensional frame shown in Fig. 2. They considered defect-free and several defective states with different levels of severities, including broken members, loss of cross-sectional area over the entire length of members, loss of area over small length of a member, and presence of one or multiple cracks in a member; Das and Haldar [13] recently extended the method to assess structural health for 3D structures.

5 Future of Structural Health Assessment

Future directions of the SHA area, as foreseen by the authors, are presented in the following sections. In the previous sections, the authors emphasized analytical concepts used for SHA and their contributions. In developing their methods, they observed many challenges yet to be resolved. Some issues are related to explicit consideration of uncertainty in describing the system and noises in the measured responses. Selection of initial state vector for large structure is also expected to be challenging. Although EKF can be used in presence of nonlinearity, the threshold nonlinearity is not known, that is, when it will fail to identify a structure. The methods proposed by the authors can identify structures with less information, but the absolute minimum number of required responses for acceptable identification needs further study. Issues related to the stability, convergence, and acceptable error in prediction need further works. Although the information of excitation is not required, characteristics of excitations need some attentions.

At the beginning of this chapter, the authors mentioned that SHA is a multidisciplinary research area. This chapter will not be complete without the discussions on sensors, intelligent sensing technologies and signal processing, next generation structural health monitoring strategies, etc. Since SHA is a relatively new area and not covered in the existing curriculum of major branches of engineering, it is necessary to emphasize education aspect of SHA. The authors are not expert in some of these areas; however, they expect that the discussions will prompt future engineers to explore them. The first author is in the process of editing a book with contributions from experts covering all these areas [20].

5.1 Transition from Present to Future: Local Level SHA Using Global Responses

SHA procedures are developed generally assuming that measured responses will be altered in presence of defects. Obviously, minor defects may not alter the responses and thus cannot be detected. A structure is expected to be inspected several times during its lifetime. Minor defects such as initiation and development of cracks and corrosion of reinforcements in concrete are expected to grow over time. The basic assumption is that when they become major, they will be detected during the periodic inspections. Environmental influences on structural behavior, for example, effect of temperature on measured responses, are not completely understood at this time. Similar comments can be made for exposure to chemicals or high-pressure gradients. Smart sensors are now being developed to detect damage for various applications. Not all sensors are equally sensitive, and noises in the measurements cannot be avoided. Depending upon to noise to signal ratios, the output of a sensor can be misleading. The discussions clearly indicate that besides analytical developments, industrial research is also going to be critical in implementing a particular health assessment strategy. Hopefully, advances in technologies, digital computing, and data processing will remove some of these hurdles.

5.2 SHA in Presence of Nonlinearity

One major assumption in most SI-based SHA procedures is that the responses are linear or mildly nonlinear. Major nonlinearities are not expected to show up in the responses during ambient excitation or when the level of excitation is relatively small. However, in real situations, the nonlinearity in the responses cannot be avoided. To understand and develop robust mathematical model of the dynamical system, the distinct effects of nonlinearities must be realistically accounted for. At the same time, it will be important to use the available resources in a very systematic manner for successful implementation of the SHA procedures. To identify a highly nonlinear structure, it is important first to identify the level of nonlinearity and establish whether available methods are appropriate or not.

Next, it will be important to determine the location, type, and form of nonlinearity and how to model them in the optimum way. Another important task will be the selection of parameters/coefficients that need to be tracked for damage assessment. The area of nonlinear system identification is still in its infancy. An extensive discussion on the related areas and future directions in nonlinear SI is discussed by Kerschen et al. [30].

5.3 Intelligent Sensing Technologies and Signal Processing

Development of new sensor technologies for various applications is expanding in an exponential scale. Use of smart wireless sensors is becoming very common. The placements of sensors, density, sources of power for their operation, calibration for maintaining them in good operating condition, acquisition of signals, advanced signal processing algorithms considering increased signal-to-noise ratio, well-developed numerical procedure for post-processing of signal, integration of software and hardware, realistic mathematical model for structures and their components, etc., are being actively studied by various researchers. Smart sensors are wireless and equipped with on-board microprocessors; they are small in size and can be procured at a lower cost. However, there are several hardware aspects such as efficient data acquisition, synchronization, limited memory, encryption and secured data transmission, and limited bandwidth that need further attention. They should be operational throughout the life of the structure, if continuous-time SHA is performed [7]. It is expected that distributed computational framework and use of agents-based architecture will expand the possibility of intelligent infrastructure maintenance in future [43].

5.4 Next Generation Structural Health Monitoring Strategy

Structural systems always change due to inevitable deterioration processes. Assessment of current state of a structure cannot be complete without taking into account the uncertainties at every step of the assessment process. Even ignoring uncertainties, monitoring a structure continuously throughout its life may not be an optimum use of available resources. Next generation structural health monitoring research needs to be performance based. Using information from the most recent assessment, mathematical model to represent the structure, placement of sensors, data collection, and interpretation methods need to be modified or updated. The integration of past and present information on structural health needs to be carried out for the cost-efficient assessment. Risk-based nondestructive evaluation concept is expected to optimize the frequency of inspection.

5.5 *SHA Curriculum in Education*

The authors sincerely hope that the previous discussions provided a flavor of multidisciplinary nature of SHA areas. Both authors are civil engineers. Their formal education in civil engineering did not train them to undertake the research discussed here. Most engineering colleges do not offer courses related to SHA. NDE mostly belongs to mechanical engineering, whereas sensors and signal processing belong to electrical engineering. So far, the SHA/SHM education for professional engineers is limited to web-based resources or short course modules. Recently, several universities in Europe are collaboratively offering an Advanced Master's in Structural Analysis of Monuments and Historical Constructions (SAMHC) funded by the European Commission. In the USA, University of California, San Diego, has started M.S. program with specialization in SHM.

There is no doubt that the SHA/SHM areas will grow exponentially in near future all over the world. Trained engineers will be essential to carry out the necessary works. A severe shortage of trained professionals is expected. It will be highly desirable if we introduce a multidisciplinary engineering discipline in SHA/SHM by integrating civil, electrical, material, and mechanical engineering departments.

6 Conclusions

Structural health assessment has become an important research topic and attracted multidisciplinary research interests. Its growth has been exponential in the recent past. Past and present developments in the related areas are briefly reviewed in this chapter. Because of their academic background, the authors emphasized the structural health assessment for civil infrastructures. Some of the future challenges are also identified. Advancements in sensor technology and signal processing techniques are also reviewed briefly. An upcoming edited book on the subject by the first author is expected to provide more information on the related areas. Because of the newness of the area, there is a major gap in current engineering curriculum. In the near future, a severe shortage is expected for experts with proper training in SHA/SHM. The authors advocate for a new multidisciplinary engineering discipline in SHA/SHM by integrating, civil, electrical, material, and mechanical engineering departments.

References

1. Aditya G, Chakraborty S (2008) Sensitivity based health monitoring of structures with static responses. Scientia Iranica 15(3):267–274
2. Anh TV (2009) Enhancements to the damage locating vector method for structural health monitoring. Ph. D. dissertation, National University of Singapore, Singapore

3. Bernal D (2002) Load vectors for damage localization. J Eng Mech ASCE 128(1):7–14
4. Bray DE (2000) Historical review of technology development in NDE. In: 15th world conference on nondestructive testing, Roma, Italy, 15–21 Oct 2000
5. Carden EP, Fanning P (2004) Vibration based condition monitoring: a review. Struct Health Monit 3(4):355–377
6. Ceravolo R (2009) Time–frequency analysis. Chapter 26. In: Boller C, Chang F-K, Fuzino Y (eds) Encyclopedia of structural health monitoring. Wiley, Chichester
7. Chang PC, Flatau A, Liu SC (2003) Review paper: health monitoring of civil infrastructure. Struct Health Monit 2(3):257–267
8. Chase JG, Begoc V, Barroso LR (2005) Efficient structural health monitoring for benchmark structure using adaptive RLS filters. Comput Struct 83:639–647
9. Chase JG, Spieth HA, Blome CF, Mandler JB (2005) LMS-based structural health monitoring of a non-linear rocking structure. Earthq Eng Struct Dyn 34:909–930
10. Choi YM, Cho HN, Kim YB, Hwang YK (2001) Structural identification with unknown input excitation. KSCE J Civil Eng 5(3):207–213
11. Das AK, Haldar A (2010) Structural integrity assessment under uncertainty for three dimensional offshore structures. Int J Terraspace Sci Eng (IJTSE) 2(2):101–111
12. Das AK, Haldar A (2010) Structural health assessment of truss-type bridges using noise-contaminated uncertain dynamic response information. Int J Eng under Uncertainity: Hazards, Assessment, and Mitigation 2(3–4):75–87
13. Das AK, Haldar A (2012) Health assessment of three dimensional large structural systems – a novel approach. Int J Life Cycle Reliab Saf Eng (in press)
14. Das AK, Haldar A, Chakraborty S (2012) Health assessment of large two dimensional structures using minimum information – recent advances. Adv Civil Eng. Article ID 582472. doi:10.1155/2012/582472
15. Doebling SW, Farrar CR, Prime MB, Shevitz DW (1996) Damage identification and health monitoring of structural and mechanical systems from changes in their vibration characteristics: a literature review, Los Alamos National Laboratory. Report no. LA-13070-MS
16. Fan W, Qiao P (2010) Vibration-based damage identification methods: a review and comparative study. Struct Health Monit 10(14):1–29
17. Garrido R, Rivero-Angeles FJ (2006) Hysteresis and parameter estimation of MDOF systems by a continuous-time least-squares method. J Earthq Eng 10(2):237–264
18. Ghanem R, Ferro G (2006) Health monitoring for strongly non-linear systems using the ensemble Kalman filter. Struct Control Health Monit 13:245–259
19. Ghosh S, Roy D, Manohar CS (2007) New forms of extended Kalman filter via transversal linearization and applications to structural system identification. Comput Methods Appl Mech Eng 196:5063–5083
20. Haldar A (ed) (2012) Health assessment of engineered structures: bridges, buildings and other infrastructures. World Scientific Publishing Co
21. Haldar A, Das AK (2010) Prognosis of structural health – nondestructive methods. Int J Perform Eng. Special Issue on Prognostics and Health Management (PHM) 6(5):487–498
22. Hellier CJ (2003) Handbook of nondestructive evaluation. The McGraw-Hill Companies, Inc, New York
23. Hoshiya M, Saito E (1984) Structural identification by extended Kalman filter. J Eng Mech ASCE 110(12):1757–1770
24. Huang NE, Huang K, Chiang W-L (2005) HHT-based bridge structural-health monitoring. Chapter 12. In: Huang NE, Shen SS (eds) Hilbert-Huang transform and its applications. World Scientific Publishing Co. Pte. Ltd., Singapore
25. Huang NE, Shen Z, Long SR, Wu MC, Shih HH, Zheng Q, Yen N-C, Tung CC, Liu HH (1998) The empirical mode decomposition and the Hilbert spectrum for nonlinear and non-stationary time series analysis. Proc R Soc Lond A 454:903–995
26. Ibanez P (1973) Identification of dynamic parameters of linear and non-linear structural models from experimental data. Nucl Eng Design 25:30–41

27. Katkhuda H (2004) In-service health assessment of real structures at the element level with unknown input and limited global responses. Ph. D. thesis, University of Arizona, Tucson, USA
28. Katkhuda H, Haldar A (2008) A novel health assessment technique with minimum information. Struct Control Health Monit 15(6):821–838
29. Katkhuda H, Martinez-Flores R, Haldar A (2005) Health assessment at local level with unknown input excitation. J Struct Eng ASCE 131(6):956–965
30. Kerschen G, Worden K, Vakakis AF, Golinval JC (2006) Past, present and future of nonlinear system identification in structural dynamics. Mech Syst Signal Process 20(3):505–592
31. Koh CG, See LM (1994) Identification and uncertainty estimation of structural parameters. J Eng Mech 120(6):1219–1236
32. Ling X, Haldar A (2004) Element level system identification with unknown input with Rayleigh damping. J Eng Mech ASCE 130(8):877–885
33. Liu X, Escamilla-Ambrosio PJ, Lieven NAJ (2009) Extended Kalman filtering for the detection of damage in linear mechanical structures. J Sound Vib 325:1023–1046
34. Martinez-Flores R, Haldar A (2007) Experimental verification of a structural health assessment method without excitation information. J Struct Eng 34(1):33–39
35. Martinez-Flores R, Katkhuda H, Haldar A (2008) A novel health assessment technique with minimum information: verification. Int J Perform Eng 4(2):121–140
36. Maybeck PS (1979) Stochastic models, estimation, and control theory. Academic, New York
37. McCaan D, Jones NP, Ellis JH (1998) Toward consideration of the value of information in structural performance assessment. Paper no. T216-6. Structural Engineering World Wide, CD-ROM
38. Montalvao M, Maia NMM, Ribeiro AMR (2006) A review of vibration-based structural health monitoring with special emphasis on composite materials. Shock Vib Dig 38(4):295–324
39. Rytter A (1993) Vibration based inspection of civil engineering structures. Ph. D. dissertation, Department of Building Technology and Structural Engineering, Aalborg University, Denmark
40. Sajjad S, Zaidi H, Zanardelli WG, Aviyente, S, Strangas EG (2007) Comparative study of time-frequency methods for the detection and categorization of intermittent fault in electrical devices. Diagnostics for electric machines, power electronics and drive, SDEMPED, IEEE symposium, 6–8 September, pp 39–45
41. Sanayei M, Imbaro GR, McClain JAS, Brown LC (1997) Structural model updating using experimental static measurements. J Struct Eng ASCE 123(6):792–798
42. Sohn H, Farrar CR, Hemez FM, Shunk DD, Stinemates DW, Nadler BR, Czarnecki JJ (2004) A review of structural health monitoring literature: 1996–2001, Los Alamos National Laboratory, LA-13976-MS
43. Spencer BF Jr, Ruiz-Sandoval ME, Kurata N (2004) Smart sensing technology: opportunities and challenges. Struct Control Health Monit 11:349–368
44. Vo PH, Haldar A (2003) Post processing of linear accelerometer data in system identification. J Struct Eng 30(2):123–130
45. Vo PH, Haldar A (2004) Health assessment of beams – theoretical and experimental investigation. J Struct Eng, Special Issue on Advances in Health Monitoring/Assessment of Structures Including Heritage and Monument Structures 31(1):23–30
46. Wang D, Haldar A (1994) An element level SI with unknown input information. J Eng Mech ASCE 120(1):159–176
47. Wang D, Haldar A (1997) System identificatin with limited observations and without input. J Eng Mech ASCE 123(5):504–511
48. Welch G, Bishop G (2006) An introduction to the Kalman filter. Technical report. TR95-041 2006, Department of Computer Science, University of North Carolina at Chapel Hill, NC
49. Yadav SK, Banerjee S, Kundu T (2011) Effective damage sensitive feature extraction methods for crack detection using flaw scattered ultrasonic wave field signal. In: Proceedings of the 8th international workshop on structural health monitoring, Stanford, USA, 13–15 September

50. Yang JN, Huang H (2007) Sequential non-linear least-square estimation for damage identification of structures with unknown inputs and unknown outputs. Int J Non-Linear Mech 42:789–801
51. Yang JN, Huang H, Lin S (2006) Sequential non-linear least-square estimation for damage identification of structures. Int J Non-Linear Mech 41:124–140
52. Yang JN, Lei Y, Lin S, Huang N (2004) Hilbert-Huang based approach for structural damage detection. J Eng Mech 130(1):85–95
53. Yang JN, Lin S (2005) Identification of parametric variations of structures based on least squares estimation and adaptive tracking technique. J Eng Mech 131(3):290–298
54. Yang JN, Lin S, Huang HW, Zhou L (2006) An adaptive extended Kalman filter for structural damage identification. J Struct Control Health Monit 13:849–867
55. Yang JN, Pan S, Lin S (2007) Least-squares estimation with unknown excitations for damage identification of structures. J Eng Mech 133(1):12–21

Characterisation of Large Fluctuations in Response Evolution of Reinforced Concrete Members

K. Balaji Rao

Abstract Large fluctuations in surface strain at the level of steel are expected in reinforced concrete flexural members at a given stage of loading due to the emergent structure (emergence of new crack patterns). Thus, there is a need to use distributions with heavy tails to model these strains. The use of alpha-stable distribution for modelling the variations in strain in reinforced concrete flexural members is proposed for the first time in the present study. The applicability of alpha-stable distribution is studied using the results of experimental investigations, carried out at CSIR-SERC and obtained from literature, on seven reinforced concrete flexural members tested under four-point bending. It is found that alpha-stable distribution performs better than normal distribution for modelling the observed surface strains in reinforced concrete flexural members at a given stage of loading.

Keywords Reinforced concrete • Surface strain • Cracking • Thermodynamics • Alpha-stable distribution

1 Introduction

Strain in steel in reinforced concrete (RC) flexural members is used in the computation of crack width. The value of strain is affected by density of cracking. It is also noted that in the context of condition assessment of existing reinforced concrete structures, measured/computed surface strains play an important role. The focus in this chapter is

More details on motivation for the proposed probabilistic model are presented in the supplementary material provided for this chapter.

K.B. Rao (✉)
Risk and Reliability of Structures, CSIR-Structural Engineering Research Centre, CSIR Campus, Taramani, Chennai 600 113, India
e-mail: balaji@serc.res.in

S. Chakraborty and G. Bhattacharya (eds.), *Proceedings of the International Symposium on Engineering under Uncertainty: Safety Assessment and Management (ISEUSAM - 2012)*,
DOI 10.1007/978-81-322-0757-3_15, © Springer India 2013

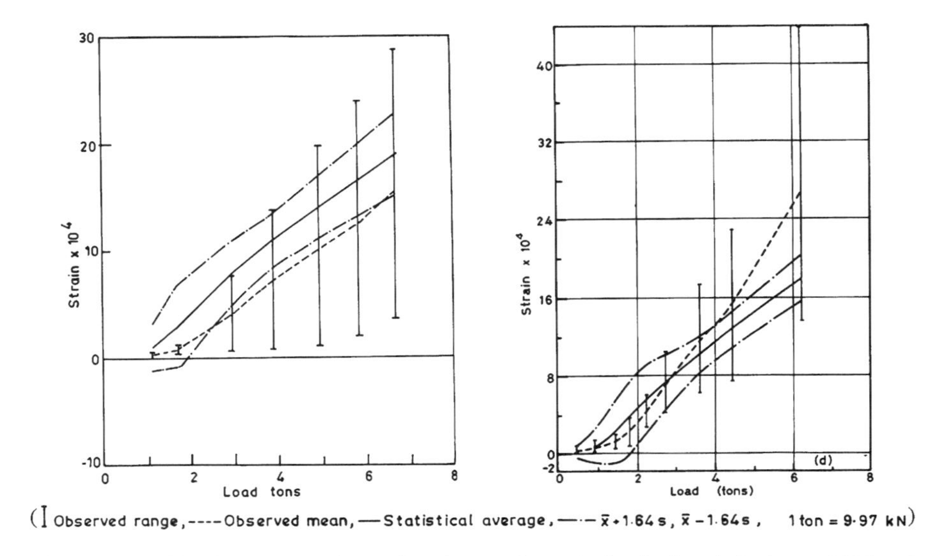

Fig. 1 Plots of load versus experimental surface strain, at the level of steel, for beams KB1 and KB2 at different stages of loading (From Desayi and Balaji Rao [17])

on identification of probabilistic models for describing the surface strain variations in reinforced concrete flexural members.

Based on an experimental investigation on RC flexural members, it has been reported that the measured strains at a given depth from extreme compression fibre, along the length of beam in pure flexure zone, exhibit large scatter [17]. Also, it has been reported that the variation of average strain across the depth is linear. Assuming linear strain variation across the depth of beam and considering various basic quantities as random variables, a probabilistic analysis of average strain, at various stages of loading, was carried out using Monte Carlo simulation. From the results of simulation (Fig. 1), it has been found that the average strain at the level of reinforcement exhibits large scatter. The probabilistic mean overestimates the experimentally observed mean strains at lower stages of loading and underestimates the same at higher stages of loading. Also, it has been found that the (mean $\pm$ 1.64 $\times$ standard deviation) limits do not enclose the observed range of strain at a given loading stage. Histogram of average strain distribution at the level of reinforcement (shown typically for the beam KB2 in Fig. 2) suggests that the distribution of average strain can be bimodal and large scatter is expected in the prediction of average strain. These observations suggest that prediction of average strain itself is beset with large uncertainty. To predict/assess the condition of a reinforced concrete member prediction of extreme (largest) value of strain is important. Hence, it is important to model the strain as a random quantity taking into account the actual mechanism of cracking and by giving due consideration to the heterogeneity of concrete. As pointed out by Bazant and Oh [7], '... This is clear even without experimental evidence, since structural analysis implies the hypothesis of smoothing of a

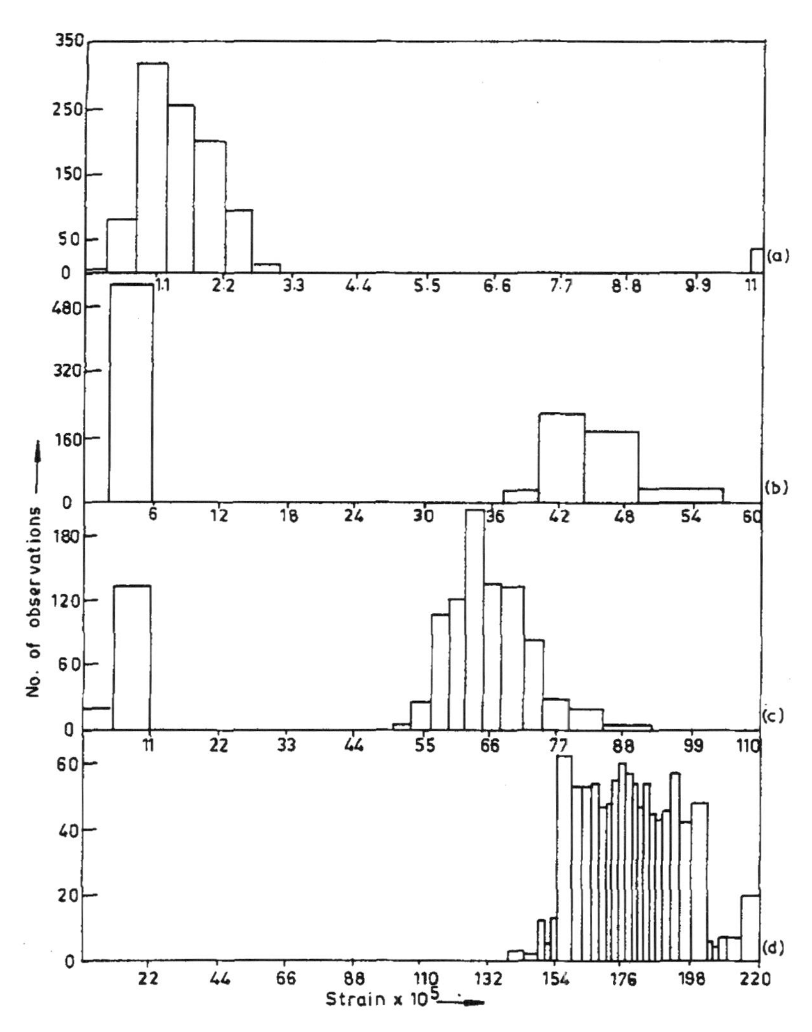

Fig. 2 Typical histograms of strains at the level of reinforcement for beam KB2 (**a**) at 1st stage of loading, (**b**) at 3rd stage of loading, (**c**) at 5th stage of loading and (**d**) at 10th stage of loading (From Desayi and Balaji Rao [17])

heterogeneous material by an equivalent homogeneous continuum in which, if one uses the language of the statistical theory of random heterogeneous materials, the stresses and strains must be understood as the averages of the actual stresses and strains in the microstructure over the so-called representative volume whose size must be taken to be at least several times the size of the heterogeneities. . .'. The authors also give an idea about the size of the representative volume element (RVE) as 10–20 times the aggregate size. They have presented, by considering the energy criterion of fracture mechanics and strength criterion, equations for crack spacing

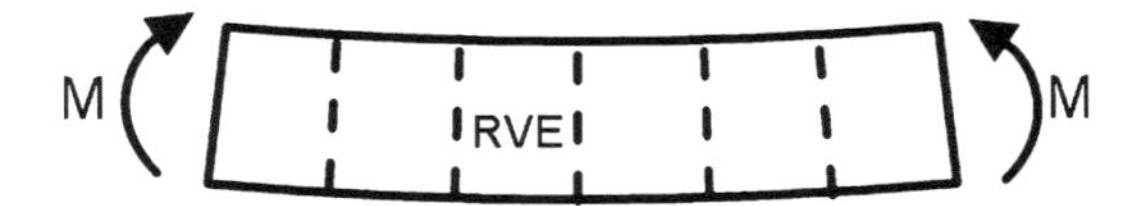

Fig. 3 The RVEs in the constant moment zone of an RC beam

and crack widths in RC members. They attribute the possible variations in cracking to the random variations in fracture energy. From this study, it is clear that when strain variation in the flexure zone is studied, the size of the RVE should be about ten times the size of aggregate.

In the present investigation, the surface strains measured over a gauge length of 200 mm are considered for further analysis satisfying the requirement of RVE. Since the RVE is statistically homogeneous and all the RVEs (Fig. 3) in the flexure zone are subjected to same moment, these elements can be considered to be identical.

As suggested by Bazant and Oh [7] and Balaji Rao and Appa Rao [2], the RC beam undergoing flexural cracking can be considered as a parallel system wherein redistribution of stresses/strains is taking place. For such a system, Bazant [5], using the fracture mechanics-based size effect theory, has recently shown that the load (or stress)-displacement (or strain) curve would exhibit jumps. However, as pointed out by Bazant [5], the size effect may not be significant in the presence of tension reinforcement. As discussed in this chapter, even the load-displacement behaviour of RC beam would exhibit multiple jumps. The points where the jumps occur are the points of bifurcation [5, 6, 15, 22]. It is known that at these points, the system which is undergoing flexural cracking would exhibit large fluctuations [33]. From the foregoing discussions, it is felt that a heavy-tailed distribution may be more appropriate to capture the large fluctuations in the concrete surface strains in reinforced concrete flexural members (at any given stage of loading, especially at the points at which an emergent crack structure forms).

Some Observations

- The RVEs located in the flexure zone of the RC member (Fig. 3) are subject to statistically similar stress/strain states. This implies that the random surface strains in the RVEs are identically distributed. While the stress/strain state in RVEs can be considered to be identical, they need not be considered to be statistically independent.
- From both the experimental and probabilistic analyses results, it is found that a random variable which exhibits large fluctuations would be preferred over those generally used (i.e. those with exponential decaying tails). Thus, it is desirable to use probability distributions that have power-law decaying tails (since they have to capture large fluctuations expected to be attendant with bifurcation phenomenon).

An experimental programme is taken up recently to check the validity of this inference. Four singly reinforced RC beams are tested under four-point bending to obtain the concrete surface strain values in the flexure zone at CSIR-SERC. Salient details of experimental investigations are presented in this chapter. An attempt is made to fit an alpha-stable distribution (which is known to have heavy tails) to the observed strains. To study further the efficacy of alpha-stable distribution for modelling the variations in strain in concrete at the level of reinforcement in RC flexural members, experimental investigations on three singly reinforced concrete beams presented by Prakash Desayi and Balaji Rao [17, 18] are also considered. Based on the results of statistical analyses, it is inferred that the alpha-stable distribution is a good choice for modelling the surface strains in concrete, at the level of reinforcement, in RC flexural members.

2 Mechanism of Cracking: A Discussion on Emergent Structure

The strain in concrete in the tension zone of a RC flexural member depends on the level of cracking in the member. To understand the strain behaviour in the tension zone, it is important to know the mechanism of cracking in RC flexural member. The mechanism of cracking is described in several references (see for instance [10, 29, 31]). Only a brief description of the same will be presented here.

When an under-reinforced concrete beam is subjected to monotonically increasing four-point bending (Fig. 4), the following points can be noted:

1. As long as the applied load is less than the first crack load of the beam, the tension forces are shared both by concrete and steel. The load-deflection curve will be essentially linear (portion A of Fig. 5). There will be internal microcracking of concrete present in the tension zone [10].
2. When the applied load is equal to the first crack load of the beam, visible crack(s) appears on the surface of the beam and the flexure zone of the beam will be divided into number of sections as shown in Fig. 6. The formation of first set of cracks will be characterised by a sudden drop in load (point B in Fig. 5). This

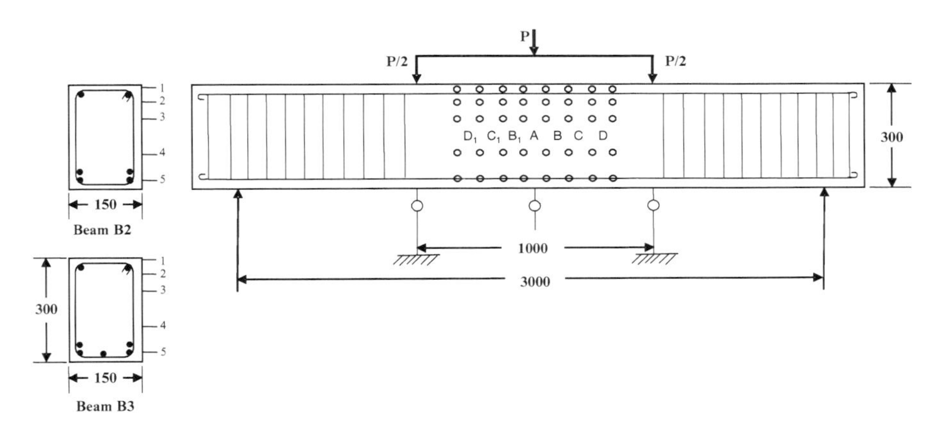

Fig. 4 Schematic representation of the test programme for beams tested at CSIR-IISc (dimensions in mm)

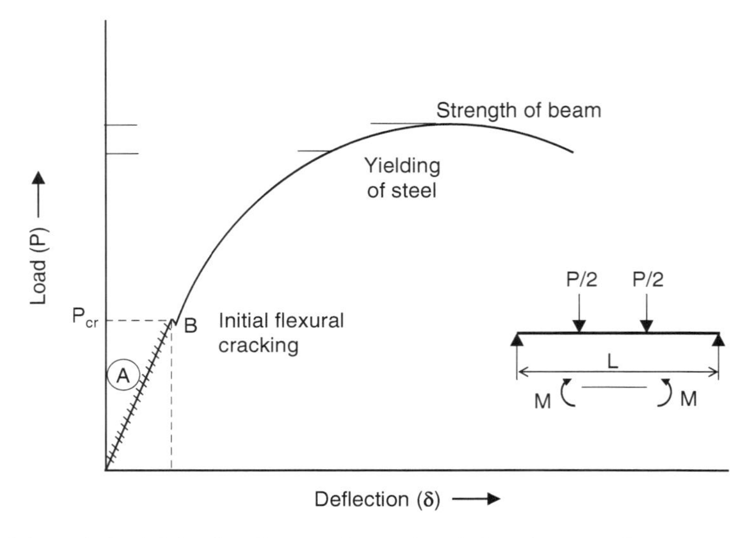

Fig. 5 Schematic load-deflection diagram of an under-reinforced concrete beam

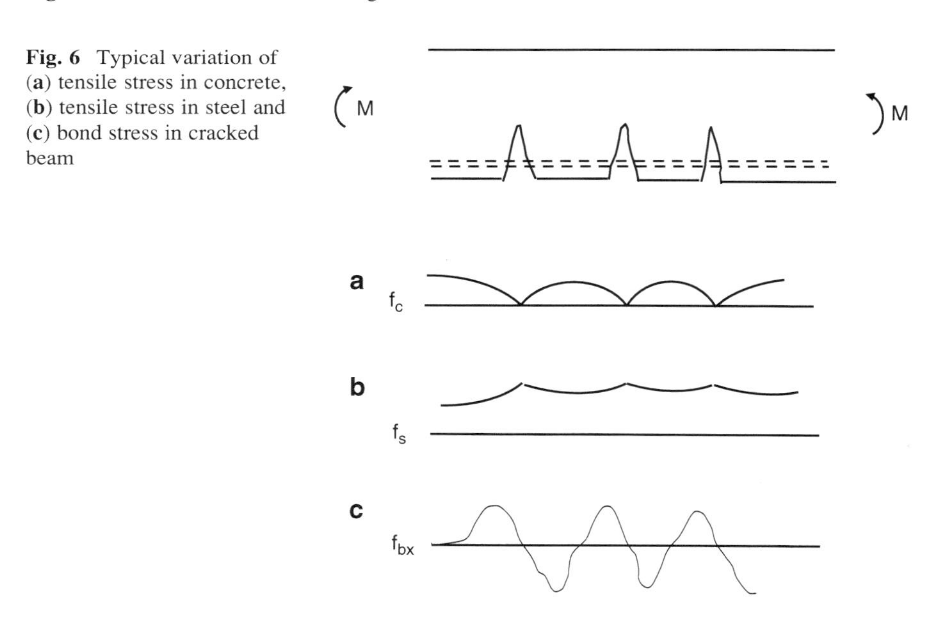

Fig. 6 Typical variation of (**a**) tensile stress in concrete, (**b**) tensile stress in steel and (**c**) bond stress in cracked beam

occurs due to sudden loss of stiffness of the beam due to cracking. Typical variations of tensile stresses in concrete and in steel and the bond stress in the flexure zone of the beam are shown in Fig. 6.

3. With the increase of load, beyond the first crack load, redistribution of stresses takes place between the cracked sections. New cracks may form in between the existing cracks, and also, the existing cracks may widen/lengthen. The formation

of new cracks results in reduction in crack spacing. The process of formation of new cracks will continue until the bottom fibre stress in concrete cannot reach a value equal to the modulus of rupture. When this condition is reached, no more new cracks form and the existing cracks will widen/lengthen with the increase of load. Thus, the spacing of cracks remains the same, and the corresponding crack spacing is called stabilised crack spacing.

2.1 Brief Review of Relevant Models for Cracking of Concrete and Reinforced Concrete

In this section, an attempt is made only to briefly review models that are relevant to the topic of this chapter, and it is considered that the *review is not exhaustive*.

From the above discussion, it is clear that the behaviour of RC flexural members under the external loading is quite complex. Also, it is clear that one of the outstanding features of the behaviour is the emergent structure at different stages of loading (in this chapter, emergent (dissipative) structure refers to formation of new cracks and/or widening and lengthening of the existing cracks on the surface of the flexural member, in the constant bending moment zone), as the loading is increased monotonically. At a given stage of loading, the emergent structure is characterised by the crack length, crack spacing and crack width. Efforts have already been made to approximately account for these observations in the estimation of crack spacing and crack widths [1, 11, 14, 19]. From the equations proposed in these codes (not presented here), it can be noted that cracking in RC flexural members depends on cross-sectional dimensions of the beam, the material strengths and the reinforcement details.

Recently, efforts are being made to develop numerical models within the framework of finite element simulation for prediction of load-deformation behaviour of RC members. The formulation, as can be expected, should consider the fact that the RC beams would have both distributed and localised damage in coupled form and also the bond-slip relation. Two types of models, namely, one which is based on characteristic length and the other based on damage evolution, have been proposed [7, 20, 22, 35]. The latter type of models recognises that there can be strong discontinuities in the displacement field at the points of bifurcation. These models are reviewed below to bring out the need for development of a probability distribution for strains at points of bifurcation. The information presented in Sects. 2.1.1 and 2.1.2 are directly obtained from Refs. [20] and [35]. These Sections are presented here for the sake of completeness, though no effort is made to use FEM in this chapter.

2.1.1 Reinforced Concrete Damage Modelling by Sluys and de Borst [35]

The authors reviewed the available constitutive models for concrete and also presented a bond-slip model that can be used in the FEM of concrete and reinforced concrete elements. The same are presented below.

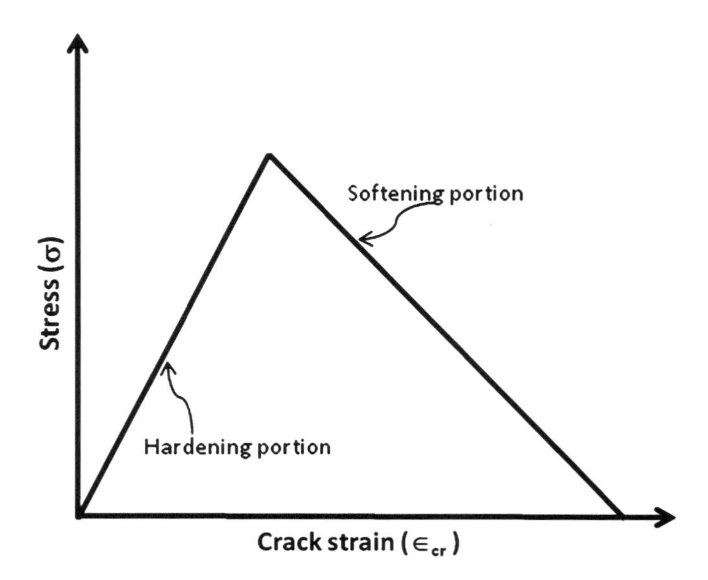

Fig. 7 Stress-strain curve for a standard crack model [35]

Concrete: The Standard Crack Model

In a standard crack model (SCM), the stress-strain curve consists of two parts, as shown in Fig. 7. The main feature of the stress-strain curve is the inclusion of the strain softening portion. The stress in softening portion is given by

$$\sigma = f_t + h\varepsilon_{cr} \tag{1}$$

where h is a constant negative softening modulus, f_t is the tensile strength and ε_{cr} is the cracking strain. The standard crack model takes into account the softening portion. However, the lack of characteristic length makes the overall response of the concrete element to be mesh sensitive, and since the problem of cracking is an ill-posed problem, the crack width is governed by the assumed size of the mesh. Also, when SCM is used for predicting the cracking in reinforced concrete members, the spacing of the cracks is more governed by the discretisation adopted. This is because by adopting finer meshes, the stress degradation occurs faster and secondary cracks form at a shorter distance (crack density increases).

Concrete: G_f-Type Crack Model

Some of the investigators have tried to overcome the lack of mesh objectivity by using higher-order terms in the displacement/strain field representation or by considering the area under the softening curve (viz. fracture energy) as a material parameter. The fracture energy depends on the size of the process/localised zone. Since G_f is a material property, the slope of the softening portion is made function of mesh size. However, while the global load-displacement response of concrete specimen will be insensitive to meshing and hence mesh objective, no refinement takes place locally. For RC members, energy released in a single primary crack can be made mesh objective, but spacing of cracks still depends on mesh configuration. This is one of the serious limitations since the crack width in reinforced concrete members depends on the spacing of cracks.

Concrete: The Rate-Dependent Crack Model

The rate-dependent models, as the name suggest, arise in cracking of concrete/ reinforced concrete due to evolution of damage/friction with time (or may be load, if it is varying). The constitutive relation in such a case is given by

$$\sigma = f_t + h\varepsilon_{cr} + m\frac{\partial \varepsilon_{cr}}{\partial t} \tag{2}$$

where m is the viscosity and all other terms are already defined. While the SCM is an ill-posed problem, rate-dependent model is well-posed initial value problem. Equation (2) caters to one experimental observation that the tensile strength increases with increase in loading rate. The length scale corresponding to this rate boundary value problem is

$$l = \frac{2m}{\sqrt{\rho E}} \tag{3}$$

Concrete: The Gradient Crack Model

One another way of solving the ill-posed problem of cracking in concrete is the use of rate of evolution of strain (i.e. rate evolution of gradients of ε_{cr}) along the spatial direction. This may be applied to static problems since the spatial variations are being dealt with. According to this model, the stress (in one dimension) is given by

$$\sigma = f_t + h\varepsilon_{cr} + \bar{c}\frac{\partial^2 \varepsilon_{cr}}{\partial x^2} \tag{4}$$

where $\bar{c}$ is the standard length parameter that takes into account interaction among the micro-defects along the length. The length scale evaluated based on solution of wave equation is given by

$$l = \sqrt{\frac{\bar{c}}{h}} \tag{5}$$

The length-scale parameter is representative of crack width in concrete and crack spacing in reinforced concrete members.

Reinforced Concrete: Bond-Slip Behaviour

It is difficult to evolve a constitutive relation for bond-slip. Recently, Dominguez et al. [20] have modelled it using the principles of thermodynamics. However, a macro-level model taking into account shear traction – slip interaction, as suggested by Dörr (in Sluys and de Borst [35]), is given by

$$\tau = \begin{cases} a\left[5\left(\frac{\delta}{\delta_0}\right) - 4.5\left(\frac{\delta}{\delta_0}\right)^2 + 1.4\left(\frac{\delta}{\delta_0}\right)^3\right] & \textit{if } 0 < \delta < \delta_0 \\ 1.9a \quad \textit{if } \delta > \delta_0 \end{cases} \tag{6}$$

where a is a constant which is taken equal to the tensile strength f_t and δ_0 is the deformation at which perfect slip occurs.

2.1.2 Strong Discontinuity Model by Dominguez et al. [20]

The authors formulate two-level damage model which works in coupled manner in a given element. The two levels identified, based on dissipation mechanism, are (1) continuum damage model and (2) discontinuity damage model. The first level considers, through continuous damage mechanics formulation, energy dissipation by formation of microcracks in the bulk material, while the second level considers the formation of macrocracks in localised zones resulting in surface dissipation of energy. The second level represents a strong discontinuity, and both levels are made to work in coupled manner by introduction of a displacement discontinuity. In the following, the formulation related to displacement discontinuity is presented:

Kinematics of Displacement Discontinuity Model

The authors introduce a surface of displacement discontinuity on which are concentrated all localised dissipative mechanisms due to formation and development of localisation zones. This is achieved by considering the domain Ω split into two sub-domains Ω^+ and Ω^- by surface of discontinuity, denoted as Γ_S.

The surface of discontinuity Γ_S is characterised 'at each point' by the normal and tangential vectors (m and n). The discontinuous displacement field can then be written as

$$u(X,t) = \bar{u}(X,t) + \bar{\bar{u}}(t)[H_{\Gamma_S}(X) - \phi(X)] \tag{7}$$

where $H_{\Gamma_S}(X)$ is a Heaviside step function and is equal to one if $X \in \Omega^+$ and zero if $X \in \Omega^-$. $\varphi(X)$ is 'at least' C_0 function with its 'boundary values' defined according to

$$\phi(X) = \begin{cases} 0 & if \quad X \in \partial\Omega \cap \Omega^- \\ 1 & if \quad X \in \partial\Omega \cap \Omega^+ \end{cases} \tag{8}$$

From Eq. (1), it is clear that $\bar{u}(X,t)$ has the same boundary values as the total displacement field $u(X,t)$. Now, the strain field can be written as

$$\varepsilon(X,t) = \nabla^S u(X,t) = \left[\nabla^S \bar{u}(X,t) + \bar{G}(X)\bar{\bar{u}}(t)\right] + \left\{(\bar{\bar{u}}(t) \otimes n)^S \delta_{\Gamma_S}(X)\right\} \tag{9}$$

In Eq. (9), the terms in square bracket represent the regular part of the strain field, while those in the curly brackets represent the singular part. Representation of the strain field in the form of two parts allows us to write the damage evolution equation as

$$\varepsilon = D : \sigma \tag{10}$$

The compliance D in Eq. (10) consists again of two parts, namely, the regular part and the other singular part. That is,

$$D = \bar{D} + \bar{\bar{D}}\delta_{\Gamma_S}(X) \tag{11}$$

It is noted that the first term in Eq. (11) corresponds to continuous damage/dissipative mechanism in the bulk continuum containing microcracks, while the second term accounts for the discrete damage/dissipative mechanism describing the localised zones. The above equation allows stress to be finite. In order to formulate the damage evolution equation, we need to identify the internal variables that can be decomposed into regular and discrete (or singular part).

Let us define the two models and their corresponding evolutionary equations:

1. *Continuum damage model*: The microcracks in the bulk material are assumed to be quasi-homogeneously distributed, and the damage is isotropic. The damage model can be formulated in the domain/framework of plasticity of concrete. Let the damage in this model be denoted by $\bar{D}$ (in the elastic stage, the damage is denoted by D^e). The variable associated with hardening is denoted by $\bar{\xi}$. The constitutive relation is obtained optimising the dissipated energy. The Lagrange multiplier used in optimisation is denoted as $\vec{\gamma}$. The main ingredients of continuum damage model are presented in Table 1 [20].
2. *Discrete damage model*: This damage model is constructed to account for steep strain gradients and localised damage due to formation of macrocracks. It is constructed in a similar way as it is done for the continuum damage model. As mentioned earlier, the discrete damage model has to be developed on the

Table 1 Main ingredients of continuum damage model [20]

Helmholtz free energy	$\bar{\psi}(\bar{\varepsilon}, \bar{\mathbf{D}}, \bar{\xi}) = \frac{1}{2}\bar{\varepsilon} : \bar{\mathbf{D}}^{-1} : \bar{\varepsilon} + \bar{\Xi}(\bar{\xi})$
Damage function	$\bar{\phi}(\sigma, \bar{q}) = \underbrace{\sqrt{\sigma : \mathbf{D}^e : \sigma}}_{\lVert\sigma\rVert_{\mathbf{D}^e}} - \frac{1}{\sqrt{E}}(\sigma_f - \bar{q})$
Constitutive equations	$\sigma = \bar{\mathbf{D}}^{-1} : \bar{\varepsilon}$ and $\bar{q} = -\frac{d}{d\bar{\xi}}\bar{\Xi}(\bar{\xi})$
Dissipation	$\bar{\mathbf{D}}(\sigma, \bar{q}) = \frac{1}{2}\sigma : \dot{\bar{\mathbf{D}}} : \sigma + \bar{q}\dot{\bar{\xi}}$
Evolution equations	$\dot{\bar{\mathbf{D}}} = \dot{\bar{\gamma}} \frac{\mathbf{D}^e}{\lVert\sigma\rVert_{\mathbf{D}^e}}$ and $\dot{\bar{\xi}} = \dot{\bar{\gamma}} \frac{1}{\sqrt{E}}$

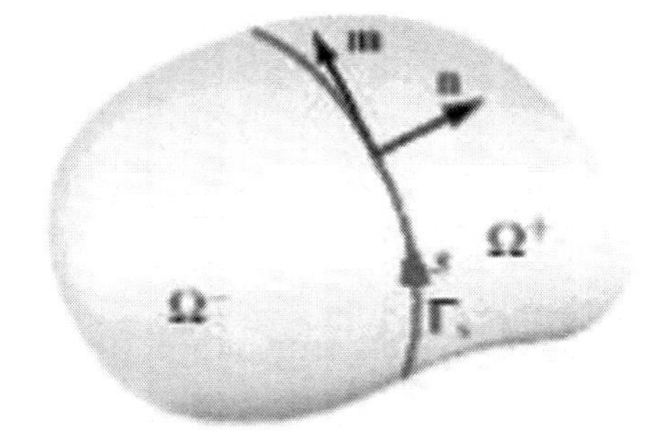

Fig. 8 Partitioning of domain Ω and the slip line Γ_S [20, 22]

Table 2 Main ingredients of discrete damage model [20]

Helmholtz free energy	$\bar{\bar{\psi}}(\bar{\bar{\mathbf{u}}}, \bar{\bar{\mathbf{Q}}}, \bar{\bar{\xi}}) = \frac{1}{2}\bar{\bar{\mathbf{u}}} : \bar{\bar{\mathbf{Q}}}^{-1} : \bar{\bar{\mathbf{u}}} + \bar{\bar{\Xi}}(\bar{\xi})$
Damage function	$\bar{\bar{\phi}}_1(t_{\Gamma_s}, \bar{\bar{q}}) = t_{\Gamma_s}.\mathbf{n} - (\bar{\bar{\sigma}}_f - \bar{\bar{q}})$
	$\bar{\bar{\phi}}_2(t_{\Gamma_s}, \bar{\bar{q}}) = \lvert t_{\Gamma_s}.\mathbf{m}\rvert - \left(\bar{\bar{\sigma}}_s - \frac{\bar{\bar{\sigma}}_s}{\bar{\bar{\sigma}}_f}\bar{\bar{q}}\right)$
Constitutive equations	$t_{\Gamma_s} = \bar{\bar{\mathbf{Q}}}^{-1} : \bar{\bar{\mathbf{u}}}$ and $\bar{\bar{q}} = -\frac{d}{d\bar{\bar{\xi}}}\bar{\bar{\Xi}}(\bar{\bar{\xi}})$
Dissipation	$\bar{\bar{\mathbf{D}}}(t_{\Gamma_s}, \bar{\bar{q}}) = \frac{1}{2}t_{\Gamma_s}.\dot{\bar{\bar{\mathbf{Q}}}}t_{\Gamma_s} + \bar{\bar{q}}\dot{\bar{\bar{\xi}}}$
Evolution equations	$\dot{\bar{\bar{\mathbf{Q}}}} = \dot{\bar{\bar{\gamma}}}_1 \frac{1}{t_{\Gamma_s},\mathrm{n}}\mathbf{n} \otimes \mathbf{n} + \dot{\bar{\bar{\gamma}}}_2 \frac{1}{\lvert t_{\Gamma_\sigma}.\mathrm{m}\rvert}\mathbf{m} \otimes \mathbf{m}$ and $\dot{\bar{\bar{\xi}}} = \dot{\bar{\bar{\gamma}}} \rightarrow + \frac{\bar{\bar{\sigma}}_s}{\bar{\bar{\sigma}}_f}\dot{\bar{\bar{\gamma}}}_2$

surface of discontinuity. As shown in Fig. 8, on the surface of discontinuity, both normal and tangential components of stress exist. The evolution of this surface is governed by a multi-surface criterion where $\dot{\bar{\bar{\gamma}}}_1$ and $\dot{\bar{\bar{\gamma}}}_2$ represent the Lagrange multipliers associated, respectively, to each damage surface defining the elastic domain. Each surface is coupled to the other via a traction like variable $\bar{\bar{q}}$. The internal variables associated to softening are denoted by $\bar{\bar{\xi}}$. The main ingredients of the discrete damage model are presented in Table 2 [20].

3. *Bond-slip modelling (at the interface)*: While the continuum and discrete damage models developed help in modelling the displacement, strain, damage and stress fields in the concrete as a material, it is important to model the same in interface between the steel and concrete in the case of reinforced concrete elements. While experimentally obtained bond-slip behaviour represents the

Table 3 Main ingredients of bond-slip model [20]

Helmholtz free energy	$\rho\psi = \frac{1}{2}[\varepsilon_N E \varepsilon_N + (1-d)\varepsilon_T G \varepsilon_T + (\varepsilon_T - \varepsilon_T^s) G d (\varepsilon_T - \varepsilon_T^s) + \gamma\alpha^2 + H(z)$
Damage function	$\phi_s(\sigma_N, \sigma_T, X) = \left\lvert \sigma_T^s - X \right\rvert - \frac{1}{3}\sigma_N \leq 0$
	$\phi_d(Y_d, Z) = Y_d - (Y_0 + Z) \leq 0$
Constitutive equations	$\sigma_N = E\varepsilon_N,$
	$\sigma_T = G(1-d)\varepsilon_T + Gd(\varepsilon_T - \varepsilon_T^s),$
	$\sigma_T^s = Gd(\varepsilon_T - \varepsilon_T^s)$
Dissipation	$Y = Y_d + Y_s = \frac{1}{2}\varepsilon_T G \varepsilon_T + \frac{1}{2}(\varepsilon_T - \varepsilon_T^s) G (\varepsilon_T - \varepsilon_T^s),$
	$X = \gamma\alpha,$
	$Z = H'(z) = \begin{cases} Z_1, & si\varepsilon_T^0 < \varepsilon_T^i \leq \varepsilon_T^2 \\ Z_1 \cdot Z_2, & si\varepsilon_T^i \geq \varepsilon_T^2 \end{cases}$
	$Z_1 = \left[\sqrt{Y_0} + \frac{1}{A_{d1}}\sqrt{\frac{G}{2}\ln\left((1+z)\frac{\varepsilon_T}{\varepsilon_T^0}\right)}\right]^2$
	$Z_2 = \left[Y_2 + \frac{1}{A_{d2}}\left(\frac{-z}{1+z}\right)\right]$
Evolution equations	$\dot{d} = \dot{\lambda}_d \frac{\partial \phi_d}{\partial Y_d}, \dot{z} = -\dot{\lambda}_d \frac{\partial \phi_d}{\partial Z},$
	$\dot{\varepsilon}_T^s = \dot{\lambda}_s \operatorname{sign}(\sigma_T^s)$, and $\dot{\alpha} = -\dot{\lambda}_s \operatorname{sign}(\sigma_T^s) + \frac{3}{2}\alpha X$

global behaviour, the local interface behaviour characterisation requires that the local friction and cracking to be modelled as a coupled phenomenon. This can be achieved using thermodynamics. In developing the thermodynamic model, the following aspects need to be considered: (a) cracking for an excessive tangential stress, $\sigma_{tt} = \sigma_T$; (b) inelastic strain due to sliding, $\in_T{}^S$; (c) coupling between tangential and normal stresses, $\sigma_{\mathrm{nn}} = \sigma_N$, in the sliding phase; and (d) hysteretic behaviour due to friction. The interface element is activated if and only if there is relative displacement between two bodies in contact. The main ingredients of the thermodynamics based bond-slip model are presented in Table 3 [20].

From the review of the above two models, it is clear that when fracture energy-based models are used in FEM of RC members, the gradient crack model proposed by Sluys and de Borst [35] with a refined bond-slip model can be used. However, it is felt that the constitutive relations proposed by Dominguez et al. [20] are best suited since they are based on strong displacement field discontinuity hypothesis. This type of model is recommended in probabilistic simulation. However, in this chapter, an attempt is made to first suggest a probabilistic model for surface strains based on phenomenological considerations and then examine its applicability based on experimental observations only. And no FE modelling and simulations are attempted. These studies are being continued at CSIR-SERC.

3 Probabilistic Model for Surface Strains in RC Members

It is known that when the principles of thermodynamics are applied to describe the cracking phenomenon of RC beams, the stage at which emergent structure forms corresponds to a transient nonequilibrium condition and subsequent formation of a metastable state(s) [3, 15]. The loading drops at the incipience of an emergent structure marking the nonequilibrium thermodynamic state. This is transient non-equilibrium thermodynamic state because further increase in load (from the load level to which it has reduced) can be achieved only with the increase in deflection of the beam. Once a particular stabilised crack pattern has formed, the beam will take further loading, which defines local equilibrium state with respect to cracking. This crack pattern will correspond to a metastable equilibrium state. The response evolution at and around the point of instability (such as the point 'B' which is in nonequilibrium state) is governed more by the fluctuations than the mean. Hence, mean field theory cannot be used to predict the behaviour around this point. However, beyond the unstable point, the response evolution can be predicted using mean field theory till another nonequilibrium point is reached (if at all possible). This behaviour is depicted in Fig. 9.

It has been pointed out in the literature [3, 15, 32] that the application of thermodynamic principles to irreversible processes (which is typically the case of cracking in reinforced concrete beams) would result in large fluctuation and

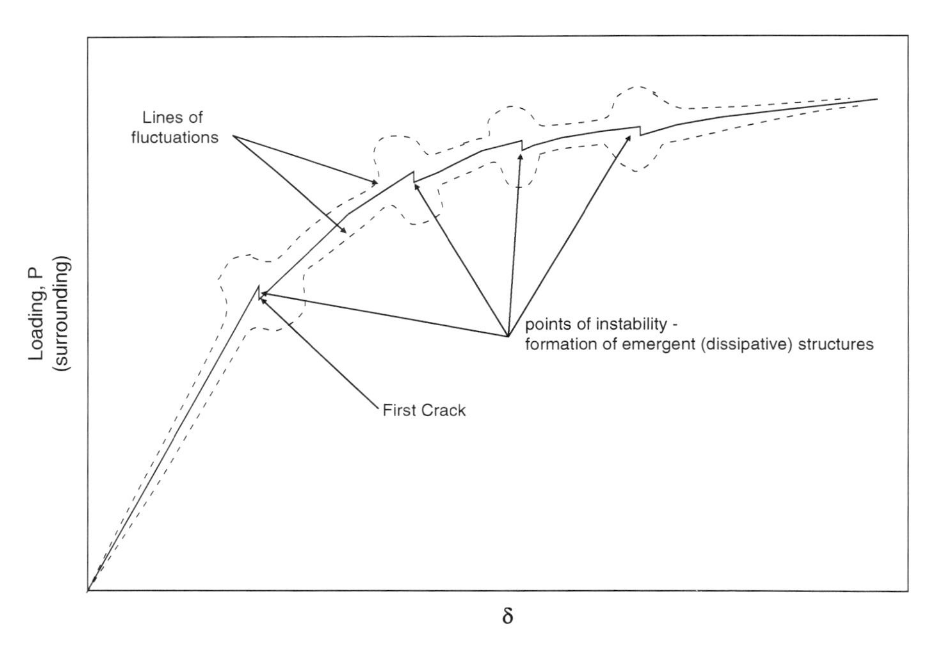

Fig. 9 Schematic of response evolution with loading showing formation of emergent (dissipative) structures

formation of a new (emergent) structure at the points of instability. This is due to the presence of wild randomness, which can simply be elucidated as follows: an environment in which a single observation or a particular number can impact the total in a disproportionate way [26]. The traditional Gaussian way of looking at the world begins by focusing on the ordinary and then deals with exceptions or so-called outliers as ancillaries. A second way, which takes the so-called exceptional as a starting point and deals with the ordinary in a subordinate manner – simply because that 'ordinary' is less consequential. These two models correspond to two mutually exclusive types of randomness: mild or Gaussian on the one hand and wild, fractal or 'scalable power laws' on the other. To characterise and quantify the fluctuations at the points of instability, a distribution with heavy tails is to be used.

3.1 Justification for Use of Alpha-Stable Distributions

One of the significant factors affecting the surface strains in concrete is the cracking in concrete. Recent developments in NDE techniques (viz. AE, GPR and other sensors) have made it possible to study the cracking process in concrete at micro-scale levels. Colombo et al. [14], with a view to identify the damage due to cracking using acoustic emission (AE) technique, conducted experimental investigations on a RC beam subjected to cyclic loading. The beam was simply supported and was subjected to two-point bending. For each loading cycle, the peak load was increased from that in the previous cycle. After the end of last (tenth) loading cycle, the RC beam was found to be severely damaged. The AE data recorded at each loading cycle was analysed to determine the b-value (the b-value is the negative gradient of the log-linear AE frequency/amplitude plot). According to Colombo et al. [14], the changes in b-value can be related to the different stages of crack growth in the RC beam. At initial stages of loading, microcracks are dominant and the macrocracks are starting to appear. The b-value corresponding to this phase is found to be greater than 1.7. In the next phase, the macrocracks are uniformly distributed along the beam and no new macrocracks are forming. The b-value corresponding to this phase is found to be between 1.2 and 1.7. In the final phase, the macrocracks are found to be opening up, as the beam is failing, and the b-value corresponding to this phase is found to be between 1.0 and 1.2. These studies thus help in locating – and assessment (qualitative) of – the damage. Carpinteri and Puzzi [12], based on both in situ field test on RC member and laboratory tests on concrete specimens, have shown that the b-value can be linked to the value of exponent of the power-law form for the tail portion of the probability distribution of crack size and that the value of exponent of the power law can be interpreted as the fractal dimension of the damage domain. Therefore, the random variables associated with crack size in RC flexural beams should have a form consistent with the power-law distribution (thus, may not have finite moments). By viewing the surface strains as a result of the indicated microscopic phenomena (such as bond-slip between steel and concrete, microcracking in concrete), the limiting distribution (attractor) is to be an alpha-stable distribution, knowing that the microscopic components may have power-law distributions.

Thermodynamics Considerations: As mentioned earlier, one of the outstanding features of the behaviour of RC flexural beams under the external loading is the emergent structure at different stages of loading. The loading drops at the incipience of an emergent structure marking the nonequilibrium thermodynamic state. This is transient nonequilibrium thermodynamic state because further increase in load (from the load level to which it has reduced) can be achieved only with the increase in deflection of the beam. Once a particular stabilised crack pattern has formed, the beam will take further loading, which defines local equilibrium state with respect to cracking. This crack pattern will correspond to a metastable equilibrium state. The response evolution at and around the point of instability is governed more by the fluctuations than the mean. Hence, to characterise and quantify the fluctuations at these points, a distribution with heavy tails is to be used.

Based on Statistical Arguments: For normal distribution, 99.74% of total probability content is contained within three times standard deviation about the mean, thus giving low values of probability to the tail regions. The symmetric nature of normal distribution also restricts its applicability to phenomena exhibiting small skewness. Hence, there is a need to use distributions with heavy tails to model the strains showing large fluctuations and to estimate the extreme values. While there are different heavy-tailed alternatives to normal distribution, like alpha-stable distribution, student's t-distribution and hyperbolic distribution, the use of alpha-stable distribution is supported by the generalised central limit theorem (see Appendix A). The use of alpha-stable distribution over normal distribution has found applications in different areas (see [30, 37]). The use of alpha-stable distribution for modelling the strain in RC flexural beams at a given loading stage is explored in the present study to account for the large fluctuations.

From the above, it is clear that to predict the extreme value of strains developed in RC flexural beams, at any stage of loading, a probability distribution with power-law tails should be used.

3.2 *Alpha-Stable Distribution*

The alpha-stable distribution is described by its characteristic function (an explicit expression for probability density function generally does not exist) given by

$$L_{\alpha,\beta}(t) = E[\exp(itX)] = \begin{cases} \exp\{-c^{\alpha}|t|^{\alpha}[1 - i\beta\,\mathrm{sign}(t)\tan(\frac{\pi\alpha}{2})] + i\delta t\}; & \text{for } \alpha \neq 1 \\ \exp\{-c|t|[1 + i\beta\,\mathrm{sign}(t)\,\frac{2}{\pi}\ln|t|] + i\delta t\}; & \text{for } \alpha = 1 \end{cases} \tag{12}$$

where X is the random variable, i is the imaginary unit, t is the argument of the characteristic function ($t \in \Re$), $E[\exp(itX)]$ denotes the expected value of $\exp(itX)$, α is an index of stability or characteristic exponent ($\alpha \in (0, 2]$), β is the skewness parameter ($\beta \in [-1, 1]$), c is a scale parameter ($c > 0$), δ is a location parameter ($\delta \in \Re$), ln denotes the natural logarithm and sign(t) is a logical function which takes values -1, 0 and 1 for $t < 0$, $t = 0$ and $t > 0$, respectively. As α approaches 2, β loses its effect and the distribution approaches the normal distribution regardless of β [9]. A stable probability density function (PDF) is symmetrical when $\beta = 0$.

3.2.1 Estimation of Parameters of Alpha-Stable Distribution

Different methods have been proposed in literature for the estimations of the parameters α, β, c and δ of the alpha-stable distribution. Fama and Roll [21] suggested a quantile-based method for estimation of characteristic exponent and scale parameter of symmetric alpha-stable distributions with $\delta = 0$. However, this method is applicable only for distributions with $\alpha > 1$. This method has been modified by McCulloch [27] to include even nonsymmetric distributions with α in the range [0.6, 2.0]. Koutrouvelis [24] proposed a characteristic function-based method involving an iterative regression procedure for estimation of the parameters of the alpha-stable distribution. Kogon and Williams [22] improved this method by eliminating the iterative procedure and simplifying the regression. Ma and Nikias [25] and Tsihrintzis and Nikias [36] proposed the use of fractional lower-order moments (FLOMs) for estimating the parameters of symmetric alpha-stable distributions. Bates and McLaughlin [4] studied the performances of the methods proposed by McCulloch [27], Kogon and Williams [22], Ma and Nikias [25], and Tsihrintzis and Nikias [36] using two real data sets. They found that there are marked differences between the results obtained using the different methods. In the present study, the parameters α, β, c and δ of the alpha-stable distribution are estimated using an optimisation procedure by minimising the sum of squares of the difference between the observed cumulative distribution function (empirical distribution function) and the cumulative distribution function (CDF) of the alpha-stable distribution.

4 Applicability of Alpha-Stable Distribution

In this section, an attempt is made to examine the applicability of alpha-stable distribution to model the surface strains in RC flexural members based on experimental observations. The data on variation of strain with loading for RC beams used in this study is based on the experimental investigations carried out at CSIR-SERC, Chennai, India, and the experimental investigations reported by Prakash Desayi and Balaji Rao [17, 18]. This data is used in the present study, since strain measurements over the entire constant bending moment region along the span at different positions for different loading stages (up to ultimate) for RC flexural beams have been taken and reported, which will be useful for studying the usefulness of alpha-stable distribution for modelling the variations in measured strain. Availability of such extensive data is scanty in literature. Salient information regarding the experimental investigations is given below.

Table 4 Details of beams tested at CSIR-SERC

Set	No. of specimens	Grade of concrete	Main reinforcement[a]	Stirrups	Hanger bars	Clear cover (mm)
B2	2 (B2-1, B2-2)	M30	2, 10-mm dia 2, 12-mm dia	6-mm dia @ 140-mm c/c	2, 6-mm dia	25
B3	2 (B3-1, B3-2)	M30	5, 10-mm dia			

[a] The main reinforcement is selected so as to have approximately the same cross-sectional area of reinforcement for all the beams

4.1 Details of Experimental Investigations

4.1.1 Experimental Investigations Carried Out at CSIR-SERC

Two sets of beams (with two beams in each set) of similar cross-sectional dimensions of 150 mm × 300 mm and 3.6 m long were cast and tested in four-point bending over an effective span of 3.0 m. Stirrups of 6-mm diameter were provided in the combined bending and shear zone to avoid shear failure, and no stirrups were provided in the constant bending moment zone. Details and properties of the beams are given in Tables 4 and 5. In the constant bending moment zone of the beams (i.e. 1 m long), seven sections (denoted as D_1, C_1, B_1, A, B, C and D on the north face and D', C', B', A', B_1', C_1' and D_1' on the south face) were identified, with each section having a gauge length of 100 mm (see Fig. 1).

In each section, demec points were fixed at five different positions across the depth on both the faces of the beam (north face and south face). As can be seen from Fig. 1, position 1 corresponds almost to the extreme compression fibre for all beams, and position 5 corresponds to position of bottom layer of main reinforcing bars. The beams were tested in four-point loading. To measure the surface strains at different positions, a Pfender gauge with least count 1/1,000 mm and gauge length of 100.1 mm was used. While the strains were also monitored using electrical strain gauges embedded on the reinforcement, for health assessment and maintenance decision-making, strain readings from surface-mounted strain gauges are more useful than point estimates of strain. Hence, the present study focuses on modelling the surface strains at different stages of loading.

At each loading stage, the strain readings are taken only after the applied load is stabilised, i.e. the loading has been increased to its original value after the drop in loading due to the incipience of emergent structure. The crack pattern observed for beams B2-1 and B3-1 at different stages of loading is shown in Fig. 10. The formation of emergent structure with loading is evident from these figures.

4.1.2 Experimental Investigations Presented by Desayi and Balaji Rao [17, 18]

Three beams of similar cross-sectional dimensions of 250 mm × 350 mm and 4.8 m long were cast and tested in two-point bending over an effective span of

Table 5 Properties of beams tested at CSIR-SERC

Beam	d_1[a] (mm)	d_2[a] (mm)	Effective depth (d) (mm)	A_{st1}[b] (mm^2)	A_{st2}[b] (mm^2)	A_{st} (mm^2)	150-mm concrete cube compressive strength (MPa)[c]	Split tensile strength (MPa)[c]	Cracking load (kN)[b]	Ultimate load (kN)[c]
B2-1	227.6	263.5	249.1	139.94	208.85	348.79	45.39	1.91	21.10[d]	77.74
B2-2	227.6	263.2	249.0	144.36	217.56	361.92	50.52	1.76	21.10[d]	82.89
B3-1	229.3	264.3	250.5	144.0	219.66	363.66	50.30	1.85	21.10[d]	83.88
B3-2	229.5	264.5	250.4	146.30	217.72	364.01	43.03	1.90	21.10[d]	82.80

Note: span (l) = 3000 mm, breadth (b) =150 mm and depth (D) = 300 mm for all the beams

[a]d_1 and d_2 are the depth from top of the beam to the centre of the main reinforcing bars in the top layer and bottom layer, respectively

[b]Based on measured diameters of the reinforcing bars

[c]Obtained from experimental investigations

[d]Cracking is initiated between applied loads of 14.72 and 27.47 kN and the average of these two loads is reported here as the cracking load

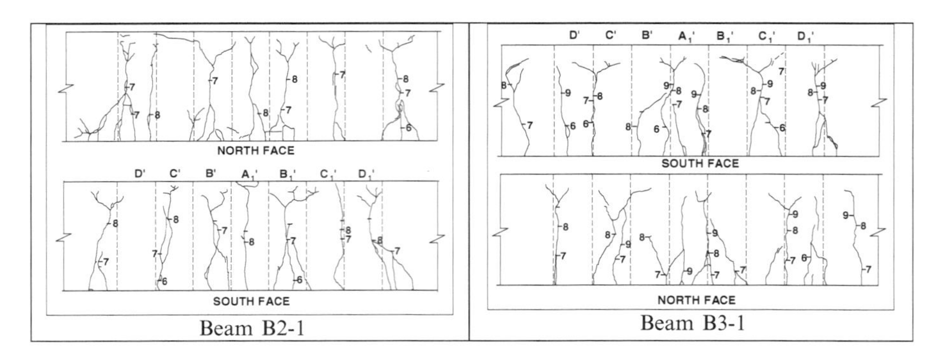

Fig. 10 Observed crack pattern for beams B2-1 and B3-1 (numbers denote the loading stage)

Table 6 Details of beams presented in Prakash Desayi and Balaji Rao [17, 18]

Beam	Effective depth (d) (mm)	A_{st} (mm^2)	150-mm concrete cube compressive strength (MPa)	Modulus of rupture (MPa)[a]	Cracking load (kN)[a]	Ultimate load (kN)[a]
KB1	311.0	402.123	33.078	4.036	23.549	95.389
KB2	305.4	437.929	40.417	3.578	14.014	104.653
KB3	303.5	529.327	22.508	2.950	8.899	84.291

Note: span (l) = 4200 mm, breadth (b) =200 mm and depth (D) = 350 mm for all three beams
[a]Obtained from experimental investigations

4.2 m. Stirrups of 6-mm diameter were provided in the combined bending and shear zone to avoid shear failure, and no stirrups were provided in the constant bending moment zone. Details of the beams are given in Table 6. The constant bending moment zone of the beams (1.4 m) was divided into eight sections (denoted as D', C', B', A', A, B, C and D on the west face and D_1, C_1, B_1, A_1, A_1', B_1', C_1' and D_1' on the east face), with each section having a gauge length of 200 mm (see Fig. 11). In each section, demec points were fixed at eight different positions on both faces of the beam (east face and west face). As can be seen from Fig. 11, position 1 corresponds almost to the extreme compression fibre for all beams, and position 7 in case of beam KB1 and positions 7 and 8 in case of beams KB2 and KB3 correspond to position of steel bars. The beams were tested in two-point loading in a 25-ton (245.25 kN) capacity testing frame. To measure the surface strains at different positions, a demec gauge with least count 1×10^{-5} and gauge length of 200.1 mm was used.

4.2 *Statistical Analysis of Strain*

Only the strains measured at the level of tension reinforcement are considered further in the analysis.

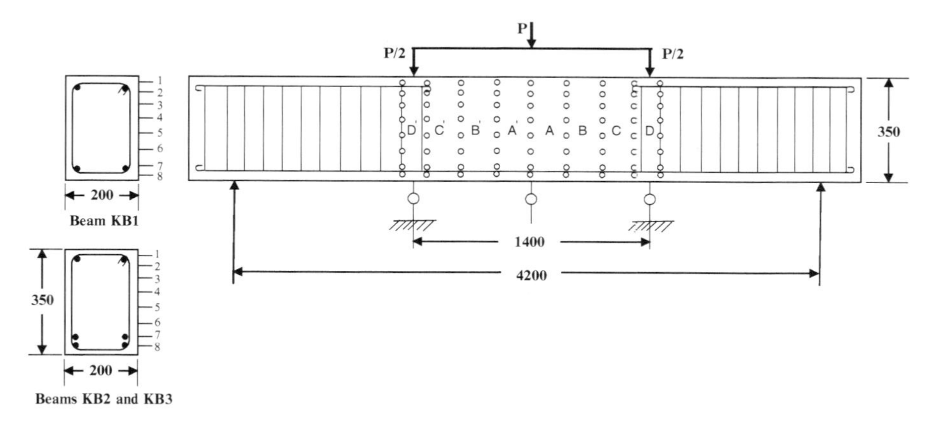

Fig. 11 Schematic representation of test programme [17, 18] (dimensions in mm)

4.2.1 Statistical Modelling of Strains for the Beams Tested at CSIR-SERC

In the present study, the aim is to model the strain in concrete at the level of bottom layer of main reinforcement (position 5 in Fig. 1) for a given loading stage. Three loading stages are considered for the sets B2 and B3, namely, those corresponding to applied loads of 27.47, 40.22 and 52.97 kN. It is known that evaluation of performance of reinforced concrete structural members under service loads is important in service life estimation. By defining the service load as approximately two-thirds of ultimate load, it is noted that the applied loads corresponding to the three loading stages considered in this study are less than the service loads, thereby helping in understanding the strain behaviour of flexural members under normal working load conditions. The typical variation of average strain across the depth for different stages of loading for the beam B2-2 is shown in Fig. 12. It is noted that the strain variation across the depth is almost linear for the loading stages considered in the present study. Similar observation is made for the other three beams also; however, the results are not presented here. This is in line with the observation regarding strain variation across the depth made by Neild et al. [28].

At any given loading stage, 14 strain readings (seven on the north face and seven on the south face) in the constant bending moment region are available for each beam in the sets B2 and B3. To enhance the sample size, the strain readings, corresponding to the same applied load, of both the beams in each set are combined together. This can be justified since the ultimate loads, cube compressive strengths and split tensile strengths for the two beams in each set are comparable with each other (see Table 5). It is also noted that the depth of neutral axis (determined using the strain gauge readings at different positions) at different loading stages for the two beams in each set is comparable, except for set B2 at an applied load of 27.47 kN. After combining the respective strain readings of beams belonging to sets B2 and B3, there are 28 strain readings at any given loading stage. These values are further processed for modelling the random variations in strain in concrete (at the level of reinforcement).

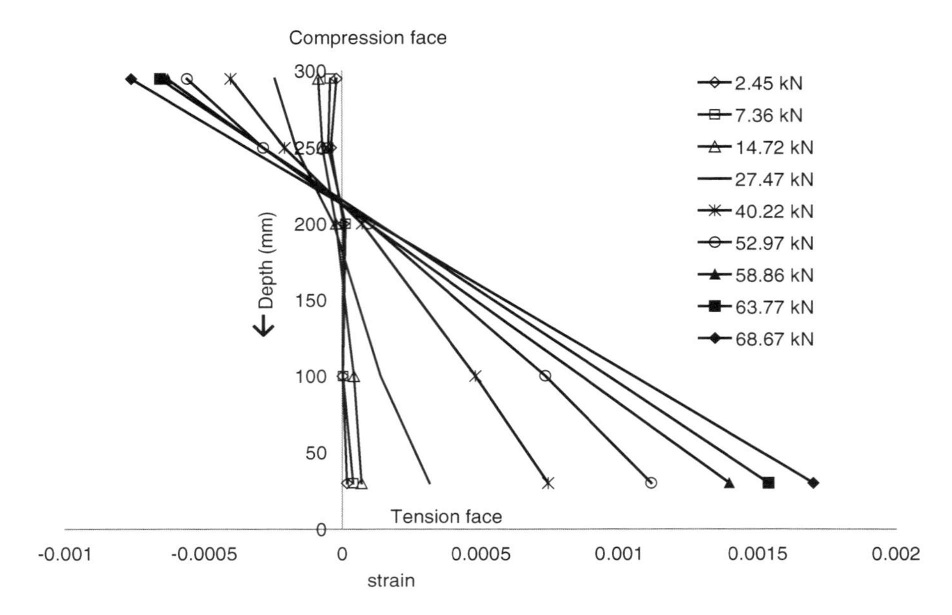

Fig. 12 Strain variation across the depth for different stages of loading for beam B2-2

4.2.2 Statistical Modelling of Strains for the Beams Presented by Desayi and Balaji Rao [17, 18]

At any given loading stage, 16 strain readings (eight on the west face and eight on the east face) in the constant bending moment region are available for the beams KB1, KB2 and KB3. These values are further processed for modelling the random variations in strain in concrete at the level of reinforcement (position 7 in case of beam KB1 and position 8 in case of beams KB2 and KB3).

4.3 Results and Discussion

The statistical properties (viz. mean, standard deviation and skewness) of the strain in concrete at the level of reinforcement have been computed based on the observed strain values. An alpha-stable distribution, $S(\alpha, \beta, c, \delta)$, is fitted to the strains in concrete at the level of reinforcement at each loading stage for the beams considered. The parameters α, β, c and δ of the alpha-stable distribution (Eq. 1) are estimated using an optimisation procedure by minimising the sum of squares of the difference between the observed cumulative distribution function (empirical distribution function) and the cumulative distribution function (CDF) of the alpha-stable distribution.

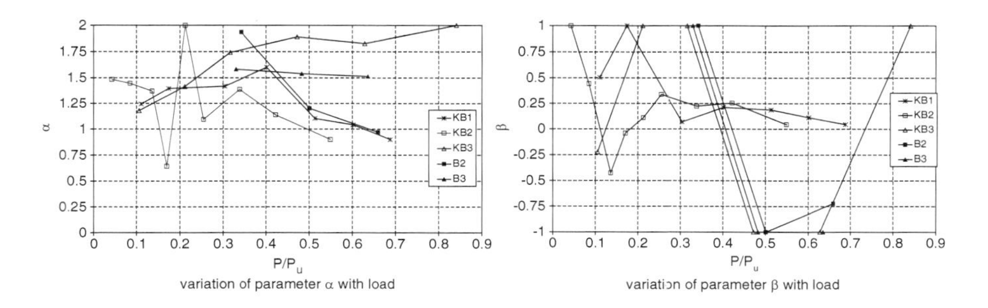

Fig. 13 Variation in parameters α and β with applied load

4.3.1 Beams Tested at CSIR-SERC

The variation in characteristic exponent, α, with applied load is shown in Fig. 13. It is noted that the values of α decrease with increase in applied load for both B2 and B3. It is known that for $\alpha = 2$, the alpha-stable distribution becomes normal distribution, and as α reduces, the tails get heavier, i.e. the tail probabilities increase [30]. This indicates that at higher stages of loading, the strain distribution deviates away from the normal distribution. This can be attributed to the formation of emergent structures and associated strain redistributions in concrete as explained in the section on mechanism of cracking. It is also noted from 13 that the reduction in α is much higher in set B2 when compared to set B3. This may be because in set B3, the cracks are more evenly distributed (lower values of crack spacing which can be attributed to the more number of smaller-diameter reinforcing bars in the beams in set B3 when compared to beams in set B2, with total area of steel in tension zone remaining approximately the same), and hence, the variability in strain is less, leading to lower values of tail probabilities.

From Fig. 13, it is noted that the value of the stability parameter (α) is almost a constant for set B3, while it shows large variation for set B2. This indicates that the strain distribution is almost stabilised for set B3. Since the strains in concrete in the tension depend on the level of cracking, a stabilised strain distribution suggests that the cracking has stabilised, i.e. no new cracks are being formed with increase in loading, rather the existing cracks are widened and extended. This is also supported by the observed crack patterns for the beams in set B3 (see Fig. 10), from which it is noted that no new major cracks are formed after the loading stage 7 (corresponding to applied load of 40.22 kN). However, from the observed crack pattern for a beam in set B2 (see Fig. 10), it is noted that a major crack has formed at the loading stage 8 (corresponding to applied load of 52.97 kN). This shows that the cracking and hence the strain distribution have not stabilised for set B2. This trend is also reflected in the values of the scale parameter (c) and the location parameter (δ). The values of c and δ increase with applied load when the strain distribution is stabilised (as is the case for set B3), indicating that the strains are increasing at an almost uniform rate in all the sections. However, when new cracks form (as is the case for set B2), there is sudden increase of strain in the section containing the crack and a decrease in strain in the adjacent sections, leading to abrupt variations in c and δ.

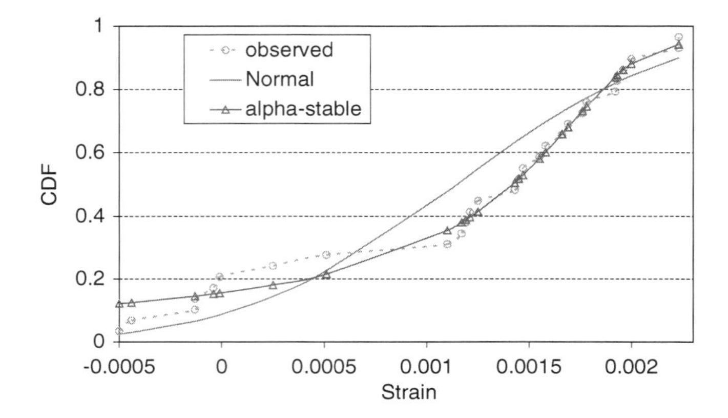

Fig. 14 Comparison of CDFs of strain for set B2 (applied load = 52.97 kN)

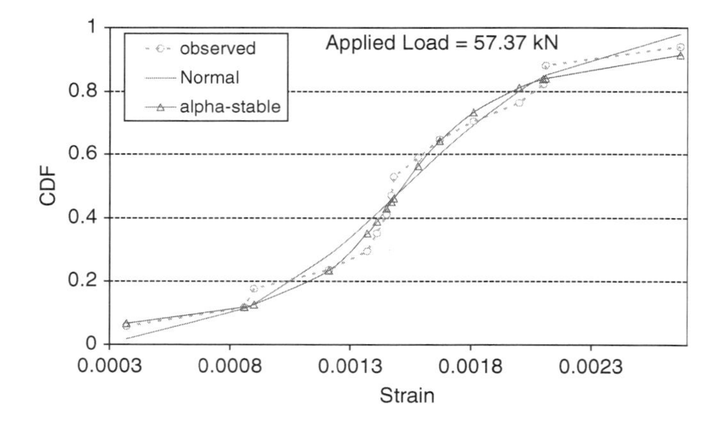

Fig. 15 Comparison of CDFs of strain for beam KB1 (applied load = 57.37 kN)

For the purpose of comparison, a normal distribution is also fitted to the observed strain data (by considering the mean and standard deviation of the normal distribution as the average and standard deviation of the observed strain data). The observed CDF, alpha-stable CDF and the normal CDF, typically for the set B2, for an applied load of 52.97 kN, are shown in Fig. 14. From this figure, it is noted that CDF corresponding to the alpha-stable distribution compares with the observed CDF better than the normal CDF (especially in the tail regions which are of interest in estimating extreme values of strains).

4.3.2 Beams Presented by Desayi and Balaji Rao [17, 18]

The observed CDF, alpha-stable CDF and the normal CDF, typically for the beam KB1, for an applied load of 57.37 kN, are shown in Fig. 15. From Fig. 15, it is noted that CDF corresponding to the alpha-stable distribution compares with the observed CDF better than the normal CDF. This observation suggests that alpha-stable distribution is a better fit to the observed strain readings (especially in the tail regions which are of interest in estimating extreme values of strains).

4.3.3 General Observations

From the variation in α with applied load (see Fig. 13) for the beams considered, it is noted that in general, α becomes close to 1.0 at loads corresponding to approximately the working loads. It may be noted that the skewness parameter β is not the same as the classical skewness parameter [30], since for non-Gaussian stable distributions, the moments do not exist. In the case of the alpha-stable distribution, the values of β indicate whether the distribution is right-skewed ($\beta > 0$), left-skewed ($\beta < 0$) or symmetric ($\beta = 0$). It is noted that the trends of skewness of the fitted alpha-stable distribution are in agreement with the trends of skewness shown by the observed strain readings. This observation also suggests the ability of the fitted alpha-stable distribution to represent the observed variations in strain. From the variation of β with applied load (Fig. 13) for the beams considered, it is noted that in general, the strain distribution becomes symmetrical ($\beta = 0.0$) at loads corresponding to approximately the working loads.

5 Conclusions

Usefulness of alpha-stable distributions for modelling the variations in surface strain in reinforced concrete flexural members, at the level of reinforcement, is studied in this chapter, using experimentally obtained strain values from reinforced concrete flexural members tested under four-point bending. From the results obtained, it is noted that alpha-stable distribution is a better fit to the observed strain readings at a given stage of loading than the normal distribution, especially in the tail regions which are of interest in estimating extreme values of strains.

6 Supplementary Material

6.1 Why Do We Need Probability Distributions with Fat Tails to Describe the Mechanism of Cracking in Reinforced Concrete Flexural Members?

The information presented in this supplementary note is based on the concepts presented by Prof. Prigogine [33]). An attempt has been made to interpret these concepts to the phenomenon of cracking in reinforced concrete flexural members.

According to Prigogine, the concept of time in the case irreversible thermodynamic systems can be replaced by associating or suitably defining a variable associated with the phenomenon of formation of dissipative structures. Through suitable thermodynamic formulations, he has addressed the problem associated with these systems both at macroscopic and microscopic levels simultaneously. However, the use of Helmoltz free energy in the thermodynamic formulations is questioned since non-equilibrium can be source of order. Therefore, in order to formulate the problem of cracking which forms dissipative structures, the concept of open system needs to be adopted. The two components, namely, internal and external, are body of the material containing microcracks and the fictitious system containing localised macrocracks which dissipate energy through mainly surface energy. These two systems are coupled and there exists a boundary between them. In this way, the thermodynamic formulations with strong discontinuity seem to show promise (*details of formulations are reviewed in the main paper*).

The second law of thermodynamics suggests that the change in entropy is equal to or greater than zero. For a closed system, the entropy production is through the irreversible damage and sets one-sidedness of time. The positive sidedness of time is associated with increase of entropy. If entropy remains constant, time will not increase! This may be a problem with thermodynamics of closed systems. Hence, more often thermodynamics principles are applied to describe the systems near the equilibrium. To extend the thermodynamics to the nonequilibrium processes, we need an explicit expression for entropy production. Progress has been achieved along this line by supposing that even outside equilibrium, entropy depends only on the same variables as at equilibrium. This assumption leads to 'local' equilibrium. This assumption enables us to use the formulations similar to equilibrium thermodynamics. Local equilibrium requires that

$$\frac{\mathrm{d}_i S}{\mathrm{d}t} = \sum_{\rho:} J_\rho X_\rho \geq 0 \tag{13}$$

The LHS of Eq. (13) gives the rate of production of internal entropy by the system due to various irreversible processes (ρ) at a macroscopic level. J_ρ and X_ρ are rate of entropy production by individual irreversible process ρ and the driving force of the process ρ. This is the basic formula of macroscopic thermodynamics of irreversible process. At thermodynamic equilibrium, we have simultaneously for all irreversible processes

$$J_\rho = 0;\ X_\rho = 0 \tag{14}$$

It is therefore natural to assume that at least near the equilibrium linear homogeneous relations between flows and forces. The assumption of thermodynamic local equilibrium for nonequilibrium systems and application of above approach, at a macroscale, allows the use of empirical laws such as Fourier's law and Fick's second law of diffusion to various phenomenon under consideration. It may be noted that we have not yet included the complex interaction that may takes place between the irreversible damage processes that are producing the entropy. The answer lies in the linearisation of the system at least near the 'local' equilibrium. This assumption enables the principle of superposition which is central to the local equilibrium thermodynamics and enables us to determine the rates of flow (flux) of a given irreversible process, taking into account the interactions among various irreversible processes contributing to the macroscopic equilibrium, using phenomenological coefficients ($L_{\rho\rho'}$) from the following relation,

$$J_\rho = \sum_{\rho'} L_{\rho\rho'} X_{\rho'} \tag{15}$$

Linear thermodynamics of irreversible processes is dominated by two important results, namely, Onsager reciprocity relations and the principle of minimum entropy production at or very near the local equilibrium point. The Onsager reciprocity relation is given by,

$$L_{\rho\rho'} = L_{\rho'\rho} \tag{16}$$

When the flow J_ρ, the flow corresponding to irreversible process ρ, is influenced by the force $X_{\rho'}$ of irreversible process ρ', then the flow $J_{\rho'}$ is also influenced by the force X_ρ through the same phenomenological coefficient. It may be seen that the Onsager's reciprocity relation is similar to the Betti's theorem in structural engineering.

We have established the 'local' equilibrium dynamics of the irreversible thermodynamic open system. The two central concepts have been explained above. The theorem dealing with minimum internal entropy production is very significant since it gives some kind of 'inertial' property to the nonequilibrium system near the local equilibrium point. When given boundary conditions prevent the system from reaching thermodynamic equilibrium (i.e. zero entropy production), the system settles down to the state of 'least dissipation'. It is to be noted that for far-from equilibrium points, the thermodynamic behaviour could be quite different. It has been proved now that the behaviour of the system can be opposite of minimum entropy production. In fact, the state of nonequilibrium (wherein there can be production of internal entropy) may be source of order at a macroscale.

It is interesting to note that Boltzmann's order principle as expressed by canonical distribution assign almost zero probability to the occurrence of Benard convection. Whenever new coherent states occur far from equilibrium, the very concept of probability, as implied in the counting of number of complexions, breaks down. In the case of Benard's convection, above a critical temperature small convection currents, appearing as fluctuations, get amplified and give rise to macroscopic current. A new supermolecular order appears which corresponds basically to a giant fluctuation stabilised by exchanges of energy with the outside world. This is the order characterised by the dissipative structures.

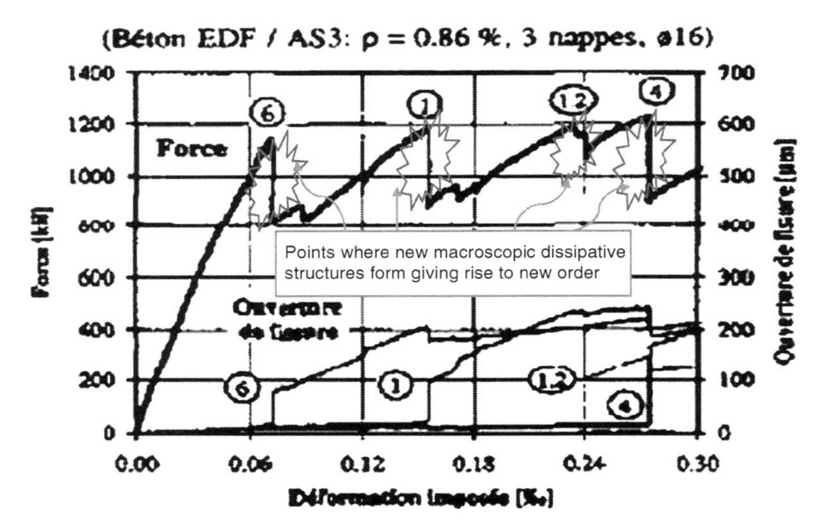

Fig. 16 Force-deformation relation of an RC member subjected to axial tension (From Ibrahimbegovic et al. [22])

Figure 16 is very important in light of the discussions presented till now. The load-deformation process in RC member under monotonically increasing loads involves irreversible damage process. But, as presented in the paper, the cracking process needs to be considered as both closed and open thermodynamic system forming dissipative structures at critical points (such as points 6, 1, 12,4 in Fig. 16). But for points at which loading drops, which are points of nonequilibrium and far from equilibrium, local equilibrium can be obtained and we can define the local stationary states of the system (this enables us to use Onsager's reciprocity relation and the minimum dissipation theorem). At the points where it has been marked as red balloons, new emergent dissipative structures form (i.e. new macrocracks form on the surface of the beam) indicating that these irreversible states correspond to the points far from equilibrium condition (something similar to Benard's convection presented by Prigogine). And, it is at these points the surface strains show large variability and the applicability of probability distribution with exponential tails is questionable. More figuratively, this is very clearly brought out in thermodynamic framework in Fig. 17.

Prigogine has shown that near critical points as well as near the coexistence curve (shown in red balloons in Fig. 16), the law of large numbers, as expressed by the expression

$$\frac{\left\langle (\delta X)^2 \right\rangle}{V} \sim \text{finite for } V \to \infty \tag{17}$$

(where X is a random variable representing an extensive quantity of thermodynamics), breaks down, as $\left\langle (\delta X)^2 \right\rangle$ becomes proportional to a higher power of volume.

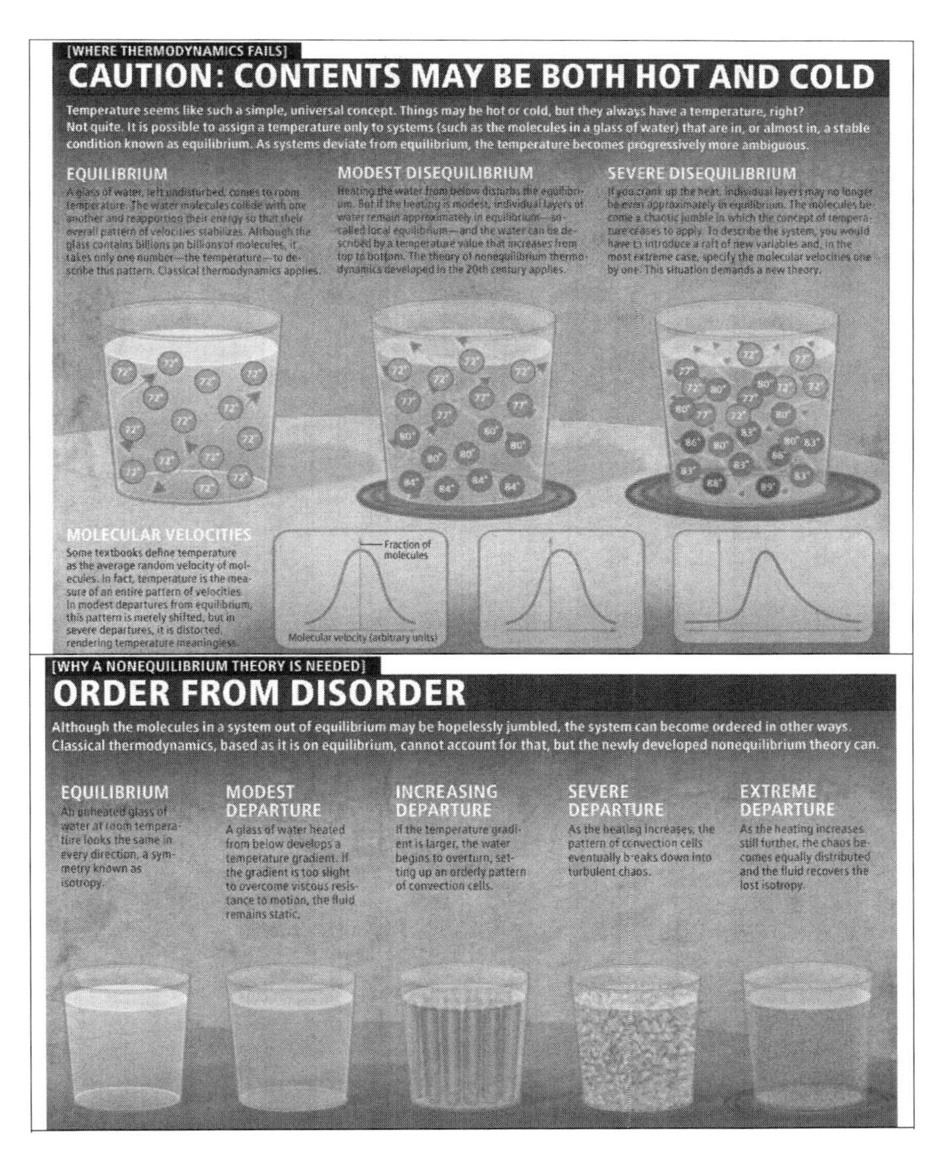

Fig. 17 Need for use of alpha-stable distribution at points of far from equilibrium due to formation of dissipative structures bringing the order (From Rubi [34])

Prigogine [33] has shown that near the critical points, probability distributions with long-range memory would be needed. It is in this context, alpha-stable distributions are proposed to describe the fluctuations in surface strains near the critical points in the flexural members.

6.1.1 Derivation of a Levy Distribution (Which Is Also Called Alpha-Stable Distribution)

The intrinsic dynamical features of the physical system need to be included in any stochastic evolution model. From the very beginning of chaotic dynamics, there were observations of intermittent character of the time/load behaviour of chaotic trajectories (as exhibited by the load-displacement curve of RC flexural member). As pointed out above, the Gaussian processes and normal diffusion are not always valid (and many situations arise when independent and weakly dependent random event models need to be abandoned).

Levy in 1937 (in Mandelbrot [25]) reconsidered law of large numbers and formulated a new approach that can also be applied to distributions with infinite second moment. The importance of approach of Levy's distributions and processes soon became clearer in many fields. Mandelbrot in [25] has presented numerous applications of the Levy's distributions and coined the notion of Levy flights.

Let $p_X(x)$ be normalised probability density function of random variable X. That is,

$$\int_{-\infty}^{\infty} p_X(x)\mathrm{d}x = 1 \tag{18}$$

with characteristic function

$$\Phi_X(t) = \int_{-\infty}^{\infty} e^{itx} p_X(x)\mathrm{d}x \tag{19}$$

Consider two random variables X_1 and X_2 and their linear combination as follows:

$$CX_3 = C_1X_1 + C_2X_2 \qquad C, C_1, C_2 > 0 \tag{20}$$

The linear combination law is stable if all $X_i s$ $(i = 1,2,3)$ are distributed according to the same $p_{X_i}(x_i)$, $(i = 1, 2, 3)$

While Gaussian distribution satisfies the stable law, another class of distributions was given by Levy (1937).

Let us write the density function of X_3 in Eq. (20).

$$p_{X_3}(x_3)\mathrm{d}x_3 = p_{X_1}(x_1).p_{X_2}(x_2).\delta\left(x_3 - \frac{C_1}{C}x_1 - \frac{C_2}{C}x_2\right).\mathrm{d}x_1\mathrm{d}x_2 \tag{21}$$

We know that the characteristic function of a random variable made up of summation of several random variables is simply multiplication of characteristic functions of individual random variables. Hence, we get

$$\Phi_{CX_3}(Ct) = \Phi_{C_1X_1}(C_1t).\Phi_{C_2X_2}(C_2t) \tag{22}$$

Taking logarithms on both the sides, we get

$$\ln \Phi_{CX_3}(Ct) = \ln \Phi_{C_1X_1}(C_1t) + \ln \Phi_{C_2X_2}(C_2t) \tag{23}$$

Equations (22) and (23) are functional ones with a solution

$$\ln \Phi_\alpha(Ct) = (Ct)^\alpha = C^\alpha e^{-\frac{i\pi\alpha}{2}(1-\text{sign}t)}|t|^\alpha \tag{24}$$

under the condition

$$\left(\frac{C_1}{C}\right)^\alpha + \left(\frac{C_2}{C}\right)^\alpha = 1 \tag{25}$$

where α is an arbitrary parameter.

Any distribution $p_X(x)$ with characteristic function

$$\Phi_\alpha(t) = e^{-C|t|^\alpha} \tag{26}$$

is known as Levy distribution with Levy index α. An important condition imposed by Levy is $0 < \alpha \leq 2$, which guarantees of positiveness of probability density function.

$$p_\alpha(x) = \int_{-\infty}^{\infty} \mathrm{d}t.e^{itx}.\Phi_\alpha(t) \tag{27}$$

The case of $\alpha = 1$ is known as Cauchy distribution

$$p_1(x) = \frac{C}{\pi}\frac{1}{\left(x^2 + C^2\right)} \tag{28}$$

An important case is the asymptotic of large $|x|$.

$$p_\alpha(x) = \frac{1}{\pi}\alpha C\,\Gamma(\alpha)\sin\frac{\pi\alpha}{2}\frac{1}{|x|^{\alpha+1}}; \quad 0 < \alpha \leq 2 \tag{29}$$

It can be shown that the moments of $p_\alpha(x)$ of order m diverge for $m \geq \alpha$.

6.2 *Implication of the Use of Alpha-Stable Distributions in Performance-Based Design of RC Structures*

The behaviour of reinforced concrete flexural members under the external loading is quite complex, as explained in the above sections. One of the outstanding features

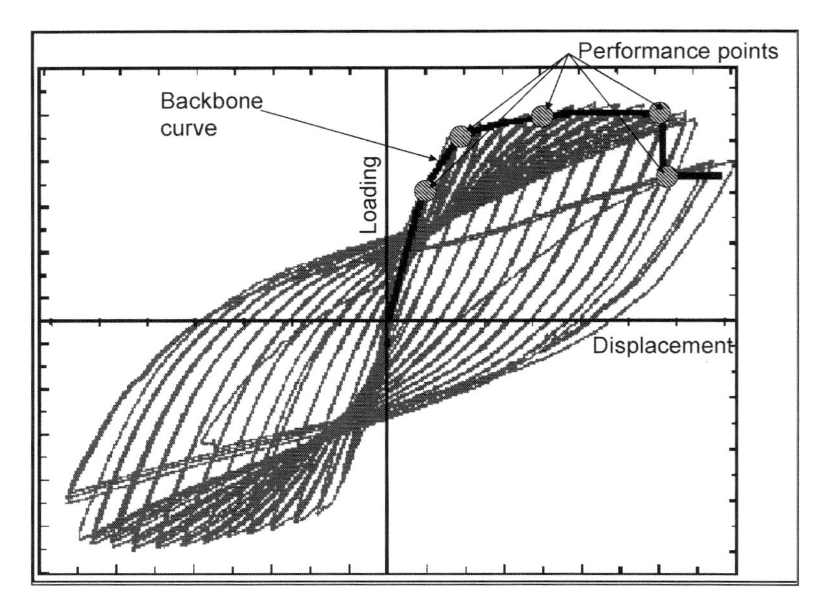

Fig. 18 Schematic of hysteretic behaviour under transient loading

of the behaviour is the emergent structures (marking the instability with respect to thermodynamic behaviour) at different stages of loading. The loading drops at the incipience of an emergent (dissipative) structure, namely, formation of new cracks and/or widening and extension of the existing cracks. Once a particular crack pattern has formed, the beam will take further loading, which defines local equilibrium state with respect to cracking. As has been explained in the earlier section on thermodynamic instability, the response evolution at and around the point of instability is governed more by the fluctuations than the mean. Hence, mean field theory fails around this point. However, beyond unstable point, the response evolution can be predicted using mean field theory.

The above discussion clearly shows that the phenomenon of cracking and deformation of reinforced concrete flexural members exhibits large scatter. To account for the inevitable large uncertainties in the behaviour, attempts have to be made to apply suitable probability distributions such as alpha-stable distributions and use convex sets to check the robustness of the system with respect to the performance limit states considered (around the points of instability shown as red balloon in Fig. 16).

While the above discussion is valid for reinforced concrete flexural members subjected to monotonic loading, this can be extended to structures/structural members subjected to transient loadings (such as earthquakes), by noting that for well designed reinforced concrete members (avoiding pinching shear failure), the load deformation response under monotonic loading forms the envelope for hysteretic behaviour (Fig. 18). The significance of this in the performance-based design is that when the reliability is sought near the point of instability, the uncertainties have to be treated properly and perhaps application of theory of convex sets is more appropriate. These suggest the need for application of hybrid reliability technique for estimating the reliability of reinforced concrete flexural members and structures than simple reliability techniques.

The research presented in this chapter has lead to the following two important observations and perhaps sets direction for future research:

- There is a need to have a relook at the probabilistic performance-based design using the capacity curve, obtained from nonlinear static pushover analysis, since at the points of bifurcation, the behaviour is to be modelled using probability distributions that account for large uncertainties. This requires a systematic approach to be followed in two steps: (a) carrying out careful experimental investigations on RC structural components and structures for characterising their load-displacement behaviour and (b) development of backbone curves that can be used in the performance-based design.
- There is also a need to examine the approximation of envelope curve as the backbone curve for representing the hysteretic behaviour of RC structural components and structures.

Acknowledgements This chapter is being published with the kind permission of Director, CSIR-SERC, Chennai, India. The author is very thankful to his colleague Dr. M. B. Anoop, Scientist, CSIR-SERC, in the preparation of this chapter. He thanks his colleagues Dr. K. Ravisankar, Chief Scientist, Head, Structural Health Monitoring Laboratory, Shri. K. Kesavan, Dr. M. B. Anoop and Shri. S. R. Balasubramanian, Scientists, CSIR-SERC, for their involvement in the experimental work. The MATLAB programmes developed by Mark Veillette, Ph.D. Scholar, Department of Mathematics and Statistics, Boston University, Boston, USA, have been used in the present study for determining the CDFs and PDFs of the alpha-stable distributions.

Appendix A

The central limit theorem states that the sum of independent, identical random variables with a finite variance converges to a normal distribution. Let $X_1, X_2, \ldots,$ $X_n$be independent identically distributed random variables. Define

$$S_n = \lim_{n \to \infty} \sum_{i=1}^{n} X_i \tag{A.1}$$

According to central limit theorem, S_n follows a normal distribution with mean and variance given by

$$\mu = \sum_{i=1}^{n} \mu_{X_i}$$
$$\sigma^2 = \sum_{i=1}^{n} \sigma_{X_i}^2$$

where μ_{X_i} and $\sigma_{X_i}^2$ are the mean and variance of X_i.

This suggests that as long as a macroscopic phenomenon is infinitely divisible into microscopic phenomena, which exhibits finite variance with exponential tails, the random variable associated with the macroscopic phenomenon can be represented using a normal distribution. However, wide variety of microscopic phenomena exhibit statistics that needs to be characterised with algebraic-tailed distributions. An example is the random variables associated with microcracks (size as well as geometry) in concrete which exhibit a power-law distribution [13] and hence may not have a finite mean and/or variance. In such cases, the random variable associated with the macroscopic phenomenon may not have a finite mean and/or variance (see Eq. A.2), and normal distribution will not be the limiting distribution of the sum $\sum_{i=1}^{\infty} X_i$.

An Important Asymptotic Property

A probability density $L(x)$ can be a limiting distribution of the sum $\sum_{i=1}^{\infty} X_i$ of independent and randomly distributed variables only if it is stable. A random variable X is stable or stable in the broad sense if for X_1 and X_2 independent copies of X and any positive constants a and b,

$$aX_1 + bX_2 \stackrel{d}{=} cX + d \tag{A.3}$$

holds for some positive c and some $d \in \Re$ [30]. The symbol $\stackrel{d}{=}$ means equality in distribution, i.e. both expressions have the same probability law. The term stable is used because the shape is stable or unchanged under sums of the type given by Eq. (A.3).

The Gaussian – and Cauchy – distributions are potential limiting distributions, depending on the physical phenomenon that is being handled. However, there are many more distributions to which the summation series $\sum_{i=1}^{\infty} X_i$ is attracted to depending on the actual behaviour. The complete sets of stable distributions have been specified by Levy and Khinchine. A probability distribution is stable if its characteristic function is of the form as given in Eq. (12).

While Eq. (12) defines the general expression for all possible stable distributions, it does not specify the conditions which the probability density function (pdf) $p(l)$ has to satisfy so that the distribution of the normalised sum $\hat{S}_n = \sum_{i=1}^{n} p_i(l)$ converges to a particular $L_{\alpha,\beta}(x)$ in the limit $n \to \infty$. If this is the case, one can say '$p(l)$ belongs to the domain of attraction of $L_{\alpha,\beta}(x)$'. This problem has been solved completely, and the answer can be summarised by the following theorem:

Theorem: *The probability density $p(l)$ belongs to the domain of attraction of a stable density $L_{\alpha,\beta}(x)$ with characteristic exponent α ($\alpha \in (0 < \alpha < 2)$) if*

$$p(l) \sim \frac{\alpha\, a^{\alpha}\, c_{\pm}}{|l|^{1+\alpha}} \quad \text{for} \quad l \to \pm\infty \tag{A.4}$$

where $c_+ \geq 0, c_- \geq 0$ and a are constants. These constants are directly related to the scale parameter c and the skewness parameter β by

$$c = \begin{cases} \dfrac{\pi(c_+ + c_-)}{2\,\alpha\,\Gamma(\alpha)\sin(\pi\,\alpha/2)} & \text{for } \alpha \neq 1 \\ \dfrac{\pi}{2}(c_+ + c_-) & \text{for } \alpha = 1 \end{cases} \tag{A.5}$$

$$\beta = \begin{cases} \dfrac{c_- - c_+}{c_+ + c_-} & \text{for } \alpha \neq 1 \\ \dfrac{c_+ - c_-}{c_+ + c_-} & \text{for } \alpha = 1 \end{cases} \tag{A.6}$$

Furthermore, if $p(l)$ belongs to the domain of attraction of a stable distribution, its absolute moments of order λ exist for $\lambda < \alpha$:

$$\left\langle |l|^{\lambda} \right\rangle = \int_{-\infty}^{\infty} \mathrm{d}l\, |l|^{\lambda}\, p(l) = \begin{cases} <\infty & \text{for } 0 \leq \lambda \leq \alpha(\alpha \leq 2) \\ \infty & \text{for } \lambda > \alpha(\alpha < 2) \end{cases} \tag{A.7}$$

The above discussion clearly brings out that the sum of independent random variables, as $n \to \infty$, may converge to an alpha-stable distribution, $\alpha = 2$ being a specific case, with $p(l) \sim 1/|l|^3$, as a Gaussian distribution. For all other values of characteristic exponent, α, $0 < \alpha < 2$, the sum would be attracted to $L_{\alpha,\beta}(x)$, and all these classes of stable distributions show the same asymptotic behaviour for large x. Thus, the central limit theorem can be generalised as follows:

The generalised central limit theorem states that the sum of a number of random variables with power-law tail distributions decreasing as $1/|x|^{\alpha+1}$ where $0 < \alpha < 2$ (and therefore having infinite variance) will tend to a stable distribution as the number of variables grows.

The characteristic exponent α and the skewness (symmetry) parameter β have to be interpreted based on physical significance. As already mentioned, α defines the shape of the distribution and decides the order of moments available for a random variable. Longer power-law tails will lead to divergence of even lower-order moments. This should not be treated as a limitation, since, in some of the physical systems, the pdf of response quantities can have power-law tails. This may also be true of nonlinear response of engineering systems, especially at bifurcation points, where the system can exhibit longer tail behaviour. Though this can be brushed aside as a transient behaviour, for seeking performance of a system, this needs to be effectively handled. Hence, it is important to understand the pdfs and associated properties so that the systems can be modelled realistically.

References

1. ACI (2002) Control of cracking in concrete structures. ACI manual of concrete practice, ACI 224. American Concrete Institute, Detroit
2. Balaji Rao K, Appa Rao TVSR (1999) Cracking in reinforced concrete flexural members – a reliability model. Int J Struct Eng Mech 7(3):303–318
3. Balaji Rao K (2009) The applied load, configuration and fluctuation in non-linear analysis of reinforced concrete structures – some issues related to performance based design. Keynote paper. In: Proceedings of international conference on advances in concrete, structural and geotechnical engineering (ACSGE 2009), BITS Pilani, India, 25–27 Oct 2009
4. Bates S, McLaughlin S (2000) The estimation of stable distribution parameters from teletraffic data. IEEE Trans Signal Process 48(3):865–870
5. Bazant ZP (2002) Scaling of structural strength, 2nd edn., 2005. Elsevier Butterworth-Heinemann, Oxford
6. Bazant ZP, Cedolin L (2010) Stability of structures – elastic, inelastic, fracture and damage theories. World Scientific, Singapore
7. Bazant ZP, Oh B-H (1983) Spacing of cracks in reinforced concrete. J Struct Eng ASCE 109:2066–2085
8. Borak S, Härdle W, Weron R (2005) Stable distributions. SFB 649 discussion paper 2005-008, SFB 649, Humboldt-Universität zu Berlin
9. Bresler B (ed) (1974) Reinforced concrete engineering, vol 1, Materials, structural elements, safety. Wiley, New York
10. BS 8110 (1997) Structural use of concrete. Code of practice for design and construction. British Standards Institution, UK
11. Carpinteri A, Puzzi S (2009) The fractal-statistical approach to the size-scale effects on material strength and toughness. Probab Eng Mech 24:75–83
12. Carpinteri A, Cornetti P, Puzzi S (2006) Scaling laws and multiscale approach in the mechanics of heterogeneous and disordered materials. Appl Mech Rev 59:283–305
13. CEB (1990) Model code for concrete structures. Euro-International Concrete Committee, Switzerland
14. Colombo IS, Main IG, Forde Mc (2003) Assessing damage of reinforced concrete beam using "*b*-value" analysis of acoustic emission signals, J Mater Civil Eng ASCE 15:280–286
15. de Borst R (1987) Computation of post-buckling and post-failure behavior of strain-softening solids. Comput Struct 25(2):211–224
16. Desayi P, Balaji Rao K (1987) Probabilistic analysis of cracking of RC beams. Mater Struct 20 (120):408–417
17. Desayi P, Balaji Rao K (1989) Reliability of reinforced concrete beams in limit state of cracking – failure rate analysis approach. Mater Struct 22:269–279
18. Desayi P, Ganesan N (1985) An investigation on spacing of cracks and maximum crack width in reinforced concrete flexural members. Mater Struct 18(104):123–133
19. Dominguez N, Brancherie D, Davenne L, Ibrahimbegovic A (2005) Prediction of crack pattern distribution in reinforced concrete by coupling a strong discontinuity model of concrete cracking and a bond-slip of reinforcement model. Eng Comput Int J Comput Aided Eng Softw 22(5/6):558–582
20. Fama EF, Roll R (1968) Some properties of symmetric stable distributions. J Am Stat Assoc 63 (323):817–836
21. Ibrahimbegovic A, Boulkertous A, Davenne L, Brancherie D (2010) Modelling of reinforced-concrete structures providing crack-spacing based on X-FEM, ED-FEM and novel operator split solution procedure. Int J Numer Methods Eng 83:452–481
22. Kogon SM, Williams DB (1995) On the characterization of impulsive noise with alpha-stable distributions using Fourier techniques. In: Proceedings of the 29th Asilomar conference on signals, systems and computers, vol. 2, pp. 787–791, Pacific Grove, CA

23. Koutrouvelis IA (1980) Regression-type estimation of the parameters of stable laws. J Am Stat Assoc 75(372):918–928
24. Ma X, Nikias CL (1995) Parameter estimation and blind channel identification in impulsive signal environments. IEEE Trans Signal Process 43:2884–2897
25. Mandelbrot BB (1982) The fractal geometry of nature – updated and augmented, edition – 1983, W.H. Freeman and Company, New York
26. Mandelbrot B, Taleb N (2006) A focus on the exceptions that prove the rule. Financial Times (3 April)
27. McCulloch JH (1986) Simple consistent estimators of stable distribution parameter. Commun Stat Simul Comput 15(4):1109–1136
28. Neild SA, Williams MS, McFadden PD (2002) Non-linear behaviour of reinforced concrete beams under low-amplitude cyclic and vibration loads. Eng Struct 24:707–718
29. Nilson AH, Winter G (1986) Design of concrete structures, 10th edn. McGraw-Hill Book Company, New York
30. Nolan JP (2009) Stable distributions: models for heavy tailed data. BirkhÄauser, Boston. Chapter 1 online at academic2.american.edu/»jpnolan. Unfinished manuscript
31. Park R, Paulay T (1975) Reinforced concrete structures. Wiley, New York
32. Prigogine I (1967) Introduction to thermodynamics of irreversible processes, 3rd edn. Interscience Publishers, New York
33. Prigogine I (1978) Time, structure, and fluctuations. Science 201(4358):777–785
34. Rubi JM (2008) The long arm of the second law. Scientific American (November):62–67
35. Sluys LJ, de Borst R (1996) Failure in plain and reinforced concrete – an analysis of crack width and crack spacing. Int J Solids Struct 33(20–22):3257–3276
36. Tsihrintzis GA, Nikias CL (1996) Fast estimation of the parameters of alpha-stable impulsive interference. IEEE Trans Signal Process 44:1492–1503
37. Yang C-Y, Hsu K-C, Chen K-C (2009) The use of the Levy-stable distribution for geophysical data analysis. Hydrogeol J 17:1265–1273

Can Fuzzy Logic Via Computing with Words Bring Complex Environmental Issues into Focus?

Ashok Deshpande and Jyoti Yadav

Abstract Information on the status and changing trends in environmental quality is necessary to formulate sound public policy and efficient implementation of environmental pollution abatement programmes. In this quest, water/air quality indices are computed using US-EPA and US-NSF proposed methods for local and regional water/air quality management in many metro cities of the world. There are different types of uncertainties while adopting the procedure in vogue in the computation of these indices. However, it does not include expert's knowledge with a view to arrive at cause–effect relationship. We believe that the development of a method to quantify association between the pollutant and air/water-borne diseases is an important step before classifying air/water quality, either in numeric or linguistic terms. There exists aleatory uncertainty in the pollution parametric data and epistemic uncertainty in describing the pollutants by the domain experts in linguistic terms such as *poor, good, and very good*. Successes of probability theory have high visibility. But what is not widely recognised is that these successes mask a fundamental limitation—the inability to operate on what may be called perception-based information. In this chapter, we describe the case study 1 that relates to fuzzy description of river water quality in River Ganga for bathing purpose, while case study 2 presents fuzzy description of air quality in Pune City.

Keywords Bathing • River water quality • Fuzzy set theory • Linguistic terms • Fuzzy number • Degree of match • Fuzzy rule base system • Degree of certainty

---

A. Deshpande (✉)
Berkeley Initiative in Soft Computing (BISC)-Special Interest Group (SIG)-Environment Management Systems (EMS), University of California, Berkeley, CA, USA

College of Engineering Pune (COEP), Pune, Maharashtra, India

National Environmental Engineering Research Institute (NEERI), Nagpur, India
e-mail: ashok_deshpande@hotmail.com

J. Yadav
Department of Computer Science, University of Pune, Pune, Maharashtra, India

S. Chakraborty and G. Bhattacharya (eds.), *Proceedings of the International Symposium on Engineering under Uncertainty: Safety Assessment and Management (ISEUSAM - 2012)*,
DOI 10.1007/978-81-322-0757-3_16,

1 Introduction

You and a friend walk outside on January morning in Pune City. You announce that the weather is *mild*. Your friend declares that it is *cold*. Who is wrong? Or are you both right?

People recognise that language can be imprecise/fuzzy and that concepts such as *cold, hot, or mild do not* have well-defined boundaries. In 1965, Prof. Lotfi Zadeh introduced fuzzy sets and thereafter fuzzy logic, a means of processing data by extending classical set theory to handle partial membership. In everyday life and in fields such as environmental health, people deal with concepts that involve factors that defy classification into crisp sets—*safe, harmful, acceptable, unacceptable,* and so on. A classic example is a regulator carefully explaining the result of a detailed quantitative risk assessment to a community group, only to be asked over and over again, *But are we safe?* In this case, *safe* defies crisp classification because it is a multivariate state with gradations that vary among different individuals and groups.

Information on the status and changing trends in environmental quality is necessary to formulate sound public policy and efficient implementation of environmental pollution abatement programmes. One of the ways of communicating the information to the policy makers and public at large is with indices. In the computation of air/water quality index (AQI/WQI), first a numerical value is computed and then the air/water quality is described in linguistic terms. There exists aleatory uncertainty in the pollution parametric data and epistemic uncertainty in describing the pollutants by the domain experts in linguistic terms such as *poor, good, and very good.* Successes of probability theory have high visibility. But what is unrecognised is that these successes mask a fundamental limitation—the inability to operate on what may be called perception-based information. In this chapter, an attempt has been made to use fuzzy-logic-based formalism in modelling these two types of uncertainties, thereby straightway describing air/water quality in linguistic terms with a degree of certainty attached to each term.

The rest of this chapter is organised as follows: Sect. 2 is a brief account of the theoretical foundation of fuzzy-logic-based method with brief description of the other mathematical framework used. While Sect. 3 relates to a case study for describing water quality, fuzzily concluding remarks and future research efforts are covered in Sect. 4.

2 Fuzzy-Logic-Based Formalism

Will I suffer from water-borne diseases (WBD) if I take a bath in polluted river water? Realising the complexity in establishing cause–effect relationship between bathing in polluted river and water-borne diseases (WBDs), an attempt has been made to present a useful method to address the issue. Some of the important facets of our approach include interviewing student community (bather/non-bather) with a

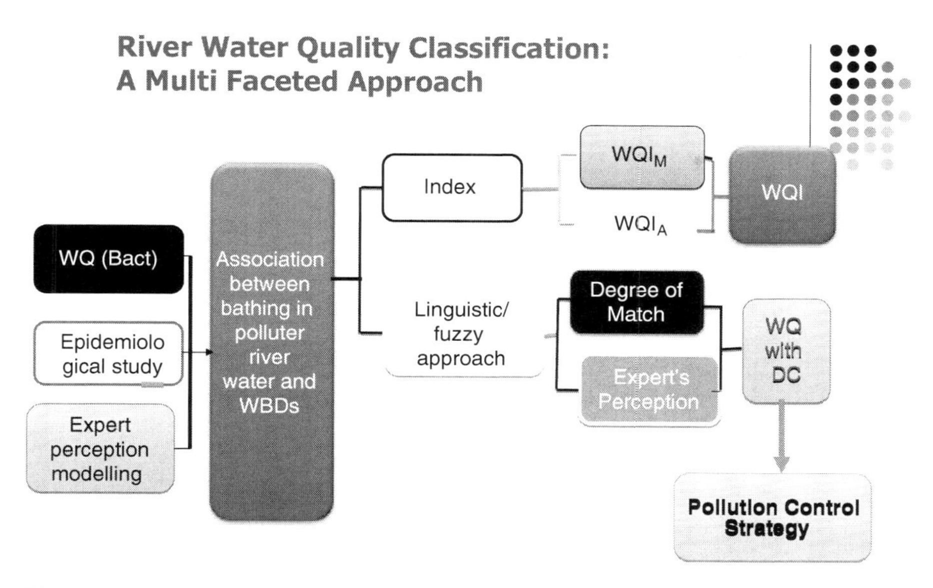

Fig. 1 Multifaceted formalism for water quality classification

structured questionnaire, collecting information on belief of the resident medical practitioners about bathing in polluted river and WBDs, and, furthermore, modelling of epistemic uncertainty in domain expert's belief in supporting their evidence for various WBDs and the like. Figure 1 presents a novel multifaceted formalism for straightway describing river water quality in linguistic terms with degree of certainty. The technique used in estimating the possible association between bathing in polluted river water and water-borne disease includes epidemiological study including case control study, river water quality analysis, perception of the resident medical professionals regarding their belief in relation to water-borne diseases, Dempster–Shafer (DS) theory of evidence, bootstrapping along with conventional statistical techniques, and the like. Some of these methods are briefly described in this section.

2.1 Fuzzy Measures and Evidence Theory

A fuzzy measure describes the vagueness or imprecision in the assignment of an element *a* to two or more crisp sets. In a fuzzy measure, the concern of attention is to describe the vagueness or imprecision in assigning the point to any of the crisp sets on the power set. Shafer developed Dempster's work and presented an important theory of evidence called Dempster–Shafer (DS) theory in which DS belief (Bel) and plausibility (Pl) are used to characterise uncertainty. A basic measure in DS theory is a basic belief assignment (BBA). The function (m) is a mapping function to express BBA for a given evidential event A, $m(A)$. BBA is a representation of partial belief not only for a single possible event but also for a set of

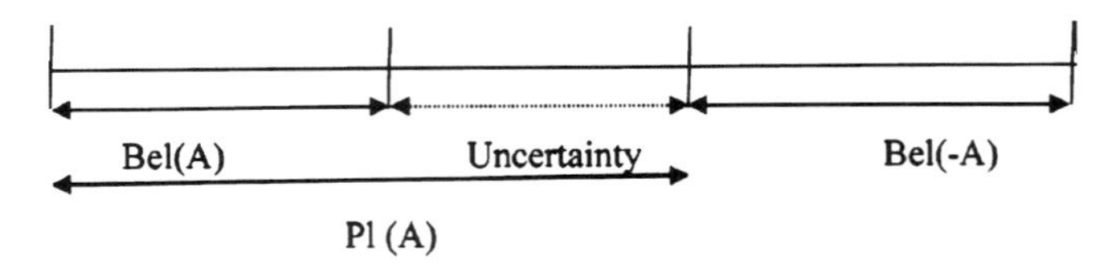

Fig. 2 Belief and uncertainty and ignorance

possible events with imprecise and incomplete knowledge and data. The main concept of evidence theory is that our knowledge of a given problem can be inherently imprecise. Hence, the bound result, which consists of both belief and plausibility, is presented (Fig. 2). BBA expresses the degree of belief in a proposition. BBA is assigned by making use of a mapping function (m) to express our belief with a number in the unit interval [0, 1]

$$m : 2x \rightarrow [0, 1]. \tag{1}$$

The number m (A) represents the portion of total belief assigned exactly to proposition A. The measure m, BBA function, must satisfy the following axioms:

$$m(A) \geq 0 \quad \text{for any } A \in 2^x \tag{2}$$

$$m(\varphi) = 0, \quad \Sigma_{\in 2x}\ mA = 1 \tag{3}$$

Though these axioms of evidence theory look similar to those of probability theory, the axioms for the BBA functions are less stringent than those for probability measure.

Dempster's Rule of Combining. The information from different sources can be aggregated by Dempster's rule of combining to make a new combined BBA structure as given in the following equation:

$$M_{12}(A) = \frac{\sum\limits_{c_i \cap c_j = A} m_1(c_i) m_2(c_j)}{1 - \sum\limits_{c_i \cap c_j = \phi} m_1(c_i) m_2(c_j)}, \quad A \neq \phi \tag{4}$$

where Ci and Cj are propositions from each sources ($m1$ and $m2$). In Eq. (4), $\sum_{c_i \cap c_i = \phi} m_1(c_i) m_2(c_j)$ can be viewed as contradiction or conflict among the information given by the independent knowledge sources (Ross 1997). Even when some conflict is found among the information, Dempster's rule disregards every contradiction by normalising with the complementary degree of contradiction to consider only consistent information. However, this normalisation can cause a counterintuitive and numerically unstable combination of information when the given information from different sources has significant contradiction or conflict. If there is a serious conflict, it is recommended to investigate the given information or to collect more information.

2.2 Belief and Plausibility Function

Owing to lack of information and various possibilities in constructing BBA structure, it is more reasonable to present a bound of the total degree of belief in a proposition, as opposed to a single value of probability given as a final result in probability theory. The total degree of belief in a proposition A is expressed within bound [bel (A), pl (A)], which lies in the unit interval [0, 1] as shown in Fig. 1, where Bel (A) and Pl (A) are given as

$$\begin{aligned} \mathrm{Bel}(A) &= \sum_{c \subset A} m(C) : \text{ Belief function} \\ \mathrm{Pl}(A) &= \sum_{c \subset A \neq \phi} m(c) : \text{ Plausibility function.} \end{aligned} \tag{5}$$

Bel (A) is obtained by the summation of BBAs for proposition, which is included in proposition A fully. Bel (A) is the total degree of belief. The degree of plausibility Pl (A) is calculated by adding BBAs of propositions whose intersection with proposition A is not an empty set. That is, every proposition consistent with proposition A at least partially is considered to imply proposition A because BBA in a proposition is not divided into its subsets. Briefly, Bel (A) is obtained by adding the BBAs of propositions that totally agree with the proposition A as a measure of belief, whereas Pl (A) plausibility is calculated by adding BBAs of propositions that correspond to the proposition A totally or partially. In a sense, these two measures consist of lower and upper probability bounds.

2.3 Fuzzy Inference System

Firstly, water/air quality experts are identified, and relevant field data is collected. Additional data generation is a logical step if the available data is inadequate for analysis. Perception of experts about the linguistic description of river water quality for bathing is obtained on interviewing or through a questionnaire. Modelling of uncertainty in the expert's perception by constructing fuzzy sets/fuzzy numbers and the uncertainty in the field data of water quality parameters using the concept of convex normalised fuzzy number is the next step. The parameters identified for defining bathing, say water quality by the experts, are faecal coliforms (*FC*), dissolved oxygen (DO), biochemical oxygen demand (*BOD*), *pH*, and *turbidity*. The relevant parameters could be considered while describing the overall air quality of a city/region.

Randomness in the air/water quality data can be transformed into a convex normalised fuzzy number A with membership grade function $\mu \mathrm{A}(x)$, thereby characterising the dynamic behaviour of the water quality parameters. We refer to

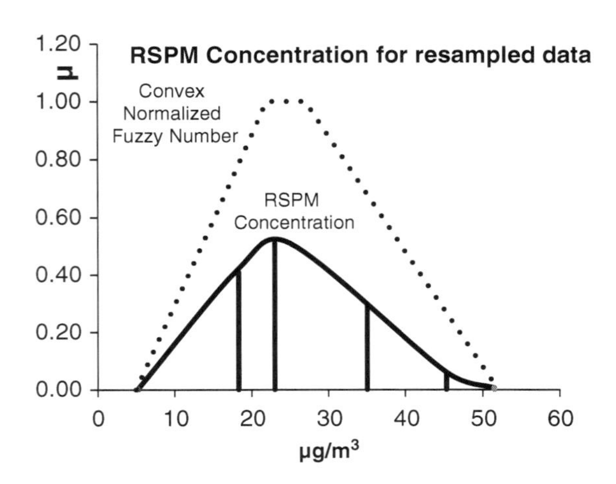

Fig. 3 Probability and possibility distribution for respirable suspended particulate matter (RSPM)

air quality parameter for illustration. If x_i is some point on the parametric domain for which $p\ (x_j)$ is maximum, then define function $\mu_A\ (x)$ (Fig. 3):

$$\mu_A(x) = p(x)/p(x_j) \tag{6}$$

Construction of fuzzy number or fuzzy sets for modelling the perception of the experts in classifying each parametric domain in linguistic terms such as *very good and good* allows for referencing all possible parametric values to be described. This transforms a random variable into a convex normalised fuzzy number A with membership grade function $\mu_A(x)$, thereby characterising the dynamic behaviour of the water quality parameter. The construction of fuzzy number or fuzzy sets for modelling the perception of the experts in classifying each parametric domain linguistically involves selection of linguistic terms such as *very good and good,* which allows for referencing all possible parametric values to be described; classification of the parametric domain and assigning linguistic terms to each class linearly by the experts reflecting the imprecision in their perception; the set of values for which all the experts assign the same linguistic term is given $\mu = 1.0$, while none of the experts assigning that term are given $\mu = 0.0$. The breakeven point membership grades 0.0 and 1.0 are connected by continuous monotonic function which presupposes that the degree of consensus among the experts goes on increasing as the parametric values approach the core of fuzzy number for the specified linguistic term.

2.4 *Matching Between Two Fuzzy Values*

The fuzzy number for field data (A) on parameters and the fuzzy number characterising linguistic terms (A') are matched together to arrive at a measure called degree of match (DM) defined by

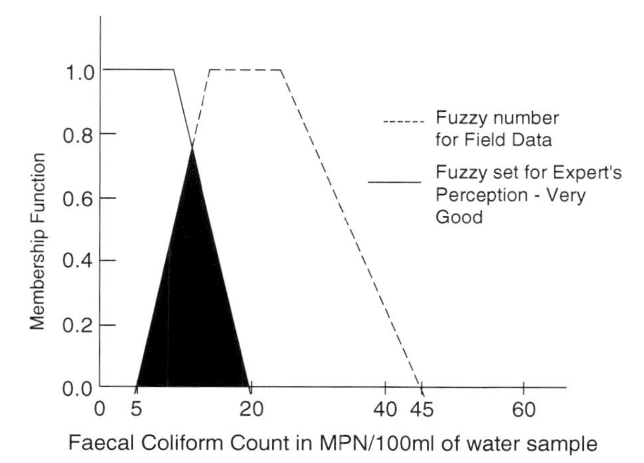

Fig. 4 Fuzzy numbers for *very good faecal coliforms*

$$\mathrm{DM_{ff}}(A, A') = \int \mu_{A \cap A'}(x)\, \mathrm{d}x / \int \mu_A(x)\, \mathrm{d}x, x \subset \mathrm{X} \quad (7)$$

in which X denotes the universe and $\mu_{\varsigma \cap \varsigma,}(x)$ is membership grade for $\varsigma \cap \varsigma,$. Furthermore, if A and A' are the discrete possibility distributions, the measure is defined as

$$\mathrm{DM_{ff}}(A, A') = \Sigma \mu_{A \cap A'}(x) / \Sigma \mu_A(x), \quad x \subset \mathrm{X} \quad (8)$$

Figure 4 shows the fuzzy number for *very good faecal coliforms* reveals that almost all the experts agree that the *faecal coliforms* count between 0 and 10 MPN/100 ml of water. The level of presumption or membership function decreases with the increasing faecal coliforms count. When the count exceeds 20, none of the experts define the parameters as *very good* for bathing purpose. This is indicated by the level of presumption $\grave{\imath} = 0$.

A set of rules is constructed for classifying air/water quality as highly acceptable, just acceptable, or not acceptable (rejected) in order to aggregate the set of attributes. Each rule has antecedent propositions connected together using AND operator, resulting in some consequences. The assertions related to its antecedent part are obtained from the users, which are imprecise or fuzzy. Thus, a fuzzy rule-based system can be developed for the knowledge representation or reasoning process. Here, the partial matching is allowed, and the analyst can estimate the extent to which the assertion satisfies the antecedent part of the rule contrary to the rule-based system which examines as to whether the antecedent part is satisfied or not [1]. A hierarchical structure for water classification resulting in a set of rules can be constructed (Fig. 5). The chemical status of water is judged in the first hierarchical level of knowledge base. The second hierarchical level characterises bacteriological, chemical, and physical status of water to arrive at the ultimate acceptable strategy of water quality for bathing purpose. If need be, a similar structure can be developed for air quality classification.

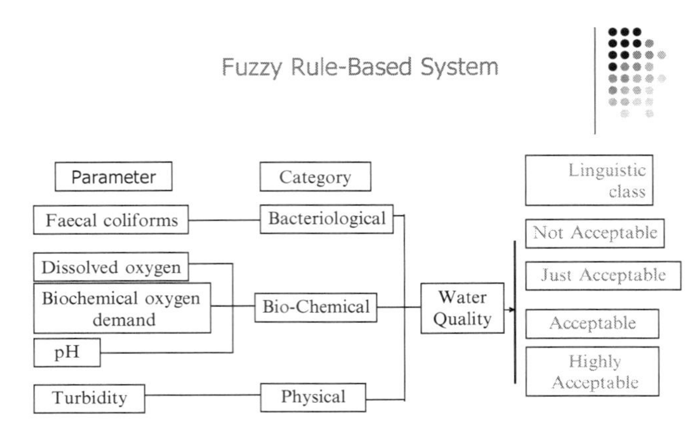

Fig. 5 Type 1 fuzzy inference system/fuzzy expert system

Following are the sample rules stored at two different hierarchical levels of the knowledge base:

Rule 1
If DO is <fair> and
BOD is <good> and
pH is <very good>,
then chemical status of water is <good>.

3 Fuzzy Description of Environment Quality

With the approach presented in this chapter, it is possible to describe any environmental quality fuzzily. We present herein water and air quality.

3.1 Case Study 1: Fuzzy Description of River Water Quality

Around 21% of communicable diseases in India are water related, and out of these, diarrheal diseases alone killed over 700,000 Indians in the year 1999. Since time immemorial, sacred bathing on the holy riverbanks (*ghats*) is practised in India and Nepal. It has been a religious belief of some of the bathers that gulping water at these holy places washes away their sins! The first study, therefore, relates to establishing the cause–effect relationship between bathing in polluted river water and water-borne diseases carried out at Alandi town near Pune situated on the

Table 1 Fuzzy description of river water quality with degree of certainty

Degree of certainty		
Water quality description	Rishikesh	Varanasi
Highly accepted	0.21	0
Accepted	**0.41**	0
Just accepted	0.36	0.01
Not accepted	0.1	**0.98**

banks of river Indrayani. We very briefly discuss the final outcome of such an investigation [2]. The analysis of the results reveals result that the combined belief of the two domain experts in identifying for the single disease *diarrhoea is 0.58*. In sum, bathing in polluted water can be a cause for the incidence of water-borne diseases in bathers [2].

Describing Ganga water quality straightway in linguistic terms, with some degree of certainty attached to each linguistic term, concludes that the Government of India should continue the efforts on the reduction of pollution levels, especially from bacteriological standpoint. Aerobic treatment is the option that should be attached great importance in future planning of GAP 2. The results depicted after Ganga Action Plan 1 infers that even at Rishikesh, the water quality of Ganga is not very good for bathing and still the pollution persists as the degree of certainty attached to fair is 0.36 while it is 0.41 for the linguistic description of water as good. Varanasi has been a serious cause of concern from the standpoint of bathing in the holy river (Table 1).

3.2 Case Study 2: Fuzzy Description of River Air Quality

The case study relates to fuzzy air quality description with the available air quality data from five monitoring stations in Pune City. These are Pimpri-Chinchwad Municipal Corporation, Karve Road, Swargate, Bhosari, and Nal Stop. In order to bring down progressive deterioration in air quality, the Government of India has enacted Air (Prevention and Control of Pollution) Act in 1981, and further stringent guidelines are promulgated in Environment (Protection) Act, 1986. The need for ambient air quality monitoring programme that is needed to determine the existing quality of air and evaluation of the effectiveness of control programme and to develop new programme was recognised. As a follow-up action, the Central Pollution Control Board (CPCB) initiated National Air Quality Monitoring (NAQM) Programme during 1984–1985 at the national level.

A well-structured air quality monitoring network involves selection of pollutants, selection of locations, frequency, duration of sampling, sampling techniques, infrastructural facilities, man power and operation, maintenance costs, and the like. The network design also depends upon the type of pollutants in the atmosphere through various common sources, called common urban air

Table 2 Comparison of conventional AQI and fuzzy description of air quality with degree of certainty

Monitoring station	Conventional AQI	AQI with degree of certainty
Karve Road	206 very poor	0.45 very poor
Bhosari	145.89 poor	0.43 very poor
Swargate	147.13 poor	0.54 poor
Nal Stop	120.2 poor	0.67 fair
PCMC	151.13 very poor	0.91 poor

pollutants, such as suspended particulate matter (SPM), respirable suspended particulate matter (RSPM), sulphur dioxide (SO_2), oxides of nitrogen (NO_x), and carbon monoxide (CO). The areas to be chosen primarily are such areas which represent high traffic density, industrial growth, human population and its distribution, emission source, public complaints if any, the land use pattern, etc. Generally, most of the times the basis of a network design is the pollution source and the pollution present.

Generation of fuzzy numbers for different linguistic hedges (*very good, good, fair, poor, very poor*) of RSPM concentration is an important issue in any FIS. According to the expert, RSPM count between 18–57 μg/m^3 is very good, 42–75 μg/m^3 is good, 67–100 μg/m^3 is fair, 90–130 μg/m^3 is poor μg/m^3, and above 120 μg/m^3 it is very poor. The level of membership function decreases with the increasing RSPM level. When it exceeds 57, the expert does not define the parameter as very good. Table 2 is the comparison between the computed AQI and the proposed fuzzy-logic-based method. It can be revealed that AQI based on the traditional method does not attach any certainty while describing the air quality. In addition, the method does not consider the aggregated effect, and the highest computed AQI is considered as the final decision on the air quality which, in our opinion, is the departure from human thinking. Alternately, using type 1 fuzzy inference system, we can describe the air quality straightway in linguistic terms with some degree of certainty attached to each term.

4 Outlook in Environmental Policy [3]

Over the past few decades, soft computing tools such as fuzzy-logic-based methods, neural networks, and genetic algorithms have had significant and growing impacts. But we have seen only limited use of these methods in environmental fields, such as risk assessment, cost-benefit analysis, and life-cycle impact assessment. Because fuzzy methods offer both new opportunities and unforeseen problems relative to current methods, it is difficult to determine how much impact such methods will have on environmental policies in the coming decades. Here, we consider some obvious advantages and limitations.

Quantitative models with explicit and crisp delineations of systems have long been the currency of discourse in engineering and the physical sciences, where basic physical laws form the foundations of analyses. These fields place high value on the causal linkages implicit in model structure and parameterization. But for problems that involve human values, language, control theory, biology, and even environmental systems, researchers have had to rely more on descriptive and empirical approaches. When the goal is to summarise the observations in an efficient and useful manner, fuzzy-logic-based methods should be further investigated as alternative—and perhaps more appropriate—methods for addressing uncertain and complex systems. For the types of complex and imprecise problems that arise in environmental policy, the ability to model complex behaviours as a collection of simple if–then rules makes fuzzy logic an appropriate modelling tool. Because fuzzy arithmetic works well for addressing linguistic variables and poorly characterised parameters, fuzzy methods offer the opportunity to evaluate and communicate assessments on the basis of linguistic terms that could possibly match those of decision makers and the public. Moreover, approximate reasoning methods such as fuzzy arithmetic do not require well-characterised statistical distributions as inputs. Another key advantage of fuzzy logic in risk assessment is the ability to merge multiple objectives with different values and meanings, for example, combining health objectives with aesthetic objectives. It also provides rules for combining qualitative and quantitative objectives [3].

But fuzzy logic has at least two limitations for expressing health risks and other environmental impacts. One problem is its strong reliance on subjective inputs. Although this is a problem in any type of assessment, fuzzy methods might provide more opportunity for the misuse of subjective inputs. Although probabilistic assessments based on tools such as Monte Carlo methods are analogous to assessments based on fuzzy logic, these two techniques differ significantly both in approach and in interpretation of results. Fuzzy logic confronts linguistic variables such as ‘safe’, ‘hazardous’, ‘acceptable’, and ‘unacceptable’, whereas Monte Carlo methods are forced to fit linguistic variables for probabilistic assessments. Fuzzy arithmetic combines outcomes from different sets in a way that is analogous to but still different from Monte Carlo methods. Possibility theory can be used as an alternative to probabilistic analysis, but this strategy creates the potential for misuse if membership functions are interpreted as probability distributions.

4.1 No More Crisp Lines?

Fuzzy logic represents a significant change in both the approach to and the outcome of environmental evaluations. Currently, risk assessment implicitly assumes that probability theory provides the necessary and sufficient tools for dealing with uncertainty and variability. The key advantage of fuzzy methods is how they reflect the human mind in its remarkable ability to store and process information that is consistently imprecise, uncertain, and resistant to classification. Our case study illustrates the ability of fuzzy logic to integrate statistical measurements with imprecise health goals. But we submit that fuzzy logic and probability theory are complementary and not competitive. In the world of soft computing, fuzzy logic has been widely used and has often been the 'smart' behind smart machines. But more effort and further case studies will be required to establish its niche in risk assessment and other types of impact assessment. Could we adapt to a system that relaxes 'crisp lines' and sharp demarcations to fuzzy gradations? Would decision makers and the public accept expressions of water- or air-quality goals in linguistic terms with computed degrees of certainty? Resistance is likely. In many regions, such as the United States and EU, both decision makers and members of the public seem more comfortable with the current system—in which government agencies avoid confronting uncertainties by setting guidelines that are crisp and often fail to communicate uncertainty. Perhaps someday, a more comprehensive approach that includes exposure surveys, toxicological data, and epidemiological studies coupled with fuzzy modelling will go a long way towards resolving some of the conflict, divisiveness, and controversy in the current regulatory paradigm.

Acknowledgement The wholehearted assistance received from Dr. D. V. Raje and Dr. Kedar Rijal for the implementation of the concept is gratefully acknowledged. My special thanks to Professor Thomas McKone who helped the author in many fuzzy ways!

References

1. Deshpande AW, Raje DV, Khanna P (1996). Fuzzy description of river water quality. Eufit 96, 2–5 September. pp 1795–1801
2. Rijal K, Deshpande A, Ghole V (2009) Bathing in polluted rivers, water-borne diseases, and fuzzy measures: a case study in India. Int. J. Environ Waste Manage 6(3–4):255–263
3. Mckone TE, Deshpande AW (2005) Can fuzzy logic bring complex environmental problems into focus? Int J Environ Sci Technol 39(2):42A–47A

Uncertainty Evaluation in Best Estimate Accident Analysis of NPPs

S.K. Gupta, S.K. Dubey, and R.S. Rao

Abstract Recent trends in carrying out deterministic safety analysis for safety assessment are attracting more attention for the use of best estimate approach. However, conservative approaches are still used in licensing safety analysis. The best estimate approach provides more realistic information with respect to conservative approach for predictions of physical behaviour and provides information about the existing safety margins and between the results of calculation and regulatory acceptance criteria, whereas conservative approach also does not give any indication about actual plant behaviour, including time scale, for preparation of emergency operating procedures. Best estimate methodology results are affected by various sources of uncertainty like code or model uncertainty, representation uncertainty, scaling uncertainty and plant uncertainty. Therefore, uncertainty in the results due to unavoidable approximation in the modelling should be quantified. Various uncertainty analysis methodologies have been emerged after the development of code scaling, applicability and uncertainty (CSAU) by US NRC in 1989, like CIAU by Italy and GRS by Germany. This chapter deals with the sources of uncertainty and its quantifications by various methodologies in terms of confidence and probability. This chapter also deals with the application of a sampling-based uncertainty evaluation in the best estimate analysis of station blackout and small break LOCA in integral test facilities carried out under the framework of IAEA CRP. Uncertainty evaluation for TMI-II accident has been carried out using this methodology.

Keywords Uncertainty • Best estimate • Accident analysis • NPPs

S.K. Gupta (✉) • S.K. Dubey • R.S. Rao
Safety Analysis and Documentation Division, Atomic Energy Regulatory Board, Anushaktinagar, Mumbai 400094, India
e-mail: guptask@aerb.gov.in

S. Chakraborty and G. Bhattacharya (eds.), *Proceedings of the International Symposium on Engineering under Uncertainty: Safety Assessment and Management (ISEUSAM - 2012)*, DOI 10.1007/978-81-322-0757-3_17, © Springer India 2013

1 Introduction

As per the AERB design code/standard requirement, deterministic safety analysis (frequently referred to as accident analysis) is an important tool for confirming the adequacy and efficiency of provisions within the defence in depth concept for the safety of nuclear power plants. The computer programmes, analytical methods and plant models used in the safety analysis shall be verified and validated, and adequate consideration shall be given to uncertainties [1]. Although the trends in accident analysis have continued to move to best estimate analysis rather than conservative analysis, conservative approaches are still use. The conservative approach does not give any indication of actual margins between the actual response and the conservatively estimated response. A conservative approach also does not give any indication about actual plant behaviour, including timescale, for preparation of emergency operating procedures (EOPs) or for use in accident management for abnormal operating conditions. In addition, a conservative approach often does not show margins that in reality could be utilised for greater operational flexibility. The concept of conservative methods was introduced at the early days (1970) of safety analyses to account for uncertainty due to the limited capability of modelling, the limited knowledge of physical phenomena and to simplify the phenomena. Use of a conservative methodology may lead to so large conservatism that important safety issues are masked. In the last four decades, thermal-hydraulic issues, intensive experimental research that has resulted in a considerable increase in knowledge and the development of computer codes have improved the ability of these codes to calculate results that agree with the experimental results. In order to overcome these deficiencies in conservative analysis, it may be preferable to use a best estimate approach together with an evaluation of the uncertainties to compare with acceptance criteria. A best estimate approach provides more realistic information about the physical behaviour, identifies the most relevant safety issues and provides information about the existing margins between the results of calculation and acceptance criteria. The best estimate approach is highly dependent upon an extensive experimental database to establish confidence in the best estimate codes and to define the uncertainties that have to be determined for the best estimate results.

The use of BE codes is generally recommended for deterministic safety analysis. Two options are offered to demonstrate sufficient safety margins in using BE codes: The first option is the use of the codes ‘in combination with a reasonably conservative selection of input data and a limited evaluation of the uncertainties of the results’. In this statement, evaluation of uncertainties is meant more in the deterministic sense: code to code comparisons, code to data comparisons and expert judgments in combination with sensitivity studies are considered as typical methods for the estimation of uncertainties. The second option is the use of the codes with realistic assumptions on initial and boundary conditions and taking credit in the analysis for availability of systems including time consideration for operator intervention. However, for this option, an approach should be based on statistically combined uncertainties for plant conditions and code models to establish, with a specified high probability that the calculated results do not exceed the acceptance criteria. Both options should be complemented by sensitivity studies, which include systematic variation of the code input variables and modelling parameters with the aim of identifying the important parameters required for the analysis and to show that there is no abrupt change in the result of the analysis for a realistic variation of inputs ('cliff edge' effects). It is necessary to quantify uncertainty in best estimate accident analysis when realistic initial and boundary conditions are used because best estimate code results are affected by various sources of uncertainty like code or model uncertainty, representation uncertainty, scaling uncertainty and plant uncertainty. This chapter deals with the sources of uncertainty, evaluation of uncertainty methodologies and quantification of uncertainty in terms of confidence and probability. In this chapter, uncertainty evaluation by sampling-based methodology in the analysis of the station blackout and small break LOCA in integral test facilities carried out under the framework of IAEA coordinated research project is reported. Application of this methodology has also been demonstrated in this chapter for actual nuclear power plant TMI-II accident.

2 Sources of Uncertainty

Uncertainty in the best estimate results due to various sources of uncertainty should be quantified. Evaluation process of various sources of uncertainties including integral effect and experimental data is reported in Fig. 1. Sources of uncertainty fall within five general categories [6]:

(a) *Code or model uncertainties*: A thermal-hydraulic system code is a computational tool that typically includes three different sets of balance equations (or of equations derived from fundamental principles), closure or constitutive equations, material and state properties, special process or component models and a numerical solution method. The three sets of balance equations deal with the fluids of the system, the solid structures including the fuel rods and the neutrons. In relation to the fluids of the system, the 1-D UVUT (unequal velocities, unequal temperatures) set of partial differential equations is part of the codes under consideration. The closure (constitutive) equations deal with the interaction between the fluid and the environment as well as with the interaction of the two phases of the fluid (i.e. the gas/vapour and the liquid phase). The interfacial drag coefficient, wall to fluid friction factor and heat transfer coefficient are typically expressed by constitutive equations. Various sets of material properties are embedded into the codes, even though the user may change these properties or add new materials. Water, nitrogen, air, uranium dioxide, stainless and carbon steel and zircaloy are materials. Different levels of sophistication usually characterise the sets of properties in the different codes. This is especially true for water (Mollier diagram quantities and

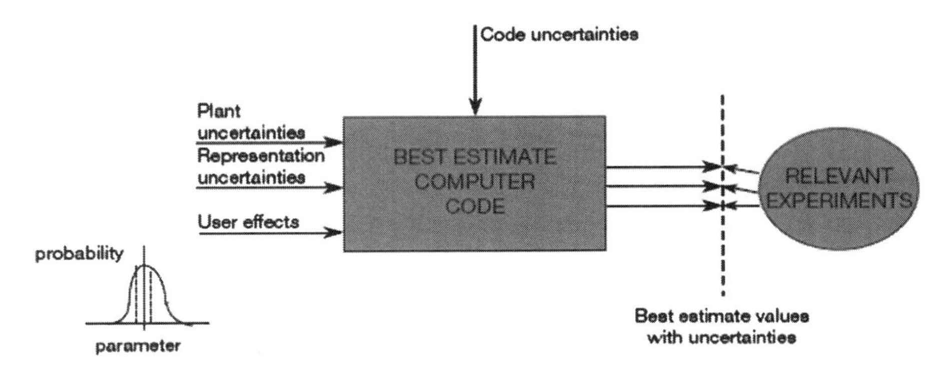

Fig. 1 Evaluation process and main sources of uncertainties [6]

related derivatives). Balance equations are not sophisticated enough for application in the modelling of special components or for the simulation of special processes. Examples of these components are the pumps and the steam separators, and examples of the special processes are the countercurrent flow limiting condition and two phase critical flow, although this is not true for all the codes. Empirical models 'substitute' the balance equations in such cases.

(b) *Representation uncertainties*: Representation uncertainty is related to the process of setting up the nodalisation (idealisation). The nodalisation constitutes the connection between the code and the 'physical reality' that is the subject of the simulation. The process for setting up the nodalisation is an activity carried out by a group of code users that aims at transferring the information from the real system (e.g. the nuclear power plant), including the related boundary and initial conditions, into a form understandable to the code. Limitation of available resources (in terms of person-months), lack of data, the target of the code application, capabilities or power of the available computational machines and expertise of the users play a role in this process. The result of the process may strongly affect the response of the code.

(c) *Scaling uncertainty*: Using data recorded in scaled experiments and the reliance on scaling laws to apply the data result to full-scale systems. Scaling is a broad term used in nuclear reactor technology as well as in basic fluid dynamics and in thermal hydraulics. In general terms, scaling indicates the need for the process of transferring information from a model to a prototype. The model and the prototype are typically characterised by different geometric dimensions, but thermal-hydraulic quantities such as pressure, temperature and velocities may be different in the model and in the prototype, as well as in the materials adopted, including working fluids. Therefore, the word 'scaling' may have different meanings in different contexts. In system thermal hydraulics, a scaling process, based upon suitable physical principles, aims at establishing a correlation between phenomena expected in a nuclear power plant transient scenario and phenomena measured in smaller scale facilities or phenomena predicted by numerical tools qualified against experiments performed in small-scale facilities.

Owing to limitations of the fundamental equations at the basis of system codes, the scaling issue may constitute an important source of uncertainties in code applications and may envelop various individual uncertainties.

(d) *Plant uncertainty*: The uncertainty bands associated with the boundary and initial conditions for the nuclear power plant condition under consideration, for example, core power, SG feed water inlet temperature. Uncertainty or limited knowledge of boundary and initial conditions and related values for a particular nuclear power plant are referred to as plant uncertainty. Typical examples are the pressuriser level at the start of the transient, the thickness of the gap of the fuel rod, the conductivity of the UO2 and the gap between the pellets and the cladding. It should be noted that quantities such as gap conductivity and thickness are relevant for the prediction of safety parameters (e.g. the PCT) and are affected by other parameters such as burn-up, knowledge about which is not as detailed as required (e.g. knowledge about each layer of a fuel element that may be part of the nodalisation). Thus, a source of error of this kind in the class of 'plant uncertainty' cannot be avoided and should be accounted for by the uncertainty method.

(e) *User effect*: The flexibility of the system codes for various volume control flags, junction control flags, solution method, etc., under consideration is a primary reason for generating a user effect. The impact of the user effect upon the final result (i.e. BE prediction plus uncertainty) may be different depending upon the selected uncertainty method. System code output results are largely affected by user's capability, qualifications and experience in the use of the code. It has been observed that results obtained by various users differ a lot for the use of same system code and same date provided for the modelling.

3 Uncertainty Methodologies

An uncertainty analysis consists of identification and characterisation of relevant input parameters (input uncertainty) as well as of the methodology to quantify the global influence of the combination of these uncertainties on selected output parameters (output uncertainty). These two main items may be treated differently by different methods. Within the uncertainty methods considered, uncertainties are evaluated using either (a) propagation of input uncertainties (Fig. 2) or (b) extrapolation of output uncertainties. For the 'propagation of input uncertainties', uncertainty is obtained following the identification of 'uncertain' input parameters with specified ranges or/and probability distributions of these parameters and performing calculations varying these parameters, schematic shown in Fig. 2. The propagation of input uncertainties can be performed by either deterministic or probabilistic methods. For the 'extrapolation of output uncertainty' approach, uncertainty is obtained from the (output) uncertainty based on comparison between calculation results and significant experimental data.

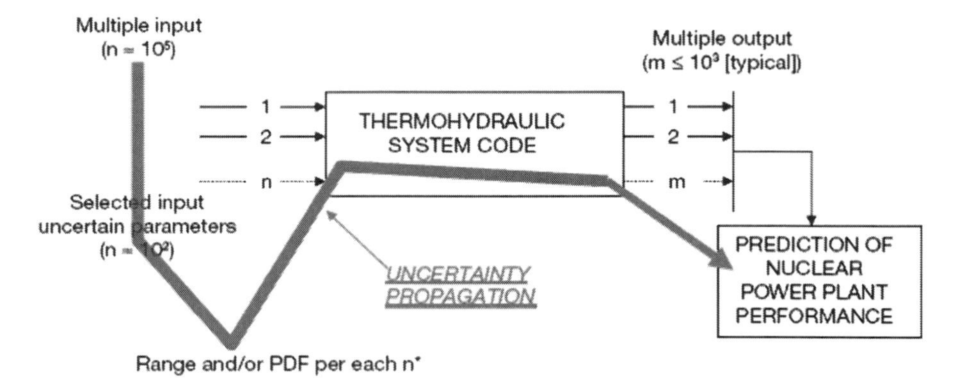

Fig. 2 Propagation of input uncertainty [6]

3.1 Propagation of Input Uncertainties

3.1.1 Probabilistic Methods

Probabilistic methods include CSAU, GRS, IPSN and the Canadian best estimate and uncertainty (BEAU) method. The probabilistic methods have the following common features:

(a) The nuclear power plant, the code and the transient to be analysed are identified.
(b) Uncertainties (plant initial and boundary conditions, fuel parameters, modelling) are identified.
(c) Some methods restrict the number of input uncertainties to be included in the calculations.

The selected input uncertainties are ranged using relevant separate effects data. The state of knowledge of each uncertain input parameter within its range is expressed by a probability distribution. Sometimes, 'state of knowledge uncertainty' is referred to as 'subjective uncertainty' to distinguish it from uncertainty due to stochastic variability. Dependency between uncertain input parameters should be identified and quantified provided that these dependencies are relevant.

3.1.2 Deterministic Methods

The deterministic methods include the Atomic Energy Authority Winfrith (AEAW) and the Electricité de France (EDF)–Framatome methods.

The deterministic methods have the following features in common with probabilistic methods:

(a) The code, nuclear power plant and transient are identified.
(b) Uncertainties (initial and boundary conditions, modelling, plant, fuel) are identified.

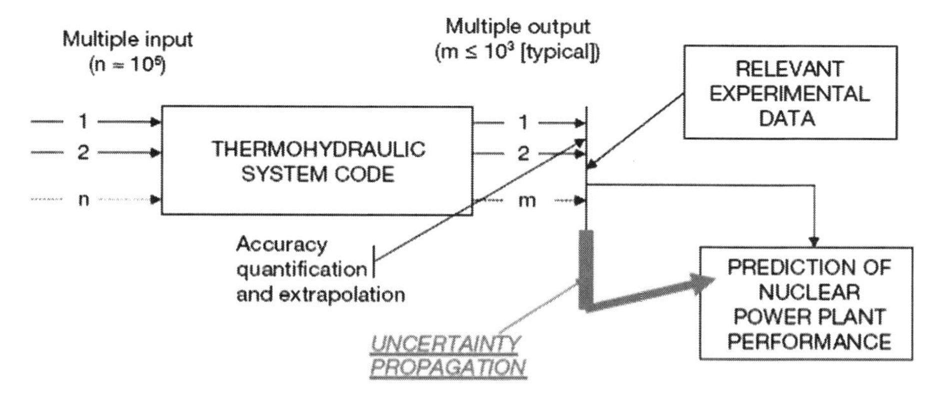

Fig. 3 Extrapolation of output uncertainty [6]

The difference with deterministic methods is in quantifying the input parameter uncertainties. No probability distributions are used; instead, reasonable uncertainty ranges or bounding values are specified that encompass, for example, available relevant experimental data. The statements of the uncertainty of code results are deterministic, not probabilistic.

3.2 Extrapolation of Output Uncertainty

The extrapolation of output uncertainty method focuses not on the evaluation of individual parameter uncertainties but on direct scaling of data from an available database, calculating the final uncertainty by extrapolating the accuracy evaluated from relevant integral experiments to full-scale nuclear power plants, known as Code with the capability to internal assessment of uncertainty (CIAU) developed by the University of Pisa, Italy. A schematic of this method is shown in Fig. 3. Considering integral test facilities (ITFs) of a reference light water reactor (LWR) and qualified computer codes based on advanced models, the method relies on code capability qualified by application to facilities of increasing scale. Direct data extrapolation from small-scale experiments to the reactor scale is difficult due to the imperfect scaling criteria adopted in the design of each scaled down facility. Only the accuracy (i.e. the difference between measured and calculated quantities) is therefore extrapolated. Experimental and calculated data in differently scaled facilities are used to demonstrate that physical phenomena, and code predictive capabilities of important phenomena do not change when increasing the dimensions of the facilities; however, available integral effect test facility scales are far from reactor scale.

4 Representation of Uncertainty

Representation of uncertainty evaluated for important parameters from the result of code output is done in many ways: confidence and probability, in terms of percentile, scatter plot and histogram, etc. The first two ways of presentation of uncertainty methodology are most commonly used in nuclear reactor technology.

4.1 *Probability and Confidence*

When regulatory acceptance criteria are checked for the calculated parameter obtained using best estimate system code, then it is to be ensured that what is the confidence with specified probability that the parameters will lie in the specified interval because many uncertainties are involved in the best estimate calculation. In many situations, a point estimate does not provide enough information about a parameter. Suppose if we want to estimate the mean peak clad temperature in case of LOCA, a single number may not be meaningful as an interval within which we would expect to find the value of this parameter as the calculated value of clad temperature strongly depends on many input parameters for which the exact real value is not known but the interval. This is an interval estimate called a confidence interval.

An interval estimate of an unknown parameter θ is an interval of the form $l \leq \theta \leq u$, where the end points l and u depend on the numerical value of the statistic of θ for a particular sample and on its sampling distribution, since different samples will produce the difference between l and u. These end points are values of random variables say L and U, respectively. From the sampling distribution of θ, the value of L and U can be determined such that the following probability statement is true:

$$P(L \leq \theta \leq U) = 1 - \alpha$$

where $0 < \alpha < 1$. Thus, there is a probability of $1 - \alpha$ of selecting a sample that will produce an interval containing true value of θ. The resulting interval $l \leq \theta \leq u$ is called a $100(1 - \alpha)$ percent confidence interval for the unknown parameter θ. The quantities l and u are called the lower and upper confidence limits, and $1 - \alpha$ is called the confidence coefficient. The interpretation of a confidence interval is that if an infinite number of random samples are collected and a $100(1 - \alpha)$ percent confidence interval for θ computed from each sample, then $100\ (1 - \alpha)$ percent of these intervals will contain the true value of θ.

Suppose that a population has unknown mean μ and known variance σ^2. A random sample of size n is taken from this population, say, $X_1, X_2, X_3, \ldots, X_n$. The sample mean $\bar{X}$ is a reasonable point estimator of the unknown mean μ. A 100 $(1 - \alpha)$ percent confidence interval on μ can be obtained by considering the sampling distribution of the sample mean $\bar{X}$. The sampling distribution of $\bar{X}$ is normal if the population is normal and approximately normal if the condition of the

Fig. 4 The distribution of Z [7]

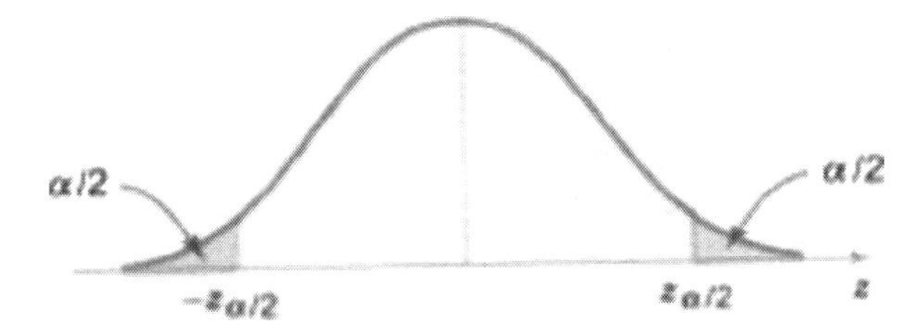

central limit theorem is met. The expected value or mean of $\bar{X}$ is μ, and the variance is σ^2/n. Therefore, the distribution of the statistic

$$Z = \frac{\bar{X} - \mu}{\sigma/\sqrt{n}}$$

is a standard normal distribution.

The distribution of $Z = \frac{\bar{X}-\mu}{\sigma/\sqrt{n}}$ is shown in Fig. 4. From the Fig. 4, it is found that

$$P\{-Z_{\alpha/2} \leq Z \leq Z_{\alpha/2}\} = 1 - \alpha$$

so that

$$P\left\{-Z_{\alpha/2} \leq \frac{\bar{X} - \mu}{\sigma/\sqrt{n}} \leq Z_{\alpha/2}\right\} = 1 - \alpha$$

This can be rearranged as

$$P\{\bar{X} - Z_{\alpha/2}\sigma/\sqrt{n} \leq \mu \leq \bar{X} - Z_{\alpha/2}\sigma/\sqrt{n}\} = 1 - \alpha$$

Therefore, it is said that if $\bar{X}$ is the sample mean of a random sample size n from a population with known variance σ^2, a 100(1 − α) percent confidence interval on μ is given by

$$\bar{X} - Z_{\alpha/2}\sigma/\sqrt{n} \leq \mu \leq \bar{X} - Z_{\alpha/2}\sigma/\sqrt{n}$$

where $Z_{\alpha/2}$ is the upper $\alpha/2$ percentage point of the standard normal distribution.

Furthermore, if $\bar{X}$ is used as an estimator of μ, then with the 100(1 − α) percent confidence that the error $|\bar{X} - \mu|$ will not exceed a specified amount E when the sample size is

$$n = \left(\frac{Z_{\alpha/2}\sigma}{E}\right)^2$$

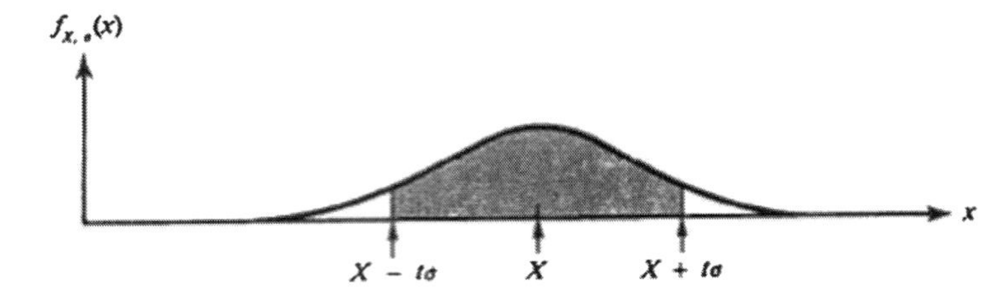

Fig. 5 Probability of a measurement with t standard deviation of X [11]

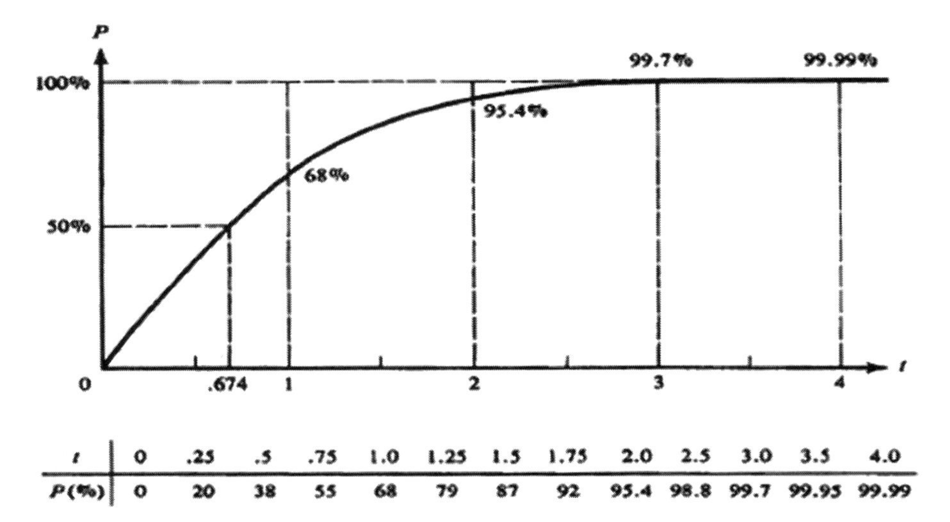

t	0	.25	.5	.75	1.0	1.25	1.5	1.75	2.0	2.5	3.0	3.5	4.0
P(%)	0	20	38	55	68	79	87	92	95.4	98.8	99.7	99.95	99.99

Fig. 6 Probability (within $t\sigma$) of a measurement that will fall within t standard deviation [11]

More generally, we could calculate $P(\text{within } t\sigma)$, which means 'the probability for an answer within $t\sigma$ of X' where t is any positive number. This probability is given by the area in Fig. 5 and calculated by

$$P(\text{within } t\sigma) = \frac{1}{\sqrt{2\pi}} \int_{-t}^{t} e^{-Z^2/2} \mathrm{d}Z$$

Probability is plotted in Fig. 6 as a function of t.

In many situations, the engineer is more concerned about where individual observations or measurements may fall than about estimating parameters. In most practical situations, μ and σ are not known, and they must be estimated from a random sample. It is possible to determine a constant k such that $\bar{X} \pm ks$ will still form a tolerance interval for normal distribution. In this case, the tolerance limits are random variables because we have replaced μ and σ by $\bar{X}$ and s and the proportion of population covered by the interval is not exact. Consequently, 100 $(1 - \alpha)$ percent confidence interval is utilised in the statement of the tolerance limit, since $\bar{X} \pm ks$ will not cover a specified proportion of the population all the time. This leads to the following conclusion.

For a normal distribution with unknown mean and variance, tolerance limits are given by $\bar{X} \pm ks$, where k is determined so that we can state with confidence 100 $(1 - \alpha)$ percent that the limit contain at least a proportion p of the population. The value of k for the required value of α and p can be obtained from Table XIV of Appendix of [7].

4.2 Percentile

For any time of a transient, if output parameters are arranged in ascending order and then rank is given to all values, from these ranks, the mean, median and the 5th and 95th percentiles are evaluated. The median value is the middle value of N ordered values (ascending order). The 5th and 95th percentiles are calculated using the following equation. The pth percentile of N ordered values is obtained by first calculating the rank (r) using the following equation and rounding to the nearest integer and taking the value which corresponds to the calculated rank:

$$r = N/100^{*}(p) + 1/2$$

And upper and lower uncertainty bands correspond to 95th and 5th value.

5 Approaches/Methodologies for Transient Analysis

The approach/methodologies to be used to qualify the nodalisation to carry out transient analysis are described here, and same nodalisation should be used for further evaluation of uncertainty. The input deck (nodalisation) is said to be qualified for particular transient when it is qualified for both 'steady-state-level' as well as 'on-transient-level' qualification. Steady-state-level qualification criteria are mainly based on D'Auria et al. [2]. These criteria consist of three parts and given in Table 1. In the first part, details of geometrical parameters from nodalisation and test facility/ NPP are compared. In the second part, boundary conditions of nodalisation and test specifications/design nominal value of NPPs are compared. In the third part, some of the important thermal-hydraulic parameters for steady state which is derived from code output are compared with the experimental data/design nominal value for NPPs. In all the case, error should be less than acceptable error. In fact, first part and second part should be checked before start of code run.

On-transient-level qualification is done to demonstrate the capability of the nodalisation to reproduce the relevant thermal-hydraulic phenomena expected in

Table 1 Acceptance criteria for steady-state-level qualification [4]

Parameter	Acceptable error (%)
Geometrical parameter	
Primary circuit volume (m^3)	1.0 %
Secondary circuit volume (m^3)	2.0 %
Active structure heat transfer area (overall) (m^2)	0.1 %
Active structure heat transfer volume (overall) (m^3)	0.2 %
Volume vs. height curve (local primary and secondary circuit volume)	10.0 %
Component relative elevation (m)	0.01 m
Flow area of components like valves, pump orifices (m^2)	1.0 %
Generic flow area (m^2)	10.0 %
Boundary conditions	
Core power, PRZ heater power, core bypass heating power, decay heat curve, axial and radial power distribution, MCP velocity, pump coast down curve, SG feed water temperature, SG feed water flow rate, SG feed water pressure, ECCS water temperature, ECCS pressure, valve opening closing time, etc.	0.0 % (Zero error means error should be within the uncertainty band in the measurement)
Thermal-hydraulic parameter (from code output) initial condition	
Primary circuit power balance (MW)	2.0 %
Secondary circuit power balance (MW)	2.0 %
Absolute Pressure, (PRZ, ACC, SG) Mpa	0.1 %
Fluid temperature (K)	0.5 %
Rod surface temperature (K)	10 K
Heat losses (kW)	10.0 %
Local pressure drop (kPa) (Close-loop cumulative differential pressure vs. length curve)	10.0 %
Mass inventory in primary circuit (kg)	2.0 %
Mass inventory in primary circuit (kg)	5.0 %
Flow rates (primary and secondary circuits) (kg/s)	2.0 %
By pass mass flow rates (kg/s)	10.0 %
PRZ level (collapsed) (m)	0.05 m
Secondary side or down comer level (m)	0.1 m

the transient. This step also makes possible to verify the correctness of some systems operating only during transient events. The nodalisation is constituted by the schematisation of a facility. It is necessary to prove the capability of the code and of the nodalisation scheme during the transient analysis. The code options selected by the user, the schematisation solutions and the logic of some systems are involved during this check.

The following steps should be performed for qualitative evaluation:

(a) List of comparison between experimental and code calculation resulting time sequence of significant events (if available).

(b) Identification/verification of CSNI phenomena validation matrix applicable to the transient: A list of phenomena for LB LOCA and SB LOCA and transients are provided by CSNI phenomena validation matrix [9]. These phenomena are for code assessment. Therefore, it is to be checked that whether code calculation is able to predict all the phenomena given in the matrix or not.
(c) Phenomenological windows: Each transient scenario, that is, measured (if available) and calculated, is divided in number of phenomenological window time spans in which a unique relevant physical process mainly occurs and a limited set of parameters control the scenario for the proper interpretation.
(d) Key phenomena and relevant thermal-hydraulic aspects (RTA). In each PhW, key phenomena and RTA must be identified. RTAs are defined for a single transient and characterised by numerical values of significant parameters: single value parameters (SVP), non-dimensional parameters (NDP), time sequence of events (TSE), integral parameters (IPA), etc. The qualitative analysis is based on five subjective judgment marks (*E*, *R*, *M*, *U*, –) the list of RTAs. *E* means excellent and a good agreement exists between code and experimental results (if available, specifically applicable for transient in integral test facility); *R* is for reasonable and means that the phenomenon is reproduced by the code, but some minor discrepancies exist; *M* is for minimal and means that a relevant discrepancy is present between the code results and the experiment, but reason for the difference is identified, and it is not caused by a nodalisation deficiency; *U* is for unqualified and means that a relevant discrepancy exists, but reasons for the difference are intrinsic to the code, and nodalisation capability are not known; and '–' means not applicable to selected test. Even if one *U* result is present during the qualitative evaluation process, then nodalisation is not said to be qualified and sensitivity analysis is recommended by suitably modifying the nodalisation till route cause is not get detected and rectified.
(e) Visual comparisons between experimental and code calculated relevant parameter time trends. Major discrepancies are noticed in the process.

All the above five steps should be performed for any transient. In case of experimental data available, it would be compared otherwise it helps in proper understanding, and major discrepancies in prediction by code are identified.

6 Sampling-Based Uncertainty Methodology

Sampling-based approaches [5] to uncertainty and sensitivity analyses are demonstrated by carrying out uncertainty evaluation in the analysis of station blackout in PSB VVER integral test facility, small break LOCA in LSTF test facility and TMI-II accident. Several sampling strategies are available, including random sampling, importance sampling and Latin hypercube sampling (LHS) [8]. LHS is very popular for use with computationally demanding models because its efficient stratification properties allow for the extraction of a large amount of uncertainty and sensitivity information with relatively small sample size. In the present study, LHS method is used. When sampling a function of N variables (input parameters), the range of each variable is divided in to M equal probable intervals. M sample points are then place to satisfy the Latin hypercube requirements; it is to be noted that it forces the number of divisions M to be equal for each variable. This sampling scheme does not require more samples for more input variables. This independence is one of the main advantages of this sampling scheme. Another advantage is that random samples can be taken one at a time, remembering which samples were taken so far. This approach involves the generation and exploration of a mapping from uncertain input parameters to uncertainty in output parameters. The underlying idea is that analysis results of output parameters (y), $y(\mathbf{x}) = [y_1(\mathbf{x}), y_2(\mathbf{x}), \ldots y_n(\mathbf{x})]$ are functions of uncertain input parameters (x), $\mathbf{x} = [x_1, x_2, \ldots x_n]$. In turn, uncertainty in input parameters (x) results in a corresponding uncertainty in output parameter (y). This leads to two questions: (1) what is the uncertainty in $\mathbf{y}$ given the uncertainty in x? and (2) how important is the individual input parameter (x) with respect to the uncertainty in output parameters (y)? The goal of uncertainty analysis is to answer the first question, and the goal of sensitivity/importance analysis is to answer the second question. In practice, the implementation of an uncertainty analysis and sensitivity analysis is very closely connected on both conceptual and computational levels. The methodology adopted for the present study consists of the following steps:

- Screening sensitivity analysis/expert judgment for the selection of uncertain input parameters
- Characterisation of uncertainty (assigning uniform distribution) to input parameters
- Calculation matrix generation using Latin hypercube sampling
- Performing best estimate thermal-hydraulic code runs
- Representation of uncertainty analysis results
- Performing importance/sensitivity analysis using linear regression (estimation of standardised rank regression coefficients, etc.)

Many authors, when referring to the degree to which an input parameter affects the model output, use words such as ‘sensitive’, ‘important’, ‘most influential’, ‘major contributor’, ‘effective’ or ‘correlated’ interchangeably. Determination of sensitivity analysis results is usually more demanding than the presentation of uncertainty analysis results due to the need to actually explore the mapping $[\mathbf{x}_i, \mathbf{y}(\mathbf{x}_i)]$, $i = 1$ to N to assess the effects of individual input parameter (x) on output parameter (y). A number of approaches to sensitivity analysis that can be used in conjunction with a sampling-based uncertainty analysis are available. The regression analysis is used to determine the importance analysis results in the present study. C_{u} standardised regression coefficients (SRCs) are obtained from the regression analysis. The SRCs provide a useful measure of variable importance with (1) the absolute values of the coefficients providing a comparative measure of variable importance (i.e. variable x_{u} is more important than variable x_{v} if $C_{\mathrm{u}} > C_{\mathrm{v}}$) and (2) the sign of C_{u} indicating whether x and y tend to move in the same direction or in the opposite direction as long as the x's are independent from other input parameter considered for uncertainty analysis.

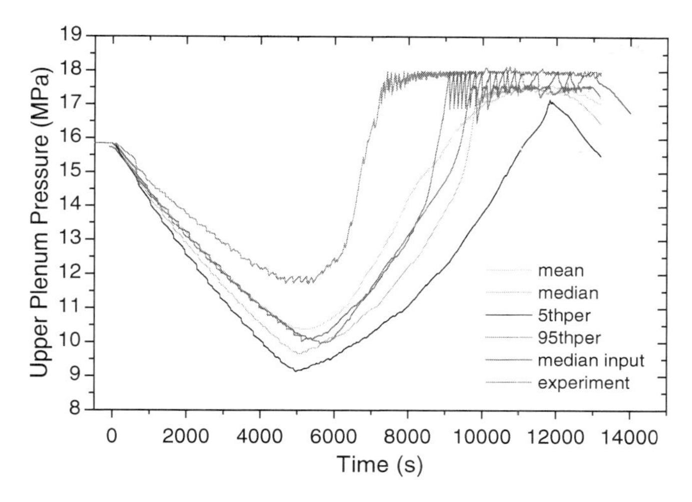

Fig. 7 Uncertainty in upper plenum pressure for SBO IN PSB VVER ITF [3]

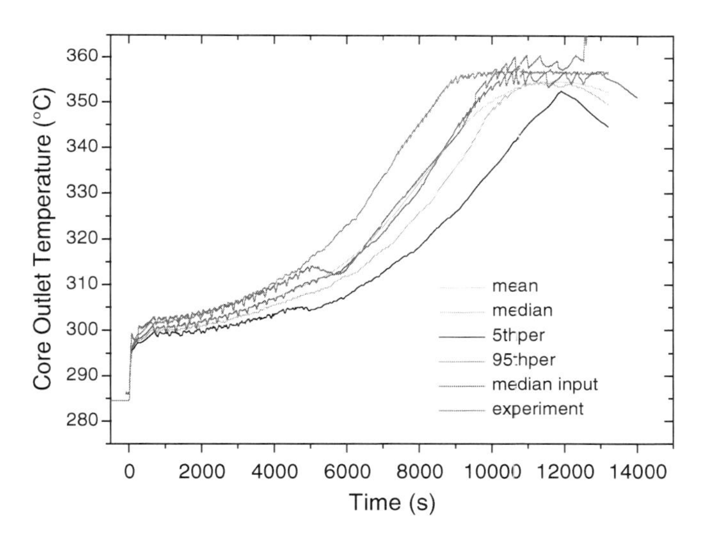

Fig. 8 Uncertainty in core outlet temperature for SBO IN PSB VVER ITF [3]

Results of uncertainty evaluation for primary pressure and core outlet temperature are shown in Fig. 7 and 8, respectively. Uncertainty in primary pressure and break discharge rate is shown in Figs. 9 and 10 for small break LOCA in LSTF. For the application of this methodology, for actual power plant uncertainty in pressuriser pressure and hot leg temperature for the analysis of TMI-II accident are shown in Figs. 11 and 12. In all the three analyses for almost all transient time experimental data and code calculated for median input are bounded by upper band (95 percentile) and lower and (5th percentile) of uncertainty.

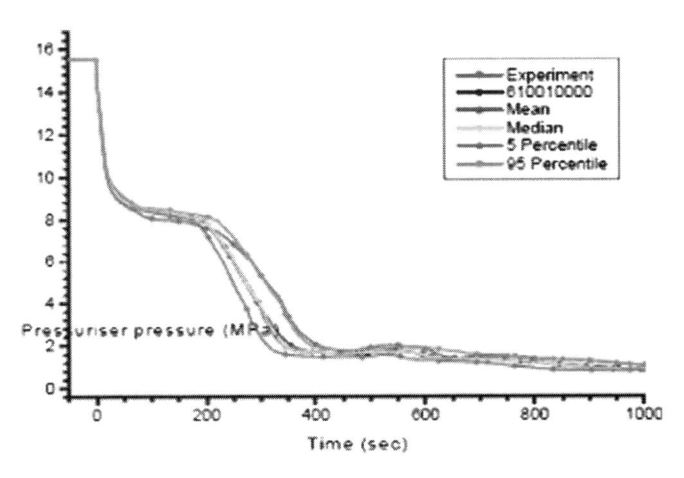

Fig. 9 Uncertainty in upper plenum pressure for small break LOCA in LSTF [10]

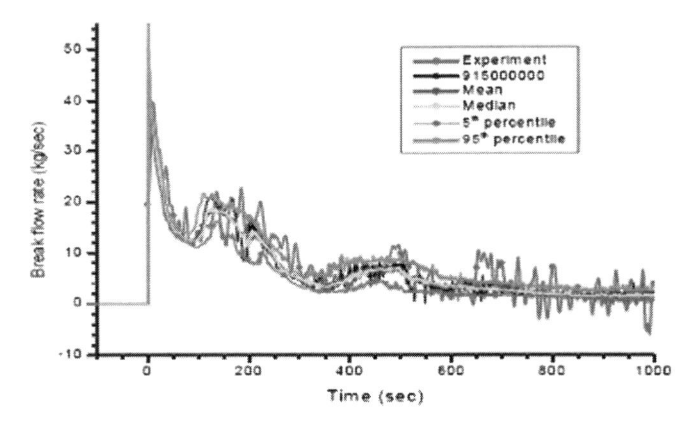

Fig. 10 Uncertainty in break flow rate for small break LOCA in LSTF [10]

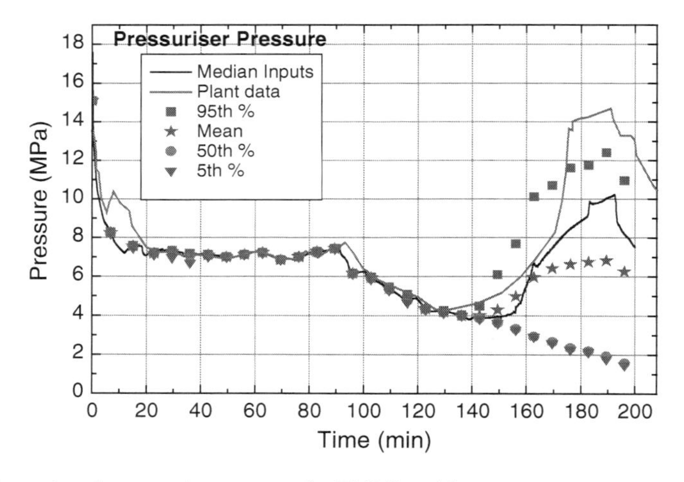

Fig. 11 Uncertainty in pressuriser pressure for TMI-II accident

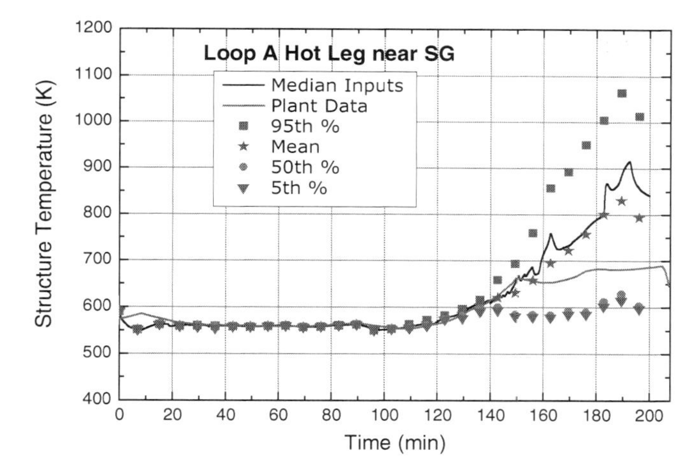

Fig. 12 Uncertainty in loop *A* hot leg temperature near SG

7 Conclusions

- Importance of conservative and best estimate accident analysis for nuclear power plants and related integral test facility are presented.
- Conservative analysis is used for licensing analysis; however, trends for deterministic safety analysis are moving towards the best estimate analysis.
- Various sources of uncertainty in carrying out best estimate accident analysis are highlighted.
- Aspects of various methodologies for uncertainty evaluation available world widely are summarised in this chapter.
- How to represent the evaluated uncertainty, that is, in terms of probability and confidence or in terms of percentile, is described in this chapter.
- Steady-state- and transient-level qualifications for the nodalisation adopted for the analysis of a transient using any system code are presented.
- Demonstration for the use of sampling-based uncertainty methodology for the evaluation of uncertainty for station blackout and small break LOCA in integral test facility is done, and application of this methodology is also extended for nuclear power plants TMI-II accident.
- This methodology is applicable to analysis of any transient using any best estimate system code.

References

1. AERB safety Code (2009) Design of pressurized heavy water reactor based nuclear power plants, no. AERB/NPP-PHWR/SC/D (Rev. 1), Mumbai
2. D'Auria F, Bousbia-Salah A, Petuzzi A, Del Nevo A (2006) State of the art in using best estimate calculation tools in nuclear technology. Nucl Eng Technol 38(1):11–32

3. Dubey SK, Rao RS, Sengupta S, Gupta SK (2011) Sampling based uncertainty analysis of station blackout in PSB VVER integral test facility. Ann Nucl Energy 38:2724–2733
4. Gupta SK (2009) Final progress report on IAEA CRP on uncertainty evaluation in best estimate accident analysis of NPPs, Mumbai
5. Helton JC (2006) Survey of sampling-based methods for uncertainty and sensitivity analysis. Reliab Eng Syst Saf 91:1175–1209
6. IAEA safety report series (2008) Best estimate safety analysis for nuclear power plants: uncertainty evaluation, IAEA, Vienna
7. Montgomery DC, Runger GC (1994) Applied statistics and probability for engineers. Wiley, New York
8. Park SR, Baek WP, Chang SH, Lee BH (1992) Development of uncertainty quantification method of the best estimate large LOCA analysis. Nucl Eng Des 135:367–378
9. Report by OECD Support Group (2001) Validation matrix for the assessment of thermal hydraulic codes for VVER LOCA and transients, NEA/CSNI/R 4
10. Sengupta S, Dubey SK, Rao RS, Gupta SK, Raina VK (2010) Sampling based uncertainty analysis of 10 % hot leg break LOCA in LSTF. Nucl Eng Technol 42(6):690–703
11. Taylor JR (1982) An introduction to error analysis-the study of uncertainty in physical measurements. Oxford University Press, Mill Valley

Failure Probability Bounds Using Multicut-High-Dimensional Model Representation

A.S. Balu and B.N. Rao

Abstract The structural reliability analysis in presence of mixed uncertain variables demands more computation as the entire configuration of fuzzy variables needs to be explored. Moreover, the existence of multiple design points plays an important role in the accuracy of results as the optimization algorithms may converge to a local design point by neglecting the main contribution from the global design point. Therefore, in this chapter, a novel uncertain analysis method for estimating the failure probability bounds of structural systems involving multiple design points in presence of mixed uncertain variables is presented. The proposed method involves weight function to identify multiple design points, multicut-high dimensional model representation technique for the limit state function approximation, transformation technique to obtain the contribution of the fuzzy variables to the convolution integral, and fast Fourier transform for solving the convolution integral. The proposed technique estimates the failure probability accurately with significantly less computational effort compared to the direct Monte Carlo simulation. The methodology developed is applicable for structural reliability analysis involving any number of fuzzy and random variables with any kind of distribution. The numerical examples presented demonstrate the accuracy and efficiency of the proposed method.

Keywords High dimensional model representation • Multiple design points • Random variables • Fuzzy variables • Convolution integral • Failure probability

---

A.S. Balu (✉) • B.N. Rao
Department of Civil Engineering, Indian Institute of Technology Madras, Chennai, Tamil Nadu, India
e-mail: arunsbalu@gmail.com; bnrao@iitm.ac.in

S. Chakraborty and G. Bhattacharya (eds.), *Proceedings of the International Symposium on Engineering under Uncertainty: Safety Assessment and Management (ISEUSAM - 2012)*,
DOI 10.1007/978-81-322-0757-3_18, © Springer India 2013

1 Introduction

Reliability analysis taking into account the uncertainties involved in a structural system plays an important role in the analysis and design of structures. Due to the complexity of structural systems, the information about the functioning of various structural components has different sources, and the failure of systems is usually governed by various uncertainties, all of which are to be taken into consideration for reliability estimation. Uncertainties present in a structural system can be classified as aleatory uncertainty and epistemic uncertainty. Aleatory uncertainty information can be obtained as a result of statistical experiments and has a probabilistic or random character. Epistemic uncertainty information can be obtained by the estimation of the experts and in most cases has an interval or fuzzy character. When aleatory uncertainty is only present in a structural system, then the reliability estimation involves determination of the probability that a structural response exceeds a threshold limit, defined by a limit state function influenced by several random parameters. Structural reliability can be computed by adopting probabilistic method involving the evaluation of multidimensional integral [1, 2].

In first- or second-order reliability method (FORM/SORM), the limit state functions need to be specified explicitly. Alternatively, the simulation-based methods such as Monte Carlo techniques require more computational effort for simulating the actual limit state function repeated times. The response surface concept was adopted to get separable and closed form expression of the implicit limit state function in order to use fast Fourier transform (FFT) to estimate the failure probability [3]. The high-dimensional model representation (HDMR) concepts were applied for the approximation of limit state function at the MPP and FFT techniques to evaluate the convolution integral for estimation of failure probability [4]. In this method, efforts are required in evaluating conditional responses at a selected input determined by sample points, as compared to full-scale simulation methods.

Further, the main contribution to the reliability integral comes from the neighborhood of design points. When multiple design points exist, available optimization algorithms may converge to a local design point and thus erroneously neglect the main contribution to the value of the reliability integral from the global design point(s). Moreover, even if a global design point is obtained, there are cases for which the contribution from other local or global design points may be significant [5]. In that case, multipoint FORM/SORM is required for improving the reliability analysis [6]. In the presence of only epistemic uncertainty in a structural system, possibilistic approaches to evaluate the minimum and maximum values of the response are available [7, 8].

All the reliability models discussed above are based on only one kind of uncertain information, either random variables or fuzzy input, but do not accommodate a combination of both types of variables. However, in some engineering problems with mixed uncertain parameters, using one kind of reliability model cannot obtain the best results. To determine the failure probability bounds of a structural system involving both random and fuzzy variables, the entire configuration of the fuzzy variables needs to be explored. Hence, the computational effort involved in estimating the bounds of the failure probability increases tremendously in the presence of multiple design points and mixed uncertain variables.

This chapter explores the potential of coupled multicut-HDMR (MHDMR)-FFT technique in evaluating the reliability of a structural system with multiple design points, for which some uncertainties can be quantified using fuzzy membership functions while some are random in nature. Comparisons of numerical results have been made with direct MCS method to evaluate the accuracy and computational efficiency of the present method.

2 High Dimensional Model Representation

High dimensional model representation (HDMR) is a general set of quantitative model assessment and analysis tools for capturing the high dimensional relationships between sets of input and output model variables [4, 9]. Let the N-dimensional vector $\boldsymbol{x} = \{x_1, x_2, \ldots, x_N\}$ represent the input variables of the model under consideration and the response function as $g(\boldsymbol{x})$. Since the influence of the input variables on the response function can be independent and/or cooperative, HDMR expresses the response $g(\boldsymbol{x})$ as a hierarchical correlated function expansion in terms of the input variables as

$$\begin{aligned} g(\boldsymbol{x}) = g_0 + \sum_{i=1}^{N} g_i(x_i) + \sum_{1 \le i_1 < i_2 \le N} g_{i_1 i_2}(x_{i_1}, x_{i_2}) + \ldots \\ + \sum_{1 \le i_1 < \ldots < i_l \le N} g_{i_1 i_2 \ldots i_l}(x_{i_1}, x_{i_2}, \ldots, x_{i_l}) + \cdots + g_{12\ldots N}(x_1, x_2, \ldots, x_N), \end{aligned} \tag{1}$$

where g_0 is a constant term representing the zeroth-order component function or the mean response of $g(\boldsymbol{x})$. The function $g_i(x_i)$ is a first-order term expressing the effect of variable x_i acting alone, although generally nonlinearly, upon the output $g(\boldsymbol{x})$. The function $g_{i_1 i_2}(x_{i_1}, x_{i_2})$ is a second-order term which describes the cooperative effects of the variables x_{i_1} and x_{i_2} upon the output $g(\boldsymbol{x})$. The higher order terms give the cooperative effects of increasing numbers of input variables acting together to influence the output $g(\boldsymbol{x})$. The last term $g_{12,\cdots,N}(x_1, x_2, \ldots, x_N)$ contains any residual dependence of all the input variables locked together in a cooperative way to influence the output $g(\boldsymbol{x})$. The expansion functions are determined by evaluating the input–output responses of the system relative to the defined reference point $\boldsymbol{c}$ along associated lines, surfaces, subvolumes, etc., in the input variable space. This process reduces to the following relationship for the component functions in Eq. (1):

$$g_0 = g(\boldsymbol{c}), \tag{2}$$

$$g_i(x_i) = g(x_i, \boldsymbol{c}^i) - g_0, \tag{3}$$

$$g_{i_1 i_2}(x_{i_1}, x_{i_2}) = g(x_{i_1}, x_{i_2}, \boldsymbol{c}^{i_1 i_2}) - g_{i_1}(x_{i_1}) - g_{i_2}(x_{i_2}) - g_0, \tag{4}$$

where the notation $g(x_i, \boldsymbol{c}^i) = g(c_1, c_2, \ldots, c_{i-1}, x_i, c_{i+1}, \ldots, c_N)$ denotes that all the input variables are at their reference point values except x_i. The g_0 term is the output response of the system evaluated at the reference point $\boldsymbol{c}$. The higher order terms are evaluated as cuts in the input variable space through the reference point. Therefore, each first-order term $g_i(x_i)$ is evaluated along its variable axis through the reference point. Each second-order term $g_{i_1 i_2}(x_{i_1}, x_{i_2})$ is evaluated in a plane defined by the binary set of input variables x_{i_1} and x_{i_2} through the reference point, etc. The first-order approximation of $g(\boldsymbol{x})$ is as follows:

$$\begin{aligned}\tilde{g}(\boldsymbol{x}) &\equiv g(x_1, x_2, \ldots, x_N) \\ &= \sum_{i=1}^{N} g(c_1, \ldots, c_{i-1}, x_i, c_{i+1}, \ldots, c_N) - (N-1) g(\boldsymbol{c}).\end{aligned} \tag{5}$$

The notion of 0th, 1st, etc., in HDMR expansion should not be confused with the terminology used either in the Taylor series or in the conventional least-squares-based regression model. It can be shown that the first-order component function $g_i(x_i)$ is the sum of all the Taylor series terms which contain and only contain variable x_i. Hence, first-order HDMR approximations should not be viewed as first-order Taylor series expansions nor do they limit the nonlinearity of $g(\boldsymbol{x})$.

3 Multicut-HDMR

The main limitation of truncated cut-HDMR expansion is that depending on the order chosen sometimes it is unable to accurately approximate $g(\boldsymbol{x})$, when multiple design points exist on the limit state function or when the problem domain is large. In this section, a new technique based on MHDMR is presented for approximation of the original implicit limit state function, when multiple design points exist. The basic principles of cut-HDMR may be extended to more general cases. MHDMR is one extension where several cut-HDMR expansions at different reference points are constructed, and the original implicit limit state function $g(\boldsymbol{x})$ is approximately represented not by one but by all cut-HDMR expansions. In the present work, weight function is adopted for identification of multiple reference points closer to the limit surface.

Let $\boldsymbol{d}^1, \boldsymbol{d}^2, \ldots, \boldsymbol{d}^{m_d}$ be the m_d identified reference points closer to the limit state function based on the weight function. MHDMR approximation of the original implicit limit state function is based on the principles of cut-HDMR expansion, where individual cut-HDMR expansions are constructed at different reference points $\boldsymbol{d}^1, \boldsymbol{d}^2, \ldots, \boldsymbol{d}^{m_d}$ by taking one at a time as follows:

$$\begin{aligned} g^k(\boldsymbol{x}) &= g_0^k + \sum_{i=1}^{N} g_i^k(x_i) + \sum_{1 \le i_1 < i_2 \le N} g_{i_1 i_2}^k(x_{i_1}, x_{i_2}) + \ldots \\ &+ \sum_{1 \le i_1 < \ldots < i_l \le N} g_{i_1 i_2 \ldots i_l}^k(x_{i_1}, x_{i_2}, \ldots, x_{i_l}) + \ldots + g_{12\ldots N}^k(x_1, x_2, \ldots, x_N); \quad k = 1, 2, \ldots, m_d \end{aligned} \quad (6)$$

The original implicit limit state function $g(\boldsymbol{x})$ is approximately represented by blending all locally constructed m_d individual cut-HDMR expansions as follows:

$$g(\boldsymbol{x}) \cong \sum_{k=1}^{m_d} \lambda_k(\boldsymbol{x}) \left[g_0^k + \sum_{i=1}^{N} g_i^k(x_i) + \ldots + g_{12\cdots N}^k(x_1, x_2, \ldots, x_N) \right]. \quad (7)$$

The coefficients $\lambda_k(\boldsymbol{x})$ possess the properties

$$\lambda_k(\boldsymbol{x}) = \begin{cases} 1 & \text{if } \boldsymbol{x} \text{ is in any cut subvolume of the } k \text{ - th reference point expansions} \\ 0 & \text{if } \boldsymbol{x} \text{ is in any cut subvolume of other reference point expansions} \end{cases} \quad (8)$$

and

$$\sum_{k=1}^{m_d} \lambda_k(\boldsymbol{x}) = 1. \quad (9)$$

There are a variety of choices to define $\lambda_k(\boldsymbol{x})$. In the present study, the metric distance $\alpha_k(\boldsymbol{x})$ from any sample point to the reference point $\boldsymbol{d}^k$; $k = 1, 2, \ldots, m_d$

$$\alpha_k(\boldsymbol{x}) = \left[\sum_{i=1}^{N} \left(x_i - d_i^k\right)^2 \right]^{\frac{1}{2}}; \quad d_i^k \equiv k\text{-th reference point} \quad (10)$$

is used to define

$$\lambda_k(\boldsymbol{x}) = \frac{\bar{\lambda}_k(\boldsymbol{x})}{\sum_{s=1}^{m_d} \bar{\lambda}_s(\boldsymbol{x})}, \quad (11)$$

where

$$\bar{\lambda}_k(\boldsymbol{x}) = \prod_{s=1;s\neq k}^{m_d} \alpha_s(\boldsymbol{x}). \quad (12)$$

The coefficients $\lambda_k(\boldsymbol{x})$ determine the contribution of each locally approximated function to the global function. The properties of the coefficients $\lambda_k(\boldsymbol{x})$ imply that the contribution of all other cut-HDMR expansions vanishes except one when $\boldsymbol{x}$ is located on any cut line, plane, or higher dimensional ($\leq l$) subvolumes through that reference point, and then the MHDMR expansion reduces to single point cut-HDMR expansion. As mentioned above, the l-th-order cut-HDMR approximation does not have error when $\boldsymbol{x}$ is located on these subvolumes. When m_d cut-HDMR expansions are used to construct an MHDMR expansion, the error-free region in input $\boldsymbol{x}$ space is m_d times that for a single reference point cut-HDMR expansion; hence, the accuracy will be improved. Therefore, first-order MHDMR approximations of the original implicit limit state function with m_d reference points can be expressed as

$$\tilde{g}(\boldsymbol{x}) \cong \sum_{k=1}^{m_d} \lambda_k(\boldsymbol{x}) \left[\sum_{i=1}^{N} g^k\left(d_1^k, \ldots, d_{i-1}^k, x_i, d_{i+1}^k, \ldots, d_N^k\right) - (N-1) g^k\left(\boldsymbol{d}^k\right) \right]. \quad (13)$$

4 Weight Function

The most important part of MHDMR approximation of the original implicit limit state function is identification of multiple reference points closer to the limit state function. The proposed weight function is similar to that used by Kaymaz and McMahon [10] for weighted regression analysis. Among the limit state function responses at all sample points, the most likelihood point is selected based on closeness to zero value, which indicates that particular sample point is close to the limit state function.

In this study, two types of procedures are adopted for identification of reference points closer to the limit state function, namely, (1) first-order method and (2) second-order method. The procedure for identification of reference points closer to the limit state function using first-order method proceeds as follows: (a) $n(=3, 5, 7 \text{ or } 9)$ equally spaced sample points $\mu_i - (n-1)\sigma_i/2$, $\mu_i - (n-3)\sigma_i/2$, ..., μ_i, ..., $\mu_i + (n-3)\sigma_i/2$, $\mu_i + (n-1)\sigma_i/2$ are deployed along each of the random variable axis x_i with mean μ_i and standard deviation σ_i, through an initial reference point. Initial reference point is taken as mean value of the random variables. (b) The limit state function is evaluated at each sample point. (c) Using the limit state function responses at all sample points, the weight corresponding to each sample point is evaluated using the following weight function:

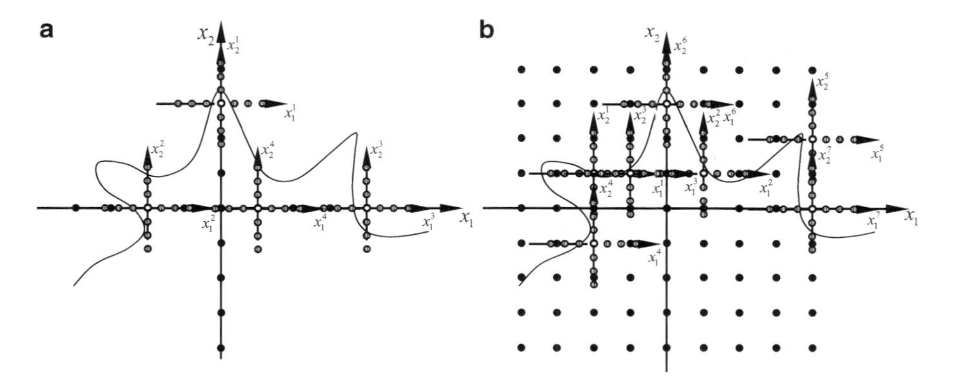

Fig. 1 MHDMR approximation of original limit state function, with (**a**) FF sampling scheme and (**b**) SF sampling scheme

$$w^I = \exp\left(-\frac{g(c_1,\ldots,c_{i-1},x_i,c_{i+1},\ldots,c_N) - g(\boldsymbol{x})|_{\min}}{\left|g(\boldsymbol{x})|_{\min}\right|}\right). \tag{14}$$

Second-order method of identification of reference points closer to the limit state function proceeds as follows: (a) A regular grid is formed by taking $n(=3,5,7$ or $9)$ equally spaced sample points $\mu_{i_1}-(n-1)\sigma_{i_1}/2$, $\mu_{i_1}-(n-3)\sigma_{i_1}/2$, ..., μ_{i_1}, ..., $\mu_{i_1}+(n-3)\sigma_{i_1}/2$, $\mu_{i_1}+(n-1)\sigma_{i_1}/2$ along the random variable x_{i_1} axis with mean μ_{i_1} and standard deviation σ_{i_1} and $n(=3,5,7$ or $9)$ equally spaced sample points $\mu_{i_2}-(n-1)\sigma_{i_2}/2$, $\mu_{i_2}-(n-3)\sigma_{i_2}/2$, ..., μ_{i_2} , ..., $\mu_{i_2}+(n-3)\sigma_{i_2}/2$, $\mu_{i_2}+(n-1)\sigma_{i_2}/2$ along the random variable x_{i_2} axis with mean μ_{i_2} and standard deviation σ_{i_2}, through an initial reference point. Initial reference point is taken as mean value of the random variables. (b) The limit state function is evaluated at each sample point. (c) Using the limit state function responses at all sample points, the weight corresponding to each sample point is evaluated using the following weight function:

$$w^{II} = \exp\left(-\frac{g(c_1,\ldots,c_{i_1-1},x_{i_1},c_{i_1+1},\ldots,c_{i_2-1},x_{i_1},c_{i_2+1},\ldots,c_N) - g(\boldsymbol{x})|_{\min}}{\left|g(\boldsymbol{x})|_{\min}\right|}\right). \tag{15}$$

Sample points $\boldsymbol{d}^1, \boldsymbol{d}^2, \ldots, \boldsymbol{d}^{m_d}$ with maximum weight are selected as reference points closer to the limit state function for construction of m_d individual cut-HDMR approximations of the original implicit limit state function locally. In this study, two types of sampling schemes, namely, FF and SF, are adopted. Figure 1a shows FF sampling scheme involving first-order method of identification of reference points and blending of locally constructed individual first-order HDMR approximations at different identified reference points using the coefficients $\lambda_k(\boldsymbol{x})$

to form MHDMR approximation $\tilde{g}(\boldsymbol{x})$. Figure 1b shows SF sampling scheme involving second-order method of identification of reference points and blending of locally constructed individual first-order HDMR approximations to form MHDMR approximation.

5 Failure Probability Bounds

Let the N-dimensional input variables vector $\boldsymbol{x} = \{x_1, x_2, \ldots, x_N\}$, which comprises of r number of random variables and f number of fuzzy variables, be divided as $\boldsymbol{x} = \{x_1, x_2, \ldots, x_r, x_{r+1}, x_{r+2}, \ldots, x_{r+f}\}$ where the subvectors $\{x_1, x_2, \ldots, x_r\}$ and $\{x_{r+1}, x_{r+2}, \ldots, x_{r+f}\}$, respectively, group the random variables and the fuzzy variables, with $N = r + f$. Then, the first-order approximation of $\tilde{g}(\boldsymbol{x})$ can be divided into three parts: the first part with only the random variables, the second part with only the fuzzy variables, and the third part is a constant which is the output response of the system evaluated at the reference point $\boldsymbol{c}$ as follows:

$$\tilde{g}(\boldsymbol{x}) = \sum_{i=1}^{r} g\left(x_i, \boldsymbol{c}^i\right) + \sum_{i=r+1}^{N} g\left(x_i, \boldsymbol{c}^i\right) - (N-1)g(\boldsymbol{c}). \tag{16}$$

The joint membership function of the fuzzy variables part is obtained using suitable transformation of the variables $\{x_{r+1}, x_{r+2}, \ldots, x_N\}$ and interval arithmetic algorithm. Using this approach, the minimum and maximum values of the fuzzy variables part are obtained at each α-cut. Using the bounds of the fuzzy variables part at each α-cut along with the constant part and the random variables part in Eq. (16), the joint density functions are obtained by performing the convolution using FFT in the rotated Gaussian space at the MPP, which upon integration yields the bounds of the failure probability.

5.1 Transformation of Fuzzy Variables

Optimization techniques are required to obtain the minimum and maximum values of a nonlinear response within the bounds of the interval variables. This procedure is computationally expensive for problems with implicit limit state functions, as optimization requires the function value and gradient information at several points in the iterative process. But, if the function is expressed as a linear combination of interval variables, then the bounds of the response can be expressed as the summation of the bounds of the individual variables. Therefore, fuzzy variables part of the nonlinear limit state function in Eq. (16) is expressed as a linear combination of intervening variables by the use of first-order HDMR approximation in order to apply an interval arithmetic algorithm as follows:

$$\sum_{i=r+1}^{N} g\left(x_i, \boldsymbol{c}^i\right) = z_1 + z_2 + \ldots + z_f, \tag{17}$$

where $z_i = (\beta_i x_i + \gamma_i)^{\kappa}$ is the relation between the intervening and the original variables with κ being order of approximation taking values $\kappa = 1$ for linear approximation, $\kappa = 2$ for quadratic approximation, $\kappa = 3$ for cubic approximation, and so on. The bounds of the intervening variables can be determined using transformations [11]. If the membership functions of the intervening variables are available, then at each α-cut, interval arithmetic techniques can be used to estimate the response bounds at that level.

5.2 Estimation of Failure Probability Using FFT

Concept of FFT can be applied to the problem if the limit state function is in the form of a linear combination of independent variables and when either the marginal density or the characteristic function of each basic random variable is known. In the present study, HDMR concepts are used to express the random variables part along with the values of the constant part and the fuzzy variables part at each α-cut as a linear combination of lower order component functions. The steps involved in the proposed method for failure probability estimation as follows:

1. If $\boldsymbol{u} = \{u_1, u_2, \ldots, u_r\}^T \in \Re^r$ is the standard Gaussian variable, let $\boldsymbol{u}^{k*} = \{u_1^{k*}, u_2^{k*}, \ldots, u_r^{k*}\}^T$ be the MPP or design point, determined by a standard nonlinear constrained optimization. The MPP has a distance β_{HL}, which is commonly referred to as the Hasofer–Lind reliability index. Construct an orthogonal matrix $\boldsymbol{R} \in \Re^{r \times r}$ whose r-th column is $\alpha^{k*} = \boldsymbol{u}^{k*}/\beta_{\mathrm{HL}}$, that is, $\boldsymbol{R} = \left[\boldsymbol{R}_1 | \alpha^{k*}\right]$, where $\mathbf{R}_1 \in \Re^{r \times r-1}$ satisfies $\alpha^{k*T}\boldsymbol{R}_1 = 0 \in \Re^{1 \times r-1}$. The matrix $\boldsymbol{R}$ can be obtained, for example, by Gram–Schmidt orthogonalization. For an orthogonal transformation, $\boldsymbol{u}=\boldsymbol{R}\,\boldsymbol{v}$.
2. Let $\boldsymbol{v} = \{v_1, v_2, \ldots, v_r\}^T \in \Re^r$ be the rotated Gaussian space with the associated MPP $\boldsymbol{v}^{k*} = \{v_1^{k*}, v_2^{k*}, \ldots, v_r^{k*}\}^T$. Note that in the rotated Gaussian space, the MPP is $\boldsymbol{v}^* = \{0, 0, \ldots, \beta_{\mathrm{HL}}\}^T$. The transformed limit state function $g(\boldsymbol{v})$ therefore maps the random variables along with the values of the constant part and the fuzzy variables part at each α-cut into rotated Gaussian space $\boldsymbol{v}$. First-order HDMR approximation of $g(\boldsymbol{v})$ in rotated Gaussian space $\boldsymbol{v}$ with $\boldsymbol{v}^{k*} = \{v_1^{k*}, v_2^{k*}, \ldots, v_r^{k*}\}^T$ as reference point can be represented as follows:

$$\begin{aligned}\tilde{g}^k(\boldsymbol{v}) &\equiv g^k(v_1, v_2, \ldots, v_r) \\ &= \sum_{i=1}^{r} g^k\left(v_1^{k*}, \ldots, v_{i-1}^{k*}, v_i, v_{i+1}^{k*}, \ldots, v_r^{k*}\right) - (r-1)g\left(\boldsymbol{v}^{k*}\right).\end{aligned} \tag{18}$$

3. In addition to the MPP as the chosen reference point, the accuracy of first-order HDMR approximation in Eq. (18) may depend on the orientation of the first $r-1$ axes. In the present work, the orientation is defined by the matrix $\boldsymbol{R}$. In Eq. (18), the terms $g^k\left(v_1^{k*},\ldots,v_{i-1}^{k*},v_i,v_{i+1}^{k*},\ldots,v_r^{k*}\right)$ are the individual component functions and are independent of each other. Equation (18) can be rewritten as

$$\tilde{g}^k(\boldsymbol{v}) = a^k + \sum_{i=1}^{r} g^k\left(v_i, \boldsymbol{v}^{k*^i}\right), \tag{19}$$

where $a^k = -(r-1)g\left(\boldsymbol{v}^{k*}\right)$.

4. New intermediate variables are defined as

$$y_i^k = g^k\left(v_i, \boldsymbol{v}^{k*^i}\right). \tag{20}$$

The purpose of these new variables is to transform the approximate function into the following form:

$$\tilde{g}^k(\boldsymbol{v}) = a^k + y_1^k + y_2^k + \cdots + y_r^k. \tag{21}$$

5. Due to rotational transformation in $\boldsymbol{v}$-space, component functions y_i^k in Eq. (21) are expected to be linear or weakly nonlinear function of random variables v_i. In this work, both linear and quadratic approximations of y_i^k are considered.
6. Let $y_i^k = b_i + c_i\,v_i$ and $y_i^k = b_i + c_i\,v_i + e_i\,v_i^2$ be the linear and quadratic approximations, where coefficients $b_i \in \Re$, $c_i \in \Re$, and $e_i \in \Re$ (nonzero) are obtained by least-squares approximation from exact or numerically simulated conditional responses $\left\{g^k\left(v_i^1, \boldsymbol{v}^{k*^i}\right), g^k\left(v_i^2, \boldsymbol{v}^{k*^i}\right), \cdots, g^k\left(v_i^n, \boldsymbol{v}^{k*^i}\right)\right\}^T$ at n sample points along the variable axis v_i. Then, Eq. (21) results in

$$\tilde{g}^k(\boldsymbol{v}) \equiv a^k + y_1^k + y_2^k + \cdots + y_r^k = a^k + \sum_{i=1}^{r} (b_i + c_i\,v_i) \tag{22}$$

and

$$\tilde{g}^k(\boldsymbol{v}) \equiv a^k + y_1^k + y_2^k + \cdots + y_r^k = a^k + \sum_{i=1}^{r} \left(b_i + c_i\,v_i + e_i\,v_i^2\right). \tag{23}$$

7. The global approximation is formed by blending of locally constructed individual first-order HDMR approximations in the rotated Gaussian space at different identified reference points using the coefficients λ_k:

$$\tilde{g}(\boldsymbol{v}) = \sum_{k=1}^{m_d} \lambda_k \tilde{g}^k(\boldsymbol{v}). \tag{24}$$

8. Since v_i follows standard Gaussian distribution, marginal density of the intermediate variables y_i can be easily obtained by simple transformation (using chain rule):

$$p_{Y_i}(y_i) = p_{V_i}(v_i)\left|\frac{dv_i}{dy_i}\right|. \tag{25}$$

9. Now, the approximation is a linear combination of the intermediate variables y_i. Therefore, the joint density of $\tilde{g}(\boldsymbol{v})$, which is the convolution of the individual marginal density of the intervening variables y_i, can be expressed as follows:

$$p_{\tilde{G}}(\tilde{g}) = p_{Y_1}(y_1) * p_{Y_2}(y_2) * \ldots * p_{Y_r}(y_r), \tag{26}$$

   where $p_{\tilde{G}}(\tilde{g})$ represents joint density of the transformed limit state function $\tilde{g}(\boldsymbol{v})$.
10. Applying FFT on both sides of Eq. (26) leads to

$$FFT\left[p_{\tilde{G}}(\tilde{g})\right] = FFT\left[p_{Y_1}(y_1)\right]FFT\left[p_{Y_2}(y_2)\right]\ldots FFT\left[p_{Y_r}(y_r)\right]. \tag{27}$$

11. By applying inverse FFT on both side of Eq. (27), joint density of $\tilde{g}(\boldsymbol{v})$ is obtained.
12. The probability of failure is given by the following equation:

$$P_{\mathrm{F}} = \int_{-\infty}^{0} p_{\tilde{G}}(\tilde{g})\mathrm{d}\tilde{g}. \tag{28}$$

13. The membership function of failure probability can be obtained by repeating the above procedure at all confidence levels of the fuzzy variables part.

6 Numerical Examples

To evaluate the accuracy and the efficiency of the present method, comparisons of the estimated failure probability bounds, both by performing the convolution using FFT in conjunction with linear and quadratic approximations and MCS on the global approximation, have been made with that obtained using direct MCS. When comparing computational efforts by various methods in evaluating the failure probability, the number of original limit state function evaluations is chosen as the primary comparison tool in this chapter. This is because of the fact that number of function evaluations indirectly indicates the CPU time usage. For direct MCS, number of original function evaluations is same as the sampling size. While evaluating the failure probability through direct MCS, CPU time is more because it involves number of repeated actual finite-element analysis.

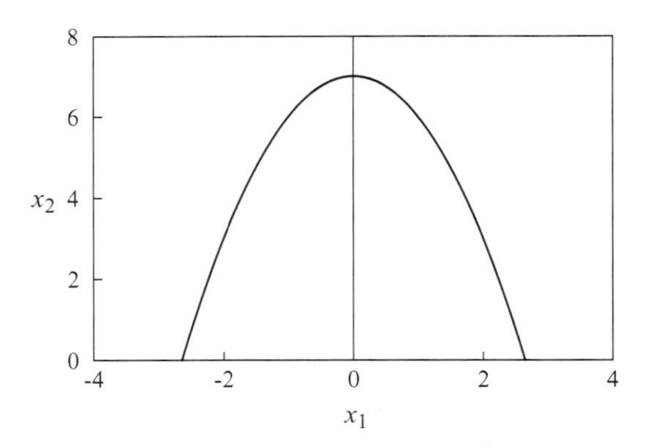

Fig. 2 Limit state function

Table 1 Identification of multiple design points with FF sampling

Sample points		$g(\boldsymbol{x})$			$g(\boldsymbol{x})\|_{\min}$			w^J		
x_1	x_2	$\alpha = 0^{(L)}$	$\alpha = 1$	$\alpha = 0^{(R)}$	$\alpha = 0^{(L)}$	$\alpha = 1$	$\alpha = 0^{(R)}$	$\alpha = 0^{(L)}$	$\alpha = 1$	$\alpha = 0^{(R)}$
−2.0	0.0	1.00	3.00	5.00	1.0	3.0	5.0	1.000	1.000	1.000
−1.0	0.0	4.00	6.00	8.00				0.050	0.368	0.549
0.0	0.0	5.00	7.00	9.00				0.018	0.264	0.449
1.0	0.0	4.00	6.00	8.00				0.050	0.368	0.549
2.0	0.0	1.00	3.00	5.00				1.000	1.000	1.000
0.0	−2.0	7.00	9.00	11.00				0.002	0.135	0.301
0.0	−1.0	6.00	8.00	10.00				0.001	0.189	0.368
0.0	0.0	5.00	7.00	9.00				0.018	0.264	0.449
0.0	1.0	4.00	6.00	8.00				0.050	0.368	0.549
0.0	2.0	3.00	5.00	7.00				0.135	0.513	0.670

6.1 Parabolic Performance Function

The limit state function considered is a parabola of the form

$$g(\boldsymbol{x}) = -x_1^2 - x_2 + x_3, \tag{29}$$

where x_1 and x_2 are assumed to be independent standard normal variables. The variable x_3 is assumed to be fuzzy with triangular membership function having the triplet [5.0, 7.0, 9.0].

The initial reference point $\boldsymbol{c}$ is taken as, respectively, the mean values and nominal values of the random and fuzzy variables. The first-order HDMR approximation, which is constructed over the initial reference point, is divided into two parts: one with only the random variables and the other with the fuzzy variables. The joint membership function of the fuzzy part of limit state function is obtained using suitable transformation of the fuzzy variables. In this example, the joint membership function is same as the membership function of the fuzzy variable x_3. As shown in Fig. 2, the limit state function given by Eq. (28) is symmetric about x_2

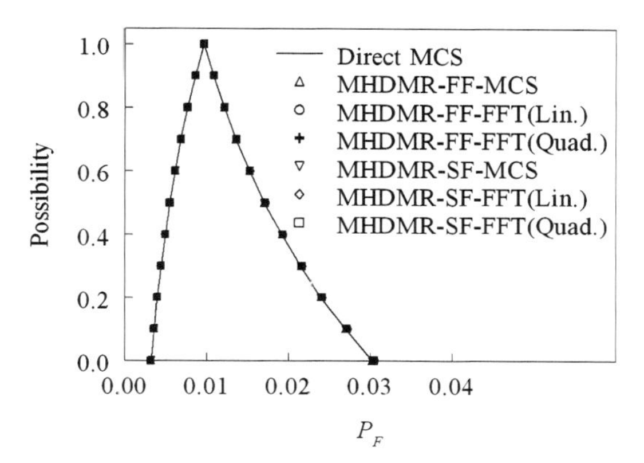

Fig. 3 Membership function of failure probability for parabolic performance function

for given value of x_3 (say the nominal value of $x_3 = 7$ at $\alpha = 1$) and has two design points. The two actual design points of the limit state function shown in Fig. 2, obtained using recursive quadratic programming (RQP) algorithm, are (2.54, 0.49) and (−2.54, 0.49) with reliability indices $\beta_1 = \beta_2 = 2.588$.

Table 1 illustrates computational details and identification of reference points $\boldsymbol{d}^1, \boldsymbol{d}^2$ using FF sampling scheme with five equally spaced sample points ($n = 5$) along each of the variable axis. In Table 1, the values corresponding to $\alpha = 0^{(L)}$ and $\alpha = 0^{(R)}$, respectively, indicate the extreme left and right values of the limit state function $g(\boldsymbol{x})$ at zero confidence level (i.e., $\alpha = 0$). Table 1 shows two reference points $\boldsymbol{d}^1 = (2, 0)$ and $\boldsymbol{d}^2 = (-2, 0)$ closer to the function. After identification of the two reference points (2, 0) and (−2, 0), local individual first-order HDMR approximations of the original limit state function are constructed at the two reference points by deploying $n = 5$ sample points along each of the variable axis. Local approximations of the original limit state function are blended together to form global approximation. The bounds of the failure probability are obtained both by performing the convolution using FFT in conjunction with linear and quadratic approximations and MCS on the global approximation.

Figure 3 shows the membership function of the failure probability P_{F} estimated both by performing the convolution using FFT and MCS on the global approximation, as well as that obtained using direct MCS.

In addition, effect of SF sampling scheme on the estimated membership function of the failure probability is studied. After identifying two reference points $\boldsymbol{d}^1 = (-2, 2)$ and $\boldsymbol{d}^2 = (2, 2)$ closer to the function producing maximum weight, the bounds of the failure probability are obtained. Figure 3 also shows the membership function of the failure probability obtained by the proposed method based on SF sampling scheme. The effect of number of sample points is studied by varying n from 3 to 9. It is observed that $n = 7$ provides the optimum number of function calls with acceptable accuracy in evaluating the failure probability with the present method.

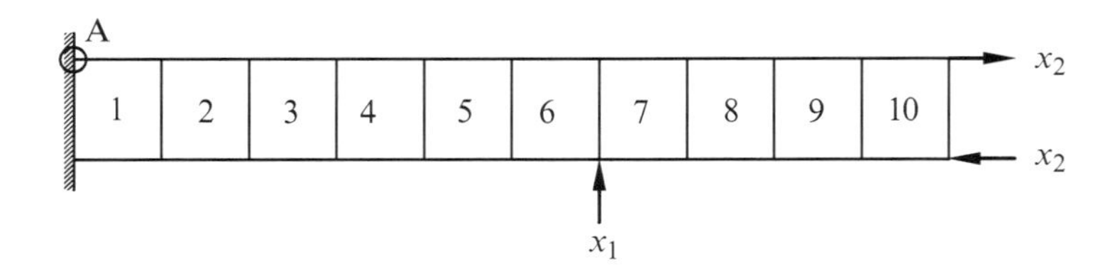

Fig. 4 Cantilever steel beam

6.2 Cantilever Steel Beam

A cantilever steel beam of 1.0 m with cross-sectional dimensions of (0.1 m $\times$ 0.01 m) is considered, as shown in Fig. 4, to examine the accuracy and efficiency of the proposed method for the membership function of failure probability estimation. The beam is subjected to an in-plane moment at the free end and a concentrated load at 0.4 m from the free end. The structure is assumed to have failed if the square of the von Mises stress at the support (at A in Fig. 4) exceeds specified threshold $V_{\max}$. Therefore, the limit state function is defined as

$$g(\boldsymbol{x}) = V_{\max} - V(\boldsymbol{x}), \tag{30}$$

where $V(\boldsymbol{x})$ is the square of the von Mises stress, expressed as a quadratic operator on the stress vector.

In this example, loads x_1 and x_2, modulus of elasticity of the beam E, and threshold quantity $V_{\max}$ are taken as uncertain variables. The variations of E and $V_{\max}$ are expressed as $E = E_0(1 + \varepsilon x_3)$ and $V_{\max} = V_{\max 0}(1 + \varepsilon x_4)$. Here, ε is small deterministic quantity representing the coefficient of variation of the random variables and are taken to equal to 0.05, $E_0 = 2 \times 10^5$ N/m^2 denotes the deterministic component of modulus of elasticity, and $V_{\max 0} = 6.15 \times 10^9$ N/m^2 denotes the deterministic component of threshold quantity. All variables are assumed to be independent. The mean values of random variables x_1 and x_2 are 1 and 0, respectively, with the standard deviation of 1. The variables x_3 and x_4 are triangular fuzzy numbers with [0.0 2.0 4.0] and [0.0, 0.1, 0.2], respectively.

The limit state function given in Eq. (30) is approximated using first-order HDMR by deploying $n = 5$ sample points along each of the variable axis and taking, respectively, the mean values and nominal values of the random and fuzzy variables as initial reference point (1.0, 0.0, 2.0, 0.1). The approximated limit state function is divided into two parts, one with only the random variables along with the value of the constant part and the other with the fuzzy variables. The joint membership function of the fuzzy part of approximated limit state function is obtained using suitable transformation of the fuzzy variables. Using FF sampling scheme, the sample point $\boldsymbol{d} = (1, -2)$ is identified as reference point closer to the limit state function producing maximum weight. In this case, since only one reference point is identified, local approximation is same as the global

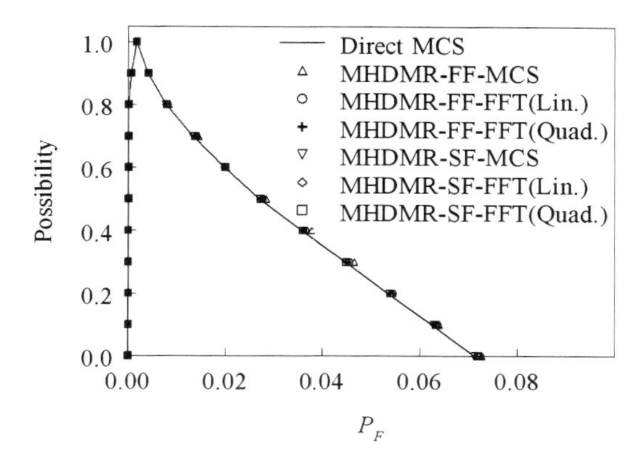

Fig. 5 Membership function of failure probability for cantilever steel beam

approximation. The bounds of the failure probability are obtained both by performing the convolution using FFT in conjunction with linear and quadratic approximations and MCS on the global approximation. Figure 5 shows the membership function of the failure probability estimated both by performing the convolution using FFT in conjunction with linear and quadratic approximations and MCS on the global approximation, as well as that obtained using direct MCS.

In addition, the membership function of the failure probability obtained by the proposed method based on SF sampling scheme is also shown in Fig. 5. The effect of number of sample points is studied by varying n from 3 to 9. It is observed that $n = 7$ provides the optimum number of function calls with acceptable accuracy in evaluating the failure probability with the present method.

6.3 80-Bar 3D Truss Structure

A 3D truss, shown in Fig. 6, is considered in this example to examine the accuracy and efficiency of the proposed method for the membership function of failure probability estimation. The loads at various levels are considered to be random, while the cross-sectional areas of the angle sections at various levels are assumed to be fuzzy as shown in Table 2.

The maximum horizontal displacement at the top of the tower is considered to be the failure criterion, as given below:

$$g(\boldsymbol{x}) = \Delta_{\text{lim}} - \Delta(\boldsymbol{x}). \tag{31}$$

The limiting deflection Δ_{lim} is assumed to be 0.15 m. The limit state function is approximated using first-order HDMR by deploying $n = 5$ sample points along each of the variable axis and taking, respectively, the mean values and nominal values of the random and fuzzy variables as initial reference point.

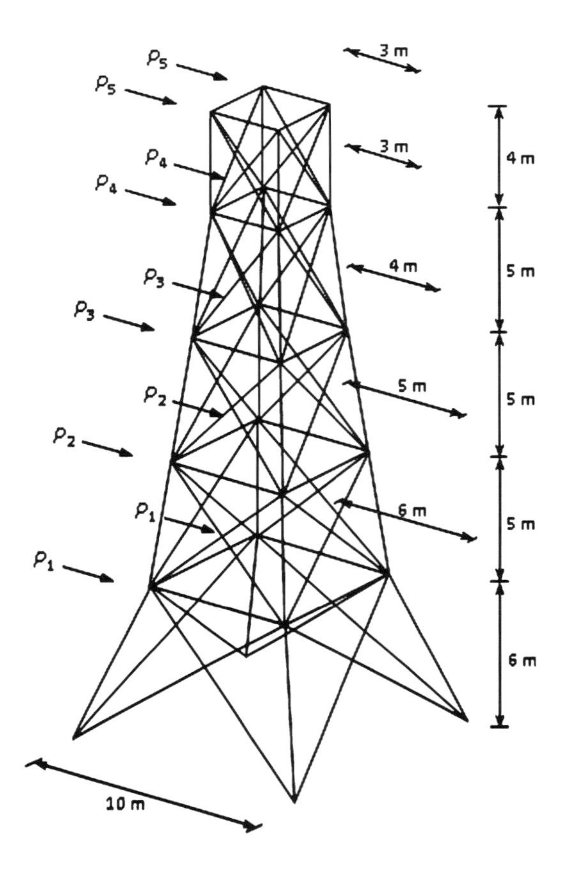

Fig. 6 3D truss structure with 80 bars

Table 2 Properties of the uncertain variables

	Random			
Uncertain variable	Mean	COV	Type	Fuzzy
P_1 (N)	1,000	0.1	Normal	
P_2 (N)	2,000	0.1	Normal	
P_3 (N)	3,000	0.1	Normal	
P_4 (N)	4,000	0.1	Normal	
P_5 (N)	5,000	0.1	Normal	
A_1 (mm^2)				[6867 7630 8393]
A_2 (mm^2)				[5571 6190 6809]
A_3 (mm^2)				[3870 4300 4730]
A_4 (mm^2)				[2088 2320 2552]
A_5 (mm^2)				[1539 1710 1881]

The approximated limit state function is divided into two parts, one with only the random variables along with the value of the constant part and the other with the fuzzy variables. The joint membership function of the fuzzy part of approximated limit state function is obtained using suitable transformation of the fuzzy variables. The two reference points closer to the function producing maximum weights,

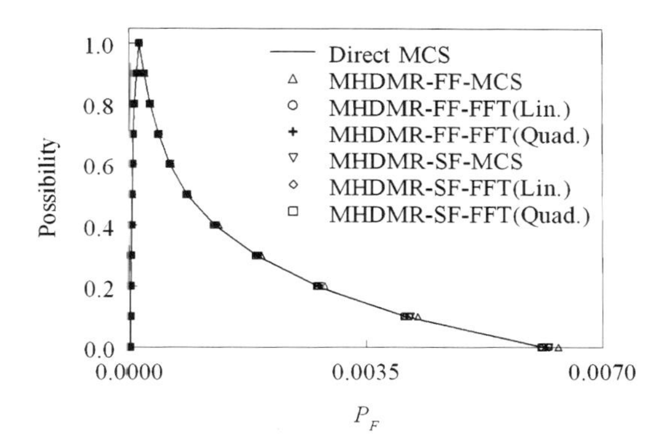

Fig. 7 Membership function of failure probability for truss structure

1.0 and 0.977, are identified. After identification of two reference points, local first-order HDMR approximations are constructed at the reference points. The bounds of the failure probability are obtained both by performing the convolution using FFT in conjunction with linear and quadratic approximations and MCS on the global approximation. Figure 7 shows the membership function of the failure probability estimated both by performing the convolution using FFT and MCS on the global approximation, as well as that obtained using direct MCS.

In addition, effects of SF sampling scheme and the number of sample points on the estimated membership function of the failure probability are studied. Figure 7 also shows the membership function of the failure probability estimate obtained by the proposed method based on SF sampling scheme.

7 Summary and Conclusions

This chapter presented a novel uncertain analysis method for estimating the membership function of the reliability of structural systems involving multiple design points in the presence of mixed uncertain variables. The method involves MHDMR technique for the limit state function approximation, transformation technique to obtain the contribution of the fuzzy variables to the convolution integral and fast Fourier transform for solving the convolution integral at all confidence levels of the fuzzy variables. Weight function is adopted for identification of multiple reference points closer to the limit surface. Using the bounds of the fuzzy variables part at each confidence level along with the constant part and the random variables part, the joint density functions are obtained by (1) identifying the reference points closer to the limit state function and (2) blending of locally constructed individual first-order HDMR approximations in the rotated Gaussian space at different identified reference points to form global approximation and (3) performing the convolution using FFT, which upon integration yields the bounds of the failure probability. As an alternative, the bounds of the failure probability are estimated by performing MCS on the global approximation in the original space, obtained by blending of locally constructed individual first-order HDMR approximations of the original limit state function at different identified reference points.

The results of the numerical examples involving explicit hypothetical mathematical function and structural/solid-mechanics problems indicate that the proposed method provides accurate and computationally efficient estimates of the membership function of the failure probability. The results obtained from the proposed method are compared with those obtained by direct MCS. The numerical results show that the present method is efficient for structural reliability estimation involving any number of fuzzy and random variables with any kind of distribution. Two types of sampling schemes, namely, FF and SF, are adopted in this study for MHDMR approximation of the original limit state function construction. A parametric study is conducted with respect to the number of sample points n used in FF and SF sampling-based MHDMR approximation, and its effect on the estimated failure probability is investigated. An optimum number of sample points n must be chosen in approximation of the original limit state function.

References

1. Breitung K (1984) Asymptotic approximations for multinormal integrals. ASCE J Eng Mech 110(3):357–366
2. Rackwitz R (2001) Reliability analysis – a review and some perspectives. Struct Saf 23(4):365–395
3. Sakamoto J, Mori Y, Sekioka T (1997) Probability analysis method using fast Fourier transform and its application. Struct Saf 19(1):21–36
4. Rao BN, Chowdhury R (2008) Probabilistic analysis using high dimensional model representation and fast Fourier transform. Int J Comput Methods Eng Sci Mech 9(6):342–357
5. Au SK, Papadimitriou C, Beck JL (1999) Reliability of uncertain dynamical systems with multiple design points. Struct Saf 21:113–133
6. Kiureghian AD, Dakessian T (1998) Multiple design points in first and second order reliability. Struct Saf 20(1):37–49
7. Briabant V, Oudshoorn A, Boyer C, Delcroix F (1999) Nondeterministic possibilistic-approaches for structural analysis and optimal design. AIAA J 37(10):1298–1303
8. Penmetsa RC, Grandhi RV (2003) Uncertainty propagation using possibility theory and function approximations. Mech Based Des Struct Mach 81(15):1567–1582
9. Rabitz H, Alis OF, Shorter J, Shim K (1999) Efficient input-output model representations. Comput Phys Commun 117(1–2):11–20
10. Kaymaz I, McMahon CA (2005) A response surface method based on weighted regression for structural reliability analysis. Probab Eng Mech 20(1):11–17
11. Adduri PR, Penmetsa RC (2008) Confidence bounds on component reliability in the presence of mixed uncertain variables. Int J Mech Sci 50(3):481–489

Reliability Considerations in Asphalt Pavement Design

Animesh Das

Abstract This chapter presents a brief overview of the concept of reliability applied to asphalt pavement design. It discusses how reliability can be estimated for a given pavement structure and how a pavement structure can be designed for a given reliability level. Reliability-based design principles for design of new asphalt pavement, as well as rehabilitation of existing pavement, have been presented.

Keywords Pavement design • Structural failure • Reliability

1 Introduction

Asphalt pavement design process involves design input as material properties, weather conditions, traffic characteristics, design period, etc. Each input is linked with a set of design parameters. For example, material property includes elastic modulus and Poisson's ratio values of asphalt, granular, cemented, subgrade layers, etc. All these parameters show significant variability [6, 11, 18, 29, 31, 32]. It would be reasonable approach to account for such variability in the pavement design process. This is done by invoking the concepts of reliability in pavement design.

Reliability issues in pavement design have been studied as early as 1970s [5, 6, 15]. Some of the initial considerations included variation of a single parameter (e.g. the subgrade strength); [28] subsequently, variability of different parameters and different modes of failures are added to the analysis.

A. Das (✉)
Department of Civil Engineering, Indian Institute of Technology Kanpur, Kanpur 208 016, India
e-mail: adas@iitk.ac.in

S. Chakraborty and G. Bhattacharya (eds.), *Proceedings of the International Symposium on Engineering under Uncertainty: Safety Assessment and Management (ISEUSAM - 2012)*,
DOI 10.1007/978-81-322-0757-3_19,

2 Concept of Reliability

Reliability is the probability of not having a failure. Failure does not happen if the number of repetitions of expected traffic (*T*) does not exceed the number of repetitions that the pavement can sustain (*N*). Since it is assumed that there are variabilities associated with the pavement design parameters, *T* and *N* can be represented in the form of probability distributions.

Figure 1 shows a schematic diagram with hypothetical distributions of *T* and *N* for any given pavement section. These distributions are shown to be nonintersecting in the present case. In case 1, the pavement would definitely fail, because the expected traffic is always greater than the allowable traffic, indicating the reliability value as 0%. Following similar logic, it can be said that the reliability of pavement for case 2 is 100%.

Reliability values of 0 and 100% (in other scale, the reliability values of 0 and 1), are two extreme cases, and generally, the reliability value of any given pavement lies somewhere in between. Such a case is represented in Fig. 2. Thus, reliability of a pavement (*R*) can be defined as:

Probability (number of repetitions a pavement can sustain is greater than number of traffic repetitions expected to occur on the pavement), that is

$$R = P(N>T) \tag{1}$$

If a parameter 'safety margin' (*S*) is defined as $S = T - N$, reliability can be defined as $R = P(S<0)$. If the probability density function (*pdf*) of *S* is given as $f_S(s)$, then *R* can be calculated as follows:

$$R = \int_{-\infty}^{0} f_S(s)\, ds \tag{2}$$

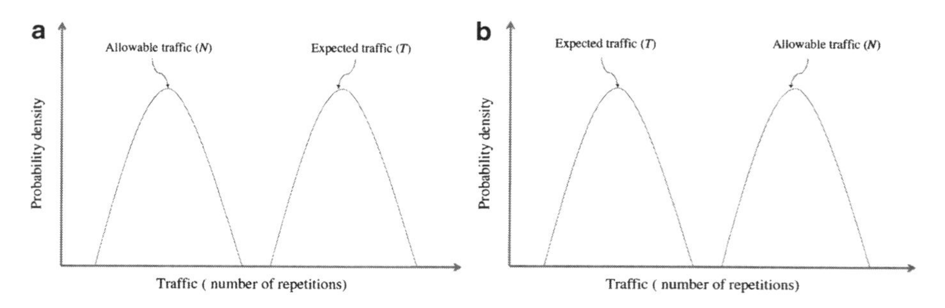

Fig. 1 Pavement reliability as 0 and 100%. (**a**) Case 1: reliability = 0%, (**b**) Case 2: reliability = 100%

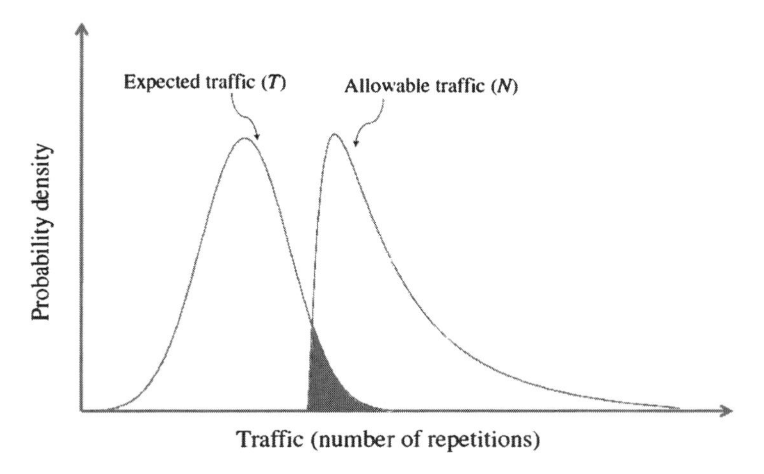

Fig. 2 Schematic diagram explaining the definition of reliability

Reliability is sometimes expressed using 'damage factor', D, defined as $D = \frac{T}{N}$. In that case, reliability can be defined as $R = P(D<1)$. If the *pdf* of D is known as $f_D(d)$, then R can be calculated as follows:

$$R = \int_0^1 f_D(d')dd' \tag{3}$$

If the distributions of T and N are known, the reliability value (R) can obtained using either Eqs. (2) or (3) [4, 14, 17, 23]. Knowing the distributions of T and N sometimes becomes a complex task.

T is dependent on traffic volume, axle load, traffic growth rate, design period, etc. The design period is generally prespecified and therefore has a fixed value. N is dependent on material properties (e.g. elastic moduli and Poisson's ratio values of the individual layers), layer thicknesses, loading configuration, performance equations, etc. The performance equation relates the critical stress/strain values to the life (in terms of traffic repetitions) of the pavement for a given mode of failure. These equations are generally developed empirically through calibration of pavement performance data. Figure 3 schematically shows various parameters that influence the distributions of T and N and, in turn, the reliability of pavement.

To estimate the pavement reliability, the distributions of these basic parameters (i.e. traffic volume, axle load, traffic growth rate, material properties, thicknesses, loading configuration, performance equation coefficients) are to be known first. Substantial literature is available which contains information on the variabilities of these parameters [4, 6, 11–13, 21, 29, 31–33]. Various analytical and numerical methods (e.g. point estimate method, first- or second-order reliability method, simulation method [3, 9, 35]) have been used [6, 14, 17] to obtain the parameter

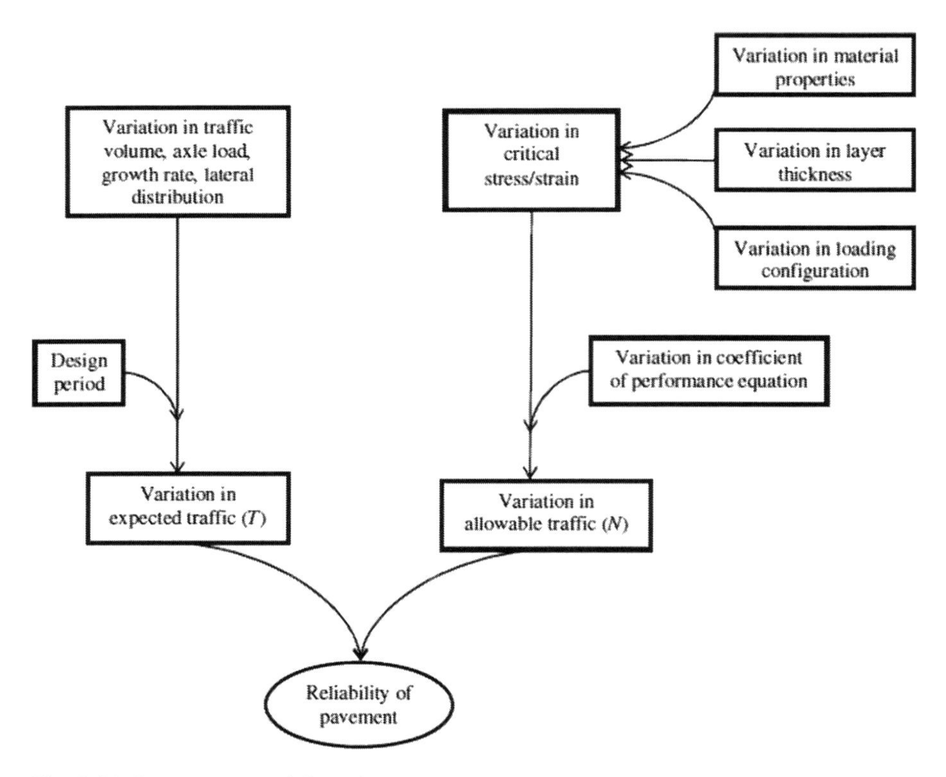

Fig. 3 Various parameters influencing reliability of pavement

related to possible distribution of T and N, when the distribution of these basic parameters are assumed to be known.

2.1 Reliability for a Single or Multiple Mode of Failure

The structural failure of asphalt pavement may occur in various modes, for example, load fatigue, thermal fatigue, rutting, thermal shrinkage and top-down cracking. The calculation of overall reliability of the pavement would depend on how the failure is defined. One may define the failure of a pavement when (1) the pavement fails due to all the failure modes or (2) the pavement fails due to any one of the failure modes [18]. These can be conceptually thought as failure modes linked in parallel (refer Fig. 4a) or failure modes linked in series (refer Fig. 4b), respectively.

If the failure probability for the ith mode is represented as F_i, then the overall reliability (R_o) can be calculated as follows:

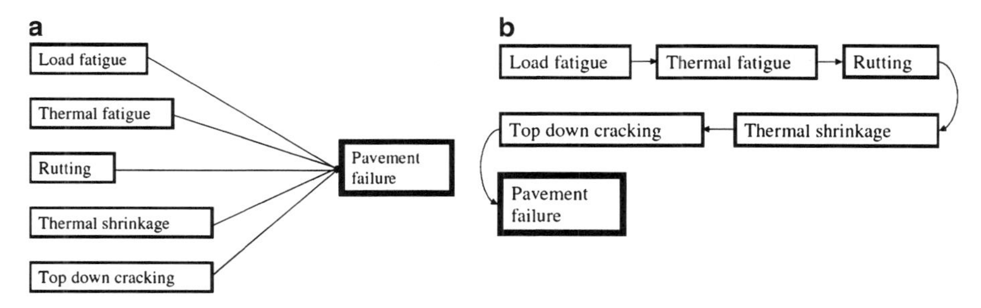

Fig. 4 Series or parallel linkage of failure modes. (**a**) Failure modes are linked in parallel, (**b**) Failure modes are linked in series

If the pavement is thought to be failed when it fails due to all the failure modes (refer Fig. 4a), then

$$R_{\mathrm{o}} = 1 - \bigcap_{i=1}^{n} F_i \tag{4}$$

If the pavement is thought to be failed when it fails due to either of the failure modes (refer Fig. 4b), then

$$R_{\mathrm{o}} = 1 - \bigcup_{i=1}^{n} F_i \tag{5}$$

One may even further consider the failure modes as mixed combinations of parallel and series connections, and accordingly, different expressions for the overall reliability (R_{o}) can be obtained. It is difficult to know the mutual dependency between the individual failure modes; the calculation of R_{o}, therefore, becomes quite simple when the failure modes are assumed to be mutually independent.

3 Estimation of Reliability

It is a difficult task to derive a closed-form analytical solution for estimating the critical stress/strain parameters (refer Fig. 3) of a multilayered asphalt pavement structure. Thus, Monte Carlo simulation method has been quite popularly [4, 13, 32, 33] used to estimate the reliability of a given pavement section. Simulation studies show that T generally follows a normal distribution, whereas the fatigue or rutting life of a given pavement generally follows a log-normal distribution [5, 18, 33]. Sensitivity studies have been conducted by various researches [3, 6, 13, 18, 30, 32] to study the effect of various parameters on reliability, and it is generally observed that the thicknesses and stiffness moduli of pavement layers significantly affect the fatigue and rutting reliabilities [3, 18, 32].

4 Design of Pavement for a Given Reliability

The effect of thickness on reliability for low-temperature shrinkage cracking and reliability of thermal cracking is generally not very significant [24, 26]. The design reliability levels of these failure modes can be achieved by adjusting at the asphalt mix design stage. Thus, the asphalt pavement thickness design becomes primarily governed by the fatigue and rutting considerations. Reliability-based design of new asphalt pavement and rehabilitation of existing asphalt pavement are discussed in the following.

4.1 Design of New Pavement

Design of a pavement for a given reliability level is an iterative process. For a given traffic data, the distribution of T is fixed, and the distribution of N value changes once the trial thickness values are changed. The iteration is continued until the reliability levels (of the fatigue and rutting failure modes, since these are strongly affected by thickness) satisfy the respective design reliability levels. Similar concept is used in the AASHTO [1] guidelines, where difference between the expected traffic and allowable traffic is adjusted to achieve a reliability level. However, the recent NCHRP guidelines [20] suggest use of reliability in terms of a system of pavements (i.e. the probability that a pavement section survives out of a number of pavement sections under similar conditions) for design purpose. In this approach, empirical equations are developed from performance data (of a number of pavement sections) to predict individual pavement distresses.

Figure 5 presents a schematic pavement design chart. The asphalt pavement is assumed to be made up of three basic layers, asphalt layer, unbound granular layer and the soil subgrade. Thus, the thicknesses of the asphalt layer (h_1) and the granular layer (h_2) are the two variables to be designed. From Fig. 5, it can be seen that for a given value, h_2, the reliability level increases as h_1 increases. Similarly, for a given value of h_1, the reliability value is higher if larger h_2 is used. Different sets of curves can be obtained for various failure modes (say, fatigue and rutting). Thus, from this design chart, for given reliability levels of fatigue and rutting failure modes, one can choose suitable values of h_1 and h_2.

As seen from Fig. 5, for a given pavement design problem, a number of alternative design solutions are possible – all of them may satisfy the reliability requirements, but their costs may differ [25]. The final design solution would be the one whose cost, C (total construction cost including the cost of materials), is the

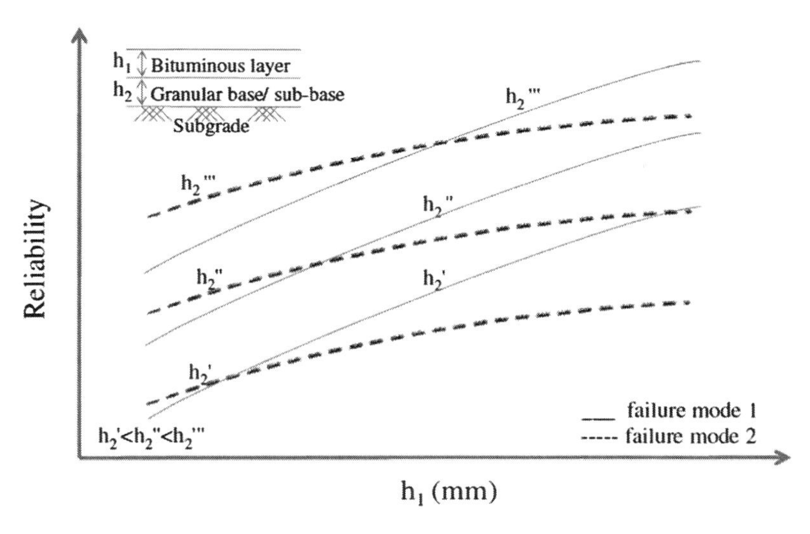

Fig. 5 Schematic diagram of a pavement design chart for two design layers and two failure modes

least. Thus, a reliability-based pavement design problem can be formulated as an optimization scheme as follows:

Minimize C

Subject to:

$$\forall_i R_i(\forall_j h_j) \geqslant R_i^{\mathrm{d}} \tag{6}$$

where j is the number of design layers, i is the number of modes for failure, $R_i(h_j)$ is the reliability of ith failure mode which is a dependent on the thickness of all the design layers (h_j) and R_i^{d} is the design reliability (i.e. target reliability) level for the ith failure mode.

4.2 *Design for Rehabilitation*

A newly constructed pavement undergoes deterioration with the passage of time. This deterioration is due to traffic and environmental factors. An appropriate structural design prevents a pavement to undergo premature failure, but the pavement would finally fail after the expiry of its design life. The pavement performance is however stochastic in nature [10, 16, 27, 34], and the purpose of using reliability approach is to take into account this stochasticity in the design process.

A pavement, before it undergoes complete failure, needs to be rehabilitated. Determining the optimal rehabilitation timing and the extent of rehabilitation to be applied is an optimization problem. A number of formulations for estimation of

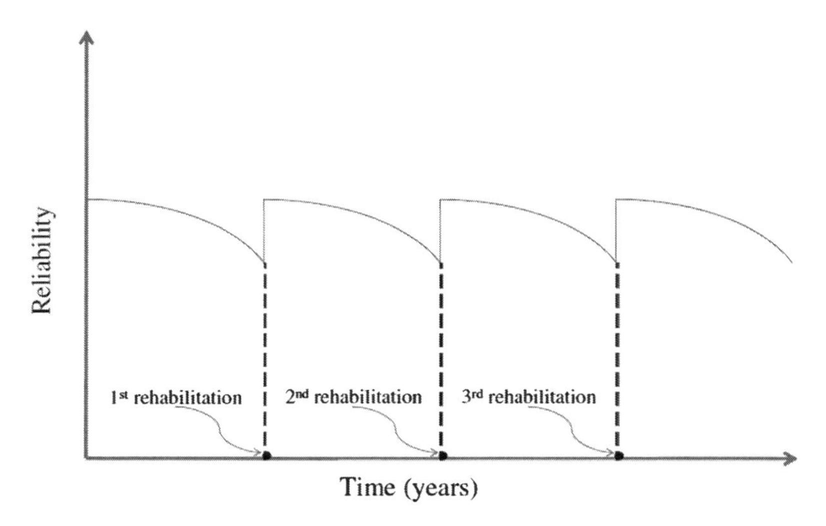

Fig. 6 Variation of reliability with time over the life cycle of the pavement

optimal rehabilitation timing are available in literature [2, 19, 22]. The scope of these formulations include minimization of total cost of construction and rehabilitation as well as the cost incurred by the road users, over the entire life cycle of the pavement.

The reliability of a pavement structure discussed above indicates the reliability value at the end of the design period, when all the expected traffic repetitions have taken place. The reliability value of a newly constructed pavement is therefore higher than the value for which it is designed, and it gradually decreases as the cumulative traffic repetitions increase [7]. Once the reliability value reaches a prespecified threshold value, a rehabilitation (say, an overlay) may be needed, which would further shift the allowable traffic (i.e. improve the pavement capacity). This is shown schematically in Fig. 6. A suitable optimization formulation would be able to estimate the optimal rehabilitation timing so that reliability value does not fall below the threshold level and at the same time the total cost is minimized [8]. This forms the basis of reliability-based rehabilitation design scheme.

5 Summary

Significant level of variabilities present in the pavement design parameters necessitates use of reliability-based approach in pavement design. This chapter has discussed the reliability principles in use for the design of asphalt pavements.

Acknowledgement The author wishes to thank Mr. Sudhir N. Varma, former master's student, Department of Civil Engineering, IIT Kanpur, for helping in drawing some of the schematic diagrams presented in this chapter.

References

1. AASHTO (1993) Guide for design of pavement structure. American Association of State Highway and Transportation Officials (AASHTO), Washington, DC
2. Abaza KA, Abu-Eisheh SA (2003) An optimum design approach for flexible pavements. Int J Pavement Eng 4(1):1–11
3. Chou YT (1990) Reliability design procedures for flexible pavements. J Trans Eng 116 (5):602–614
4. Chua KH, Kiureghian AD, Monismith CL (1992) Stochastic model for pavement design. J Trans Eng 118(6):769–786
5. Darter MI, McCullough BF, Brown JL (1972) Reliability concepts applied to the Texas flexible pavement system. Highw Res Rec HRB 407:146–161
6. Darter MI, Hudson WR, Brown JL (1973) Statistical variation of flexible pavement properties and their consideration in design. Proc Assoc Asphalt Paving Technol 42:589–615
7. Despande VP, Damnjanvic ID, Gardoni P (2010) Modeling pavement fragility. J Trans Eng 136(6):592–596
8. Despande VP, Damnjanvic ID, Gardoni P (2010b) Reliability-based optimization models for scheduling pavement rehabilitation. Comput-Aided Civ Infrastruct Eng 25:227–237
9. Harr ME (1987) Reliability based design in civil engineering. McGraw-Hill Book Company, New York
10. Hong HP, Wang SS (2003) Stochastic modeling of pavement performance. Int J Pavement Eng 4(4):235–243
11. Jiang YJ, Selezneva O, Mladenovic G, Aref S, Darter MI (2003) Estimation of pavement layer thickness variability for reliability-based design. Trans Res Rec 1849:156–165, TRB, National Research Council, Washington, DC
12. Kim HB, Lee SH (2002) Reliability-based design model applied to mechanistic empirical pavement design. KSCE J Civ Eng 6(3):263–272
13. Kim HB, Buch N (2003) Reliability-based pavement design model accounting for inherent variability of design parameters. TRB 82nd annual meeting, CD-ROM, Washington, DC
14. Kulkarni RB (1994) Rational approach in applying reliability theory to pavement structural design. Trans Res Rec 1449:13–17, TRB, National Research Council, Washington, DC
15. Lamer AC, Moavenzadeh F (1971) Reliability of highway pavements. Highw Res Rec HRB 362:1–8
16. Li N, Haas R, Xie W (1996) Reliability-based processing of Markov chains for modeling pavement network deterioration. Trans Res Board 1524:203–213
17. Lytton RL, Zollinger D (1993) Modelling reliability in pavement. TRB 72nd annual meeting, Washington, DC
18. Maji A, Das A (2008) Reliability considerations of bituminous pavement design by mechanistic-empirical approach. Int J Pavement Eng 9(1):19–31
19. Mamlouk MS, Zaniewski JP, He W (2000) Analysis and design optimization of flexible pavement. J Trans Eng 126(2):161–167
20. Mechanistic-Empirical Design of New & Rehabilitated Pavement Structures, Appendix B: Design reliability (2003) Final document, NCHRP project 1-37A, National Research Council, Washington, DC, 2003. http://www.trb.org/mepdg/guide.htm. Accessed Aug 2010
21. Noureldin SA, Sharaf E, Arafah A, Al-Sugair F (1994) Estimation of standard deviation of predicted performance of flexible pavements using AASHTO model. Trans Res Rec 1449:46–56, Transportation Research Record, TRB, National Research Council, Washington, DC
22. Ouyang Y, Madanat S (2004) Optimal scheduling of rehabilitation activities for multiple pavement facilities: exact and approximate solutions. Trans Res A Policy Pract 38:347–365

23. Rajbongshi P, Das A (2008) Estimation of structural reliability of asphalt pavement for mixed axle loading conditions. In: Proceedings of the 6th International Conference of Roads and Airfield Pavement Technology (ICPT), Sapporo, Japan, pp 35–42
24. Rajbongshi P, Das A (2008) Thermal fatigue considerations in asphalt pavement design. Int J Pavement Res Technol 1(4):129–134
25. Rajbongshi P, Das A (2008) Optimal asphalt pavement design considering cost and reliability. J Trans Eng 134(6):255–261
26. Rajbongshi P, Das A (2009) Estimation of temperature stress and low-temperature crack spacing in asphalt pavements. J Trans Eng 135(10):745–752
27. Sanchez-Silva M, Arroyo O, Junca M, Caro S, Caicedo B (2005) Reliability based design optimization of asphalt pavements. Int J Pavement Eng 6(4):281–294
28. Shell International Petroleum Company Limited (1978) Shell pavement design manual – asphalt pavements and overlays for road traffic, Shell International Petroleum Company Limited, London
29. Stubstad RN, Tayabji SD, Lukanen EO (2002) LTPP data analysis: variation in pavement design inputs. Final report, NCHRP web document 48, TRB, National Research Council, Washington, DC. http://gulliver.trb.org/publications/nchrp/nchrp_w48.pdf. Accessed Aug 2010
30. Tarefder RA, Saha N, Stormont JC (2010) Evaluation of subgrade strength and pavement designs for reliability. J Trans Eng 136(4):379–391
31. Timm DH, Briggison B, Newcomb DE (1998) Variability of mechanistic-empirical flexible pavement design parameters. In: Proceedings of the 5th international conference on the bearing capacity of roads and airfields, vol 1, pp 629–638
32. Timm DH, Newcomb DE, Briggison B, Galambos TV (1999) Incorporation of reliability into the Minnesota mechanistic-empirical pavement design method. Final report, submitted to Minnesota Department of Transportation, Department of Civil Engineering, Minnesota University, Minneapolis
33. Timm DH, Newcomb DE, Galambos TV (2000) Incorporation of reliability into mechanistic-empirical pavement design. Trans Res Rec 1730:73–80, TRB, National Research Council, Washington, DC
34. Wang KCP, Zaniewski J, Way G (1994) Probabilistic behavior of pavements. J Trans Eng 120 (3):358–375
35. Zhang Z, Damnjanović I (2006) Applying method of moments to model reliability of pavements infrastructure. J Trans Eng 132(5):416–424

Structural Reliability Analysis of Composite Wing Subjected to Gust Loads

D.K. Maiti and Anil Kumar Ammina

Abstract The design of any engineering system is a process of decision-making, under constraints of uncertainty. The uncertainty in the design process results from the lack of deterministic knowledge of different physical parameters and the uncertainty in the models with which the design is performed. In this study, the reliability analysis is conducted for composite wing subject to gust loads. For this, the probability distribution function of bending and shear stresses from random gust is calculated by power spectral analysis, and the material properties of composite skin are assumed to be normal random variables to consider uncertainty. With these distributions of random variables, the probability of failure of the wing structure is calculated by Monte Carlo simulation. The necessary modification is carried out, and it is found that the suggested modification improves the reliability of the design.

Keywords Uncertainty • Random gust load • Probability of failure • Uncertainty • Reliability analysis

1 Introduction

The response of an airplane in flight due to gust is one of the most important dynamic response problems from the structural design considerations. Gusts are the result of atmospheric turbulence. They can be categorised into two types: [1, 2] (1) vertical/lateral gusts, wherein a component of the gust velocity is at right angles

D.K. Maiti (✉)
Department of Aerospace Engineering, Indian Institute of Technology, Kharagpur, Kharagpur 721302, WB, India
e-mail: dkmaiti@aero.iitkgp.ernet.in

A.K. Ammina
Reliability Engineering Centre, Indian Institute of Technology, Kharagpur, Kharagpur 721302, India

S. Chakraborty and G. Bhattacharya (eds.), *Proceedings of the International Symposium on Engineering under Uncertainty: Safety Assessment and Management (ISEUSAM - 2012)*, DOI 10.1007/978-81-322-0757-3_20, © Springer India 2013

to the flight path, and (2) head-on or longitudinal gusts, wherein the gust velocity is parallel to the flight path. Vertical/lateral gust causes a change in angle of attack/angle of sideslip, which is equal to the gust velocity divided by the forward speed. On the other hand, the head-on (longitudinal) gust produces only a change in the dynamic pressure. The change in lift force produced by a head-on gust is negligible compared to that of a vertical/lateral gust. Hence, a vertical/lateral gust is more critical from design considerations. There are two approaches for solving the problem of gust response. One is called as discrete gust approach which is relatively easy to handle; on the other hand, the second approach is called as turbulence gust approach. The second approach considers the nature of gust is random in nature and responses are calculated based on the statistical approach considering the uncertainty in the gust model.

The uncertainties that occur in the design process are employed for analysis of the loads and, in the geometric parameters of structure, have been dealt with for generations by experience and safety factors. The main sources of uncertainties in structural analysis and design are (a) uncertainties in the determination of the physical and mathematical model used for analysis, including uncertainties in the failure criteria (model); (b) uncertainties in the determination of the magnitudes, locations, frequency content and correlations of the external loads (either static or dynamic); and (c) uncertainties in various structural parameters such as geometries, dimensions, material properties and allowable stochastic structure. These three categories do not include other more subjective uncertainties such as human errors in the design and production.

Aerospace structures are excited by aerodynamic loads, which are usually random in nature. Flow around the wing structure creates pressure fluctuations, which have a wide range of frequency and amplitude content. The same phenomena are caused by acoustic noise created by rocket and jet outlet flows. Rotating elements such as engines and rotors create excitations with a better defined frequency content that are in many cases random in amplitudes. The use of composite materials reduced the weight and increases the payload capacity, but the flexibility and high aspect ratio become a concern particularly under gust conditions. Enough research has also been done on design and optimisation of composite structures under different loads like flutter, flight loads and natural frequency but all in connection with stability not dealt the reliability of structure.

In the past, the simple discrete gust type, for example, a one-minus-cosine pulse, was used to model the atmospheric turbulence. But, in natural, a gust profile is continuous and irregular. So the continuous gust profile, which can be idealised as a stationary Gaussian random process, is widely used for gust load analysis recently. By considering the atmospheric turbulence as a stationary random process, the power spectral methods are used for finding the root mean square (RMS) values of bending stresses. In FAR-25, two basic types of power spectral gust loads criteria are described, which are mission analysis and design envelope, respectively.

Advanced composite materials are widely used in modern aircraft structural design mainly to achieve the weight-efficient structure. Thus, the research efforts have been devoted to the optimal design of wing structures in connections of various objectives and constraints [3–7]. Penmetsa and Grandhi [8] calculated the failure probability using interval analysis where the wing skin thickness and the loading were considered to be available as an interval. Mahadevan and Liu [9] developed the system reliability analysis procedure and applied it to the analysis of composite wing structure. In their work, material properties, ply thicknesses and orientations and pressure loads were assumed to be random variables. The aeroelastic response to time-dependent external excitation of a two-dimensional rigid/elastic-lifting surface in incompressible flow field featuring plunging--pitching-coupled motion is addressed by Marzocca et al. [10].

In this chapter, the reliability analysis of composite wing subject to continuous random gust is conducted based on mission analysis criteria. To evaluate the probability of failure of wing structure, the wing root bending and shear stresses induced by gust loads are analysed by power spectral method, and then the probability distribution function of bending and shear stresses is calculated. The material properties of composite wing are assumed to be random normal variable to consider uncertainties. With the probability distribution functions for bending and shear stresses and material strength properties, the failure probability of wing structure can be evaluated by Monte Carlo simulation.

2 Analysis Procedure

The aeroelastic response analysis is carried out using the university version of FE package MD.NASTRAN [11]. The modal domain formulation is used for the random gust response analysis. The details of analysis procedure are given below. The basic equation in modal coordinates is

$$\left[-M_{\mathrm{hh}}\omega^2 + iB_{\mathrm{hh}}\omega + (1+ig)K_{\mathrm{hh}} - \frac{1}{2}\rho V^2 Q_{\mathrm{hh}}(m,k)\right]\{u_{\mathrm{h}}\} = \{P(\omega)\} \tag{1}$$

where M_{hh} is modal mass, B_{hh} modal damping and K_{hh} is modal stiffness. Q_{hh} is modal aerodynamic load in terms of Mach number (m) and reduced frequency (k). The external modal load, $P(\omega)$, is expressed as a function of frequency (ω) which can be aerodynamic or non-aerodynamic in nature and is a function of the frequency. The generalised load due to gust is expressed as

$$P(\omega) = qw_{\mathrm{g}}PP(\omega)\left[Q_{ij}\right]\left\{w_j(\omega)\right\} \tag{2}$$

where PP (ω) is the user supplied frequency or time variation of the gust (if the applied loading is in time domain, Fourier transform techniques are used to convert the loading into the frequency domain). q is dynamic pressure, w_{g} is gust scale factor and Q_{ij} is the aerodynamic influence coefficients. The gust downwash matrix is a function of frequency and the geometry of the aerodynamic model:

$$w_j(\omega_j) = \cos\gamma_j e^{-i\omega_j(x_j - x_0)/V} \tag{3}$$

Equation (1) is solved, and the frequency response, $H_{ja}(\omega)$, of any physical variable, u_j, due to some excitation source, $Q_a(t)$, is obtained. Then the power spectral density of the response, $\Phi_j(\omega)$, is related to the power spectral density of the gust load, $\Phi_a(\omega)$, by

$$\Phi_j(\omega) = \left|H_{ja}(\omega)\right|^2 \Phi_a(\omega) \tag{4}$$

The expected value of the number of zero crossings with positive slope per unit time, or mean frequency, is another quantity of interest from fatigue analysis and design of aircraft for gusts. This mean frequency, N_0, can be found from the spectral density:

$$N_0{}^2 = \frac{\int_0^\infty (\omega/2\pi)^2 \Phi_j(\omega) d\omega}{\int_0^\infty \Phi_j(\omega) d\omega} \tag{5}$$

3 Modelling of Gust Load

The gust velocity (V_{gust}) is small compared to forward velocity (V_∞). Otherwise, it would cause such a large change in angle of attack that the wing would stall. For a relatively small V_{gust}, the magnitude of the change in angle of attack, $\Delta\alpha$, is given by $\Delta\alpha = \tan^{-1}\left(V_{\text{gust}}/V_\infty\right) \approx V_{\text{gust}}/V_\infty$, and the change in lift coefficient is $\Delta C_{\text{L}} = C_{\text{L}_\alpha}$ $\Delta\alpha = C_{\text{L}_\alpha}\left(V_{\text{gust}}/V_\infty\right)$. The change in lift is $\Delta L = \Delta C_{\text{L}} qS = \left(C_{\text{L}_\alpha} \rho V_{\text{gust}} V_\infty S/2\right)$, and the change in load factor is $\Delta n = (\Delta L/W) = \left(C_{\text{L}_\alpha} \rho V_{\text{gust}} V_\infty / 2(W/S)\right)$. Assuming the load factor prior to encountering the gust is 1, the maximum load factor during the encounter is

$$n_{\text{gust}} = 1 + \Delta n = 1 + \frac{C_{\text{L}_\alpha} \rho V_{\text{gust}} V_\infty}{2(W/S)} \tag{6}$$

An aircraft is just as likely to encounter a *downdraft* as an updraft when flying through turbulent air. A downdraft is a vertical air current like an updraft, except that the direction of the air flow is downwards. The reaction of an aircraft to a downdraft is similar to an updraft, but since V_{gust} is negative, the second term in (6) is negative as well. The first term in (6) remains positive, so the magnitude of n_{gust} is less. However, since most aircraft have lower negative structural limits, encountering a downdraft could still be a problem.

3.1 Concepts of Random Gust

The gust profiles typically tend to be continuous and irregular in nature. Such a profile can be modelled as a "stationary Gaussian random process". A stationary Gaussian random process can be considered to be generated by the superposition of an infinite number of sinusoidal components. These components differ infinitesimally in frequency from one to the next. Each component is of prescribed infinitesimal amplitude, and each is randomly phased relative to the others. The profile or time history thus idealised is stationary in that it is considered to be of infinite duration and its statistical properties are the same whenever it may be sampled. The magnitude of a stationary random process is statistically defined by its RMS (root mean square) value and its probability distribution.

Two shapes of power spectral density (PSD), $\Phi(\omega)$, function for atmospheric turbulence have been widely used, the von Kármán and the Dryden. These are defined by mathematical expressions as follows:

$$\Phi(\omega) = \frac{2{\sigma_{\mathrm{w}}}^2 (L_{\mathrm{g}}/V_{\mathrm{gust}})[1 + 2(p+1)(\bar{k}L_{\mathrm{g}}\omega/V_{\mathrm{gust}})^2]}{[1 + (\bar{k}L_{\mathrm{g}}\omega/V_{\mathrm{gust}})^2]^{p+3/2}} \tag{7}$$

where L_{g} is scale of turbulence and σ_{w} RMS gust velocity. Values of the parameters k and p are given in the following table:

	Dryden	Von Kármán
$\bar{k}$	1.0	1.339
P	0.5	0.333

The von Kármán spectrum gives a better fit to observed experimental data and is normally used for design purposes.

4 Results and Discussion

4.1 Wing Geometry and Properties

A simple wing model as shown in Fig. 1 is used for gust response analysis. The aircraft wing panel is idealised as a cantilevered composite plate made of graphite/epoxy with four-noded quadrilateral elements as shown in Fig. 2. The stacking sequence is $[90/\pm45/0]_{2s}$, and each layer has uniform thickness of 0.018 cm. An orthotropic material is a homogeneous linear elastic material having two planes of symmetry in terms of mechanical properties, and these two planes are being perpendicular to each other. Then one can show that the number of independent

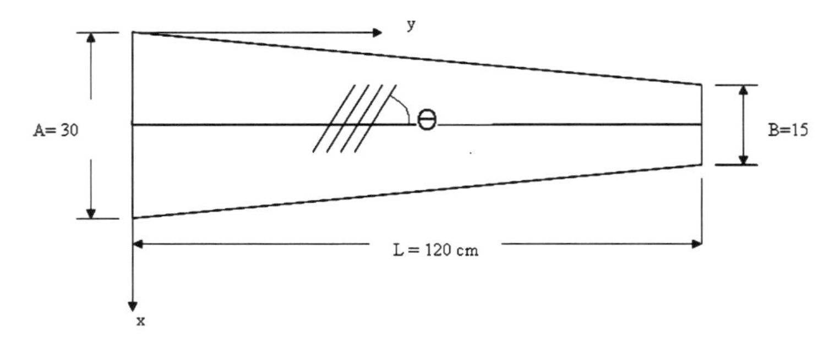

Fig. 1 Geometry and coordinate system of composite wing

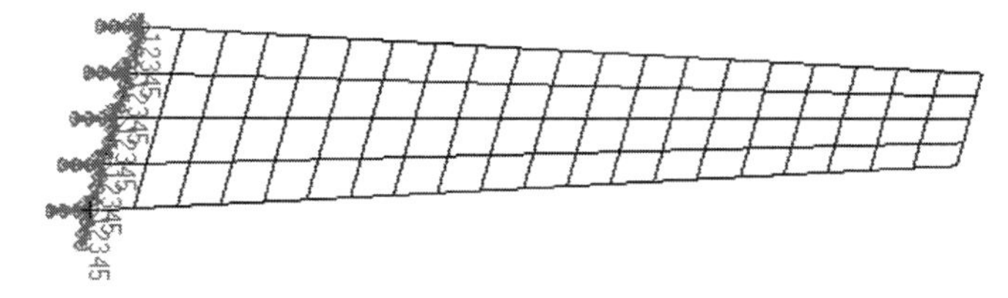

Fig. 2 Finite element model of the wing

elastic constants is nine. The constitutive relation expressed in the so-called "orthotropic" axes, defined by three axes constructed on the two orthogonal planes and their intersection line, can be written in the following form, called the engineering notation because it utilises the elastic modulus and Poisson's ratio. The material properties of the graphite/epoxy are as follows:

$E_1 = 132.16$ GPa, $E_2 = 8.65$ GPa, $\upsilon_{12} = 0.3$, $\rho = 1.59$ g/cm^3, $G_{12} = G_{13} = G_{23} = 4.12$ GPa

Thickness of the each layer = 0.018 cm.

Sequence of each stacking = $[90/\pm45/0]_{2s}$

4.2 Modal Analysis

To investigate the dynamic characteristics of wing model, normal mode analysis is performed. The first four natural frequencies and mode shapes are shown in Fig. 3. It is evident from the statistical gust profile that the intensity of gust power spectral density is weak with the higher frequency ranges. In this study, first six natural modes which are less than 100 Hz are considered for gust response analysis. The summary of the first six natural frequencies and the description of the mode shapes are presented in Table 1.

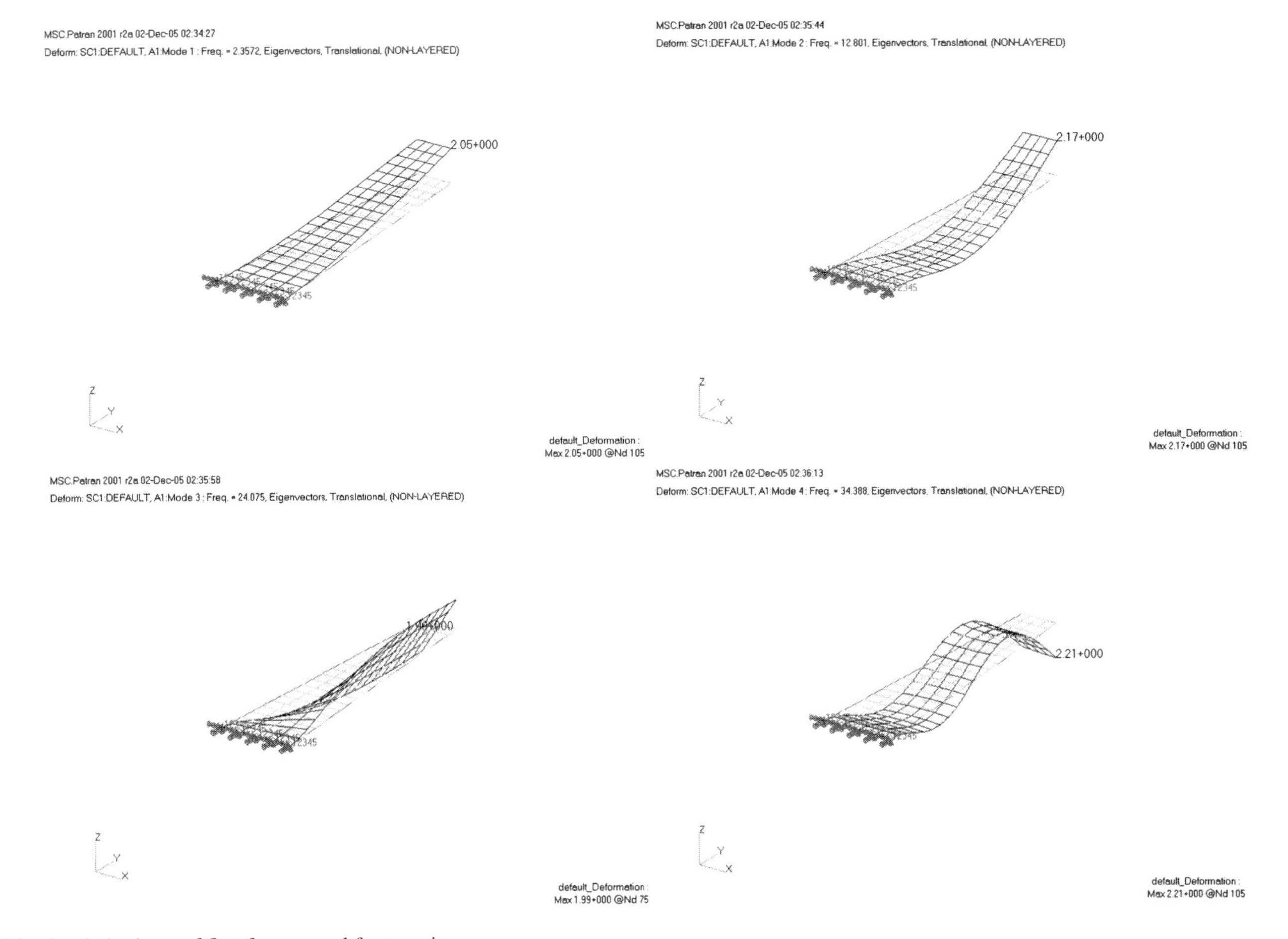

Fig. 3 Mode shape of first four natural frequencies

Table 1 Normal modes

Mode no.	Frequency (Hz)	Description of mode
1	2.45	First spanwise bending
2	13.30	Second spanwise bending
3	24.20	First chordwise bending
4	35.70	Third spanwise bending
5	61.40	Second spanwise bending and first twisting mode
6	69.20	Fourth spanwise bending

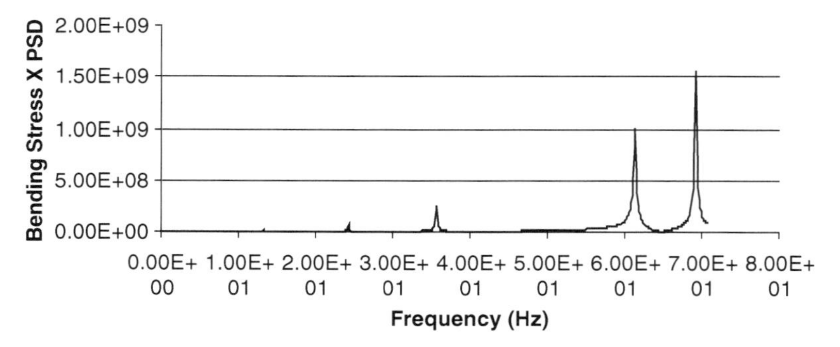

Fig. 4 PSD for wing root bending stress in X direction

4.3 Random Response Analysis Using MD.NASTRAN

The FEM structural model used in this work was considered as a semi-span model made of graphite/epoxy composite. The mesh divisions are as shown in Fig. 2. The aerodynamic mesh is also created to perform the unsteady aerodynamic computation in Mach (m) and reduced frequency (k) pair.

A corresponding full-span model possessing identical stiffness and mass properties was used in the two-dimensional PSD gust analysis. MD.NASTRAN is used for obtaining the bending moment PSD at the wing root. The von Karman gust PSD of the gust is considered for gust response analysis with the following input information:

$$V_{\text{gust}} = 15 \text{ m/s}, \quad \sigma_{\text{W}} = 0.3048 \text{ m/s}, \quad L_{\text{g}} = 762 \text{ m}$$

After running the random gust response analysis in MD.Nastran, it generates the required response quantities. These statistical quantities are important for failure and fatigue analyses. Here, the responses are wing root bending and shear stresses along the fibre, across the fibre and shear. The response parameters are calculated as PSD function of frequency. The required statistical quantities are calculated from these responses. The wing root bending stresses are plotted in Figs. 4, 5 and 6. It is observed from the figure that stresses are maximum at natural frequencies of the structure.

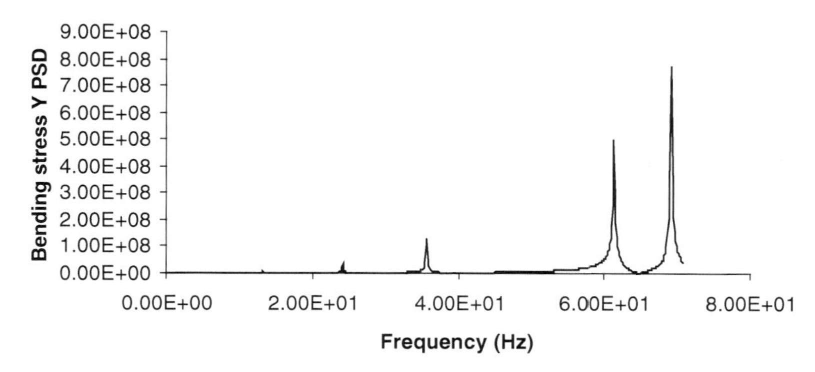

Fig. 5 PSD for wing root bending stress in *Y* direction

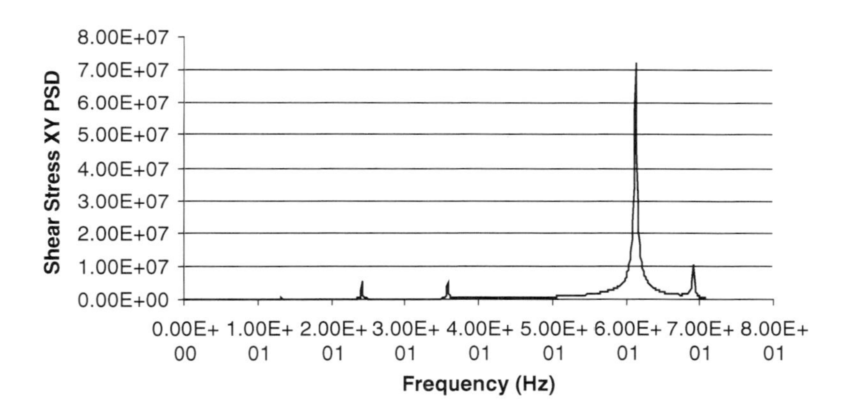

Fig. 6 PSD for wing root shear stress

4.4 Probability of Exceedance

The PSD of bending stress is calculated by power spectral analysis with gust PSD as an input. Then, the number of crossings of a given level y per unit time can be obtained by the following formula:

$$N(y) = N_0 \left[P_1 e^{-\frac{(y/\bar{A})}{b_1}} + P_2 e^{-\frac{(y/\bar{A})}{b_2}} \right] \tag{8}$$

N_0 is the number of crossing rate of level 0 as defined earlier, and $\bar{A}$ is the ratio of the RMS of output to that of the gust, which is given by

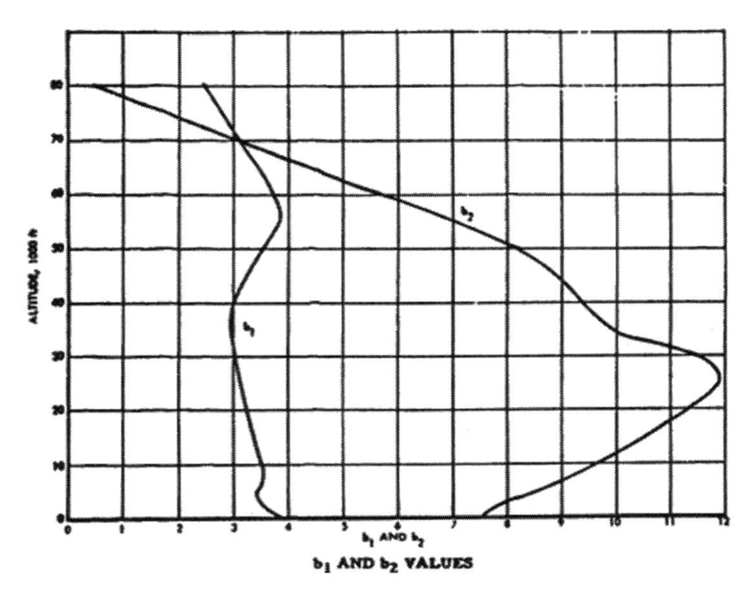

Fig. 7 Plot for b_1 and b_2 at different altitudes

$$\bar{A}^2 = \frac{\int_0^\infty \Phi_\psi(\omega)d\omega}{\int_0^\infty \Phi_w(\omega)d\omega} \tag{9}$$

In practice, the integrals defining $\bar{A}$ and N_0 are evaluated only up to a reasonable upper limit, beyond which the contribution to the integrals for computing $\bar{A}$ is negligible. According to FAR 25, P_1, P_2, b_1 and b_2 (Figs. 7 and 8 and Table 2) are constants describing the probability distribution of σ_W and depend only on altitude. P_1 and P_2 are fractions of time in non-storm and storm turbulence, respectively, and b_1 and b_2 are constants indicative of probable intensities. The statistical model of Eq. 9 is valid for an airplane flying at a constant speed and altitude. For a mission profile composed of segments that are representative of the aircraft usage, the exceedance expression given by Eq. 9 is superimposed as given below:

$$N(y) = \sum_{i}^{n_p} t_i N_{0i} \left[P_{1i} e^{\left(-\frac{(y/\bar{A}_i)}{b_{1i}}\right)} + P_{2i} e^{\left(-\frac{(y/\bar{A}_i)}{b_{2i}}\right)} \right] \tag{10}$$

where n_p is the number of segments in the mission profile being analysed, t_i is the fraction of time in segment i relative to the sum of all other segments, and N_{0i} and $\bar{A}_i$ are obtained by dynamic analysis for each segment. Once $N(y)$ in Eq. 10 is obtained, it is easy to calculate the probability of exceedance. However, for the frequency of

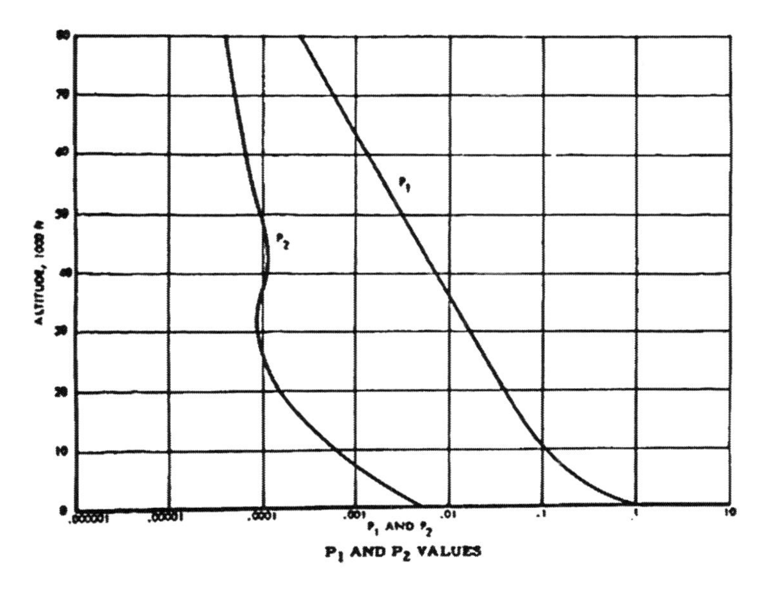

Fig. 8 Plot for P_1 and P_2 for different altitudes

Table 2 Sample values of P_1, P_2, b_1 and b_2 at different altitudes

S. No.	Altitude (ft)	P_1	P_2	b_1	b_2
1	0 (sea level)	1	0.0017	3.932	7.547
2	1,000	0.19	0.00163	3.760	7.600
3	5,000	0.1348	0.00119	3.412	8.530
4	10,000	0.112	6.5E-4	3.519	9.641
5	20,000	0.041	1.566E-4	3.25	11.354
6	30,000	0.017	8.642E-4	3.05	11.523
7	40,000	0.0008	1.265E-4	3.04	9.341
8	60,000	0.0012	8.2E-5	3.715	5.675
9	80,000	3.04E-4	6.01E-5	2.316	0.51

exceedance of very much lower value of y, the probability of exceedance obtained by multiplying frequency by flight time can give the value greater than unity. So, in this chapter, the following equation is used for relating frequency of exceedance and probability:

$$P = 1 - e^{-\lambda t} \tag{11}$$

where t is the period to which the probability applies and λ is the frequency of exceedance.

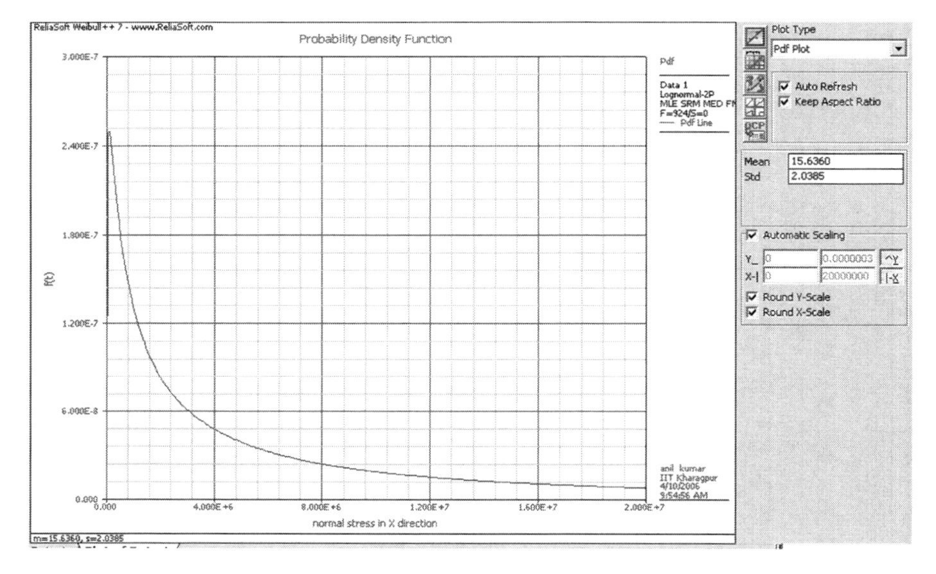

Fig. 9 PDF of normal stress (σ_x)

4.5 Probability Distribution of Responses

Once the frequency of exceedance is determined, it can be easily converted to probability of exceedance by Eq. 12. Namely, the probability that y is lower than y_1 is

$$F(y) = P(y<y_1) = 1 - (1 - e^{-\lambda t}) = e^{-\lambda t} \tag{12}$$

Equation 12 is the cumulative distribution function (CDF) for bending moment/ stress, so probability density function (PDF) can be obtained by differentiating CDF and expressed as

$$f(y) = \frac{dF(y)}{dy} = -te^{-\lambda t}N_0\left[-\frac{P_1}{Ab_1}e^{\left(-\frac{(y/A)}{b_1}\right)} - \frac{P_2}{Ab_2}e^{\left(-\frac{(y/A)}{b_2}\right)}\right] \tag{13}$$

The PDF for wing root bending stresses, obtained by Eq. 13, are shown in Figs. 9, 10 and 11. The values for parameters P_1, P_2, b_1 and b_2 can be determined from the Table 2, while $\bar{A}$ and N_0 are obtained by power spectral analysis, and corresponding flight time t is selected as 100 h. The values of N_0 and $\bar{A}$ for different altitudes or segments in the mission profile for all responses, namely, Stress X, Stress Y and Shear XY, are calculated.

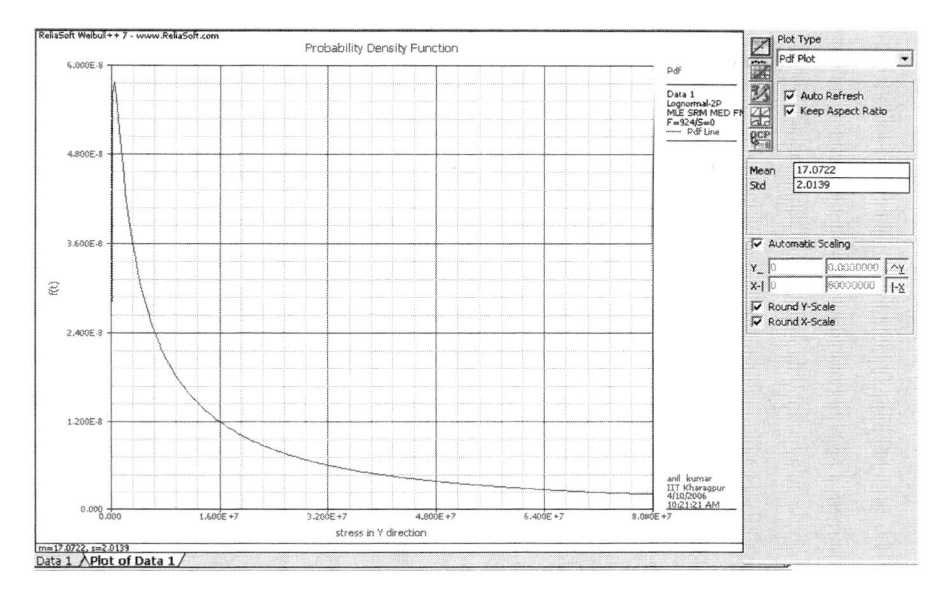

Fig. 10 PDF of normal stress (σ_y)

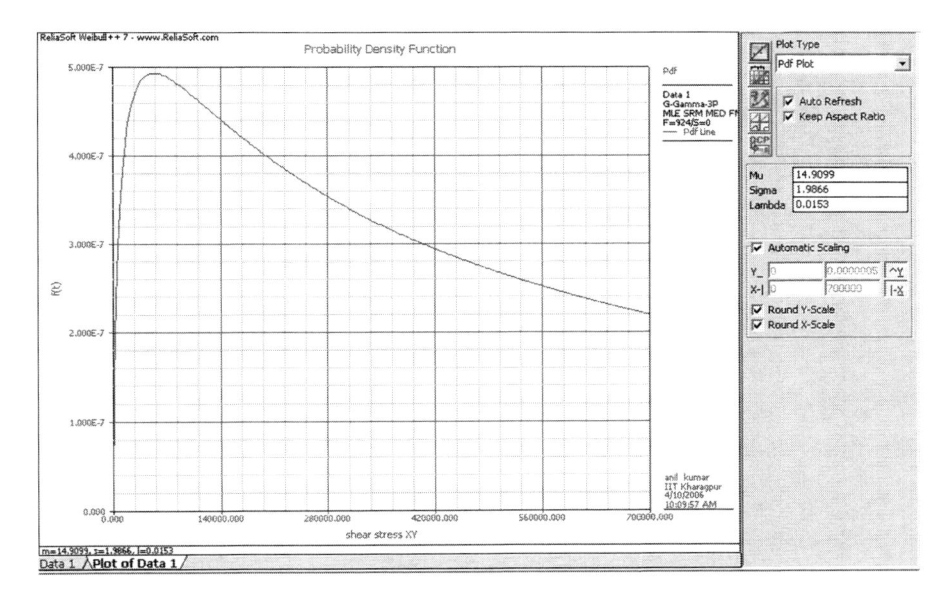

Fig. 11 PDF of shear stress (τ_{xy})

4.6 Probability of Failure

To check the failure due to wing root bending moment, failure criterion is calculated at each ply of wing root. According to Tsai–Hill criterion [12], failure occurs when

$$g(\sigma) = \frac{\sigma_1^2}{X^2} + \frac{\sigma_2^2}{Y^2} - \frac{\sigma_1\sigma_2}{X^2} + \frac{\tau_{12}^2}{S^2} - 1 \geq 0 \tag{14}$$

where X, Y and S are lamina strengths and σ_1, σ_2 and τ_{12} are stress components in the principal material axes.

Also, material properties of the composite wing, X, Y and S, are modelled by normal random variables to consider uncertainty. Standard deviations of the distributions are assumed to be 10% of mean values. The other properties are assumed to be deterministic. With 2,000 sampling points for each random variable, an inequality of Tsai–Hill equation is examined for failure check. A sample failure table is presented in Table A.1. N_f is the number of simulation cycles; when $g(\sigma)$ is greater than zero and N is the total number of simulation cycles, the probability of failure can be expressed as

$$P_\mathrm{f} = \frac{N_\mathrm{f}}{N} \tag{15}$$

The above equation would give the value of failure probability, and one minus of that will give the reliability of the composite wing. This method is very efficient way to check the safety and reliability of aircraft structure subjected to gust. For $\sigma_\mathrm{W} = 0.3048$ m/s, the failure probability, calculated using Eq. 15, was found to be 0.06. Therefore, the composite wing's success probability is 0.94 under the gust condition, but this is very low, and it is a matter of concern.

5 Modification in Wing Structure

The success probability of the previous design is very low and needed to do certain changes to improve the reliability. Methods to improve reliability are as follows:

- Optimising the mechanical properties of the wing by improving the strength in the week direction; this can be done by changing the orientation of the plies.
- Increasing the no. of laminas.
- Stiffening the structure with *stringers or longerons.*
- Going for high-strength composite.

In the present study, the modification is considered to remove two 0° and 90° layers and is replaced by two ±45° layers. The modified lamination sequence is

[90/±45/0/ ∓ 45/90/±45/90/±45/0/ ∓ 45/90]. Then the above procedure is repeated to get the failure probability. This time, the failure probability is decreased to 0.015.

N_f =30 and N = 2,000;

Therefore, success probability is 0.985.

6 Conclusion

In this study, the FE method has been established to perform the dynamic gust response analysis of a simplified aircraft wing model. First six natural frequencies are considered for the gust response study as the power spectral density diminishes with higher frequency range. Failure analysis of composite structure is performed to assess the failure probability of the base model. A reliability analysis was conducted for a composite wing subjected to continuous random gust. For this, gust load was represented by wing root bending stresses, and its probability distribution function was obtained. Monte Carlo simulation was used to handle random variables, and numerical results show that failure probability increases nonlinearly with the growth of RMS gust velocity.

It is observed that the success rate is below the acceptable range for the base model. The modification of the lamination sequence is considered, and failure probability is calculated again. It is observed that with the modification of the lamination sequence, the failure probability reduced and it is within the acceptable range.

Appendix

Table A.1 Sample failure index

1	Stress X	Stress Y	Shear X	Strength	Strength	Strength	*g*
2	401,151.3	86,797.84	41,013.67	1.23E + 09	90,337,605	66,178,034	−1
3	138,012.4	109,750.3	58,837.44	1.25E + 09	84,377,789	80,236,318	−1
4	112,563.3	73,538.05	103,408.7	1.04E + 09	79,070,863	73,021,476	−1
5	951,786.2	160,006.2	30,854.64	1.22E + 09	81,725,809	67,688,895	−1
6	23,128.1	49,263.65	168,206.2	1.03E + 09	84,451,323	81,083,256	−1
7	941,522.6	159,833.3	31,898.26	1.2E + 09	77,231,433	82,033,068	−1
8	1,506,422	160,806	134,387.1	1.39E + 09	87,221,197	82,911,923	−0.99999
9	2,217,938	40,466.99	144,768.8	1.14E + 09	61,389,979	68,970,014	−0.99999
10	1,105,908	45,493.47	213,074.1	1.1E + 09	87,425,271	77,149,897	−0.99999
11	1,540,061	241,163.3	38,051.89	1.2E + 09	86,792,604	54,968,889	−0.99999
12	2,356,498	29,929.41	187,735.7	1.06E + 09	91,187,309	81,251,512	−0.99999
13	2,436,160	150,225.9	74,459.79	9.73E + 08	79,874,621	72,569,404	−0.99999
14	1,775,837	205,306.9	121,056.8	1.11E + 09	80,983,458	87,974,349	−0.99999
15	3,346,433	192,095.2	42,981.48	1.12E + 09	90,144,578	72,829,447	−0.99999

(continued)

Table A.1 (continued)

1	Stress X	Stress Y	Shear X	Strength	Strength	Strength	g
16	1,237,790	72,521.88	223,634.9	1.18E + 09	85,450,125	62,238,419	−0.99999
17	546,138.1	86,031.75	281,387.2	1.18E + 09	91,798,601	76,211,431	−0.99999
18	3,493,641	42,323.16	171,631.3	1.18E + 09	77,783,132	68,977,825	−0.99998
19	262,642.5	75,834.34	298,013.9	1.13E + 09	83,062,112	77,773,890	−0.99998
20	2,354,829	229,962.5	24,160	1.13E + 09	64,372,400	85,230,015	−0.99998
21	3,689,102	173,391.6	127,277.1	1.01E + 09	80,294,048	73,735,580	−0.99998
22	4,978,842	123,911.8	181,979.8	1.3E + 09	77,079,460	86,167,184	−0.99998
23	80,245.52	171,473.3	339,163.3	1.09E + 09	79,240,045	82,956,734	−0.99998
24	3,921,881	163,553.1	201,569.9	1.14E + 09	84,551,745	75,967,414	−0.99998
25	1,265,606	163,452.7	363,795.4	1.21E + 09	81,435,284	85,102,869	−0.99998
26	434,054.5	96,452.5	376,276	1.28E + 09	78,872,300	80,120,513	−0.99998
27	5,081,898	9,256.32	164,585.8	1.14E + 09	88,853,717	76,372,852	−0.99998
28	1,580,984	61,176.75	389,969.7	1.07E + 09	83,822,140	80,917,604	−0.99997
29	469,398.1	39,938.21	395,661.3	1.14E + 09	74,587,737	76,865,455	−0.99997
30	613,988.5	288,373.6	316,655.4	1.01E + 09	79,824,018	82,825,087	−0.99997
31	4,745,057	59,674.38	242,368.2	1.35E + 09	87,934,583	61,958,646	−0.99997
32	4,790,091	224,434.9	134,304.5	1.11E + 09	82,586,012	71,723,634	−0.99997
33	98,208.7	272,921.3	366,708.5	1.13E + 09	81,049,631	83,968,725	−0.99997
34	5,103,056	232,005.5	222,548.9	1.22E + 09	80,178,208	95,259,879	−0.99997
35	4,384,522	246,126.8	228,573	1.19E + 09	83,412,499	68,326,241	−0.99997

References

1. Hoblit FM (1988) Gust loads on aircraft: concepts and applications, AIAA education series. AIAA, Washington, DC
2. Bisplinghoff RL, Ashley H, Halfman RL (1955) Aeroelasticity. Addison-Wesley Publishing, Cambridge, MA
3. Starnes JH, Haftka RT (1979) Preliminary design of composite wings for buckling, strength and displacement constraints. J Aircraft 16:564–570
4. Liu IW, Lin CC (1991) Optimum design of composite wing structures by a refined optimality criterion. Compos Struct 17:51–65
5. Lernet E, Markowitz J (1979) An efficient structural resizing procedure for meeting static aeroelastic design of objective. J Aircraft 16:65–71
6. Liu B, Haftka RT, Akgun MA (2000) Two-level composite wing structural optimization using response surface. Struct Multidiscip Optim 20:87–96
7. Eastep FE, Tischler VA, Venkayya VB, Knot NS (1999) Aeroelastic tailoring of composite structures. J Aircraft 36:1041–1047
8. Penmetsa RC, Grandhi RV (2002) Efficient estimation of structural reliability for problems with uncertain intervals. Comput Struct 80:1103–1112
9. Mahadevan S, Liu X (2002) Probabilistic analysis of composite structure ultimate strength. Am Inst Aeronaut Astronaut J 40:1408–1414
10. Marzocca P, Librescu L, Chiocchia G (2001) Aeroelastic response of 2-D lifting surfaces to gust and arbitrary explosive loading signatures. Int J Impact Eng 25:41–65
11. Aeroelastic Analysis User's guide, MSC.Nastran Version 68, MSC.Software Corporation, 2004.
12. Sinha PK (2006) Composite materials and structures. Department of Aerospace Engineering, IIT Kharagpur, Kharagpur (Web publication)

Seismic Fragility Analysis of a Primary Containment Structure Using IDA

Tushar K. Mandal, Siddhartha Ghosh, and Ajai S. Pisharady

Abstract The seismic fragility of a structure is the probability of exceeding certain limit state of performance given a specific level of hazard. This fragility is typically estimated for multiple hazard levels considering monotonically increasing intensity measures, such as peak ground acceleration (PGA). The seismic safety of the primary/inner containment structure, which is the most important civil engineering structure in a nuclear power plant (NPP) housing the reactor and other major safety related components, is of utmost concern for both old and new NPP. This chapter presents a novel approach of obtaining the seismic fragility curves for a primary containment structure using incremental dynamic analysis (IDA). The limit state of performance selected for these fragility estimations is based on the collapse of the structure. In order to reduce the computation involved, a simple 'stick model' of the containment structure is used for the nonlinear response-history analyses in the multi-earthquake IDA. The seismic fragility curves obtained using the proposed approaches are compared with those obtained using the conventional approach considering an elastic response spectrum and a linear elastic seismic analysis of the structure. The IDA-based fragilities are found to be more realistic than those obtained using conventional methods.

Keywords Fragility analysis • Probabilistic seismic risk analysis • Nuclear containment • Inner containment • Incremental dynamic analysis

---

T.K. Mandal • S. Ghosh (✉)
Department of Civil Engineering, Indian Institute of Technology Bombay, Mumbai, India
e-mail: sghosh@civil.iitb.ac.in

A.S. Pisharady
Siting & Structural Engineering Division, Atomic Energy Regulatory Board, Mumbai, India

S. Chakraborty and G. Bhattacharya (eds.), *Proceedings of the International Symposium on Engineering under Uncertainty: Safety Assessment and Management (ISEUSAM - 2012)*,
DOI 10.1007/978-81-322-0757-3_21,

1 Introduction

The objective of seismic probabilistic safety assessments (PSA) for nuclear power plants is to examine the existence of vulnerabilities against postulated earthquake hazards [7]. It involves assessing the plant's (or its components') safety numerically, in a probabilistic framework, so that appropriate measures can be taken to enhance an NPP's safety level, if needed. One of the major components in the seismic PSA of an NPP is the seismic fragility evaluation. Seismic fragility is defined as the conditional probability of failure for a given seismic intensity level. These fragilities are typically expressed using fragility plots, where these conditional probabilities are plotted against varying values of seismic intensity. Seismic fragility can be defined both at the component level and at the system level in an NPP. Fragility definitions also depend on how failure is defined while estimating the probability of failure.

India has 20 operational nuclear reactor units, 18 of which are pressurized heavy water reactor (PHWR) with the earliest dating back to 1973. All of these are located in moderate seismic zones (Zones 2 and 3 as per the current seismological intensity map of India), except for those in Narora, UP, which is in Zone 4 (IS 1893–2002). Seismic re-evaluation of these reactors, including those in moderate seismic zones, is an extremely important task, considering several factors, such as:

1. A change in the seismicity of the site based on newer information
2. Requirement of checking the safety level for greater seismic hazard than the original design basis
3. Lack of seismic design or, more commonly, poor seismic design and detailing not meeting current standards
4. Low-level analysis adopted in the original qualification (many a times owing to a lack of computational tools necessary to perform high-level analyses)

2 Conventional Seismic Fragility Analysis

Seismic fragility analyses of nuclear power plant structures and other critical components typically adopt the method proposed by Kennedy and Ravindra [8]. In their pioneering work on seismic fragility analysis, they stated that the objectives of a seismic PSA were to estimate the frequencies of occurrence of earthquake-induced accidents and to identify the key risk contributors so that necessary risk reductions could be achieved. They identified the component fragility analysis to be a major part of the seismic PSA (other parts being seismic hazard analysis, system-level analysis, accident sequence identification, etc.). Among many others, two major achievements of this work were in the identification of different levels of damage and in the treatment of system-level fragilities separately from component-level fragilities.

In this fragility analysis approach, the conditional probability of failure is computed as [1, 2]

$$P_{\mathrm{f}} = \Phi\left(\frac{\ln\left(\frac{a}{A_{\mathrm{m}}}\right) + \beta_{\mathrm{U}}\Phi^{-1}(Q)}{\beta_{\mathrm{R}}}\right) \tag{1}$$

where P_{f} is the conditional probability of failure for an earthquake intensity given by its peak ground acceleration (PGA) $= a$, $A_{\mathrm{m}} =$ median ground acceleration capacity of the structure/component and $Q =$ confidence level in terms of non-exceedance probability. A_{m} is related to the actual ground acceleration capacity parameter, $A = A_{\mathrm{m}}\varepsilon_{\mathrm{R}}\varepsilon_{\mathrm{U}}$, where $A_{\mathrm{m}} = A_{\mathrm{RBGM}}F_{\mathrm{m}}$.

A_{RBGM} is the PGA of review basis ground motion (RBGM) or review level earthquake (RLE). ε_{R} is the random variable representing the aleatory uncertainties, i.e. inherent randomness associated with ground acceleration capacity. ε_{U} is the random variable representing epistemic uncertainty in the determination of median value, A_{m}, i.e. the uncertainty associated with data, modelling, methodology, etc. F_{m} is the median value of factor of safety, F. ε_{R} and ε_{U} are taken as lognormally distributed random variables, with logarithmic standard deviations of β_{U} and β_{R}, respectively, and both having unit median.

Once A_{RBGM} is known, determination of median factor of safety, F_{m}, is the key to derive A_{m}. Generic expression of F can be written as [10]

$$F = F_1F_2F_3 \tag{2}$$

F_1 is a factor representing ratio of capacity to demand and is a strength factor. F_2 corresponds to the level of conservatism in assessing the capacity; it depends primarily on the energy absorption capacity of structure, system or component (SSC) beyond elastic limit. F_3 represents the conservatism associated with calculating demand. Different methodologies for fragility analysis are all about determining the median values of F_1, F_2 and F_3 and selection of corresponding β.

There are numbers of components of an NPP which should remain functional during a seismic event, so it is very difficult to check the seismic qualification of each component individually. Components of a NPP are grouped in a number of categories, and different methods are recommended for their seismic qualification [10].

2.1 Drawbacks of Conventional Methods

The conventional methodologies for seismic fragility analysis have the following drawbacks [5,6]:

1. Though these methods are easy to implement, they require considerable engineering judgement especially in case of selecting of parameters for aleatory and epistemic uncertainties.

2. The use of a double lognormal model is mathematically feasible but doesn't have too strong theoretical basis/background.
3. The use of response spectrum-based methods introduces epistemic uncertainties:

   (a) Response contribution from different modes depends on the accuracy of different modal combination rules.
   (b) In case of nonlinear system, modal analysis is not applicable.

Thus, fragility estimated using conventional methods is not very realistic specifically for nonlinear response of structure.

3 The Proposed Method of Fragility Analysis

This chapter presents a novel approach of obtaining the seismic fragility curves for a primary containment structure using incremental dynamic analysis (IDA). The response of the IC structure, modelled as a 2-D stick, is studied for different types of ground motions.

The important assumptions made for fragility analysis are:

1. Randomness associated with the seismic forces is much more compared to that for structural parameters [3]. Hence, the randomness in seismic forces is considered only, and structural parameters are considered to be deterministic.
2. The structure will behave linearly in shear even while it behaves nonlinearly in flexure. It is done as we do not want the stick model to be failed in shear.
3. Prestressing forces are considered as uniaxial compression for the simplified stick model.
4. Reinforcement orientation in the dome portion is taken vertically for each element though it is not the case in actual, assuming that this will not affect too much in final result of the analysis as maximum strain and interstorey drift ratio usually occurs at the base of the cantilever kind of structure.
5. Openings are not considered for the stick model though it can be done by manually selecting some equivalent reduced sectional properties at the location opening. The reason behind that procedure of taking reduced section will work well during linear behaviour of structure, but it will not work beyond that as nonlinear behaviour depends largely on the actual section geometry which cannot be resembled using equivalent reduced section, and our main motive is to do nonlinear response-history analysis (NLRHA).

The basic steps of fragility analysis are:

1. Choose the ground motion data comparing its response spectrum with the design spectrum of the site and the seismological location of the site, i.e. either intraplate or at plate boundary.
2. Prepare the mathematical model appropriate for NLRHA.

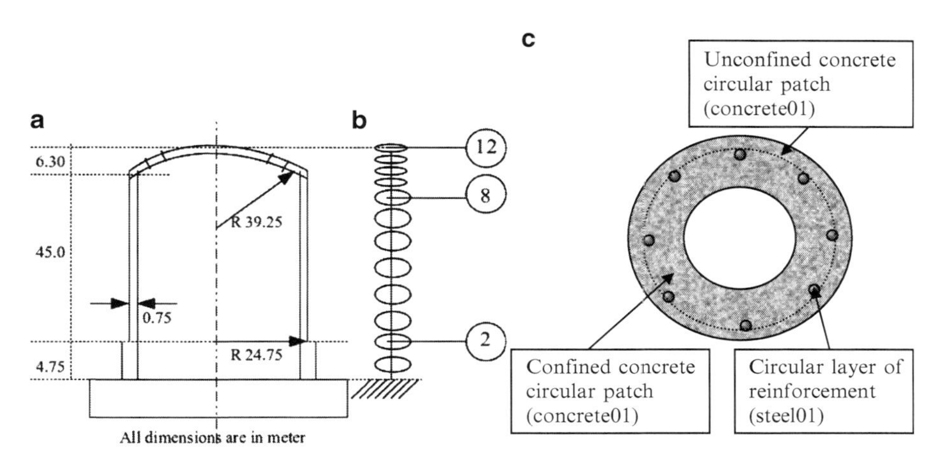

Fig. 1 (**a**) Containment structure and (**b**) its stick model, (**c**) sample fibre section (not to scale)

3. Do the incremental dynamic analysis which is basically a set of NLRHA for all ground motion data.
4. Calculate the probability of exceedance of particular limit states from the output of IDAs, and plot it to obtain fragility plot.

4 Modelling of the Structure

4.1 Description of Structure

The IC structure considered for this study consists of a prestressed concrete cylindrical wall capped by a segmental prestressed concrete dome through a massive ring beam. The containment shell is supported on a circular raft. The typical containment structure considered for the study is depicted in Fig. 1a. The containment structure responds to seismic excitation like a cantilever beam with a circular cross section. The segmental dome along with the ring beam acts to stiffen the circular cross section and also adds to the mass of the system.

4.2 Mathematical Model

The inner containment structure is idealized as a system of lumped masses at elevations of mass concentrations, connected by two-dimensional BeamColumn elements with actual section geometry other than the zone of openings. The structure is assumed to be fixed at the top of the raft foundation. The earthquake excitation is constrained along a single horizontal direction only. The stick model of a containment structure so developed is also shown in Fig. 1b. Nonlinearity of the system is modelled in the programme *OpenSees* [9] using *NonLinearBeamColumn* element. Concrete is modelled using *Concrete01* and reinforcing steel as *steel01* in *OpenSees*. Sections at different levels are defined as *fibre* sections (Fig. 1c) with circular concrete patch and circular layer of reinforcements. As stick model made of single *fibre* section at each level cannot directly take the shear deformation, the shear-deformation behaviour is incorporated using *SectionAggregator* command. A different uniaxial elastic material is defined as the slope of stress-strain plot as GA_s, where G is the shear modulus of concrete and A_s is the shear area of the containment cross section. The *SectionAggregator* command is used to combine this material with *fibre* section previously defined with actual geometrical properties. This section is used to model the element at that level. The gravity load is calculated for each element and applied as nodal load on the upper node of each two-noded element. The average prestress is taken as 10 MPa. Total prestressing force is calculated by multiplying the prestress with the average cross section area, and it is applied as compressive force at top-most and bottom-most node.

5 Incremental Dynamic Analysis

Incremental dynamic analysis [12] is an emerging analysis method offering detailed seismic demand and capacity prediction capability through a series of NLRHA for multiple scaled ground motion. Results of IDA are presented as IDA plots. A single IDA plot is basically the variation of maximum structural response at different intensity of a scaled ground motion. Maximum structural response is known as damage measure (DM), and intensity of ground motion is known as intensity measure (IM). After analysing the structure for multiple ground motion data, the results are plotted on a single paper and the generated plot is known as multi-IDA plot.

5.1 Ground Motions Considered

The containment is assumed to be in the stable continental region of Indian peninsula. Considering this, ground motion records selected for performing a multi-IDA of this containment structure are sourced from recorded earthquakes in similar seismic regions across the world. 5% damped elastic response spectra of these records are compared with the design spectrum of the site, and those varying significantly from this design spectrum are filtered out. Details of

Table 1 Summary of ground motion data considered

Record name	No. of records	Event	Component	Epicentral distance, km	PGA-range, g
GM-1 to 2	2	Bhuj, 2001	Horizontal	Unknown	0.08–0.08
GM-3	1	Koyna, 1967	Horizontal	Unknown	0.474
GM-4 to 25	22	Saguenay, 1988	Horizontal	45–167	0.002–0.174
GM-26 to 32	7	Miramichi, 1982	Horizontal	11–23	0.125–0.575

Table 2 Limit states

Damage measure	LS-1	LS-2	LS-3	Reference
Drift	0.004	0.006	0.0075	FEMA-356 [4]
Plastic rotation (θ_P)	0.0015	0.005	0.005	FEMA-356 [4]
Curvature (obtained from θ_P)	0.00015	0.00025	0.00025	FEMA-356 [4] and Priestley [11]
Compressive strain	0.002	0.0035	0.005	IS-456 and Priestley [11]
Tensile strain	0	0.00014	0.013	OpenSees Concrete02 model

ground motions considered for study are provided in Table 1. The PGA of a ground motion data is adopted to be the intensity measure (IM) for the IDA. Based on previous literature on fragility analysis of nuclear containment structures, a maximum PGA of 5.0 g is considered. For successive NLRHA, the PGA is incremented by 0.075 g on a trial basis. For numerical convergence, this increment is modified as discussed later.

5.2 *Structural Limit States Considered*

The limit states of the containment structure subjected to seismic loading considered in the study are tensile cracking of concrete, crushing of concrete in compression, interstorey drift ratio (IDR) and plastic rotation of section.

1. The tensile and compressive strains developed at the innermost and the outermost fibre of each section are stored for each scale factor of ground motion and compared with limiting valued specified by FEMA-356 [4]. As FEMA-356 does not consider any tensile strength of concrete, the limiting values for tensile strain are adopted from the material model *Concrete02*.
2. The allowable total (elastic and plastic) IDR values are adopted from Table 6-19 of FEMA-356.
3. The allowable plastic rotation values are adopted from Table 6-7 of FEMA-356. Using Preistley's equation [11] limiting values in terms of curvature is calculated since *OpenSees* provides curvature as output not plastic rotation.

5.3 Numerical Convergence of IDA

As stated by Vamvatsikos and Cornell [13,14], numerical convergence is a very critical issue in performing IDA especially in the zone of higher PGA levels when the structure may reach a state of global dynamic instability. To deal with this, following steps are adopted:

1. The structure is modelled with *NonLinearBeamColumn* element which has capability to track the distributed plasticity across the member section and the length of the member.
2. Instead of single algorithm for solving nonlinear equations, a number of algorithms in sequence are tried. The following algorithms are used in sequence if one fails next one is used:

   (a) Newton-Raphson
   (b) Modified Newton-Raphson
   (c) KrylovNewton – with different combinations of tangents: initial and current
   (d) Broyden
   (e) Newton with line search (for different constants)

3. For a specific ground motion, if all algorithm fails at a particular PGA, it may be due to (1) global dynamic instability or (2) numerical failure. Here, if all the algorithm fails at any particular scale factor, it is assumed that it is global dynamic instability situation not only numerical failure, and the control is sent back to the previous scale factor with updated increment of scale factor equal to 1/10 of the regular increment of scale factor. The procedure of obtaining multi-IDA is illustrated in detail in Fig. 2.

6 Results and Discussion

Eigen value analysis is done only to get an overview of the dynamic properties of the primary containment structure (Table 3).

6.1 IDA Plots from Raw Data and Its Correction

If the raw data obtained from an IDA are plotted, it is sometimes found that there are unexpected large rebounds at high PGA level. Such a rebound is due to single data point being located far off from the other data points of the same IDA (Fig. 3c). This is inconsistent even with the previous and the next data points from structural perspective. Such problematic data points are considered to be points of numerical failure, which has not been detected by the algorithm. The IDA is replotted by

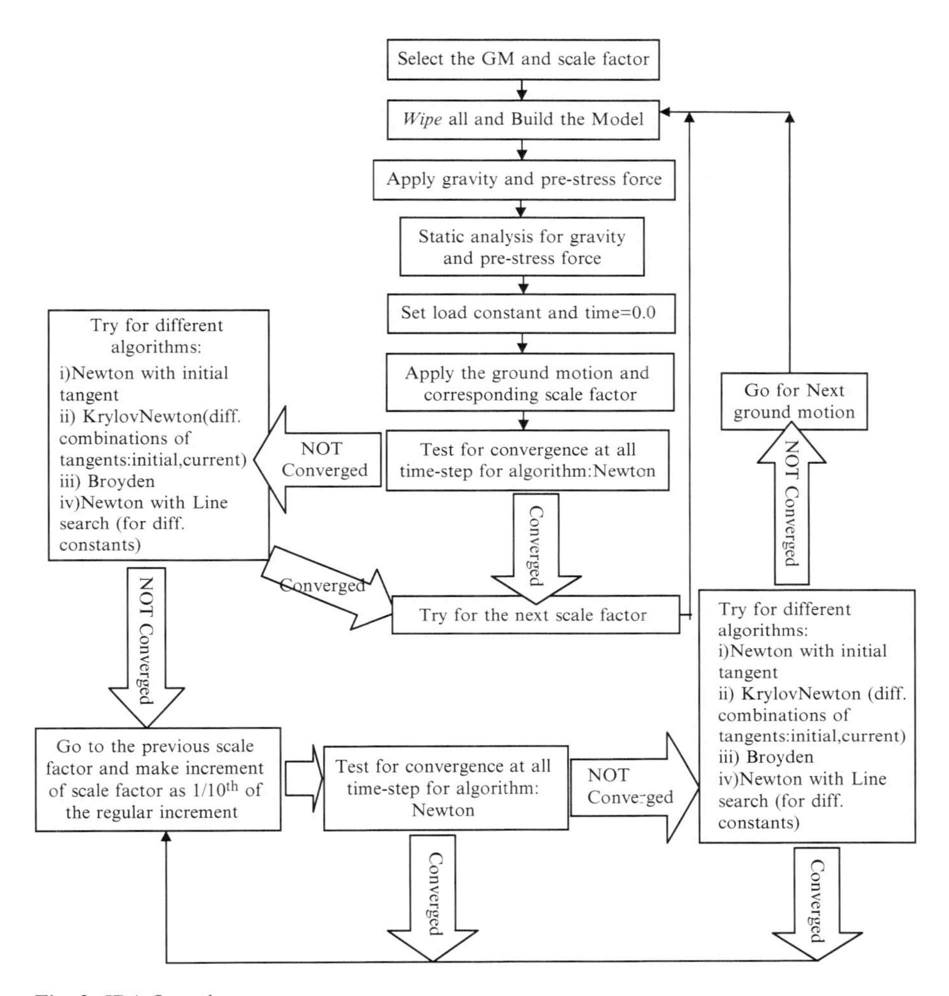

Fig. 2 IDA flow chart

Table 3 Natural period and frequencies of different modes

Mode	*T*, s	*f*, Hz
Mode-1	0.1826	5.48
Mode-2	0.0988	10.12
Mode-3	0.0595	16.82
Mode-4	0.0473	21.13
Mode-5	0.0385	25.96

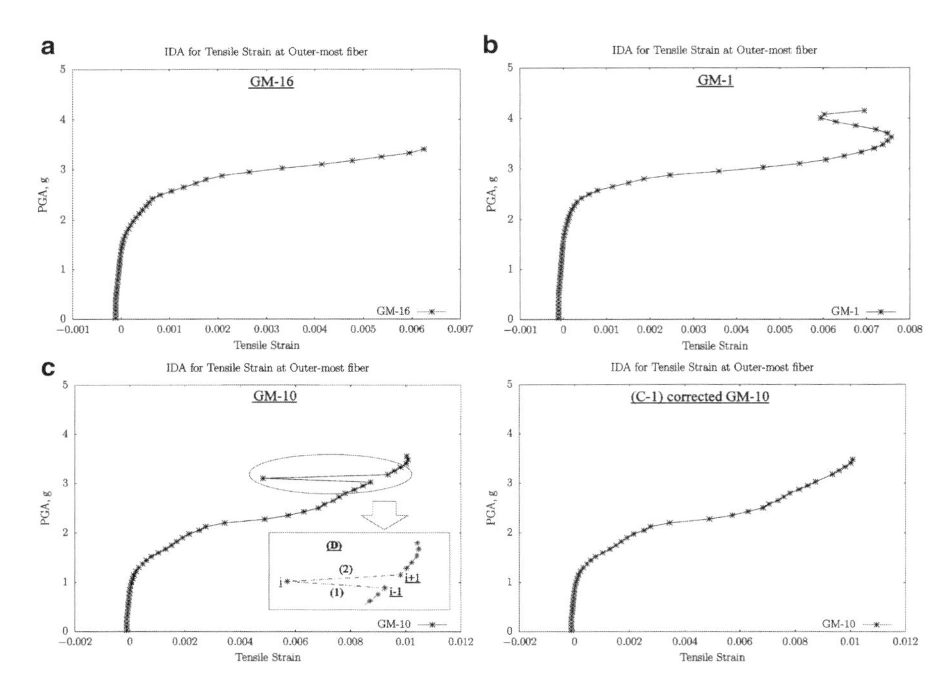

Fig. 3 Sample IDA plots: (**a**) without any rebound (**b**) with realistic rebound (**c**) with unrealistic sharp rebound (C-1) corrected IDA plot with rebound removed

removing these points causing large rebounds by placing some limiting conditions on rate of change in slope of the IDA plot (Fig. 3C-1). The other plots (Fig. 3a, b) are quiet normal as per literature review.

6.2 Fragility Plots

Seismic fragility of a structure, which is the probability of a predefined damage measure (DM) exceeding certain predefined limit states (LS) for a given intensity measure (IM), is calculated as the fraction of IDA curves exceeding the LS at the selected PGA. Fragility analysis results in a set of $P(\text{DM} > \text{LS}|\text{IM})$ vs. IM plots. All the fragility plots so obtained are stepped since a discrete number of ground motion data are used for multi-IDA. The stepped plots are smoothened using weighted cubic spline approximation by increasing the number of ground motion records and reducing the PGA increment for each IDA.

Figure 4 shows fragility plots based on various performance limit states as mentioned earlier in Table 2. Figure 4b, d also show fragility curves based on Eq. (1). Some major observations from these fragility plots are:

- Based on IDR, the fragility is zero up to PGA = 5g (Fig. 4a).
- The fragility, based on plastic rotation (or curvature) limits, is zero up to 2g for LS-1 and up to 3g for LS-2 and LS-3 (Fig. 4b).

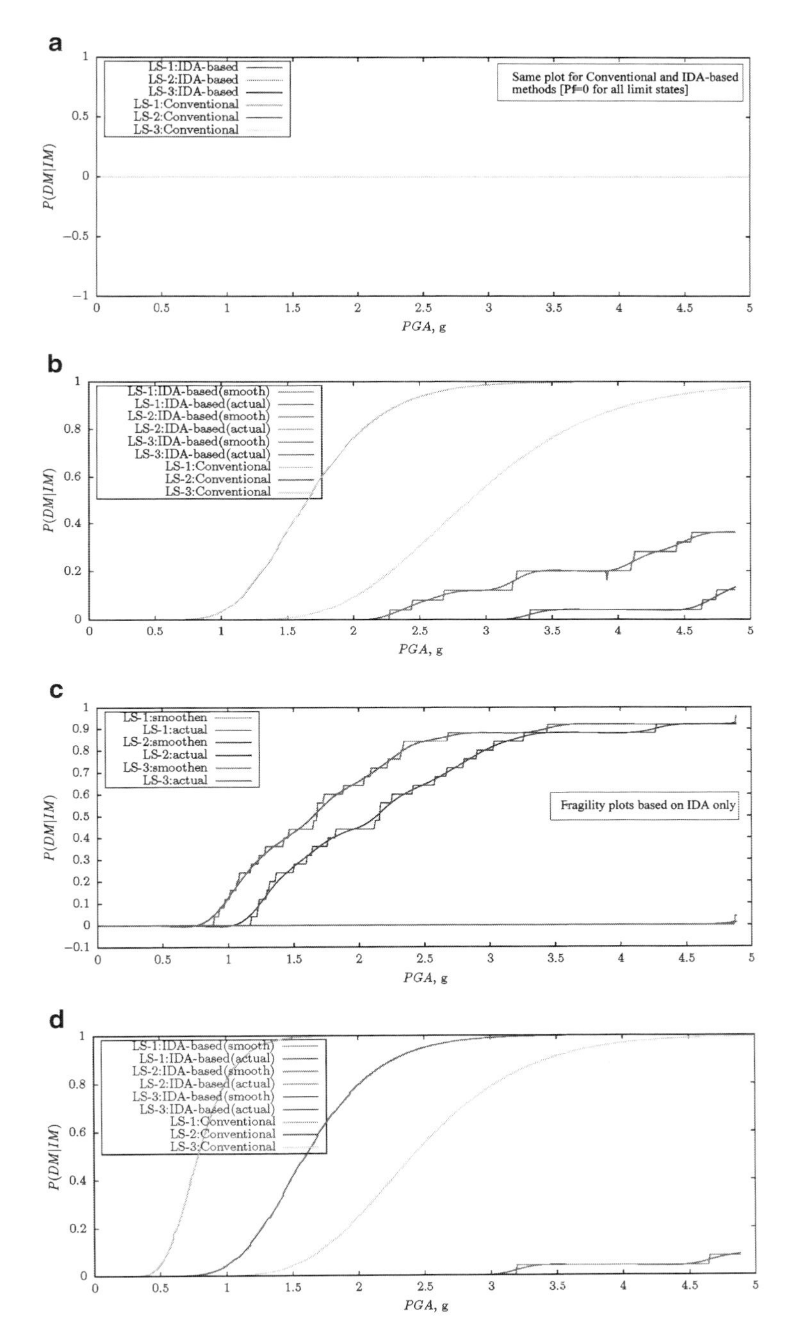

Fig. 4 Fragility plots (IDA-based and conventional methods) (**a**) Fragility for inter storey drift ratio, (**b**) Fragility for curvaturem, (**c**) Fragility for tensile strain at inner-most fiber, (**d**) Fragility for compressive strain at outer-most fiber

- The fragility based on tensile strain at the innermost fibre, which indicates a through crack along the thickness, is almost zero up to PGA = 0.5g for LS-1 (zero tensile strain) and LS-2 (cracking strain) and up to PGA = 5g for LS-3. It reaches almost 0.95 for LS-1 and LS-2 at PGA = 5g (Fig. 4c).
- Fragility curve for concrete crushing is almost zero up to PGA = 5g for all limit states.

Fragilities are also computed for the same containment using Eq. (1). A_{RBGM} is adopted as 0.214g for this PHWR. The median value of F_1 is obtained from a linear elastic analysis. The standard deviations for uncertainties and randomness are based on the recommendations of Pisharady and Basu [10]. The fragility plots obtained this way are significantly different from those obtained using multi-IDA (e.g. Fig. 4b, d). Since the linear elastic analysis shows no tensile strain at the RBGM or design PGA level, the through crack fragility based on conventional method is zero even at PGA = 5g. This shows the unrealistic nature of the fragility curves using the conventional method.

7 Conclusions

Primary containment structures are found to have almost zero fragility considering limit states based on IDR and crushing of concrete. But these structures have very high probability of failure in terms of through crack formation, which results in radiation leakage. The IDA-based fragility curves are found to be more realistic than fragility curve obtained using conventional method. This is primarily due to the fact that the effect of nonlinearities is directly incorporated in IDA-based estimation of fragility. However, it should be noted that the IDA-based fragilities shown here do not include uncertainties associated with structural modelling. Future works in this area should focus on reducing the model uncertainty by using detailed structural model, soil-structure interaction and larger number of earthquake records.

References

1. Bhargava K, Ghosh AK, Agrawal MK, Patnaik R, Ramanujam S, Kushwaha HS (2002) Evaluation of seismic fragility of structures – a case study. Nucl Eng Des 212(1–3):253–272
2. Bhargava K, Ghosh AK, Ramanujam S (2005) Seismic response and failure modes for a water storage structure – a case study. Struct Eng Mech 20(1):1–20
3. Ellingwood BR, Celik OC, Kinali K (2007) Fragility assessment of building structural systems in mid-America. Earthq Eng Struct Dyn 36(3–5):1935–1952
4. FEMA-356 (2000) Prestandard and commentary for the seismic rehabilitation of buildings. Federal Emergency Management Agency, Washington, DC
5. Gupta S, Manohar CS (2006) Reliability analysis of randomly vibrating structures with parameter uncertainties. Ph.D. thesis, Indian Institute of Science, Bangalore, pp 1–407

6. Gupta S, Manohar CS (2006) Reliability analysis of randomly vibrating structures with parameter uncertainties. J Sound Vib 297(3–5):1000–1024
7. Hari Prasad M, Dubey PN, Reddy GR, Saraf RK, Ghosh AK (2006) Seismic PSA of nuclear power plants: a case study, BARC/2006/015, BARC
8. Kennedy RP, Ravindra MK (1984) Seismic fragilities for nuclear power plant risk studies. Nucl Eng Des 79(1):47–68, 38
9. McKenna F, Fenves GL (2001) Opensees command language manual, version 1.2, Pacific Earthquake Engineering Research Center, University of California, Berkeley, USA.
10. Pisharady AS, Basu PC (2010) Methods to derive seismic fragility of npp components: a summary. Nucl Eng Des 240(11):3878–3887
11. Priestley MJN (1997) Displacement-based seismic assessment of reinforced concrete buildings. J Earthq Eng 1(1):157–192, Cited By (since 1996): 62
12. Rizkalla SH, Lau BL, Simmonds SH (1984) Air leakage characteristics in reinforced concrete. J Struct Eng 110(5):1149–1162
13. Vamvatsikos D, Cornell CA (2002) Incremental dynamic analysis. Earthq Eng Struct Dyn 31 (3):491–514
14. Vamvatsikos D, Cornell CA (2004) Applied incremental dynamic analysis. Earthq Spectra 20 (2):523–553

Nanotoxicology: A Threat to the Environment and to Human Beings

D. Dutta Majumder, Sankar Karan, A. Goswami, and N Banerjee

Abstract Nanotechnology research and development is directed towards the understanding and control of matter at dimension of roughly 1–100 nm. At this size, the physical, chemical, and biological properties of materials differ in fundamental and potentially useful ways from the properties of individual's atoms and molecules. Its applications advanced very quickly while very little has been done to measure and assess the risks of nanoparticles (NPs) to biological systems and to the ecosystems. In the year 2000, the National Nanotechnology Initiative (NNI) was formed to ensure public confidence in the field of nanotechnology research, engineering and manufacturing of nanoscale product. In this chapter, we present different kind of nanotoxicology studies to examine the environment and health risks associated with nanoparticles exposure. We also point the penetration of nanoparticles into human body via various routes and interact with the system with their toxic properties that depends on surface chemistry, particle size, surface

---

D. Dutta Majumder (✉)
ECSU, Indian Statistical Institute, 203, B.T. Road, Kolkata 700108, India

Institute of Cybernetics Systems and Information Technology, 155, Ashoke Garh, Kolkata 700108, India
e-mail: duttamajumder.isi@gmail.com

S. Karan
Institute of Cybernetics Systems and Information Technology, 155, Ashoke Garh, Kolkata 700108, India

Institute of Radiophysics and Electronics, C.U., 92, APC Road, Kolkata 700006, India
e-mail: sankar.karan@gmail.com

A. Goswami
Biological Science Division, ISI, 203, B T Road, Kolkata 700108, India
e-mail: agoswami@isical.ac.in

N. Banerjee
Department of Health and Family Welfare, Sastha Bhawan, Kolkata 700091, India
e-mail: nupur.special.eng@gmail.com

S. Chakraborty and G. Bhattacharya (eds.), *Proceedings of the International Symposium on Engineering under Uncertainty: Safety Assessment and Management (ISEUSAM - 2012)*,
DOI 10.1007/978-81-322-0757-3_22, © Springer India 2013

charge, and surface area. Nanoparticles of oxides like Sio_2 produced and characterised in our laboratory were tested against insect's pests and pathogens. Nano silica against insect pests shows nearly 100% mortality.

Keywords Nanotechnology • Risk assessment • Capped nanoparticles • Nanotoxicology • Environmental risk and hazards • Occupational safety • Environmental risk management

1 Introduction

Nanotechnology, the science and technology of controlling matter and energy at the nanoscale (1–100 nm), promises to have far reaching impacts on the science and technology industry in areas ranging from consumer products to health care to transportation [1]. The present challenge facing nanotechnology is to manipulate matter at dimension of roughly 1–100 nm (10^{-9} m) in a controlled way to create new substances with very special properties. Industry thinks the technology holds promise to change every fact of life in some way. Substances at nanoscale, or nanoparticles, demonstrate novel physiochemical properties compared to large particles of the same substance. Their use, thus, helps to improve products. But there has not been any significant contribution as far as our knowledge goes to measure and assess the risk to human health [2] and the environment [3]. The Royal Society of London [4] appointed a committee to make a study on the subject, the report of which indicated the importance and need for extensive study on the subject. The National Nanotechnology Initiative (NNI), USA has been taken the initiative in the year 2000 for protecting human health, and the environment [5, 6] from risk and hazards generated due to nano research, engineering and manufacturing.

1.1 Background and Perspective

In the evolution of nanotechnology, manufactured nanomaterials are an important step towards a long-term vision of building objects atom-by-atom and molecule-by-molecule with processes such as self-assembly [7] or molecular assemblers [8]. Innovations in analytical and imaging technologies first paved the way for perceiving, measuring, and manipulating nanoscale objects [9, 10], typically defined as those having a characteristic dimension <100 nm. The ability to design materials at the nanoscale is now leading to the rapid development of an industry that provides nanomaterials for a range of industrial and consumer products. Overall nanotechnology has revolutionised the industry. At present, there are more than 200 nanoparticle-based products worldwide. Separate researches on the ill effects of nanoparticles are being carried out. But most are inconclusive

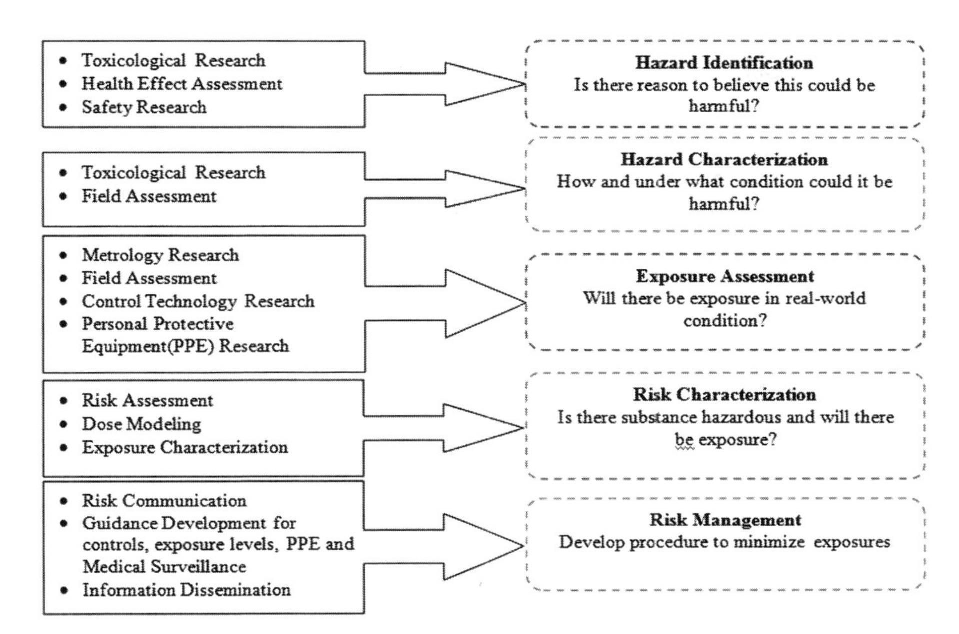

Fig. 1 Important steps to management of risk involved with nanotechnology

still. However, there is evidence that some nanoparticles are harmful [11] and regulations on their use are essential. In 2009, the Drug Controller General of India (DCGI) recalled Albupax, a nano-based medicine for breast cancer. US firm Abraxis BioScience had complained against the maker that the drug damaged the liver [12]. The generic failed the DCGI test. In medicines, nanoparticles are so tiny that they can pass through the blood-brain barrier [13]. They may be toxic, but this is little understood. Carbon as graphite used in pencils is innocuous, but carbon nanotubes show alarming similarities to asbestos [14]. Nanoparticles of zinc and titanium oxide damage DNA in cell lines [15], found IITR. The two are important constituents of sunscreen lotions. Nanosilver has antibacterial properties [16], but its unrestricted use weakens children's ability to fight infections. Nanoparticles accumulate in the environment and kill the beneficial bacteria [17]. To understand the impact of nanoparticles on human health, the U. S. Food and Drug Administration (FDA) established a nanotechnology task force in 2006 [18]. In January this year, it issued draft guidelines on safe use of nanoparticles in consumer products. The US EPA, through its Toxic Substances Control Act, keeps a tab on the manufacturing of nano-based products in the country. European Commission's REACH, a regulatory framework for chemicals, has not suggested a policy that covers nanomaterial, but it does call them substances that need to be tested (Fig. 1).

2 Materials and Methods

Silica nanoparticles were synthesised [19] and characterised in the laboratory and tested against insect's pests and pathogens. Although there are not yet international definitions of NMs and their sizes, aggregates with overall dimension in the μm range but made of primary particles of <100 nm would be regarded as nanoparticles. From the toxicological point of view, particles with diameter 10–2.5 μm are defined as coarse particles, for diameter of 2.5 μm or less are fine particles, and for less than 0.1 μm (<0.1 μm) as ultrafine particles. NPs are particles with sizes between 0.1 and 100 nm (diameter) but provide a very large surface to volume ratio [20]. The surface properties of NPs and their biocompatibility depend on the charges carried by the particle and its chemical reactivity. It has been shown that polycationic macromolecules have a strong interaction with cell membranes in vitro. Also, the interaction of NPs with the surface lining layers of biological tissues is determined by their surface chemistry and reactivity.

2.1 Mechanisms Behind Toxicity of Nanomaterial

Nano-sized particles can penetrate the human body via various routes and could persist in the system because of the incapability of the macrophages to phagocytose [21] them. Whether these persisting nanomaterial react with the body, stay inert, or interact with the system will govern their toxic properties and is primarily dependent on their surface properties.

2.1.1 Surface Chemistry

The role of surface chemistry had been underemphasized in the present research of nanotoxicity. Whether the particle remains suspended as an individual particle or as an aggregate depends upon its surface chemistry [22]. A small aggregate or single particle is presumed to be more toxic than an aggregate of nano-sized particle(NSPs) as the relative surface area could change, determining whether the material has a good wetting characteristic or has a surface characteristic that catalyses specific chemical reactions or remains passive and allows fibrous tissue to grow on its surface. It was shown that the rats, when treated to polymeric vapours of polytetrafluoroethylene (PTFE) [23] having a diameter of 18 nm, suffered severe lung injury with high mortality rate of within 4 h after a 15-min inhalation exposure to 50 $\mu g/m^3$.

2.1.2 Particle Size

Nanoparticles may be surrounded by water molecules and may or may not get agglomerated when present in fluid medium, which, in turn, will govern the diffusion of species. The diffusion coefficient of the particles can be derived using the well-known Stokes–Einstein equation as

$$D = \frac{K_{\mathrm{B}}T}{6\pi\eta R} \tag{1}$$

Equation 1, where R is the hydrodynamic radius, T is the absolute temperature, η is the viscosity, and K_{B} is the Boltzmann constant. The aggregation of the particles can also be quantified using the following simple formulation:

$$\frac{dn_t}{dt} = \frac{1}{2}\frac{\beta}{W}n_t^2 \tag{2}$$

If dn_t/dt is the mono disperse particle population and β is the aggregation rate constant, then Eqs. 2 and 3 can be used to derive the characteristic time of a doublet formation τ from initial particle concentration (Eq. 2), where W is the stability ratio relating the steric and electronic hindrance towards the aggregation and gives the ratio of the aggregation constant of diffusion limited cluster aggregation and the observed aggregation constant. The time for doublet formation which gives a measure of the agglomeration is given in Eq. 3:

$$\tau = \frac{2W}{\beta} = \frac{3\eta W}{4K_{\mathrm{B}}Tn_0} \tag{3}$$

2.1.3 Surface Charge

It can be noted that the particle size has an effect on the stability constant, and this can be altered by small changes on the particle surface, such as charge distribution. The surface potential (ΔV) of a monolayer spread at the air/water interface can be interpreted using the Helmholtz Eq. 4

$$\Delta V = \frac{\mu_0}{\varepsilon_0 A} \tag{4}$$

$$\mu_n = \mu_0 + \mu_W \tag{5}$$

where μ_n is the effective molecular dipole moment at the interface, ε is the vacuum dielectric permittivity, and A is the surface area per molecule. It is important to study, whether NPs can cross this phosphor-lipid membrane barrier and/or interact with it will decide the level of toxicity. A particle having a high affinity for phosphates may or may not react with the phospholipid membranes. Such being the case, monolayers of particles spread at the phospholipid membrane can be treated as a Vogel Mobius two capacitor model [24, 25], an effective molecular dipole moment is given by Eq. 5, where μ_w represents the contribution of the tail group and μ_α represents the contribution from the head group of a phospholipid. These contributions can be directly influenced by the surface charge present on the molecule and the complex forming ability of the particle with the phospholipid. Zeta potential (ζ) [26] and ultraviolet-visible-spectroscopy measurements may thus provide an estimate of adsorption of proteins. The particles having mobility V_{E} in body fluid media which becomes an important parameter and is given by the Smoluchowski Equation

$$V_{\mathrm{E}} = 4\pi\varepsilon_0\varepsilon_{\mathrm{r}}\frac{\varsigma}{6\pi\eta}(1 + kr) \tag{6}$$

where ε_{r} and ε_0 are the relative dielectric constant and the electrical permittivity of a vacuum, respectively, η is the solution viscosity, r is the particle radius, and $\kappa = (2n_0z^2e^2/\varepsilon_{\mathrm{r}}\varepsilon_0k_{\mathrm{B}}T)^{1/2}$ is the Debye–Hückel parameter, n_0 is the bulk ionic concentration, z is the valence of the ion, e is the charge of an electron, k_{B} is the Boltzmann constant, and T is the absolute temperature.

2.1.4 Particle Shape

When developing NPs as catalysts, their shape is very important. For a certain volume of material, NPs make best catalysts when they have a large surface area. It is a challenge to find the shape that has the largest surface area for its volume. The perception of shape has been used for pattern recognition, computer vision, shape analysis [27], and image registration. Here, we used Dutta Majumder's generalised method of shape analysis and shape-based similarity measures, shape distance, and shape metric to measure the NPs shape.
The shape of an object can be defined as a subset X in R^2 if

1. X is closed and bounded.
2. Interior of X is nonempty and connected.
3. Closure property holds on interior of X.

This representation of shape remains invariant with respect to translation, rotation, and scaling. Moreover another object Y in R^2 is of same shape to object $X \in R^2$ if it preserves translation, rotation, and scaling invariance. In terms of set, these three transformations can be represented as

$$\text{Translation: } Y = \{(x + a), (y + b) : x, y \in X\} \tag{7}$$

$$\text{Rotation: } Y = \{P1(\alpha).P2(\beta)X\} \quad \text{where } P1 \,\&\, P2 \text{ are rotation around } x \text{ and } y \text{ axes} \tag{8}$$

$$\text{Scaling: } Y = \{(kx, ky) : x, y \in X\} \tag{9}$$

Distance d_1 between shape X and Y in F is defined as follows:

$$d_1(X, Y) = m_2[(X - Y) \cup (Y - X)] \tag{10}$$

where m_2 is Lebesgue measure in R^2 and d_1 satisfies following rules:

1. $d_1(X,Y) \geq 0$
2. $d_1(X,Y) = 0$ if and only if $X = Y$
3. $d_1(X,Y) = d_1(Y,X)$
4. $d_1(X,Y) + d_1(Y,Z) \geq d_1(X,Z.)$

We consider that two nanoparticles are of same shape if and only if one of the images is translation, scaling, and rotation of other.

2.2 *Exposures to Nanomaterial: Most Likely Routes*

All substances in the world are toxic at some exposure levels. Currently, very little is known about the interaction of nanoparticles in biological systems [28, 29]. Initial toxicology experiments have shown that some types of nanoparticles can enter the body, affect organ function, and possibly lead to health problems. But these experiments must be repeated many times under precisely controlled conditions using specific types and concentrations of nanoparticles to identify short-term and long-term health effects of different kinds of nanoparticles on human health. This research should study the response of living organisms to the many different kinds of nanoparticles that have different chemical compositions, sizes, shapes, and surface areas [20] (Fig. 2).

2.3 *Life Cycle Analysis (LCA) Modelling: Potential Impact Indicators*

Goal and scope definition defines the goal and intended use of the LCA (life cycle analysis) [30–33] and scopes the assessment concerning system boundaries, function and flow, required data quality, technology, and assessment parameters. We define the following parameter for analysis.

2.3.1 Life Cycle Inventory Analysis (LCI)

We define LCI as an activity for collecting data on inputs (resources and intermediate products) and outputs (emissions, wastes) for all the processes in the product system.

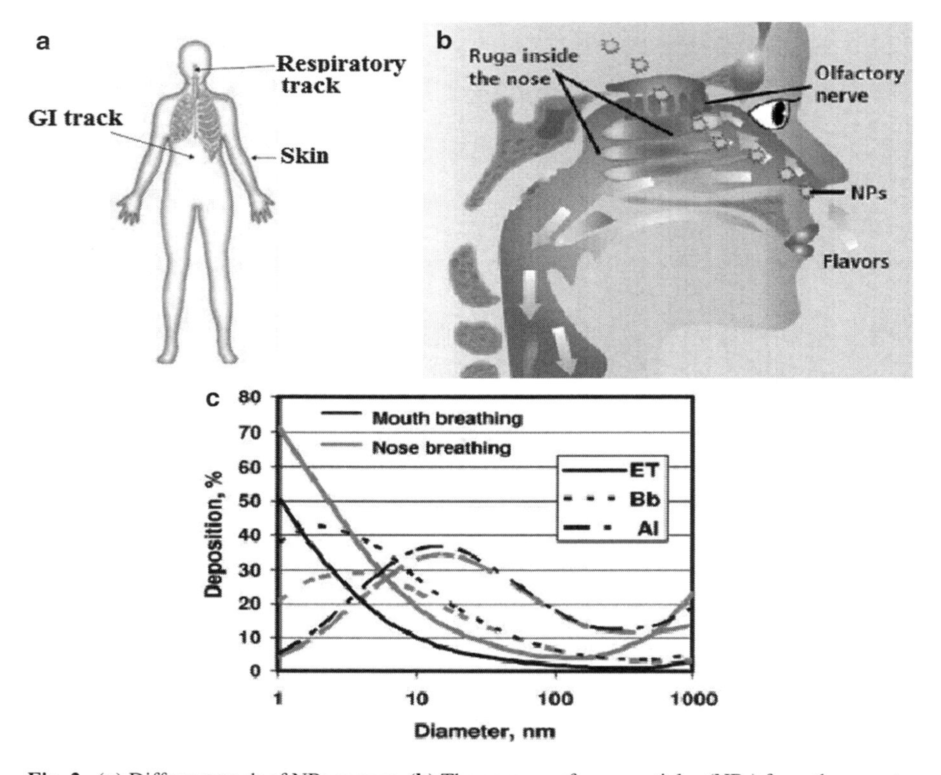

Fig. 2 (**a**) Different track of NPs expose. (**b**) The passage of nanoparticles (NPs) from the nose to the cerebral system via the cibriform plate, which separates the nasal sinus from the brain and protects the nasal nerves and nervous receptors. (**c**) Deposition of NPs in mouth and nose breathing with diameter

2.3.2 Life Cycle Impact Assessment (LCIA)

We define LCIA [34, 35] as the phase of the LCA where inventory data on inputs and outputs are translated into indicators about the product system's potential impacts on the environment, on human health, and on the availability of natural resources.

Interpretation is the phase where the results of the LCI and LCIA are interpreted according to the goal of the study and where sensitivity and uncertainty analysis are performed to qualify the results and the conclusions (Fig. 3).

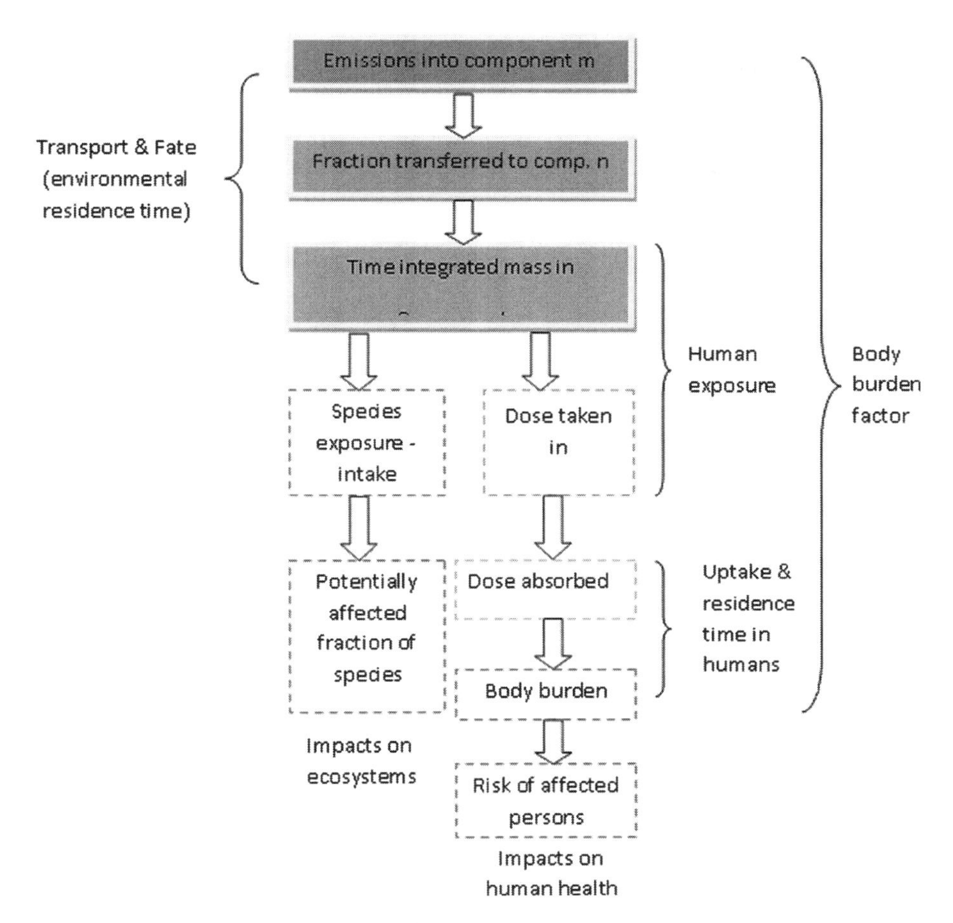

Fig. 3 Framework for assessment of toxic impacts in LCA

2.4 Damage Amount Analysis: A Case Study of SWNT (Fig. 4)

2.5 The Risk Model: Evaluation, Characterisation, and Reduction Using Genetic Algorithm (Fig. 5)

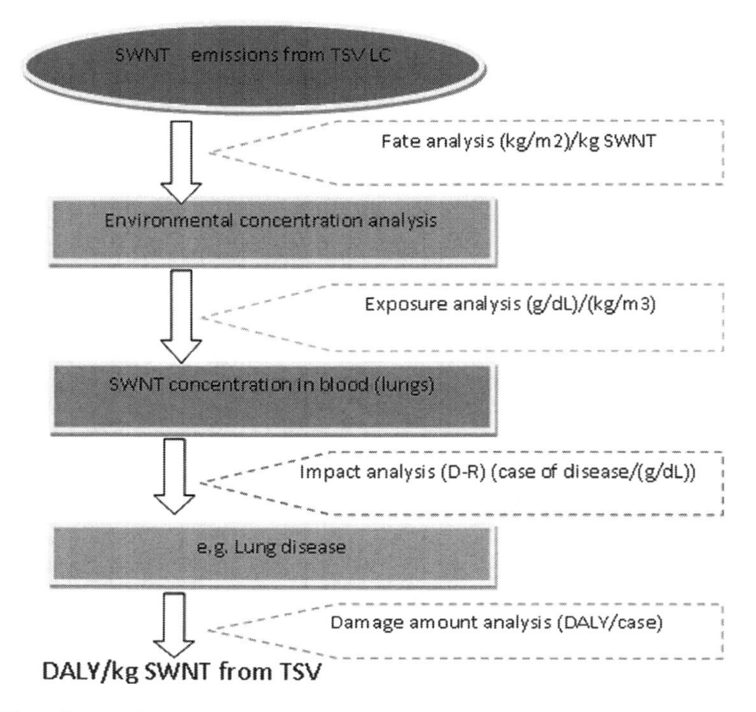

Fig. 4 Flow diagram for emitted SWNT from TSV (DALY/kg)

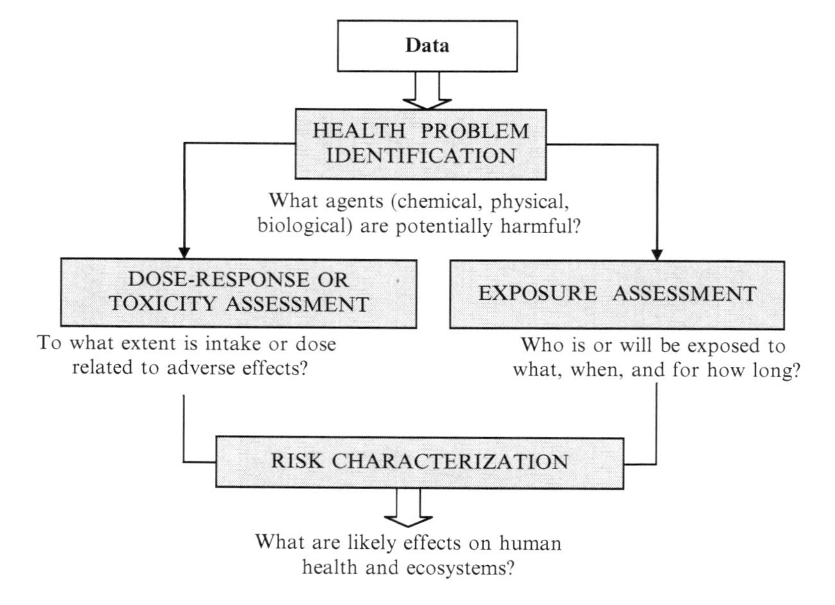

Fig. 5 Data analysis flow chart for risk evaluation for human health and ecosystem

3 Experimental Details

In this study, diatomaceous earth (DE) was used to design amorphous nano-sized hydrophilic, hydrophobic, and lipophilic silica in 15–30 nm size range. Surface-functionalized silica nanoparticles (SNP) are a viable alternative to conventional pesticides. Entomotoxicity of SNP was tested against rice weevil, *Sitophilus oryzae*, and its efficiency was compared with bulk-sized silica (individual particle larger than 1 μm). Amorphous SNP was found to be highly effective against this insect pest causing more than 90% mortality, indicating the effectiveness of SNP to control insect pests. On one hand, we studied the health and environmental hazards of NPs, and parallelly, we established the entomotoxic effect of silica NPs against *Sitophilus oryzae*. Insects *S. oryzae* were reared on whole rice grain (IR64) at 30C±1C, 75±5% r.h. in continuous darkness (insects were inbred in our laboratory by sib-mating for 20 generation). The r.h. was maintained by using saturated solution of sodium chloride. Adults less than 2 weeks old were used for the experiments.

Hydrophilic spherical SNPs of different size were synthesised in the laboratory by sol-gel method from aqueous alcohol solution of silicon alkoxide. First, aqueous ethanolic solution of silicon alkoxide was subjected to ultrasonication for 10 min. Next, a known volume of TEOS was added and the mixture was sonicated for another 10 min. At the end of the process, 24% NH_4OH was added as a catalyst to promote the condensation reaction. Sonication was continued for a further 30 min to get a white turbid suspension. The reaction was performed at room temperature. The size of the spherical monodispersed SNP could be varied by varying the concentration of the reactants. The size of the SNP obtained depends on the chain length of the alcohol used. The SNPs were lyophilized after synthesis and washed several times with double distilled water to remove all trace of ammonium hydroxide used in the process of synthesis as catalyst. Particle size of SNP synthesised in our laboratory was measured by field emission scanning electron microscope (FE-SEM, FEI Quanta 200 F, FEI, USA) in the central instrumental facility of Indian Institute of Technology (IIT), Roorkee, India (Fig. 6).

The bioassay on *S. oryzae* was performed in small plastic screw-capped jars. Twenty grams of rice (IR64) was placed in each jar. Rice in each jar was treated individually with custom-made SNPs (hydrophilic, hydrophobic, and lipophilic), laboratory-made hydrophilic modified stober SNP (average particle size 30 nm), and bulk size silica at three doses rate 0.5, 1, and 2 g kg^{-1} rice. The jars were kept for 24 h before 20 unsexed adults of *S. oryzae* were introduced into each jar. All bioassays were performed at 30°C±1°C, 75±5% r.h. Insect mortality was checked after 1, 2, 3, 7, and 14 days. Data analysis has been performed as per Table 1, 2, 3, 4, and 5 where insect mortality as the response variable and treatment, dose, and exposure interval were the main effects.

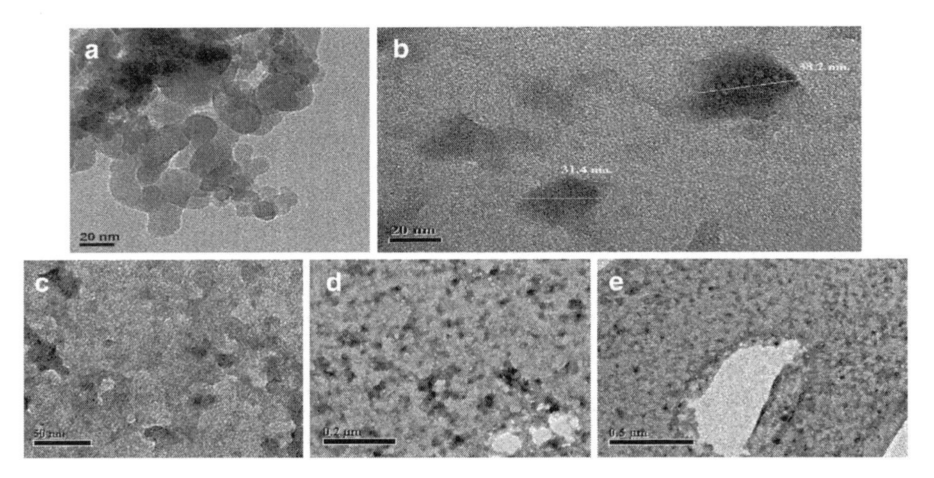

Fig. 6 Electron micrographs of custom-made silica nanoparticles. (**a**) Average diameter 20 nm. (**b**) Individual silica NPs 31.4 and 38.2 nm. (**c**) Average diameter 50 nm. (**d**) Average diameter 0.2 µm. (**e**) Average diameter 0.5 µm

Nano silica in insect pests shows nearly 100% mortality. The nanoparticle surface is proposed to be modified by coating the particle with a chemical. These surface changes can totally alter the toxicity of a specific product and have a major impact on nanoparticles' toxicity or safety.

4 Discussion

Nanotechnology is a double-edge sword [36], the same novel properties making nanoparticles attractive, which makes them potentially toxic. In the risk assessment study, our hypothesis at this point is that increase in surface area is the main reason that is associated with nanoparticle exposure hazards [20] since this increase in surface area causes nanoparticles to become more active and show toxic properties that we did not anticipate for such materials, and this surface increase also makes nanoparticles more flammable, increasing the risk of fire and explosion in workplaces. So determining the amount of size and shape [19] of nanoparticle (called critical size and critical shape) existing in a workplace is a necessary hazard prevention step. By identifying leakage sources and applying essential critical steps, we can control and reduce exposure risks associated with nanoparticles. Our occupational safety guidelines can minimise exposure, but until more is known about the potential hazards of nanomaterial, safe-handling practices may be inadequate. There has to be compromise between testing all the possible scenarios for each nanoparticle and creation of standards to unify tests.

Table 1 ANOVA parameters for main effects and their associated interaction

Source	df	Day1		Day2		Day3		Day7		Day14	
		F	P	F	P	F	P	F	P	F	P
Treatment	4	78.15	<0.001	88.07	<0.001	117.21	0.001	94.24	<0.001	121.7	<0.001
Dose	3	282.61	<0.001	593.98	<0.001	786.19	<0.001	667.39	<0.001	809.98	<0.001
Treatment × dose	12	16.87	<0.001	23.74	<0.001	25.69	<0.001	21.25	<0.001	24.04	<0.001

Table 2 Mean mortality (±SE.) of *Sitophilus oryzae* adults exposed for 1 day on rice treated with bulk and nano silica at three dose rates with control

Nanoparticle	0 g/kg	0.5 g/kg	1 g/kg	2 g/kg
SiO_2 – hydrophilic	0.0 ± 0.0 Aa	14.2 ± 6.6 Ba	67.0 ± 8.4 Ca	86.0 ± 7.4 Da
SiO_2 – hydrophobic	0.0 ± 0.0 Aa	6.0 ± 8.2 Aa	7.0 ± 7.6 Ab	42.0 ± 7.6 Bb
SiO_2 – lipophilic	0.0 ± 0.0 Aa	7.9 ± 4.2 Aa	7.0 ± 5.7 Ab	34.0 ± 8.2 Bb
SiO_2 – (modified Stober)	0.0 ± 0.0 Aa	13.2 ± 5.8 Aa	61.0 ± 8.2 Ba	81.0 ± 4.2 Ca
SiO_2 – bulk	0.0 ± 0.0 Aa	3.0 ± 4.5 Aa	7.0 ± 5.7 Ab	17.0 ± 2.7 Bc

Within each column, means followed by the same lower case letter are not significantly different, within each row means followed by the same upper case letter are not significantly different; Tukey–Kramer HSD test; $P = 0.05$

Table 3 Mean mortality (±SE) of *Sitophilus oryzae* adults exposed for 2 days on rice treated with bulk and nano silica at three dose rates with control

Nanoparticle	0 g/kg	0.5 g/kg	1 g/kg	2 g/kg
SiO_2 – hydrophilic	1.0 ± 2.3 Aa	23.3 ± 7.7 Ba	89.0 ± 4.2 Ca	95.0 ± 3.5 Da
SiO_2 – hydrophobic	1.0 ± 2.3 Aa	34.0 ± 6.5 Ba	49.0 ± 9.6 Cb	97.0 ± 2.7 Da
SiO_2 – lipophilic	1.0 ± 2.3 Aa	49.5 ± 8.4 Bb	48.0 ± 5.7 Bb	97.0 ± 4.5 Ca
SiO_2 – (modified Stober)	1.0 ± 2.3 Aa	22.3 ± 5.9 Ba	82.0 ± 5.7 Ca	92.0 ± 7.6 Ca
SiO_2 – bulk	1.0 ± 2.3 Aa	7.0 ± 4.5 Ac	11.0 ± 4.2 Ac	23.0 ± 4.5 Bb

Within each column, means followed by the same lower case letter are not significantly different, within each row means followed by the same upper case letter are not significantly different; Tukey–Kramer HSD test; $P = 0.05$

Table 4 Mean mortality (±SE) of *Sitophilus oryzae* adults exposed for 7 days on rice treated with bulk and nano silica at three dose rates with control

Nanoparticle	0 g/kg	0.5 g/kg	1 g/kg	2 g/kg
SiO_2 – hydrophilic	2.0 ± 2.7 Aa	35.4 ± 8.3 Ba	95.0 ± 5.0 Ca	97.0 ± 2.7 Ca
SiO_2 – hydrophobic	2.0 ± 2.7 Aa	62.0 ± 9.1 Bb	86.0 ± 8.2 Ca	100.0 ± 0.0 Da
SiO_2 – lipophilic	2.0 ± 2.7 Aa	62.4 ± 5.6 Bb	71.0 ± 8.9 Bb	100.0 ± 0.0 Da
SiO_2 – (modified Stober)	2.0 ± 2.7 Aa	35.4 ± 8.3 Ba	94.0 ± 4.2 Cab	97.0 ± 2.7 Ca
SiO_2 – bulk	2.0 ± 2.7 Aa	16.5 ± 5.5 Bc	21.9 ± 6.5 Bc	34.0 ± 5.5 Cb

Within each column, means followed by the same lower case letter are not significantly different, within each row means followed by the same upper case letter are not significantly different; Tukey–Kramer HSD test; $P = 0.05$

Table 5 Mean mortality (±SE) of *Sitophilus oryzae* adults exposed for 14 days on rice treated with bulk and nano silica at three dose rates with control

Nanoparticle	0 g/kg	0.5 g/kg	1 g/kg	2 g/kg
SiO_2 – hydrophilic	4.1 ± 2.3 Aa	42.5 ± 9.1 Ba	96.0 ± 4.2 Ca	100.0 ± 0.0 Ca
SiO_2 – hydrophobic	4.1 ± 2.3 Aa	69.0 ± 9.6 Bb	92.0 ± 6.7 Ca	100.0 ± 0.0 Ca
SiO_2 – lipophilic	4.1 ± 2.3 Aa	69.2 ± 5.8 Bb	89.0 ± 2.2 Ca	100.0 ± 0.0 Ca
SiO_2 – (modified Stober)	4.1 ± 2.3 Aa	41.5 ± 9.7 Ba	95.0 ± 3.5 Ca	99.0 ± 2.2 Ca
SiO_2 – bulk	4.1 ± 2.3 Aa	23.0 ± 5.7 Bc	25.0 ± 6.1 Cb	40.0 ± 6.1 Db

Within each column, means followed by the same lower case letter are not significantly different, within each row means followed by the same upper case letter are not significantly different; Tukey–Kramer HSD test; $P = 0.05$

Although progress has recently been made towards understanding the health and environmental consequences of these materials, challenges remain for future research.

Acknowledgements We are thankful particularly to all engineers and staff of Institute of Cybernetics Systems and Information Technology (ICSIT), all our colleagues at the Biological Science Division, Indian Statistical Institute, Kolkata, ECSU, Indian Statistical Institute, Kolkata, and Department of Radiophysics and Electronics, Calcutta University, Kolkata.

References

1. Wagner V, Dullaart A, Bock AK, Zweck A (2006) The emerging nanomedicine landscape. Nat Biotechnol 24(10):1211–1217
2. Mansson A, Sundberg M, Bunk R, Balaz M, Nicholls IA, Omling P, Tegenfeldt JO, Tagerud S, Montelius L (2005) Actin-based molecular motors for cargo transportation in nanotechnology – potentials and challenges. IEEE Trans Adv Packag 28(4):547–555
3. Matsui Y (2010) Cross-sectional risk assessment of various nano materials and production stages. In: Nanotechnology (IEEE-NANO), 2010 10th IEEE conference, pp 188–191
4. Royal Society, Royal Academy of Engineering (2004) Nanoscience and nanotechnologies: opportunities and uncertainties. The Royal Society and The Royal Academy of Engineering, London
5. Marra J, Den Brink W, Goossens H, Kessels S, Philips Res. Labs, Eindhoven (2009) Nanoparticle monitoring for exposure assessment. IEEE Nanotechnol Mag 3(2):6–37
6. Maynard AD, Pui DYH (2007) Nanoparticles and occupational health. Springer, Dordrecht
7. Vauthey S, Santoso S, Gong H, Watson N, Zhang S (2002) Molecular self-assembly of surfactant-like peptides to form nanotubes and nanovesicles. Proc Natl Acad Sci USA 99 (8):5355–5360
8. Drexler E, Smalley R (1993) Nanotechnology: Drexler and Smalley make the case for and against 'molecular assemblers'. Chem Eng News 81(48):37–42
9. Drexler KE (1985) Engines of creation: the coming era of nanotechnology. Forth Estate, London
10. Drexler KE (1992) Nanosystems: molecular machinery, manufacturing, and computation. Wiley, New York
11. Moore MN (2006) Do nanoparticles present ecotoxicological risks for the health of the aquatic environment? Environ Int 32(8):967–976
12. Arora S, Jain J, Rajwade JM, Paknikar KM (2009) Interactions of silver nanoparticles with primary mouse fibroblasts and liver cells. Toxicol Appl Pharmacol 236(3):310–318
13. Schroeder U, Sommerfeld P, Ulrich S, Sabel BA (1998) Nanoparticle technology for delivery of drugs across the blood–brain barrier. J Pharm Sci 87(11):1305–1307, Wiley
14. Greenemeier L (2008) Study says carbon nanotubes as dangerous as asbestos. Scientific American, 20 May 2008. http://www.sciam.com/article.cfm?id = carbon-nanotube-danger
15. Marissa D, Newman MD, Mira Stotland MD, Jeffrey I, Ellis MD (2009) The safety of nanosized particles in titanium dioxide– and zinc oxide–based sunscreens. J Am Acad Dermatol Sci Direct 61(4):685–692
16. Marambio-Jones C, Hoek EMV (2008) A review of the antibacterial effects of silver nanomaterials and potential implications for human health and the environment. J Nanopart Res 12(5):1531–1551
17. Choi O, Hu Z (2008) Size dependent and reactive oxygen species related nanosilver toxicity to nitrifying bacteria. Environ Sci Technol ACS 42(12):4583–4588

18. Wilson RF (2006) Nanotechnology: the challenge of regulating known unknowns. J Law Med Ethics 34(4):704–713, Wiley
19. Dutta Majumder D, Karan S, Goswami A (2011) Characterization of gold and silver nanoparticles using it's color image segmentation and feature extraction using fuzzy C-means clustering and generalized shape theory. In: Proceeding of the IEEE international conference on communications and signal processing 2011, pp 70–74
20. Duttamajumder D, Karan S, Goswami A (2011) Synthesis and characterization of gold nanoparticle – a fuzzy mathematical approach. In: PReMI2011, Springer, LNCS 6744, pp 324–332
21. Fadok VA, Laszlo DJ, Noble PW, Weinstein L, Riches DW, Henson PM (1993) Particle digestibility is required for induction of the phosphatidylserine recognition mechanism used by murine macrophages to phagocytose apoptotic cells. J Immunol Am Assoc Immunol 151 (8):4274–4285
22. Clift MJD, Rothen-Rutishauser B, Brown DM, Duffin R, Donaldson K, Proudfoot L, Guy K, Stone V (2008) The impact of different nanoparticle surface chemistry and size on uptake and toxicity in a murine macrophage cell line ScienceDirect. Toxicol Appl Pharm 232(3):418–427
23. Lee KP, Seidel WC (1991) Pulmonary response of rats exposed to polytetrafluoroethylene and tetrafluoroethylene hexafluoropropylene copolymer fume and isolated particles. Inhal Toxicol 3(3):237–264
24. Taylor DM, De Oliveira ON Jr, Morgan H (1990) Models for interpreting surface potential measurements and their application to phospholipid monolayers. J Colloid Interface Sci 139 (2):508–518, Elsevier
25. Karakoti S, Hench LL, Seal S (2006) The potential toxicity of nanomaterials—the role of surfaces. JOM J Miner, Met Mater Soc 58(7):77–82
26. Alice S, David E, Liqing R, Dongqing L (2003) Zeta-potential measurement using the Smoluchowski equation and the slope of the current–time relationship in electroosmotic flow. J Colloid Interface Sci 261(2):402–410, Elsevier
27. Dutta Majumder D (1995) A study on a mathematical theory of shapes in relation to pattern recognition and computer vision. Indian J Theor Phys 43(4):19–30
28. Yuliang Zhao, Hari Singh Nalwa (2006) Nanotoxicology – interactions of nanomaterials with biological systems. American Scientific Publishers, Stevenson Ranch
29. Stark WJ (2011) Nanoparticles in biological systems. Angew Chem Int Ed 50(6):1242–1258, General & Introductory Chemistry, Wiley
30. Hellweg S, Hofstetter TB, Hungerbuhler K (2005) Time-dependent life-cycle assessment of emissions from slag landfills with the help of scenario analysis. J Clean Prod 13(3):301–320
31. Hellweg S, Fischer U, Scheringer M, Hungerbuhler K (2004) Environmental assessment of chemicals: methods and application to a case study of organic solvents. Green Chem 6 (8):418–427
32. Ciambrone DF (1997) Environmental life cycle analysis. Lewis Publishers, Inc., Boca Raton
33. Keoleian GA, Menerey D (1994) Product life cycle assessment to reduce health risks and environmental impacts. Noyes Publications, Boca Raton
34. Seppala J (1997) Decision analysis as a tool for life cycle impact assessment. Finnish Environment Institute, Helsinki
35. Sonnemann G (2003) Integrated life-cycle and risk assessment for industrial processes. CRC Press, London
36. Dreher KL (2004) Health and environmental impact of nanotechnology: toxicological assessment of manufactured nanoparticles. Toxicol Sci 77:3–5

Probabilistic Assessment of Strengths of Corrosion-Affected RC Beams

Kapilesh Bhargava, Yasuhiro Mori, and A.K. Ghosh

Abstract Corrosion of reinforcement causes premature deterioration in reinforced concrete (RC) structures and reduces their intended residual service life. Damages to RC structures due to reinforcement corrosion generally manifest in the form of expansion, cracking and eventual spalling of the cover concrete, loss of steel cross-sectional area and loss of bond between corroded reinforcement and surrounding cracked concrete. These damages may sometime result in structural failure. This chapter initially presents predictive models for time-dependent damages in corrosion-affected RC beams, recognized as loss of mass and cross-sectional area of reinforcing bar and loss of concrete section owing to the peeling of cover concrete. Then these models have been used to present analytical formulations for evaluating time-dependent flexural and shear strengths for corroded RC beams based on the standard composite mechanics expressions for RC sections. Further, by considering variability in the identified basic variables that could affect the time-dependent strengths of corrosion-affected RC beams, an attempt is made in this chapter to present simple estimations for the time-dependent mean strengths and time-dependent coefficient of variation (c.o.v.) associated with the strengths for a typical simply supported RC beam. Comparison of presented simple estimations of mean strengths and c.o.v. associated with strengths has been made with those obtained using Monte Carlo simulation.

K. Bhargava (✉)
Architecture & Civil Engineering Division, Bhabha Atomic Research Center, Mumbai, India
e-mail: kapilesh_66@yahoo.co.uk; kapil_66@barc.gov.in

Y. Mori
Graduate School of Environmental Studies, Nagoya University, Nagoya, Japan
e-mail: yasu@sharaku.nuac.nagoya-u.ac.jp

A.K. Ghosh
Health Safety and Environment Group, Bhabha Atomic Research Center, Mumbai, India
e-mail: ccss@barc.gov.in

S. Chakraborty and G. Bhattacharya (eds.), *Proceedings of the International Symposium on Engineering under Uncertainty: Safety Assessment and Management (ISEUSAM - 2012)*, DOI 10.1007/978-81-322-0757-3_23, © Springer India 2013

Keywords Reinforcement corrosion • Annual mean corrosion rate • Time-dependent flexural strength • Time-dependent shear strength • Time-dependent c.o.v. Latin hypercube sampling

1 Introduction

The basic safety requirements of a nuclear power plant (NPP) include the safe shutdown of the reactor, to remove decay heat and to limit the release of radioactivity to the environment. Safety evaluation of the structures is an important issue for any NPP, which has to be carried out to take care of number of factors including the ageing effects, if any. As NPP structures age, a number of degradation mechanisms start affecting the load carrying capacity and serviceability of these structures. Some of the degradation mechanisms for RC structures include corrosion of steel reinforcement, alkali-silica reaction, freeze-thaw damage, sulphate attack, etc. Out of these mechanisms, corrosion of steel has been identified as being the most widespread and predominant mechanism responsible for the deterioration of RC structures. Corrosion causes the reduction of reinforcement cross-sectional area which in extreme forms, can be significant enough to reduce the strength of structural members below the minimum requirements. It also results in cracking and spalling of cover concrete due to formation of expansive of corrosion products, and reduction of bond between the corroded reinforcement and concrete, thereby resulting in further structural damage. By proper control and monitoring of the reinforcement corrosion, premature failure of RC structures can be prevented. Also the assessment of performance of corrosion-damaged RC structures to withstand extreme events during their anticipated service life would help in arriving decisions pertaining to the inspection, repair, strengthening, replacement and demolition of such structures.

In this chapter initially, predictive models are presented for the quantitative assessment of time-dependent damages in RC beams, recognized as loss of concrete section owing to the peeling of cover concrete and loss of mass and cross-sectional area of reinforcing bar. Then these models have been used to present analytical formulations for evaluating time-dependent flexural and shear strengths of corroded RC beams based on the standard composite mechanics expressions for RC sections. For the corroded RC beams, loss of flexural and shear strengths would be mainly due to loss of cross section for concrete and reinforcing steel. The scope of flexural strength estimation has been limited to either by the yielding of tensile reinforcement or by the crushing of concrete in compression zone. Although the continued rebar corrosion would also affect the composite action of concrete and reinforcing steel due to bond deterioration between them, the evaluation of time-dependent flexural strength due to loss of bond has not been considered in the present study. The performance of the presented formulations has been evaluated through their ability to reproduce the available experimental trends. This chapter further presents probabilistic assessment of time-dependent strengths for a typical simply supported corroded RC beam. The basic variables that can affect time-dependent strengths for a corroded RC beam are identified as material strengths of concrete and reinforcement, modulus of elasticity of reinforcing steel plus expansive corrosion products combine, creep coefficient for concrete, dimensions of the beams and annual mean corrosion rate. By considering variability in these variables, simple estimations of following are presented: (1) time-dependent mean strength and c.o.v. associated with the strength and (2) time-dependent mean degradation function and c.o.v. associated with the degradation function, wherein, the degradation function is defined as the ratio of strength at time, t, to the initial strength for an un-corroded RC beam. The estimation of time-dependent strengths and degradation functions are carried out for two limit states: (a) flexural failure and (b) shear failure. An attempt has also been made to present analytical models for estimating time-dependent c.o.v. associated with the degradation functions. The performance of simple estimations of mean strengths and c.o.v. associated with strengths and degradation functions has been evaluated by comparing their results with those obtained using Monte Carlo simulation.

2 Corrosion Propagation in RC Beams

A quantitative description of corrosion propagation is generally given in terms of the loss of metal per unit surface area per unit time, and this can be obtained by measuring the mass differences in the reinforcing steel with reference to its surface area exposed to corrosion. Most of the non-destructive techniques used for the corrosion monitoring are based on the electrochemical measurements, in which the annual mean corrosion rate is estimated in terms of the corrosion current density, i_{COR} [39]. This i_{COR} can be transformed into the loss of metal by using the diffusion law related to the growth of expansive corrosion products [6–8, 26]. Reduction in cross-sectional area of reinforcing bars shall result in the reduced flexural and shear strengths of the corrosion-damaged RC beams [5]. Corrosion of reinforcement causes cracking, and eventually spalling and/or peeling of the cover concrete. Reduction in concrete section owing to the peeling of cover concrete shall also result in the reduced flexural and shear strengths of the corrosion-damaged RC beams [34]. Following subsections present estimation of time-dependent loss of concrete and steel sections for the purpose of corrosion propagation in RC beams.

2.1 Loss of Concrete Section

Reduction in concrete section occurs due to peeling of bottom, top and side covers to the reinforcements. Methodology for estimating the time required for peeling of cover concrete is adopted from Bhargava [5]. Figure 1 shows the crack propagating

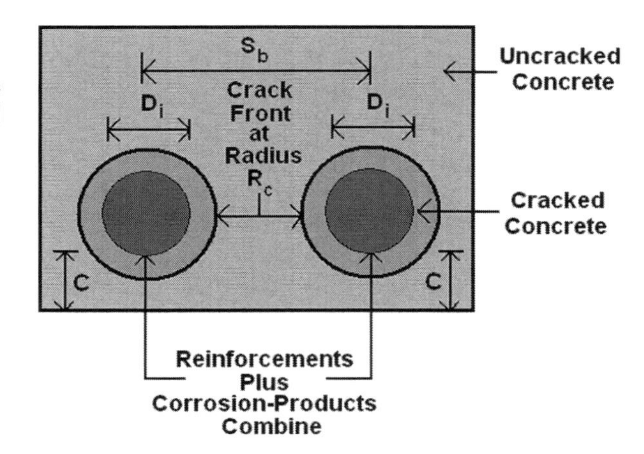

Fig. 1 Crack propagating condition for concrete block with two reinforcing bars due to reinforcement corrosion [5]

condition for the concrete block with two reinforcing bars of initial diameter D_i. The bars have clear cover, C, and the centre line spacing, S_b.

Due to the formation of corrosion products, the propagation of radial splitting cracks shall take place in all the directions to the same distance R_c, i.e. the radius of crack front. With $R_i = (D_i/2)$ and $R_o = (R_i + C)$, the cover concrete is assumed to be fully cracked when R_c becomes equal to R_o [8]. Bazant [3] reported that failure may occur in two different modes in case of R_c becoming equal to R_o: (1) when $C > (S_b - D_i)/2$, then the failure shall consist of peeling of cover concrete, and (2) when the spacing of bars, S_b, is large (say $S_b > 6 \cdot D_i$), the failure shall consist of inclined cracking. Based on the suggested failure philosophy of Bazant [3], the cover peeling time for top, bottom and side covers are evaluated using corrosion cracking model of Bhargava et al. [8].

2.2 *Loss of Reinforcing Steel*

Corrosion process is a dynamic process; growth of expansive corrosion products is given by Eq. (1) [6–8, 26].

$$\frac{dW_r}{dt} = \frac{k_p}{W_r} \tag{1}$$

where W_r = mass of expansive corrosion products (mg/mm), t = corrosion time (years) and k_p = function of rate of metal loss. k_p is expressed by Eq. (2) [8].

$$k_p = 2.48614 \cdot \pi \cdot D_i \cdot i_{\mathrm{COR}} \tag{2}$$

where D_i = initial diameter of reinforcement (mm) and i_{COR} = annual mean corrosion rate ($\mu A/\mathrm{cm}^2$). Various parameters associated with the loss of reinforcing steel at time, t, since initiation of corrosion are evaluated by Eqs. (3, 4, 5 and 6) [5].

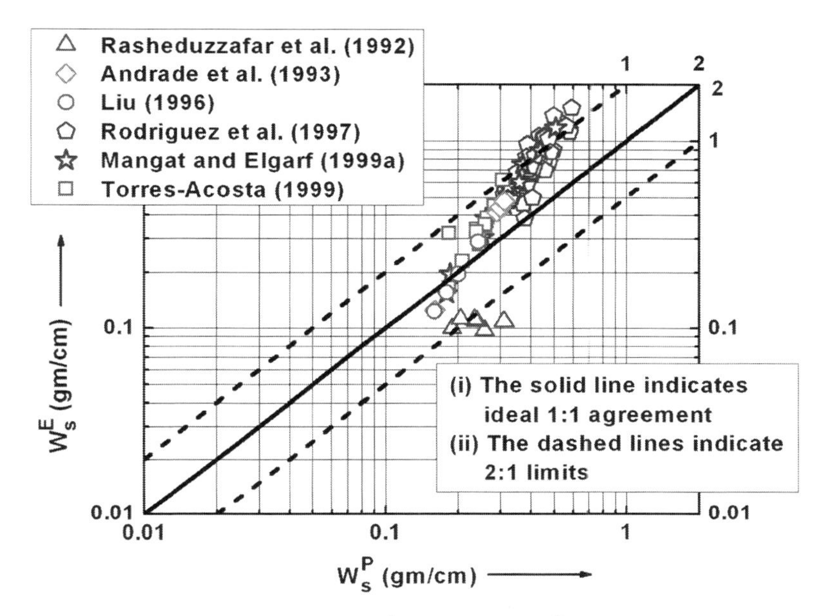

Fig. 2 Comparison between experimental W_s^E and predicted W_s^P using Eq. (3)

$$W_s(t) = 2.42362 \cdot \sqrt{D_i \cdot i_{\mathrm{COR}} \cdot t} \tag{3}$$

$$D_r(t) = \sqrt{D_i^2 - 0.39245 \cdot \sqrt{D_i \cdot i_{COR} \cdot t}} \tag{4}$$

$$A_{cor}(t) = 0.30835 \cdot \sqrt{D_i \cdot i_{\mathrm{COR}} \cdot t} \tag{5}$$

$$X(t) = \frac{1}{2} \cdot [D_i - D_r(t)] \tag{6}$$

where $W_s(t)$= mass of steel per unit length of the reinforcement (mg/mm) getting consumed by corrosion process, $D_r(t)$= reduced bar diameter (mm), $A_{\mathrm{cor}}(t)$= loss of cross-sectional area of steel (mm^2) and $X(t)$ = corrosion penetration depth (mm).

2.3 *Predictions for Loss of Reinforcing Steel: Comparison with Experimental Results*

Analytical predictions for W_s and X are made for the available experimental data by using Eqs. (3 and 6) [2, 25, 27, 33, 34, 38]. Figure 2 presents the comparison between W_s^P and W_s^E for the available experimental data. Similarly, Fig. 3 presents the comparison between X^P and X^E for the same experimental data. The superscripts

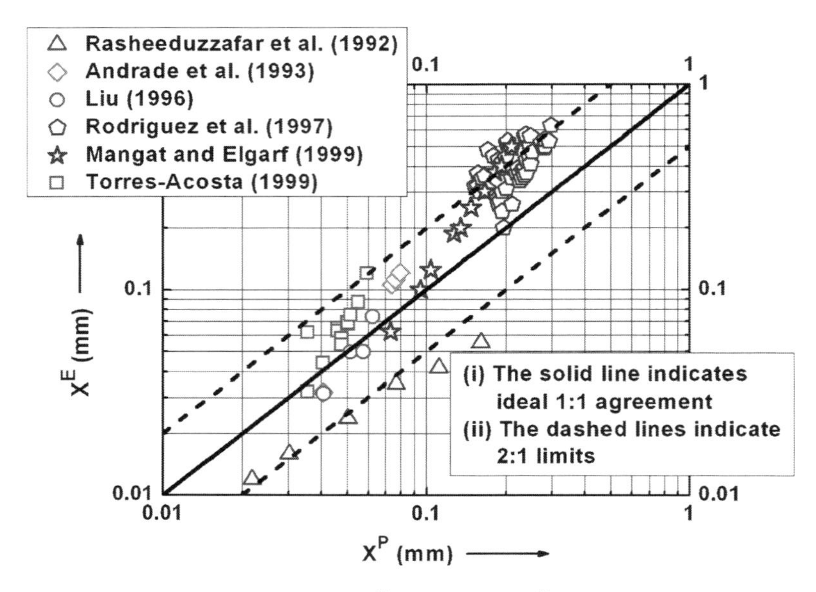

Fig. 3 Comparison between experimental X^{E} and predicted X^{P} using Eq. (6)

P and *E* correspond to the analytically predicted and the experimental observed values, respectively. The data in both figures are presented by different symbols to represent the analytical predictions made for different experimental data. It is clear from the same figures that the deviation between the analytically predicted and the experimentally observed values is generally less than by a factor of two, and this is a considerably good agreement in view of the large variability associated with the corrosion phenomena.

To test the goodness of Eqs. (3 and 6), the correlation between the predicted and experimental values is estimated. Assuming that $x = W_s^P$ (the independent variable), and $y = W_s^E$ (the dependent variable), the values of both $r^2{}_{xy}$ and $s^2{}_{yx}$ are estimated as 0.895 and 0.126 gm^2/cm, respectively, for Eq. (3), wherein r is the coefficient of correlation between x and y, and s is the root mean square error of estimate of y on x. Similarly, by assuming $x = X^P$ and $y = X^E$, the values of both $r^2{}_{xy}$ and $s^2{}_{yx}$ are estimated as 0.851 and 0.028 mm^2, respectively, for Eq. (6). The quite high values associated with $r^2{}_{xy}$ in both the predictions suggest that Eqs. (3 and 6) can be effectively used for estimating the values of W_s and X for the reinforced concrete members exposed to the corrosive environment.

3 Time-Dependent Strengths of Corroded RC Beams

Figure 4 shows typical cross section of a simply supported un-corroded RC beam, which is subjected to flexure and shear under loads. This doubly reinforced beam has b and D as its width and depth, respectively. The beam is reinforced with

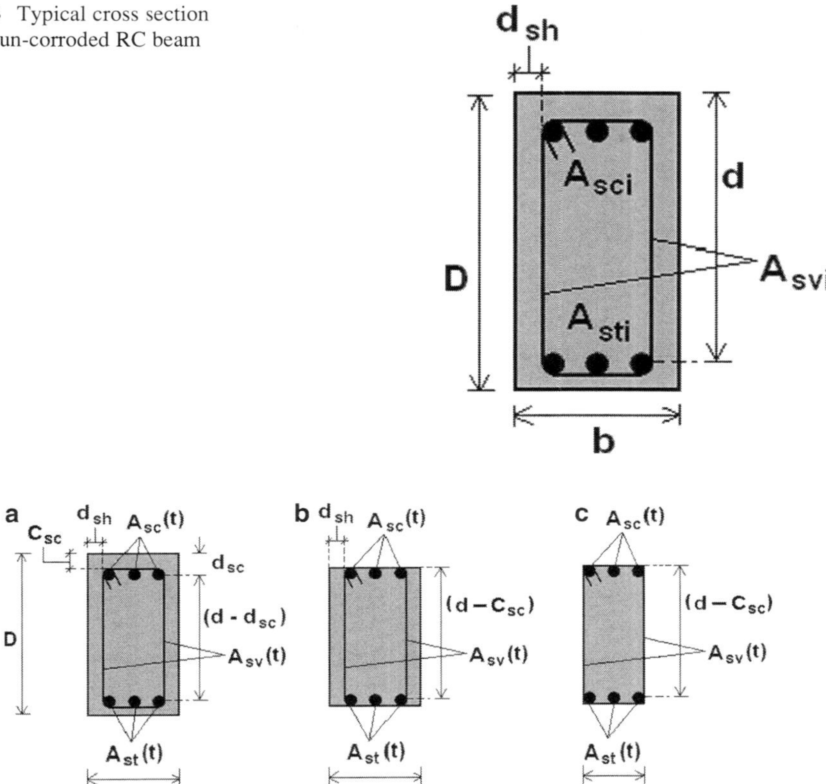

Fig. 4 Typical cross section of an un-corroded RC beam

Fig. 5 Different schemes of deteriorated reinforced concrete sections for corroded RC beams (**a**) Section 1, (**b**) Section 2, (**c**) Section 3 [34]

bottom tensile reinforcing steel bars having initial area, A_{sti}, and top compressive reinforcing steel bars having initial area, A_{sci}. The distance between the centroid of tensile steel and the edge of the compression zone is d (also known as effective depth). The shear stirrups are having an initial area, A_{svi}, and are provided at spacing, S_v.

Figure 5 shows three representative schemes of deteriorated reinforced concrete sections for a corrosion affected RC beam [34]. The Section 1 indicates the intact concrete section, wherein the peeling of covers is yet to occur. The Section 2 indicates the reduced concrete section due to the peeling of top and bottom covers. In Section 2, effective depth to tensile reinforcement of the beam is reduced to $(d - C_{sc})$, wherein C_{sc} is clear cover to the compression steel. The Section 3 indicates the reduced concrete section due to the peeling of all the top, bottom and side covers. In Section 3, effective depth to tensile reinforcement and effective width of

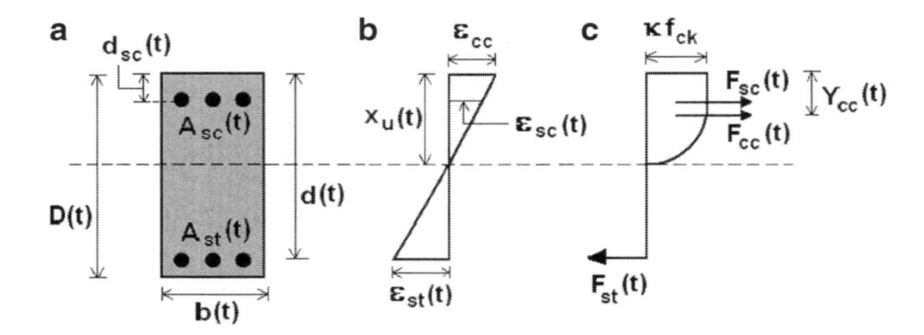

Fig. 6 Formulation of time-dependent flexural strength of corroded RC beams: (**a**) beam section, (**b**) strain distribution [9] and (**c**) stress distribution [9]

the beam are reduced to $(d - C_{sc})$ and$(b - 2 \cdot d_{sh})$, respectively, wherein d_{sh} is the clear cover to the shear stirrups. All the three sections also indicate the reduced sections of main bars and shear stirrups at time, t. It is very important to point out that the reduced concrete section for a deteriorated corroded beam at time, t, is governed by the individual cover peeling time for top, bottom and side covers, and in some cases, it may be different than those shown in Fig. 5. Formulations for time-dependent flexural and shear strengths of corroded RC beams are proposed with the considerations that the loss of strengths is mainly due to reduction in cross-sectional areas of reinforcing steel and concrete.

3.1 Time-Dependent Flexural Strength

Figure 6a shows the typical beam section for a doubly reinforced corroded beam, wherein the different notations pertaining to dimensions and reinforcing steels have their usual meanings at time, t, from the initiation of corrosion. Figure 6b and c present the strain and stress distribution across the cross section of the beam, respectively, wherein ε_{cc} is the ultimate strain in concrete and is taken as 0.0035 [9].

In the same figures, $\varepsilon_{st}(t)$ and $\varepsilon_{sc}(t)$ are the strains in tensile and compressive reinforcements, respectively; $x_u(t)$ is the height of compression zone; $F_{cc}(t)$ and $F_{sc}(t)$ are the forces of compression in concrete and compressive steel, respectively; $Y_{cc}(t)$ is the distance of point of application of $F_{cc}(t)$ from the edge of compression zone; $F_{st}(t)$ is the force of tension in tensile steel and f_{ck} is the 28-day characteristic cube compressive strength of concrete [9]. Uniform corrosion around the surface and along the length of the bar is assumed. Considering the simple bending theory, $\varepsilon_{st}(t)$ and $\varepsilon_{sc}(t)$are given by Eq. (7).

$$\varepsilon_{st}(t) = \left[\frac{d(t) - x_u(t)}{x_u(t)}\right] \cdot \varepsilon_{cc}; \varepsilon_{sc}(t) = \left[\frac{x_u(t) - d_{sc}(t)}{x_u(t)}\right] \cdot \varepsilon_{cc} \tag{7}$$

The total force of compression, $F_c(t)$, is given by Eq. (8).

$$F_c(t) = F_{cc}(t) + F_{sc}(t) \quad (8)$$

$F_{cc}(t)$ and $Y_{cc}(t)$ are given by Eq. (9).

$$F_{cc}(t) = \kappa \cdot f_{ck} \cdot b(t) \cdot x_u(t) \cdot \left[\frac{3 \cdot \varepsilon_{cc} - 0.002}{3 \cdot \varepsilon_{cc}}\right]; \quad Y_{cc}(t) = x_u(t) - x_u(t) \cdot \left[\frac{6 \cdot \varepsilon_{cc} - \left(\frac{0.000004}{\varepsilon_{cc}}\right)}{12 \cdot \varepsilon_{cc} - 0.008}\right] \quad (9)$$

where $\kappa = a$ factor which is decided based on the design compressive strength of the concrete in the structures and the partial safety factor appropriate to the material strength of concrete [9]. The force of compression in compressive steel, $F_{sc}(t)$, is given by Eq. (10).

$$F_{sc}(t) = \varepsilon_{sc}(t) \cdot E_{st} \cdot A_{sc}(t); \text{ for } \varepsilon_{sc}(t) \le \varepsilon_{sy}, \text{ and, } F_{sc}(t) = \eta \cdot f_y \cdot A_{sc}(t); \text{ for } \varepsilon_{sc}(t) > \varepsilon_{sy} \quad (10)$$

where f_y = yield strength of reinforcing steels, η = a factor which is decided based on the partial safety factor appropriate to material strength of reinforcing steels [9], E_{st} = modulus of elasticity of reinforcing steel and ε_{sy} = yield strain for the reinforcing steels = $(\eta \cdot f_y / E_{st})$. In the present study, both κ and η are considered as 1.0. The distance, $Y_c(t)$, of point of application of total force of compression, $F_c(t)$, from the edge of compression zone is given by Eq. (11).

$$Y_c(t) = \frac{F_{cc}(t) \cdot Y_{cc}(t) + F_{sc}(t) \cdot d_{sc}(t)}{F_{cc}(t) + F_{sc}(t)} \quad (11)$$

The force of tension, $F_{st}(t)$, in tensile steel is given by Eq. (12).

$$F_{st}(t) = \varepsilon_{st}(t) \cdot E_{st} \cdot A_{st}(t); \text{ for } \varepsilon_{st}(t) \le \varepsilon_{sy}, \text{ and, } F_{st}(t) = \eta \cdot f_y \cdot A_{st}(t); \text{ for } \varepsilon_{st}(t) > \varepsilon_{sy} \quad (12)$$

By equating the force of compression given by Eq. (8) and the force of tension given by Eq. (12), the height of compression zone $x_u(t)$ is evaluated. The flexural strength, $Mu(t)$, at time, t, is then determined by Eq. (13).

$$Mu(t) = F_{st}(t) \cdot [d(t) - Y_c(t)] \quad (13)$$

3.2 Time-Dependent Shear Strength

The permissible shear stress of concrete at time, t, from the initiation of corrosion is given by Eq. (14) [9, 35].

$$\tau_c(t) = \frac{0.85 \cdot \sqrt{0.8 \cdot f_{ck}} \cdot \left(\sqrt{1 + 5 \cdot \beta(t)} - 1\right)}{6 \cdot \beta(t)}; \beta(t) = \frac{0.8 \cdot f_{ck} \cdot b(t) \cdot d(t)}{689 \cdot A_{st}(t)} \quad (14)$$

The shear strength, $Vu(t)$, at time, t, is determined by Eq. (15).

$$Vu(t) = \tau_c(t) \cdot b(t) \cdot d(t) + \frac{\eta \cdot f_y \cdot A_{sv}(t) \cdot d(t)}{S_v} \quad (15)$$

The maximum value of the shear strength of the corroded RC beam, $Vu_{max}(t)$, is given by Eq. (16) [9].

$$Vu_{max}(t) = \tau_{c\,max} \cdot b(t) \cdot d(t) \quad (16)$$

where τ_{cmax} = maximum shear stress of concrete for a given value of f_{ck} [9]. If calculated $Vu(t)$ is more than $Vu_{max}(t)$, then $Vu(t)$ is limited to $Vu_{max}(t)$ [9].

3.3 Predictions for Time-Dependent Flexural and Shear Strengths of Corroded RC Beams: Comparison with Experimental Results

Predictions have been made for the residual flexural and shear capacity of the corrosion-degraded RC beams for which the experimental results are available [34]. Rodriguez et al. [34] tested six different types of RC beams of sections 150 × 200 mm with spans ranging from 2,050 to 2,300 mm. The beams were provided with different ratios of tensile and compressive reinforcement, different spacing of shear reinforcement and different locations of curtailment of tensile reinforcing bars. The various reinforcing bars were corroded to different degree of corrosion in terms of attack penetration, i.e. reduction in bar radius. After having corroded the reinforcement, the beams were tested up to the failure.

Figures 7 and 8 present the comparison of experimentally observed and analytically predicted values for time-dependent flexural and shear strengths, respectively, for type 13 beams of Rodriguez et al. [34]. Analytical predictions are presented for all the three deteriorated schemes for RC sections (Fig. 5) for comparison purposes. However, for these beams, Section 2 deteriorated scheme (Fig. 5) is expected at the end of the corrosion period. Analytical predictions are found to agree within 17% of the experimentally observed values for flexural and shear strengths for Section 2

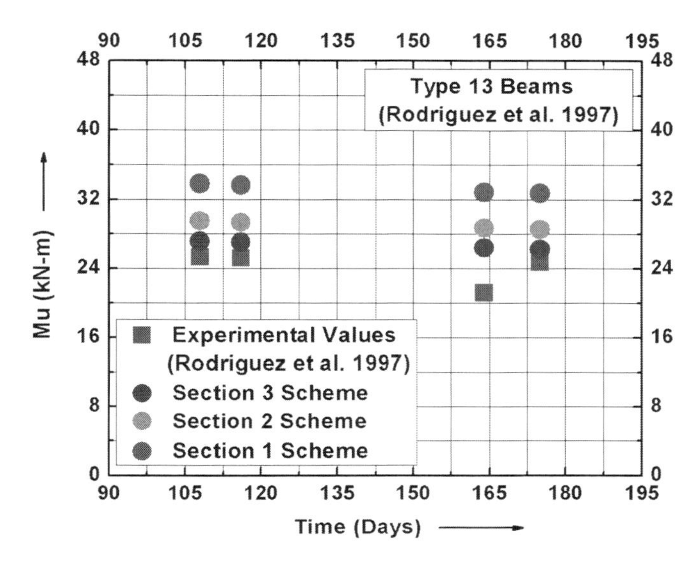

Fig. 7 Time-dependent flexural strength for type 13 corroded beams of Rodriguez et al. [34]

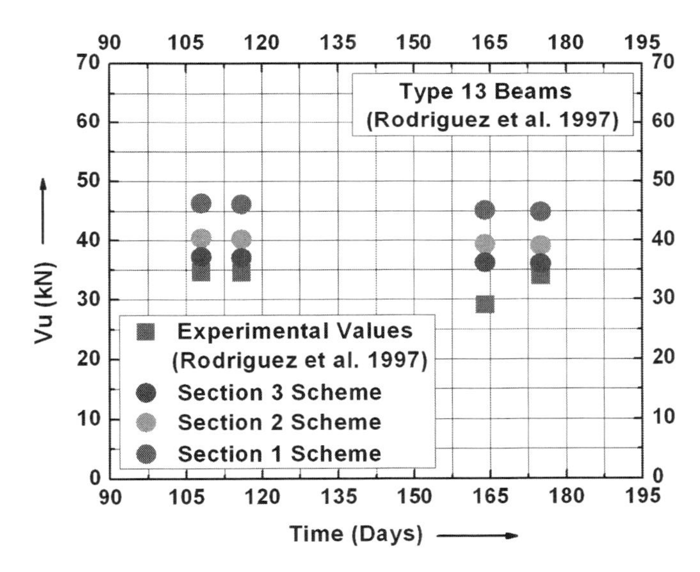

Fig. 8 Time-dependent shear strength for type 13 corroded beams of Rodriguez et al. [34]

deteriorated scheme; this is a considerably good agreement in view of the large variability associated with the corrosion phenomena. Therefore, the proposed analytical formulations for time-dependent flexural and shear strengths predict the analytical trends which are in considerably good agreement with those of the observed experimental trends.

4 Illustration of Time-Dependent Strengths of Corroded RC Beams: Probabilistic Approach

For the purpose of illustration, a simply supported RC beam with its reinforcement details as shown in Fig. 9 is considered. The span of the beam is considered as 4.0 m, and it is subjected to corrosion attack. The statistical parameters for the basic variables for material strengths, dimensions and annual mean corrosion rate appropriate for the RC beam are given in Table 1. These statistical parameters are similar to the ones suggested by many researchers [1, 9–24, 29–32, 36, 37, 39–41].

Time-dependent flexural and shear strengths are estimated using the formulations given in the preceding sections. Since the formulations are provided based on BIS [9]

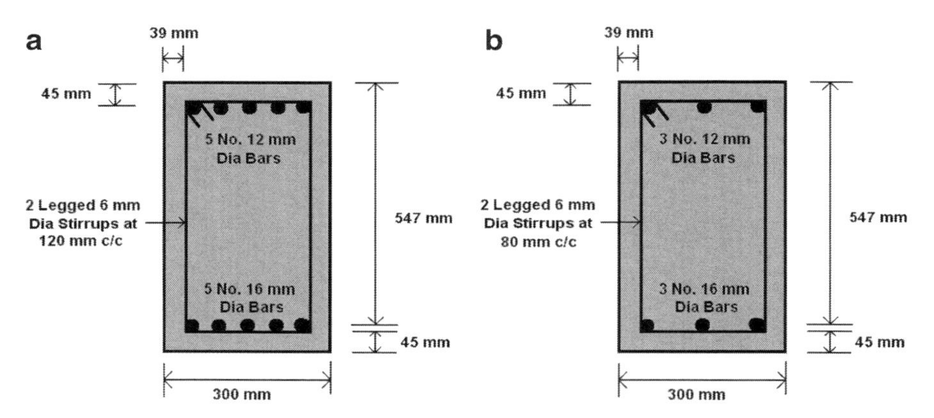

Fig. 9 Typical RC beam cross section: (**a**) at mid span and (**b**) at supports

Table 1 Statistical parameters for the basic variables for material strengths, dimensions and annual mean corrosion rate

Variables		Mean	c.o.v.	Distribution
Material strengths	f_{cm}	25.8 MPa	0.18	Normal
	τ_C	0.421 MPa	0.18	Normal
	θ	2.0	0.20	Normal
	f_y	466.88 MPa	0.11	Lognormal
	E_s	200,000 MPa	0.051	Lognormal
Dimensions	b	310.3 mm	0.033	Normal
	D	614.4 mm	0.017	Normal
	C_B	46.6 mm	0.123	Normal
	C_T	48.2 mm	0.105	Normal
	C_S	40.6 mm	0.099	Normal
Annual mean corrosion rate	i_{COR}	1, 3, 5 μA/cm^2	0.1, 0.2 and 0.3	Normal

Notations: f_{cm}: compressive strength of concrete, θ; creep coefficient, E_s; modulus of elasticity of steel plus corrosion products combine, C_B, C_T *and* C_S: clear covers to bottom, top and side reinforcements, respectively

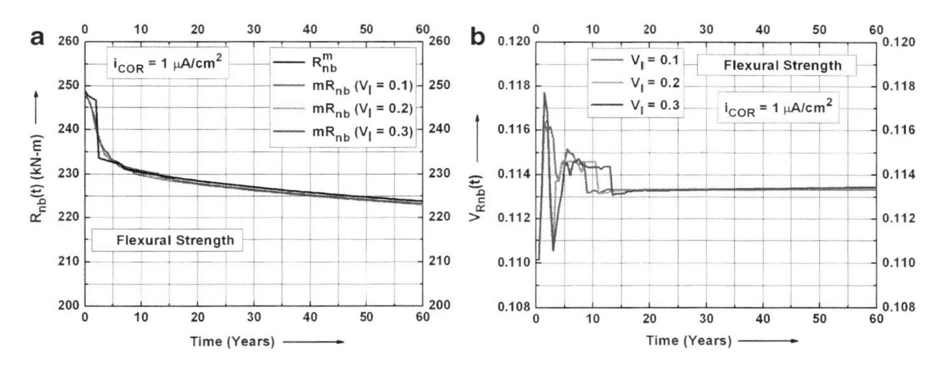

Fig. 10 (**a**): Time-dependent flexural strength (i_{COR} = 1 μA/cm²) (**b**): Time-dependent c.o.v. for flexural strength (i_{COR} = 1 μA/cm²)

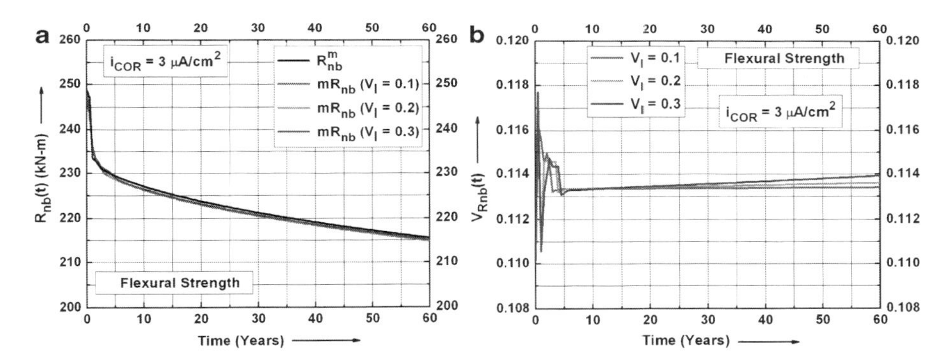

Fig. 11 (**a**): Time-dependent flexural strength (i_{COR} = 3 μA/cm²) (**b**): Time-dependent c.o.v. for flexural strength (i_{COR} = 3 μA/cm²)

and SP16 [35], following is adopted in the present study: (1) time-dependent flexural strength is calculated by using f_{cm} instead of $\kappa.f_{ck}$ and by using E_s instead of E_{st}, and (2) time-dependent shear strength is calculated by using f_{cm} instead of f_{ck}.

Monte Carlo simulation is used for evaluating time-dependent mean strengths and c.o.v. associated with the strengths. Here, the "Latin hypercube sampling (LHS)" technique with 40 samples [28] is used for an efficient sampling considering the variability in the variables given in Table 1. The results are presented in Figs. 10, 11, 12, 13, 14 and 15 and are discussed in the following sections. The 40 random samples of the basic variables and their random combinations obtained through LHS technique are used for evaluating the sample mean of strengths and c.o.v. associated with the strengths in Figs. 10, 11, 12, 13, 14 and 15. The approximated mean for strengths is evaluated by considering all the basic variables to be at their mean values. In all the figures, V_I stands for the c.o.v. of i_{COR}. Results are presented typically for a corrosion period of 60 years.

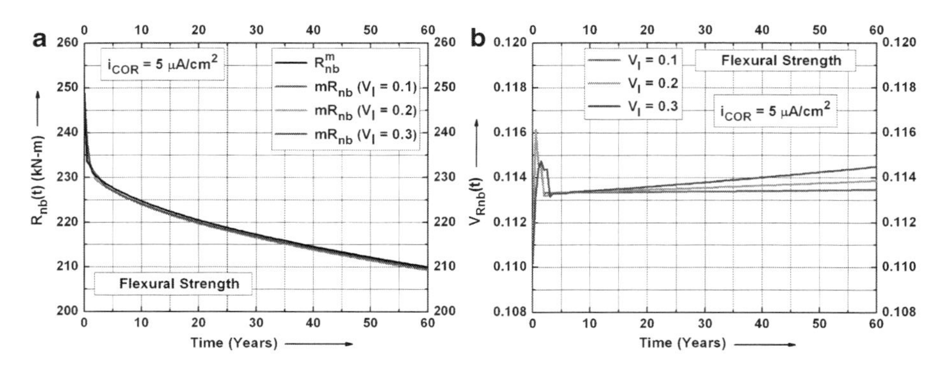

Fig. 12 (**a**): Time-dependent flexural strength (i_{COR} = 5 μA/cm^2) (**b**): Time-dependent c.o.v. for flexural strength (i_{COR} = 5 μA/cm^2)

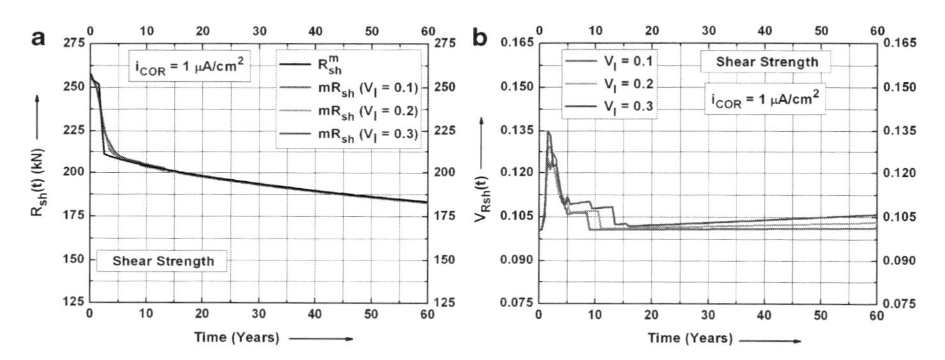

Fig. 13 (**a**): Time-dependent shear strength (i_{COR} = 1 μA/cm^2) (**b**): Time-dependent c.o.v. for shear strength (i_{COR} = 1 μA/cm^2)

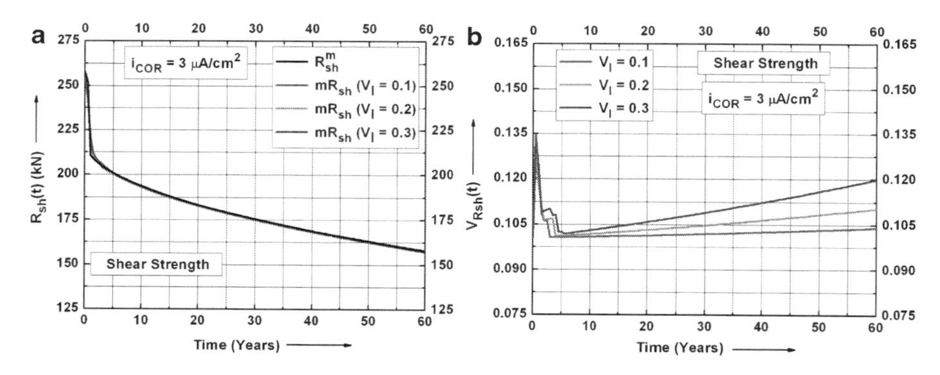

Fig. 14 (**a**): Time-dependent shear strength (i_{COR} = 3 μA/cm^2) (**b**): Time-dependent c.o.v. for shear strength (i_{COR} = 3 μA/cm^2)

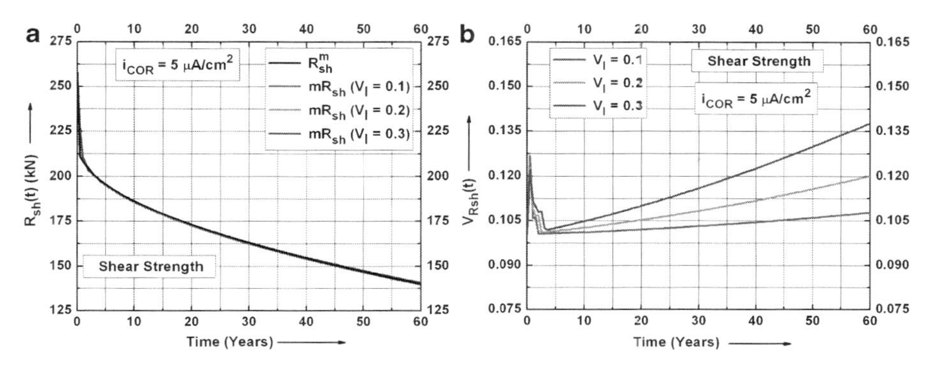

Fig. 15 (**a**): Time-dependent shear strength (i_{COR} = 5 μA/cm²) (**b**): Time-dependent c.o.v. for shear strength (i_{COR} = 5 μA/cm²)

4.1 Time-Dependent Flexural Strength

Figures 10a, 11a and 12a present the time-dependent flexural strength, $R_{nb}(t)$, at time t (years) from the initiation of corrosion, for different mean values of i_{COR}. In the same figures, mR_{nb} stands for the sample mean flexural strength, and R^m_{nb} stands for the approximated mean flexural strength. Figures 10b, 11b, and 12b present the time-dependent c.o.v., $V_{Rnb}(t)$, associated with $R_{nb}(t)$, at time t (years) from the initiation of corrosion, for different mean values of i_{COR}. These figures depict the following:

1. For a given i_{COR}, R^m_{nb} agrees well with mR_{nb} except in the time interval of 0–5 years, where a slight difference between them is observed. This difference is mainly attributed to the randomness associated with the time of cover peeling at bottom, top, and side resulting in the change of cross section for the concrete. Flexural strength is mainly governed by all the basic variables (except τ_C) given in Table 1 and is a non-linear function of those variables. Figure 6 also shows that, at time t, $R_{nb}(t)$ is evaluated after estimating various parameters, such as beam dimensions, cross-sectional area of steels, neutral axis depth, forces of compression in concrete and compression steel and their points of application and force of tension in tension steel and its point of application. Good agreement between the sample mean and approximated mean values of $R_{nb}(t)$ shows that the time-dependent mean flexural strength can be approximated by considering the linear terms in Taylor series expansion of performance functions (Eqs. (5, 7, 8, 9, 10, 11, 12, and 13) for the aforementioned parameters needed to evaluate $R_{nb}(t)$.
2. For a given i_{COR}, an increase in $V_{Rnb}(t)$ is observed in the time interval of 0–5 years for all V_I values. This is attributed to the randomness associated with the time of cover peeling at the bottom, top and side resulting in the change of cross section for the concrete.

3. For a given i_{COR}, V_I has negligible effect on the mean values of the $R_{\mathrm{nb}}(t)$; however, it affects $V_{\mathrm{Rnb}}(t)$. As V_I increases, $V_{\mathrm{Rnb}}(t)$ also increases. This is because $R_{\mathrm{nb}}(t)$ is a function of i_{COR}, and variation in i_{COR} values will result in the variation of corresponding $R_{\mathrm{nb}}(t)$ values.
4. i_{COR} affects $R_{\mathrm{nb}}(t)$ and $V_{\mathrm{Rnb}}(t)$. As i_{COR} increases, $As_{\mathrm{COR}}(t)$ for tension steel also increases, thus further resulting in reduced $R_{\mathrm{nb}}(t)$ and more variability for $As_{\mathrm{COR}}(t)$. Increase in variability for $As_{\mathrm{COR}}(t)$ will result in the increase in $V_{\mathrm{Rnb}}(t)$ values. Here, As_{COR} is defined as the ratio of loss of cross-sectional area to the initial un-corroded area of reinforcement.

4.2 Time-Dependent Shear Strength

Figures 13a, 14a and 15a present the time-dependent shear strength, $R_{\mathrm{sh}}(t)$, at time t (years) from the initiation of corrosion, for different mean values of i_{COR}. In the same figures, mR_{sh} stands for the sample mean shear strength, and $R_{\mathrm{sh}}^{\mathrm{m}}$ stands for the approximated mean shear strength. Figures 13b, 14b and 15b present the time-dependent c.o.v., $V_{\mathrm{Rsh}}(t)$, associated with $R_{\mathrm{sh}}(t)$, at time t (years) from the initiation of corrosion, for different mean values of i_{COR}. These figures depict the following:

1. For a given i_{COR}, $R_{\mathrm{sh}}^{\mathrm{m}}$ agrees well with mR_{sh} except in the time interval of 0–5 years, where a slight difference between them is observed. This difference is mainly attributed to the randomness associated with the time of cover peeling at the bottom, top, and side resulting in the change of cross section for the concrete. Shear strength is governed by all the basic variables given in Table 1 and is a non-linear function of those variables. At time t, $R_{\mathrm{sh}}(t)$ is evaluated after estimating various parameters, such as beam dimensions, cross-sectional area of steels and shear strength of concrete and stirrups. Good agreement between the sample mean and approximated mean values of $R_{\mathrm{sh}}(t)$ shows that the time-dependent mean shear strength can be approximated by considering the linear terms in Taylor series expansion of performance functions (Eqs. 5, 14, 15 and 16) for the aforementioned parameters needed to evaluate $R_{\mathrm{sh}}(t)$.
2. For a given i_{COR}, an increase in $V_{\mathrm{Rsh}}(t)$ is observed in the time interval of 0–5 years. This is attributed to the randomness associated with the time of cover peeling at the bottom, top and side resulting in the change of cross section for the concrete.
3. For a given i_{COR}, V_I has negligible effect on the mean values of the $R_{\mathrm{sh}}(t)$; however, it affects $V_{\mathrm{Rsh}}(t)$. As V_I increases, $V_{\mathrm{Rsh}}(t)$ also increases. This is because $R_{\mathrm{sh}}(t)$ is a function of i_{COR}, and variation in i_{COR} values will result in the variation of corresponding $R_{\mathrm{sh}}(t)$ values.
4. i_{COR} affects $R_{\mathrm{sh}}(t)$ and $V_{\mathrm{Rsh}}(t)$. As i_{COR} increases, $As_{\mathrm{COR}}(t)$ for shear stirrups also increases, thus further resulting in reduced $R_{\mathrm{sh}}(t)$ and more variability for $As_{\mathrm{COR}}(t)$. Increase in variability for $As_{\mathrm{COR}}(t)$ will result in the increase in $V_{\mathrm{Rsh}}(t)$ values.

5. At time $t = 0$, $V_{\mathrm{Rsh}}(\text{at } t = 0)$ is slightly lower than $V_{\mathrm{Rnb}}(\text{at } t = 0)$. It is mentioned that shear strength of beam is evaluated as sum of the shear strength contributions of concrete section and stirrups. Shear strengths of concrete section is evaluated as a product of beam dimensions and permissible shear stress of concrete, while that of stirrups is evaluated as a product of effective depth of beam, yield strength and cross-sectional area of stirrups (Eq. 14, 15 and 16). Flexural strength for the given beam is mainly governed by the yielding of tensile reinforcement and is evaluated as a product of lever arm (which is a function of effective depth of beam), yield strength and cross-sectional area of tensile reinforcement (Eq. 13). Since the considered variability in beam dimensions are smaller as compared to those for material strengths of concrete and steel, it may result in $V_{\mathrm{Rsh}}(\text{at } t = 0)$ slightly lower than $V_{\mathrm{Rnb}}(\text{at } t = 0)$.

5 Time-Dependent Degradation Functions for Time-Dependent Strengths of Corroded RC Beams

Degradation function is defined as the ratio of strength at time, t, to the initial strength for an un-corroded RC beam. Time-dependent mean degradation functions for the time-dependent flexural and shear strengths, for the considered RC beam are expressed by Eq. (17).

$$g_{\mathrm{nb}}(t) = \frac{R_{\mathrm{nb}}(t)}{R_{\mathrm{nb0}}}; \quad g_{\mathrm{sh}}(t) = \frac{R_{\mathrm{sh}}(t)}{R_{\mathrm{sh0}}} \tag{17}$$

where $R_{\mathrm{nb}}(t)$and $R_{\mathrm{sh}}(t)$ = flexural strength and shear strength, respectively, at time t; R_{nb0}andR_{sh0} = flexural strength and shear strength, respectively, at time $t = 0$; and $g_{\mathrm{nb}}(t)$and$g_{\mathrm{sh}}(t)$ = mean degradation functions for flexural strength and shear strength, respectively, at time t.

Figures 16 and 17 present the time-dependent mean degradation functions, $g_{\mathrm{nb}}(t)$ and $g_{\mathrm{sh}}(t)$, for flexural strength and shear strength, respectively, for different mean values of i_{COR}. With the increase in i_{COR}, reduction in $g_{\mathrm{nb}}(t)$ and $g_{\mathrm{sh}}(t)$ is observed. This is attributed to the reduction in $R^{\mathrm{m}}_{\mathrm{nb}}$ and $R^{\mathrm{m}}_{\mathrm{sh}}$, respectively, due to increase in i_{COR}.

6 Analytical Estimation of Time-Dependent c.o.v. Associated with Degradation Function

Since time-dependent strengths and degradation functions for corroded RC beam are related to each other by Eq. (17), the time-dependent c.o.v. associated with degradation functions is evaluated by Eq. (18) [4].

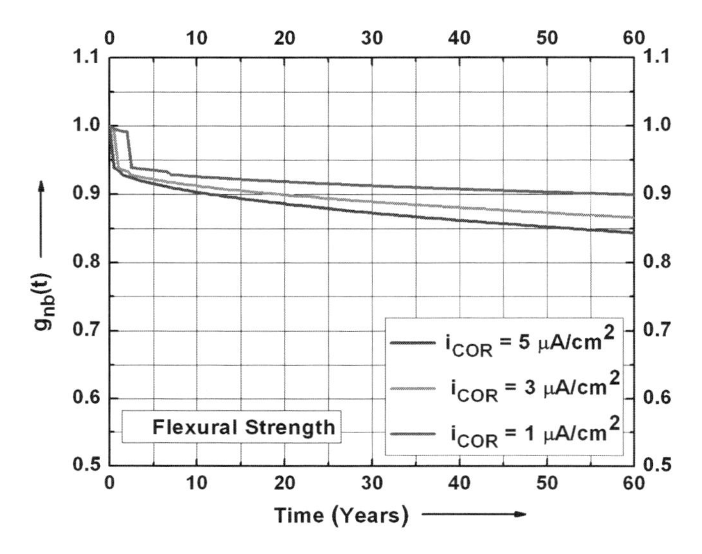

Fig. 16 Time-dependent mean degradation function for flexural strength

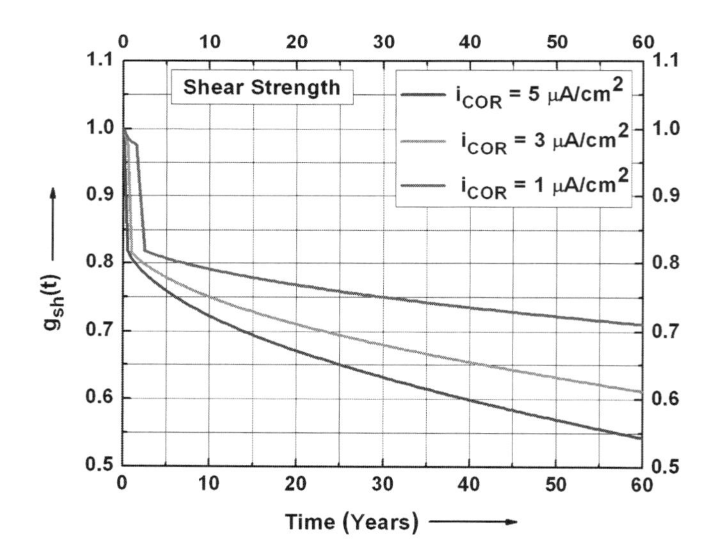

Fig. 17 Time-dependent mean degradation function for shear strength

$$V_{\mathrm{Gnb}}(t) = \sqrt{\frac{V_{\mathrm{Rnb}}^2(t) - V_{\mathrm{Rnb0}}^2}{1 + V_{\mathrm{Rnb0}}^2}}; \quad V_{\mathrm{Gsh}}(t) = \sqrt{\frac{V_{\mathrm{Rsh}}^2(t) - V_{\mathrm{Rsh0}}^2}{1 + V_{\mathrm{Rsh0}}^2}} \tag{18}$$

where $V_{\mathrm{Gnb}}(t)$ and $V_{\mathrm{Gsh}}(t)$ = c.o.v. associated with degradation functions for flexural strength and shear strength, respectively, at time t; $V_{\mathrm{Rnb}}(t)$ and

$V_{\text{Rsh}}(t)$ = c.o.v. associated with flexural strength and shear strength, respectively, at time t; and V_{Rnb0} and V_{Rsh0} = c.o.v. associated with initial flexural strength and initial shear strength, respectively, at time $t = 0$. The estimation of $V_{\text{Rnb}}(t)$, $V_{\text{Rsh}}(t)$, $V_{\text{Gnb}}(t)$ and $V_{\text{Gsh}}(t)$is addressed in the following sections.

6.1 Analytical Estimation of $V_{Rnb}(t)$ and $V_{Gnb}(t)$

At time t, the flexural strength, $R_{\text{nb}}(t)$, is estimated by Eq. (13) (performance functions for $R_{\text{nb}}(t)$) for tension failure. $R_{\text{nb}}(t)$ is a function of $F_{\text{st}}(t)$, $Y_{\text{c}}(t)$ and $d(t)$. At time t, if σ_{Rnbt} is the standard deviation associated with $R_{\text{nb}}(t)$, then the first-order approximation of variance of $R_{\text{nb}}(t)$ is given by Eq. (19) [18].

$$\sigma_{\text{Rnbt}}^2 = \left(\frac{\partial R_{\text{nb}}(t)}{\partial F_{\text{st}}(t)}\right)^2 \cdot \sigma_{\text{Fstt}}^2 + \left(\frac{\partial R_{\text{nb}}(t)}{\partial Y_c(t)}\right)^2 \cdot \sigma_{\text{Yct}}^2 + \left(\frac{\partial R_{\text{nb}}(t)}{\partial d(t)}\right)^2 \cdot \sigma_{\text{dt}}^2 \tag{19}$$

where σ_{Fstt} = standard deviation associated with $F_{\text{st}}(t)$, σ_{Yct} = standard deviation associated with $Y_c(t)$ and σ_{dt} = standard deviation associated with $d(t)$. First-order approximation for σ_{Fstt}, σ_{Yct} and σ_{dt} are given by Eqs. (20a, 20b, 21, 22a and 22b).

$$\sigma_{\text{Fstt}}^2 = \left(\frac{\partial F_{\text{st}}(t)}{\partial E_s}\right)^2 \cdot \sigma_{Es}^2 + \left(\frac{\partial F_{\text{st}}(t)}{\partial X_u(t)}\right)^2 \cdot \sigma_{\text{Xut}}^2 + \left(\frac{\partial F_{\text{st}}(t)}{\partial d(t)}\right)^2 \cdot \sigma_{dt}^2 + \left(\frac{\partial F_{\text{st}}(t)}{\partial i_{\text{COR}}}\right)^2 \cdot \sigma_{\text{icor}}^2 \quad \text{for } \varepsilon_{\text{st}}(t) \le \varepsilon_{\text{sy}} \tag{20a}$$

$$\sigma_{\text{Fstt}}^2 = \left(\frac{\partial F_{\text{st}}(t)}{\partial f_y}\right)^2 \cdot \sigma_{fy}^2 + \left(\frac{\partial F_{\text{st}}(t)}{\partial i_{\text{COR}}}\right)^2 \cdot \sigma_{\text{icor}}^2; \text{ for } \varepsilon_{st}(t) > \varepsilon_{sy} \tag{20b}$$

$$\sigma_{\text{Yct}}^2 = \left(\frac{\partial Y_c(t)}{\partial F_{\text{cc}}(t)}\right)^2 \cdot \sigma_{\text{Fcct}}^2 + \left(\frac{\partial Y_c(t)}{\partial Y_{\text{cc}}(t)}\right)^2 \cdot \sigma_{\text{Ycct}}^2 + \left(\frac{\partial Y_c(t)}{\partial F_{\text{sc}}(t)}\right)^2 \cdot \sigma_{\text{Fsct}}^2 + \left(\frac{\partial Y_c(t)}{\partial d_{\text{sc}}(t)}\right)^2 \cdot \sigma_{\text{dsct}}^2 \tag{21}$$

$$\sigma_{dt} = \sqrt{\sigma_D^2 + \sigma_{CB}^2}\,; \text{(before peeling of top cover)} \tag{22a}$$

$$\sigma_{dt} = \sqrt{\sigma_D^2 + \sigma_{CB}^2 + \sigma_{CT}^2}\,; \text{(after peeling of top cover)} \tag{22b}$$

where σ_{Es}, σ_{fy}, σ_{icor}, σ_D, σ_{CB} and σ_{CT} = standard deviation associated with E_s, f_y, i_{COR}, D, C_B and C_T, respectively, and are evaluated based on the statistical parameters for these variables given in Table 1. σ_{Xut} is the standard deviation associated with $X_u(t)$, and the first-order approximation of variance of $X_u(t)$ for different strain conditions in reinforcing steels shall give its value. σ_{Fcct}, σ_{Ycct}, σ_{Fsct} and σ_{dsct} are the standard deviations associated with $F_{\mathrm{cc}}(t)$, $Y_{\mathrm{cc}}(t)$, $F_{\mathrm{sc}}(t)$ and $d_{\mathrm{sc}}(t)$, respectively, and the first-order approximation of variances of $F_{\mathrm{cc}}(t)$, $Y_{\mathrm{cc}}(t)$ $F_{\mathrm{sc}}(t)$ and $d_{\mathrm{sc}}(t)$ shall give their respective values. The time-dependent c.o.v., $V_{\mathrm{Rnb}}(t)$, associated with $R_{\mathrm{nb}}(t)$ is estimated from Eq. (23).

$$V_{\mathrm{Rnb}}(t) = \frac{\sigma_{\mathrm{Rnbt}}}{R_{\mathrm{nb}}(t)} \tag{23}$$

At time $t = 0$, $V_{\mathrm{Rnb}}(t = 0) = V_{\mathrm{Rnb0}}$. Once V_{Rnb0} and $V_{\mathrm{Rnb}}(t)$ are known, $V_{\mathrm{Gnb}}(t)$ is evaluated from Eq. (18).

6.2 Analytical Estimation of $V_{Rsh}(t)$ and $V_{Gsh}(t)$

The shear strength, $R_{\mathrm{sh}}(t)$, at time t, is estimated by Eq. (15) (performance functions for $R_{\mathrm{sh}}(t)$). $R_{\mathrm{sh}}(t)$ is a function of $\tau_C(t)$, $b(t)$, $d(t)$, f_y and i_{COR}. For the considered corroded RC beam (Fig. 9), $\tau_C(t)$ is not calculated by Eq. (14), rather τ_C is considered as a separate random variable with statistical parameters provided in Table 1. At time t, if σ_{Rsht} is the standard deviation associated with $R_{\mathrm{sh}}(t)$, then the first-order approximation of variance of $R_{\mathrm{sh}}(t)$ is given by Eq. (24) [18].

$$\begin{aligned}\sigma_{\mathrm{Rsht}}^2 = & \left(\frac{\partial R_{\mathrm{sh}}(t)}{\partial \tau_C}\right)^2 \cdot \sigma_{TC}^2 + \left(\frac{\partial R_{\mathrm{sh}}(t)}{\partial b(t)}\right)^2 \cdot \sigma_{bt}^2 + \left(\frac{\partial R_{\mathrm{sh}}(t)}{\partial d(t)}\right)^2 \cdot \sigma_{dt}^2 \\ & + \left(\frac{\partial R_{\mathrm{sh}}(t)}{\partial f_y}\right)^2 \cdot \sigma_{fy}^2 + \left(\frac{\partial R_{\mathrm{sh}}(t)}{\partial i_{\mathrm{COR}}}\right)^2 \cdot \sigma_{\mathrm{icor}}^2\end{aligned} \tag{24}$$

where $\sigma_{\tau c}$ = standard deviations associated with τ_C and is evaluated based on the statistical parameters for τ_C given in Table 1; σ_{bt} = standard deviation associated with $b(t)$ and first-order approximation of variance of $b(t)$ shall give its value. The time-dependent c.o.v., $V_{\mathrm{Rsh}}(t)$, associated with $R_{\mathrm{sh}}(t)$ is estimated from Eq. (25).

$$V_{\mathrm{Rsh}}(t) = \frac{\sigma_{\mathrm{Rsht}}}{R_{\mathrm{sh}}(t)} \tag{25}$$

At time $t = 0$, $V_{\mathrm{Rsh}}(t = 0) = V_{\mathrm{Rsh0}}$. Once V_{Rsh0} and $V_{\mathrm{Rsh}}(t)$ are known, $V_{\mathrm{Gsh}}(t)$ is evaluated from Eq. (18).

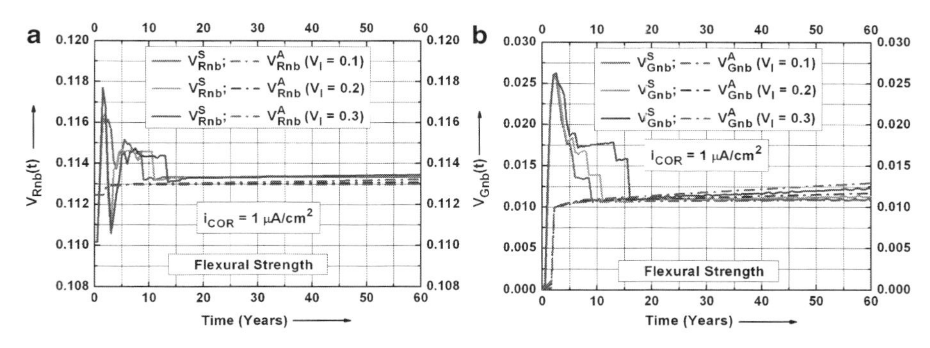

Fig. 18 (**a**): Comparison of time-dependent c.o.v. V^A_{Rnb} and V^S_{Rnb} for flexural strength (i_{COR} = 1 μA/cm²) (**b**): Comparison of time-dependent c.o.v. V^A_{Gnb} and V^S_{Gnb} for degradation function for flexural strength (i_{COR} = 1 μA/cm²)

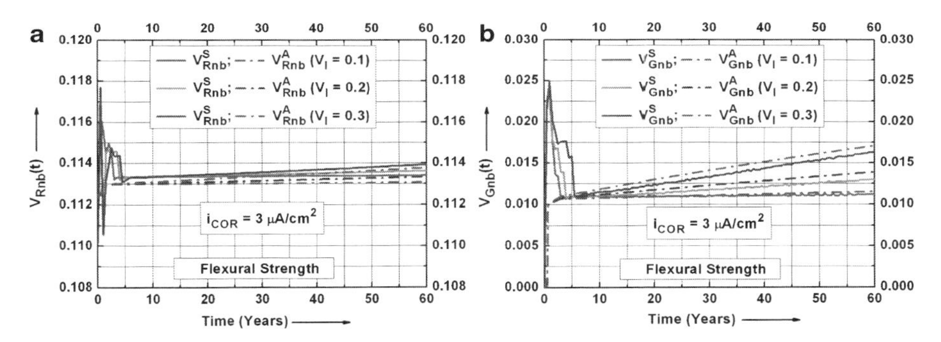

Fig. 19 (**a**): Comparison of time-dependent c.o.v. V^A_{Rnb} and V^S_{Rnb} for flexural strength (i_{COR} = 3 μA/cm²) (**b**): Comparison of time-dependent c.o.v. V^A_{Gnb} and V^S_{Gnb} for degradation function for flexural strength (i_{COR} = 3 μA/cm²)

6.3 Discussion of Results for Analytical Estimation of Time-Dependent c.o.v. Associated with Time-Dependent Strengths and Time-Dependent Degradation Functions

For different mean values of i_{COR}, at time t (years) from the initiation of corrosion, (1) Figures 18a, 19a and 20a present the time-dependent c.o.v., $V_{Rnb}(t)$, associated with $R_{nb}(t)$; (2) Figures 18b, 19b and 20b present the time-dependent c.o.v., $V_{Gnb}(t)$, associated with $G_{nb}(t)$; (3) Figures 21a, 22a, and 23a present the time-dependent c.o.v., $V_{\mathrm{Rsh}}(t)$, associated with $R_{\mathrm{sh}}(t)$; and (4) Figures 21b, 22b, and 23b present the time-dependent c.o.v., $V_{\mathrm{Gsh}}(t)$, associated with $G_{\mathrm{sh}}(t)$. In the same figures, $V^{\mathrm{S}}_{\mathrm{Rnb}}$ and $V^{\mathrm{S}}_{\mathrm{Gnb}}$ and $V^{\mathrm{S}}_{\mathrm{Rsh}}$ and $V^{\mathrm{S}}_{\mathrm{Gsh}}$ are evaluated using 40 random samples, $V^{\mathrm{A}}_{\mathrm{Rnb}}$ and $V^{\mathrm{A}}_{\mathrm{Rsh}}$ are evaluated using the analytical formulations presented in the preceding sections and

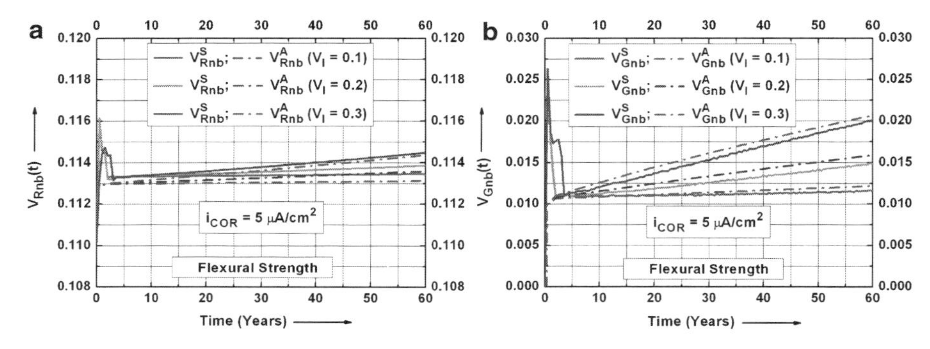

Fig. 20 (**a**): Comparison of time-dependent c.o.v. V^A_{Rnb} and V^S_{Rnb} for flexural strength (i_{COR} = 5 μA/cm^2) (**b**): Comparison of time-dependent c.o.v. V^A_{Gnb} and V^S_{Gnb} for degradation function for flexural strength (i_{COR} = 5 μA/cm^2)

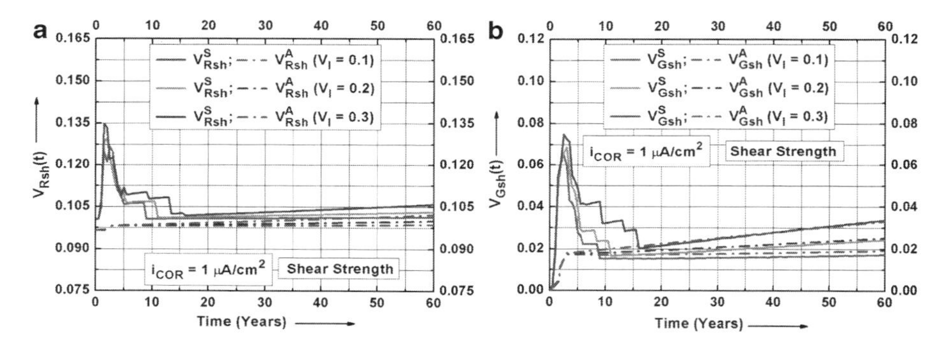

Fig. 21 (**a**): Comparison of time-dependent c.o.v. $V^A_{Rsh}$and V^S_{Rsh} for shear strength (i_{COR} = 1 μA/cm^2) (**b**): Comparison of time-dependent c.o.v. V^A_{Gsh} and V^S_{Gsh} for degradation function for shear strength (i_{COR} = 1 μA/cm^2)

$V^{\rm A}_{\rm Gnb}$ and $V^{\rm A}_{\rm Gsh}$ are evaluated using Eq. 18. For the purpose of comparison, (1) in Figs. 18a, 19a and 20a, $V^{\rm S}_{\rm Rnb}$ are reproduced from Figs. 10b, 11b and 12b, and (2) in Figs. 21a, 22a and 23a, $V^{\rm S}_{\rm Rsh}$ are reproduced from Figs. 13b, 14b and 15b. For a given $i_{\rm COR}$, good agreement is observed between $V^{\rm A}_{\rm Rnb}$ and $V^{\rm S}_{\rm Rnb}$, $V^{\rm A}_{\rm Gnb}$ and $V^{\rm S}_{\rm Gnb}$, $V^{\rm A}_{\rm Rsh}$ and $V^{\rm S}_{\rm Rsh}$ and $V^{\rm A}_{\rm Gsh}$ and $V^{\rm S}_{\rm Gsh}$, except in the time interval of about (a) 0–15 years, for $i_{\rm COR}$ = 1 μA/cm^2, (b) 0–5 years for $i_{\rm COR}$ = 3 μA/cm^2 and (c) 0–3 years for $i_{\rm COR}$ = 5 μA/cm^2. The difference in the aforementioned time intervals is mainly attributed to the randomness associated with the time of cover peeling at the bottom, top and side resulting in the change of cross section for the concrete. Good agreement between the analytically estimated and simulated values for the c.o.v. in the remaining time intervals shows that the first-order approximation of mean and variance of strengths can be used for estimating c.o.v. for both strengths and degradation functions.

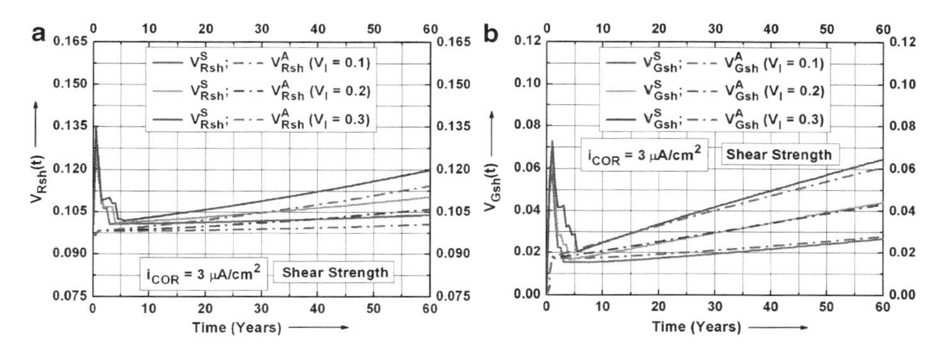

Fig. 22 (**a**): Comparison of time-dependent c.o.v. V^A_{Rsh} and V^S_{Rsh} for shear strength (i_{COR} = 3 μA/cm²) (**b**): Comparison of time-dependent c.o.v. V^A_{Gsh} and V^S_{Gsh} for degradation function for shear strength (i_{COR} = 3 μA/cm²)

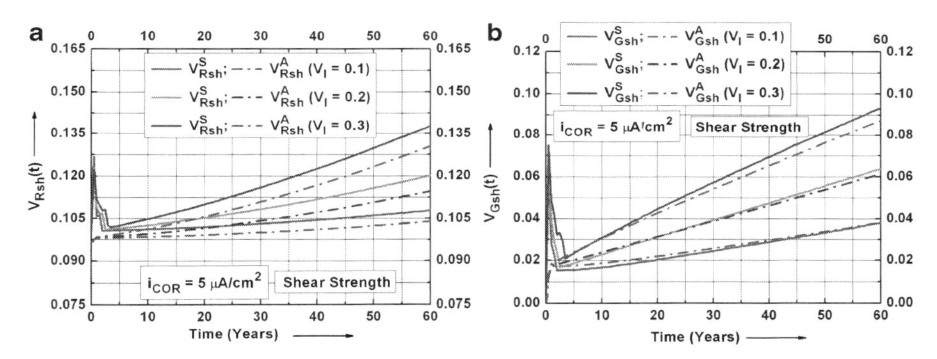

Fig. 23 (**a**): Comparison of time-dependent c.o.v. V^A_{Rsh} and V^S_{Rsh} for shear strength (i_{COR} = 5 μA/cm²) (**b**): Comparison of time-dependent c.o.v. V^A_{Gsh} and V^S_{Gsh} for degradation function for shear strength (i_{COR} = 5 μA/cm²)

7 Conclusions

The following conclusions are drawn from the present study:

1. Correlations with the experimental results indicate that the equations proposed for estimating the time-dependent mass loss of the reinforcement, W_s; reduced diameter of reinforcement, D_r; loss of cross-sectional area of reinforcement, A_{cor}; and corrosion penetration depth, X, are capable of providing their reasonable estimates that are in line with the available experimental trends.
2. Correlations with the experimental results indicate that good agreement is observed between analytical predictions and experimental results, for ultimate flexural and shear strengths for corrosion-affected RC beams. These findings also highlight the fair estimation of (1) time to peeling of cover concrete using

the proposed methodology to arrive at the time-dependent reduced concrete section and (2) W_s to arrive at the time-dependent reduced cross section of reinforcing steel.

3. A methodology to evaluate time-dependent strengths and c.o.v. associated with strengths is presented for corrosion-affected RC beams, by using LHS technique.
4. For corrosion-affected RC beams, it is shown that time-dependent mean flexural and shear strengths can be approximated by considering the linear terms in Taylor series expansion of their performance functions.
5. For corrosion-affected RC beams, time-dependent c.o.v. associated with the strength is influenced by the c.o.v. associated with annual mean corrosion rate, i_{COR}, at late stages of degradation (say 40–60 years).
6. Good agreement is observed between the analytical and simulated estimations for the time-dependent c.o.v. associated with flexural and shear strengths.
7. Good agreement is observed between the analytical and simulated estimations for the time-dependent c.o.v. associated with the degradation functions for the flexural and shear strengths.
8. Analytical estimation of c.o.v. associated with strengths and degradation functions substantially reduces the computational efforts involved in their estimation using LHS technique.

Acknowledgement The first author gratefully acknowledges the financial support provided by Japan Society for the Promotion of Science under JSPS RONPAKU Dissertation Ph.D. programme for pursuing Ph.D. at Nagoya University, Nagoya, Japan.

References

1. ACI (1995) Building code requirement for structural concrete (ACI 318-95) and commentary (ACI 318R-95). American Concrete Institute, Farmington Hills
2. Andrade C, Alonso C, Molina FJ (1993) Cover cracking as a function of rebar corrosion: part I – experimental test. Mater Struct 26:453–464
3. Bazant ZP (1979) Physical model for steel corrosion in sea structures – theory. J Struct Div ASCE 105(6):1137–1153
4. Benjamin JR, Cornell CA (1970) Probability, statistics, and decision for civil engineering. McGraw-Hill, New York
5. Bhargava K (2008) Time-dependent degradation and reliability assessment of RC structures subjected to reinforcement corrosion, Doctor of Engineering dissertation, Graduate School of Environmental Studies, Nagoya University, Nagoya, Japan
6. Bhargava K, Ghosh AK, Mori Y, Ramanujam S (2003) Analytical model of corrosion-induced cracking of concrete considering the stiffness of reinforcement. Struct Eng Mech Int J 16(6): 749–769
7. Bhargava K, Ghosh AK, Mori Y, Ramanujam S (2005) Modeling of time to corrosion-induced cover cracking in reinforced concrete structures. Cem Concr Res 35(11):2203–2218
8. Bhargava K, Ghosh AK, Mori Y, Ramanujam S (2006) Model for cover cracking due to rebar corrosion in RC structures. Eng Struct 28(8):1093–1109
9. BIS (2000) IS 456: 2000, Indian standard, plain and reinforced concrete – code of practice, 4th Rev, Bureau of Indian Standards, New Delhi, India

10. CEB–FIP (1990) CEB-FIP model code 1990, Comite Euro-International du Beton-Federation International de la Precontrainte. Thomas Telford, London
11. Ellingwood B (1977) Statistical analysis of RC beam-column interaction. J Struct Div ASCE 103(ST7):1377–1388
12. Ellingwood B (1982) Safety checking formats for limit states design. J Struct Div ASCE 108 (ST7):1481–1493
13. Ellingwood BR, Ang AHS (1974) Risk-based evaluation of design criteria. J Struct Div ASCE 100(ST9):1771–1788
14. Ellingwood B, Hwang H (1985) Probabilistic descriptions of resistance of safety-related structures in nuclear plants. Nucl Eng Des 88:169–178
15. Enright MP, Frangopol DM (1998) Service-life prediction of deteriorating concrete structures. J Struct Eng ASCE 124(3):309–317
16. Enright MP, Frangopol DM (1998) Probabilistic analysis of resistance degradation of reinforced concrete bridge beams under corrosion. Eng Struct 20(11):960–971
17. Frangopol DM, Lin KY, Estes AC (1997) Reliability of reinforced concrete girders under corrosion attack. J Struct Eng ASCE 123(3):286–297
18. Haldar A, Mahadevan S (2000) Reliability assessment using stochastic finite element analysis, 1st edn. Wiley, New York
19. Hong HP (2000) Assessment of reliability of aging reinforced concrete structures. J Struct Eng ASCE 126(12):1458–1465
20. Hong HP, Zhou W (1999) Reliability evaluation of RC columns. J Struct Eng ASCE 125 (7):784–790
21. Hwang H, Ellingwood B, Shinozuka M, Reich M (1987) Probability-based design criteria for nuclear plant structures. J Struct Eng ASCE 113(5):925–942
22. Israel M, Ellingwood B, Corotis R (1987) Reliability-based code formulations for reinforced concrete buildings. J Struct Eng ASCE 113(10):2235–2252
23. Li CQ (2003) Life cycle modeling of corrosion affected concrete structures – propagation. J Struct Eng ASCE 129(6):753–761
24. Li CQ, Lawanwisut W, Zheng JJ (2005) Time-dependent reliability method to assess the serviceability of corrosion-affected concrete structures. J Struct Eng ASCE 131(11):1674–1680
25. Liu Y (1996) Modeling the time to corrosion cracking of the cover concrete in chloride contaminated reinforced concrete structures, Ph.D. dissertation, Virginia Polytechnic Institute and State University, Blacksburg, Virginia, USA
26. Liu Y, Weyers RE (1998) Modelling the time-to-corrosion cracking in chloride contaminated reinforced concrete structures. ACI Mater J 95(6):675–681
27. Mangat PS, Elgarf MS (1999) Flexural strength of concrete beams with corroding reinforcement. ACI Struct J 96(1):149–158
28. Mckay MD, Bechman RJ, Conover WJ (1979) A comparison of three methods for selecting values of input variables in the analysis of output from a computer code. Technometrics 21(2):239–245
29. Mirza SA, MacGregor JG (1979) Variations in dimensions of reinforced concrete members. J Struct Div ASCE 105(ST4):751–766
30. Mirza SA, MacGregor JG (1979) Variability of mechanical properties of reinforcing bars. J Struct Div ASCE 105(ST5):921–937
31. Mirza SA, Hatzinikolas M, MacGregor JG (1979) Statistical description of strength of concrete. J Struct Div ASCE 105(ST6):1021–1037
32. Ranganathan R (2000) Structural reliability analysis and design, second Jaico impression. Jaico Publishing House, Mumbai
33. Rasheeduzzafar ASM, Al-Saadoun SS, Al-Gahtani AS (1992) Corrosion cracking in relation to bar diameter, cover and concrete quality. J Mater Civ Eng ASCE 4(4):327–343
34. Rodriguez J, Ortega LM, Casal J (1997) Load carrying capacity of concrete structures with corroded reinforcement. Constr Build Mater 11(4):239–248

35. SP16 (1980) Design aids for reinforced concrete to IS: 456-1978. Bureau of Indian Standards, New Delhi
36. Stewart MG, Rosowsky DV (1998) Time-dependent reliability of deteriorating reinforced concrete bridge decks. Struct Saf 20:91–109
37. Thoft-Christensen P (1998) Assessment of the reliability profiles for concrete bridges. Eng Struct 20(11):1004–1009
38. Torres-Acosta AA (1999) Cracking induced by localized corrosion of reinforcement in chloride contaminated concrete, Ph.D. dissertation, Department of Civil and Environmental Engineering, University of South Florida, Tampa, Florida, USA
39. Val DV, Stewart MG, Melchers RE (1998) Effect of reinforcement corrosion on reliability of highway bridges. Eng Struct 20(11):1010–1019
40. Vu KAT, Stewart MG (2000) Structural reliability of concrete bridges including improved chloride-induced corrosion models. Struct Saf 22:313–333
41. Vu KAT, Stewart MG (2005) Predicting the likelihood and extent of reinforced concrete corrosion-induced cracking. J Struct Eng ASCE 131(11):1681–1689

Refined Modeling of Crack Tortuousness to Predict Pressurized Airflow Through Concrete Cracks

L.R. Bishnoi, R.P. Vedula, and S.K. Gupta

Abstract Cracks may appear in the pressurized concrete containment of a nuclear power plant during a severe accident and provide leak paths for release of radioactive aerosols dispersed in the contained air. In this study, numerical results for air leakage through concrete cracks are reported for a range of pressure gradients and crack widths relevant to containment atmosphere during a severe accident scenario. Crack geometry in 2D is generated using statistical crack model to account for crack tortuousness. While airflow predictions through such models provide good agreement with experimental results reported in literature, the computational results generally provide over-prediction of airflow. The statistical crack models account for the gross tortuousness of the concrete cracks based on experimental studies. The local tortuousness of the order of grain size that can deflect the crack from straight path is not accounted in these models. In this study, fractal geometry-based curves are used to introduce the local tortuousness within the global crack segments represented by straight lines in the statistical models. Comparison of pressurized airflow rates obtained from such refined crack model with the experimental values reported in literature for plain concrete shows very good agreement. The effect of local tortuousness on the pressurized airflow rates was accounted indirectly in 3D crack models for reinforced concrete with modified crack morphology due to reinforcing steel. The computational results with corrections due to local tortuousness compared well with the experimental values

L.R. Bishnoi (✉)
Siting & Structural Engineering Division, Atomic Energy Regulatory Board, Mumbai, India
e-mail: lr.bishnoi@gmail.com

R.P. Vedula
Mechanical Engineering Department, Indian Institute of Technology Bombay, Mumbai, India
e-mail: rpv@iitb.ac.in

S.K. Gupta
Safety Analysis & Documentation Division, Atomic Energy Regulatory Board, Mumbai, India
e-mail: guptask@aerb.gov.in

S. Chakraborty and G. Bhattacharya (eds.), *Proceedings of the International Symposium on Engineering under Uncertainty: Safety Assessment and Management (ISEUSAM - 2012)*,
DOI 10.1007/978-81-322-0757-3_24,

for pressurized airflow through cracks in reinforced concrete panels reported in literature.

Keywords Severe accident • Containment • Concrete cracks • Tortuousness • Fractal geometry • Airflow

1 Introduction

A major fraction of the airborne radioactivity within nuclear reactor containment, consequent to a postulated severe accident involving reactor core meltdown, consists of aerosols generated by condensation of volatile fission products. The containment envelope becomes pressurized during a severe accident, and there is a possibility of cracks through concrete shell of the containment, which can provide leak paths for pressurized air and aerosol release to the outside atmosphere.

Rizkalla et al. [1] reported experimental data for leakage rate through cracks in reinforced concrete test panels and suggested correlations. Riva et al. [2] performed finite element analysis of reinforced concrete test panels to calculate equivalent average single crack width for evaluating leakage rate with different correlations and reported good match of the test leakage rate with the calculated rate using correlation of Rizkalla et al. [1]. Gelain and Vendel [3] performed experiments on plain concrete panels and computed crack geometry as an equivalent rectangular channel, which would have the same flow rate as the experimental data. Boussa et al. [4] generated cracks in a large number of test specimens of different concrete grades, modelled the crack profile in terms of statistical parameters and reported good agreement of the crack profile obtained from the statistical model with the experimental data. Bishnoi et al. [5] conducted computational studies for airflow and aerosol transport through cracks in plain concrete using statistical crack model of Boussa et al. [4] and reported good match between the computational and experimental results. Bishnoi et al. [6] studied effect of reinforcing steel on airflow through cracks in reinforced concrete using computational models incorporating modified crack morphology due to reinforcing steel derived from stress analysis of reinforced concrete panels.

In this study, numerical results for air leakage through concrete cracks are reported for a range of pressure gradients and crack widths relevant to containment atmosphere during a severe accident scenario. Crack geometry is generated using statistical crack model of Boussa et al. [4] to account for crack tortuousness. While airflow predictions using such computational model provide good agreement with experimental results reported in literature, the numerical results generally provide over-prediction of airflow. The Boussa crack model defines gross tortuousness of the concrete cracks in statistical terms derived from experimental studies. Local tortuousness of the order of grain size that can deflect the crack from straight path is not accounted in these models.

The Boussa crack model is refined in this study by using fractal geometry-based curves to introduce the local tortuousness within the global crack segments represented by straight lines. The airflow study results from such refined model are compared again with the experimental values. The comparison shows very good agreement between computational values from the refined crack model and the experimental values of Gelain and Vendel [3] for cracks in plain concrete. The computational airflow studies through cracks in reinforced concrete also shows very good agreement with experimental results of Rizkalla et al. [1] when the effect of local tortuousness is accounted along with the effect of reinforcing steel on the crack morphology.

Statistical crack model including refinement of the crack morphology to represent the local tortuousness using fractal geometry methods was implemented in MATLAB. Fluent (version 6.2.16, 2005) was used for airflow computations through the crack models.

2 Numerical Procedure

The crack model parameters were chosen to represent concrete grade in the range of M50–M60, typically used to construct containment structures. The range of pressure gradients and crack widths for the study has been reported in Bishnoi et al. [5]. These parameters were chosen to be representative of air leakage through containment leak paths, which are likely to exist under postulated severe accident conditions in water-cooled reactor-based nuclear power plants.

The 2D crack profile is defined in terms of straight segments and deviation of these segments from horizontal line by specifying mean value and standard deviation separately for these two parameters as per the approach suggested by Boussa et al. [4]. Two identical profiles placed at a constant spacing equal to the crack width represent two lips of the crack in 2D. Details of statistical crack model in 2D are reported in Bishnoi et al. [5]. Refinement of the 2D crack model was done to represent roughness due to local tortuousness, introduced by crack tip deflecting material grains using fractal geometry, for each of the straight segments. Random midpoint displacement method [7] was used to approximate local deflection of the crack due to micrometer size concrete material grains as approximate fractal Brownian motion representation.

A typical statistical crack model geometry representing global tortuousness of the crack is shown in Fig. 1. Typical global tortuousness and local tortuousness in portion ‘A’ of the global crack model geometry, derived using random midpoint displacement method, are depicted in Fig. 2.

To construct the fractal geometry representing local tortuousness in a straight segment of the crack, displaced y-value of the midpoint (based on x-coordinates of the end points) of the straight line is calculated as the average of the y-values

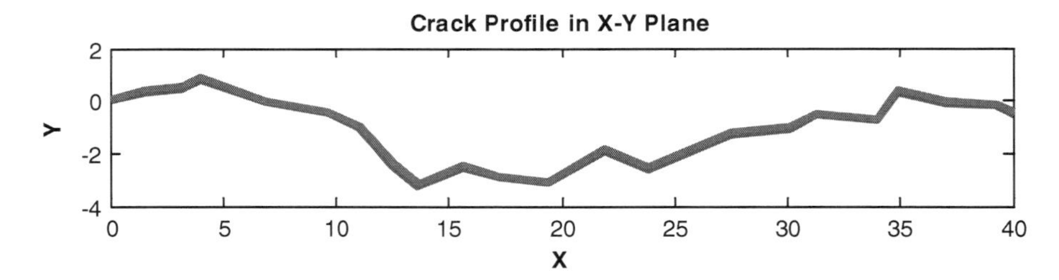

Fig. 1 Typical 2D crack model representing global tortuousness (all dimensions in mm)

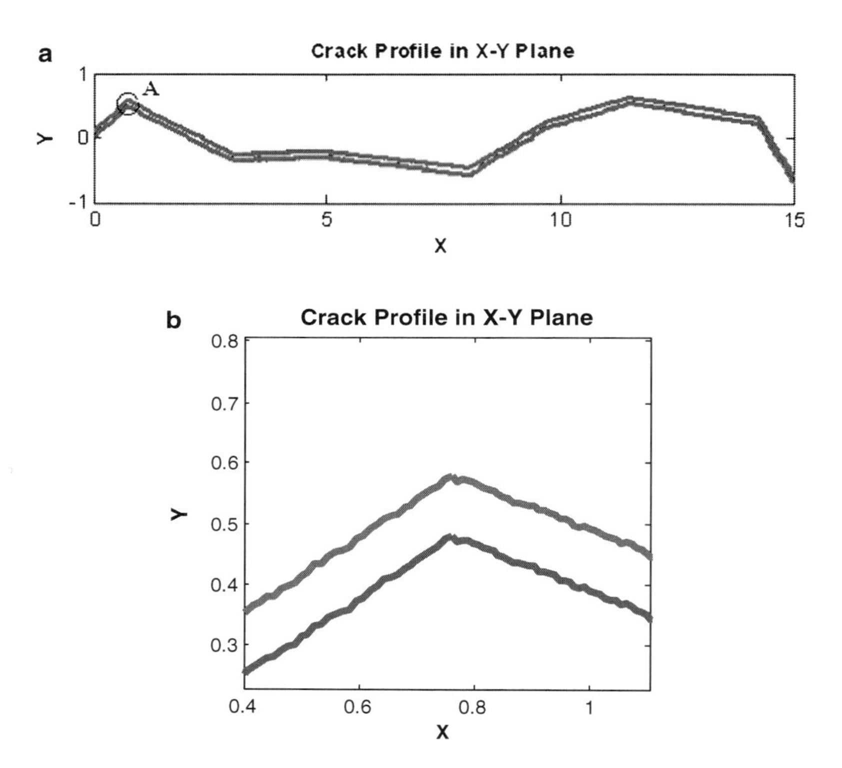

Fig. 2 Depiction of typical global and local tortuousness (all dimensions in mm). (**a**) Global tortuousness, (**b**) enlargement of portion 'A' in (**a**) to view local tortuousness

of the endpoints plus a random offset. The process is repeated by calculating a displaced y-value for the midpoint of each half of the subdivided line. The subdivision is continued until the subdivided line segments are less than a preset value, s_{lim}.

The random offset, r, is calculated as follows:

$$r = sr_g\langle L\rangle \tag{1}$$

where s is a selected 'surface roughness' factor, r_g is a Gaussian random value with mean zero and variance one and L is the length of the straight line.

Nominal diameter of the capillary pores in cement paste varies from about 0.3 to 3 μm [8]. Assuming water cement ratio of 0.45, porosity of cement paste works out to be about 11.5% [8]. Considering average size of capillary pores in the range of 1–1.5 μm, the material inhomogeneity affecting crack path (i.e. effective grain size for crack deflection) was taken in the range of 7.5–15 μm. The surface roughness (i.e. deviation from straight path) was considered to be in the range of ¼ to ½ of the grain size, and the actual variation was approximated to be 2.5–5 μm. The values adopted in the current study are 3 and 10 μm for s and s_{lim}, respectively, which were arrived at from parametric airflow computations on crack models with varying values of s and s_{lim} in the size ranges stated above.

Stress analyses were performed on finite element (FE) model of a reinforced concrete experimental test specimen (L4) from Rizkalla et al. [1] to derive morphology and extent of crack surfaces to be used for airflow calculations. Specimen L4 was selected because complete details of geometry, reinforcement, cracking and airflow results are reported for this specimen only. Uniaxial tensile load was applied to reinforcing steel bars in the stress analysis models as was done in the experimental study. Two non-linear analyses were conducted: one with reinforcing steel and another with plain concrete without reinforcement to see the effect of reinforcing steel on crack surface pattern growth compared to the plain concrete. These analyses indicated sudden spread of the concrete damage in plain concrete from exterior faces to the entire cross section, whereas the reinforcement does not allow spread of the damage over the entire cross section even at ultimate load, though the extent of damage keep increasing with load increments. Typical damage spread over the cross section of the test specimen for plain concrete and reinforced concrete are shown in Fig. 3a, b, respectively.

Linear stress analysis was conducted with a priori crack in the form of a slit of finite width to obtain the likely morphology of the crack around reinforcing steel bars under uniaxial tensile load. Typical crack surface contours due to the effect of reinforcing steel are depicted in Fig. 4. Details of these analyses and the modified 3D crack geometry are reported in Bishnoi et al. [6].

The standard Navier–Stokes and continuity equations are solved for the flow domain defined by the crack morphology using the finite volume method (FVM) with SIMPLEC algorithm for pressure–velocity coupling and second-order upwind scheme with under relaxation factor of 0.5 for discretization of momentum equations. Atmospheric pressure was assumed at outlet, and inlet pressure was calculated according to the specified pressure gradient. No-slip condition was imposed at crack wall boundaries. All case studies were conducted for a constant temperature of 300 K. Since the pressure drop across the crack is not large, the flow is assumed incompressible and confirmed by checking the flow Mach numbers, which remained much below unity for all the cases. The constant air density was assumed to be 1.225 kg/m^3. Details of optimized computational grid and validation of the crack model for 2D airflow computations are reported in Bishnoi et al. [5]. Typical grids for numerical computation of pressurized airflow in the crack model

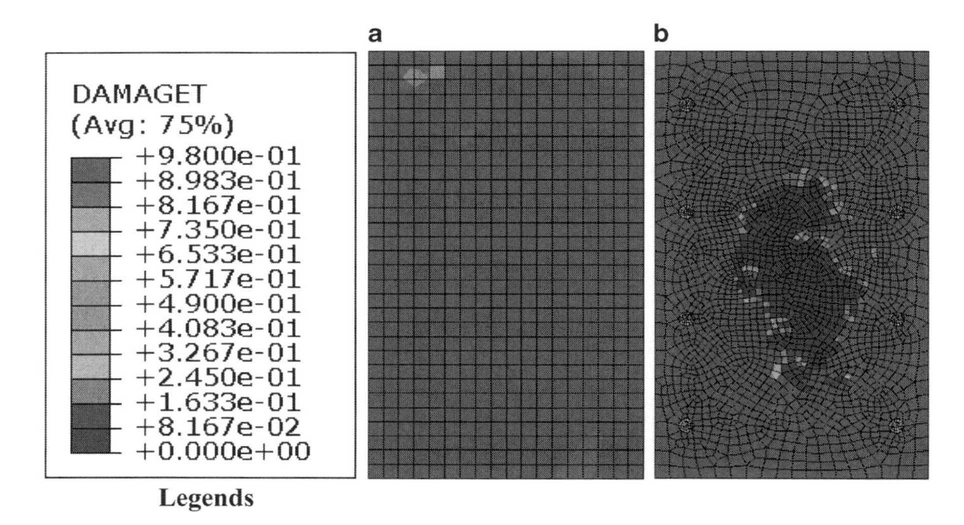

Fig. 3 Concrete damage spread over the X-sections of specimen models at ultimate load, (**a**) plain concrete, (**b**) reinforced concrete

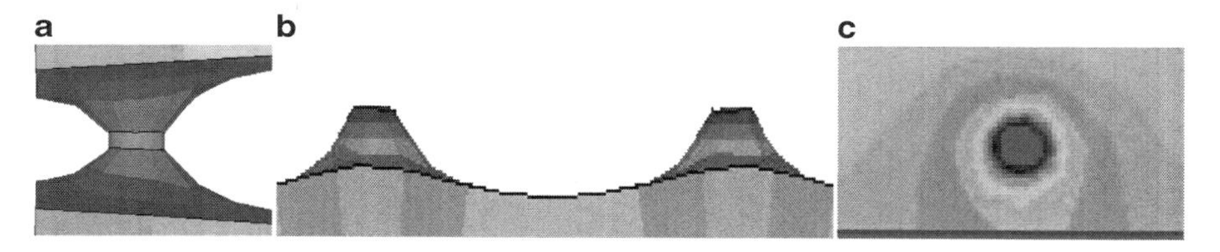

Fig. 4 Crack surface contours around rebar, (**a**) cross section along crack path through rebar, (**b**) typical crack profile across the flow path between rebar at the edge and (**c**) typical crack surface contour in plan around a rebar

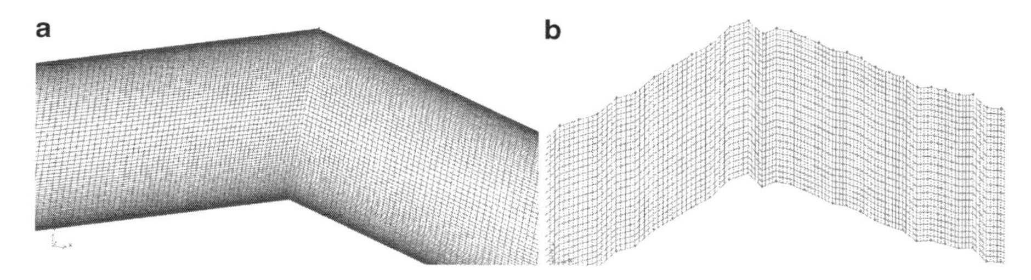

Fig. 5 Typical grids used for numerical computation of pressurized airflow using FVM, (**a**) crack model with global tortuousness alone, (**b**) refined crack model with both global and local tortuousness

with (a) the global tortuousness alone and (b) the refined crack model incorporating the local tortuousness are depicted in Fig. 5.

Numerical procedure established for airflow through 2D crack path was extended to 3D crack surfaces for airflow computations through cracks in reinforced concrete.

3 Results and Discussion

3.1 *Airflow Through Cracks in Plain Concrete*

Considering enormous computational resources and convergence issues associated with large-size models, a comparative study of flow rate was conducted for crack lengths of 15, 40 and 70 mm and crack widths of 0.1, 0.2 and 0.3 mm with same model parameters to explore feasibility of restricting crack length for computational model. The study confirms that a crack length as small as 15 mm could also be considered as a representative sample for airflow studies. In view of this, an intermediate crack length of 40 mm was chosen for airflow computations with the global tortuousness. The crack length was restricted to 15 mm for the refined model incorporating the local tortuousness because of computational constraints arising due to model size.

The results in terms of friction factor (f) as a function of Reynolds number (Re) for the Boussa crack model representing the global tortuousness of the crack are shown in Fig. 6a. Several different pressure gradients (0.25–1 bar/m) were used in the computations to generate the plot. The results reported by Gelain and Vendel [3] for plain concrete are superposed in the figure and the comparison is noticed to be within 25% for majority of the data points. Figure 6b depicts f versus Re along with a curve fit.

The computational study was repeated with the refined crack model incorporating the local tortuousness besides the global tortuousness. The results are shown in Fig. 7a along with the experimental results of Gelain and Vendel [3].

While most of the data points from refined model computation and experimental values are within 9%, the maximum difference in friction factor (f) values is within 16%. Figure 7b depicts f versus Re along with a curve fit for the results from the refined model.

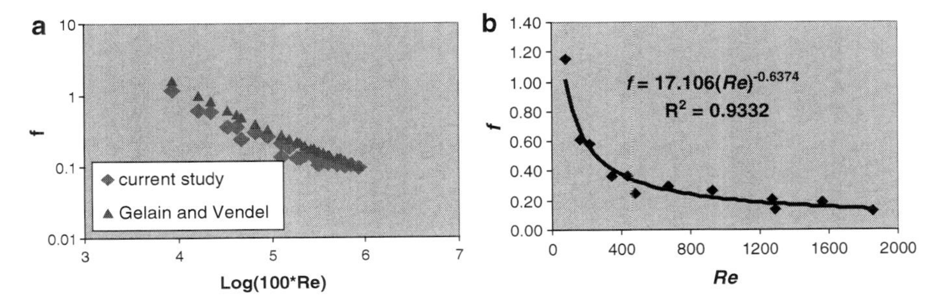

Fig. 6 Plot of f versus Re for the crack model with global tortuousness alone, (**a**) comparison of computational and experimental results, (**b**) curve-fit for computational results

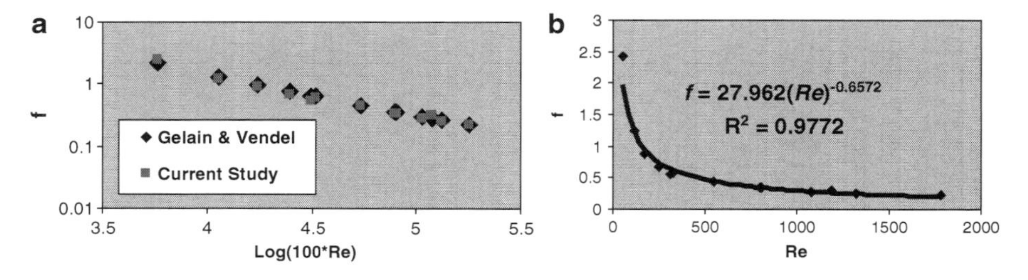

Fig. 7 Plot of f versus Re for the refined crack model incorporating local tortuousness, (**a**) comparison of computational and experimental results, (**b**) curve-fit for computational results

Table 1 Experimental and computational airflow rates through cracks in RC

Measured flow through reinforced concrete test specimen L4 [1]	Flow calculated from refined 2D crack model for plain concrete	3D crack model for plain concrete		3D crack model with modified morphology due to rebar and flow area constricted in the central region of the test specimen			
		Calculated flow	Diff. w.r.t. (2) (%)	Calculated flow	Diff. w.r.t. (1) (%)	Diff. w.r.t. (2) (%)	Diff. w.r.t. (3) (%)
(1)	(2)	(3)	(4)	(5)	(6)	(7)	(8)
5.18e-4	7.084e-4	8.288e-4	+17	5.305e-4	+2.4	−25.11	−36

3.2 *Airflow Through Cracks in Reinforced Concrete*

Airflow calculations were performed for specified pressure difference of 106 kPa across the specimen thickness of 178 mm (i.e. pressure gradient of 5.96 bar/m) and specified average crack width of 0.06 mm using the computational model generated for 3D crack morphology of the test specimen, as described earlier in numerical procedure section, as well as with refined 2D crack model. The airflow computations were repeated with 3D crack model modified by restricting the crack flow surface area in the central region of the model as observed in the non-linear stress analysis. Details of the airflow computational models are reported in Bishnoi et al. [6].

The airflow rates through the two 3D crack models were compared with the flow rate through a 3D plane crack in which two plane surfaces were separated by the average crack width value instead of the tortuous crack morphology. All the 3D crack morphology computations for flow rate were adjusted to account for the correction due to local tortuousness, which could not be accounted directly in these models. A detailed comparison of flow rates (m^3/s) from different 2D and 3D models is reported in Bishnoi et al. [6]. Results relevant to the scope of this chapter are provided in Table 1.

From the above comparison, it is seen that the 3D crack model, incorporating the crack morphology derived by considering the local effect of rebar including the restricted flow surface area in the central region of the test specimen, a phenomenon revealed in RC members through non-linear stress analysis [6], as well as the effect of local tortuousness, provide excellent comparison with the flow obtained experimentally. The difference between the flow rate calculated from 2D crack model for plain concrete and 3D model for plain crack is attributed to the global crack tortuousness, which has been incorporated in the 2D model only and not in 3D model for plain concrete.

4 Conclusions

Results of computational studies for pressurized airflow through cracks in concrete are reported. The statistical crack model available in the literature has been refined to incorporate local tortuousness of the crack due to the smallest material inhomogeneity that can affect the airflow rates. The local tortuousness is introduced using fractal geometry-based curves. The computational results for pressurized airflow through refined crack models are compared with the experimental values reported in the literature for cracks in plain concrete and reinforced concrete. The comparison shows very good agreement between the computational results and the experimental values.

These studies have certain limitations. The parameters considered for fractal curve approximation of the crack tortuousness are the surface roughness and the limiting value of straight crack segment, which are considered to be functions of the level of material inhomogeneity affecting the crack propagation and its deflection from the straight path. These parameters have been approximated from the knowledge of cement paste properties and parametric studies for airflow were conducted to arrive at the most appropriate values. Further studies are required to establish these parameters for different grades of concrete and sensitivity of the airflow results to these parameters. In case of reinforced concrete, establishing the size of uncracked zone in the interiors of the concrete structural elements and sensitivity of the airflow results to this parameter requires further studies.

References

1. Rizkalla SH, Lau BL, Simmonds SH (1984) Air leakage characteristics in reinforced concrete. J Struct Eng (ASCE) 110(5):1149–1162
2. Riva P, Brusa L, Contri P, Imperato L (1999) Prediction of air and steam leak rate through cracked reinforced concrete panels. Nucl Eng Des 192:13–30
3. Gelain T, Vendel J (2008) Research works on contamination transfers through cracked concrete walls. Nucl Eng Des 238:1159–1165
4. Boussa H, Tognazzi-Lawrence C, La Borderie C (2001) A model for computation of leakage through damaged concrete structures. Cem Concr Compos 23:279–287

5. Bishnoi LR, Vedula RP, Gupta SK (2010) Characterization of pressurized air leakage and aerosol transport through cracks in concrete. In: Proceedings of the 37th international and 4th national conference on fluid mechanics and fluid power (FMFP2010). Paper ID: 311, Indian Institute of Technology Madras, Chennai, India, 16–18 December
6. Bishnoi LR, Vedula RP, Gupta SK (2010) Effect of reinforcing steel on pressurized air leakage through cracks in concrete, transactions, SMiRT21. Div-III: Paper ID# 496, New Delhi, India, 6–11 November 2011
7. Hearn D, Baker M (1994) Computer graphics, 2nd edn. Prentice Hall, Englewood Cliffs
8. Neville AM (2000) Properties of concrete. Pearson Education Asia Pte. Ltd., Harlow

Experiences in Subsurface Investigations Using GPR

G. Venkatachalam, N. Muniappan, and A. Hebsur

Abstract The primary source of information in a geotechnical engineering exploration programme is boreholes. However, in many projects, the investigation is inadequate, and uncertainties about the subsurface prevail and adversely affect the cost-effectiveness and reliability of the design process. Safe designs are adopted to overcome the lack of information. Since collecting additional information through boreholes may be difficult, a non-invasive technique could help. Ground-penetrating radar (GPR) is one such non-invasive technique suitable for shallow and deep subsurface mapping. It has, in recent times, emerged as a viable tool for subsurface mapping. With the choice of GPR antennae of appropriate frequencies, one can explore to depths varying from 3 to 100 m. This chapter describes the principle and practice of operation of a GPR and presents four instances where uncertainties and inadequate knowledge of subsurface were alleviated through GPR surveys and critical decisions regarding foundations could be taken confidently.

Keywords Uncertainties • GPR • Non-invasive technique • Radargram • Subsurface mapping • Foundation

1 Introduction

Geotechnical exploration is an expensive activity in any project and, not infrequently, accorded inadequate importance. Therefore, a geotechnical engineer has to frequently deal with situations involving limited in situ investigations and

G. Venkatachalam (✉)
Emeritus Fellow, Department of Civil Engineering, Indian Institute of Technology Bombay, Mumbai, India
e-mail: gveecivil@gmail.com

N. Muniappan • A. Hebsur
Research Scholar, Department of Civil Engineering, Indian Institute of Technology Bombay, Mumbai, India
e-mail: muniappan19@gmail.com; almelu84@gmail.com

S. Chakraborty and G. Bhattacharya (eds.), *Proceedings of the International Symposium on Engineering under Uncertainty: Safety Assessment and Management (ISEUSAM - 2012)*,
DOI 10.1007/978-81-322-0757-3_25,

inadequate data. Uncertainties, therefore, are not new in geotechnical engineering. The errors and uncertainties associated with the data make the decision making process difficult and of questionable reliability.

1.1 Uncertainties

There are several perceptions regarding uncertainties. An excellent idea about uncertainties can be had from Juang and Elton [11], Dodagoudar and Venkatachalam [6]. Very broadly, they are classified as aleatory and epistemic [2]. Comparative studies of different methods may be seen in Bhattacharya et al. [3] and Venkatachalam [19]. Uncertainties in geotechnical engineering arise at the exploration stage itself and propagate right up to the performance stage owing to a number of subjective factors as listed below:

1. Exploration stage:
   - (a) Field description of subsurface strata in borehole logs leaves scope for interpretation, since it is not always done by a domain specialist.
   - (b) Descriptions of strata even at the same location could vary when exploration is done by different agencies.
   - (c) Delineating the boundaries between strata involves subjective averaging.
2. Sampling and testing stage:
   - (a) Uncertainties due to spatial variability in properties.
   - (b) Uncertainties due to statistically inadequate sampling.
   - (c) Errors due to testing techniques, testing equipment and test conditions not simulating field conditions – (1) systematic errors and (2) random errors.
   - (d) Errors due to skill of the testing personnel.
   - (e) Errors due to scale of the problem – local or regional; some errors of a spatial nature get averaged out when the region is large.
3. Analysis stage:
   - (a) Level of abstraction/idealization of problem – in this case, limit equilibrium method and its attendant assumptions and idealizations
   - (b) Accuracy of computational models used/performance function used
   - (c) Uncertainties in estimating triggering factors
   - (d) Modelling support measures, e.g. reinforcement
4. Design stage:
   - (a) Simplified design philosophies
   - (b) Subjective judgement exercised during choice of design parameters
5. Performance stage:
   - (a) Intermittent slope modification
   - (b) Use of same factor of safety for short- and long-term performance

Due consideration must be given for the uncertainties arising out of these variabilities. And that is what reliability analysis does. It may be said that reliability analysis is all about the confidence that can be reposed on the deterministic analysis, and the two are to be looked upon and used as complementary techniques. Reliability engineering helps to address these uncertainties in the data. However, a way to supplement the data with more information and remove the uncertainties to an extent possible is a better alternative. Fortunately, a trend towards conducting a GPR survey as a prelude to geotechnical exploration is now evident. This is a welcome trend because it gives an opportunity to plan the geotechnical investigations economically and optimally so that no surprises are thrown up during actual foundation excavation and construction. Ground penetrating radar (GPR) is one such non-invasive technique suitable for shallow and deep subsurface mapping.

Since collecting more information through additional boreholes may be difficult, a non-invasive technique could help. GPR has, in recent times, emerged as a viable tool for subsurface mapping. With the choice of GPR antennae of appropriate frequencies, one can explore to depths varying from 3 to 100 m. This chapter describes the principle and practice of operation of a GPR and presents four instances where uncertainties and inadequate knowledge of subsurface were alleviated through GPR surveys and critical decisions regarding foundations could be taken confidently.

2 Ground Penetrating Radar

Ground penetrating radar, also referred to as georadar or GPR, is a tool for subsurface imaging, which is a form of close-range remote sensing. It is a type of radar which can be used to detect objects buried underground, in contrast with the radar used to identify features on the land or in the ocean or sky. This method uses the radio waves to probe the 'ground', which simply means any low-loss dielectric material (Jol [12]). It operates in the frequency range from 10 MHz to 2 GHz (microwave band), depending on the device. It has an antenna emitting electromagnetic energy and another that receives the reflected energy from the surfaces as well as from the inner layers. It works on the basis of contrasts in dielectric constant in the subsurface. The energy reflected is transformed into visual images, which provide extensive data on the subsurface materials and objects, when interpreted properly.

Ground penetrating radar (GPR) is perhaps the only geophysical technique, which detects buried objects and helps to get continuous, real-time profiles of the subsurface, through imaging in an appropriate frequency of the electromagnetic spectrum.

The degree of information obtainable about the buried objects or subsurface layers depends on the GPR antenna parameters, mainly central frequency, radial and lateral resolutions. At lower frequencies, GPR pulses can get information for higher depths but at lower resolution; similarly at higher frequencies, GPR pulses have lesser penetrability but much better resolution.

Subsurface mapping applications are reported in archaeology, foundation engineering and seepage. Conyers et al. [4] have described archaeological applications, wherein successive layers of excavation are carefully planned after ascertaining the presence of buried artefacts. Also, Nobes and Lintott [16] have successfully mapped the foundation of Victorian-age lecture hall using GPR, and Nobes and Sikma [17] tested stability of the foundations of the Cathedral of the Blessed Sacrament church. Successful identification of subsurface oil leaks by GPR is given by King [13], and accurate determination of the failing concrete floor slab which is 50 m in diameter and 8 m deep which is 7 m below the ground level was carried out by King et al. [14]. GPR has been used successfully mainly for mapping underground utilities such as pipes, conduits and cables. Underground features which have been successfully detected include buried land mines (Daniels [5]), pipes (Shihab and Al-Nuaimy [18]) and erosional voids (Anchuela et al. [1]). Gomez-Ortiz et al. [8] have reported the use of a number of frequencies, namely, 100, 200, 250 and 500 MHz for imaging the shallow substructure in coastal Spain. Whereas the 100-MHz antenna was useful in locating the water table at 5–6 m depth, the 200- to 500-MHz antennae allowed the determination of the internal structure of the coastal sand deposits helping to infer their progradation dynamics.

An interesting application of GPR is in detecting clandestine mass burial sites during criminal investigations. Fiedler et al. [7] and Koppenjan et al. [15] have reported successfully tracing buried bodies of humans, pigs and buffaloes using 200- to 900-MHz frequencies. The higher frequency gave slightly better penetration below the concrete slab covering the cadavers and the effectiveness of degradation of the body also could be judged.

3 Applications

In this chapter, four instances are illustrated, wherein GPR has been successfully used to gather useful information not available in boreholes. In these instances, the main uncertainty was regarding the lateral or vertical extension of a specific subsurface layer, which was to be the founding layer for either a shallow or a deep foundation. The details of the studies are given below.

4 Study 1

This was a site where a 16-storey residential building was being planned. Figure 1 shows the site map along with the locations of bore holes BH1, BH2, BH3 and BH4. The issue here was the uncertainty concerning the continuity of the breccia layer between the boreholes as the boreholes were far from each other, especially boreholes 2 and 4. In view of this, the usual linear interpolation did not appear reasonable. This had implications for deciding the level at which the pile foundations of the proposed building should be terminated.

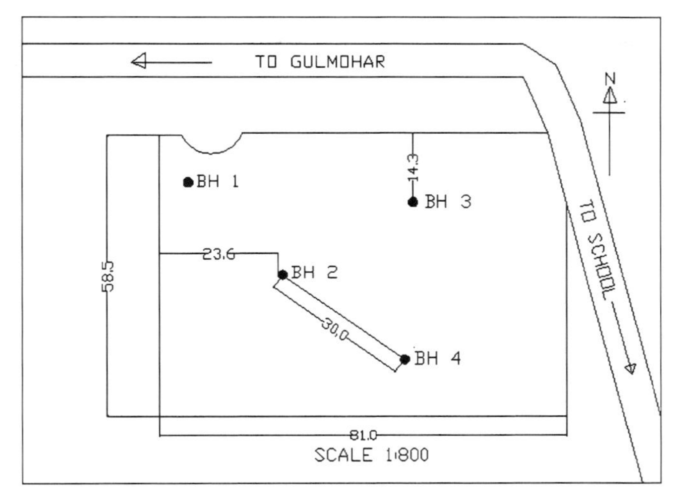

Fig. 1 Site map along with the line of traverse study area 1

Table 1 Initial settings made during data collection

Antenna used (MHz)	80
Mode of data collection	2D (point mode)
Mode of polarization	Co-polarized
Transmitter and receiver spacing (m)	1
Length or area of survey	30 m
Dielectric constant	30
Range (ns)	370
Scans/s	64
Scans/m	5
Vertical IIR filter range (MHz)	60–100
Horizontal IIR stack (no. of scans)	64

GPR data was collected along several traverses, including along the line from BH2 to BH4. Initial settings used are given in Table 1.

The borehole data BH2 and BH4 along with the core recovery values are as shown in Fig. 2. According to BH2, the location where it was driven has four layers. BH4, on the other hand, shows only 2 layers. Though the core recovery values are changing, there is no change in the layers. It is also quite possible that the layers are not captured properly as these stratifications are done manually on the basis of visual interpretation. This necessitated verification of the thicknesses of the layers, particularly, 3 and 4.

The GPR scan data is processed for position correction, data filtering using spatial FFT and stretching, (GSSI 2007) [9]. Spatial FFT is applied with low- and high-pass filters of 120 and 40 MHz, respectively, to remove the noise. For clarity, the filtered data is stretched four times using horizontal scaling. Figure 3a, b show the raw data and the corrected and stretched data.

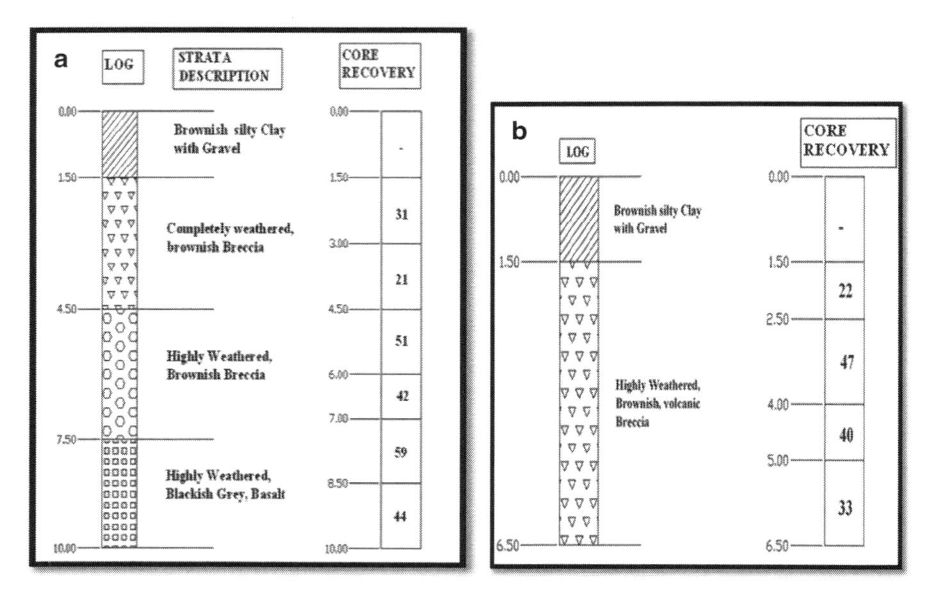

Fig. 2 Borehole data (**a**) BH2 data with core recovery values (**b**) BH4 data with core recovery values

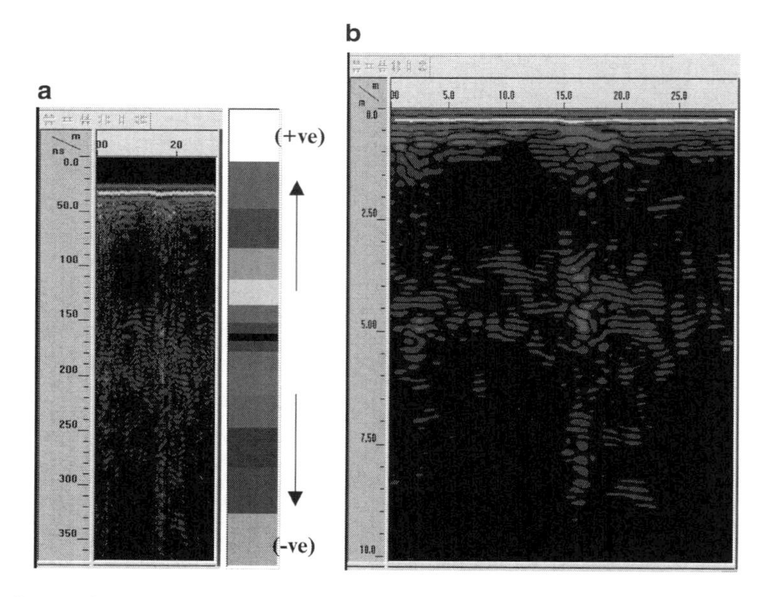

Fig. 3 Raw and post-processed data. (**a**) Raw data (**b**) corrected and stretched data

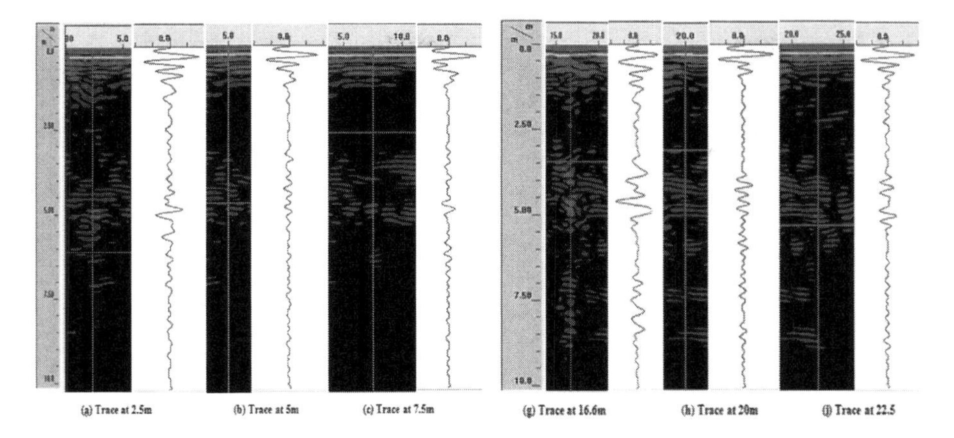

Fig. 4 Traces along the survey traverse

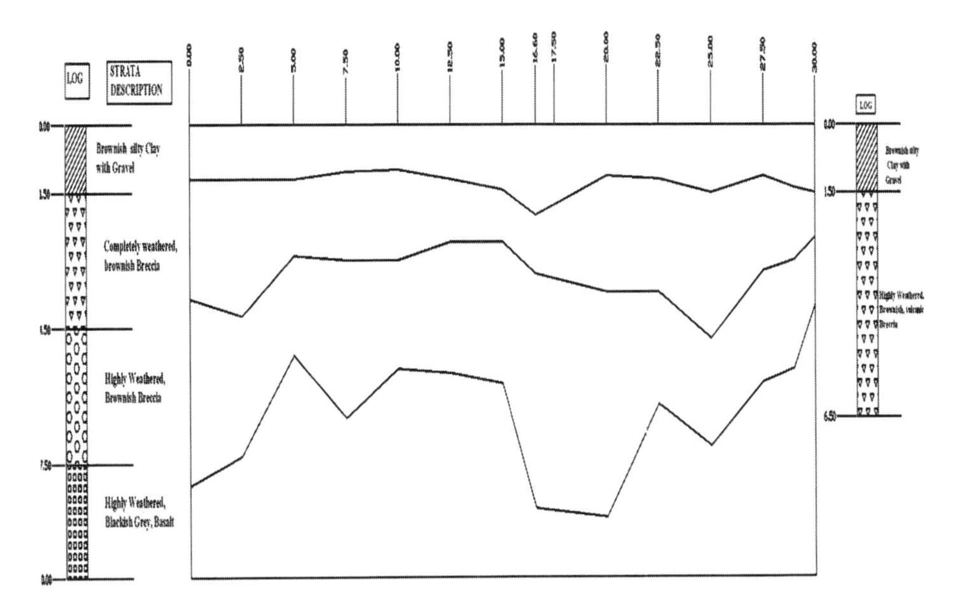

Fig. 5 Variation of subsurface layers from BH2 to BH4

From the above stretched data, at every interval of 2.5 m, wiggle traces have been taken and those up to 7.5 m from either side are presented in Fig. 4. Based on the wiggle traces and the radargram, a subsurface section is drawn as depicted in Fig. 5. The figure shows anomalies in third layer thickness at two places –2.5–7.5 m and 15–22.5 m. This was useful in deciding the founding depth of the foundations at this location. This also raised doubts about the log of BH4. A second look at the core recovery values in BH4 showed that, in fact, there are two breccia layers of different degrees of weathering.

5 Study 2

This study also relates to GPR subsurface profiling at a construction site. A four-storey building was planned at this site. The issue here was that only three boreholes were drilled almost parallel to the eastern side of the plot thus leaving the central and western sides unexplored (Fig. 6). In order to get this missing information, data was collected with 400, 200, 80 and 40 MHz antennae, with penetrabilities ranging from 3 to 15 m. GPR data was collected along 2D grids and linear traverses. A single traverse was taken from borehole location BH3 to borehole location BH1. The results of the single traverse are presented in this chapter. This was also useful to check the adequacy of the borehole investigation.

Figure 6 shows the site map along with the locations of boreholes BH1, BH2 and BH3. GPR data was collected along the line of traverse AB stretching from near BH3 to a point close to BH1.

The data acquisition settings are given below in Table 2.

The data obtained for traverse AB for 200 and 400 MHz, after post-processing are presented in Figs. 7 and 8. For 80 MHz, the results are shown in Fig. 9 in the form of wiggle traces for the first 10 m and last 10 m.

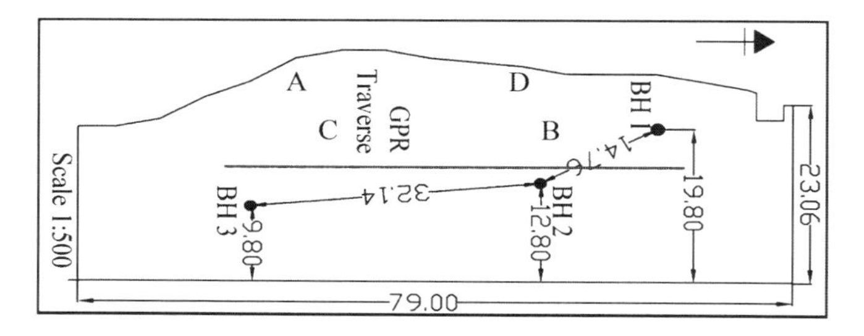

Fig. 6 Site map along with the line of traverse study area 2

Table 2 Initial settings made during data collection

Antenna used (MHz)	200, 400, 80, 40
Mode of data collection: 200, 400	Dist. mode
80, 40	Point mode
Mode of polarization	Co-polarized
Transmitter and receiver spacing (m)	1
Length or area of survey	50 m
Dielectric constant	6
Range (ns)	370
Scans/s	64
Scans/m	5
Vertical IIR filter range (MHz)	60–100
Horizontal IIR stack (no. of scans)	64

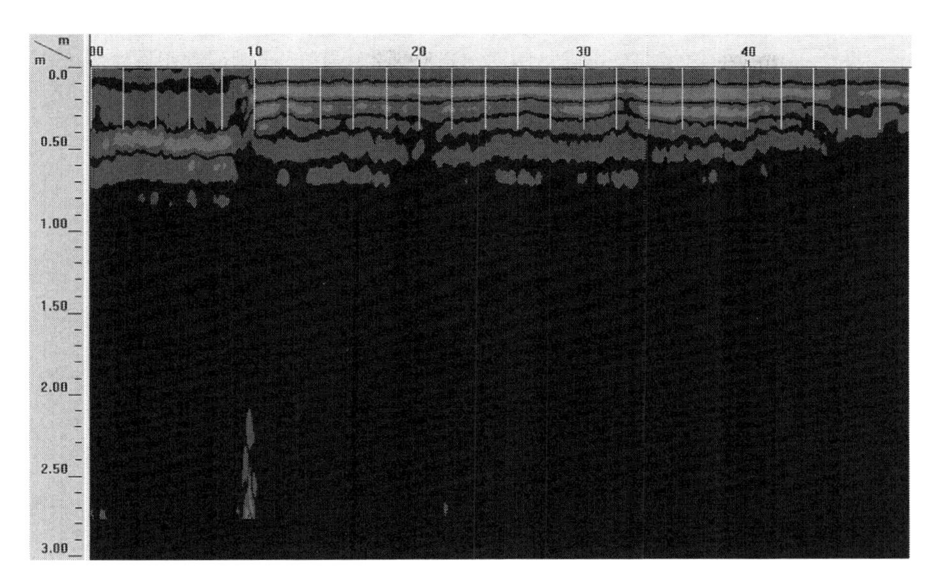

Fig. 7 Post-processed radargram for 200 MHz

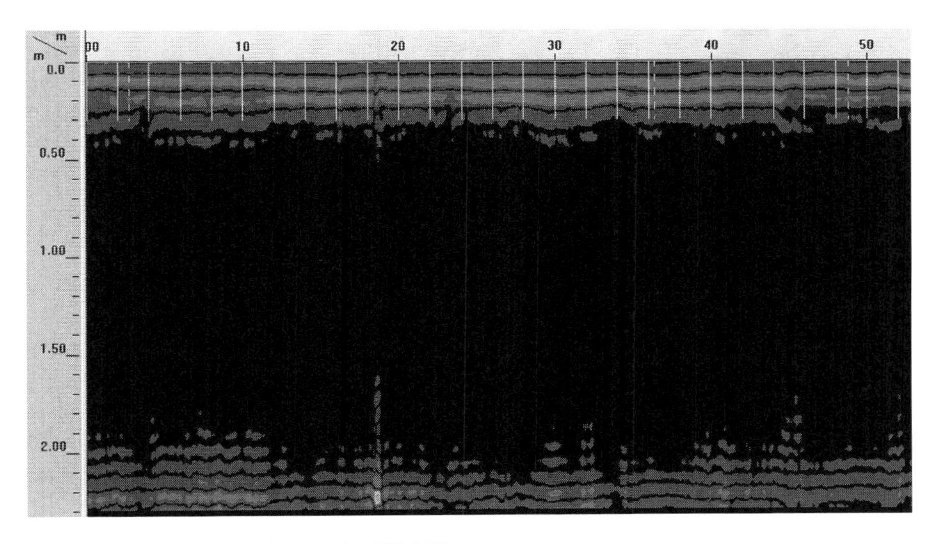

Fig. 8 Post-processed radargram for 400 MHz

Some obvious observations based on the radargrams are the following: the 200- and 400-MHz antennae show the presence of a uniform layer up to a depth of 0.5 m followed by another more reflecting material up to about 2.5 m. However, 80-MHz data shows that the second layer is extending beyond 2.5 m, possibly up to about 7 m. Beyond 7 m, 80 MHz loses the signal strength and is not sensitive enough for

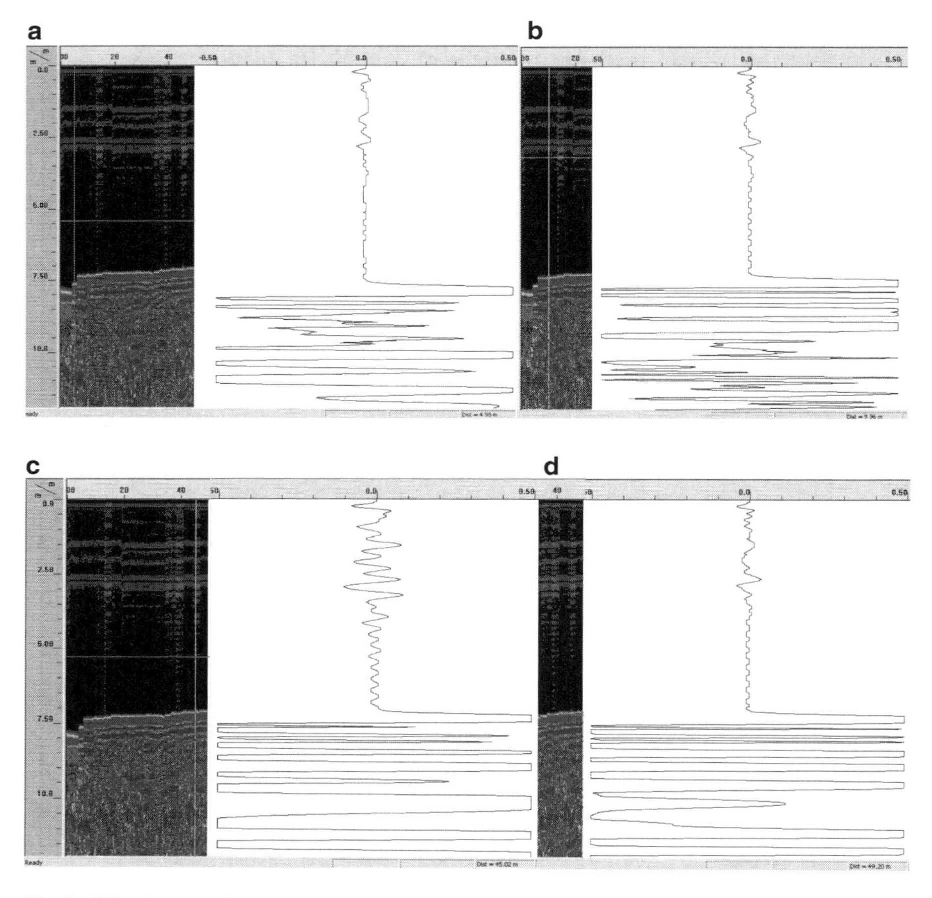

Fig. 9 Wiggle traces from 80 MHz data at 5 m interval (**a**) Trace at 5 m (**b**) trace at 10 m, (**c**) trace at 45 m (**d**) trace at 50 m

subsurface profiling. A closer study shows that there is a small zone of non-reflecting layer at around 5–7 m. This is typical of a highly weathered material in which signals are completely diffracted.

However, a more appropriate method of delineation of the layers would be through digital techniques of signal processing. In this case, the Hilbert transform is useful because it helps to separate the real and the imaginary parts and hence, the magnitude and the phase and frequency components as shown in Fig. 10. The layers of soil and rock are clearly brought out in these figures, especially in the phase components.

The GPR results are compared in Fig. 11 with the available borehole data. Several anomalies were detected at the unexplored locations of the site. One such instance at Point C (Fig. 6) is shown in Fig. 12 along with the profile as obtained

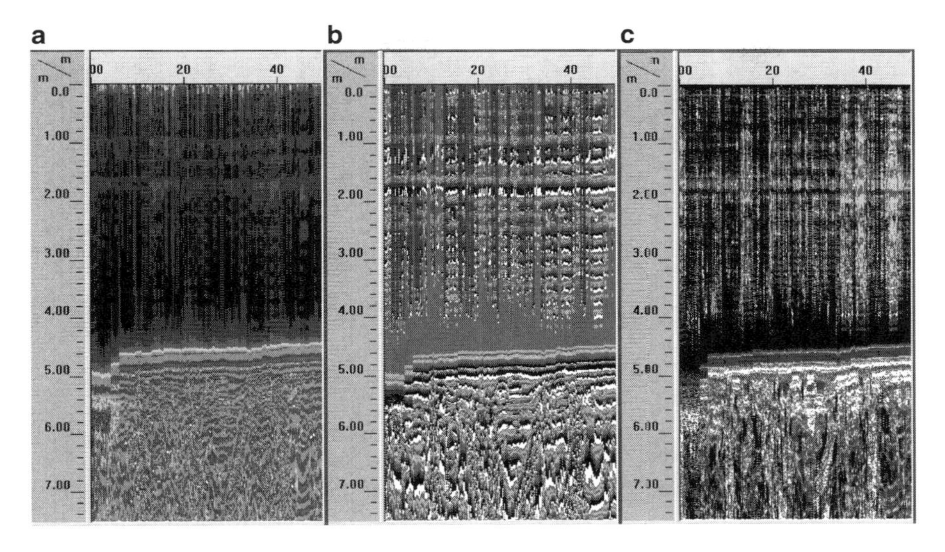

Fig. 10 HT components for 80-MHz data (a) Magnitude (b) phase (c) frequency

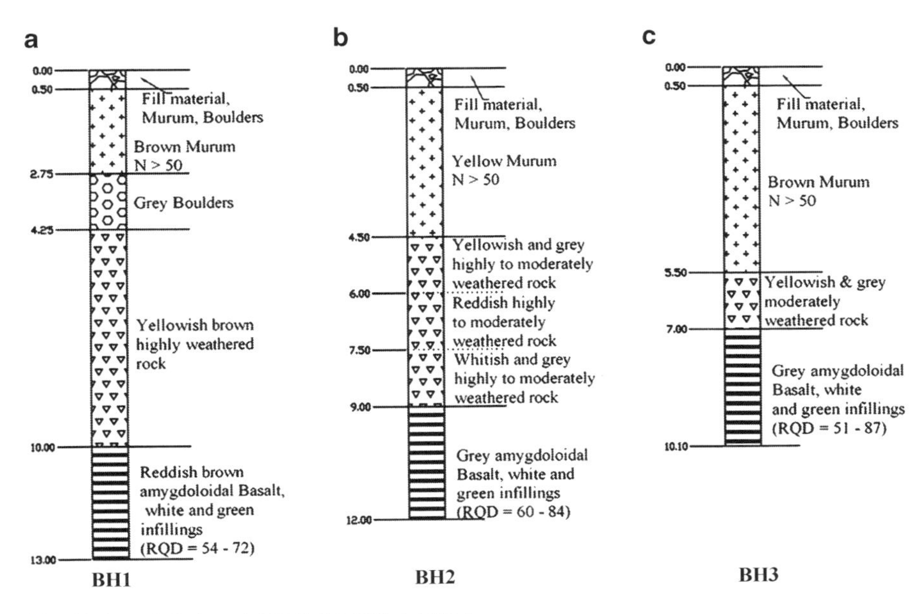

Fig. 11 Borehole logs (a)BH1 (b) BH2 (c) BH3

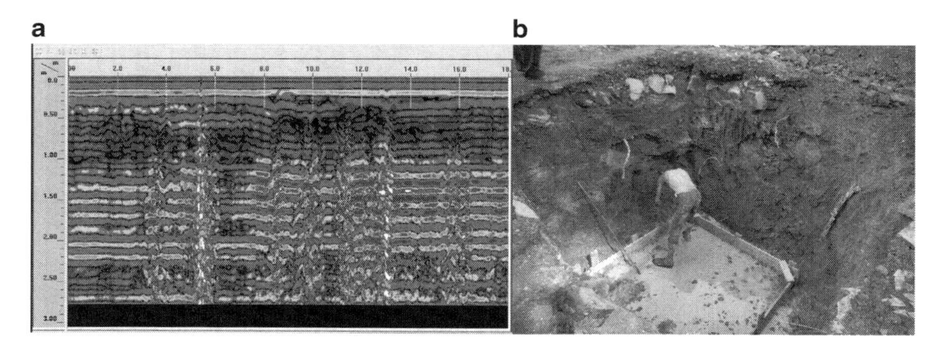

Fig. 12 Anomaly at location C (**a**) 400 MHz profile, (**b**) Photograph of rock layer

after foundation excavation was carried out. A similar situation prevailed at Point D as well. This indicated that extrapolation from borehole data to the western side of the plot would have been grossly in error. GPR studies helped detecting the anomalies in a cost-effective manner and be forewarned. However, in this case, the error was on the safer side as the rock layer was closer to the surface than predicted from boreholes.

6 Study 3

This was carried out at a construction site sandwiched between two existing buildings (Hebsur et al. [10]). A building existed at this site earlier and was subsequently demolished. The issue here was to detect buried utilities which connect the two flanking buildings and other objects, if any. This was necessary to avoid damage to buried facilities during excavation for foundation for the proposed six-storey building at the same site. The layout of the site is shown in Fig. 13.

The GPR studies included scans using 200- and 80-MHz antennae. The initial settings are shown in Table 3. The 21 m $\times$ 30 m grid used for the 200-MHz survey is shown in Fig. 13. There were 7 X traverses at 5-m interval and 8 Y traverses at 3-m interval. The data was collected in zigzag mode. The 80-MHz data was collected along three transects. Along one of the transects, both XX and YY polarizations were used.

The post-processed data is presented in 3D view (Fig. 14). One of the transects, namely, X1, is also presented as a B scan in Fig. 15.

The study yielded very valuable information about the subsurface. It was found that there were no buried utility lines to be taken care of. More importantly, it showed that there were several shallow footings which had to be taken cognizance of in order to excavate without damage to the excavating equipment. Thus, in this study, the uncertainties relating to the nature and position of buried objects were resolved without doubt.

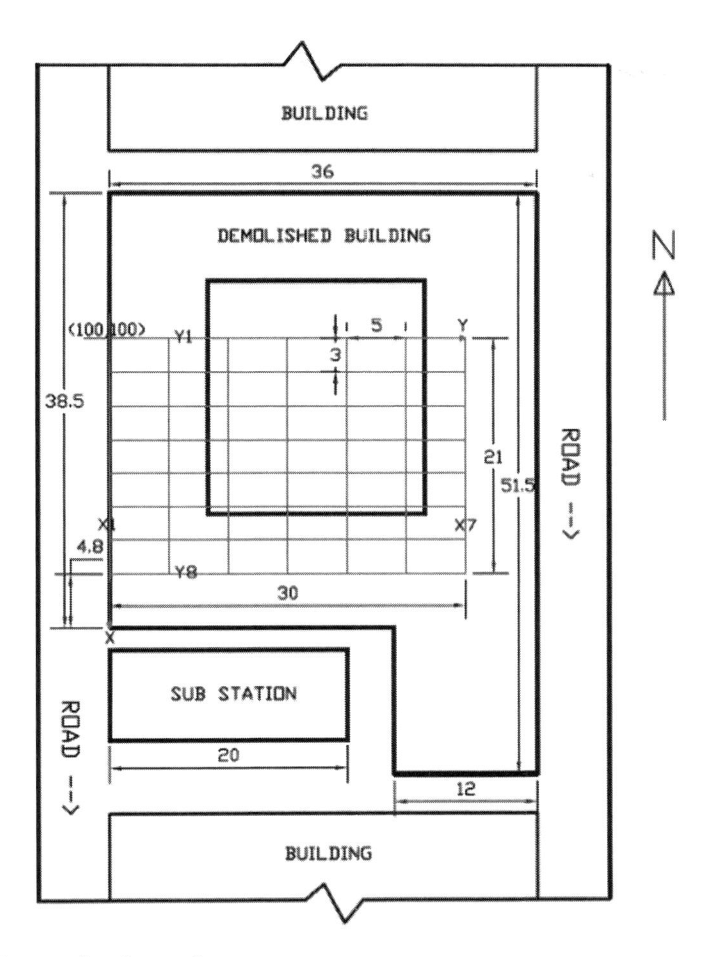

Fig. 13 Layout of study area 3

Table 3 Initial settings made during data collection

Antenna used (MHz)	200	80	
Mode of data collection	3D (time mode)	2D (point mode)	
Mode of polarization	XX & YY	XX	YY
Transmitter and receiver spacing(m)	(Fixed) 0.332	1	2
Area (m^2) or length of survey (m)	21×30	21	21
Dielectric constant	4.5	5	5
Range (ns)	80	200	200
Scans/s	64	64	64
Scans/m	50	5	5
Vertical IIR filter range (MHz)	180–220	60–100	60–100
Horizontal IIR stack (no. of scans)	3	64	64

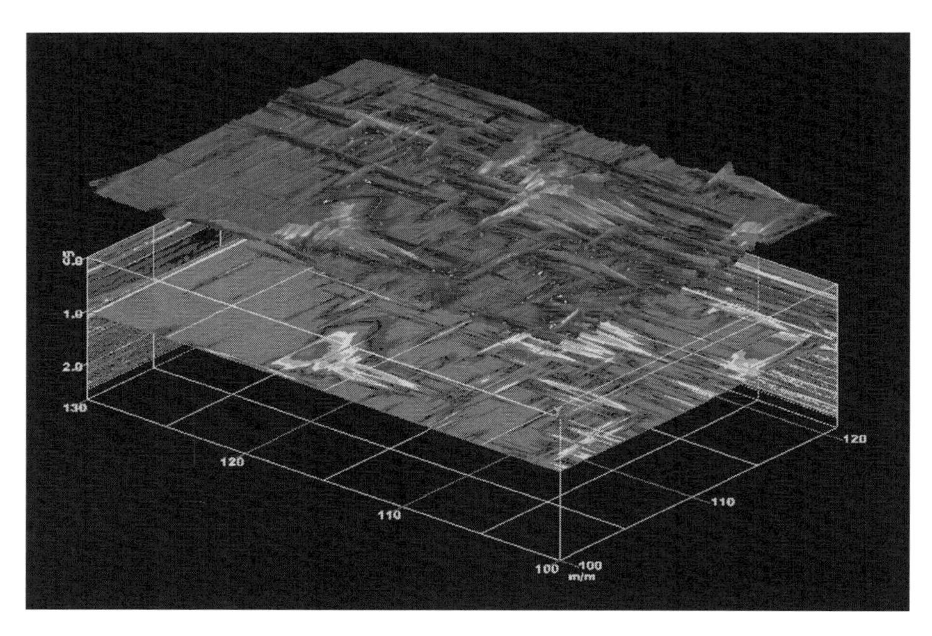

Fig. 14 200-MHz data in 3D mode at 1.4 m depth

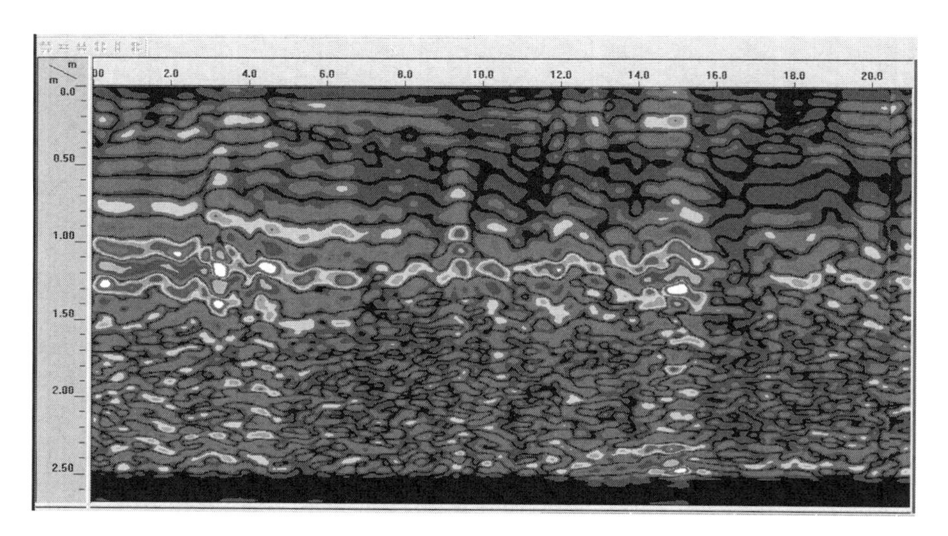

Fig. 15 Post-processed transect X1 data

7 Study 4

This study relates to the site of a tall tower of over 60 storeys planned to be erected on a slope (Fig. 16). As per prevailing construction practice, it was planned to use touch piles to stabilize the ground before excavation. The obvious choice of foundation type was piles. Thus, from considerations of slope stabilization and pile foundation, it was necessary to explore the depth and extent of suitable founding layers. The study reported here pertains to a small segment between two boreholes, 1 and 2. The issue here was that, the rock layer in one borehole was highly weathered at the top in BH2 and at the other it was not. Therefore, it was necessary to determine the depth and lateral spread of the weathered rock layer.

GPR survey using 80 and 40 MHz antennae was carried out close to the borehole which did not show weathering. The initial settings used are given in Table 4. The post-processed radargram for 40 MHz study is shown below in Fig. 17. The weathered rock layer can be seen to exist at a depth of 20 m and varying in

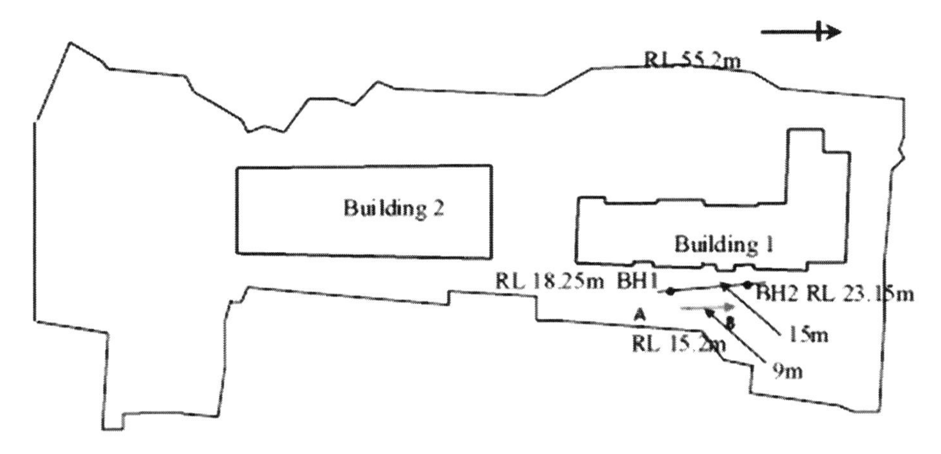

Fig. 16 Layout of study area 4

Table 4 Initial settings made during data collection

Antenna used (MHz)	80 and 40	
Mode of data collection	2D (point mode)	
Mode of polarization	XX	YY
Transmitter and receiver spacing(m)	1	2
Length of survey (m)	9	9
Dielectric constant	5,8,10	5,8,10
Range (ns)	1,000	1,000
Scans/s	64	64
Scans/m	5	5
Vertical IIR filter range (MHz)	10–90	10–90
Horizontal IIR stack (no. of scans)	4	4

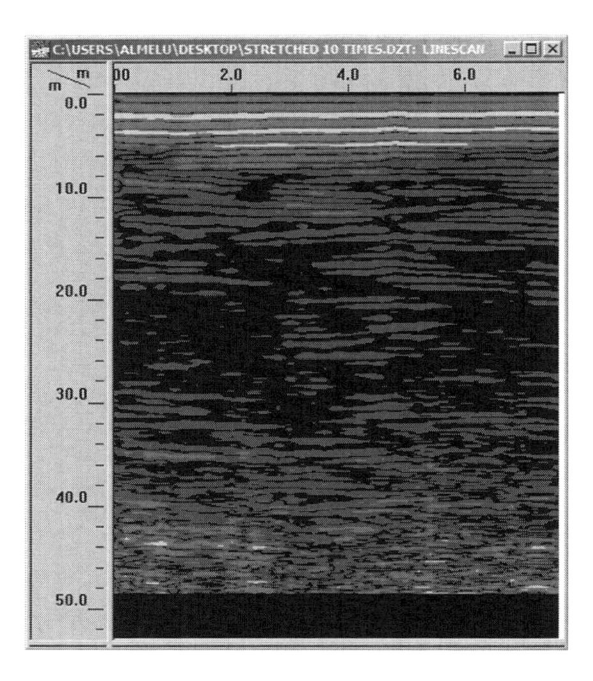

Fig. 17 Post-processed stretched radargram for 40 MHz

thickness in the lateral direction from the first borehole. The requirement at site was to embed the piles in the unweathered rock. The GPR study helped resolve the uncertainty regarding the founding stratum and depth of piles.

8 Other Uncertainties

Perhaps, the most important application of GPR is in the area of utility mapping, where determination of depth, diameter and dielectric constant of the material of the pipe are beset with uncertainties. This requires controlled basic studies on the response of buried utilities under laboratory or field conditions. But, since this is prohibitively expensive, a viable alternative is a simulation through wave propagation modelling. This is an active area of current research.

9 Conclusions

The subsurface profile obtained from GPR is in close agreement with that obtained from boreholes. Hence, GPR profiles can be used for locations within a site which are far from the boreholes and for detecting anomalies, if any. This is particularly useful when questions arise regarding the founding stratum where foundations should rest.

Acknowledgement The help, advice and critical comments and suggestions of Prof. E.P. Rao are gratefully acknowledged.

References

1. Anchuela OP, Casas-Sainz AM, Soriano MA, Pocoví-Juan A (2009) Mapping subsurface karst features with GPR: results and limitations. Environ Geol 58(2):391–399
2. Baecher GB, Christian JT (2003) Reliability and statistics in geotechnical engineering. Wiley, Hoboken
3. Bhattacharya G, Chowdhury SS, Mukherjee S, Chakraborty S (2004) Slope reliability – effect of deterministic and probabilistic methods. In: Babu GLS, Phoon KK (eds) Proceedings of the international workshop on risk assessment in site characterization and geotechnical design (GEORISK – 2004), Bangalore, pp 140–149
4. Conyers LB, Ernenwein EG, Bedal LA (2002) Ground penetrating radar (GPR) mapping as a method for planning excavation strategies. Petra, Jordan. Report of Society for American Archaeology. www.du.edu/lconyer/~petra/petra.07.html. Accessed 2 Apr 2011
5. Daniels DJ (2006) A review of GPR for landmine detection. Sens Imaging 7(3):90–123
6. Dodagoudar GR, Venkatachalam G (2000) Reliability analysis of slopes using fuzzy sets theory. Comput Geotech 27:101–115
7. Fiedler S, Illich B, Berger J, Graw M (2009) The effectiveness of ground-penetrating radar surveys in the location of unmarked burial sites in modern cemeteries. J Appl Geophys 68(3):380–385
8. Gomez-Ortiz D, Pereira M, Martin-Crespo T, Rial FI, Novo A, Lorenzo H, Vidal JR (2009) Joint use of GPR and ERI to image the subsoil structure in a sandy coastal environment. J Coast Res 56:956–960
9. GSSI (2007) RADAN 6.5 User's manual. Geophysical Survey Systems Inc., Salem, p 34
10. Hebsur A, Muniappan N, Rao EP, Venkatachalam G (2010) A methodology for detecting buried solids in second-use sites using GPR. In: Proceedings of Indian geotechnical conference – 2010, Geotrendz, IIT Bombay
11. Juang CH, Elton DJ (1996) A practical approach to uncertainty modelling in geotechnical engineering, ASCE geotechnical special publication no. 58. Uncertainty in the geologic environment: from theory to practice, ASCE, New York, pp 1269–1283
12. Jol HM (2009) Ground penetrating radar: theory and applications, 1st edn. Elsevier Sc., Killington
13. King ML (2000) Locating a subsurface oil leak using ground penetrating radar. In: GPR 2000: Proceedings of the 8th international conference on ground penetrating radar, SPIE, Gold Coast, vol 4084, pp 346–350
14. King ML, Wu DP, Nobes DC (2003) Non-invasive ground penetrating radar investigation of a failing concrete floor slab. In: International symposium (NDT-CE 2003) on non-destructive testing in civil engineering, Berlin
15. Koppenjan SK, Schultz JJ, Falsetti AB, Collins ME, Ono S, Lee H (2003) The application of GPR in Florida for detecting forensic burials. In: Proceedings of SAGEEP, San Antonio, TX, USA
16. Nobes DC, Lintott B (2000) Rutherford's "old tin shed": mapping the foundations of a victorian-age lecture hall. In: Proceedings of the 8th international conference on ground penetrating radar, SPIE, Gold Coast, vol 4084, p 887. doi:10.1117/12.383534, 22–25 May 2000
17. Nobes DC, Sikma TJ (2003) Non-invasive ground penetrating radar in the NDT of the foundations of the Cathedral of the Blessed Sacrament, Christchurch, New Zealand. In: International symposium (NDT-CE 2003) on non-destructive testing in civil engineering, Berlin.
18. Shihab S, Al-Nuaimy W (2005) Radius estimation for cylindrical objects detected by ground penetrating radar. Subsurf Sens Technol Appl 6(2):151–166
19. Venkatachalam G (2006) Reliability analysis of slopes and tunnels. Indian Geotech J 36(1):1–66

A Robust Controller with Active Tendons for Vibration Mitigation in Wind Turbine Rotor Blades

Andrea Staino and Biswajit Basu

Abstract A new robust active controller design to suppress flapwise vibrations in wind turbine rotor blades is presented in this chapter. The control is based on active tendons mounted inside the blades of a horizontal-axis wind turbine (HAWT). The multimodal model proposed includes the effects of centrifugal stiffening, gravity, and aerodynamic loading. Dynamic interaction between the blades and the tower has been included, and variable mass and stiffness per unit length of the blade have been also taken into account. A robust model predictive control (MPC) algorithm has been implemented to study the effectiveness of the proposed active control system. Due to its high complexity and to the variable nature of its operating environment, a wind turbine is subjected to changes in operating condition. As a consequence, significant variations may occur in certain parameters of the turbine. Therefore, robustness is of particular concern for control design purposes. The main advantage of the proposed method is to explicitly incorporate plant model uncertainty in designing the controller. Numerical simulations have been carried out by using data describing aerodynamic and structural properties for a 5-MW wind turbine.

Keywords Wind turbine • Active vibration control • Robust control • Model predictive control • Blade-tower interaction • Blade element momentum theory

1 Introduction

The increased flexibility of the blades (and the tower) in large multi-megawatt wind turbines entails a higher sensitivity to induce mechanical vibrations. The uncontrolled vibrations might significantly shorten components' lifetime, and large amplitude oscillations might even compromise safe operation of the power plant.

A. Staino • B. Basu (✉)
School of Engineering, Trinity College Dublin, College Green, Dublin, Ireland
e-mail: stainoa@tcd.ie; basub@tcd.ie

S. Chakraborty and G. Bhattacharya (eds.), *Proceedings of the International Symposium on Engineering under Uncertainty: Safety Assessment and Management (ISEUSAM - 2012)*, DOI 10.1007/978-81-322-0757-3_26,

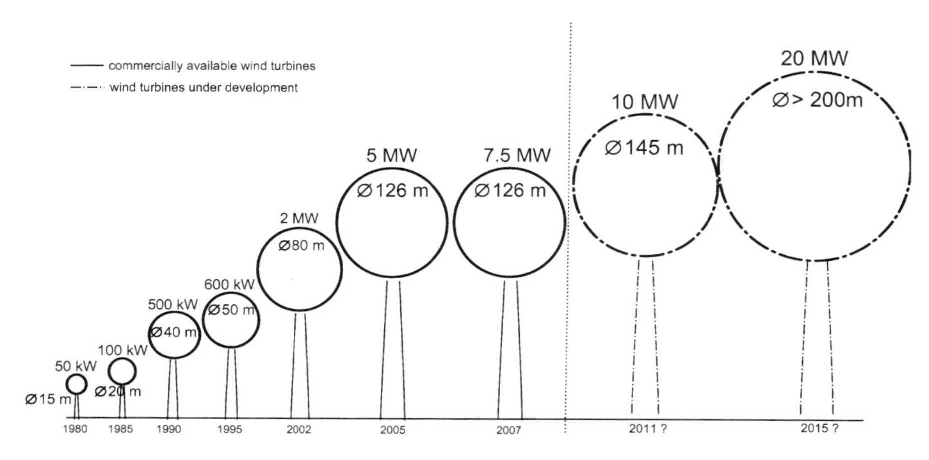

Fig. 1 Growth in size of commercial wind turbine designs

Moreover, by mitigating vibration effects, a more efficient design of the structure can be achieved, as lighter materials and components can be used (Fig. 1).

Different approaches have been investigated in the literature in order to tackle problems associated with aerodynamic loads in wind turbines. Dueñas-Osorio and Basu [3] carried out a probabilistic study on the impact of acceleration-induced damages in the wind turbine components leading to a loss of availability and reduction of power generation. Murtagh et al. [13] developed a dynamic model describing the vibration response of a wind turbine consisting of three flexible rotating blades connected to a flexible supporting tower. In Murtagh et al. [14], the authors extended their work by studying the use of a passive control device (i.e., a tuned mass damper, TMD) for reducing the wind-induced vibrations experienced by the wind turbine structure. Colwell and Basu [2] simulated the response of an offshore wind turbine subjected to wind and wave loadings and found that a considerable reduction in the peak response may be achieved by equipping the plant with a tuned liquid mass damper (TLCD). A semi-active control algorithm for the control of flapwise vibrations in wind turbine blades by means of semi-active tuned mass dampers has been proposed by Arrigan et al. [1]. Recently, significant research has also been carried out into the development of active vibration control in wind turbine blades [6, 11, 12]. An innovative structural control scheme based on active controllers has been proposed by Staino et al. [15].

In this chapter, we illustrate the use of a recently proposed active controller for suppressing the flapwise vibration in blades and mitigate their damaging effects. The effectiveness of the proposed control architecture is tested by simulating the application of a robust model predictive control (MPC) strategy [9], whose main advantage is to explicitly incorporate plant model uncertainty in designing the controller. The MPC algorithm implemented in this study explicitly takes into account physical constraints on the evolution of the variables involved in the process to be controlled.

2 HAWT Model with Controller

A modern multimegawatt wind turbine is a highly complex mechano-electrical system consisting of several components, including structural elements like tower, rotor (consisting of nacelle and blades), and other mechanical and electrical elements such as gears, converters, and transformers as well as a high number of different sensors, actuators, and controllers. It follows that the modeling of large wind turbines is also complex and challenging, and getting accurate models entails studying the dynamics of many degrees of freedom (DOFs), leading to a high-dimensional set of equations. Because we are interested in studying the flapwise dynamics of rotor vibrations in a wind turbine, here we formulate a mathematical model that takes into account only the relevant states or degrees of freedom, representing the flapwise vibration responses and the associated coupling of the blade with the tower/nacelle motion [13]. A schematic representation of a three-bladed HAWT is shown in Fig. 2. The blades are modeled as Bernoulli-Euler cantilever beams of length "*L*," with variable bending stiffness *EI*(*x*) and variable mass per unit length $\mu(x)$ along the length. The blades rotate at a constant

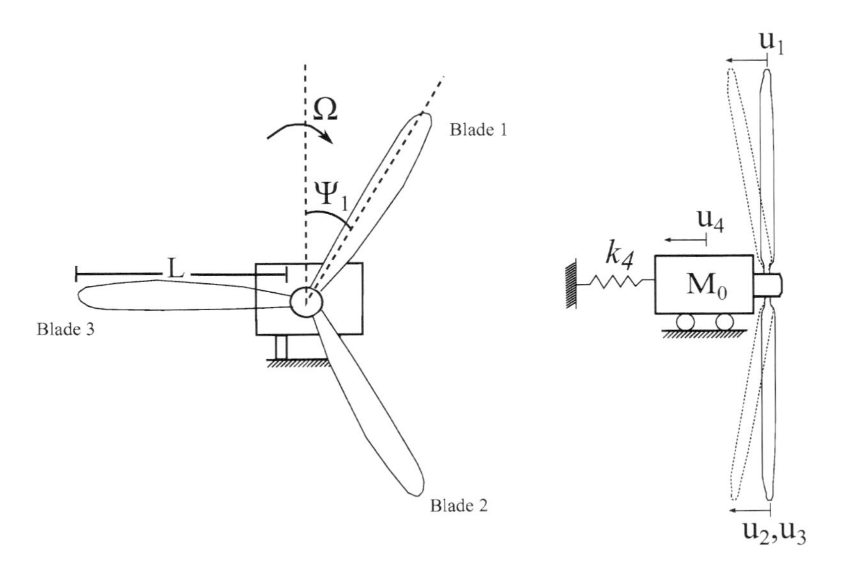

Fig. 2 Flapwise model with nacelle coupling

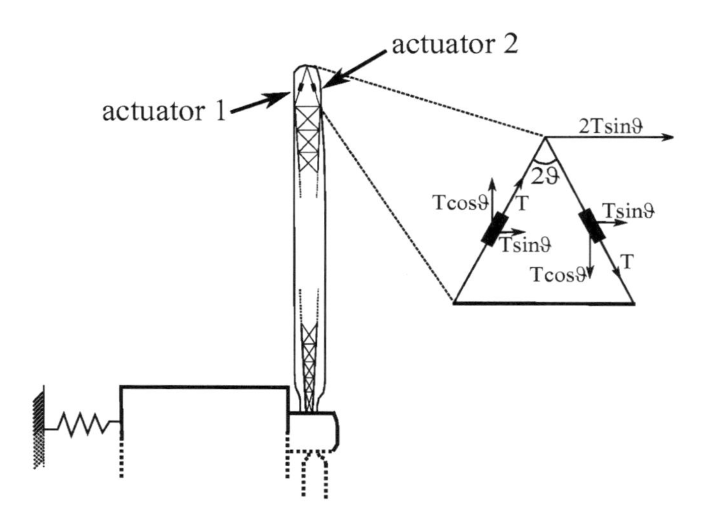

Fig. 3 Flapwise model with the controllers

rotational speed Ω rad s^{-1}, and the azimuthal angle $\Psi_j(t)$ of blade "j" at the time instant "t" is given by

$$\Psi_j(t) = \Psi_1(t) + (j-1)\frac{2\pi}{3}, \ \Psi_1(t) = \Omega t; \ j = 1, 2, 3 \tag{1}$$

The dynamic coupling between the blade and the tower has been included through the fore-aft motion of the nacelle. The tower is modeled as a single degree of freedom system with the mass M_0, which represents the modal mass of the tower and the mass of the nacelle. The variables $u_j(x, t)$, $j = 1;2;3$, and $u_4(t)$ denote the flapwise blade and the nacelle displacements, respectively. The generalized (or modal) stiffness of the tower is represented by k_4.

The control is implemented by means of two linear actuators located inside the blade (Fig. 3). The actuators, each exerting a controlled force $T_i(t)$, are mounted on a frame supported from the nacelle. For the ith blade, the net control force acting on the blade tip in the flapwise direction is denoted by $f_{ci}(t)$. This force is proportional to $T_i(t)$ and to the sine of the angle ϑ and represents the control input to the system.

The reaction forces are transmitted along the supporting structure finally to the nacelle. The support structure for applying the control forces has to satisfy the requirement of transferring the force to the hub. This has to be accomplished ideally by avoiding the generation of a reaction force in the flapwise direction of the blade or practically by eliminating the possibility of any reaction force in the close to medium spatial proximity of the tip. This design condition can be achieved by introducing active elements in the support structure (such as active braces or active tendons) as is typically used in large engineering structures for protection against wind or earthquake loads. A Lagrangian approach has been used in order to derive the equations of motion for the system considered.

2.1 Flapwise Model Formulation

A generalized flexible model of the blade with N modes of vibration is formulated. In the generalized representation of the wind turbine, each mode of vibration is associated to the corresponding modeshape $\Phi_i(x)$, for which an appropriate function approximation can be computed from the eigenanalysis of the blade structural data. The system is therefore described by 3 N +1 generalized coordinates that provides an accurate description of the flexible blade behavior. Let $\underline{q}(t)$ be the vector of the generalized coordinates of the system defined as

$$\underline{q}(t) = \langle\, q_{11}(t) \quad q_{12}(t) \quad \ldots \quad q_{1N}(t) \quad q_{21}(t) \quad \ldots \quad q_{ji}(t) \quad \ldots \quad q_4(t) \,\rangle^{\mathrm{T}} \in R^{3N+1} \tag{2}$$

The degree of freedom $q_{ji}(t)$; $j = 1, 2, 3$, $i = 1, \ldots, N$ denotes the ith flapwise mode for the blade "j." The variable $q_4(t) = u_4(t)$ represents the motion of the nacelle. The total flapwise displacement along the blade is given by

$$u_j(x,t) = \sum_{i=1}^{N} \Phi_i(x) q_{ji}(t) \tag{3}$$

A set of differential equations, describing the dynamics of the flapwise vibrations for the system considered, has been obtained by applying the Euler-Lagrange method:

$$\frac{d}{dt}\left(\frac{\partial T}{\partial \dot{q}_i}\right) - \frac{\partial T}{\partial q_i} + \frac{\partial V}{\partial q_i} = Q_{ext,i} \tag{4}$$

where $Q_{ext,i}$ is the generalized nonconservative load for the degree of freedom i, and the terms T and V are the total kinetic energy and the total potential energy of the system, respectively, given by

$$T = \frac{1}{2}\sum_{j=1}^{3} \int_0^L \mu v_{b,j}^2 dx + \frac{1}{2} M_0 \dot{q}_4^2 \tag{5a}$$

and

$$V = \frac{1}{2}\sum_{j=1}^{3} \int_0^L EI\left(\frac{\partial^2 u_j}{\partial x^2}\right)^2 dx + \frac{1}{2} k_4 q_4^2 \tag{5b}$$

The term $v_{b,j}(x,t)$ in Eq. (5a) denotes the total velocity of the blade in the flapwise direction, i.e., it is calculated by taking into account the nacelle motion. The inclusion of the generalized coordinate $q_4(t)$ into the Lagrangian formulation has allowed modeling of the coupling between the blade and the tower. As found by Hansen [4], the effect of centrifugal stiffening has been added to the model by considering the additional potential energy V_c which is given by:

$$V_c = \frac{1}{2}\Omega^2 \sum_{j=1}^{3} \int_0^L \left[\left(\frac{\partial u_j}{\partial x} \right)^2 \int_x^L \mu(\xi)\xi d\xi \right] dx \tag{6}$$

The contribution given from the component of the gravity acting along the blade axis has also been considered. For the jth blade, the component of the gravitational force at a distance "x" from the blade root is

$$F_{g,j}(x) = -\int_x^L \mu(\xi) g \cos(\psi_j) d\xi = -g\cos(\psi_j) \int_x^L \mu(\xi) d\xi \tag{7}$$

and the potential energy associated is

$$V_g = -\frac{g}{2} \sum_{j=1}^{3} \int_0^L \left[\left(\frac{\partial u_j}{\partial x} \right)^2 \int_x^L \mu(\xi) d\xi \right] dx \cos(\psi_j) \tag{8}$$

The final equations of motion of the HAWT flapwise model with active controllers are expressed in matrix form as

$$\mathbf{M}\ddot{\underline{q}}(t) + \mathbf{C}\dot{\underline{q}}(t) + K\underline{q}(t) = \underline{\boldsymbol{Q}}_{ext}(t), \tag{9}$$

where **M**, **C**, and **K** represent the mass, damping, and stiffness matrices of the system, $q(t)$ is the vector of generalized coordinates, and $\underline{Q}_{ext}(t)$ the generalized loading due to the aerodynamic/gravity loads and the active control forces. Wind excitation is modeled as an external modal load applied to the blade in the flapwise direction. The generalized aerodynamic load on the blade"j" for the ith mode is computed as

$$Q_{ji} = \int_0^L p_j(x,t)\Phi_i(x)dx \tag{10}$$

with $p_j(x,t)$ representing the variable wind load intensity along the blade length in the flapwise direction.

The generalized load on the nacelle corresponds to

$$Q_4 = \sum_{j=1}^{3} \int_0^L p_j(x,t)dx \tag{11}$$

Quasi-static aerodynamic wind loading time series are computed by applying the corrected blade element momentum (BEM) method.

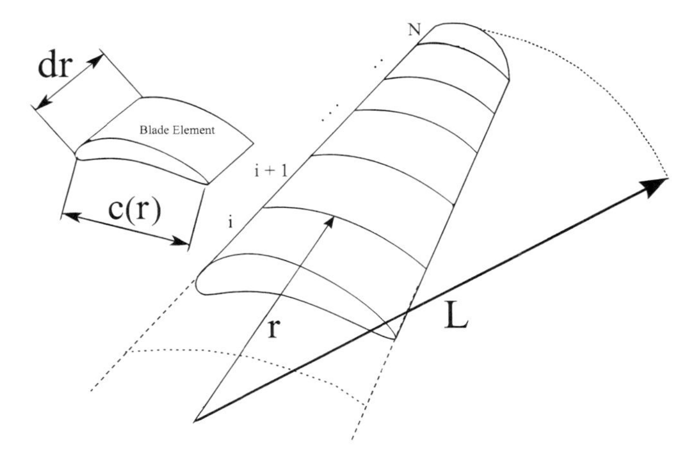

Fig. 4 Blade model according to the BEM theory approach

2.2 *Wind Loading: Blade Element Momentum Theory*

In this chapter, in order to have a realistic estimate of the wind loading to which the rotor is subjected to, models based on the blade element momentum (BEM) theory have been adopted [5]. These models allow to obtain a detailed quantitative description of the wind turbine rotor behavior, which is based on the aerodynamic properties of the blade section airfoils, the geometrical characteristics of the rotor, as well as the wind speed and the rotational velocity of the blades. BEM analysis is carried out by combining momentum theory and blade element theory.

The blade is assumed to be discretized into N sections (elements). Each element is located at a radial distance r from the hub (Fig. 4), and it has chord length $c = c(r)$ and width dr. Assuming no radial dependency for the annular sections, i.e., no aerodynamic interactions between different elements, and assuming that the forces on the blade elements depend only on the lift and drag characteristics of the airfoil shape of the blades, the BEM theory provides a method to estimate the axial and tangential induction factors, a and a', respectively. Once these parameters are known, local loads on each segment can be determined. The total forces acting on the blade can then be computed by performing numerical integration along the blade span. In order to describe the BEM algorithm for calculating quasi-static aerodynamic wind loads, the following quantities are defined:

$$
\begin{aligned}
V_{rel}(r,t) &= \sqrt{(V_0(r,t)(1-a))^2 + \Omega^2 r^2 (1+a')^2} \\
\phi(r,t) &= \tan^{-1}\left(\frac{(1-a)V_0(r,t)}{(1+a')\Omega r}\right) \\
\alpha(r,t) &= \phi(r,t) - \beta(t) - \kappa(r)
\end{aligned}
\tag{12}
$$

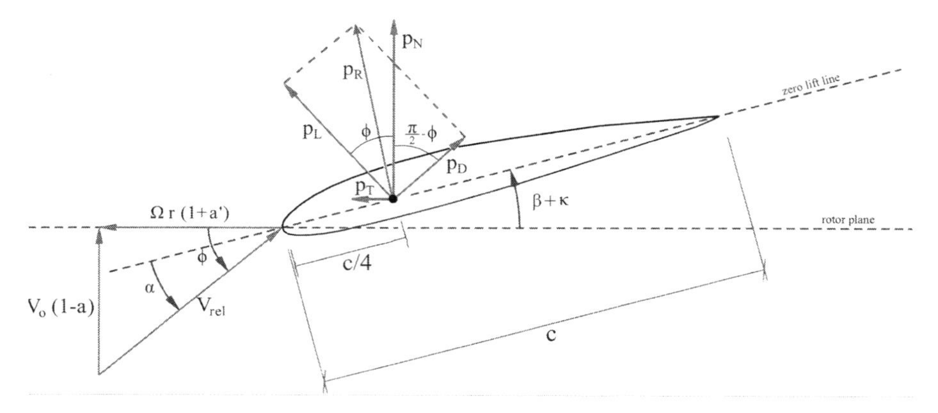

Fig. 5 Local forces and velocities in the BEM model of the blade

where V_{rel} and V_O denote the relative and the instantaneous wind speed, respectively, ϕ is the flow angle, α the instantaneous local angle of attack, β the pitch angle, and κ the local pre-twist of the blade (Fig. 5).

The local lift and drag forces can be respectively computed as

$$
\begin{aligned}
p_L(r,t) &= \frac{1}{2}\rho V_{rel}^2(r,t)c(r)C_l(\alpha) \\
p_D(r,t) &= \frac{1}{2}\rho V_{rel}^2(r,t)c(r)C_d(\alpha)
\end{aligned}
\tag{13}
$$

where ρ is the density of air and $C_l(\alpha)$ and $C_d(\alpha)$ represent the lift and drag coefficients, respectively, whose values depend on the local angle of attack. Finally, the aerodynamic forces normal to the rotor plane (corresponding to the aerodynamic loads in the flapwise direction) can be obtained by projecting the lift and the drag along the normal plane, as shown in Fig. 5. Therefore, the local flapwise load is given by

$$
p_N(r,t) = p_L(r,t)\cos(\phi) + p_D(r,t)\sin(\phi) \tag{14}
$$

As suggested in Hansen [5], in order to improve the accuracy of the model, Prandtl tip loss factor and Glauert correction have been applied. The former corrects the assumption, used in the classical blade element momentum theory, of an infinite number of blades, while the latter has been applied in order to compute the induced velocities more accurately when the induction factor a is greater than *a* critical value a_c.

Once the local loads on the blade elements have been calculated, by integrating Eq. (14) along the blade length and considering the appropriate modeshape of the blade, the generalized flapwise load can be calculated using Eqs. (10) and (11). To account for the variation in the vertical wind shear due to the rotation of the blade, the term V_O in Eq. (12) can be approximately assumed as a constant wind speed linearly varying with height.

2.3 Model with Polytopic Uncertainty

For the purpose of designing a controller, a number of simplifying hypothesis and approximations are assumed in the formulation of a mathematical model of the physical system. This is done in order to obtain a set of "tractable" equations, on which the design of the control law will be based. This simplified mathematical representation, in some cases, can be inaccurate or even inappropriate to adequately describe the dynamics of the phenomenon of interest. In particular, considering all parameters as constant is a highly restrictive assumption. In fact, this does not allow to take into account the change in the physical properties of the system due to fatigue and cyclic loading in time or due to change in environmental conditions. Therefore, it is reasonable to deem that the model is affected by a degree of uncertainty, which expresses the fact that many physical and geometric properties of the system are not precisely known a priori.

In this chapter, a polytopic (or multimodel) representation of the uncertainty has been adopted; [9] this accounts for errors in the identification and parametric variations in the stiffnesses of blades and nacelle/tower. Underlying the multimodel paradigm is a linear time-varying (LTV) representation of the system:

$$\begin{cases} x(k+1) = A(p(k))x(k) + B(p(k))u(k) \\ y(k) \quad\quad\;\; = Cx(k) \end{cases} \qquad (15)$$
$$[A(p(k)) \quad B(p(k))] \in \Pi$$

where $x(k) \in \mathrm{R}^{n_x}$ is the state of the system at the time instant "k," $u(k) \in R^{n_u}$is the control input, and $y(k) \in R^{n_y}$ is the output of the plant. The system matrices A and B depend on time through the vector of time-varying parameters $p(k) \in R^p$. By assuming bounded parametric variations, i.e., $p_{\min} \leq p(k) \leq p_{\max}\ \forall k$, l linear time invariant (LTI) systems can be obtained, and the uncertainty set (polytope) Π can be expressed as

$$\Pi = C_0\{[A_1 \quad B_1], \ldots, [A_l \quad B_l]\} \qquad (16)$$

where Co(·) denotes the convex hull of the vertices $[A_1\ B_1], \ldots, [A_l\ B_l]$. For each parametric variation inside the given bounds, every possible dynamic realization of the system is included in the polytope Π; moreover, according to the definition, Eq. (16), every possible realization $\left[\bar{A} \quad \bar{B}\right]$ of Eq. (15) can be obtained as

$$[\bar{A} \quad \bar{B}] = \sum_{i}^{L} \lambda_i [A_i \quad B_i], \quad \sum_{i=1}^{l} \lambda_i = 1 \quad 0 \leq \lambda_i \leq 1 \qquad (17)$$

The uncertain dynamics of Eq. (15) can hence be described by the variation of λ_i in Eq. (16).

In this study, parametric uncertainties in the fundamental natural frequencies of the blade ω_b and the tower ω_n have been modeled. The vector of uncertain parameters corresponds to

$$p(k) = \begin{bmatrix} \omega_b(k) \\ \omega_n(k) \end{bmatrix}, \quad \omega_b(k) \in [\omega_{b_{\min}}, \omega_{b_{\max}}], \qquad \omega_n(k) \in [\omega_{n_{\min}}, \omega_{n_{\max}}] \quad \forall k \quad (18)$$

By evaluating the matrices of the multimodel representation of the system, Eq. (15) in the extreme values of the uncertain parameters, four different operating points are obtained ($l = 4$), representing the vertex of the uncertainty polytope Π for the case considered.

2.4 *Robust-Constrained MPC*

The purpose of the robust controller is to guarantee that the closed loop system is stable and the requirements on input and output variables are met for every possible realization of the system in the uncertainty polytope Π. The synthesis of the robust control law, based on a linear model with polytopic uncertainty, is performed by minimizing an appropriate upper bound on a robust (*worst-case*) quadratic objective function:

$$\begin{array}{lll} \min & \max & J_\infty(k) \\ u(k+i|k) & [A(k+i)B(k+i)] & \text{belongs to the polytope PI cfr. Eq. 16} \\ i = 0, \ldots, m-1 & i \geq 0 & \end{array}$$

$$J_\infty(k) = \sum_{i=0}^{\infty} \left[x^T(k+i|k) R_x x(k+i|k) + u^T(k+i|k) R_u u(k+i|k) \right] \quad (19)$$

where $R_x \in R^{n_x \times n_x}$ is the weight on the state and $R_u \in \mathrm{R}^{n_u \times n_u}$ is the weight on the input variables in the optimization process. The problem of determining the sequence of inputs that solves the given control problem, ensuring the fulfillment of constraints on input and output variables (the former referred as "*hard constraints*"), is reformulated as a convex optimization problem. Given the state $x(k) = x(k|k)$ of the uncertain system (15) at the time instant k, the robust-constrained MPC algorithm is implemented by computing the state feedback gain $F_k = YQ^{-1}$, where $Q = Q^{\mathrm{T}} \geq 0$ and Y are obtained by solving the following minimization problem with linear matrix inequalities (LMIs) constraints:

$$\min_{\gamma, Y, Q} \quad \gamma \quad (20a)$$

$$\begin{bmatrix} 1 & x^T(k|k) \\ x(k|k) & Q \end{bmatrix} \geq 0 \quad (20b)$$

$$\begin{bmatrix} Q & QA_j^T + Y^T B_j & QR_x^{\frac{1}{2}} & Y^T R_u^{\frac{1}{2}} \\ A_j Q + B_j Y & Q & 0 & 0 \\ R_x^{\frac{1}{2}} Q & 0 & \gamma I & 0 \\ R_u^{\frac{1}{2}} Y & 0 & 0 & \gamma I \end{bmatrix}_{j=1\ldots l} \geq 0 \tag{20c}$$

$$\begin{bmatrix} X & Y \\ Y^T & Q \end{bmatrix} \geq 0, \quad X_{ii} \leq u_{i,\max}^2, i = 1\ldots n_u \tag{20d}$$

$$\begin{bmatrix} Q & (A_j Q + B_j Y)^T C_i^T \\ (A_j Q + B_j Y) C_i & y_{i,\max}^2 \end{bmatrix}_{\substack{j=1\ldots l \\ i=1\ldots ny}} \geq 0 \tag{20e}$$

where $u_{i,\max}$ and $y_{i,\max}$ denote the constraint on the ith input and output components, respectively. At the time instant "k," if a solution to Eqs. (20a, 20b, 20c, 20d, 20e) exists, the control input $u(k|k) = F_k\, x(k|k)$ is implemented. At the next sampling time, the state $x(k + 1)$ is measured, and the optimization problem is solved in order to compute F_{k+1}. The details of the control algorithm used in this study are provided in Kothare et al. [9].

3 Results

The robust MPC algorithm (20) has been simulated in Matlab® by using the LMI parser "YALMIP" and the SDP solver "SeDuMi" provided in the "Multi-Parametric Toolbox" [10]. A reduced order model has been derived for the system under consideration in order to reduce the number of states required for implementing the control and hence to decrease the computational cost associated with the calculation of the control law. In particular, for the design of the controller, each beam is assumed to be vibrating in its fundamental mode. This leads to a reduced order model with 4DOF. Also, the effect of gravity has been ignored, as it may not have a significant impact on the flapwise vibration. The proposed control scheme has been tested using data relative to the NREL 5-MW baseline offshore wind turbine, which is a three-bladed upwind turbine with a rotor diameter of about 126 m and hub height of 90 m. The blade considered (LM 61.5 P2) is 61.5 m long with an overall mass of 17,740 kg. A damping ratio of 1% in the blade flapwise mode has been specified. Full details of the NREL 5-MW baseline wind turbine are provided in Jonkman et al. [7].

Aerodynamic loads in the flapwise direction have been calculated using the blade element momentum (BEM) theory, according to the algorithm outlined by Hansen [5]. The BEM method allows to compute an estimate of the loads induced by the wind by taking into account the aerodynamic properties of the blade section

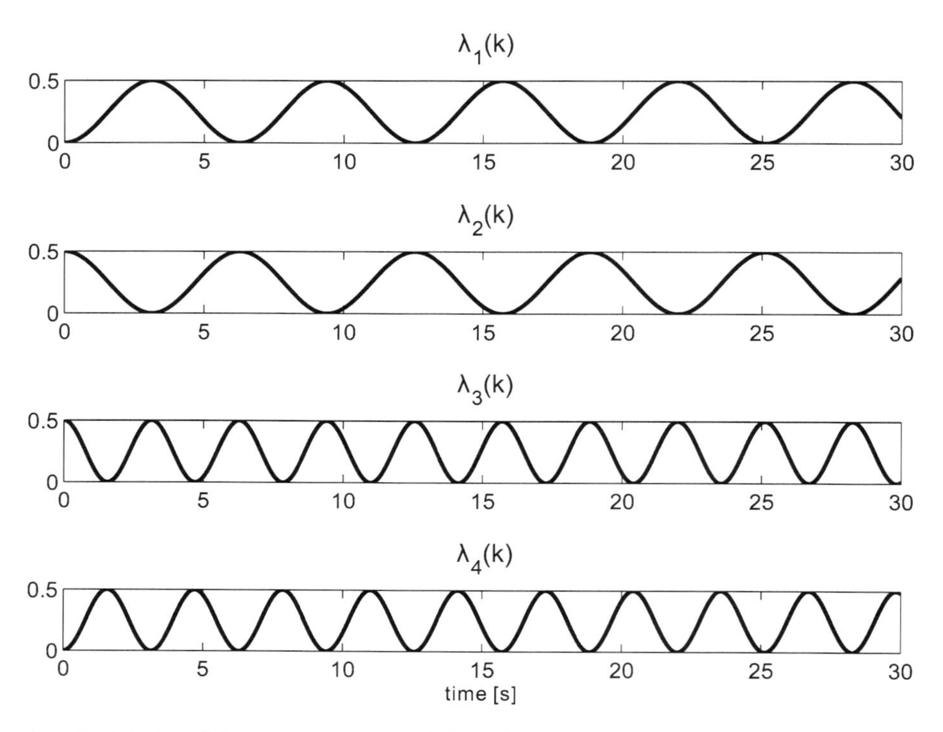

Fig. 6 Variation of the system parameters λ_i (case I)

airfoils and the structural properties of the rotor. In this chapter, a steady wind load with a mean speed value of 12 ms^{-1} has been considered, in addition to a linear wind shear in the vertical direction producing a periodic loading variation.

For studying the effect of uncertainties in the dynamic parameter variation in a wind turbine, a few different cases have been considered. Since the main aim is to analyze the robustness of the controller in presence of uncertainty, contributions from higher modes and gravity have been excluded in the numerical simulations. Instead, uncertainties in the fundamental natural frequencies of the blade and the tower as in Eq. (18) have been investigated, and different scenarios have been analyzed by setting various input and output constraints, as well as different forms of variation in the parameters of the model during the simulation.

3.1 Case I: Sinusoidal Variation

For the first numerical experiment, the bounds on the uncertain parameters assumed for the design of the robust control law are $\omega_{b_{\min}} = 2\,\mathrm{rad/s}$, $\omega_{b_{\max}} = 15\,\mathrm{rad/s}$, $\omega_{n_{\min}} = 1\,\mathrm{rad/s}$, and $\omega_{n_{\max}} = 12\,\mathrm{rad/s}$. During the simulation, the parameters λ_i in Eq. (17) are varied according to a sinusoidal pattern as shown in Fig. 6. The control

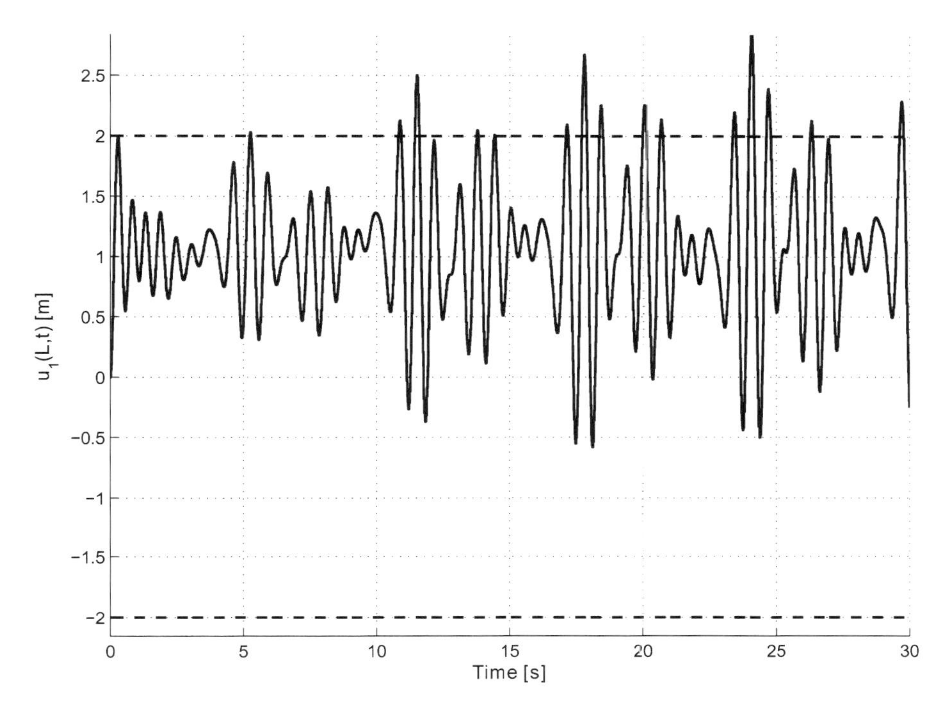

Fig. 7 Blade 1 tip displacement with sinusoidal parameter variation

law is synthesized in order to allow for each actuator a maximum control force corresponding to 15% of the total weight of the blade. Figure 7 shows the time history of the displacement response of blade 1 with the constraint limits imposed on the response.

It is observed that the constraints are violated a few times. This can be attributed to the fact that the constraints imposed are soft constraints and a penalty is imposed when violation occurs. A closer look at the response reveals that the constraints are satisfied initially (about 10 s). Subsequently, the excursion of the constraint occurs a few times. The reason for this may be because of the nature of the excitation which is persistent and decaying. The steady-state loading condition has not been accounted for in the formulation of the set of conditions for the LMI (20). In spite of this, it may be noted that the control force is limited to the constrained values once the limit on the displacement is reached. This is evident from Fig. 8 where the control force time history on the blade 1 has been plotted. It is also worth noting that the violation of the displacement limit on one blade may result in limiting the control force on another blade if the control force on that blade is close to the limit.

Figure 9 compares the controlled displacement response of blade 1 for the robust MPC algorithm with the uncontrolled response. It is observed that there is a significant reduction in the displacement response and the robust MPC is successful

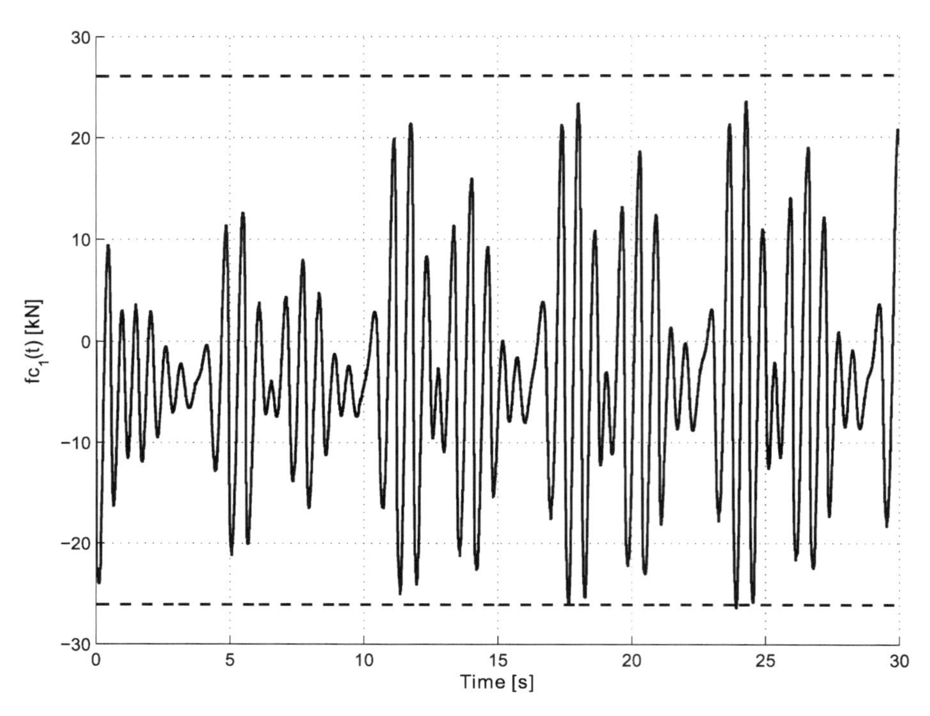

Fig. 8 Control force on blade 1 with sinusoidal parameter variation

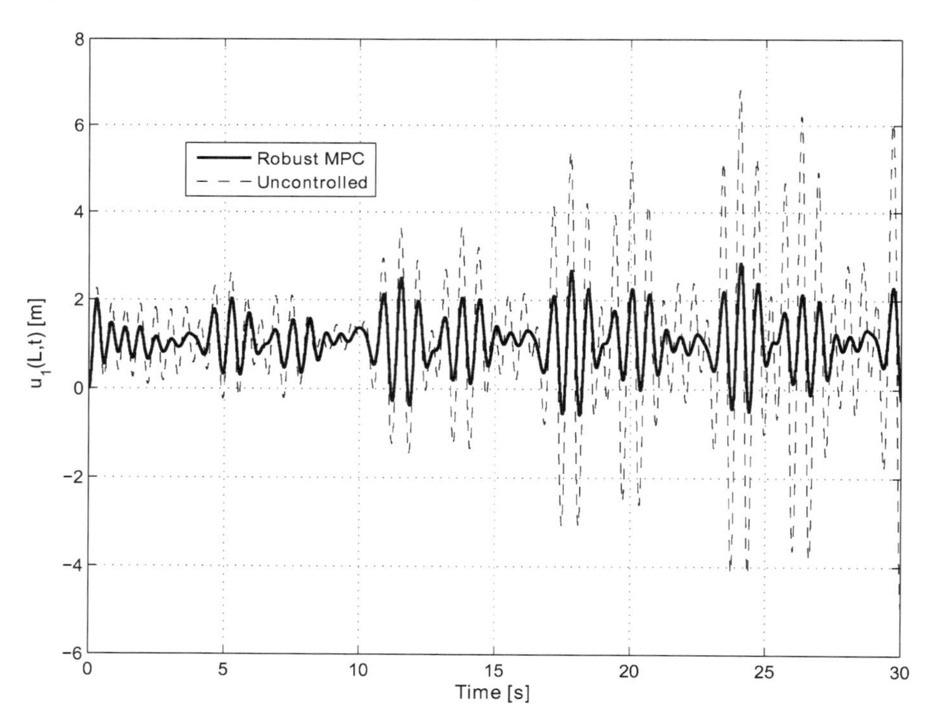

Fig. 9 Uncontrolled (*dashed*) and robust MPC-controlled (*solid*) displacement response (sinusoidal parameter variation)

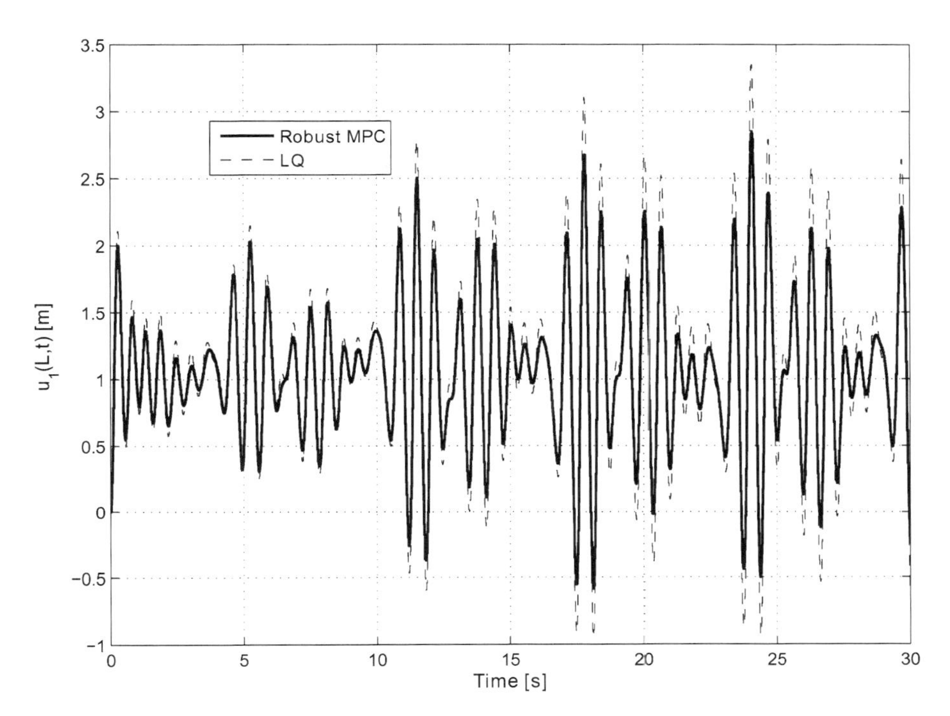

Fig. 10 LQ-controlled (*dashed*) and robust MPC-controlled (*solid*) displacement response (sinusoidal parameters variation)

in taking into account the uncertainties in the system and suppressing the vibrations while ensuring the fulfillment of hard constraints. To examine the performance of the robust MPC with the LQ regulator algorithm, the displacement response of blade 1 has been plotted in Fig. 10 for the two algorithms. It is concluded that the robust MPC is slightly better in controlling vibration response of the blade under uncertain parametric conditions.

Finally, the Fourier spectra of the controlled response (applying robust MPC) and the uncontrolled response has been plotted in Fig. 11. It is seen that the parametric uncertainties in the system have the impact of introducing a large number of high-frequency components in the response, and the robust MPC algorithm suppresses those frequency components.

3.2 Case II: Extreme Vertex

To consider the effect of an extreme variation in the parameter due to uncertainty in the system, the natural frequency variation has been assumed to be the vertex of the parameter polygon space with maximum values. Assuming that the

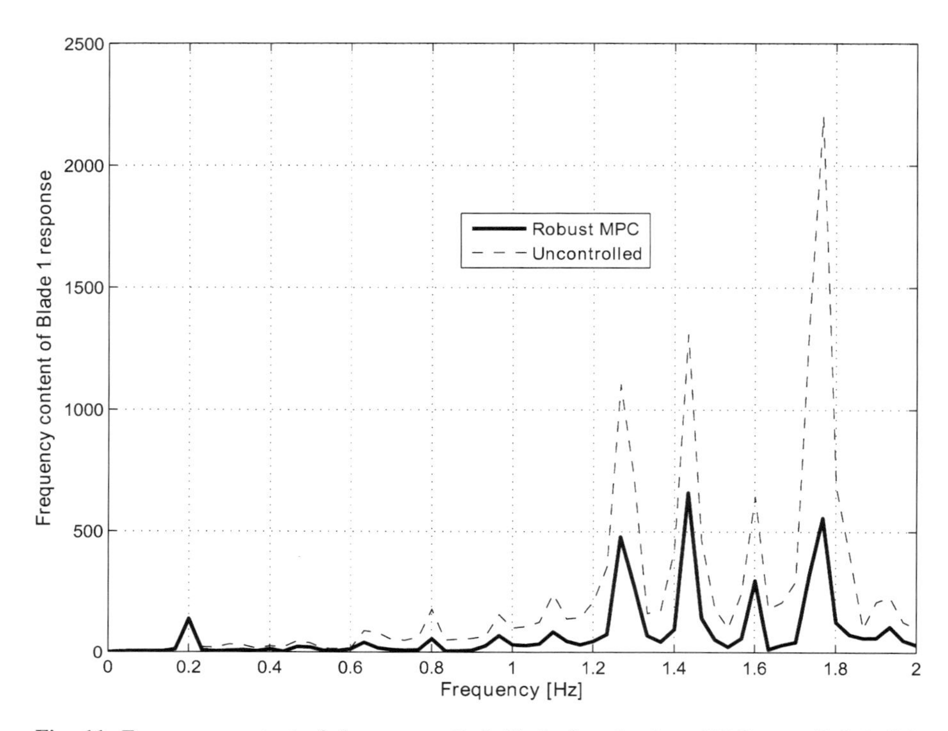

Fig. 11 Frequency content of the uncontrolled (*dashed*) and robust MPC-controlled (*solid*) displacement response (sinusoidal parameter variation)

bounds for the design of the control law $\omega_{b_{\min}}$, $\omega_{b_{\max}}$, $\omega_{n_{\min}}$, and $\omega_{n_{\max}}$ are 2, 6, 1, and 4 rad/s, respectively, the flapwise vibration model used during the simulation is instantiated with $\omega_b = \omega_{b_{\max}}$, $\omega_n = \omega_{n_{\max}} \forall k$. This corresponds to setting $\lambda_i = 0$, $i = 1, 2, 3$, $\lambda_4 = 1$. Therefore, the dynamic realization of the system used in the test is one of the vertexes of the uncertainty polytope on which the synthesis of the robust control law is based. In this respect, this corresponds to a worst-case scenario since the controller has to cope with extreme values for the modeled uncertainties. The results for this case using a maximum control force of 20% of the total blade weight and an output constraint of 5 m have been plotted in Figs. 12, 13, 14 and 15. As in the previous case, no major qualitative deviation in conclusion has been found.

Also for this case, the controlled displacement responses based on the robust MPC algorithm have been compared with the case with nominal MPC (i.e., the optimization problem in the MPC algorithm has been solved for the case with the nominal natural frequency parameters). The results indicate that the application of the nominal MPC controller with the given constraints and designed for the case considered leads to unstable displacement response while the robust MPC has again shown excellent performance (Fig. 16).

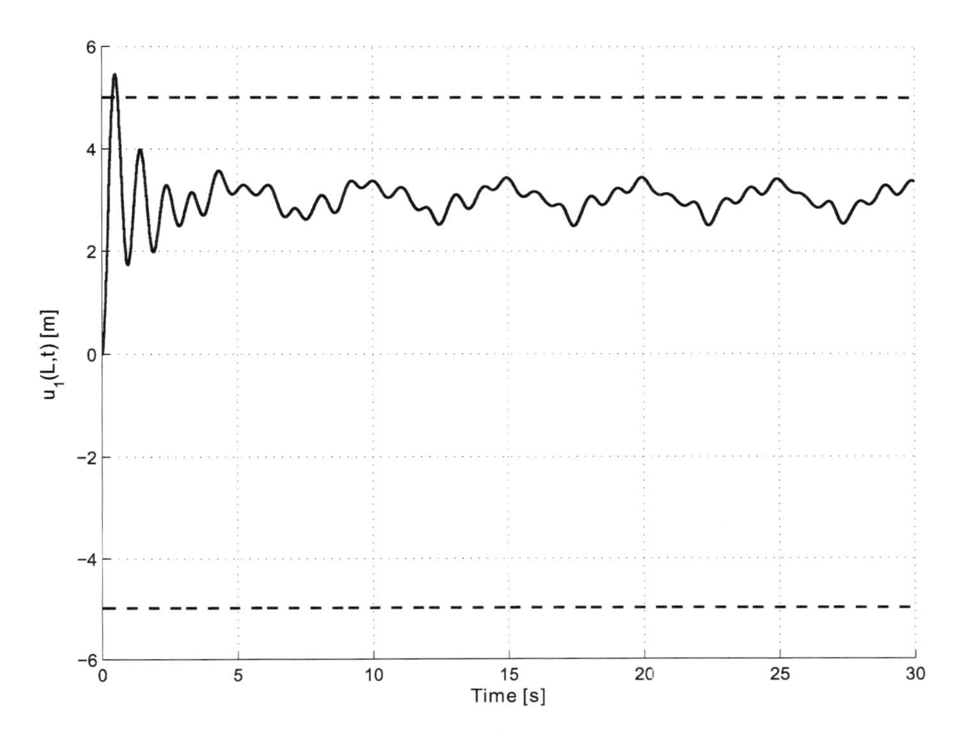

Fig. 12 Blade 1 tip displacement (extreme vertex)

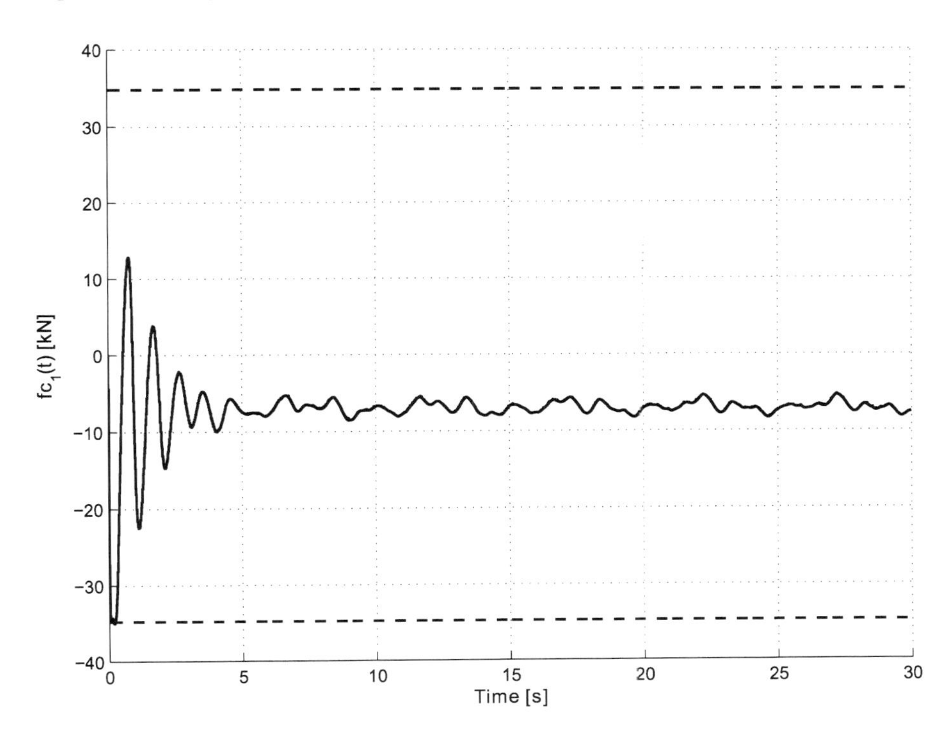

Fig. 13 Control force on blade 1 (extreme vertex)

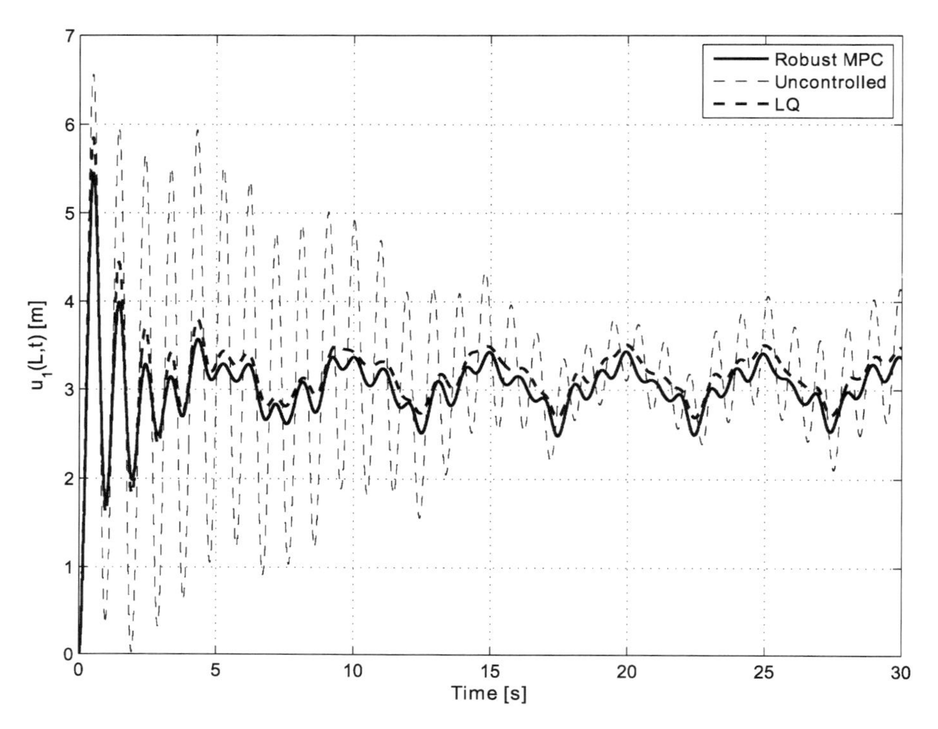

Fig. 14 Uncontrolled (*thin dashed*), LQ-controlled (*thick dashed*), and robust MPC-controlled (*solid*) displacement response (extreme vertex)

3.3 Output-Only Feedback with Robust State Observer

The robust state feedback MPC algorithm can be extended in order to implement an output-only feedback MPC controller [8]. This is obtained by designing a robust state observer, which provides an estimate $\hat{x}(k|k)$ of the state of the system based on the output measurements. By assuming that the system parameters (i.e., λ_i) are measurable at each sampling time, the state update equation for the observer is given by

$$\hat{x}(k|k) = \hat{x}(k|k-1) + GCA(k)e(k-1) \tag{21}$$

$e(k) = x(k) - \hat{x}(k|k)$ is the state estimation error. The observer gain G is designed to robustly stabilize $e(k)$ and is computed as $G = P^{-1}Y$, where $P = P^{\mathrm{T}} > 0$ and Y are obtained by solving the following set of linear matrix inequalities:

$$\begin{bmatrix} P & PA_j + YCA_j \\ A_j^T P + A_j^T C^T Y^T & P \end{bmatrix} \geq 0, \quad j = 1 \ldots l \tag{22}$$

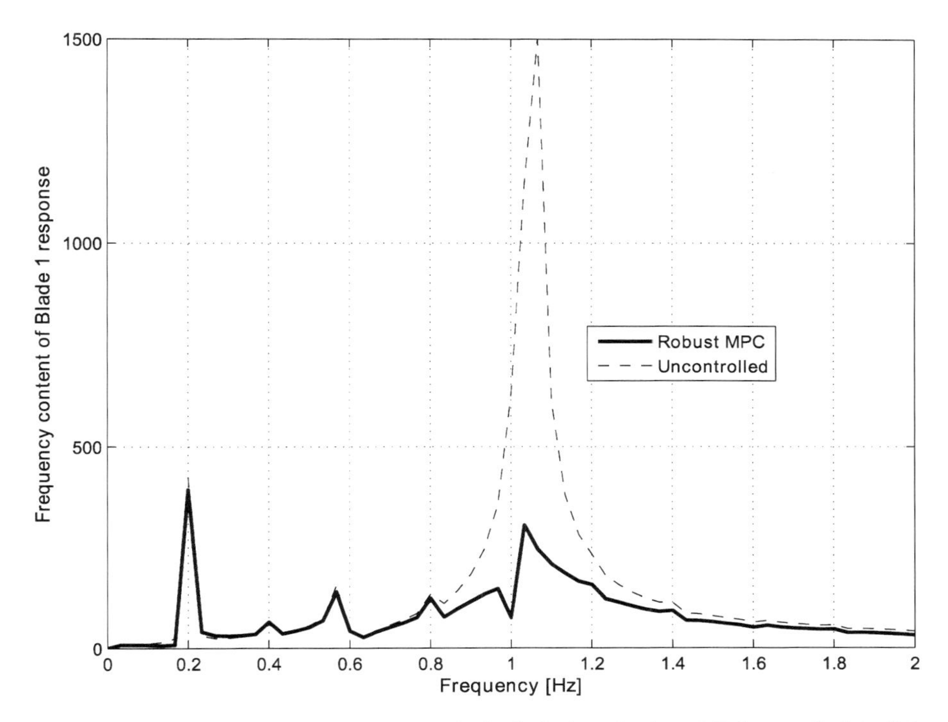

Fig. 15 Frequency content of the uncontrolled (*dashed*) and robust MPC-controlled (*solid*) displacement response (extreme vertex)

By replacing the state system $x(k|k)$ in Eq. (20a, 20b, 20c, 20d, 20e) with the estimate $\hat{x}(k|k)$ reconstructed by feeding back output variables only, the robust-constrained MPC control law is implemented as $u(k|k) = F_k\hat{x}(k|k)$. In this study, displacements have been assumed as measured outputs, and the performance of the full-state feedback controller has been compared to the output feedback case as shown in Fig. 17.

The numerical experiments have been carried out by assuming the bounds in case II and a maximum-controlled force of 15% of the total blade weight. The initial state of the observer has been set zero, while for the system state a randomly generated initial condition has been chosen:

$$x(0) = [0.1219 \quad 0.5221 \quad 0.1171 \quad 0.7699 \quad 0.3751 \quad 0.8234 \quad 0.0466 \quad 0.5979]^T \tag{23}$$

It is interesting to note that after an initial mismatch (as shown in the zoomed plot in Fig. 17) due to different initial condition, the observer converges to the system state even in the presence of persistent disturbances, as there is no appreciable difference between the output-only feedback and state feedback-controlled blade response.

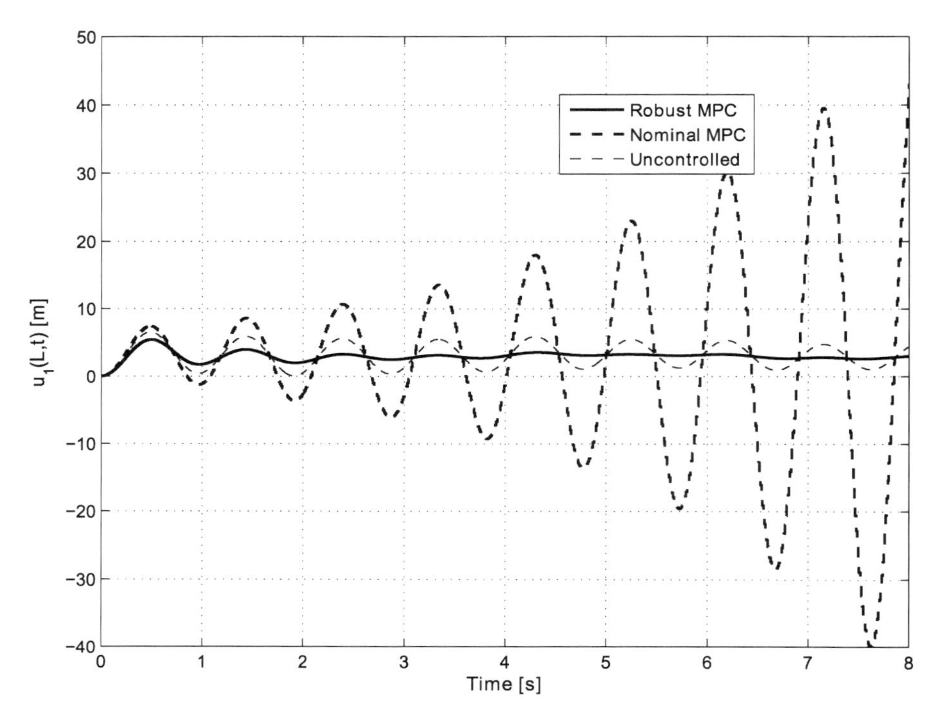

Fig. 16 Uncontrolled (*thin dashed*), nominal MPC-controlled (*thick dashed*), and robust MPC-controlled (*solid*) displacement response (extreme vertex)

4 Conclusions

In this chapter, a new control scheme for mitigating flapwise vibrations in wind turbine blades has been presented. The potential use of a new active control configuration located inside the blades has been considered. The study has been carried out by developing a mathematical model focused on the dynamics of flapwise vibrations, including the interaction between the blades and the tower and the active controllers. Steady wind loading conditions, including linear wind shear, have been considered. Uncertainties in the fundamental natural frequencies of the blade and the tower have been modeled in the framework of structured polytopic uncertainty. The robust-constrained MPC algorithm proposed by Kothare et al. [9] has been implemented in Matlab in order to ascertain the effectiveness of the proposed control strategy in presence of time-varying parametric uncertainties and hard constraints. Simulation results show that in different scenarios analyzed, the robust MPC is effective in reducing the response of the blades even when variations in the considered parameters occur and the control input is limited

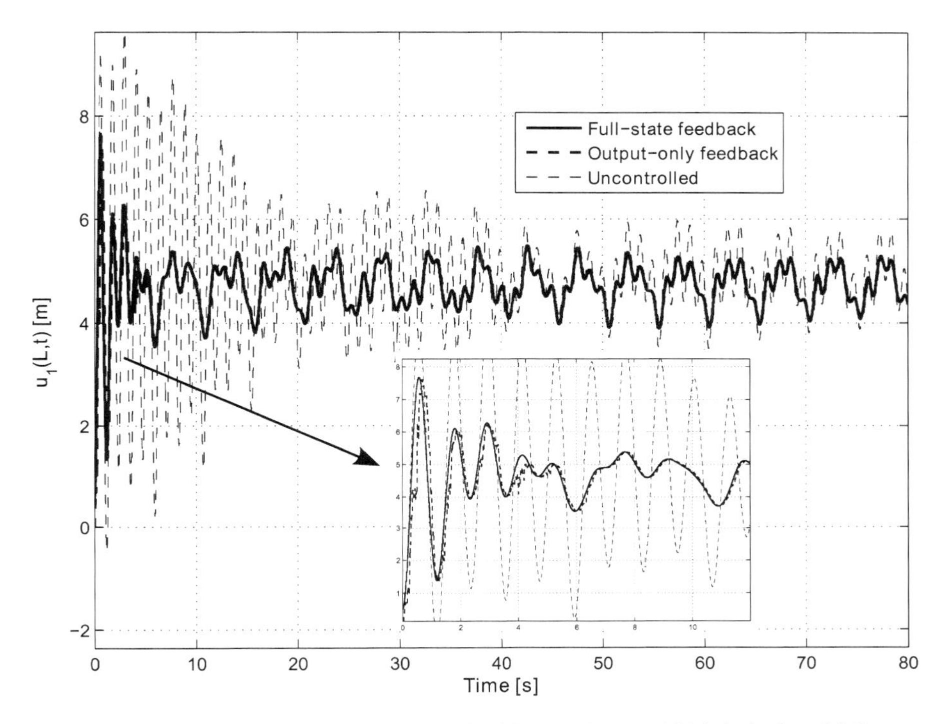

Fig. 17 Uncontrolled (*thin dashed*), robust MPC with state observer (*thick dashed*), and full-state feedback robust MPC-controlled (*solid*) displacement response (output feedback case)

to a prescribed value. The proposed active tendon controller has been also used for investigating the control of edgewise vibration of wind turbine rotor blades (which is a time-varying system), and encouraging results have been reported [15].

Acknowledgments This research is carried out under the EU FP7 funding for the Marie Curie ITN project SYSWIND (Grant No. PITN-GA- 2009-238325).

References

1. Arrigan J, Pakrashi V, Basu B, Nagarajaiah S (2010) Control of flapwise vibration in wind turbine blades using semi-active tuned mass dampers. Struct Control Health Monit 18(8): 840–851 (in press)
2. Colwell S, Basu B (2009) Tuned liquid column dampers in offshore wind turbines for structural control. Eng Struct 31(2):358–368
3. Dueñas-Osorio L, Basu B (2008) Unavailability of wind turbines from wind induced accelerations. Eng Struct 30(4):885–893
4. Hansen MH (2003) Improved modal dynamics of wind turbines to avoid stall-induced vibrations. Wind Energy 6(2):179–195
5. Hansen MOL (2008) Aerodynamics of wind turbines. Earthscan, Sterling

6. Johnson SJ, Baker JP, van Dam CP, Berg D (2010) An overview of active load control techniques for wind turbines with an emphasis on microtabs. Wind Energy 13(2–3):239–253
7. Jonkman JM, Butterfield S, Musial W, Scott G (2009) Definition of a 5-MW reference wind turbine for offshore system development. National Renewable Energy Laboratory, Technical report NREL/TP-500-38060
8. Kothare M (1997) Control of systems subject to constraints, Ph.D. thesis, California Institute of Technology
9. Kothare M, Balakrishnan V, Morari M (1996) Robust constrained model predictive control using linear matrix inequalities. Automatica 32(10):1361–1379
10. Kvasnica M, Grieder P, Baotíc M (2004) Multi-Parametric Toolbox (MPT). http://control.ee.ethz.ch/mpt/
11. Lackner MA, van Kuik G (2010) A comparison of smart rotor control approaches using trailing edge flaps and individual pitch control. Wind Energy 13(2–3):117–134
12. Maldonado V, Farnsworth J, Gressick W, Amitay M (2010) Active control of flow separation and structural vibrations of wind turbine blades. Wind Energy 13(2–3):221–237
13. Murtagh PJ, Basu B, Broderick BM (2005) Along-wind response of a wind turbine tower with blade coupling subjected to rotationally sampled wind loading. Eng Struct 27(8):1209–1219
14. Murtagh PJ, Ghosh A, Basu B, Broderick B (2008) Passive control of wind turbine vibrations including blade/tower interaction and rotationally sampled turbulence. Wind Energy 11(4): 305–317
15. Staino A, Basu B, Nielsen SRK (2011) Actuator control of edgewise vibrations in wind turbine blades. J Sound Vib. doi:10.1016/j.jsv.2011.11.003

Disaster Mitigation of Large Infrastructure Systems

Baidurya Bhattacharya

Abstract Modern engineering has become a hugely complex and demanding endeavour – better and cheaper products must be produced at an ever increasing pace while they must continue to be safe and reliable. But, systems can fail and big systems fail in big ways, causing large losses. Failure, if it occurs, often happens in hitherto unknown ways. It is imperative that large infrastructure systems be designed not only to provide full functionality under normal conditions; they must also be able to absorb limited damages without tripping, be able to provide essential services after a major strike and be able to have the ability to come back up online within a reasonable time after being hit by a disaster. These requirements can be met by (1) spelling out in precise measurable terms the system performance expectations both in intact and damaged conditions, (2) clearly understanding the hazards for each performance level and (3) specifying the reliability or confidence with which these performance expectations must be met. The challenges in meeting these tasks can be grouped into three categories: (1) uncertainty quantification, (2) system level modelling and (3) risk communication.

Keywords Large infrastructure • Disaster mitigation • Uncertainty quantification • System level modelling • Risk communication

1 Introduction

Infrastructure refers to the basic framework that underlies and holds together a complex system. A country or a region depends on its civil, communication, military, financial and other infrastructure to function and serve its citizens. This chapter is

B. Bhattacharya (✉)
Department of Civil Engineering, Indian Institute of Technology Kharagpur,
Kharagpur, West Bengal, India
e-mail: baidurya@civil.iitkgp.ernet.in

S. Chakraborty and G. Bhattacharya (eds.), *Proceedings of the International Symposium on Engineering under Uncertainty: Safety Assessment and Management (ISEUSAM - 2012)*,
DOI 10.1007/978-81-322-0757-3_27,

about civil infrastructure systems, i.e., the built environment which includes the transportation, power, water, etc., infrastructure systems. An infrastructure has many interconnected parts, working together to provide a desired solution to society. For example, the transportation infrastructure of a nation is composed of its port, airport, road, bridge, air, rail, etc., systems. Large-scale industrialization started in the nineteenth century, and the word infrastructure came into English less than a 100 years ago. With the harnessing of steam, electricity and explosives, together with the ability to make new and better materials, humans had the ability to span distance and reduce communication time at a scale not possible before.

2 Evolution of Engineering Design

Building complex systems does not happen by chance or in a vacuum. Up until the middle of the nineteenth century, engineering, be it making a castle, a bridge or a watch, was mostly an art – it was conceived by intuition, designed by experience, performed often by one very talented individual, expected to last long and put up by factor of safety. Testing, repeatability, collaboration or optimization was not of primary concern. The art aspect is still very much central to engineering, for one cannot build something if one cannot imagine it, but rigorous science has become the bedrock of modern engineering. The modern engineer has made society more democratic: more people have access to what once belonged only to kings – be it indoor plumbing, indoor illumination, high-speed travel or instant communication. In the process, modern engineering has also become a hugely complex and demanding endeavour – better and cheaper products must be produced at an ever increasing pace while they must continue to be safe and reliable. Such demands require constant innovation and teamwork involving hundreds or thousands of professionals often spread over a large geographical area [1, 2].

When building something, the engineer always knows that something might go wrong with it, and his/her solution might not work the way it is supposed to. Failure might mean economic/human/environmental loss to the owner and society on the one hand and, on the other, loss of business and reputation, penalty and, in ancient times, even death for the engineer. Factors of safety have been the traditional means to prevent such undesirable occurrences.

Factors of safety work well when the system is mature, pace of innovation is slow, overdesign is not a deal-breaker and knowledge of system performance – particularly under trying conditions – is limited. Traditional design is component based. The engineer designs the system component by component for ordinary demands, then makes each component safer by a comfortable factor and hopes that the system will hold good under extraordinary demands [3]. As stated above, this strategy works when the technology is mature, the cost is not much of a concern and the system is not expected to perform under extraordinary or exceptional conditions.

3 Managing Failure

But, systems do fail. And big systems fail in big ways, causing big losses. And if they fail, they often fail in hitherto unknown ways. In 2004, the worldwide loss estimate from natural disasters was USD 120 billion [4] much of which resulted from infrastructure damage. If the country's financial system is too big to fail, and it is necessary that important elements of that infrastructure be propped up to prevent collapse of the entire system, so is any one among the country's civil infrastructures – its water supply, power, transportation, building and other infrastructure systems. The difference is that important elements of the civil infrastructure system, such as a large bridge, a nuclear power plant or an airport, if they fail, cannot be propped up or replaced immediately. It can take months or even years for the system to come back up to full functionality. Remember the still unfinished levee system of New Orleans after Katrina in 2005, the unfolding of Fukushima-Daiichi nuclear meltdown of 2011, the cleanup after the Deep Horizon blow-up and oil spill in Gulf of Mexico in 2010, the aftermath of Hurricane Aila in West Bengal and Bangladesh in 2009, the destruction after the Indian Ocean tsunami of 2004, etc.

So it is imperative that large infrastructure systems be designed not only to provide full functionality under normal conditions; they must also be able to absorb limited damages without tripping, be able to provide essential services after a major strike and have the ability to come back up online within a reasonable time after being hit by a disaster. These are very demanding requirements and rather idealistic in nature. However, we must evaluate our existing as well as our upcoming infrastructure systems against these expansive desiderata in precise measurable ways, if we wish to have our engineering infrastructure serve us in normal times as well as in times of crises.

4 Performance Expectations and How to Achieve Them

The first task in designing an infrastructure system, then, is to spell out our expectations – its performance requirements – under a range of system conditions, e.g., normal, partially damaged and severely damaged [5–10]. The damage states must be defined in precise measurable terms.

Once the performance requirements are understood, the designer must make a comprehensive survey of the hazards that are likely to befall the system during its design life [11]. Different performance levels should generally be evaluated against different types and/or magnitudes of hazard [12, 13]. Man-made hazards are different from natural hazards in that the former are inflicted by an intelligent agent to cause harm and thus may cause damage disproportionate to the extent and scale of attack [14]. The engineer also has to define the so-called design envelope in order to admit that it is either too costly or technically impossible or both to meet hazard scenarios beyond this envelop.

The third task is to define the confidence or reliability with which the system must perform its intended functions subjected to the appropriate hazards [15–18]. Uncertainties abound in any engineering activity, and the uncertainties about a large infrastructural system are significant indeed. There are uncertainties about the occurrence and magnitudes of the hazards, the loads they cause on the system, the strength of each element of the system [19], the manner in which these elements influence and interact with each other, and finally in the mathematical models with which we evaluate the hazards and system performance [20]. Under such myriad uncertainties, it is clear that the system can meet its requirements not every time; the frequency or confidence with which it does so must be evaluated probabilistically and compared to a predefined target. These target reliabilities/availabilities are not in the purview of the engineer alone; they need to be set by engineers, economists and policymakers, and must take into account the consequences of failure, the cost of mitigation measures [21] and the perception of risk from the failed infrastructure by members of the public.

5 Current Challenges

Once these three tasks are in place, design of the infrastructure system can proceed in the usual iterative manner, and it is the responsibility of the designer to provide the most economical solution for the design. It can so happen that in case of severe system damage under an extreme hazard, the system can meet its performance requirements with the required reliability only if adequate post-disaster management activities are factored into the design.

At the current state of the art, the impediments to realizing the ideal solution described above relate to three major aspects: the first to do with uncertainty, the second in regard to modelling of the system and the third to do with risk communication.

1. *Uncertainty quantification.* There is lack of complete knowledge about the input, i.e., the future hazards and the future demands, to the system. For hazards that arise out of extremes of geophysical processes, how does one reconcile their spatio-temporal scales that are orders of magnitude larger than those of the engineering systems? How does uncertainty in the input propagate through a complex system? How accurately is it possible to predict the state/output of a complex system in the face of significant uncertainty in the input and the model? How are uncertainties arising from human intervention, human error and public behaviour going to affect the response of the system when disaster strikes?
2. *System level model.* It is comparatively easy to model a system in its intact form operating under normal conditions. The model of the system in severely damaged or in near failure conditions becomes inaccurate and cannot be verified against experimental data. How much is the error in the system model itself? Important system failure modes and weak progressive failure sequences may be

missed. It is relatively easy to model dependence among events if they are causally related, but associative dependence is more difficult and easy to miss which might give a false sense of safety through redundancy. If the system is instrumented, how can the sensed data under normal conditions, and those under damaged conditions, be used to estimate the extent of damage and to direct disaster response operations?

3. *Risk communication.* How much risk to life, property and the environment is society willing to accept for the benefits that it gets from the infrastructure if it fails? How much money is it willing to spend to mitigate an additional unit of risk? What failure costs are to be taken into account, and which are to be kept out? What is the value of natural beauty that is threatened by a disaster? These questions directly affect the reliability/availability to which the infrastructure system needs to be designed. There may be a large difference between the actual risk of failure of a system and the risk perceived by the public. How is the proper risk to be communicated? Society's tolerable risk to an activity may change with time: how is one to keep up with it?

References

1. Longstaff TA, Haimes YY (2002) A holistic roadmap for survivable infrastructure systems. IEEE Trans Syst Man Cybern A Syst Humans 32(2):260–268
2. Woo G (2005) Topical issues in infrastructure risk management. In: ICOSSAR 2005, Millpress, Rotterdam
3. Galambos TV (1992) Design codes. In: Blockley D (ed) Engineering safety. McGraw Hill, London/New York, pp 47–69
4. Kunreuther H (2005) Catastrophe modeling: a new approach to managing risk. In: ICOSSAR 2005. Millpress, Rotterdam
5. Augusti G, Ciampoli M (2008) Performance-based design in risk assessment and reduction. Probab Eng Mech 23:496–508
6. Collins KR et al (1996) Dual-level seismic design: a reliability-based methodology. Earthq Eng Struct Dyn 25:1433–1467
7. FEMA (1997) FEMA-273 NEHRP guidelines for the seismic rehabilitation of buildings. Federal Emergency Management Agency, Washington, DC
8. FEMA (2000) FEMA-350 Recommended seismic design criteria for new steel moment-frame buildings. Federal Emergency Management Agency, Washington, DC
9. SEAOC (1995) Vision 2000, Performance based seismic engineering of buildings. Structural Engineers Association of California
10. Wen Y-K (2001) Minimum lifecycle cost design under multiple hazards. Reliab Eng Syst Saf 73:223–231
11. Wen Y-K (1990) Structural load modeling and combination for performance and safety evaluation, Developments in civil engineering. Elsevier, Amsterdam
12. Ghobarah A (2001) Performance based design in earthquake engineering: state of development. Eng Struct 23:878–884
13. Kinali K, Ellingwood BR (2007) Performance of non-seismically designed PR frames under earthquake loading. In: International conference on applications of statistics and probability (ICASP 10), Tokyo, Japan

14. AASHTO (2002) A guide to highway vulnerability assessment for critical asset identification and protection [cited 9 Sept 2005]. Available from: http://security.transportation.org/sites/security/docs/guide-VA_FinalReport.pdf
15. Bhattacharya B et al (2001) Developing target reliability for novel structures: the case of the Mobile Offshore Base. Mar Struct 14(12):37–58
16. Ditlevsen O (2003) Decision modeling and acceptance criteria. Struct Saf 25:165–191
17. ISO (1998) ISO 2394 General principles on reliability for structures, 2nd edn. International Organization for Standardization
18. Wen Y-K (2001) Reliability and performance based design. Struct Saf 23:407–428
19. Ellingwood BR et al (1980) Development of a probability based load criterion for American National Standard A58. NBS Special Publication 577. 1980, U.S. Department of Commerce, National Bureau of Standards, Washington, DC
20. Ditlevsen O (1982) Model uncertainty in structural reliability. Struct Saf 1:73–86
21. Frangopol DM et al (1997) Life-cycle cost design of deteriorating structures. J Struct Eng ASCE 123(10):1390–1401

An Overview of Application of Nanotechnology in Construction Materials

A.K. Tiwari and Subrato Chowdhury

Abstract Nanotechnology has changed the way material and process are being today, in a used number of applications. For civil engineering applications, however, the effect is not so visible, though a couple of applications are available. One of the biggest issues of concern for civil engineers is the amount of materials being used for various developmental projects around the world. This large consumption is mostly exhausting the natural materials, which are non-reclaimable and hence the present use is unsustainable. Nanotechnology can help reduce uses of these natural materials without losing their optimum applications. Construction being the single largest industry today in the world would certainly benefit with this application. Nanotechnology has the potential to make construction faster, cheaper, safer, and more varied. Automation of nanotechnology construction can allow for the creation of structures from advanced homes to massive skyscrapers much more quickly and at much lower cost. An overview of application and opportunity of nanotechnology in construction materials is briefly introduced in the present chapter with critical insight.

Keywords Nanotechnology • Construction materials • Cement-based composites • Admixtures • Coatings • Steel • Glass

1 Introduction

Nanotechnology has changed the way material and process are being done today, in a number of applications. For civil engineering applications, however, the effect is not so visible, though a couple of applications are available. One of the biggest

A.K. Tiwari
UltraTech Cement Limited, Mumbai, India

S. Chowdhury (✉)
Research & Development, UltraTech Cement Limited, Mumbai, India
e-mail: subrato.chowdhury@adityabirla.com

S. Chakraborty and G. Bhattacharya (eds.), *Proceedings of the International Symposium on Engineering under Uncertainty: Safety Assessment and Management (ISEUSAM - 2012)*,
DOI 10.1007/978-81-322-0757-3_28,

issues of concern for civil engineers is the amount of materials being used for various developmental projects around the world. This large consumption is mostly exhausting the natural materials, which are non-reclaimable and hence the present use is unsustainable. Nanotechnology can help reduced uses of these natural materials without losing their optimum applications. Construction being the single largest industry today in the world would certainly benefit with this application. Nanotechnology has the potential to make construction faster, cheaper, safer, and more varied. Automation of nanotechnology construction can allow for the creation of structures from advanced homes to massive skyscrapers much more quickly and at much lower cost.

Nanotechnology is one of the most active research areas that encompass a number of disciplines such as electronics, biomechanics, and coatings including civil engineering and construction materials. The use of nanotechnology in construction involves the development of new concept and understanding of the hydration of cement particles and the use of nano-size ingredients such as alumina and silica and other nanoparticles. The manufacturers are also investigating the methods of manufacturing of nano-cement. If cement with nano-size particles can be manufactured and processed, it will open up a large number of opportunities in the fields of ceramics, high-strength composites, and electronic applications. At the nanoscale the properties of the material are different from that of their bulk counterparts. When materials become nano-sized, the proportion of atoms on the surface increases relative to those inside, and this leads to novel properties.

Use of nanomaterials can improve fluidity, strength, and durability of the concrete. Nanomaterials also have the potential to be used to improve the reinforcement qualities like anticorrosion. Nano-enabled coating of construction materials is going to constitute the largest application of nanotechnology in construction. Nano-products like architectural paints, water sealers, and deck treatments, treatments applied during fabrication, such as scratch-resistant coatings on vinyl or wood flooring, insulation coatings, etc., offer immense market opportunities for nanomaterials. Nanotech products and applications, among other benefits, may enhance the performance with regard to blocking of the ultraviolet rays, transparency of the structures, photo reactivity, and resistance to stain and odor. Moreover, nanotechnology-based coatings can enable creating self-cleaning surfaces. Many of these are already being embedded into window glasses and plumbing fixtures. Nanomaterials and nanotechnology-based applications will thus take the construction industry much beyond bricks and mortar.

2 Opportunities in the Fields of Cement-Based Composites

Nanotechnology is being used for the creation of new materials, devices, and systems at molecular, nano-, and microlevel [1–3]. Interest in nanotechnology concept for Portland cement composites is steadily growing. The most reported

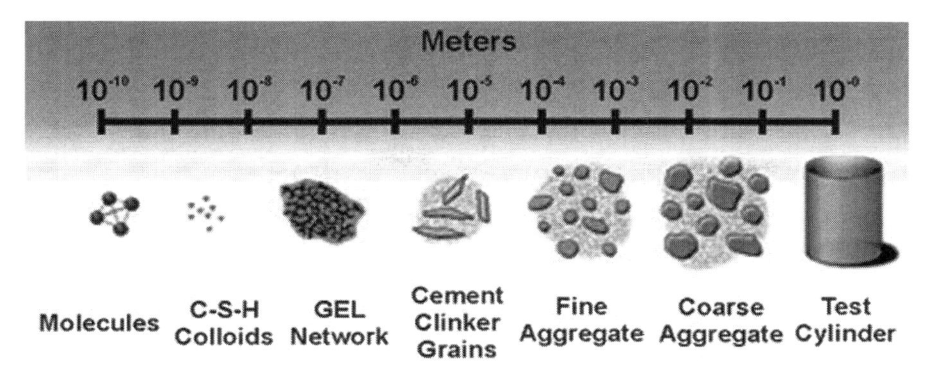

Fig. 1 Scales of various constituents of concrete [4]

research work regarding application of nanotechnology in cement-based materials is either related to coating or enhancement of mechanical and electrical properties. Some of the widely reported nanoparticles in cement concrete industries are titanium dioxide (TiO_2), nano-silica (SiO_2), alumina (Al_2O_3), carbon nanotube (CNT), etc. Currently, the most active research areas dealing with cement and concrete are the following: understanding of the hydration of cement particles and the use of nano-size ingredients such as alumina and silica particles [1–3]. A typical scale of various constituent of a normal concrete is given in Fig. 1.

Average size of Portland cement particle is about 50 μm. In applications that require thinner final products and faster setting time, micro-cement with a maximum particle size of about 5 μm is being used. Knowledge at the nanoscale of the structure and characteristics of materials will promote the development of new applications and new products to repair or improve the properties of construction materials. For example, the structure of the fundamental calcium-silicate-hydrate (C–S–H) gel which is responsible for the mechanical and physical properties of cement pastes, including shrinkage, creep, porosity, permeability, and elasticity. C–S–H gel can be modified to obtain better durability. Cement-based materials containing carbon nanotubes can be used for both strengthening and enhancing electrical and electronic properties of the concrete besides their mechanical properties. Development of smart concrete using carbon nanotubes would be easier. If nano-cement particles can be processed with nanotubes and nano-size silica particles, conductive, strong, tough, and more flexible, cement-based composites can be developed with enhanced properties, for electronic applications and coatings.

3 Nano-concrete and Nano-ingredients

Nano-concrete is defined as a concrete made with Portland cement particles with sizes ranging from a few nanometers to a maximum of about 100 μm. Nano-ingredients are ingredients with at least one dimension of nanometer size.

Therefore, the particle size has to be reduced in order to obtain nano-Portland cement. If these nano-cement particles can be processed with nanotubes and reactive nano-size silica particles, conductive, strong, tough, more flexible, and cement-based composites can be developed with enhanced properties, for electronic applications and coatings. There is also limited information dealing with the manufacture of nano-cement. If cement with nano-size particles can be manufactured and processed, it will open up a large number of opportunities in the fields of cement-based composites. Current research activity in concrete using nano-cement and nano-silica includes the following:

4 Carbon Nanotubes (CNTs)

Carbon nanotubes are among the most extensively researched nanomaterials today. CNTs are tubular structures of nanometer diameter with large aspect ratio. These tubes have attracted much attention in recent years not only for their small dimensions but also for their potential applications in various fields. A single sheet of graphite is called grapheme. A CNT can be produced by curling a graphite sheet. Carbon sheets can also curl in a number of ways. CNT can be considered as the most superior carbon fiber ever made. Addition of small amount (1% by wt) of CNT can improve the mechanical properties consisting of the main Portland cement phase and water. A CNT can be singled or multiwalled. CNTs are the strongest and most flexible molecular material with Young's modulus of over 1 TPa. The approximate diameter is 1 nm with length to micron order. CNTs have excellent flexibility. These are essentially free from defects. Nanotubes are highly resistant to chemical attack and have a high strength to weight ratio (1.8 g/cm^3 for MWNTs and 0.8 G/cm^3 for SWNTs). CNT has maximum strain of about 10% which is higher than any other material. Figure 2 shows the flexible behavior of CNTs. Electrical conductivity of CNTs are six orders of magnitude higher than copper, hence, have very high current-carrying capacity. Hence, carbon nanotubes have excellent potential for use in the cement composites (Fig. 3).

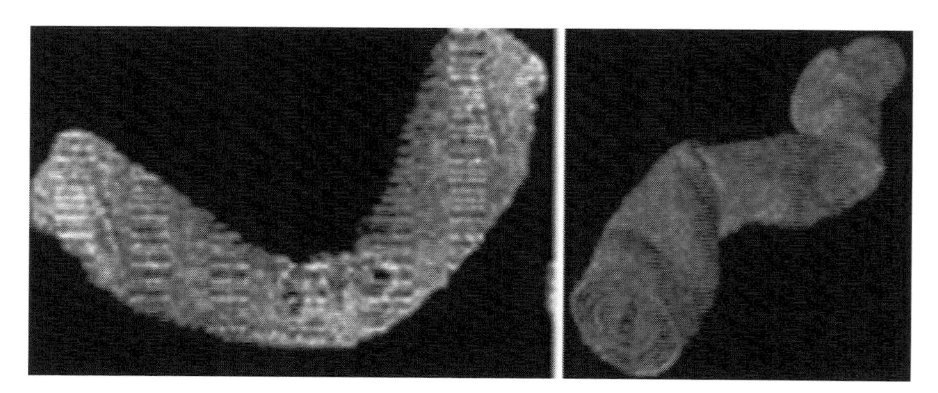

Fig. 2 Flexible behavior of CNTs [5]

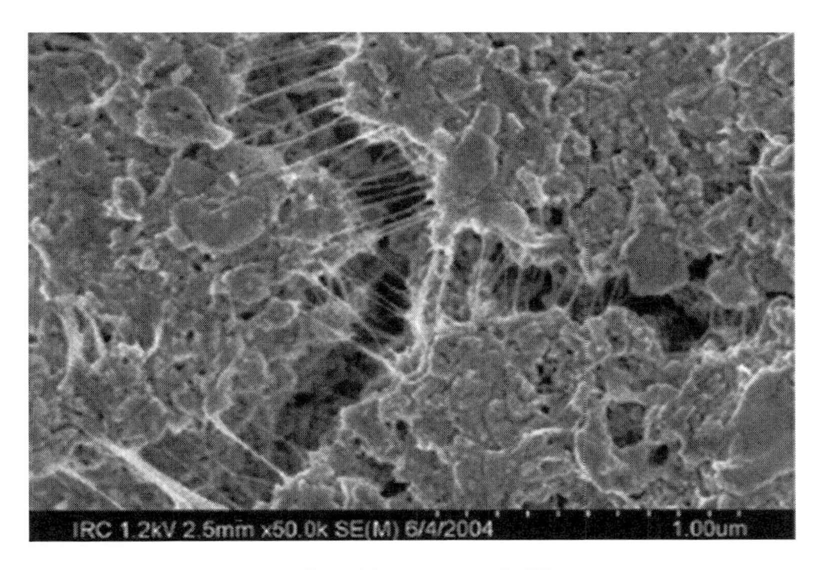

Fig. 3 Crack bridging by CNT bundles with cement matrix [5]

5 Nano-silica Fume for Improving Concrete Performance

Nano-silica is most common nano-additive to concrete. It is reported that nano-silica was found to be much effective than micron-sized silica for improving the performance such as permeability and, subsequently, durability. In addition, reduced amount of about 15–20 kg of nano-silica was found to provide same strength as 60 kg of regular or micro-silica (Fig. 4).

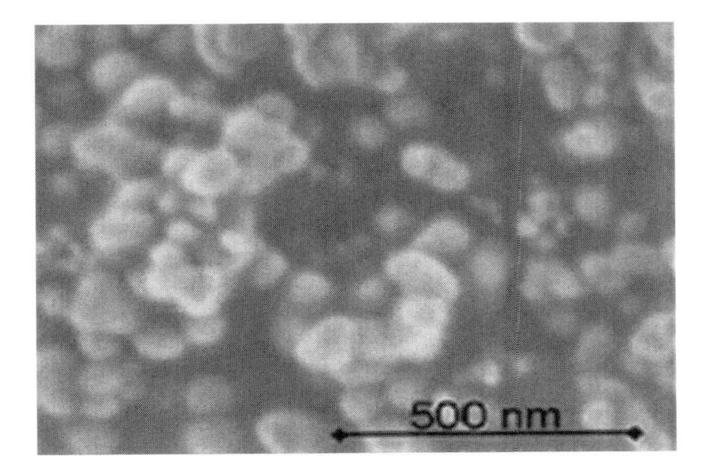

Fig. 4 A typical SEM of nano-silica particles [6]

Nano-silica is an effective additive to polymers and concrete, a development realized in high-performance and self-compacting concrete with improved workability and strength. Nano-silica addition to cement-based materials controls the degradation of the fundamental C–S–H (calcium-silicate-hydrate) reaction in water as well as blocks water penetration and leads to improvement in durability. The addition of nano-SiO_2 particles enhances the density and strength of concrete. The results indicate that nano-silica behaves not only as a filler to improve microstructure but also as an activator to promote pozzolanic reaction for fly ash concrete; as a result strength of the fly ash concrete improves particularly in the early stages.

6 Coatings for Concrete

Another major large volume application of nano-powder in cement-based materials is the area of coatings. The attractive coloring on ancient Czech glasses is found to contain nanoparticles. This shows that nanotechnology was used for coating surfaces, that is, spraying and making a product look attractive from ancient time. Nano-powders have a remarkable surface area. The surface area imparts a serious change of surface energy and surface morphology. The change in properties causes improved catalytic ability, tunable wavelength-sensing ability, and better designed pigments and paints with self-cleaning and self-healing features. One promising area of application of nanoparticle for cement-based materials is development of self-cleaning coating. Titanium oxide is commonly used for this purpose. It is incorporated, as nanoparticles to block UV light. It is added to paints, cements, and windows for its sterilizing properties as TiO_2 breaks down organic pollutants, volatile organic compounds, and bacterial membrane through powerful catalyst reactions and can reduce airborne pollutants applying to outer surfaces. Additionally, it is hydrophilic and therefore gives self-cleaning properties to surface to which it is applied.

7 Controlled Release of Admixtures

Currently, there is an extensive use of chemical admixtures mainly to control/modify the fresh and hardened properties of concrete. The most common admixtures for cement and concrete include accelerators, set retarders, air entraining agents, and superplasticizers. Their successful use requires a basic knowledge of concrete technology, standard construction procedures, and familiarity with cement-admixture interactions. A particular challenge of interest to the authors is to optimize the use of dispersing agents such as superplasticizers in high-performance concretes containing high volumes of supplementary cementing materials (SCMs). Dispersing agents such as superplasticizers are commonly used in these concretes. There are,

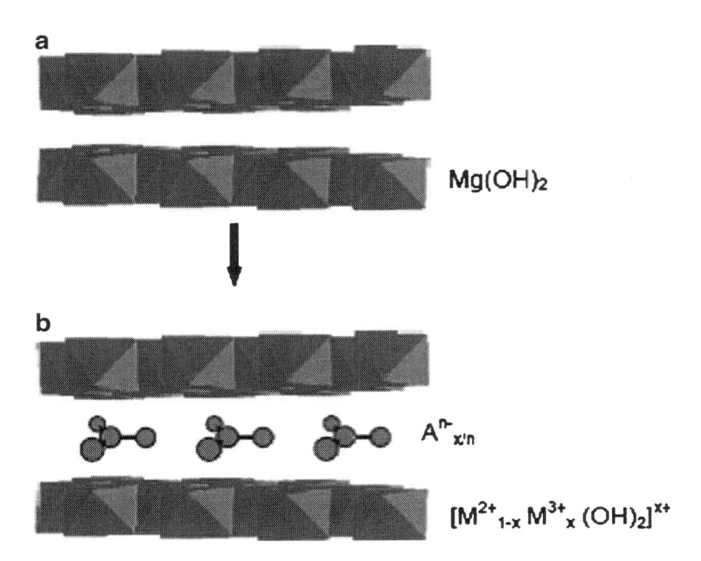

Fig. 5 Crystal structure of brucite (**a**) and LDH (**b**) [3]

however, practical problems such as loss of workability with time that are controlled by interactions with cement components. Controlling the timing of the availability of an admixture in cement systems is essential for its optimal performance. Control release technology provides a route to prolonged delivery of chemicals while maintaining their concentration over a specific time period. Here, a nanotechnology-based approach for controlled release of admixtures in cement systems using layered double hydroxides can be used.

There have been a number of applications in cement and concrete where different means of controlling the effect of admixtures via a controlled release technique were used. A number of patents and research articles describe "encapsulation" procedures for delivery of liquids and solids. A corrosion inhibitor, such as calcium nitrite, was dispersed by encapsulation in coated hollow polypropylene fibers. This anticorrosion system was activated automatically when conditions would allow corrosion to initiate in a steel-reinforced concrete. Porous aggregates were also used to encapsulate antifreezing agents. Porous solid materials (e.g., metal oxides) have also been used as absorbing matrices to encapsulate chemical additives (e.g., accelerators, retarders, and dispersants) and to release them at a slower rate when combined with oil well-treating fluids (Figs. 5 and 6).

Another method to control the release of chemicals in cement-based materials is by "intercalation/de-intercalation." A cement additive for inhibiting concrete deterioration was developed with a mixture of an inorganic cationic exchanger, a calcium zeolite capable of absorbing alkali ions (sodium, potassium, *etc.*), and an inorganic anionic exchanger, hydrocalumite capable of exchanging anions (chlorides, nitrates, sulfates, *etc.*). The results of their tests showed the potential of increasing concrete durability by exchange of alkali and chloride ions to inhibit alkali-aggregate reaction and corrosion of rebar.

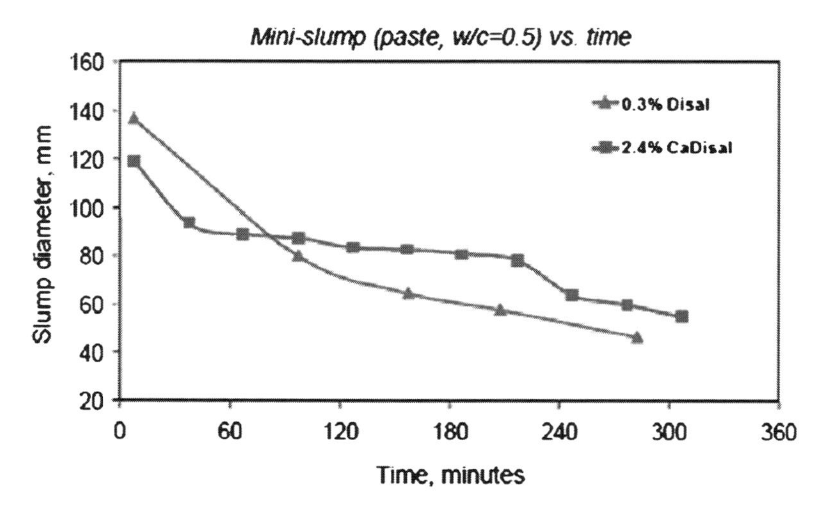

Fig. 6 Slump retention with LDH technology

More recently, work examined means to control the timing of the release of chemical admixtures through their incorporation in nanoscale composite materials. More specifically, the technique consisted of intercalating an admixture into a hydrocalumite-like material, a calcium-based LDH derivative, and adding this composite to a cement-based mix. De-intercalation of the admixture can be actively programmed through controlled chemistry involving, for example, type of layered inorganic material, charge density, concentration, and/or pH. A sulfonated naphthalene formaldehyde-based superplasticizer, called Disal™, was used to produce the controlled release formulation (CaDisal).

The effectiveness of Disal™ alone in controlling the slump-loss *versus* time characteristic was compared to that of the controlled release formulation CaDisal.

8 Nanoparticles and Steel

Steel has been a widely available material and has a major role in the construction industry. The use of nanotechnology in steel helps to improve the properties of steel. The fatigue led to the structural failure of steel due to cyclic loading, such as in bridges or towers. The current steel designs are based on the reduction in the allowable stress, service life, or regular inspection regime. This has a significant impact on the life cycle costs of structures and limits the effective use of resources. The stress risers are responsible for initiating cracks from which fatigue failure results. The addition of copper nanoparticles reduces the surface unevenness of steel which then limits the number of stress risers and hence fatigue cracking. Advancements in this technology using nanoparticles would lead to increased safety, less need for regular inspection regime, and more efficient materials free from fatigue issues for construction.

The nano-size steel produces stronger steel cables which can be in bridge construction. Also, these stronger cable materials would reduce the costs and period of construction, especially in suspension bridges as the cables are run from end to end of the span. This would require high-strength joints which lead to the need for high-strength bolts. The capacity of high-strength bolts is obtained through quenching and tempering. The microstructures of such products consist of tempered martensite. When the tensile strength of tempered martensite steel exceeds 1,200 MPa, even a very small amount of hydrogen embrittles the grain boundaries, and the steel material may fail during use. This phenomenon is known as delayed fracture, which hindered the strengthening of steel bolts, and their highest strength is limited to only around 1,000–1,200 MPa.

The use of vanadium and molybdenum nanoparticles improves the delayed fracture problems associated with high-strength bolts reducing the effects of hydrogen embrittlement and improving the steel microstructure through reducing the effects of the intergranular cementite phase.

Welds and the heat-affected zone (HAZ) adjacent to welds can be brittle and fail without warning when subjected to sudden dynamic loading. The addition of nanoparticles of magnesium and calcium makes the HAZ grains finer in plate steel, and this leads to an increase in weld toughness. The increase in toughness at would result in a smaller resource requirement because less material is required in order to keep stresses within allowable limits. The carbon nanotubes are exciting material with tremendous properties of strength and stiffness; they have found little application as compared to steel, because it is difficult to bind them with bulk material, and they pull out easily, which make them ineffective in construction materials.

9 Nanoparticles in Glass

The glass is also an important material in construction. There is a lot of research being carried out on the application of nanotechnology to glass. Titanium dioxide (TiO_2) nanoparticles are used to coat glazing since it has sterilizing and antifouling properties. The particles catalyze powerful reactions which break down organic pollutants, volatile organic compounds, and bacterial membranes.

The TiO_2 is hydrophilic (attraction to water) which can attract raindrops which then wash off the dirt particles. Thus, the introduction of nanotechnology in the glass industry incorporates the self-cleaning property of glass. Fire-protective glass is another application of nanotechnology. This is achieved by using a clear intumescent layer sandwiched between glass panels (an interlayer) formed of silica nanoparticles (SiO_2) which turns into a rigid and opaque fire shield when heated. Most of glass in construction is on the exterior surface of buildings. So the light and heat entering the building through glass has to be prevented. The nanotechnology can provide a better solution to block light and heat coming through windows.

10 Nanoparticles in Coatings

Coatings is important element in construction and are extensively used to paint the walls, doors, and windows. Coatings should provide a protective layer which is bound to the base material to produce a surface of the desired protective or functional properties. The coatings should have self-healing capabilities through a process of "self-assembly." Nanotechnology is being applied to paints to obtain the coatings having self-healing capabilities and corrosion protection under insulation. These coatings are hydrophobic and repel water from the metal pipe and can also protect metal from saltwater attack. Nanoparticle-based systems can provide better adhesion and transparency. The TiO_2 coating captures and breaks down organic and inorganic air pollutants by a photocatalytic process, which leads to putting roads to good environmental use.

11 Conclusion

Nanotechnology can change the way we construct our structure today. It can help us to utilize the natural resources to optimum level and make them sustainable. This will also allow use of resources for present development and leave the same for the use of future generations also. Though presently at research level, scope is enormous, and engineers and scientist need to increase their effort on the directions to overcome challenges.

References

1. Patel-Predd P (2007) The Nano secret to concrete, MIT Technology review
2. Ge Z (2008) Applications of nanotechnology and nano-materials in construction. In: First International Conference on Construction In Developing Countries (ICCIDC–I) 2008, Karachi, Pakistan
3. Raki L et al (2010) Cement and concrete nano-science and nanotechnology. Materials 3:918–942
4. Porro P (2005) Nano science and nanotechnology in construction Materials. In: 2nd international symposium on nanotechnology in construction 2005, Bilbao, Spain
5. Mann, S (2006) Nanotechnology and construction. Report of Nano forum
6. Ji T (2005) Preliminary study on the water permeability and microstructure of concrete incorporating nano-SiO_2. Cement Concr Res 35:1943–1947

Sensor Network Design for Monitoring a Historic Swing Bridge

Giuseppe C. Marano, Giuseppe Quaranta, Rita Greco, and Giorgio Monti

Abstract Significant advances in the development and customization of various sophisticated technologies for structural monitoring have emerged during the last decade. Technologies for instrumentation, monitoring, load testing, nondestructive evaluation and/or characterization, three-dimensional finite element modeling, and various types of analyses have now become available at a reasonable cost. Within this framework, this chapter focuses on the issues addressed in designing a sensor network for dynamic monitoring of a historic swing bridge in Taranto (Italy).

Keywords Dynamic monitoring • Movable bridge • Sensor network

1 Introduction

Bridge structures are very critical elements within a complex transportation system, and movable bridges are especially important because they allow traffic across active waterways, thus granting passage to ships that would otherwise be blocked by the structure. Therefore, reliability assessment as well as health monitoring of movable bridge structures are challenging issues that deserve significant attention

---

G.C. Marano (✉) • R. Greco
Department of Environmental Engineering and Sustainable Development, Technical University of Bari, Viale del Turismo, 10, 74100 Taranto, Italy
e-mail: g.marano@poliba.it

G. Quaranta
Department of Civil and Environmental Engineering, University of California, Davis, One Shields Avenue, 95616 Davis, CA, USA
e-mail: gquaranta@ucdavis.edu

G. Monti
Department of Structural and Geotechnical Engineering, Sapienza University of Rome, Via A. Gramsci 53, 00197 Rome, Italy
e-mail: giorgio.monti@uniroma1.it

S. Chakraborty and G. Bhattacharya (eds.), *Proceedings of the International Symposium on Engineering under Uncertainty: Safety Assessment and Management (ISEUSAM - 2012)*, DOI 10.1007/978-81-322-0757-3_29,

Fig. 1 Location of the "St. Francesco da Paola" bridge (*left*) and a photograph (*right*)

because a structural failure and/or a temporary lack of service may have a tremendous socioeconomic impact. Developing a reliable network for monitoring these infrastructures is, therefore, an efficient way for supporting numerical studies by making available reference experimental data to be used, for example, in finite element model elaborations.

Having this in mind, we are developing a sensor network for monitoring the "St. Francesco da Paola" bridge (Fig. 1), a historic swing bridge located in Taranto (Italy).

Taranto is a coastal city located in the southern part of Italy, in the Apulia region, in front of the Ionian Sea. The city is characterized by a hydrologic system based on two basins close to the Ionic coast. An inner, semi-enclosed basin with lagoon features – named "Mar Piccolo" – is connected with the outer basin (named "Mar Grande") through two canals, namely, the "Navigabile" canal and the "Porta Napoli" canal. The inner basin "Mar Piccolo" is divided into two inlets – named first and second inlet – which have a maximum depth of 13 and 8 m, respectively (Fig. 1). The actual swing bridge over the "Navigabile" canal – the "St. Francesco da Paola" bridge – was opened to traffic more than fifty years ago, on March 10, 1958. This swing bridge was built with two equally armed movable portions, and each of them rotates about one vertical axis. The structural health and the functionality of the opening system are crucial, because of the bridge's central role within the local transport network. The current structural reliability of the bridge is substantially unknown because it has not been investigated previously by using modern simulation-based numerical techniques. Significant experimental data are not available to date. There is no digitalized information about the bridge and only historic documents were found. Therefore, a numerical model calibrated on experimentally recorded data is an important first step for a reliable condition assessment of the bridge. For instance, records obtained from a dynamic test may be used to identify natural frequencies, mode shapes, and damping characteristics of the structure for the purpose of model updating.

In this perspective, this chapter provides an overview about our work in developing numerical models as well as experimental- and information-based technologies for monitoring the "St. Francesco da Paola" swing bridge.

2 Structural Monitoring of Bridge Structures

The evaluation of existing bridges has become an increasingly important topic in the effort to deal with deteriorating infrastructures. This is because a considerable ratio of historic bridges may be classified as deficient or in need of rehabilitation. The most critical issue for historic bridges is typically due to their seismic reliability. However, historic bridges located within relatively low-seismic regions may also have an insufficient reliability. For instance, this is due either to their decreased capacity due to decay or to an increased demand with respect to when they were built (i.e., the loads and the traffic flows are higher, fewer disturbances due to traffic-induced vibrations are accepted). Thus, performing a reliable structural assessment of a bridge becomes essential to avoid its disposal, which involves seldom-acceptable economical and safety implications, not to mention the irreparable loss of a cultural heritage artifact. The accuracy of bridge evaluation can be improved by using recent developments in the fields of bridge diagnostics, structural tests, and material tests. Advanced diagnostic procedures can be applied to the evaluation of the current capacity of the structure, monitoring of load as well as resistance histories, and evaluation of the accumulated damages. On the contrary, traditional visual-inspection-based condition assessment of bridges cost significantly while restricting operations for many months. Moreover, since visual inspections are conducted by trained and experienced bridge engineers, and/or inspectors, according to some (standardized) procedures, many have pointed out the limitations and shortcomings associated with evaluating and managing bridges primarily on the basis of essentially subjective data. As a consequence, the use of numerical models calibrated on experimental data is a more reliable and appropriate way for bridge condition assessment. The available technologies of structural monitoring tools can be classified as experimental, analytical, and information technologies. Experimental technologies are further classified as:

- Geometry monitoring (to track changes in the geometry, such as geometry changes in cable systems)
- Controlled testing (which should be static or dynamic, nondestructive, destructive or "localized" nondestructive evaluations, and continuous monitoring)

Dynamic testing of bridges, sometimes termed as "vibration analysis" by civil engineers, is an exceptionally powerful experimental technique. For instance, records obtained from a dynamic test may be used to assess the comfort on pedestrian-accessible bridges and to identify relevant modal features. Dynamic tests of bridges and civil constructions are sometimes a necessity as there is no other test technique available that provides a direct measurement of the global dynamic properties of a structure and without the need for any external measurement reference frame.

Analytical technologies that are used for bridge health monitoring have been classified as:

- CAD and reverse CAD
- Analytical modeling based on a macroscopic, element-level, microscopic, or mixed approaches
- Linear analysis under static, moving, or dynamic loads
- Nonlinear analysis incorporating material nonlinearity, geometric nonlinearity, or both types

Information technologies cover the entire spectrum of efforts related to the acquisition, processing, and interpretation of data. This includes sensing, data acquisition, preprocessing, communication and control, transmission and synchronization, quality testing, post-processing, analysis, display and visualization, database archival and management, and interpretation for decision-making. Given the advances in experimental and analytical technologies, an extensive level of expertise is needed to take full advantage of advanced information technology tools. Teams of computer scientists and structural and electrical engineers have to be brought together so that all the necessary ingredients of know-how may be integrated into meaningful structural monitoring applications.

3 CAD and FE Model

The geometrical model of the bridge was undertaken using CAD techniques (see Figs. 2 and 3). The "St. Francesco da Paola" bridge over the "Navigabile" canal has a span equal to 89.52 m and accommodates two traffic lanes with a total 6.00 m roadway. Two pedestrian lanes are on both sides of the bridge, and the sidewalk span is equal to 1.50 m for each lane. The two movable portions of the bridge are

Fig. 2 CAD model: general view of the bridge

Fig. 3 CAD model: view on the steel truss structure

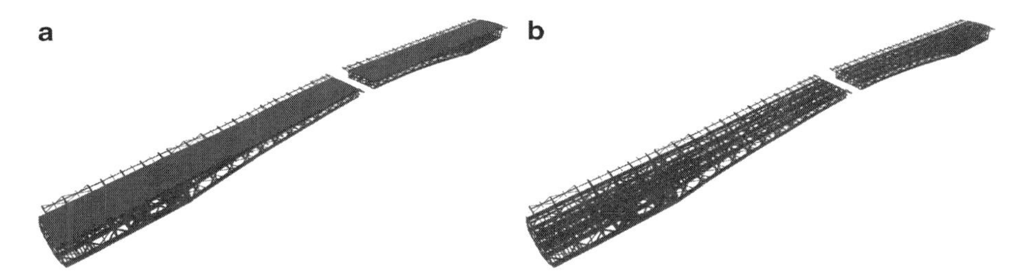

Fig. 4 FE model with (**a**) and without (**b**) steel orthotropic deck

steel truss structures. One rim bearing for each movable portion of the bridge is placed on the corresponding abutment. Each rim bearing consists of a series of conical rollers which are loaded during the bridge movement. The two movable parts of the bridge are connected to each other in four points located in the transversal section of the free ends. The steel truss structure of each movable portion of the bridge was realized as follows: 4 longitudinal truss beams, 19 transverse braces, and horizontal braces on both lower and upper chord of the truss structure.

A three-dimensional finite element (FE) model of the bridge was built by using all the available information collected into the above-illustrated CAD model. The construction of the FE model was undertaken using SAP2000© (version 12), and it is shown in Fig. 4. Only structural elements were included in the model, whereas nonstructural elements are considered as extra masses. All structural elements are made of steel with the following properties: elastic modulus 200,000 N/mm^2, Poisson's ratio 0.30, and mass density 7,850 kg/m^3. Soil-structure interaction is not considered for the purpose of the modal analysis, and the rim bearing is replaced with an appropriate set of constraints. Beam-type elements were adopted to model the element of the truss structure. Thin shell elements were used to model steel plates (including the orthotropic plate-type deck).

4 Hardware Components of the Sensor Network

Developing experimental and information technologies that meet internationally recognized guidelines and standards was one of the starting points for our project. Experimental and information technologies that conform to international standards and guidelines were the priorities in order to ensure the reliability of the final results, which the correctness of the analytical strategies depends upon. Technologies that are described in this chapter meet the requirements provided by the following standards and guidelines:

- Guideline for ANSS seismic monitoring of engineered civil systems [1]
- IEEE standards regarding hardware and software implementation

The developed system for structural dynamic monitoring was named "THOR" (Fig. 5).

THOR consists of three main components: *ThorSensors*, *ThorAgents*, and *ThorServers*. THOR is able to manage more distributed sensor networks with real-time acquisition and data processing for structural analyses. The single THOR network is able to monitor a zone with one *ThorAgent* linked to a number of *ThorSensors*. Different and also geographically distributed networks can be managed by one *ThorServer*. *ThorAgents* and *ThorSensors* cooperate with *ThorServers* to

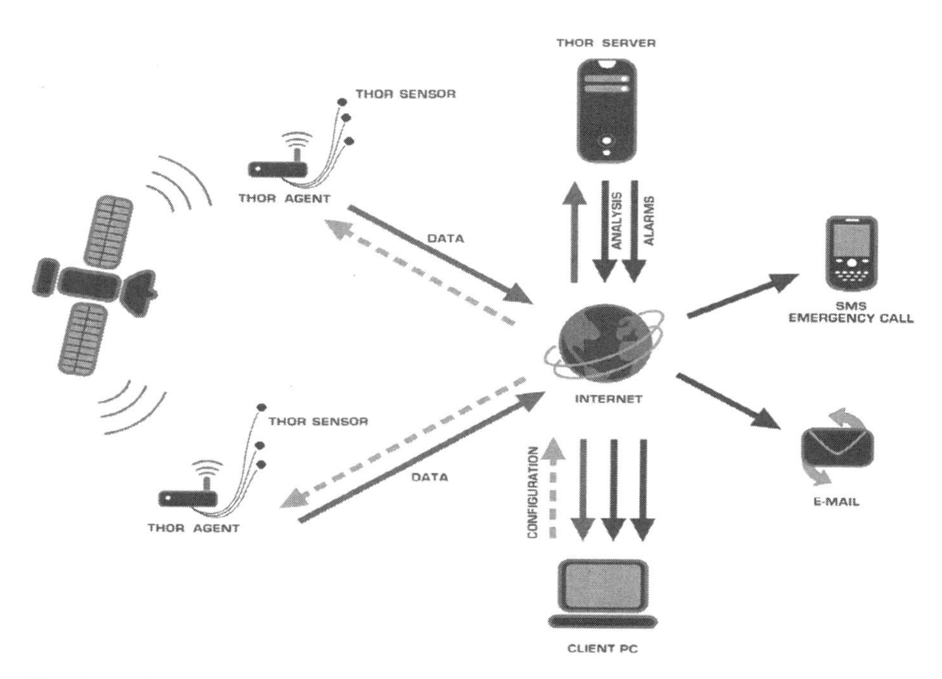

Fig. 5 "THOR" working scheme

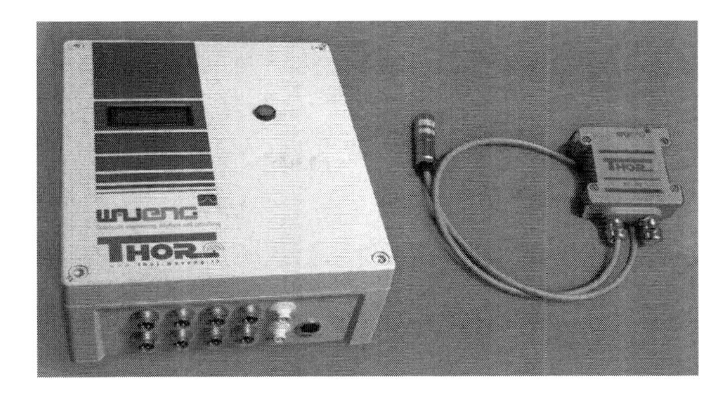

Fig. 6 Some THOR components

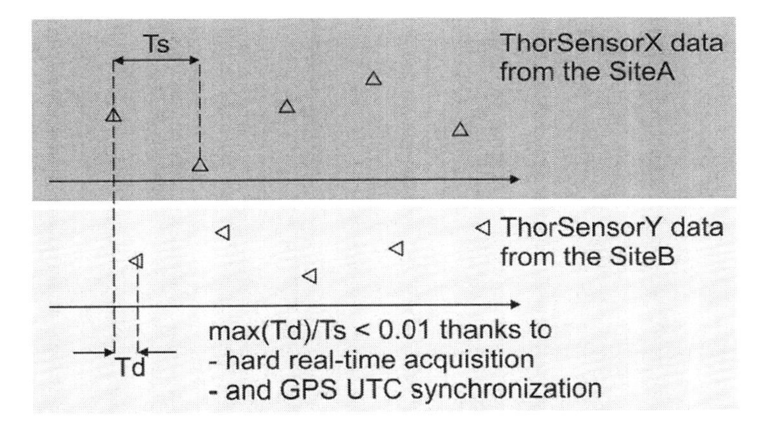

Fig. 7 GPS synchronization

monitor more sites. Some network components are shown in Fig. 6. The key point of this system is that – although each network works alone – all the data collected from different networks can be analyzed and correlated, because the system keeps hard real-time acquisition and GPS synchronization, see Fig. 7. In this way, *ThorServer* can execute intelligent algorithms for data processing and detection of critical conditions. If dangerous situations occur, *ThorServer* is able to send alarms via e-mail, GSM calls, and SMS. The entire system can be remotely accessed via web. Main features of THOR are:

- Rugged design for operation in hostile environments
- Modular design to ensure scalability of the sensor networks
- Hard real-time data acquisition with high-performance sensors
- Capability to manage more geographically distributed sensor networks

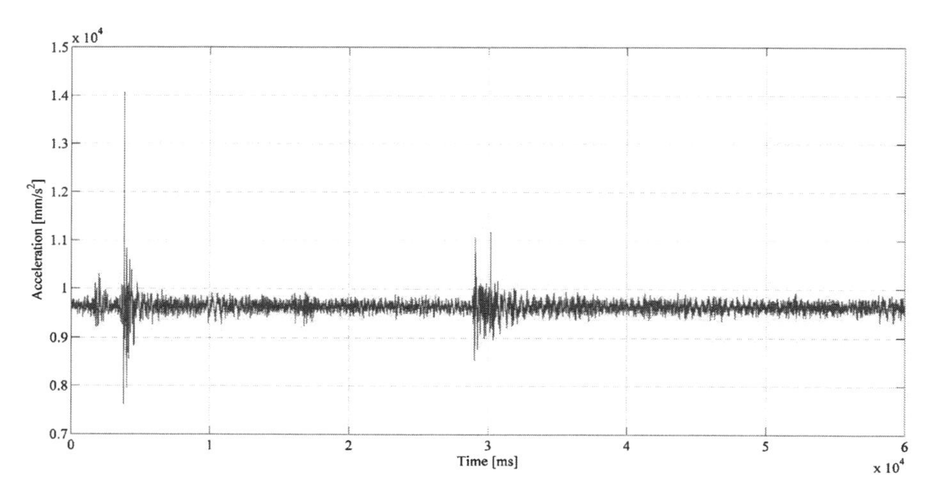

Fig. 8 Data streaming example samples analyzed online

- Capability to synchronize all the data from different networks via GPS UTC synchronization
- Capability to analyze and process real-time data from in one base time line
- Comprehensive tools for online/offline data management

These goals are typically achieved by means of expensive hardware and software technologies with lab-oriented design. On the contrary, THOR integrates the features above with reasonable costs and rugged hardware ready to use in real-world applications.

ThorSensors are force balance accelerometers that meet the following specifications: dynamic range 140 dB, bandwidth DC to 200 Hz, user selectable full-scale range up to ± 4 g, linearity $< 1{,}000$ μg/g^2, hysteresis $< 0.1\%$ of full scale, cross-axis sensitivity $< 1\%$ (including misalignment), operating temperature from -20 to 70°C, and weight 0.35 kg. *ThorAgent* specifications are real-time acquisition with GPS time stamping, data samples on all channels taken simultaneously within 1% of the sample interval, clock accuracy to UTC $\pm$ 50 ns, sample rate 1 kHz, GPRS/UMTS remote connection, Wi-Fi interface, rugged case (IP67), local storage 8 GB, and operating temperature from -20 to +60°C. An example of acquisition is reported in order to demonstrate how the developed structural monitoring network works (Fig. 8). *ThorServer* is able to show a user-defined period for each sensor acquisition via web. This allows speeding up the analysis by the user who can have a glance of a long data streaming and then can directly zoom on the interested part without any download and extra tool. Moreover, the web server implemented on *ThorServer* is also able to support the analysis with online elaboration, which results very comfortable during on-site operations where it is hard to have the office facilities.

5 Sensors Placement

Because of the existence of budgetary and practical constraints, civil engineered systems are typically monitored under natural dynamic loads (i.e., wind or traffic-induced loads) available at no cost. As historic infrastructures are concerned, output-only techniques are also preferred because of cheaper and faster test execution and minimum interference with the use and preservation of the structure. Based on these considerations, output-only measurements will be considered for this case study. This implies that the optimal sensor placement (OSP) problem is of importance for this application. It should be remarked that most of the applications regarding OSP techniques are in the field of aeronautic, aerospace, and mechanical engineering. On the contrary, few applications deal with civil engineered systems. In this field, the selection of the best DOFs to be monitored seems to be mostly based on engineering judgments only. However, numerical techniques for the OSP may provide a valuable support in order to design a sensor network for civil structural dynamic monitoring as well [2–4].

Therefore, a preliminary study was performed in order to look for the most significant DOFs to be monitored by using the available sensors and data acquisition system. To this end, the effective independence method [5] was used with the aim to look for sensor positions that maximize both the spatial independence and the signal strength of the target mode shapes by maximizing the determinant of the associated Fisher information matrix. Results for 8, 12, and 16 sensors are shown in Figs. 9, 10, and 11, respectively.

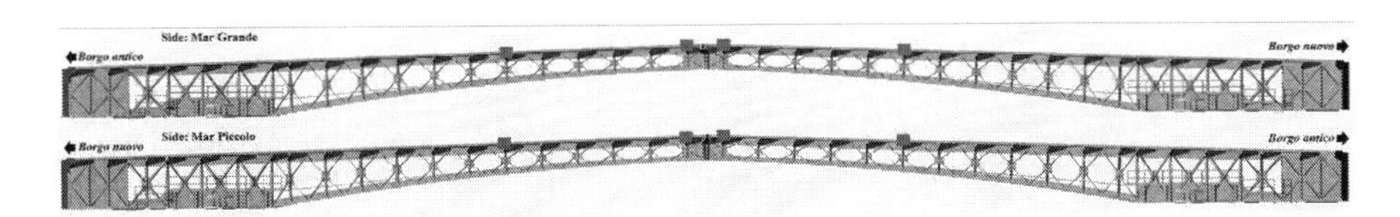

Fig. 9 Optimal sensors placement according to EFI method (8 sensors)

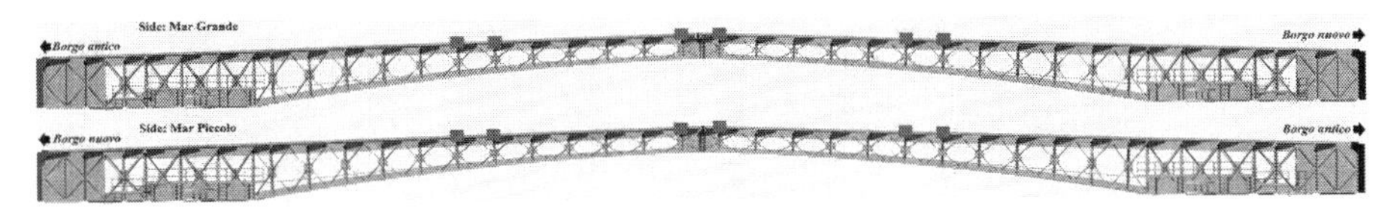

Fig. 10 Optimal sensors placement according to EFI method (12 sensors)

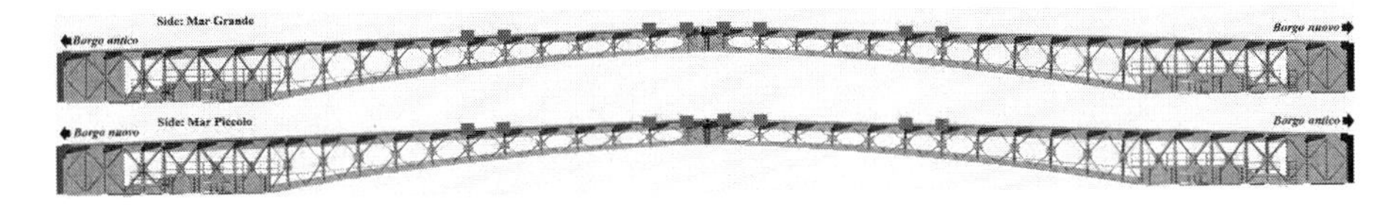

Fig. 11 Optimal sensors placement according to EFI method (16 sensors)

6 Conclusions

This chapter provided an overview about the most important steps that we are addressing for dynamic structural monitoring applications, with emphasis on the case study regarding the "St. Francesco da Paola" swing bridge in Taranto (Italy). The developed system for structural dynamic monitoring has been named as "THOR," and it fulfills the most important international standards and guidelines in this field. First, both CAD and FE model are briefly presented. Therefore, hardware specifications of the most important system components are listed. Finally, a preliminary study about the optimal sensors placement via effective independence method is presented. Results in this chapter will be useful for supporting further studies in order to design a reliable structural identification instrumentation for this bridge. Although this chapter deals with a particular bridge structure, considerations and results herein presented may provide a constructive framework for similar engineering applications.

Acknowledgements The support provided by Giuseppe Leonardo Cascella, Ph.D., and Davide Cascella, Ph.D. (Department of Electrotechnic and Electronic, Technical University of Bari) is appreciated. The assistance provided by Waveng srl (via Robert Schuman 14, 70126 Bari, Italy) is also acknowledged. Finally, we wish to thank Eng. Domenico Antonio Ricci, Eng. Erika Mastromarino, and Marco Denitto for their assistance.

References

1. ANSS (2005) Guideline for ANSS seismic monitoring of engineered civil systems. U.S. Dept. of the Interior, U.S. Geological Survey, Reston
2. Meo M, Zumpano G (2005) On the optimal sensor placement for a bridge structure. Eng Struct 27:1488–1497
3. Marano GC, Monti G, Quaranta G (2011) Comparison of different optimum criteria for sensor placement in lattice towers. Struct Des Tall Spec Build 20:1048–1056. doi:10.1002/tal.605
4. Monti G, Quaranta G, Marano GC (2010) Robustness against the noise in sensors network design for heritage structures: the case study of the Colosseum. In: 4th international workshop on Reliable Engineering Computation (REC2010), 3–5 March 2010, Singapore, pp 241–254
5. Kammer DC, Brillhart RD (1996) Optimal sensor placement for modal identification using system-realization methods. J Guid Control Dyn 19:729–731

Sensor Applications for Structural Diagnostics and Prognostics

Anindya Ghoshal

Abstract This chapter examines emerging sensor technologies in aerospace structural prognostics health management. A review of existing and emerging in situ sensor technologies for structural health monitoring for aerospace applications has been discussed in details. Details of the sensor selection criteria for the sensor technologies have been stated. For successful implementation of condition-based maintenance of aerospace vehicles, such emerging sensors are key technologies that would be required.

Keywords Condition-based structural health monitoring • In situ sensor technologies • Structural sensing • Diagnostics and prognostics

1 Introduction

Considerable advancement has been made in the sensor technology development for in situ sensor technologies for structural health diagnostics and prognostics [1]. This chapter is done with the objective of defining and selecting appropriate damage detection sensor(s) for direct monitoring of subcritical fatigue cracks in airframe primary structural elements. The sensor hardware should also provide reliable detection in representative airframe joints and/or attachments.

A. Ghoshal (✉)
ARL, Towson, MD, USA
e-mail: anindo_ghoshal@yahoo.com

S. Chakraborty and G. Bhattacharya (eds.), *Proceedings of the International Symposium on Engineering under Uncertainty: Safety Assessment and Management (ISEUSAM - 2012)*,
DOI 10.1007/978-81-322-0757-3_30,

2 Sensor Selection Criteria

The following parameters are used in sensor selection and sensor evaluation:

- Functionality of structural damage
  - Minimal detectable size of damage
  - Probability of detection (POD)
  - Sensitivity variation to size, orientation, and location of cracks
  - Boundary conditions, presence of joints, loads, and structural layers.
- Technology maturity
  - Commercial off-the-shelf with minimal customization
  - Demonstrated capability for aircraft applications or on similar products
- Sensor durability, reliability, and false alarm rate
  - Long-term stability, repeatability, and low drift (including bonding durability)
  - Sensitivity to environmental variation and normal workload
  - Temperature, vibration, and dynamic loading/static loading/structural deformation
  - Built-in smartness (through software) to reject noises or disturbance
- Structural embeddability
  - Be able to permanently mount on or bond to surface of structure for real-time monitoring or periodical scanning
  - Minimal intrusive to the structure being monitored—low profile and lightweight

3 Physical Sensor Review

This chapter further elaborates few promising damage-sensor technologies and associated vendors among a dozen of potential candidates.

Local crack monitoring sensors

- MWM-Array eddy current sensors: JENTEK Sensors, MA
- Active current potential drop sensors: Matelect Ltd, UK
- Comparative vacuum sensors: SMS Systems, Australia

Global damage sensors

- Piezoelectric acoustic sensors, Acellent, CA
- Fiber-optic sensors (fiber Bragg grating): Luna Innovations Inc, VA; Micron Optics, WA; and Insensys Inc
- Time-domain reflectometry, Material Sensing & Instrumentation, Inc., PA

- Magnetostrictive sensors: Southwest Research Institute, TX
- Carbon nanotube and graphene-based sensors

The six technologies that have high potential for the structural health monitoring application are evaluated in details as follows.

4 Eddy Current Sensors

4.1 *Principle of Operation*

An eddy current sensing system makes its measurement by measuring the electrical impedance change of the eddy current probe. The probe consists of coils that carry high-frequency current and generates an electromagnetic field. When the probe is placed near a metallic structure, the EM field penetrates the conductive surface and creates an eddy current within the structure. The intensity of the current or the electric impedance of the coil is a function of the material properties such as electric conductivity and permeability which is sensitive to the local structure damage or defects. This material property variation around the measurement point can then be translated into the structure defects or damage information via either a mathematic model or calibration against empirical database. The eddy current sensors can be made on thin polymer film with electric coil printed on it so they can be customizable in shapes, conformable to the surface of the structure, and cannot be easily mounted onto any complex surfaces permanently.

JENTEK Inc is today a major player in the technology of crack detection by eddy current sensing [2]. Its Meandering Winding Magnetometer Array (MWM-Array) system features high-resolution multiple-channel impedance measurement instrumentation and high-resolution imaging for crack detection. The sensing element configurations provide improved detection performance along with reduced calibration requirements and setup time.

4.2 *MWM-Sensor Array*

A MWM-Array system has a single period spatial mode drive with a linear array of sensing elements. The MWM-Array provides images of electrical conductivity and is suitable for crack detection with high special resolution. The MWM-Array sensors have a primary winding that is driven with a high-frequency current to produce a time-varying magnetic field with a spatial wavelength that is determined by the physical spacing between drive winding segments. The MWM-Arrays typically operate at frequencies from 10 kHz to 15 MHz. At these frequencies, the wavelength of traveling waves is long compared to the dimensions of the sensor, so the distance between the drive winding segments defines the shape of the applied magnetic field. The magnetic field produced by the winding induces eddy currents in the material being tested. These eddy currents create their own magnetic fields that oppose the applied field. At low frequencies, these eddy currents are distributed well into the material under test; at high frequencies, these induced eddy currents are concentrated on a thin layer near the surface of the test specimen. A surface-breaking crack interferes with the flow of these eddy currents or the impedance of the winding. When MWM-Array is scanned across, an image of the impedance of the sensor array is generated which maps the structural or material abnormality of the test specimen.

Comparing with conventional eddy current sensor, MWM-Array features:

- Absolute sensing configurations (as opposed to differential sensing element designs) capable of inspecting regions likely to have cracks forming from micro-cracks into larger cracks.
- Calibration is performed on-site using either "air" or uniform reference parts without cracks, reducing calibration and training requirements.
- A crack signature is extracted off-site, only once, using either real cracks, EDM notch standards, or a simulated crack signature. This is an advantage because it eliminates dependence on crack standards and avoids potential errors encountered during calibration on such standards.

4.3 Technology and Product Maturity

JENTEK's MWM-Array sensor system has been under development and improvement since its inception in 1996 and has reached a certain level of product maturity. It currently offers a line of products typically off the shelf. The system is built around the following components: (1) a parallel architecture impedance instrument, (2) magnetic field (MWM) sensor arrays, and (3) Grid Station software environment, application modules, and tools. JENTEK is currently delivering two versions of imaging sensor array systems. The 39-channel system is a high-resolution imaging system with comprehensive imaging, decision support, and procedure development tools. The 7-channel system is designed for less image-intensive applications. These systems are supported by a wide selection of MWM-Array sensor configurations.

5 Active Current Potential Drop Sensors

5.1 Principle of Operation

Alternating current potential drop (ACPD) or direct current potential drop (DCPD) is an electrical resistance measurement technique for sizing surface-breaking defects in metals [3]. ACPD works by inputting an alternating current into the electric conductive object. At the points (I, I′), a constant direct current is supplied. An increase in crack length produces an increase of the potential drop measured between the potential leads (V, V′). Presence of defect or crack in the material between these two points will result in a local resistance larger than that of its vicinity. By comparing potential differences with a reference value, calibrating against empirical database or FEM modeling results, crack depth and size can be estimated.

The reference measurement of potential drop is usually required to provide comparison and needs to be as close to the crack as possible. Because of the skin effect, ACPD system is more capable of measuring surface crack while DCPD can measure in-depth crack but with lower sensitivity. The crack size can be estimated as Crack Depth=$\Delta/2\ (V_c/V_r-1)$, where Δ is the probe separation and V_c is the crack voltage. V_r is the reference voltage. The techniques are available for both thick and thin structures. Custom-made ACPD sensing probe can be either hand-holding scan probe or wire spot-weld in structure. Similar to MWM-Array sensors, these sensing wires need to be permanently mounted near the cracks or where cracks would potentially develop for detection. This means it is only a local crack detection system.

5.2 *Technology and Product Maturity*

Among a few vendors, Matelect Ltd of UK has been selected for investigation of product availability and specification. Matelect has a line of commercial off-the-shelf ACPD or DCPD products. Their most popular crack detection is CGM-7 microprocessor-based crack growth monitor system with operating frequency ranging from 0.3 to 100 kHz and current up to 2 A. The system is able to detect cracks of 0.02–10 mil on lab specimen and 40 mil on aircraft components. For permanent crack monitoring, the potential measuring probes are usually spot-weld into structure and probe can withstand 600°C temperature. The products have been used for Rolls-Royce engine turbine disk dovetail crack inspection, and it is recently being tested on CRJ aircraft structure for crack monitoring. The vendor claimed POD is 85% on 0.004 crack, sensitive to crack orientation.

The summary of pros and cons for use of ACPD/DCPD is as follows:

- High sensitivity to incipient crack and has long history of industrial application.
- Detestability is sensitive to the orientation of crack.
- Electrode sensor needs to be spot-weld into structure for permanent monitoring and is less suitable for retrofit application.
- Need complex in field calibration.
- Vulnerable to electromagnetic interference.

6 Comparative Vacuum Sensors

6.1 Principle of Operation

Comparative vacuum monitoring (CVM) offers a novel method for in situ, real-time monitoring of structural crack initiation and propagation [4]. CVM makes use of the principle that a steady-state vacuum, maintained within a small volume, is extremely sensitive to any leakage of air. It measures the differential pressure between fine galleries containing a low vacuum alternating with galleries at atmosphere in a simple manifold. The manifold is directly mounted on structure surface being monitored for crack. If no flaw is present, the vacuum in galleries will remain at a stable level. If a crack develops and creates a passage between vacuum and atmosphere galleries, air will flow through the passage created from the atmosphere to the vacuum galleries. Sensors may either take the form of self-adhesive polymer "pads" or may form part of the component. A transducer measures the fluid flow or pressure difference between the galleries.

CVM has been developed primarily as a tool to detect crack initiation. Once a sensor has been installed, a base line reading is made. Generally, the differential pressure between the reference vacuum and the sensor will be approximately 0 Pa. However, if there is a known existing flaw or crack beneath the sensor, or the permeability of the test material is high, the fluid flow meter will measure a nonzero base value. If this nonzero value is constant, it will not affect the ability of the CVM system to detect an increase in total crack length.

When a vacuum gallery is breached by a crack, molecules of air will begin to flow through the path created by the crack. Once the system has reached an equilibrium flow rate, the volume of air passing through the crack is equal to the volume of air passing through the flowmeter, and the measured differential pressure will become constant at a higher value. Therefore, the system is very sensitive to any changes in the total crack size. The CVM method is unable to differentiate between a single large crack and several smaller flaws, but is sensitive to any increase in the total crack length. The sensitivity of the sensor is determined by the gallery wall thickness.

6.2 Technology and Product Maturity

Structural Monitoring System Inc of Australia has developed this sensor technology. A variety of sensor types have been developed. These include self-adhesive elastomer sensors for the measurement of surface crack initiation or propagation and sensors integral within structure, for example, permeable fiber within a composite. The sensors are produced from a variety of materials. The accuracy of the crack propagation sensor is governed by the accuracy of the galleries, measured optically at better than 10 mm. Once the sensor has been installed, the leading edge of the first gallery is determined optically, and from this initial measure, all subsequent gallery positions are determined.

CVM sensors have been applied to a variety of aerospace structures for crack detection. Sandia National Lab, in conjunction with Boeing, Northwest Airlines, Delta Airlines, Structural Monitoring Systems, the University of Arizona, and the FAA, has conducted validation testing on the CVM system in an effort to adopt comparative vacuum monitoring as a standard NDI practice. The system has been tested by on Boeing 737, DC 9, C130 aircrafts, and Blackhawk helicopter by Australian air force. According to the SMS Inc., the sensor pad adhesive can hold the vacuum in the galleries for as long as 18 month. Ninety percent probability of detection with 0.020–0.025″ crack on 2024 aluminum structure has been reported.

The summary of pros and cons for use of CVM is as follows:

- High sensitivity to 0.020 mils.
- The system is lightweight, inert (safe), and less vulnerable to EMI than any electric-based system.
- Elastomeric sensor is low cost and conformable to any curved surfaces and can be easily integrated to complex structural surface.
- Easy to operate and calibrate.
- The system is capable of detecting surface break cracks only.

Overall, CVM is a good candidate technology for local crack detection

7 Networked Piezoelectric Sensor

Piezoelectric (PZT) transducer can be used to monitor structures for internal flaws when it is embedded in or surface mounted to structures. PZT transducer can act as both transmitters and sensors due to its direct and reverse piezoelectric effect. As transmitters, piezoelectric sensors generate elastic waves in the surrounding material driven by alternating electric field. As acoustic sensor, they receive elastic waves and transform them into electric signals. It is conceivable to imagine arrays of active sensors, in which each element would take, in turn, the role of transmitter and acoustic sensor and thus scan large structural areas with high-frequency acoustic waves. As global damage sensing, two PZT sensor network-based approaches are commonly used for structural health monitoring:

- Self-electromechanical (E/M) impedance method for flaw detection in local area using effect of structural damage on EM impedance spectrum
- Lamb wave propagation method for large area of detection using acoustic wave propagation and interception by presence of structural damage on the acoustic path

In the self-E/M impedance approach, pattern recognition methods are used to compare frequency domain impedance signatures and to identify damage presence and progression from the change in these signatures. In the cross impedance approach, the acousto-ultrasonic methods identifying changes in transmission velocity, phase, and additional reflections generated from the damage site are used. Both approaches can benefit from the use of artificial intelligence neural network algorithms that can extract damage features based on a learning process.

The Acellent Technologies of Mountain View, California, offers a PZT acoustic-based structural health monitoring system commercially off-the-shelf [5]. The system comprises of SMART Layer sensors, SMART Suitcase, and ACESS Software®. This structural health monitoring solution is capable of monitoring both metallic and composite structures. The SMART Layer consists of multiple piezoelectric sensing elements with wires lithographically imprinted on the Kapton film. These sensors operate based upon the principles of piezoelectricity and its converse effects. The sensor suite uses both pulse echo and transmission mode for operability, making it capable for deployment as distributed sensor network for the global damage monitoring sensing system (GDMS). The Smart Suitcase includes the portable diagnostic hardware and customized form factor. The hardware is capable of monitoring up to 64 sensor channels simultaneously.

Under active interrogation mode, the system has the capability of generating a 50–500-kHz input excitation with a maximum of 50 V peak-to-peak amplitude in the form a single-cycle or multiple-cycle pulse through one of the transducer. The adjacent sensors are used to detect the transmitted signals generated by the traveling stress waves. This is then repeated sequentially to cover map the whole structure which is under the sensor coverage. The signals are then compared with historical data. Both the transmitted signals and the pulse echo signals are used for analysis. Through transmission, the system bandwidth is 10 kHz to 1 MHz and the pulse echo system bandwidth mode is 10 kHz to 5 MHz.

Acellent is currently flight-testing its system by deploying the sensor on an F-16 test aircraft landing gear door to monitor the edge crack growth. The system had been demonstrated to detect 0.531″ crack in 500 cycles in metallic components. For a flawed composite doubler, the Sandia National Labs tested the system's capability to monitor crack length greater than 1 in. Acellent has done considerable testing on coupons and components under laboratory conditions. This system needs calibration of crack size. Herein, it should be noted that crack size determination has not yet been proven along with its robustness. Currently under a separate program, Acellent is preparing data for flight certification (salt fog, moisture, etc.).

As this chapter is primarily on its global sensing capability, this evaluation of the Acellent SHM system focuses on its capability of crack locating and sizing. The conclusions were:

- The hardware system is basically a multiple-channel high-speed data acquisition with a set of function generating capability typically used in acoustic wave-based flaw detection.
- Passive and active mode hardware systems are separate systems as they require different DAQ boards with different sampling frequencies.
- Its standard software, "Access," comes with a diagnostic imaging plug-in that enables raw acoustic data imaging and interpreting, but not damage locating.

With a network of sensors and current version of Access software, the system is able to indicate the structural changes in between the sensors due to damage.
- To achieve the damage localization, new algorithm is to be developed and the Access software must be customized and field calibrated with sufficient number of tests on a particular specimen.
- The system is currently unable to perform detection on a 3D geometry without further customization.

8 Fiber-Optic Sensors

There are two types of FO sensors available in the market: fiber Bragg sensors and Fabry-Perot interferometry sensors (extrinsic and intrinsic) [6, 7]. The following section discusses them in some details.

8.1 *Fiber Bragg Grating*

Fiber-optic Bragg gratings utilize a photo- or heat-induced periodicity in the fiber core refractive index to create a sensor whose reflected or transmitted wavelength is a function of this periodicity. The biggest advantage of fiber Bragg grating sensors (FBG) is that they can be easily multiplexed to enable multiple measurements along a single fiber. One approach for multiplexing Bragg gratings is to place gratings of different wavelength in a single fiber and utilize wavelength division multiplexing (WDM). However, the limited bandwidth of the source, as well as that supported by the fiber, and the range over which the physical parameter of interest is being measured provide practical limitations on the number of gratings that can be multiplexed in a single fiber with WDM approaches. The system, based on the principle of optical frequency domain reflectometry (OFDR), enables the interrogation of hundreds or thousands of Bragg gratings in a single fiber. OFDR essentially eliminates the bandwidth limitations imposed by the WDM technique as all of the gratings are of nominally the same wavelength. Very low reflectivity gratings are utilized, which allow reflections from large numbers of gratings to be recorded and analyzed. By tracking wavelength changes in individual gratings, one is able to measure mechanical- or thermal-induced strain in the grating. United Technologies Research Center is the original developer of the fiber Bragg sensors for strain measurements. The recent developments in multi-axis FBG strain sensor technology offer a distinct advantage in creating a series of fiber sensor types that are extremely compatible with one another, allowing the usage of similar readout equipment for a variety of applications. It is possible to configure a fiber grating system so that each fiber grating is sensitive to different frequency bands. This could be done in an array that measures multi-axis strain and other key parameters. This sensor array/neural network is nominally intended to predict and/or last the lifetime of the structure or component regarding mechanical damage tolerance. "Pre-assembly"of the fiber array could be affected in appliqué coatings. In all these potential arrangements, the simplest attachment methods will be developed capable for re-hooking the fiber array to the readout equipment. FBGs have minimal risk of electromagnetic interference and high bandwidth/sensitivity and can noninvasively inquire (passively or actively) into the health of a structure. The disadvantages of the FBG sensors are the uncertainties in the long-term durability of the sensors, sensor bonding to the airframe structure, and the fragility of the quartz elements, especially when the fiber is turned around. Several organizations, for example, NASA Langley and companies (e.g., Blue Road Research, Luna Innovations Inc., Micron Optics Inc., New Focus Inc.), are in the development of FBG demodulators for potential structural health monitoring applications, and the sensors are mainly developed. The main fiber-optic sensor manufacturers are Canadian-based companies like LXSix and FISO technologies. The normal FBG sensors are 150–250 μm in diameter and approximately 5 mm long.

8.2 Fabry-Perot Interferometry

Fiber-optic sensors can be separated into two classes for discrete strain and temperature measurement: cavity-based designs and grating-based designs. Cavity-based designs utilize an interferometric cavity in the fiber to create the sensor. Examples include the extrinsic Fabry-Perot interferometer (EFPI), the intrinsic or fiber Fabry-Perot interferometer (IFPI or FFPI), and all other etalon-type devices. Although such sensor designs have been utilized in a wide variety of applications such as in high temperature and EMI environments, they do not allow for multiplexing capability in a single fiber and thus may be limited for applications requiring large number of sensors. Originally developed by a team of scientists at Virginia Tech, this has seen widespread applications into different areas. To measure both strain and temperature, a broadband light source is transmitted to the cleaved end of a single-mode fiber. To perform the strain measurements, upon reaching the end of the fiber, the light is partially reflected while the remaining light travels past the end of the fiber and is reflected off a secondary reflector. The reflectors (also fibers) are aligned with the main fiber in a capillary tube and attached to a substrate. The two reflected light signals interfere with each other forming a fringe pattern. As the structural substrate strains, the distance between the two fiber end-faces vary, causing the fringe pattern to change. Using a spectrometer, the changing gap is measured to obtain the strain. The sensor is less prone to failure because the fiber itself is not being strained by the substrate. The temperature sensor has a small, single-crystal chip on the end of the fiber. The two faces of the chip are reflectors. Precise temperature can be obtained by measuring the temperature-dependent optical path length through the chip. Different types of fibers are used for making the FO sensors. For temperature less than 700°C, silica fibers are used for FO sensor, and for temperature greater than 900°C, applications currently single-crystal sapphire fibers are being investigated. New fiber-optic sensor materials being developed in this area include photonic crystal fibers and wholly fibers.

8.3 Technology and Products Maturity

Several vendors of fiber-optic sensing have been evaluated for global damage and load sensing. Luna Innovations of Blacksburg, VA, has two types of commercially available solution for fiber-optic sensing—(1) Distributed Sensing System and (2) FiberPro2 (an older version is also marketed known as FiberScan). The Distributed Sensing System is able to monitor several (10,000 s) FBG sensor nodes on a single fiber, which gives the ability to measure strain at several locations on a single fiber. However, the laser scan rate is limited to 10 Hz limiting its utility for application at 8P load ranges (40 Hz). For the current rotorcraft airframe application, this system is not useful for either damage monitoring or load monitoring. FiberPro2 is the newer version of FiberScan which can monitor both FBG and Fabry-Perot Interferometer sensor. It can be connected to "MU8" which is a multiplexer, allowing the ability to individually monitor eight single sensors on eight different fibers (channels) at the same time. The demodulator can monitor both fiber Bragg sensor and Fabry-Perot Interferometry sensor. The single FPI sensors can be used to monitor strain at a higher loading frequency levels. This is suitable if we want to monitor single optic fiber sensor mounted on separate fibers at the same time at different locations. However, this is limited to eight channels (i.e., eight sensors mounted on eight fibers), which makes application of FiberPro2 also limited with regards to multisensor array on a single fiber. One of the issues that is significant is that vibration in connecting fiber can cause drifting in the sensor readings. This is significant in terms of accuracy in the measurement during flight and the airframe undergoing variable load history. The system should be able to zero out the effects of sensor drift.

Alternatives to Luna's systems are Insensys and Micron *Optics. Both of them have systems which can monitor dynamic loads over 40 Hz. Invensys did demonstrate the capability of monitoring continuous dynamic loading on a cantilever* I-beam at the AIAA SDM Conference held at Newport, April 2006. A significant innovation was the designed brackets used for mounting the fiber-optic sensors onto the airframe. Attaching the fiber-optic sensors to airframes and long-term durability of such mounts are significant challenges. Also the results from a successful observation of a bird impact strike using the FO system on a winglet spar were presented. Unlike an electric-based sensing system, FO can actually detect a lightning strike because of its electromagnetic immunity. The team is talking to Insensys, which is based in United Kingdom for a follow-up demo in Connecticut.

Micron Optics developed several optical sensing interrogators which are wavelength division multiplexing (WDM) based. The WDM interrogators work with both FPI and fiber Bragg grating sensors. The si425-500 combines a PC with a high power, low-noise laser source. It is a stand-alone system, which can provide optical power and rapid measurement of 512 FBG sensors mounted on 4 separate optical fibers (128 sensors on a single fiber). The sensors can be placed as close as 1 cm apart. The system is expandable to 8–12 channels and customizable. The scan rate frequency is 250 Hz, and the wavelength is 1,520–1,570 nm. The sm130 can provide power and rapid measurements of several hundred sensors mounted on four separate optical fibers up to a scan rate of 1-kHz range. This unit is more applicable for rugged and harsh environment deployment.

The fiber-optic Bragg sensors that would be tested during sensor characterization part are going to be obtained from LXSix or FISO companies, which are main FO suppliers of Micron Optics. Clearly for the applications as load monitoring and damage monitoring sensors, the fiber Bragg sensors are more applicable than FPI sensors; as for FBG, we have the capability of monitoring several points for strain using a single fiber, whereas the FPI is a single discrete sensor at a single point location. Because of the limitation of the scan rate to 1 kHz, it is envisioned that the FBG sensors are more applicable for load (strain) monitoring at several locations on the component rather than using them as damage monitoring sensors.

8.4 *Time-Domain Reflectometry*

Time-domain reflectometry (TDR) is a method of sending a fast pulse down a controlled-impedance transmission line and detecting reflections returning from impedance and geometric discontinuities along the line. Time scales are fast, so reflections occurring at different positions in the line are separated by time-of-flight, forming a “closed-circuit radar.”

TDR can potentially be used as structural global damage sensing due to its distributed nature. TDR has gained popularity in recent years in infrastructure applications. The transmission line is embedded in a bridge or highway structure, such that a flaw in the surrounding structure causes a mechanical distortion in the line, which produces an impedance discontinuity, which is located by time-of-flight.

It has been investigated for composite parts defect detection after instrumentation high spatial location resolution becomes available. TDR structural health monitoring probes the structural health of a composite part by propagating a fast electrical pulse along a distributed linear sensor which has been fabricated directly in the laminate. The sensor is formed from the native graphite fibers already used in composite manufacture and constitutes zero defects. Fibers are patterned into a microwave waveguide geometry, or transmission line, and interrogated by a rapid pulse as shown below. Structural faults along the line cause distortions in waveguide geometry, producing reflected pulses similar to radar. Cracking, delamination, disbonds, moisture penetration, marcelling, and strain can be detected by propagation delay for sensor lengths up to several meters. These features make

TDR appropriate for the permanently embedded and distributed monitoring of the structural characteristics variation.

Material Sensing & Instrumentation, Inc (MSI) of Lancaster, PA [8, 9], is one of the major players specializing in TDR concrete and composite cure monitoring. It also performs R&D in the area of composite debond or delamination detection. TDR provides a new approach to cure monitoring of advanced polymer composites fabrication process. Using a high-speed pulse and inexpensive microwave sensor, TDR cure monitoring provides an alternative between high-frequency fiber-optic methods and low-frequency dielectric methods, combining optical-style precision and miniaturization with electrode-based simplicity and robustness.

MSI is a small company which has successfully executed several government SBIRs. Its primary technical expertise lies on the composite curing monitoring. The company typically does not/is not able to provide COTS equipment to customers. The structural health monitoring using TDR, especially metallic structures, is still under preliminary development. Due to the non-dielectric nature of metallic structure, further development of this technology is currently necessary.

9 Magnetostrictive Sensors

This sensing approach is based on a novel thin-film magnetostrictive sensor material that has recently been developed by Southwest Research Institute (SwRI) for turbine engine applications [10, 11]. This thin-film is 4 μm thick and achieves high activation efficiency, as well as temperature stability, using alternate crystalline and amorphous nano-layers. Defect detection is accomplished by activating the magnetostrictive thin-film causing emission of ultrasonic guided waves into the component that are subsequently backscattered and detected by the same sensor in "pitch-catch" fashion mostly done in a pulse echo mode. Energy harvesting and radio frequency (RF) communication enable multiple, individually addressable sensors to detect and monitor damage in structural airframe components. This sensing system provides a low mass sensing system, which does not affect the dynamic response of the component and high-efficiency sensor in power density requirements for electromechanical conversion. The robustness, durability, and the accuracy level of the magnetostrictive sensors for the rotorcraft airframe component structure need to be reviewed under rotorcraft loading environment. The thin-film magnetostrictive sensor is still at R&D stage and is not commercially available yet for the time being. The currently available sensor system is handheld scanning type. The sensor-related hardware is bulky and not suited for in situ crack monitoring. However, under the DARPA SIPS program, SwRI have made considerable technological improvements in terms of the sensor hardware readiness levels for deployment. Currently, a small company located in Colorado is trying to commercialize the magnetostrictive sensor technology.

10 Conclusions

The critical conclusion drawn out of this study is that for the time being, although there are some vendors/technologies of the structural damage sensing system commercially available on the market, there are some technology gaps required to be covered to reach a sufficient maturity to be able to claim commercial off-the-shelf (COTS). Substantial customization to both the "standard hardware and software" has to be made for each particular application. The damage detection algorithms/ software is still semiempirical and lacking of generality, and the sensor system will need in situ calibration if the structure being monitored is slightly different in detection ability.

References

1. Fu-Kuo Chang (ed) (2001) Structural health monitoring—the demands and challenges. In: Proceedings of the 3rd international workshop on structural health monitoring, held at Stanford University, 12–14 Sept 2001. CRC Press, Boca Raton
2. Zilberstein V, Schlicker D, Walrath K, Weiss V, Goldfine N (2001) MWM eddy current sensors for monitoring crack initiation and growth during fatigue tests and in service. Int J Fatigue 23(Suppl 1):477–485
3. Merah N (2003) Detecting and measuring flaws using electric potential techniques. J Qual Maint Eng 9(2):160–175. ISSN: 1355–2511
4. Wheatley G (2003) Comparative vacuum monitoring as an alternate means of compliance. In: Structural health monitoring 2003: from diagnostics & prognostics to structural health management: proceedings of the 4th international workshop on structural health monitoring, Stanford University, Stanford, CA, 15–17 Sept 2003, pp 1358–1365
5. Beard S, Qing PX, Hamilton M, Zhang DC (2004) Multifunctional software suite for structural health monitoring using SMART technology. In: Proceedings of the 2nd European workshop on structural health monitoring, Germany, July 2004
6. Silva JMA, Devezas TC, Silva AP, Ferreira JAM (2005) Mechanical characterization of composites with embedded optical fibres. J Compos Mater 39(14):1261–1281
7. Stewart A, Carman G, Richards L (2005) Health monitoring technique for composite materials utilizing embedded thermal fibre optic sensors. J Compos Mater 39(3):199–213
8. Kwun H, Light GM, Kim SY, Spinks RL (2002) Magnetostrictive sensor for active health monitoring in structures. In: Proceedings of SPIE, the International Society for Optical Engineering SPIE, vol 4702, pp 282–288
9. Kwun H, Kim SY, Light GM (2001) Long-range guided wave inspection of structures using the magnetostrictive sensor. J Korean Soc NDT 21:383–390
10. Hager NE III, Domszy RC (2001) Time-domain-reflectometry cure monitoring. In: SAMPE proceedings, Long Beach, CA, 6–10 May 2001, pp 2252–2264
11. Hager NE III, Domszy RC (1998) Time-domain-reflectometry cure monitoring. In: American Helicopter Society—affordable composite structures proceedings, Oct 1998

Application of Artificial Neural Network (ANN) Technique to Reduce Uncertainty on Corrosion Assessment of Rebars in Concrete by NDT Method

M. Bal and A.K. Chakraborty

Abstract The basic objective of this study is to assess corrosion behavior of steel bars in concrete members by nondestructive method of testing. Corrosion possibility is assessed by half-cell potential method (using CANIN) while resistivity meter (RESI) is used to estimate risk of corrosion of rebars in concrete members. Interestingly, higher half-cell potential indicates more possibility of corrosion, but higher value of resistivity indicates lower probability of corrosion. An extensive research program was undertaken in order to assess risk of corrosion using type 1 (Fe415:TATA-TISCON), type 2 (Fe500:TISCON-CRS), type 3 (Grade Fe415: ELEGANT steel), type 4 (Fe415:VIZAG steel), and type 5 (Fe500:SRMB steel) TMT steel bars of 16 mm diameter with M20, M40, and M60 grade concrete samples, prepared with OPC exposed both in *AIR* and *NaCl* for a period of 900 days. Taking only average values of experimental data, initially huge uncertainties were found. Secondly, after applying standard statistical method, uncertainties were slightly reduced. Third analysis was then taken up by modifying standard statistical process; satisfactory results were still not obtained. Then, fourth analysis was carried out with optimum values to minimize the uncertainties. Three-dimensional graphs for each case were plotted using MATLAB, an ANN-based software. More appropriate ANN-based software is required for better correlation.

Keywords Corrosion of rebar in concrete • Half-cell potential • Concrete resistivity • Artificial neural network

---

M. Bal • A.K. Chakraborty (✉)
Department of Civil Engineering, Bengal Engineering and Science University, Shibpur, Howrah 711 103, India
e-mail: manabendra_ball@yahoo.com; arunchakraborty@mailcity.com

S. Chakraborty and G. Bhattacharya (eds.), *Proceedings of the International Symposium on Engineering under Uncertainty: Safety Assessment and Management (ISEUSAM - 2012)*, DOI 10.1007/978-81-322-0757-3_31, © Springer India 2013

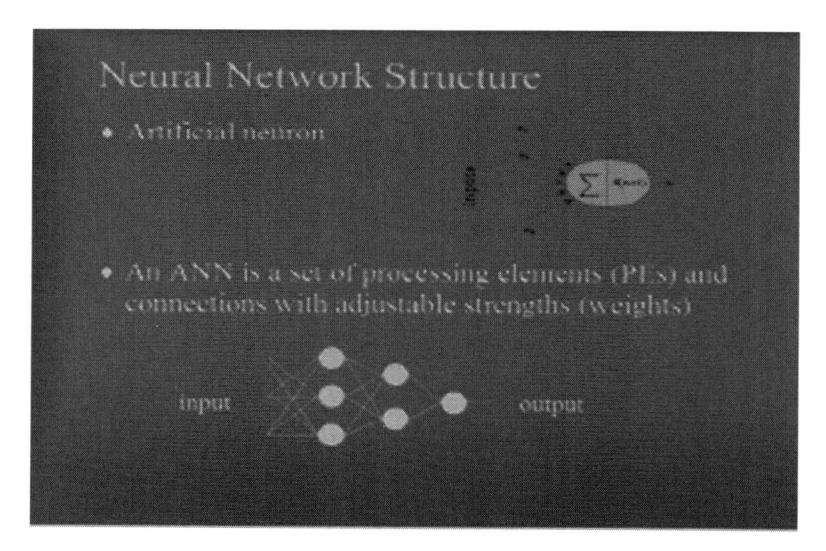

Fig. 1 Typical structure of neural network

1 Introduction

Neural network is a powerful data modeling tool, which is able to capture and represent complex input/output relationships. The motivation for the development of neural network technology stemmed from the desire to develop an artificial system that could perform "intelligent" tasks similar to those performed by our human brain. Neural networks resemble the human brain in the following two ways (Fig. 1):

1. A neural network acquires knowledge through learning.
2. A neural network's knowledge is stored within interneuron connection strengths known as synaptic weights.

True power and advantage of neural networks lies in their ability to represent both linear and nonlinear relationships and in their ability to learn these relationships directly from the data being modeled. A graphical representation of an MLP is shown below (Fig. 2).

Inputs are fed into input layer and get multiplied by interconnection weights as they are passed from the input layer to first hidden layer. Within the first hidden layer, they get summed and then processed by a nonlinear function (usually the hyperbolic tangent). As the processed data leaves the first hidden layer, again it gets multiplied by interconnection weights and then summed and processed by the second hidden layer. Finally, the data is multiplied by interconnection weights and then processed one last time within the output layer to produce the neural network output. MLP and many other neural networks learn using an algorithm called back propagation. With back propagation, the input data is repeatedly presented to the neural network. With each presentation, the output of the neural

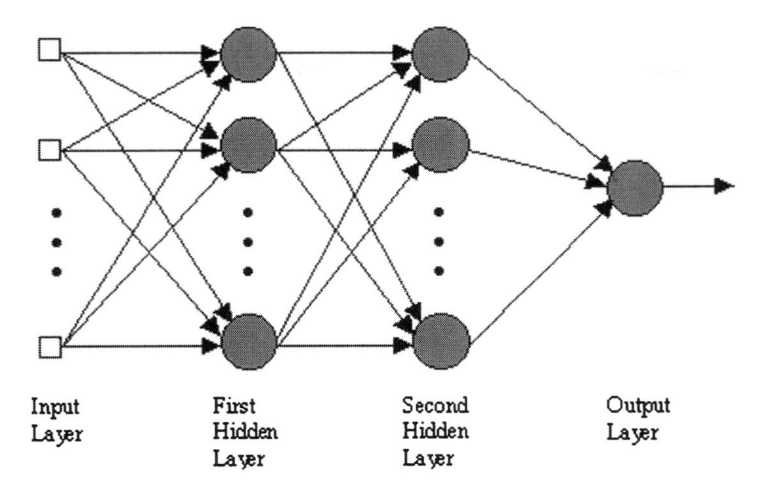

Fig. 2 Graphical representation of MLP

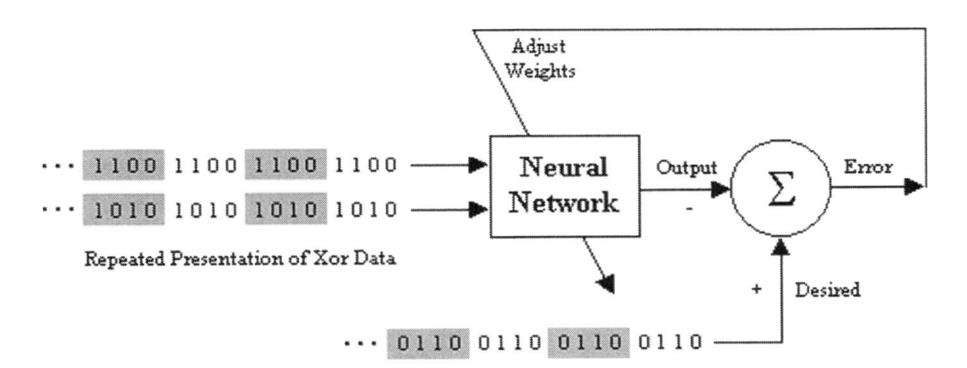

Fig. 3 Schematic diagram of "training" process

network is compared to the desired output and an error is computed. This error is then fed back (back propagated) to the neural network and used to adjust the weights such that the error decreases with each iteration and the neural model gets closer and closer to producing the desired output. This process is known as "training" (Fig. 3).

Demonstration of neural network learning is to model the exclusive-or (Xor) data. The Xor data is repeatedly presented to the neural network. With each presentation, the error between the network output and the desired output is computed and fed back to the neural network. The neural network uses this error to adjust its weights such that the error will be decreased. This sequence of events is usually repeated until an acceptable error has been reached or until the network no longer appears to be learning.

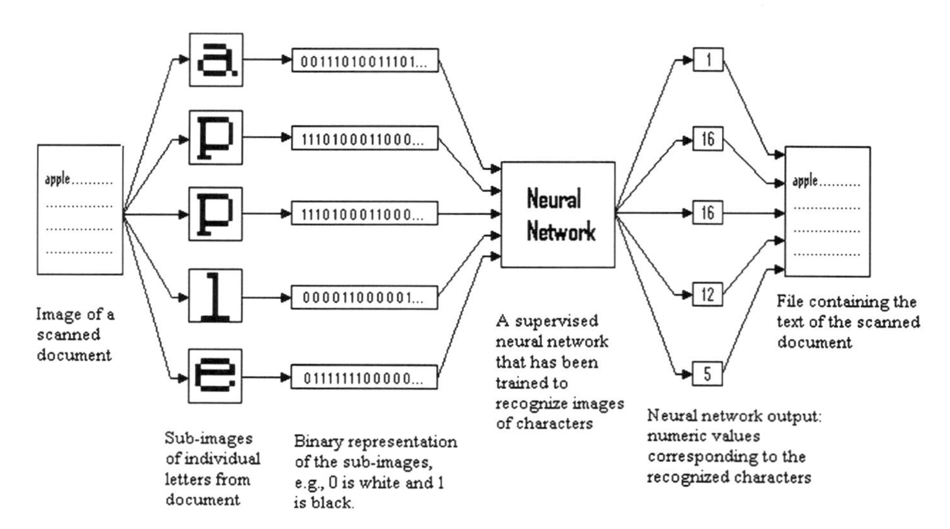

Fig. 4 Demonstration of optical character recognition (OCR)

A good way to introduce the topic is to take a look at a typical application of neural networks. Many of today's document scanners for the PC come with software that performs a task known as optical character recognition (OCR). OCR software allows you to scan a printed document and then convert the scanned image into to an electronic text format such as a Word document, enabling you to manipulate the text. In order to perform this conversion, the software must analyze each group of pixels (0 and 1s) that form a letter and produce a value that corresponds to that letter. Some of the OCR software on the market uses a neural network as the classification engine.

Demonstration of a neural network used within an optical character recognition (OCR) application is shown in Fig. 4. Original document is scanned into a computer and saved as an image. OCR software breaks image into sub-images, each containing a single character. Sub-images are then translated from an image format into a binary format, where each 0 and 1 represents an individual pixel of the sub-image. Binary data is then fed into a neural network that has been trained to make association between the character image data and a numeric value that corresponds to the character (Fig. 4).

Of course character recognition is not the only problem that neural networks can solve. Neural networks have been successfully applied to broad spectrum of data-intensive applications, such as:

- *Machine diagnostics* – Detect when a machine has failed so that the system can automatically shut down the machine when this occurs.
- *Portfolio management* – Allocate the assets in a portfolio in a way that maximizes return and minimizes risk.
- *Target recognition* – Military application which uses video and/or infrared image data to determine if an enemy target is present.

- *Medical diagnosis* – Assisting doctors with their diagnosis by analyzing the reported symptoms and/or image data such as MRIs or X-rays.
- *Engineering and technological events* – To create a network model for developing relationship among different properties of engineering materials, which creates physical or mathematical modeling between input and output data/information.
- *Credit rating* – Automatically assigning a company's or individual's credit rating based on their financial condition.
- *Targeted marketing* – Finding the set of demographics which have the highest response rate for a particular marketing campaign.
- *Voice recognition* – Transcribing spoken words into ASCII text.
- *Financial forecasting* – Using the historical data of a security to predict the future movement of that security.
- *Quality control* – Attaching a camera or sensor to the end of a production process to automatically inspect for defects.
- *Intelligent searching* – An internet search engine that provides the most relevant content and banner ads based on the users' past behavior.
- *Fraud detection* – Detect fraudulent credit card transactions and automatically decline the charge.

NeuroSolutions is one of the leading edge neural network development software that combines a modular, icon-based network design interface with an implementation of advanced learning procedures, such as Levenberg-Marquardt and back-propagation through time. Some other notable features include C++ source code generation, customized components through DLLs, neuro-fuzzy architectures, and programmatic control from Visual Basic using OLE Automation. This can be used through softwares that are conveniently available in present market. It can be tried for building and training of a neural network with available data.

2 Possibility for Application of Artificial Neural Network Technique

Like traditional use in biological events, artificial neural networks are also presently used in solving different problems of civil engineering in order to achieve particular targeted tasks by developing some mathematical models in the field of several engineering and technological events. *Artificial neural network (ANN)* provides its inbuilt properties, which facilitates to build up interconnections between artificial neurons, which may be considered as "nodal points" of a structure in civil engineering analysis. Artificial neural network (ANN) is one of the very important tools toward processing a large number of data collected from site. Practically ANN can accept both linear and nonlinear engineering, and statistical back-propagation learning algorithm is presented. For example, problem involves pattern recognition;

otherwise, it could be difficult to code in a conventional program. TMT steel bars are normally used as reinforcement in cement concrete as per design requirements. Besides various advantages, about 80% of damages occur in RCC members due to the attack of corrosion that takes place in steel bars embedded into it. RCC member as a whole loses its strength gradually and fails to survive its designed lifetime. It is similar to that of cancer infection in a body. Sometimes the effect of corrosion becomes so dangerous that a structure finally reaches to its final state of collapse quickly. It is therefore very important to understand the behavior of TMT steel bars especially in terms of susceptibility to attack of corrosion.

Factors responsible for attack of corrosion in RCC member are:

1. Thickness of cover on steel reinforcement bars in concrete
2. Type of cement used in concrete like OPC (ordinary Portland cement), blended cements like PPC (Portland pozzolana cement) and PSC (Portland slag cement), etc.
3. Grade of concrete mix like M20, M40, and M60
4. Grade of steel used like Fe415, Fe500, and CRS (corrosion resistant steel)
5. Diameter of steel bars used like 8.0, 16.0, and 25.0 mm Dia
6. Permeability of concrete
7. Electrical resistively of concrete
8. Type of exposure condition such as in NaCl and in natural air
9. Degree of carbonation.
10. Chloride ingress

Following principles of ANN, different softwares that are recently developed and available in the market can be used in computer. In this exercise, one of such computer software, *MATLAB*, having version **7.6.0.324 (R2008a),** was used as a tool to express results graphically.

Vulnerability of seismically deficient older buildings, risk parameters, significant threat to life safety, and its survivability can be assessed before by ANN [1]. Chloride-induced corrosion of steel reinforcement bars embedded in reinforced cement concrete can be enumerated before by applying neural network techniques [2]. Ductility performance of hybrid fiber reinforced concrete can be predicted by applying neural network techniques [3]. Prediction of density and compressive strength of concrete cement paste containing silica fume can be predicted using artificial neural networks [4]. Application of neural network in predicting damage of concrete structures caused by chlorides [5]. Prediction of stress-strain relationship for reinforced concrete sections by implementing neural network technique [6]. Strengthening of corrosion-damaged reinforced concrete beams can be done with glass fiber reinforced polymer laminates [7]. Corrosion mitigation in mature reinforced concrete using nanoscale pozzolan deposition [8]. The cathodic protection of reinforcing steel bars using platinized-type materials [9]. Corrosion protection of steel rebar in concrete using migrating corrosion inhibitors MCI 2021 and 2022 [10].

3 Data Processing on Experimental Results and Correlation

Data obtained from samples are at *stressed* condition in *atmospheric AIR* and *in NaCl* solution,[*NaCl at 5%* by weight of water] *for 900 days,* respectively. Readings are as follows:

1. *Readings* of *half-cell potential values* of concrete samples in *mVolt*
2. *Readings* of electrical *resistivity* of concrete samples in *kΩ–cm*
3. *Readings* of *ultrasonic pulse velocity* of concrete samples in *km/s*
4. *Readings* of *rebound hammer* with the help of *Schmidt hammer* on concrete samples in *numbers* to obtain *strength of concrete* to determine *loss of strength of concrete* in %
5. *Readings* of *clear cover* in *mm of concrete* samples

4 Processing of Supplied Data and Readings

A. *Initial Analysis:* Steps are as follows:

   1. Average values have been calculated for each set of observations to minimize error in taking readings indicated as *initial averages.*
   2. Three-dimensional graphs have been plotted with *initial average* values.

B. *Second Analysis:* Initial analysis is a most generalized form and does not represent true key area because average value is influenced by abnormally extreme higher and lower readings as all readings have been added altogether. To avoid this, *statistical analysis* has been done. Steps are:

   1. Extreme abnormally higher and lower values lying *above* (+) *15%* and *below* (−) *15%* of *initial average* have been omitted by *standard statistical screening process.*
   2. Again, average values have been recalculated with reasonable readings and tabulated.
   3. Three-dimensional graphs have been plotted with those *average* values.

C. *Third Analysis:* Average obtained from *standard statistical screening process* also does not represent truly because it does not obey characteristics of test parameters. Potential difference increases while resistivity decreases with the increase of degree of corrosion; hence, both extreme higher and lower values are required to preserve. Then, statistical analysis has been modified a little as per requirement. Steps are:

   1. All higher readings of potential difference and all lower values of resistivity had been preserved. Readings lying *below* (−) *15%* and *above* (+) *15%* of the *initial average*, respectively, have been omitted.
   2. Average values have been recalculated and three-dimensional graphs have been plotted.

D. *Fourth Analysis:* Analysis has been carried out with readings of optimum magnitudes for correlation with different parameters, considering the following assumptions:

   1. Corrosion in concrete develops maximum at where the thickness of cover is minimum.
   2. Potential difference readings exhibit maximum at where corrosion is maximum.
   3. Resistivity of concrete readings shows minimum at where corrosion is maximum.
   4. Ultrasonic pulse runs slower at where corrosion is maximum.
   5. Percentage loss of strength of an RCC shall obviously be more at where corrosion is maximum.

On the basis of above mentioned assumptions, data have been reanalyzed. Steps are:

1. All readings of maximum potential difference have been preserved and others have been omitted.
2. All readings of minimum resistivity have been preserved and others have been omitted.
3. Average values have been recalculated with optimum readings and tabulated.
4. Three-dimensional graphs have been plotted with average values for study on correlation.

All three-dimensional graphs are shown as Graphs 1, 2, 3, 4, 5, and 6.

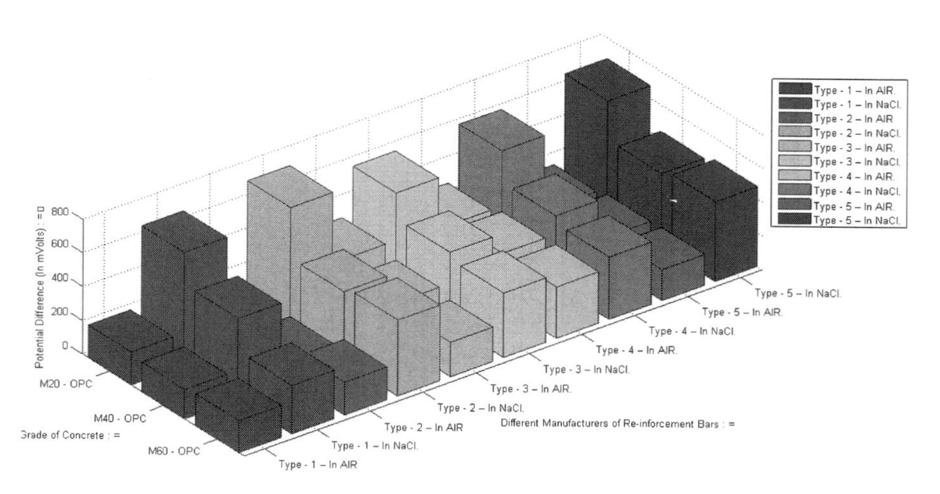

Graph 1 3D bar diagram: potential difference (initial average) vs. grade of concrete vs. different manufacturer of steel re-inforement bars with opc, using 16 mm dia bars both in "Air & Nacl" for 900 days

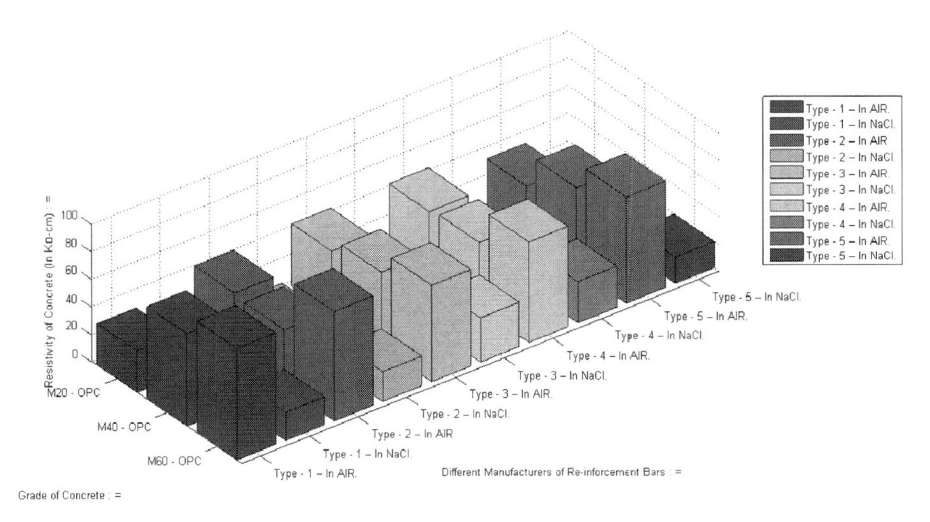

Graph 2 3D bar diagram: resistivity (initial average) vs. grade of concrete vs. different manufacturer of steel re-inforement bars with opc, using 16 mm dia bars both in "Air & Nacl" for 900 days

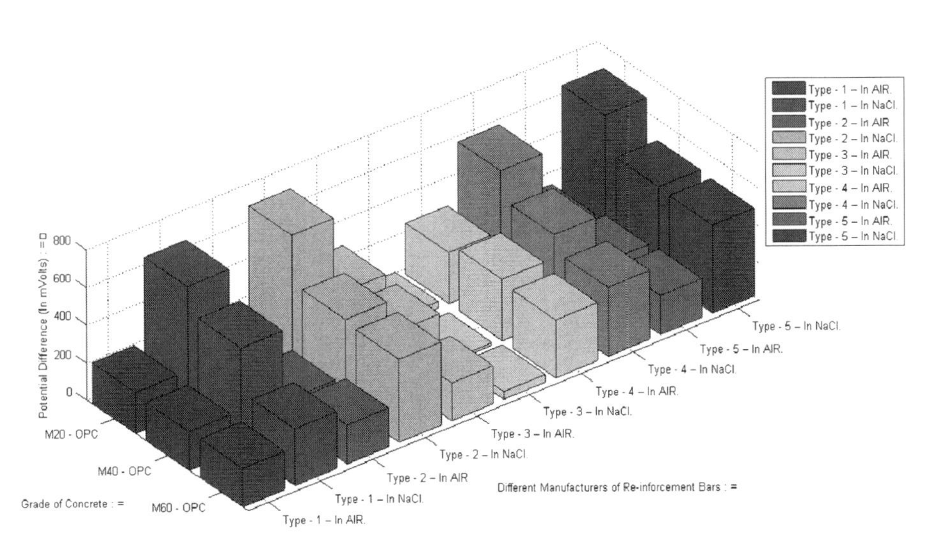

Graph 3 3D bar diagram: potential difference vs. grade of concrete vs. different types of steel re-inforement bars with opc, using 16 mm dia bars both in "Air & Nacl" for 900 days

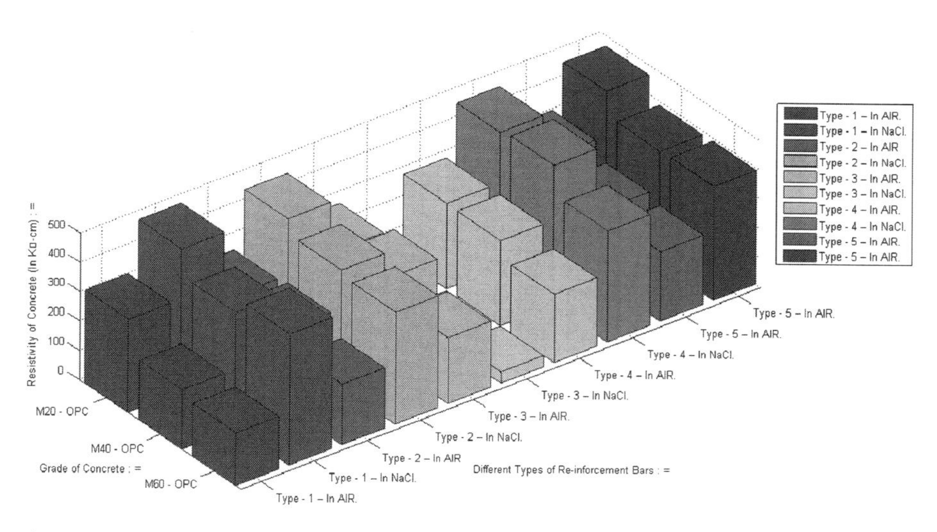

Graph 4 3D bar diagram: resistivity of concrete vs. grade of concrete vs. different types of steel re-inforement bars with opc, using 16 mm dia bars both in "Air & Nacl" for 900 days

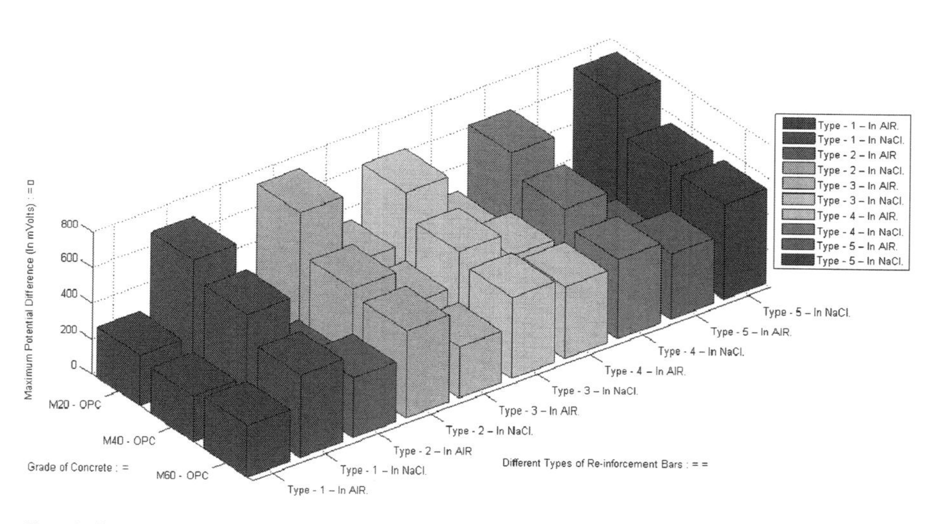

Graph 5 3D bar diagram: maximum potential difference vs. grade of concrete vs. different types of steel re-inforement bars with opc, using 16 mm dia bars both in "Air & Nacl" for 900 days

5 Conclusion

As MATLAB 7.6.0.324 (R2008a) is the most available ANN-based software, hence it is used in this exercise for the purpose of correlation of different parameters of corrosion of rebars in concrete to reduce the uncertainties. Out of all, only 3 (three) of any parameters can be plotted at a time in MATLAB against 3 (three) mutually

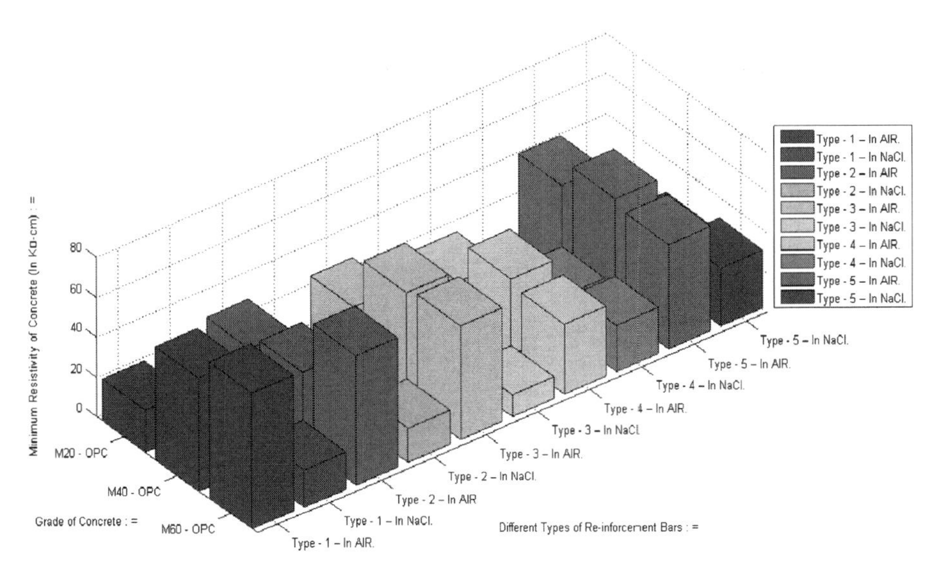

Graph 6 3D bar diagram: minimum resistivity of concrete vs. grade of concrete vs. different types of steel re-inforement bars with opc, using 16 mm dia bars both in "Air & Nacl" for 900 days

perpendicular axes. So, a number of 3- (three-) dimensional graphical representations had been obtained with different possible combinations. But more important factor was noticed at that time, 3- (three-) dimensional graphs are not clearly decipherable because normally we looked upon a 2- (two-) dimensional figure. Therefore, only 2- (two-) dimensional graphs are more convenient and easily readable. Since, corrosion is one of the most complicated phenomenon and involves so many uncertainties; hence, it was felt that a more powerful and suitable ANN-based software is necessary so that at least 10–12 parameters could be plotted simultaneously at a time in order to correlate different parameters related to corrosion of rebars in concrete assessed by nondestructive method of testing and could also be represented graphically in order to reduce uncertainties more accurately. Knowing characteristics of each parameter, mathematical expressions could be derived correspondingly. Substituting certain known values, other unknown variables/parameters could be obtained analytically for better correlations, and thus, uncertainties could also be minimized.

References

1. Solomon T, EERI M, Murat S (2008) Vulnerability of seismically deficient older buildings, risk parameters, significant threat to life safety and its survivability. Earthq Spectra 24 (3):795–821
2. Glass GK, Buenfeld NR (2001) Chloride-induced corrosion of steel reinforcement bars embedded in reinforced cement concrete by applying neural network techniques. Prog Struct Eng (Struct Control Health Monit) 2(4):448–458

3. Eswari S, Raghunath PN, Suguna K (2008) Ductility performance of hybrid fibre reinforced concrete by applying neural network techniques. Am J Appl Sci 5(9):1257–1262, ISSN 1546–9239
4. Rasa E, Ketabchi H, Afshar MH (2009) Prediction on density and compressive strength of concrete cement paste containing silica fume using artificial neural networks. Trans A Civ Eng 16(1):33–42
5. Neven U, Ivana BP Velimir U (2004) Application of neural network in predicting damage of concrete structures caused by chlorides. Published in proceedings of international symposium ASFCACT, pp 187–194
6. Mansour NJ (1996) Prediction of stress-strain relationship for reinforced concrete sections by implementing neural network technique. J King Saud Univ Eng Sci 9(2):169–189
7. Rose AL, Suguna K, Ragunath PN (2009) Strengthening of corrosion-damaged reinforced concrete beams with glass fiber reinforced polymer laminates. J Comput Sci 5(6):1549–3636, ISSN 1549–3636
8. Cardenas H, Kupwade-Patil K, Eklund S (2011) Corrosion mitigation in mature reinforced concrete using nanoscale pozzolan deposition. J Mater Civil Eng 23(6):752–760
9. Hayfield PCS (1986) The cathodic protection of reinforcing steel bars using platinised-type materials. Platin Metals Rev 30(4):158–166
10. Behzad B, Lisa R (2002) Corrosion protection of steel rebar in concrete using migrating corrosion inhibitors MCI 2021 & 2022. A report published by College of Engineering and Computer Science, California State University, Northridge, CA, pp 1–10

Effect of Very Mild Random Tremors on Saturated Sub-surface Flow

Amartya Kumar Bhattacharya and Debasish Kumar

Abstract Very mild random tremors in a saturated soil are not uncommon. Mild earthquakes, nearby piling, and the passage of underground trains all lead to vibrations in a manner that cannot be exactly predicted. High pore pressures and low effective stresses in cohesionless soil can lead to soil liquefaction and complete loss of bearing capacity of the soil. As long as the saturated sub-surface water flow is Darcian, the governing partial differential equation is elliptic in nature at all points in the flow domain. While analytical methods can be used in limited cases, numerical methods are universally available to find out the velocity components and pressure at all points in the flow domain. Tremors disrupt the steady-state flow of saturated sub-surface water. Velocity and pressure patterns vary in a random way. The possibility of saturated sub-surface water pressure mounting followed by soil liquefaction arises. In this chapter, canonical equations related to the finite element method have been considered, and the problem has been analysed.

Keywords Mild • Random • Tremors • Sub-surface • Flow

1 Introduction

Very mild random tremors in a saturated soil are not uncommon. Mild earthquakes, nearby piling, and the passage of underground trains all lead to vibrations in a manner that cannot be exactly predicted. High pore pressures and low effective stresses in cohesionless soil can lead to soil liquefaction and complete loss of bearing capacity of the soil. As long as the saturated sub-surface water flow is Darcian, the governing partial differential equation is elliptic in nature at all points

---

A.K. Bhattacharya • D. Kumar (✉)
Department of Applied Mechanics, Bengal Engineering and Science University, Shibpur, Howrah 711103, West Bengal, India
e-mail: dramartyakumar@gmail.com; debasishkumar0@gmail.com

S. Chakraborty and G. Bhattacharya (eds.), *Proceedings of the International Symposium on Engineering under Uncertainty: Safety Assessment and Management (ISEUSAM - 2012)*, DOI 10.1007/978-81-322-0757-3_32, © Springer India 2013

in the flow domain. While analytical methods can be used in limited cases, numerical methods are universally available to find out the velocity components and pressure at all points in the flow domain. Tremors disrupt the steady-state flow of saturated sub-surface water. Velocity and pressure patterns vary in a random way. The possibility of saturated sub-surface water pressure mounting followed by soil liquefaction arises. In this chapter, canonical equations related to the finite element method have been considered, and the problem has been analysed.

A two-dimensional vertical slice through a soil stratum is taken. The superficial seepage velocity components in the vertical plane are v_x and v_y, where v_x and v_y are the velocity components in the horizontal and vertical directions, respectively. If the velocity potential is denoted by ϕ, then

$$v_x = \frac{\partial \phi}{\partial x} \text{ and } v_y = \frac{\partial \phi}{\partial y} \tag{1}$$

satisfies Laplace's equation,

$$\frac{\partial^2 \phi}{\partial x^2} + \frac{\partial^2 \phi}{\partial y^2} = 0 \tag{2}$$

In a static homogeneous aquifer, the hydraulic conductivity, K, is the same at all points in the flow domain. If an aquifer is subjected to very mild random tremors, the mean value of K is $\bar{K}$, and the standard deviation of K is σ_k; $\sigma_k \ll \bar{K}$.

Because it is difficult to solve for the velocity potential analytically in a vast range of situations, numerical methods like the finite element method [4, 5] have been deployed to compute the velocity potential. In the above two works, three-noded triangular finite elements with Lagrangian interpolation have been used. This has been extended to six-noded triangular finite elements with Lagrangian interpolation by Choudhury [3]. The present work follows Bhattacharya [1,2]. Whatever be the exact nature of the element being utilised, ultimately the matrix equation developed comes out to be of the form

$$\{\phi(K)\} = [G(K)]^{-1}\{P(K)\} \tag{3}$$

where $[G(K)]$ is the global constitutive matrix, $\{\phi(K)\}$ is the matrix of the nodal velocity potentials, and $\{P(K)\}$ is the equivalent of the load matrix in solid mechanics.

If the aquifer is static, a definite pattern of equipotential lines is obtained for definite flow geometry. Under the action of random tremors, the equipotential lines become stochastic. The present work addresses this situation:

$$\{\phi(K)\} = [G(K)]^{-1}\{P(K)\} \tag{3}$$

Or

$$[G(K)]\{\phi(K)\} = \{P(K)\} \tag{4}$$

Now, on differentiating Eq. (4) with respect to K, the following equation is obtained:

$$[G(K)]\frac{\partial}{\partial K}\{\phi\{K\}\} = \frac{\partial}{\partial K}\{P(K)\} - \frac{\partial}{\partial K}[G(K)]\{\phi(K)\} \tag{5}$$

or

$$\frac{\partial}{\partial K}\{\phi(K)\} = [G(K)]^{-1}\left\{\frac{\partial}{\partial K}\{P(K)\} - \{\phi(K)\}\frac{\partial}{\partial K}[G(K)]\right\} \tag{6}$$

where $\frac{\partial}{\partial K}\{\phi(K)\}$ is the sensitivity of ϕ with respect to K.

Equation (6) can be written as

$$\{\xi(K)\} = [G(K)]^{-1}\frac{\partial}{\partial K}\{P^*(K)\} \tag{7}$$

$$\{\xi(K)\} = \frac{\partial}{\partial K}\{\phi(K)\} \tag{8}$$

and

$$\{P^*(K)\} = \frac{\partial}{\partial K}\{P(K)\} - \frac{\partial}{\partial K}[G(K)]\{\phi(K)\} \tag{9}$$

Now, undertaking a Neumann expansion,

$$[G(K)] = [\bar{G}(K)] + [G'(K)] \tag{10}$$

where $[\bar{G}(K)]$ is the deterministic component of $[G(K)]$ and $[G'(K)]$ is the residual component:

$$\begin{aligned}[G(K)]^{-1} &= ([\bar{G}(K)] + [G'(K)])^{-1} = ([I] + [H])^{-1}[\bar{G}(K)]^{-1} \\ &= \left([I] - [H] + [H]^2 - [H]^3 + \ldots\right)\bar{G}(K)]^{-1}, \text{ so,}\end{aligned} \tag{11}$$

$$[G(K)]^{-1} = \left(\sum_{n=0}^{\infty}[-H]^n\right)[\bar{G}(K)]^{-1} \tag{12}$$

$$[H] = [\bar{G}(K)]^{-1}[G'(K)] \tag{13}$$

Therefore, the velocity potential matrix can be written as

$$\{\phi(K)\} = \left(\sum_{n=0}^{\infty}[-H]^n\right)[\bar{G}(K)]^{-1}\{P(K)\}$$
$$= \left([I] - [H] + [H]^2 - [H]^3 + \ldots..\right)\{\bar{\phi}(K)\} \quad (14)$$

where $\{\bar{\phi}(K)\}$ is the mean of $\{\phi\ (K)\}$.

This can be written as

$$\{\phi(K)\} = \{\bar{\phi}(K)\} - \{\phi_1(K)\} + \{\phi_2(K)\} - \{\phi_3(K)\} + \quad (15)$$

$$\{\bar{\phi}(K)\} = [\bar{G}(K)]^{-1}\ \{P(K)\} \quad (16)$$

Now, let the random tremors be considered to be very mild.

Then,

$$\lfloor\bar{G}(K)\rfloor >> [G'(K)] \quad (17)$$

and $[H]$ is small.

It is permissible, then, to write

$$[G(K)]^{-1} \approx ([I] - [H])\ [\bar{G}(K)]^{-1} \quad (18)$$

and

$$\{\phi(K)\} \approx ([I] - [H])\ \{\bar{\phi}(K)\} \quad (19)$$

Thus,

$$\{v_x\} = \frac{\partial}{\partial x}\left[([I] - [H])\ \{\bar{\phi}\ (K)\}\right] \quad (20)$$

and

$$\{v_y\} = \frac{\partial}{\partial y}\left[([I] - [H])\ \{\bar{\phi}\ (K)\}\right] \quad (21)$$

The stochastic nature of the $\{vx\}$ and $\{vy\}$ matrices is introduced through the presence of the matrix $[H]$. However, the two velocity matrices are almost deterministic as $[H]$ is small.

2 Conclusions

It can be concluded that very mild random tremors do not appreciably alter the velocity distribution of the saturated sub-surface water. Localised changes in point velocity do occur, but these are unlikely to have a major impact on, for example, the discharge into a pumping well. The situation can change totally if the random tremors no longer remain very mild.

References

1. Bhattacharya AK (2005) Groundwater flow in a weakly-random hydraulic conductivity field. Electron J Geotech Eng 10, Bundle – E, EJGE Paper 2005–0582
2. Bhattacharya AK, Choudhury S (2001) Uncertainty analysis of groundwater flow during earthquakes. In: Proceedings, all India seminar on disaster management and civil engineering solutions, Calcutta, India, 7–8 Dec 2001
3. Choudhury S (2002) Evaluation of seepage under sheet-pile using the finite element method with six-noded triangular elements. ME thesis, Bengal Engineering College (Deemed University), Howrah, India
4. Maji S, Mondal S, Bhattacharya AK, Manna MC, Choudhury S (2002) Finite element analysis of flow under a sheet pile. In: Proceedings, international conference on water resource management in arid regions, Kuwait, 23–27 Mar 2002
5. Mondal S (2001) Finite element analysis of seepage flow under dam. ME thesis, Bengal Engineering College (Deemed University), Howrah, India

Slope Reliability Analysis Using the First-Order Reliability Method

Subhadeep Metya and Gautam Bhattacharya

Abstract This chapter pertains to a study on the reliability evaluation of earth slopes under a probabilistic framework. This study is concerned in the first phase with the determination of reliability index and the corresponding probability of failure associated with a given slip surface and then in the second phase with the determination of the critical probabilistic slip surface and the associated minimum reliability index and the corresponding probability of failure. The geomechanical parameters of the slope system have been treated as random variables for which different probability distributions have been assumed. The reliability analyses have been carried out using two methods, namely, the approximate yet simple mean-value first-order second-moment (MVFOSM) method and the rigorous first-order reliability method (FORM). Based on a benchmark illustrative example of a simple slope in homogeneous soil with uncertain strength parameters along a slip circle, an effort has been made to numerically demonstrate the nature and level of errors introduced by adopting the MVFOSM method for reliability analysis of earth slopes still widely used in the geotechnical engineering practice, vis-à-vis a more accurate method such as the FORM.

Keywords Slope stability • Slip surface • Uncertainty • Random variable • Probability distribution • Reliability analysis

S. Metya (✉)
Department of Civil Engineering, Budge Budge Institute of Technology, Kolkata 700137, India
e-mail: subhadeep.metya@gmail.com

G. Bhattacharya
Department of Civil Engineering, Bengal Engineering and Science University, Shibpur, Howrah 711103, India
e-mail: gautam@civil.becs.ac.in

S. Chakraborty and G. Bhattacharya (eds.), *Proceedings of the International Symposium on Engineering under Uncertainty: Safety Assessment and Management (ISEUSAM - 2012)*, DOI 10.1007/978-81-322-0757-3_33,

1 Introduction

In recent years, there has been a growing appreciation among the researchers in the field of geotechnical engineering of the fact that geotechnical parameters, especially the strength parameters including pore water pressure, are highly uncertain or random. Conventional deterministic approach is, therefore, being increasingly replaced with probabilistic approach or reliability analysis within a probabilistic framework. Slope stability analysis is one of the important areas where the recent trend is to determine the probability of failure of slopes instead of, or complementary to, the conventional factor of safety.

During the last decade, quite a few studies on reliability evaluation of earth slopes have been reported in the literature. Most of these studies used the simple yet approximate reliability analysis method known as the mean-value first-order second-moment (MVFOSM) method based on a Taylor series expansion of the factor of safety. However, this method suffers from serious shortcomings such as the following: (1) The method does not use the distribution information about the variables when it is available. (2) The performance function is linearized at the mean values of the basic variables. When the performance function is nonlinear, significant errors may be introduced by neglecting higher order terms, for the reason that the corresponding ratio of mean of performance function to its standard deviation which is evaluated at the mean values may not be the distance to the nonlinear failure surface from the origin of the reduced variables. (3) Furthermore, first-order approximations evaluated at the mean values of the basic variates will give rise to the problem of invariance for mechanically equivalent limit states; that is, the result will depend on how a given limit-state event is defined.

The first-order reliability method (FORM), on the other hand, does not suffer from the above shortcomings and is, therefore, widely considered to be an accurate method. The method has been finding increasing use especially in structural engineering applications for more than a decade. More recently in the geotechnical engineering field also, there have been quite a few attempts at reliability analysis of earth slopes using the FORM method [2–4].

In this chapter, an attempt has been made to develop computational procedures for slope reliability analysis based on the first-order reliability method (FORM). Computer programs have been developed to demonstrate the application of FORM in the determination of (1) the reliability index for a given slip surface and, more importantly, in the determination of (2) the probabilistic critical slip surface and the associated minimum reliability index. Different probability distributions have been considered for the basic random variables. In determining the probabilistic critical slip surface, the basic methodology suggested by Bhattacharya et al. [1] has been adopted. The above reliability analyses have also been carried out using the approximate MVFOSM method, and the results obtained have been compared with those obtained by using the FORM to bring out the difference clearly and demonstrate numerically the shortcomings of the MVFOSM method.

2 Formulation

2.1 *Deterministic Analysis*

The conventional slope stability analysis follows a deterministic approach wherein out of a number of candidate potential slip surfaces, the one with the least value of factor of safety is searched out and is termed the critical slip surface. It has now been widely appreciated that the slope stability analysis is essentially a problem of optimization wherein the coordinates defining the shape and location of the slip surface are the design variables and the factor of safety functional expressed as a function of the design variables is the objective function to be minimized subject to the constraints that the obtained critical slip surface should be kinematically admissible and physically acceptable. In practice, analysis is often done based on the assumption that the slip surface is an arc of a circle, as it greatly simplifies the problem. The ordinary method of slices (OMS) [5] is the simplest and the earliest method of slices that assumes a circular slip surface geometry.

The factor of safety functional (FS) for the ordinary method of slices (OMS) is given by the following expression [Eq. (1)], where the notations have their usual meaning. Specifically, c' and ϕ' denote the effective cohesion and effective angle of shearing resistance, respectively; W_i and u_i are the weight and the pore water pressure at the base of the ith slice, respectively; θ_i is the base inclination of the ith slice; and Δl_i and $\widehat{L}$ are the base length of the ith slice and the total arc length of the slip circle, respectively:

$$\mathrm{FS} = \frac{c' \widehat{L} + \tan\phi' \sum_{i=1}^{i=n} (W_i \cos\theta_i - u_i \Delta l_i)}{\sum_{i=1}^{i=n} (W_i \sin\theta_i)} \tag{1}$$

Substituting $W_i = \gamma b h_i$, and $u_i = r_u \gamma h_i$, where γ and b are the unit weight of soil and the common width of slice, respectively, h_i is the mean height of the ith slice, and r_u is the pore pressure ratio, Eq. (1) reduces to

$$\mathrm{FS} = \frac{c' \widehat{L} + \tan\phi' \sum_{i=1}^{i=n} (\gamma b h_i \cos\theta_i - r_u \gamma h_i \Delta l_i)}{\sum_{i=1}^{i=n} (\gamma b h_i \sin\theta_i)} \tag{2}$$

2.2 *Reliability Index β Based on the MVFOSM Method*

Taking the performance function as the expression for FS in a limit equilibrium method of slices such as Eq. (1) or (2) for analyzing slope stability and the corresponding limit-state equation as $\mathrm{FS} - 1 = 0$, the reliability index β based on the MVFOSM method is given by

$$\beta = \frac{E[\mathrm{FS}] - 1}{\sigma\ [\mathrm{FS}]}$$
$$= \frac{\mathrm{FS}(\mu_{x_i}) - 1}{\sqrt{\sum_{i=1}^{n} \left(\frac{\partial \mathrm{FS}}{\partial X_i}\right)^2 \sigma^2 [X_i] + 2 \sum_{i,j=1}^{n} \left(\frac{\partial \mathrm{FS}}{\partial X_i}\right) \left(\frac{\partial \mathrm{FS}}{\partial X_j}\right) \rho\ \sigma[X_i]\ \sigma[X_j]}} \tag{3}$$

where n is the number of soil strength parameters (c', tan ϕ', r_u, γ, etc.) taken as random variables; E[FS], the expected value of FS; σ[FS], the standard deviation of FS; μ_{xi}, the mean value of random variable X_i; $\sigma[X_i]$, the standard deviation of X_i; and ρ, correlation coefficient between X_i and X_j.

2.2.1 Mechanically Equivalent Limit State

When using MVFOSM method, it is of interest to study how the results of reliability analysis differ when other mechanically equivalent limit states are adopted. A limit state equivalent to FS−1 = 0 mentioned above is given by ln (FS) = 0. For such a limit state, the reliability index is given by

$$\beta = \frac{E[ln\ \mathrm{FS}]}{\sigma_{ln\,\mathrm{FS}}} \tag{4a}$$

where

$$E[ln\ \mathrm{FS}] = ln(E[\mathrm{FS}]) - \frac{\sigma^2_{ln\,\mathrm{FS}}}{2} \tag{4b}$$

and

$$\sigma_{ln\,\mathrm{FS}} = \sqrt{ln\left(1 + \left(\frac{\sigma_{\mathrm{FS}}}{E[\mathrm{FS}]}\right)^2\right)} \tag{4c}$$

2.3 *Reliability Index* β *Based on the FORM Method*

In this method, the reliability index (β) is defined as the minimum distance ($D_{\min}$) from the failure surface [$g(X') = 0$] to the origin of the reduced variates, as originally proposed by Hasofer and Lind [7]. For general nonlinear limit states, the computation of the minimum distance ($D_{\min}$) becomes an optimization problem as stated below:

Minimize $D = \sqrt{X'^{t}X'}$
Subject to the constraint $g(X') = 0$

where X' represents the coordinates of the checking point on the limit-state equation in the reduced coordinates system.

Two optimization algorithms are commonly used to solve the above minimization problem to obtain the design point on the failure surface and the corresponding reliability index β [6]. In the first method [10] referred to as FORM method I by Haldar and Mahadevan [6], it is required to solve the limit-state equation during the iteration. The second method [11] referred to as FORM method II by Haldar and Mahadevan [6] does not require solution of the limit-state equation. It uses a Newton-type recursive formula to find the design point. The FORM method II is particularly useful when the performance function is implicit, that is, when it cannot be written as a closed-form expression in terms of the random variables. The FORM method, however, is applicable only for normal random variables. For non-normal variables, it is necessary to transform them into equivalent normal variables. This is usually done following the well-known Rackwitz–-Fiessler method [6].

2.4 Probability of Failure

Once the value of the reliability index β is determined by any of the methods discussed above, the probability of failure p_F is then obtained as

$$p_F = \Phi(-\beta) \tag{5}$$

where $\Phi(.)$ is the standard normal cumulative probability distribution function, values of which are tabulated in standard texts.

2.5 Determination of Probabilistic Critical Slip Surface

Bhattacharya et al. [1] proposed a procedure for locating the surface of minimum reliability index, $\beta_{\min}$, for earth slopes. The procedure is based on a formulation similar to that used to search for the surface of minimum factor of safety, $FS_{\min}$, in a conventional slope stability analysis. The advantage of such a formulation lies in enabling a direct search for the critical probabilistic surface by utilizing an existing deterministic slope stability algorithm or software with the addition of a simple module for the calculation of the reliability index β. This is definitely an improvement over the indirect search procedure proposed earlier by Hassan and Wolff [8].

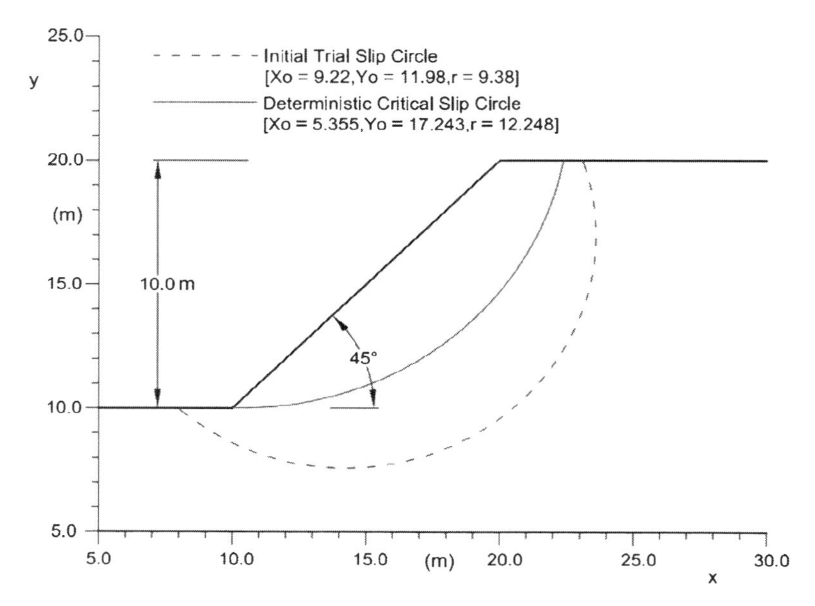

Fig. 1 Slope section and the deterministic critical slip circle in the illustrative example

Table 1 Statistical properties of soil parameters

Parameter (1)	Mean (2)	Standard deviation (3)	Coefficient of variation (4)
c′	18.0 kN/m^2	3.6 kN/m^2	0.20
tan ϕ'	tan 30°	0.0577	0.10
γ	18.0 kN/m^3	0.9 kN/m^3	0.05
r_u	0.2	0.02	0.10

3 Illustrative Example

Figure 1 shows a section of a simple slope of inclination 45° and height 10 m in a homogeneous c- ϕ soil. Previous reliability analyses of this slope under a probabilistic framework include those reported by Li and Lumb [9], Hassan and Wolff [8], and Bhattacharya et al. [1] using different methods of analysis. Thus, this example can well be regarded as a benchmark example problem. In all the previous investigations, all four geotechnical parameters, namely, the effective cohesion c', the effective angle of shearing resistance ϕ', the pore pressure ratio r_u, and the unit weight γ, were treated as random variables, and their statistical properties (mean, standard deviation, and coefficient of variation) are as in Table 1.

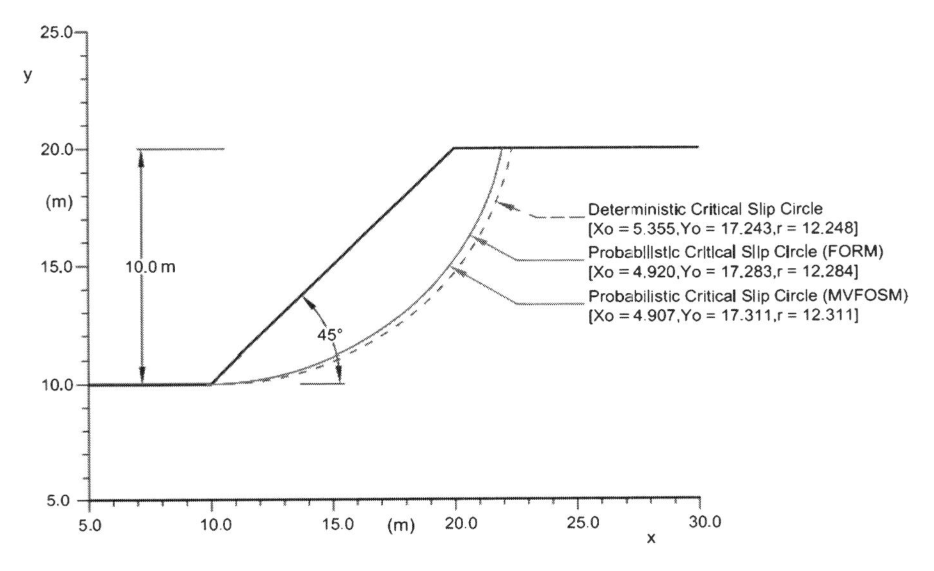

Fig. 2 Probabilistic and deterministic critical slip surfaces

4 Results and Discussion

4.1 Deterministic Analysis

For the purpose of determination of the critical slip circle, a trial slip circle ($x_o = 9.22$ m, $y_o = 11.98$ m, $r = 9.38$ m with reference to the axis system shown in Fig. 1) has been arbitrarily selected. Using Eq. (1) or (2) for the ordinary method of slices, its factor of safety (FS) is obtained as 1.70, when the parameters c', tan ϕ', γ, and r_u are assumed constant at their mean values (Table 1). With this slip circle as the initial slip surface, the developed computer program based on the sequential unconstrained minimization technique (SUMT) of nonlinear optimization coupled with the ordinary method of slices (OMS) yields a critical slip circle ($x_c = 5.355$ m, $y_c = 17.243$ m, $r_c = 12.248$ m) which passes through the toe, as shown in Fig. 2. The associated minimum factor of safety (F_{min}) is obtained as 1.26.

4.2 Reliability Analysis

Reliability analysis of this slope was attempted using two methods, namely, the mean-value first-order second-moment (MVFOSM) method and the first-order reliability method (FORM), with a view to compare the two sets of results. All four parameters c', tan ϕ', γ, and r_u are assumed to be normally distributed and uncorrelated. However, reliability analyses were carried out in three phases: In phase I only two parameters, namely, the cohesion c' and the effective angle of shearing resistance in the form of tan ϕ', were treated as random variables, while the other two parameters γ and r_u were assumed as constants at their mean values. In phase II, three parameters, namely, the cohesion c', the effective angle of shearing resistance tan ϕ', and r_u, were treated as random variables, while the parameter γ was assumed as constant at its mean value, while in phase III all four parameters were assumed as random variables.

4.3 Reliability Analysis for a Given Slip Surface Using MVFOSM

For the two slip surfaces shown in Fig. 1, namely, (1) the initial slip circle ($x_o = 9.22$, $y_o = 11.98$, $r = 9.38$) and (2) the deterministic critical slip circle ($x_o = 5.355$, $y_o = 17.243$, $r = 12.248$), the reliability indices were determined by MVFOSM method by taking two mechanically equivalent limit states: $\text{FS}-1 = 0$ and $\ln(\text{FS}) = 0$ using Eqs. (3) and (4a, 4b, 4c), respectively, for phase I, phase II, and phase III described above. The reliability index values for the different cases are summarized in Table 2.

4.4 Reliability Analysis for a Given Slip Surface Using FORM

Reliability index values have also been determined for the above mentioned slip surfaces using FORM. In particular, the algorithm for FORM method I [6] has been used in this case. All three phases mentioned above have been analyzed. For the sake of comparison as well as numerical demonstration, both the limit states considered in the analyses by MVFOSM have also been used here. The results are summarized in Table 2 again, alongside those obtained by using MVFOSM.

From Table 2, the following observations are made:

1. For the same slip surface and the same set of random variables (in phases I, II, and III), values of reliability index obtained for different mechanically equivalent limit states are markedly different when MVFOSM is used as the method of reliability analysis, whereas these values are identical when analyzed by the FORM method. This observation clearly demonstrates that unlike the FORM method, the MVFOSM method suffers from the "problem of invariance," that is, the result depends on how a given limit-state event is defined. In this respect, another observation from Table 2 is that when the limit state is taken as $\ln(\text{FS}) = 0$, the reliability index values are higher in all three phases of analysis.
2. It may be noted from Eqs. (1) and (2) that the performance function FS is linear when only c' and tan ϕ' are treated as random variables as in phase I of reliability analysis. However, when c', tan ϕ', and r_u are treated as random variables as in

Table 2 Summary of results of reliability analyses for given slip surfaces

Method of reliability analysis	Slip surface	Values of reliability index					
		Limit state: FS − 1 = 0			Limit state: ln (FS) = 0		
		Phase I	Phase II	Phase III	Phase I	Phase II	Phase III
		(c', tan ϕ')	(c', tan ϕ', r_u)	(c', tan ϕ', r_u, γ)	(c', tan ϕ')	(c', tan ϕ', r_u)	(c', tan ϕ', r_u, γ)
MVFOSM	Initial slip circle	3.955 (3.83×10^{-5})	3.812 (6.89×10^{-5})	3.736 (9.35×10^{-5})	5.059 (2.10×10^{-7})	4.874 (5.46×10^{-7})	4.775 (8.98×10^{-7})
	Deterministic critical slip circle	1.671 (4.73×10^{-2})	1.643 (5.02×10^{-2})	1.601 (5.47×10^{-2})	1.815 (3.47×10^{-2})	1.783 (3.73×10^{-2})	1.734 (4.14×10^{-2})
FORM (all random variables are normally distributed)	Initial slip circle	3.955 (3.83×10^{-5})	3.862 (5.63×10^{-5})	3.851 (5.88×10^{-5})	3.955 (3.83×10^{-5})	3.862 (5.63×10^{-5})	3.851 (5.88×10^{-5})
	Deterministic critical slip circle	1.671 (4.73×10^{-2})	1.646 (4.99×10^{-2})	1.625 (5.21×10^{-2})	1.671 (4.73×10^{-2})	1.646 (4.99×10^{-2})	1.625 (5.21×10^{-2})

Note: Figures in the parentheses indicate the values of the probability of failure for the respective slip surface

phase II or when c', tan ϕ', r_u, and γ are treated as random variables as in phase III, the performance function FS becomes nonlinear, and the degree of nonlinearity increases from phase II to phase III. Now, from Table 2, it is seen that for phase I, the values of reliability index yielded by MVFOSM and FORM are exactly the same, whereas they are different for phases II and III. Further, this difference in case of phase III is more than in case of phase II. This observation clearly demonstrates that in those situations where the performance function is linear and all the variables are normally distributed and statistically independent, the values of reliability index by MVFOSM method agree with those given by the FORM, and the error associated with MVFOSM method increases as the degree of nonlinearity of the performance function (or limit-state equation) increases.

3. Another important observation from Table 2 is that when the number of random variables increases, the value of reliability index (β) decreases and probability of failure increases.

4.5 Reliability Analysis for Given Slip Surfaces: Effect of Probability Distributions of the Basic Variates

As already stated, the MVFOSM method does not use the information on probability distribution of the basic random variables. In the FORM method, on the other hand, this information can be incorporated in the analysis. In the present analysis, the effect of variation of probability distributions has been studied using FORM method I. Only two distributions have been considered, namely, the normal distribution and the lognormal distribution. Results have been obtained for phase I only, that is, when only two parameters c'. and tan ϕ' are treated as random variables. Table 3 summarizes the results. It can be observed that there are substantial differences in the values of the reliability index obtained by using FORM when different combinations of probability distributions for the random variates c'. and tan ϕ' are considered. It is further observed that the β values from MVFOSM agree with those from FORM only when both the random variables are assumed to be normally distributed. Thus, it can be said that the MVFOSM method, though does not make use of any such knowledge regarding distribution of variates, implicitly assumes that all variables are normally distributed.

4.6 Probabilistic Critical Slip Surface and the Associated β_{min}

The probabilistic critical slip surface (surface of minimum β) has been determined following the same computational procedure as used for the determination of the deterministic critical slip surface, simply by replacing the objective function FS with

Table 3 Variation of reliability index with different probability distributions for the basic variates

			Values of reliability index for			
	Probability distribution		Initial trial slip circle		Deterministic critical slip circle	
Limit state surface	c'	$\tan\phi'$	MVFOSM	FORM	MVFOSM	FORM
$FS-1=0$	Lognormal	Normal	3.955	4.806	1.671	1.859
	Normal	Lognormal		4.038		1.661
	Lognormal	Lognormal		5.318		1.854
	Normal	Normal		3.955		1.671
$\ln FS=0$	Lognormal	Normal	5.059	4.806	1.815	1.859
	Normal	Lognormal		4.038		1.661
	Lognormal	Lognormal		5.318		1.854
	Normal	Normal		3.955		1.671

Note: These results correspond to the phase I analysis, that is, when only c'. and $\tan\phi'$ are treated as random variables

β [1]. A computer program was developed based on the sequential unconstrained minimization technique (SUMT) of nonlinear optimization coupled with a method of reliability analysis, MVFOSM or FORM, as the case may be.

For this search, the deterministic critical slip surface shown in Fig. 1 has been used as the initial slip surface. Several such probabilistic critical slip surfaces have been determined, and the associated minimum reliability index (β_{min}) values are summarized in Table 4. For the sake of clarity, only two of these critical surfaces are plotted in Fig. 2: the probabilistic critical slip surface for phase III analysis using MVFOSM ($x_c = 4.907$ m, $y_c = 17.311$ m, $r_c = 12.311$ m) and the probabilistic critical slip surface for phase III analysis using FORM with all four random variables normally distributed ($x_c = 4.920$ m, $y_c = 17.283$ m, $r_c = 12.284$ m). For the sake of comparison, the deterministic critical slip surface ($x_c = 5.355$ m, $y_c = 17.243$ m, $r_c = 12.248$ m) has also been plotted in Fig. 2.

From Fig. 2, as well as from the magnitudes of the coordinates of centers and radii, it is seen that the two probabilistic critical slip surfaces are located very close to each other while the deterministic critical slip circle is somewhat apart. The closeness of the deterministic and the probabilistic critical slip surfaces for the case of simple homogeneous slopes is in agreement with those reported by earlier investigators. The detailed results presented in Table 4 generally corroborate the observations made earlier from Table 2 with reference to the reliability analyses of the given slip surfaces.

5 Summary and Conclusions

In view of the growing appreciation of the uncertainty associated with the geotechnical parameters, especially, the strength parameters including the pore water pressure, the conventional deterministic approach of analysis is increasingly

Table 4 Summary of results of the minimum reliability analyses associated with the probabilistic critical slip surface

	Values of minimum reliability index					
	Limit state: FS−1 = 0			Limit state: ln (FS) = 0		
	Phase I	Phase II	Phase III	Phase I	Phase II	Phase III
Method of reliability analysis	(c', tan ϕ')	(c', tan ϕ', r_u)	(c', tan ϕ', r_u, γ)	(c', tan ϕ')	(c', tan ϕ', r_u)	(c', tan ϕ', r_u, γ)
MVFOSM	1.643 (1.671)	1.618 (1.643)	1.576 (1.601)	1.785 (1.815)	1.756 (1.783)	1.707 (1.734)
FORM (all random variables are normally distributed)	1.643 (1.671)	1.620 (1.646)	1.599 (1.625)	1.643 (1.671)	1.620 (1.646)	1.599 (1.625)

Note: Figures in the parentheses indicate the values of reliability index for the deterministic critical slip surface

being replaced by probabilistic approach of analysis or reliability analysis under a probabilistic framework. The mean-value first-order second-moment (MVFOSM) method based on a Taylor series expansion is rather widely used by the practitioners in the geotechnical engineering field mainly due to the simplicity and early origin of the method. However, in other fields of engineering, for example, in the structural engineering field, it is an established fact for quite some time that the MVFOSM method suffers from serious shortcomings such as the problem of invariance, as mentioned in an earlier section of this chapter.

This chapter concerns a study on the reliability analysis of earth slopes with uncertain soil strength parameters under a probabilistic framework. Reliability analyses have been carried out using a rigorous method, namely, the first-order reliability method (FORM) in conjunction with a simple slope stability model, namely, the ordinary method of slices (OMS). For the sake of comparison, results in the form of the reliability index and probability of failure have also been obtained using the MVFOSM method. Computer programs have been developed for the determination of reliability index based on both FORM and MVFOSM method for a given slip surface and then for the optimization-based determination of the probabilistic critical slip surface and the associated minimum reliability index. The developed programs have been applied to a benchmark example problem concerning a simple slope in homogeneous soil in which the geotechnical parameters are treated as random variables with given values of statistical moments. The differences between the two sets of results have been brought out for the cases of an arbitrarily selected given slip surface, the deterministic critical slip surface, and also the probabilistic critical slip surface. The study has been successfully used to demonstrate numerically all the major shortcomings of the approximate MVFOSM method and the error involved vis-à-vis the more accurate FORM method.

References

1. Bhattacharya G, Jana D, Ojha S, Chakraborty S (2003) Direct search for minimum reliability index of earth slopes. Comput Geotech 30(6):455–462
2. Cho SE (2009) Probabilistic stability analyses of slopes using the ANN-based response surface. Comput Geotech 36(5):787–97
3. Chowdhury R, Rao BN (2010) Probabilistic stability assessment of slopes using high dimensional model representation. Comput Geotech 37(7–8):876–884
4. Das SK, Das MR (2010) Discussion of "Reliability-based economic design optimization of spread foundations- by Y. Wang". J Geotech Geoenviron Eng 136(11):1587–1588
5. Fellenious W (1936) Calculation of stability of earth dams transaction. In: Second Congress on large dams, Washington, vol 4, p 445
6. Haldar A, Mahadevan S (2000) Probability, reliability, and statistical methods in engineering design. Wiley, New York
7. Hasofer AA, Lind AM (1974) Exact and invariant second moment code format. J Geotech Eng Div ASCE 100(1):111–121
8. Hassan AM, Wolff TF (1999) Search algorithm for minimum reliability index of earth slopes. J Geotech Geoenviron Eng 125(4):301–308
9. Li KS, Lumb P (1987) Probabilistic design of slopes. Can Geotech J 24(4):520–535
10. Rackwitz R, Fiessler B (1976) Note on discrete safety checking when using non-normal stochastic models for basic variables. Load project working session. MIT, Cambridge
11. Rackwitz R, Fiessler B (1978) Structural reliability under combined random load sequences. Comput Struct 9(5):484–494

Design of a Tuned Liquid Damper System for Seismic Vibration Control of Elevated Water Tanks

Anuja Roy and Aparna (Dey) Ghosh

Abstract In this paper, a Tuned Liquid Damper (TLD) system is proposed for the response mitigation of an elevated water tank structure under seismic excitation. First, a study on the performance of the TLD system attached to an elevated water tank structure modelled as a single-degree-of-freedom (SDOF) system is carried out in the frequency domain. Fluid-structure interaction is not considered in the present study. A realistic example of an elevated water tank structure is considered. The performance of the TLD system is examined on the basis of reduction in the root mean square (rms) value of the structural displacement. The sensitivity of the performance of the TLD system to change in the structural frequency due to fluctuation in the amount of water in the water tank container is examined. A time domain study is also carried out with the recorded accelerogram of the El Centro earthquake as the base input. The procedure to obtain a practical configuration of the TLD units is outlined. The performance of the designed TLD system indicates that the TLD is a promising device for the seismic vibration control of elevated water tanks.

Keywords TLD • Elevated water tank • Seismic excitation • Passive control

1 Introduction

Elevated water tanks are heavy mass structures that are vulnerable to seismic forces. Again, these are critical facilities, and it is essential that they remain serviceable even in the regions of maximum shaking intensity following a design level earthquake.

A. Roy (✉) • A.D. Ghosh
Department of Civil Engineering, Bengal Engineering and Science University, Shibpur, Howrah, India
e-mail: anuja.civil@gmail.com; aparnadeyghosh@gmail.com

S. Chakraborty and G. Bhattacharya (eds.), *Proceedings of the International Symposium on Engineering under Uncertainty: Safety Assessment and Management (ISEUSAM - 2012)*,
DOI 10.1007/978-81-322-0757-3_34,

Rai [6] in his study on the performance of elevated tanks in the Bhuj 2001 earthquake noted that the elevated water tank structure does not have much redundancy, and this is especially true for tanks with circular shaft-type stagings. Further, thin sections of shaft-type stagings also have low ductility. Thus, apart from having stringent design criteria and high-quality construction practices for these structures, it is also judicious to provide them with seismic protection devices.

Studies on the passive control of the seismic response of elevated tanks by base isolation were carried out by several researchers. Shenton and Hampton [7] evaluated the seismic response of an isolated elevated water tank structure. They compared the results with those of the corresponding fixed base tank and concluded that seismic isolation is efficient in reducing the tower drift, base shear, overturning moment and tank wall pressures for the practical range of tank capacities and height-to-diameter ratios. Shrimali and Jangid [8] examined the application of linear elastomeric bearings in the seismic response mitigation of elevated liquid storage steel tanks. Shrimali and Jangid [9] applied resilient-friction base isolator and friction pendulum system in their study of earthquake response of elevated cylindrical liquid storage tanks. Shrimali [10] investigated the earthquake response of elevated cylindrical liquid storage tanks isolated by friction pendulum system under bidirectional excitation. These studies yielded promising results in the field of seismic isolation of elevated liquid storage tanks.

In this chapter, the popular passive control device, the TLD, is studied for the seismic vibration control of elevated water tanks. The main advantages of this kind of damper over mass dampers are low installation, operational and maintenance costs, ease of installation in case of existing structures, easy adjustment of natural frequency and effectiveness even against small-amplitude vibrations.

The TLD was first proposed by Frahm in the early 1900s in which the frequency of motion of the water in two interconnected tanks was tuned to the fundamental rolling frequency of a ship so that this component of motion was successfully reduced [11]. The concept of applying TLDs in civil engineering structures as a passive vibration control device was initiated in the mid-1980s. Welt and Modi [15, 16] suggested the use of TLD in buildings to reduce structural response under wind or earthquake forces. Housner [3] took into consideration both the impulsive and convective liquid pressures in the analysis of the hydrodynamic pressures generated in a fluid container when it was subjected to horizontal acceleration. Fujino et al. [2] concluded that the fundamental sloshing frequency of TLD must be tuned to the natural frequency of the structure to obtain a higher damping value of the TLD under small-amplitude structural vibration. Koh et al. [5] reported that the liquid motion is highly sensitive to the natural frequency of the TLD as well as the amplitude and frequency content of the excitation spectrum. Yu et al. [17] presented an equivalent Tuned Mass Damper (TMD) model of the TLD and illustrated its satisfactory performance over a wide range of values of the excitation amplitude. The study by Banerji et al. [1] revealed the need for greater mass ratio of the TLD to be more efficient for larger structural damping. Kim et al. [4] performed shaking table experiments on the TLD that revealed the dependence of liquid sloshing on the amplitude of vibration and on the frequency of the damper. Tait et al. [13] studied the performance of unidirectional and bidirectional TLD for random vibrations. Tait and Deng [12] compared the performance of TLD with different tank geometries.

In this work, a transfer function formulation for the elevated water tank structure, modelled as a SDOF system, with TLD is developed. The input excitation is characterized by a white noise power spectral density function (PSDF). A sensitivity study on the performance of the TLD is included. It is followed by an investigation on the optimum tuning ratio. A simulation study using a recorded earthquake excitation is also carried out. The design of the TLD for a realistic example elevated water tank structure is performed.

2 Modelling of the Elevated Water Tank-TLD System

The TLD is a right circular cylindrical container of radius a (Fig. 1). It is filled with water upto a height h. It has a free upper surface. The thickness of the container wall is assumed to be uniform. The water in the damper container is assumed to respond in impulsive and convective modes under horizontal base excitation. Though in theory, there are several convective modes, consideration of only a few modes is enough for practical purpose. Here, only the first mode is considered. The liquid mass associated with the first convective mode, m_d, is assumed to be attached to the SDOF system representing the elevated water tank structure (Fig. 2) by a linear spring with stiffness k_d and a linear viscous damper with damping coefficient c_d. The fluid-structure interaction in the water tank is not being taken into account. The weight of the damper container is neglected. The damping ratio of the convective mode of vibration of the damper liquid is ζ_d, while ω_d denotes the natural frequency of the fundamental sloshing mode of vibration of the liquid. The expressions for m_d and ω_d are given by Veletsos and Tang [14]. The mass, stiffness and damping of the

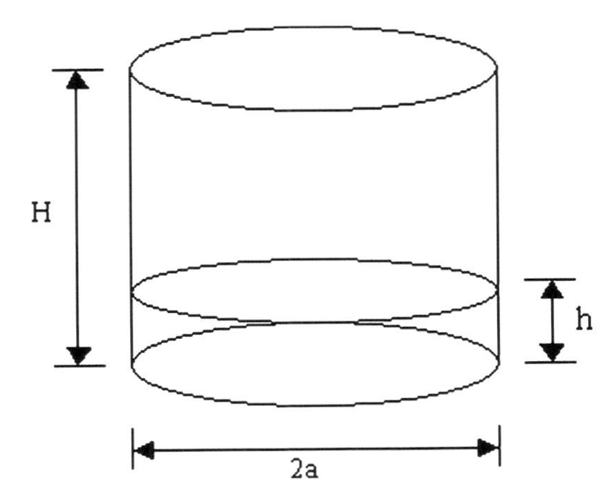

Fig. 1 Definition sketch for circular cylindrical Tuned Liquid Damper (TLD)

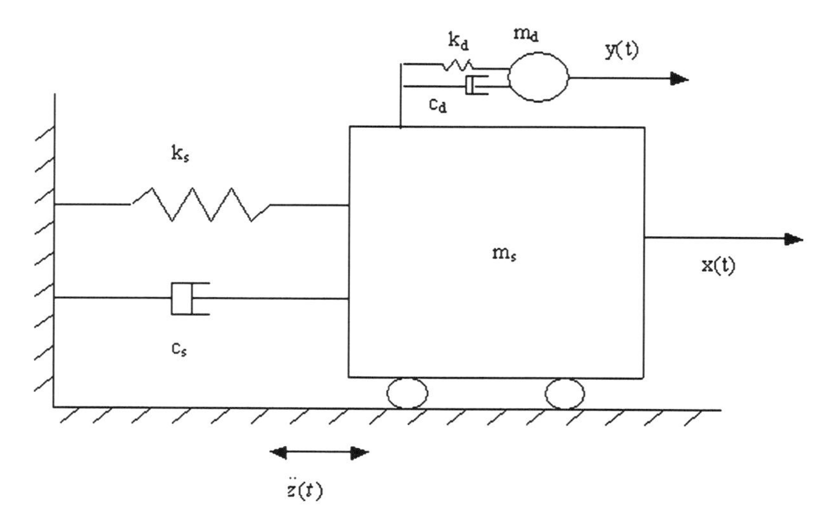

Fig. 2 Model of TLD-elevated water tank system

SDOF system representing the elevated water tank are m_s, k_s and c_s respectively. The damping ratio and the natural frequency of the same are ζ_s and ω_s.

3 Formulation of Transfer Function

3.1 *Equations of Motion*

Let $x(t)$ denote the horizontal displacement of the SDOF structural system relative to ground motion at a time instant t. Further, $y(t)$ represents the displacement of the convective water mass of the TLD relative to the structure at time instant t.

The equation of motion of m_d can be written as

$$\ddot{y}(t) + \ddot{x}(t) + 2\zeta_d\omega_d\dot{y}(t) + \omega_d^2 y(t) = -\ddot{z}(t). \tag{1}$$

Consideration of the dynamic equilibrium of the TLD-elevated water tank system gives

$$\ddot{x}(t) + 2\zeta_s\omega_s\dot{x}(t) + \omega_s^2 x(t) - 2\zeta_d\omega_d\mu\dot{y}(t) - \omega_d^2\mu y(t) = -\ddot{z}(t), \tag{2}$$

where $\mu = m_d/m_s$ is the ratio of the convective liquid mass within the TLD to the mass of the structural mass.

On Fourier transforming Eqs. (1) and (2) and by appropriate substitution, the following transfer function relating the structural displacement to the input ground acceleration is obtained in frequency domain

$$X(\omega) = \frac{\{\mu[2\zeta_d\omega_d i\omega + \omega_d^2]H_1(\omega) - 1\}H_2(\omega)}{\{\mu[2\zeta_d\omega_d i\omega + \omega_d^2]\omega^2 H_1(\omega)H_2(\omega) - 1\}}\ddot{Z}(\omega),$$

where $H_1(\omega) = -\frac{1}{\omega_d^2 - \omega^2 + 2\zeta_d\omega_d i\omega}$, $H_2(\omega) = -\frac{1}{\omega_s^2 - \omega^2 + 2\zeta_s\omega_s i\omega}$ and $X(\omega)$ and $\ddot{Z}(\omega)$ are the Fourier transforms of the corresponding time-dependent variables, $x(t)$ and $\ddot{z}(t)$ respectively.

4 Numerical Study in Frequency Domain

For numerical investigation, an example elevated water tank structure is considered. To examine the performance of TLD for the passive control of elevated water reservoir, the following realistic reinforced concrete tank structure with flat top cylindrical container and shaft-type staging is taken as the example structure. The relevant data that are assumed to define the structure are listed below:

Height of shaft support = 30 m; mean diameter of shaft = 3 m; thickness of shaft wall = 0.125 m; inner height of the tank container = 4 m; mean diameter of the tank container = 10 m; thickness of the container wall = 0.2 m; thickness of the container bottom = 0.2 m; thickness of the top cover of the container = 0.1 m; height of water in the tank container = 3.8 m.

Based on the above assumed data, the three-dimensional modelling of the example tank structure is analysed using the software package Staad.Pro2004. The fundamental natural frequency of the structure is found to be 4.5432 rad/s (1.383 s) for empty condition of the water tank and 3.625 rad/s (1.733 s) for full condition of the water tank. The value of the structural damping is assumed to be 1%.

The performance criterion of the TLD is the reduction in the rms value of the displacement of the elevated water tank system. To investigate the influence of mass ratio (the ratio of the convective water mass in the damper to the mass of the structure) in the structural response reduction, the values of mass ratio considered, are 1%, 2% and 5%. The value of ζ_d is taken as 0.01 as per Veletsos and Tang [14]. The whole system is subjected to a white noise PSDF input with $S_o = 100\ \text{cm}^2/\text{s}^3$.

To investigate the effect of the TLD, the same structure with a TLD located at the top is considered. The natural frequency of the structural system varies between the values corresponding to the full condition to that for the empty condition of the tank as the water mass in the water tank is a variable quantity. Thus, if the TLD is tuned to a particular frequency of the tank, there will be some amount of detuning as the water content of the tank changes. The effect of this detuning on the performance of the damper is studied for two cases, viz., when the damper is tuned to the structural frequency corresponding to the full condition of

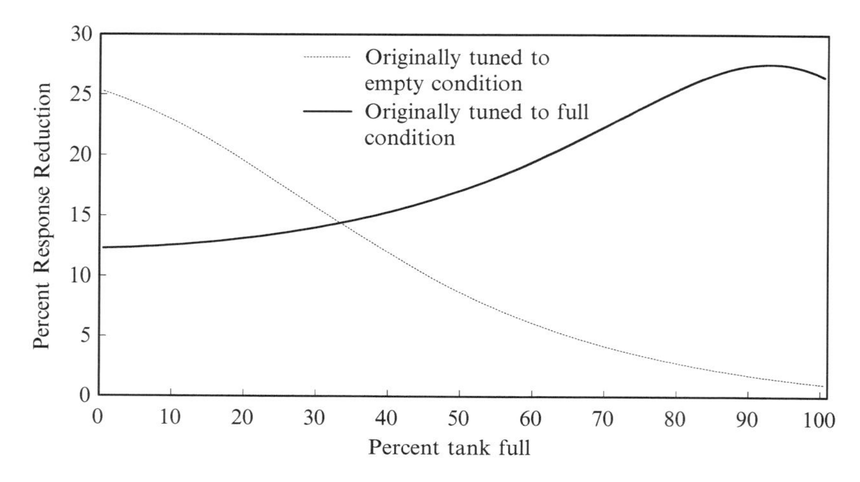

Fig. 3 Variation of percent response reduction with percent tank full

the tank and when the damper is tuned to the structural frequency corresponding to the empty condition of the tank. The response reduction varies from 12 to 26% when the damper is tuned to the structural frequency corresponding to the full condition of the tank (Fig. 3). The response reduction varies from 1 to 25% when the damper is tuned to the structural frequency corresponding to the empty condition of the tank. Thus, the results of this particular case study with mass ratio of 1% and tuning ratio (the ratio of the frequency of the TLD to the structural frequency) of unity indicated in Fig. 3 show that the performance of TLD is better if it is designed to be tuned to the frequency corresponding to the full condition of the tank. Thus, in this study, the TLD is kept tuned, with a tuning ratio of unity, to the frequency corresponding to the full condition of the tank.

Figure 4 shows the displacement transfer function curve for the structure in full condition of the tank without and with damper, subjected to ground acceleration process with white noise PSDF. Figure 4 also shows the displacement transfer function curves in full condition of the tank for different mass ratios, and it is observed that as compared to the without damper curve, now there are two peaks in each transfer function curve. Since the damper is tuned to the structural frequency, the tuning effect can be understood by the presence of the two peaks of reduced amplitude on either side of the peak of the 'without damper' curve. The nature of the curves is similar for the cases of 1%, 2% and 5% mass ratio. However, the reduction in the amplitude of the peaks due to the presence of the damper is dependent on the mass ratio. The greatest reduction for full condition of water tank is obtained for the case of 1% mass ratio. Here, the modelling of the TLD is very similar to a TMD. Generally, for a TMD, the response reduction increases with increase in mass ratio. To study the converse behaviour in the present case, a study is made on the response reduction achieved by a TMD with

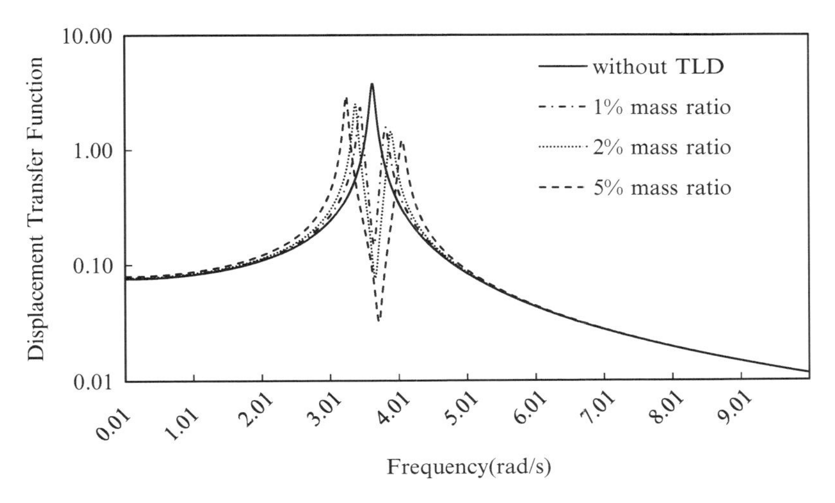

Fig. 4 Displacement transfer function of structure in full condition of the tank

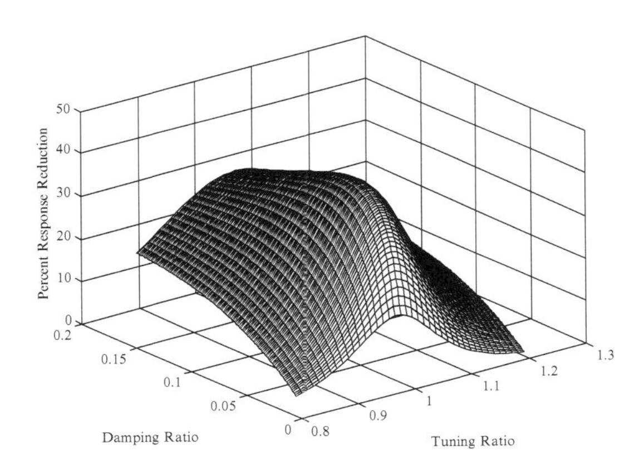

Fig. 5 Variation of percent response reduction with tuning ratio and damping ratio for 1% mass ratio

variation in tuning ratio and damping ratio for 1% and 5% mass ratio as shown in Figs. 5 and 6, respectively. A numerical illustration of the percent response reduction achieved by the damper for the full and empty conditions of the tank for the three different mass ratios is presented in Table 1.

Thus, from Table 1, it is seen that the response reduction varies in the range 6%–26% for 1% mass ratio, 10%–25% for 2% mass ratio and 19%–21% for 5% mass ratio. Hence, significant response reduction is achieved by the TLD.

Further, the sensitivity of the performance of the TLD to tuning ratio is examined. The variation in the response reduction achieved by the TLD system,

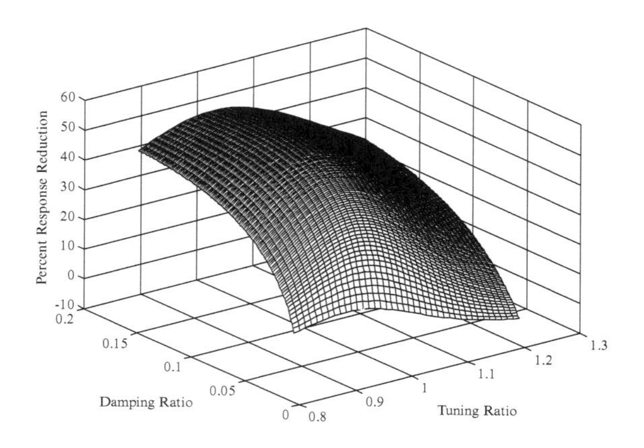

Fig. 6 Variation of percent response reduction with tuning ratio and damping ratio for 5% mass ratio

Table 1 Comparison of the percent response reduction by the TLD for different mass ratios with white noise PSDF input

Mass ratio (%)		Percentage reduction in RMS value of structural displacement	Tuning ratio
1	Empty condition	6.22	0.7979
	Full condition	26.48	1.0
2	Empty condition	10.59	0.7979
	Full condition	25.66	1.0
5	Empty condition	19.59	0.7979
	Full condition	21.73	1.0

in terms of the rms displacement of the structure, over a range of tuning ratio values, for the different values of mass ratios is observed (Fig. 7). From that study, it is found that optimal tuning ratios are 0.98, 0.97 and 0.92 for mass ratio equal to 1%, 2% and 5% respectively.

5 Simulation Study

The fourth-order Runge-Kutta method is being employed for time history analysis. The example structure is subjected to the recorded accelerogram of the El Centro earthquake. The TLD is tuned, with a tuning ratio of unity, to the frequency corresponding to the full condition of the tank, and mass ratio of 1%

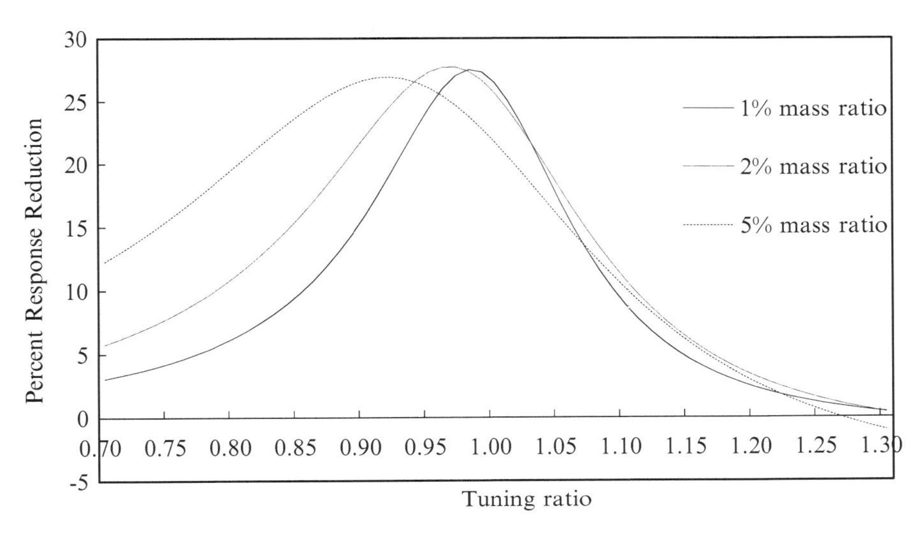

Fig. 7 Percent response reduction with tuning ratio for different mass ratios for full condition of the tank

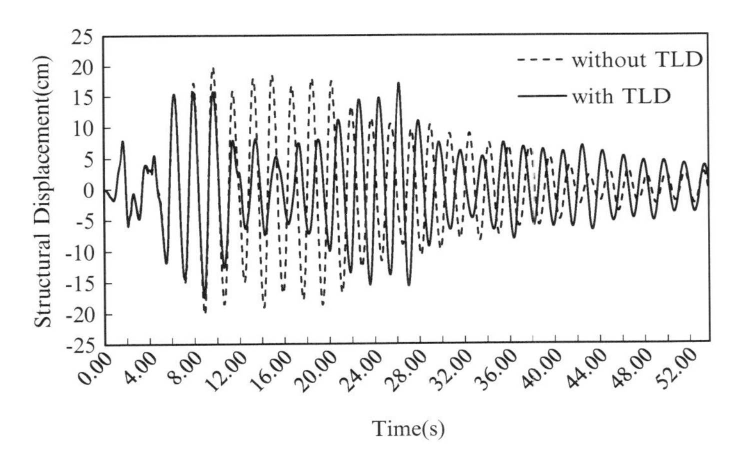

Fig. 8 Displacement Time history of SDOF structural system to E1 Centro earthquake when the tank is full

is considered. The displacement time history of the structure alone and with the TLD system for full condition of the tank is presented in Fig. 8. The response reduction achieved by the TLD system in terms of the rms displacement of the structure is 18.46% and that in terms of the peak displacement of the structure is 13.34% for full condition of the tank. The effect of detuning that occurs when

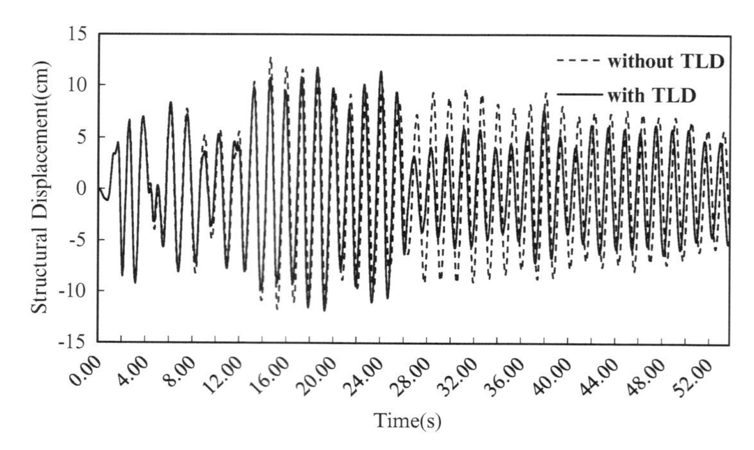

Fig. 9 Displacement Time history of SDOF structural system to E1 Centro earthquake when the tank is empty

the tank is empty is shown in Fig. 9. In this case, the response reduction achieved by the TLD system in terms of the rms displacement of the structure is 14.94% and that in terms of the peak displacement of the structure is 6.93% for empty condition.

6 Design of TLD System

The design parameters of TLD system (Fig. 1) are (1) mass ratio, (2) tuning ratio, (3) no. of TLD, (4) radius of TLD container (a), (5) height of liquid in TLD (h), (6) height of TLD container (H) and (7) h/a ratio.

The parameters should be chosen in such a way that they are compatible with the practical situation, and the TLD system should give sufficient response reduction. The studies carried out show that about 26.48% reduction in the RMS value of the displacement of the structure may be achieved by the TLD when tuned to the full condition of the tank with a tuning ratio (the ratio of the frequency of the TLD to the structural frequency) of unity for a corresponding mass ratio (convective water mass in the TLD/total mass of the structure) of about 1%. As it is not practicable to go for a very high mass ratio, 1% mass ratio is adopted. The optimum tuning ratio for 1% mass ratio is obtained as 0.98. So, a tuning ratio of 0.98 is adopted. The TLD is tuned to the frequency corresponding to the full condition of the tank. The ratio of depth of water to radius of damper container (h/a ratio) is chosen as 0.3. The radius of the TLD container (a), the height of liquid in the TLD container (h) and the height of the TLD container (H) for a single unit of damper as in Fig. 1 are designed to be 0.7 m, 0.21 m and 0.4 m, respectively. The number of TLD required to provide the total water mass in the TLD system is 20. So 20 dampers are to be accommodated on the top of the elevated water tank container. These dampers are to be stacked. The total area required to accommodate dampers is available on the top of the elevated water tank container. The design frequency of the damper is 3.599 rad/s. The design tuning ratios are 0.993 and 0.792 for full and empty condition of the elevated water tank container respectively.

7 Performance of the Designed TLD System

The performance is observed through reduction in rms displacement of the structure in time domain and in frequency domain. In time domain, the reduction in peak displacement of the structure is also evaluated. In frequency domain study, white noise PSDF with intensity of 0.01 m^2/s^3 is chosen as the base input excitation. In case of time history analysis, the recorded accelerogram of the El Centro earthquake is taken as the input excitation.

Representative set of transfer function curve for the full condition of the water tank is given in Fig. 10 and is compared with that of the structure without damper. The variation of percent response reduction with percent content of water in the water tank is presented in Fig. 11.

The numerical value of the response reduction achieved by the TLD system varies between 11% and 27% from empty to full condition of the water tank for the designed tuned condition for white noise PSDF. The displacement time history of the TLD-structural system for full condition is presented in Fig. 12.

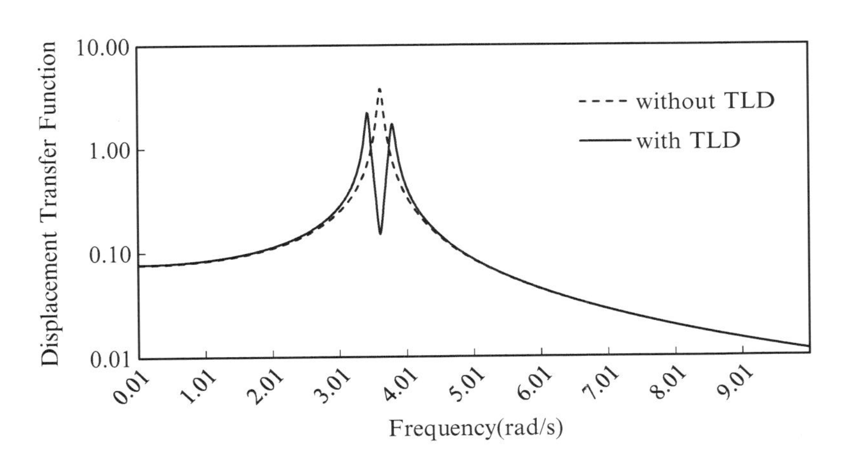

Fig. 10 Displacement transfer function of structure in full condition of the tank with designed TLD

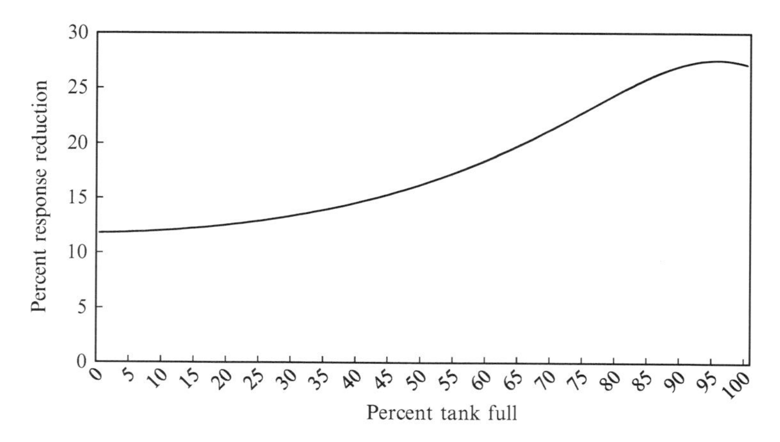

Fig. 11 Variation of percent response reduction with percent tank full

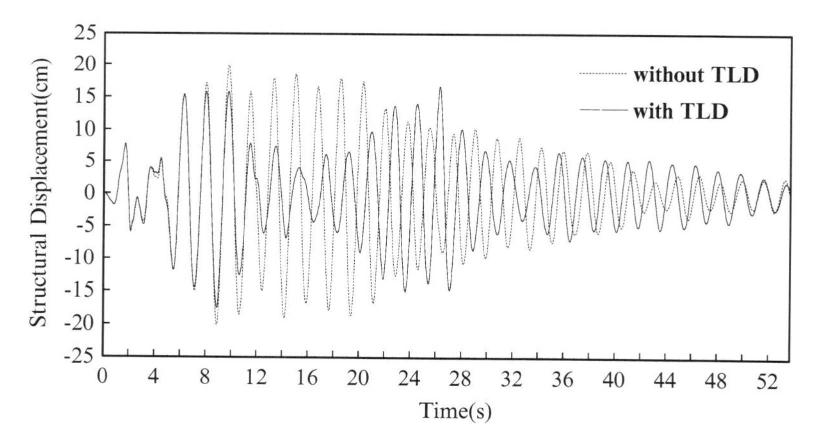

Fig. 12 Displacement Time history of SDOF structural system to E1 Centro earthquake when the tank is full

The response reduction achieved by the TLD system in terms of the RMS displacement of the structure is 22.94% and that in terms of the peak displacement of the structure is 12.89% for full condition of the tank. The effect of detuning that occurs when the tank is empty is clear from Fig. 13 where the response reduction achieved by the TLD system in the RMS displacement of the structure is 14.11%, while that in the peak displacement of the structure is 8.16%.

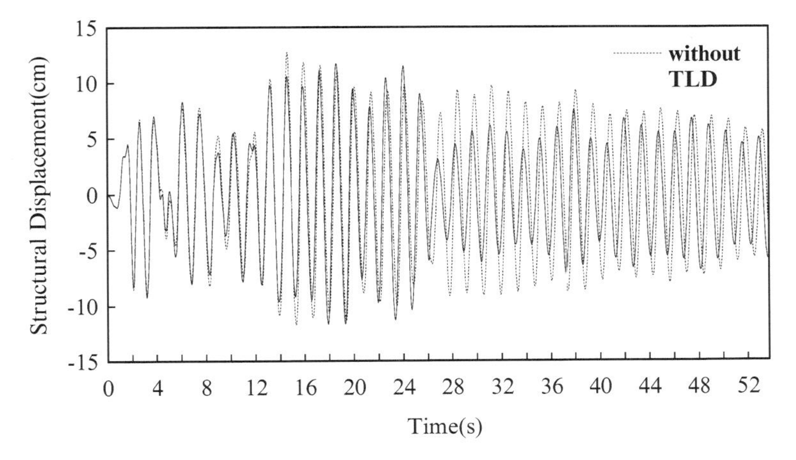

Fig. 13 Displacement Time history of SDOF structural system to E1 Centro earthquake when the tank is empty

8 Conclusions

Through this work, the applicability of TLD as a seismic vibration response reduction device for an elevated water tank is investigated. An attempt is also made to design a TLD system for a realistic elevated water tank structure. The performance of the damper system is evaluated both in frequency domain as well as in time domain. Based on the observations during the study, the following conclusions are drawn about the TLD system:

- The numerical study indicates that a TLD can be used as a promising control device for the seismic vibration suppression of elevated water tank structures. The results in frequency domain study show that substantial reduction in rms value of structural response upto 27% may be achieved by the TLD in the tuned condition, while this response reduction may be 18% as illustrated by the results of the simulation study. The reduction in case of peak displacement as obtained in time domain is 13%.
- The performance of the TLD system is better if it is tuned to the frequency corresponding to the frequency of structure in tank full condition out of the two possibilities of tuning to the frequency corresponding to the frequency of structure in empty and full condition of the tank. The reduction in the rms value of the structural response varies from 12% to 27% over the range of structural frequencies from empty condition to full condition of the tank.
- In this study, the percent response reduction is evaluated in the possible range of detuning of the damper system for the different mass ratios of the TLD system. The results indicate that the optimum tuning ratio is very close to unity for a lower mass ratio of 1%.

- The design example of the TLD system for realistic elevated water tank reveals that it will be possible to get a feasible configuration of the damper. In frequency domain, a maximum reduction of about 27% is achieved by the designed damper system even with a relatively lower mass ratio. In the extreme condition of detuning, at least 11% of response reduction is obtained. Using recorded accelerogram as input, the reduction in rms displacement of structure is 22% when the tank is full. When the tank becomes empty, it becomes 14%. The reduction in peak displacement of structure, as obtained, is 12% for the tank full condition, while that for the tank empty condition is 8%.

References

1. Banerji P, Murudi M, Shah AH, Popplewell N (2000) Tuned liquid dampers for controlling earthquake response of structures. Earthq Eng Struct Dyn 29:587–602
2. Fujino Y, Pacheco BM, Chaiseri P, Sun LM (1988) Parametric studies on tuned liquid damper (TLD) using circular containers by free-oscillation experiments. Struct Eng/Earthq Eng JSCE 5(2):381–391
3. Housner GW (1957) Dynamic pressures on accelerated fluid containers. Seismol Soc Am 47 (1):15–35
4. Kim YM, You KP, Cho JE, Hone DP (2006) The vibration performance experiment of tuned liquid damper and tuned liquid column damper. J Mech Sci Technol 20(6):795–805
5. Koh CG, Mahatma S, Wang CM (1994) Theoretical and experimental studies on rectangular liquid dampers under arbitrary excitations. Earthq Eng Struct Dyn 23:17–31
6. Rai DC (2002) Review of code design forces for shaft supports of elevated water tanks. In: Proceedings of 12th symposium on earthquake engineering, IIT Roorkee, 16–18 Dec, pp 1407–1418
7. Shenton HW III, Hampton FP (1999) Seismic response of isolated elevated water tanks. J Struct Eng ASCE 125(9):965–976
8. Shrimali MK, Jangid RS (2003) Earthquake response of isolated elevated liquid storage steel tanks. J Constr Steel Res 59:1267–1288
9. Shrimali MK, Jangid RS (2006) Seismic performance of elevated liquid tanks isolated by sliding systems. In: Proceedings of first European conference on earthquake engineering and seismology, Geneva, Switzerland, 3–8 Sept 2006
10. Shrimali MK (2008) Earthquake response of elevated liquid storage tanks isolated by FPS under bi-direction excitation. Adv Vib Eng 7(4):389–405
11. Soong TT, Dargush GF (1997) Passive energy dissipation systems in structural engineering. Wiley, UK
12. Tait MJ, Deng X (2010) The performance of structure – tuned liquid damper systems with different tank geometries. Struct Control Health Monit 17:254–277
13. Tait MJ, Isyumov N, El Damatty AA (2008) Performance of tuned liquid dampers. J Eng Mech 134(5):417–427
14. Veletsos AS, Tang Y (1990) Soil-structure interaction effects for laterally excited liquid storage tanks. Earthq Eng Struct Dyn 19:473–496
15. Welt F, Modi VJ (1989a) Vibration damping through liquid sloshing: part I – a nonlinear analysis. Proceedings of diagnostics, vehicle dynamics and special topics. ASME, Design Engineering Division (DE) 18(5):149–156.
16. Welt F, Modi VJ (1989b) Vibration damping through liquid sloshing: part II –experimental results. Proceedings of diagnostics, vehicle dynamics and special topics. ASME, Design Engineering Division (DE) 18(5):157–165
17. Yu JK, Wakahara T, Reed DA (1999) A non-linear numerical model of the tuned liquid damper. Earthq Eng Struct Dyn 28:671–686

Structural Reliability Evaluation and Optimization of a Pressure Vessel Using Nonlinear Performance Functions

P. Bhattacharjee, K. Ramesh Kumar, and T.A. Janardhan Reddy

Abstract Structural safety is one of the most important factors of any aerospace product. Until recently, a design is considered to be robust if all the variables that affect its life has been accounted for and brought under control. The meaning of robustness is changing. Designer and engineers have traditionally handled variability with safety factors. In this chapter, in the first phase, a pressure vessel made of titanium alloy is considered for safety index (structural reliability) study. The safety index is evaluated based on the data collected during manufacturing and operation. Various methods like mean value and moment methods are used for safety evaluation and same have been discussed. In the second phase of this chapter, an attempt has been made to carry out multi-objective design analysis taking into account the effect of variation of design parameters. Multiple objective of interests include structural weight, load-induced stress, deflection and structural reliability. The design problem is formulated under nonlinear constrained optimization and has been solved. Nonlinear regression relations are used for various performance functions. Nonlinear regression model is validated and found to be in good agreement with experimental results. Finally, optimum design parameters are suggested for design operating conditions.

Keywords Structural Reliability • Pressure Vessel • Safety Index and Nonlinear Constrained Optimization

---

P. Bhattacharjee (✉)
Reliability Engineering Division, DRDL, Hyderabad, India
e-mail: pradeep9_rqa@yahoo.com

K.R. Kumar
Production Planning Division, DRDL, Hyderabad, India
e-mail: rkkatta@rediffmail.co

T.A.J. Reddy
Mechanical Engineering, Osmania University, Hyderabad, India
e-mail: thanam.engineer@gmail.com

S. Chakraborty and G. Bhattacharya (eds.), *Proceedings of the International Symposium on Engineering under Uncertainty: Safety Assessment and Management (ISEUSAM - 2012)*,
DOI 10.1007/978-81-322-0757-3_35,

1 Introduction

Probabilistic structural design evaluation method is fast growing in aerospace engineering. In this method, all uncertainties like variability in material properties, geometry and loads are considered during design which enables a product to have better reliability compared to deterministic design. The study of reliability engineering is also developing very rapidly. The desire to develop and manufacture a product with superior performance and reliability than its predecessor is a major driving force in engineering design. The design of any engineering system requires the assurance of its reliability and quality. Traditional deterministic method has accounted for uncertainties through empirical safety factor. Such safety factors do not provide a quantitative measure of safety margin in design and are not quantitatively linked to influence different design variables and their uncertainties on overall system performance. In this chapter, a titanium air bottle (pressure vessel) is identified for structural safety index study. These air bottles are extensively used in aerospace and ground operations. The detailed safety index evaluation is discussed in this chapter.

Generally, the objective and constraint functions, load conditions, failure modes, structural parameters and design variables are treated in a deterministic manner. The problem with this approach is that, in many cases, deterministic optimization gives designs with higher failure probability than optimized structures. Therefore, since uncertainties are always present in the design for engineering structure, it is necessary to introduce reliability theory in order to achieve a balance between cost and safety for optimal design. A straightforward approach for the modelling and analysis of uncertainties is to introduce probabilistic models in which structural parameters and/or design variables are considered stochastic in nature. Then, by the combination of reliability-based design procedures and optimization technique, it is possible to devise a tool to obtain optimal designs.

The aim of this work is to establish a simple methodology, which will be useful to pressure vessel designer during design and development. This chapter deals with analysis of a typical titanium pressure vessel that is used to store high-pressure air, nitrogen or inert gas to run turbine to generate power as well as for pneumatic actuation for control system. These bottles have to be safe and reliable and shall be of lower weight. The optimum design problem formulated under nonlinear constrained optimization using nonlinear regression has been solved as 'Nonlinear Constrained Minimization' optimization. Optimum parameters of air bottle are suggested for design operating pressure.

2 Nomenclature

a, b, c: Regression coefficient
$g(x)$: Performance function
P: Pressure

p_f: Probability of failure
R: Reliability
R^2_{adj}: Adjusted R-square
R^2: Regression square
R_i: Internal radius
R_0: External radius
s, M: Material strength
t: Thickness
U, u, Z: A vector of statistically independent random variables with zero mean and unit standard deviation
V: Volume
W: Weight
β: Safety index
β_0: Design safety index requirement
γ: Poisson's ratio
δ: Deformation
μ_M: Mean material strength
μ_σ: Mean induced stress
ρ: Density
σ: Stress/standard deviation
σ_M: Standard deviation of material strength
σ_σ: Standard deviation of induced stress
Φ: Normal cdf

3 Problem Statement and Methodology Adapted

Weight optimization is one of the prime requirements of any aerospace product. Aerospace product has to be optimum in respect of weight, size, volume, cost, etc. Any weight-saving results in increase in payload capacity. Similarly, packaging density can be increased with volume optimization. But all these design optimizations should not be at the cost of safety and reliability. In this chapter, we have identified a pressure vessel which has already been designed, developed and successfully used in various grounds and space vehicles applications. These air bottles operate at very high pressure and are filled with dry air, nitrogen or inert gas. Due to its higher operating pressure, it has to be totally safe as ground personnel handle these bottles in fully charged condition.

In the first phase, safety index (structural reliability) is evaluated using mean value method (MVM) and advanced first-order second-moment (AFOSM) method using data collected during manufacturing and testing. In the second phase, an attempt is made to optimize the weight of the air bottle to meet the target reliability. Finite element analysis (FEA) is carried out to generate maximum stress, strain and deformation for various design parameters and operating conditions. Nonlinear performance functions (regression relations) are established for stress and deflection.

4 Safety Index Evaluation

4.1 Mean Value Method

This method [1–3] is commonly referred to as the mean value first-order second-moment (MVFOSM or simply MV) method since it involves a first-order expansion about the mean to estimate the first and second moments. MV method involves developing the Taylor series expansion of $g(x)$ about the nominal or mean value of the individual random variables. The moments of the resulting approximating function are found, using which approximate statements can be made regarding the probability of failure.

$$g(x) = g(x_1, x_2, \ldots x_n); \quad \text{safety index } \beta = \frac{\mu_g}{\sigma_g}$$

where μ_g and σ_g are the mean and standard deviation of performance function.

4.2 Advanced First-Order Second-Moment

Hasofer and Lind (HL) [4–6] proposed a method for evaluating the safety index (β). According to HL approach, the constraint is linearized by using Taylor series expansion retaining up to the first-order terms. The linearization point selected is that of maximum likelihood of occurrence and is known as the most probable failure point. This method is called advanced first-order second-moment (AFOSM) method. The most probable failure point is determined by transforming original random variables to normalized and independent set of reduced variables as shown in Fig. 1.

$$U = \frac{x - x_\mu}{x_\sigma} \tag{1}$$

The failure surface is mapped onto the corresponding failure surface in the reduced space. The point on this surface with minimum distance from the origin is the most probable failure point, and the geometric distance to the origin is equal

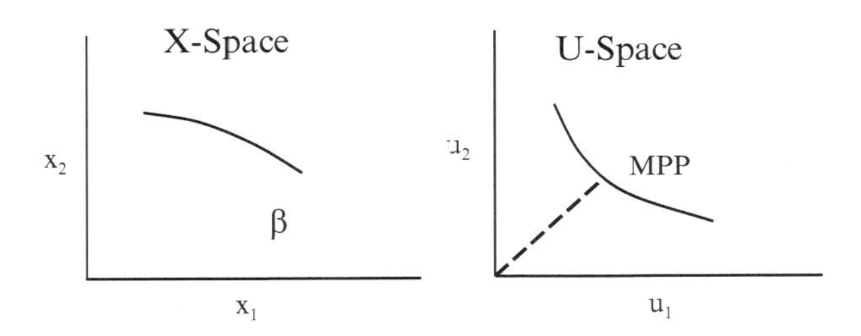

Fig. 1 Transformation of coordinate into standard space

to the safety index β. The failure surface is generally a nonlinear function, and the point with minimum distance to the origin can be evaluated by solving the following optimization problem, that is,

$$\text{Minimize } \left(U^T U\right)^{\frac{1}{2}} = \beta \tag{2}$$

$$\text{Subject to } g(U) = 0$$

where $g(U)$ is the failure function or limit state equation in reduced space. Using AFOSM, this optimization can be solved by using nonlinear optimization or iterative algorithms.

5 Optimization Formulation

This chapter deals with a practical problem of high-pressure air bottles. In the first phase of the study, we have evaluated the safety index (structural reliability) of these air bottles. After confirming that these air bottles have enough margins of safety, an attempt has been made to optimize the weight of the air bottles. These air bottles are at present under serial production and are being used in various ground and space applications. Static linear strength, deformation and safety index have been considered as design constraints. The brief optimization formulation of this problem is discussed below.

5.1 Nonlinear Performance Functions

5.1.1 Nonlinear Regression Relations

The regression relations [7] are established for induced stress and deformation due to internal load. In many engineering situations, the relationship between dependent and independent variable may not be linear but exponential, Weibull, logarithmic or inverse type. These types of equations generally fall under the category of having

Table 1 ANSYS output

		Design load 40 MPa	
Thickness (t)	Outer radius (R_0)	Hoop stress (σ) MPa	Deformation (δ) mm
2.5	152.0	1,233	1.0555
3.0	152.5	1,007	0.8819
3.5	153.0	885	0.7567
4.0	153.5	746	0.6644
4.5	154.0	665	0.5927
5.0	154.5	599	0.5353
5.5	155.0	544	0.4883
6.0	155.5	500	0.4492
6.5	156.0	461	0.4161
7.0	156.5	428	0.3877
7.5	157.0	399	0.3632
8.0	157.5	375	0.3417
8.5	158.0	358	0.3227

nonlinear parameter. In order to apply the principles of least squares, the equation should be reduced to linear form. The process of transforming the nonlinear equation into a linear form is called linear transformation. It is observed from ANSYS output data, Table 1, that the hoop stress and deflection have nonlinear relation with its wall thickness and outer radius. A logarithmic relation has been established and tested statistically for significance. The regression relations considered are as given below:

$$\log \sigma = a_1 + b_1 \log t + c_1 \log R_0 \tag{3}$$

$$\log \delta = a_2 + b_2 \log t + c_2 \log R_0 \tag{4}$$

The wall thickness of the hemispherical shell and outer radius of the air bottle are the main design geometrical parameters which account for weight. Hence, these parameters are considered for weight optimization. The nonlinear performance function relations for stress and deformation are as given below:

$$\sigma = a_1 t^{b_1} R_0^{c_1} \tag{5}$$

$$\delta = a_2 t^{b_2} R_0^{c_2} \tag{6}$$

Load-induced stress and deformation under various operating load (pressures) is evaluated using finite element analysis (ANSYS), where a_1, a_2, b_1, b_2, c_1 and c_2 are the coefficients of above regression equations. The nonlinear regressions established are given in Table 2.

Table 2 Nonlinear regression equations

Design load	$\sigma = a_1 t^{b_1} R_0^{c_1}$	$\delta = a_2 t^{b_2} R_0^{c_2}$
40 (MPa)	$\sigma = 0.0059 t^{-1.0967} R_0^{2.6386}$	$\delta = 0.00037 t^{-1.0242} R_0^{1.7701}$

5.2 *Optimization Constraints*

5.2.1 Static Linear Strength

The static linear strength [8] design criteria are based on load-induced stress at the critical location in the spherical shell should be less than or equal to maximum allowable stress. In other words, the maximum stress generated due to internal pressure should not exceed the material strength as defined below:

$$\frac{\sigma}{\sigma_{\max}} - 1 \leq 0 \tag{7}$$

5.2.2 Deformation

The spherical air bottle will dilate radially outward due to internal pressure. In no case the air bottle shall deform beyond maximum allowable deflection, that is,

$$\frac{\delta}{\delta_{\max}} - 1 \leq 0 \tag{8}$$

5.2.3 Safety Index (Structural Reliability)

The safety index β is defined as

$$\beta = -\frac{\mu_M - \mu_\sigma}{\left(\sigma_M^2 + \sigma_\sigma^2\right)^{\frac{1}{2}}} \tag{9}$$

$$\beta \geq \beta_0; 1 - \frac{\beta}{\beta_0} \leq 0$$

where β_0 is design safety index requirement.

'β' is achieved safety index and is evaluated using strength-stress interference model. In this chapter, both material strength and induced stress are considered normally distributed to be random variable.

5.3 Nonlinear Constrained Optimization

The main objective is to minimize the weight of the air bottle, without compromising the safety requirement. A nonlinear single objective constrained optimization method is adapted. The nonlinear regression equations as given in Table 2 are used for optimization formulation [9–11]. The optimization problem formulation is as given below:

Objective function

Minimize $W = \rho V$

Subject to

$$\frac{\sigma}{\sigma_{\max}} - 1 \leq 0; \quad \frac{\delta}{\delta_{\max}} - 1 \leq 0; \quad 1 - \frac{\beta}{\beta_0} \leq 0$$

where $\sigma = a_1 t^{b_1} R_0^{c_1}$; $\delta = a_2 t^{b_2} R_0^{c_2}$; $\beta = \frac{\mu_M - \mu_\sigma}{\left(\sigma_M^2 + \sigma_\sigma^2\right)^{\frac{1}{2}}}$; $V = \frac{4}{3}\pi\left[3R_0^2 t - 3R_0 t^2 + t^3\right]$

Bound constraints $t_l < t < t_u$ and $R_{0\,l} < R_0 < R_{0\,u}$

l and u are lower and upper design limit.

6 Nonlinear Performance Function Generation

6.1 Finite Element Analysis

To establish the performance function relations, the ANSYS output is generated for each set of design input parameters, that is, wall thickness and outer radius of spherical shell. The details of ANSYS output are given in Table 1.

6.2 Nonlinear Regression and Regression Statistics

The nonlinear regression of induced stress (σ) and deformation (δ) versus wall thickness (t) and outer radius (R_0) is established using ANSYS output as given in Table 1. These regression equations are used for optimization formulation. The nonlinear regression and regression statistics are given in Tables 2 and 3.

R^2_{adj}: Adjusted regression square; rmse: root mean square error; mse: mean square error as R^2_{adj} values are close to one, and 'rmse' value is low indicating that the regression relations are fitting well.

Table 3 Regression statistics

Design load	$\sigma = a_1 t^{b_1} R_0^{c_1}$	$\delta = a_2 t^{b_2} R_0^{c_2}$
40 (MPa)	$R^2_{\text{adj}} = 0.9999$ rmse = 0.0020 mse = $3.9 * 10^{-6}$	$R^2_{\text{adj}} = 1.0000$ rmse = 0.0010 mse = $1.0377 * 10^{-6}$

7 Design Data

To evaluate safety index, initially design data are collected from design document. The important design parameters of the air bottle are given in Table 4.

A systematic data collection is carried out during manufacturing at production centres starting from raw material to final product. Material mechanical properties are taken from test certificates provided by the suppliers which cover chemical compositions, heat treatment details, tensile strength and percentage elongation. Similarly, thickness mapping for wall thickness and internal radius is carried out before joining of each spherical shell. The statistical dispersion of various parameters is given in Table 5.

8 Design Safety Analysis

8.1 *Safety Index (Structural Reliability)*

The structural safety [1] of the air bottle is evaluated considering the statistical variability from the data collected during manufacturing, testing and operation. The safety index (β) evaluated using mean value method and advanced first-order second-moment method as discussed in safety index evaluation section is as follows:

8.1.1 Mean Value Method

$$g = \left(s - \frac{PRi}{2t}\right) \tag{10}$$

$$\mu_g = 484.84$$

$$\sigma_g = \left[\left(\frac{\partial g}{\partial s}\sigma_s\right)^2 + \left(\frac{\partial g}{\partial P}\sigma_P\right)^2 + \left(\frac{\partial g}{\partial R}\sigma_R\right)^2 + \left(\frac{\partial g}{\partial t}\sigma_t\right)^2\right]^{\frac{1}{2}} \sigma_g = 23.2585$$

$$\beta = \frac{\mu_g}{\sigma_g}; \quad \beta = 20.8$$

Table 4 Design parameter

Internal radius (R_i)	:	149.5 mm
Wall thickness (t)	:	8.0 mm
Design pressure (P)	:	40 MPa
Material	:	Titanium alloy
Construction	:	Welded
Type of welding	:	Electron beam
Weight	:	10.5 kg

Table 5 Variability observed

Parameter	Mean (μ)	Standard deviation (σ)
Operating pressure (P)	36 (MPa)	1.98 (MPa)
Internal radius (R_i)	149.5 (mm)	0.5 (mm)
Wall thickness (t)	7.91 (mm)	0.25 (mm)
Material strength (M)	860 (MPa)	8.6 (MPa)

8.1.2 Advanced First-Order Second-Moment Method

Data from Table 5 is transformed to normal space

$$Z_1 = \frac{s - s_\mu}{\sigma_s} = \frac{s - 860}{8.6}$$

$$Z_2 = \frac{P - P_\mu}{\sigma_P} = \frac{P - 40}{1.98}$$

$$Z_3 = \frac{R_i - R_\mu}{\sigma_R} = \frac{R_i - 149.5}{0.5}$$

$$Z_4 = \frac{t - t_\mu}{\sigma_t} = \frac{t - 7.91}{0.25}$$

$$s = 8.6Z_1 + 860$$

$$P = 1.98Z_2 + 40$$

$$R_i = 0.5Z_3 + 149.5$$

$$t = 0.25Z_4 + 7.91$$

Substituting s, P, R_i and t in Eq. 10

$$g(Z) = (8.6Z_1 + 860) - \left[\frac{(1.98Z_2 + 40)\,(0.5Z_3 + 149.5)}{2(0.25Z_4 + 7.91)}\right]$$

Minimize $\beta = \sqrt{Z^T Z}$
Such that $g(Z) = 0$
Solving the above optimization problem, we get $\beta = 14.39$

9 Reliability Optimization

9.1 Nonlinear Constrained Optimization

The main objective of this chapter is to minimize the weight of the air bottle as thickness, and outer radius are the design parameters which contribute to the weight of titanium material. Hence, these parameters are optimized so as to minimize the weight of the air bottle. The nonlinear constrained optimization [10–12] is formulated as given below:

Objective function

Minimize weight (W) $= V * \rho$

Subject to

$$\frac{\sigma}{\sigma_{\max}} - 1 \leq 0, \quad \frac{\delta}{\delta_{\max}} - 1 \leq 0, \quad 1 - \frac{\beta}{\beta_0} \leq 0$$

where $V(\text{volume}) = \frac{4}{3}\pi\left[3R_0^2 t - 3R_0 t^2 + t^3\right]$

$$\rho(\text{density}) = 4.46 * 10^{-6}\ \text{kg/mm}^3; \quad \sigma_{\max} = 860\ \text{MPa}; \quad \delta_{\max} = 0.5\ \text{mm}$$

β_0 (Target safety index) ≥ 4; R (Structural reliability) ≥ 0.99996; R_i (Internal radius) $= 149.5$;

σ (Induced stress) $= a_1 t^{b_1} R_0^{c_1}$; δ (Deformation) $= a_2 t^{b_2} R_0^{c_2}$

$\beta = -\frac{\mu_M - \mu_\sigma}{\left(\sigma_M^2 + \sigma_\sigma^2\right)^{\frac{1}{2}}}$; $\sigma_\sigma = 25.91$ (experimentally observed induced stress standard deviation)

$\mu_\sigma = \sigma$, $\mu_M = \sigma_{\max}$, $\sigma_M = 8.6$ MPa and P (design load) $= 40$ MPa

Rewriting the above optimization formulation,

Objective function

Minimize weight (W) $= \left[55.67R_0^2 t - 55.67R_0 t^2 + 18.5\, t^3\right]$

Subject to

$$6.8604 * 10^{-6} t^{-1.0967} R_0^{2.6386} - 1 \leq 0 \tag{11}$$

$$7.4 * 10^{-4} t^{-1.0242} R_0^{1.7701} - 1 \leq 0 \tag{12}$$

$$5.4 * 10^{-5} t^{-1.0967} R_0^{2.6386} - 6.875 \leq 0 \tag{13}$$

Inequality constraints

$3 < t < 8$ and $152.5 < R_0 < 157.5$

Solving the above optimization problem using nonlinear programming technique, the optimized design parameters obtained are given in Table 6.

Table 6 Optimized parameters

Parameter	Design	Optimized
Inner radius 'R_i'	149.5 mm	149.5 mm
Thickness 't'	8.00 mm	5.3453 mm
Outer radius 'R_0'	157.5 mm	154.8453 mm
Weight 'W'	10.5 kg	6.8919 kg
Safety index 'β'	14.39	6.0949

10 Results and Discussion

The structural reliability (safety index) study has been carried out using moment methods; the safety index 'β' is found to be high. This indicates that titanium air bottle is over designed, and these bottles are very safe at design operating pressure. Hence, this gives scope for weight optimization.

The nonlinear regression relation of performance functions established using ANSYS output is found to be useful for prediction of stress, strain and deformation. These relations are simple and help in optimization formulation. Many complex performance functions can be written in simple linear, nonlinear or polynomial forms.

Using the above findings, the weight of the existing air bottle is optimized without compromising quality, design and safety requirements. A net weight reduction of 3.5 kg is possible which in turn helps in increasing payload capacity for aerospace mission. The optimized safety index is now $\beta = 6.09$, that is, the probability of failure of this optimized air bottle is $p_f = 0.522\text{E-}09$.

11 Conclusion

A nonlinear constraint optimization for air bottle using regression model has been developed and found to be useful in their domain validity. This method accounts for effect of design variability while providing a realistic design model where conflicting and multi-objectives, namely, structural weight, operating load, induced stress, strain, deformation and structural reliability (safety index), exist. Suggested nonlinear regressions (nonlinear performance functions) are validated. In addition to design parameters, fatigue and fracture can also be considered in safety and optimization studies.

Acknowledgment The constant encouragement and support extended by Director DRDL and help rendered by Director R&QA, DRDL are gratefully acknowledged.

References

1. Bhattacharjee P (2009) Structural reliability assessment of pressure vessel. J Aerosp Qual Reliab 5:159–163
2. Robinson DG (1998) A survey of probabilistic methods used in reliability, risk and uncertainty analysis analytical techniques −1. Sandia report SAND 98:1189–1998
3. Wong FS (1985) First order second moment methods. Comput Struct 20(4):779–791
4. Melchers RE (1987) Structural reliability analysis and prediction. Ellis Harwood Limited, Chichester, pp 104–141
5. Shu HD, Wang MO (1992) Reliability analysis in engineering applications. Van Nostrand Reinhold, New York, pp 61–132
6. Sorensen JD (2004) Notes in structural reliability theory and risk analysis. Aalborg University, Aalborg
7. Montgomery DC (2004) Design and analysis of experiments. Wiley, New York
8. El-Sayed ME (1999) Structural optimization for reliability using nonlinear goal programming. NAGI - 1837
9. Kyumchoi S, Grandhi RV, Canfield RA (2006) Reliability – based structural design. Springer, London
10. Grandhi RV, Wang L (1999) Structural reliability analysis and optimization: use of approximations. Wright State University, Ohio. NASA/CR – 209154
11. Bhattacharjee P, Ramesh Kumar K, Janardhan Reddy TA (2010) Reliability design evaluation and optimization of a nitrogen gas bottle using response surface method. Int J ReliabQual Safety Eng 17(2):119–132
12. Bhattacharjee P, Janardhan Reddy TA, Ramesh Kumar K (2009) Structural reliability evaluation using response surface method. In: Proceedings of international conference on reliability, maintainability and safety. IEEE ICRMS, pp 972–977

Dynamic Response of Rectangular Bunker Walls Considering Earthquake Force

Indrajit Chowdhury and Jitendra Pratap Singh

Abstract In prevalent design of rectangular reinforced concrete bunkers, seismic effect on the wall is usually ignored. The stored material including the container is usually considered as a rigid body whose mass is lumped to the supporting frame by a rigid link. The seismic force induced therein is considered in the frame while walls of the bunker are designed for static pressure only. In this chapter, a method is proposed to estimate the dynamic pressures induced on the wall due to earthquake force, and when in the process, the dynamic amplification to the static pressure due to vibration of the frame is also induced in it. Finally, this pressure is utilized to determine the modal response of the bunker wall and estimate its coupled (wall + frame) response considering appropriate boundary condition.

Keywords Eigenvalues • Plate element • Galerkin's method • Rankine's formula • Dynamic pressure

1 Introduction

In different industries like power, oil and gas, and steel plants, rectangular bunkers are often deployed to store materials like coal, sulfur pellets, and coke either as an input or an output product. In many cases, it becomes essential that these storage vessels remain operable after a major earthquake, for their failures could lead to severe functional problem for the plant which is expected to remain operational even after such calamities.

The present state of art does not furnish any guidelines as to how to cater to the seismic force on such bunkers (especially the walls) and are left to the structural designer's personal judgment. Common methodology that is adapted is to assume

I. Chowdhury • J.P. Singh (✉)
Petrofac International Limited, Sharjah, UAE
e-mail: Indrajit.Chowdary@petrofac.com; jitendra.singh@petrofac.com

S. Chakraborty and G. Bhattacharya (eds.), *Proceedings of the International Symposium on Engineering under Uncertainty: Safety Assessment and Management (ISEUSAM - 2012)*,
DOI 10.1007/978-81-322-0757-3_36,

the bunker and its content as a lumped mass, and the force induced in it due to seismic vibration is transferred to the frame supporting the container [3, 5]. No procedure exists to assess the dynamic pressure that is induced on the wall due to seismic force and its amplification due to primary vibration of the supporting frame.

This chapter makes an attempt to develop a procedure based on which this dynamic pressure on the wall can be estimated and be catered for – a phenomenon that has been ignored till date

2 Proposed Method

The proposed method consists of two parts – (1) analysis of the bunker frame and (2) finally the dynamic response of the bunker wall – that are elaborated hereafter.

2.1 Analysis of Bunker Frame

The mathematical model as perceived for dynamic analysis of the bunker frame is a shown in Fig. 1.

Considering stiffness matrix $[K_F]$ and lumped mass matrix $[M_F]$ of the frame are given as

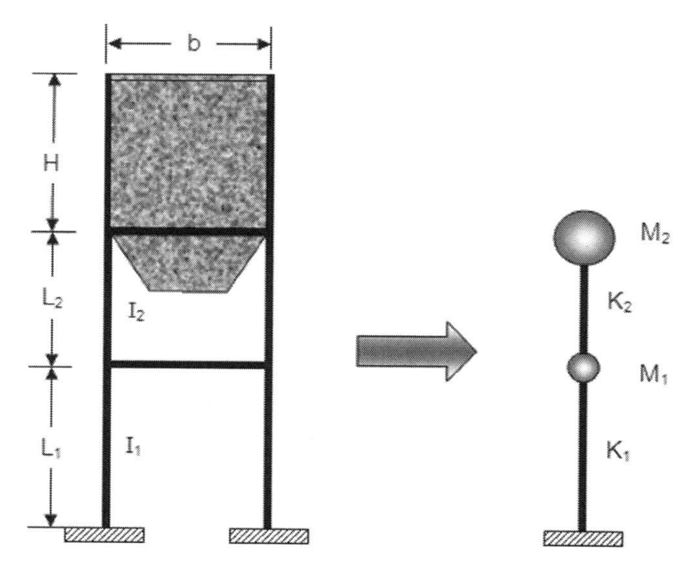

Fig. 1 Mathematical model of frame of rectangular bunker

$$[K_F] = \begin{bmatrix} K_1 + K_2 & -K_2 \\ -K_2 & K_2 \end{bmatrix} \text{ and } [M_F] = \begin{bmatrix} M_1 & 0 \\ 0 & M_2 \end{bmatrix} \quad (1)$$

where $K_1 = n\frac{12EI_1}{L_1^3}$ and $K_2 = n\frac{12EI_2}{L_2^3}$ are flexural stiffness of frame below and above tie level.

n = number of columns in the frame.
M_2 = mass of container and fill material.
M_1 = effective mass of column and beam at tie level.

The eigenvalue of the problem is given by

$$[\lambda] = \begin{bmatrix} K_1 + K_2 - M_1\lambda & -K_2 \\ -K_2 & K_2 - M_2\lambda \end{bmatrix} \quad (2)$$

Let λ_1 and λ_2 be the eigenvalues and let the corresponding eigenvectors be expressed as

$$[\phi] = \begin{bmatrix} \phi_{11} & \phi_{12} \\ \phi_{21} & \phi_{22} \end{bmatrix} \quad (3)$$

Normalized eigenvectors for first and second modes of vibration are given by

$$\begin{bmatrix} \phi_{11}^n \\ \phi_{21}^n \end{bmatrix} = \frac{1}{\sqrt{M_1\phi_{11}^2 + M_2\phi_{21}^2}} \begin{bmatrix} \phi_{11} \\ \phi_{21} \end{bmatrix} \text{ and}$$
$$\begin{bmatrix} \phi_{12}^n \\ \phi_{22}^n \end{bmatrix} = \frac{1}{\sqrt{M_1\phi_{12}^2 + M_2\phi_{22}^2}} \begin{bmatrix} \phi_{12} \\ \phi_{22} \end{bmatrix} \quad (4)$$

Modal participation factors are expressed as

$$\kappa_1 = \frac{M_1\phi_{11}^n + M_2\phi_{21}^n}{M_1\left(\phi_{11}^n\right)^2 + M_2\left(\phi_{21}^n\right)^2} \text{ and } \kappa_2 = \frac{M_1\phi_{12}^n + M_2\phi_{22}^n}{M_1\left(\phi_{12}^n\right)^2 + M_2\left(\phi_{22}^n\right)^2} \quad (5)$$

Let T_1 and T_2 be the time period of the frame corresponding to eigenvalues λ_1 and λ_2. Let $\frac{S_{a1}}{g}$ and $\frac{S_{a2}}{g}$ be the spectral acceleration coefficients from code for T_1 and T_2, respectively.

Design spectral acceleration coefficients at top of frame (node 2 of lumped mass model) for two modes of vibration are expressed as

$$\frac{D_{a1}}{g} = \kappa_1\beta\frac{S_{a1}}{g}\varphi_{21}^n \text{ and } \frac{D_{a2}}{g} = \kappa_2\beta\frac{S_{a1}}{g}\varphi_{22}^n \quad (6)$$

Fig. 2 Variation of pressure on vertical wall of bunker

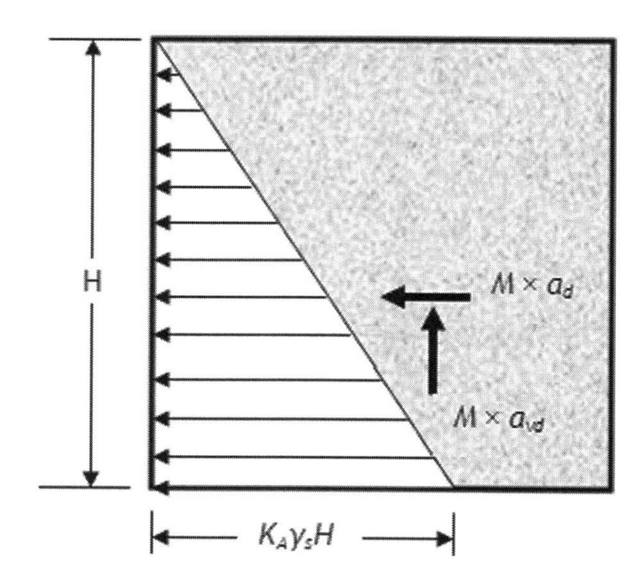

where $\beta = \frac{ZI}{2R}$ is a code factor for seismic zone, importance, and response reduction factor [4].

The above accelerations will work on the container as shown Fig. 2.

During earthquake, pressure at any depth z from top is expressed as

$$p_h = K_A\gamma_s z + \gamma_s z\left(\frac{D_{ai}}{g}\right) + K_A\gamma_s z\left(\frac{2}{3}\frac{D_{ai}}{g}\right) \tag{7}$$

The last expression in Eq. (7) represents the effect of vertical acceleration on horizontal pressure on the bunker wall:

$$\text{or } p_h = K_A\gamma_s z\left[1 + \frac{1}{K_A}\frac{D_{ai}}{g} + \frac{2}{3}\frac{D_{ai}}{g}\right] \tag{8}$$

$$\text{or } p_h = K_A\gamma_s z\Lambda_i \tag{9}$$

$$\text{where } \Lambda_i = 1 + \frac{1}{K_A}\frac{D_{ai}}{g} + \frac{2}{3}\frac{D_{ai}}{g} \tag{10}$$

Here Λ is a dimensionless amplification factor for two modes of vibration ($i = 1, 2$) contributing to enhance the static wall pressure and shall always be ≥ 1.0.

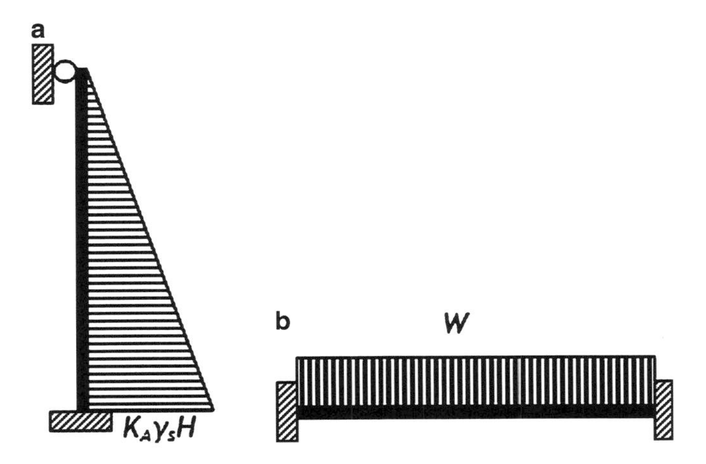

Fig. 3 (**a**) Propped cantilever and (**b**) fixed supported strips

2.2 Analysis of Bunker Wall

For analysis of the bunker wall, the wall is usually considered as three sides fixed and one side (the vertical top edge) as hinged – as in most of the cases, the roof is covered by removable precast slabs or steel checkered plates or gratings. Now considering a strip of the wall in vertical and horizontal direction, the boundary conditions and the loading on these strips are as shown in Fig. 3.

2.2.1 Shape Function of Plate

For propped cantilever strip as shown in Fig. 3a, displacement at depth z from the top is expressed as

$$\delta_z = \frac{K_A \gamma_s H^5 \Lambda_i}{30EI} \left[\frac{1}{4} \left(\frac{z}{H} \right)^5 - \frac{1}{2} \left(\frac{z}{H} \right)^3 + \frac{1}{4} \left(\frac{z}{H} \right) \right] \tag{11}$$

$$\text{or } \delta_z = \delta_{\text{static}} \times f(\xi) \tag{12}$$

where $\xi = z/H$ is a nondimensional term that varies between 0 and 1, and $f(\xi)$ is the generic shape function of propped cantilever beam given as

$$f(\xi) = \frac{\xi^5}{4} - \frac{\xi^3}{2} + \frac{\xi}{4} \tag{13}$$

Similarly, for a strip which is fixed at both ends in horizontal direction, the generic shape function of displacement can be expressed as

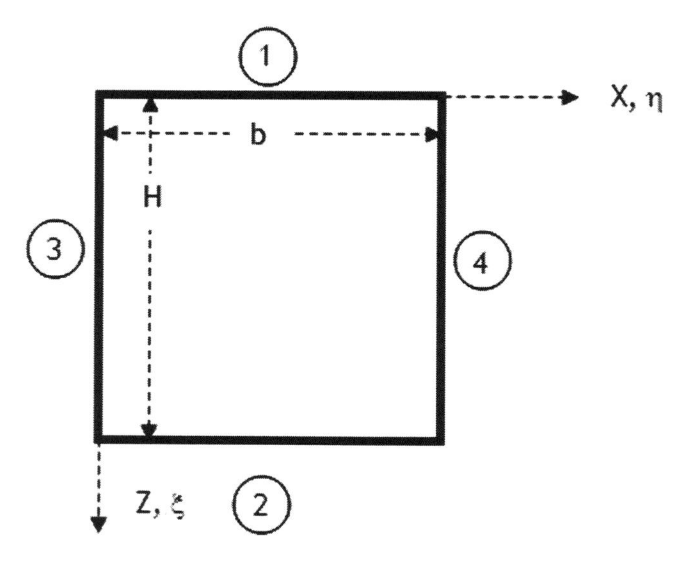

Fig. 4 Coordinate system of Bunker wall

$$f(\eta) = 16\eta^4 - 32\eta^3 + 16\eta^2 \tag{14}$$

where $\eta = x/b$ is a nondimensional term that varies between 0 and 1.

The shape functions derived in Eqs. (13) and (14) will generically satisfy plate equilibrium equation having boundary conditions of three sides fixed and one side propped but will have residual error (R_e) as they are not derived from exact analysis of the fourth order partial differential equation of an isotropic plate expressed as [6]

$$\frac{\partial^4 u}{\partial x^4} + 2\frac{\partial^4 u}{\partial x^2 \partial z^2} + \frac{\partial^4 u}{\partial z^4} = \frac{q}{D} \tag{15}$$

where u is displacement of the plate under a pressure load q. D is flexural stiffness of plate expressed as $D = Et^3/12(1 - v^2)$, and v is the Poisson ratio of bunker wall material (Fig. 4).

In natural coordinate, as expressed in Eqs. (13) and (14), Eq. (15) can be expressed as

$$\frac{1}{b^4}\frac{\partial^4 u}{\partial \eta^4} + \frac{2}{b^2H^2}\frac{\partial^4 u}{\partial \eta^2 \partial \xi^2} + \frac{1}{H^4}\frac{\partial^4 u}{\partial \xi^4} = \frac{q}{D} \tag{16}$$

$$\text{where displacement } u \text{ is expressed as } u = \Delta.f(\xi).f(\eta) \tag{17}$$

Substituting Eq. (17) in (16), we have

$$\frac{\Delta}{b^4}\frac{\partial^4}{\partial\eta^4}f(\xi).f(\eta)+\frac{2\Delta}{b^2H^2}\frac{\partial^4}{\partial\eta^2\partial\xi^2}f(\xi).f(\eta)+\frac{\Delta}{H^4}\frac{\partial^4}{\partial\xi^4}f(\xi).f(\eta)-\frac{q}{D}=0 \quad (18)$$

The shape functions $f(\xi)$ and $f(\eta)$ will generically satisfy Eq. (18) as they conform to boundary condition of the given plate, however will have residual error R_e which may be expressed as

$$R_e=\frac{\Delta}{b^4}\frac{\partial^4}{\partial\eta^4}f(\xi).f(\eta)+\frac{2\Delta}{b^2H^2}\frac{\partial^4}{\partial\eta^2\partial\xi^2}f(\xi).f(\eta)+\frac{\Delta}{H^4}\frac{\partial^4}{\partial\xi^4}f(\xi).f(\eta)-\frac{q}{D} \quad (19)$$

The residual error R_e is now minimized by Galerkin's [1] weighted residual method based on which

$$\int_0^1\int_0^1 [R_e].f(\xi).f(\eta).d\xi.d\eta=0 \quad (20)$$

Equation (20) can thus be expressed as

$$\frac{\Delta}{b^4}\int_0^1\int_0^1 f(\xi)^2.f''''(\eta).f(\eta).d\xi.d\eta+\frac{2\Delta}{b^2H^2}\int_0^1\int_0^1 f''(\xi)f''(\eta)f(\xi)f(\eta)d\xi.d\eta$$
$$+\frac{\Delta}{H^4}\int_0^1\int_0^1 f''''(\xi)f(\xi)f(\eta)^2d\xi d\eta-\frac{1}{D}\int_0^1\int_0^1 qf(\xi)f(\eta)d\xi.d\eta=0 \quad (21)$$

Considering $q=K_A.\gamma_s.z.\Lambda_i$, Eq. (21) can be expressed in a simplified form as

$$\Delta\left[X_1+\frac{2H^2}{b^2}X_2+\frac{H^4}{b^4}X_3\right]=\frac{K_A.\gamma_s.\Lambda_i.H^5}{D}X_4 \quad (22)$$

Considering $r=H/b$, the aspect ratio of the plate, Eq. (22) can be expressed as

$$\Delta=\frac{K_A.\gamma_s.\Lambda_i.H^5}{D(X_1+2r^2X_2+r^4X_3)}X_4 \quad (23)$$

where X_1, X_2, X_3, and X_4 are integral functions that can be solved numerically or explicitly and are as expressed in Table 1.

The maximum static displacement enhanced by the vibration of the frame for two modes of vibration of bunker wall is calculated using Eq. (23) as

$$\Delta_1=\frac{K_A.\gamma_s.\Lambda_1.H^5}{D(X_1+2r^2X_2+r^4X_3)}X_4 \quad (24)$$

and

Table 1 Values of integral functions X_1 to X_4

Integral functions	Values
$X_1 = \int_0^1 \int_0^1 f(\xi)^2 . \left[\frac{\partial^4}{\partial \eta^4} f(\eta)\right] . f(\eta).d\xi.d\eta$	0.473
$X_2 = \int_0^1 \int_0^1 \left[\frac{\partial^2}{\partial \xi^2} f(\xi)\right] \left[\frac{\partial^2}{\partial \eta^2} f(\eta)\right] f(\xi).f(\eta).d\xi.d\eta$	0.124
$X_3 = \int_0^1 \int_0^1 \left[\frac{\partial^4}{\partial \xi^4} f(\xi)\right] . f(\xi).f(\eta)^2.d\xi.d\eta$	0.232
$X_4 = \int_0^1 \int_0^1 \xi.f(\eta).f(\xi).d\xi.d\eta$	0.010

$$\Delta_2 = \frac{K_A.\gamma_s.\Lambda_2.H^5}{D(X_1 + 2r^2X_2 + r^4X_3)} X_4 \tag{25}$$

The displacement profile u of container wall for first two modes of vibration can thus be expressed as

$$u_1(\xi,\eta) = \Delta_1.f(\xi).f(\eta) \text{ and } u_2(\xi,\eta) = \Delta_2.f(\xi).f(\eta) \tag{26}$$

Here $f(\xi)$ and $f(\eta)$ are as expressed in Eqs. (13) and (14), respectively.

The maximum deflection will occur at $\xi = 0.447$ and $\eta = 0.5$ which gives

$$\begin{aligned} u_1(0.447, 0.5) &= \Delta_1.f(0.447).f(0.5) \text{ and} \\ u_2(0.447, 0.5) &= \Delta_2.f(0.447).f(0.5) \end{aligned} \tag{27}$$

Time period of the wall for first two modes of vibration can be expressed as

$$T_{w1} = 2\pi\sqrt{\frac{u_1(0.447, 0.5)}{g}} \text{ and } T_{w2} = 2\pi\sqrt{\frac{u_2(0.447, 0.5)}{g}} \tag{28}$$

Let S_{aw1} and S_{aw2}be the spectral acceleration coefficients from code for T_{w1} and T_{w2}, respectively.

For modal analysis, maximum amplitudes (S_{wd}) in terms of code for first two modes of vibration are expressed as [2]

$$S_{dw1} = \kappa_{w1}.\beta.\frac{S_{aw1}.g}{\omega_1^2} \text{ and } S_{dw2} = \kappa_{w2}.\beta.\frac{S_{aw2}.g}{\omega_2^2} \tag{29}$$

where $\omega_1 = 2\pi/T_{w1}$ and $\omega_2 = 2\pi/T_{w2}$ are natural frequency of the wall, β = code factor given earlier, and κ_{w1} and κ_{w2} = mode participation factors for first two modes of wall given in Eq. (30)

$$\kappa_{w1} = \kappa_{w2} = \frac{\int_0^1 \int_0^1 \xi.f(\xi).f(\eta).d\xi.d\eta}{\int_0^1 \int_0^1 \xi.(f(\xi).f(\eta))^2.d\xi.d\eta} \tag{30}$$

Equation (30) on computation gives a value of $\kappa_{w1} = 24$.

Thus, based on Eq. (29), the combined spectral displacement of the wall can be calculated using SRSS method as

$$S_{dw} = \sqrt{(S_{dw1} + \Delta_1)^2 + (S_{dw2} + \Delta_2)^2} \tag{31}$$

Finally, dynamic SRSS displacement of wall can be expressed as

$$u(\xi, \eta) = S_{dw}.f(\xi).f(\eta) \tag{32}$$

Equation (32) gives a complete dynamic displacement profile of the wall between limits 0 and 1 in both X and Z directions for first two modes.

2.2.2 Dynamic Bending Moment and Shear Force

Bending moment and shear force induced in the wall can be expressed as [6]

$$M_x = -D\left[\frac{\partial^2 u}{\partial x^2} + v\frac{\partial^2 u}{\partial z^2}\right] \tag{33}$$

$$M_z = -D\left[v\frac{\partial^2 u}{\partial x^2} + \frac{\partial^2 u}{\partial z^2}\right] \tag{34}$$

$$Q_x = -D\left[\frac{\partial^3 u}{\partial x^3} + \frac{\partial^3 u}{\partial x.\partial z^2}\right] \tag{35}$$

$$Q_z = -D\left[\frac{\partial^3 u}{\partial x^2 \partial z} + \frac{\partial^3 u}{\partial z^3}\right] \tag{36}$$

Transferring the above Eqs. (33), (34), (35), and (36) in natural coordinate and substituting in Eq. (32), dynamic moments and shears can be calculated using following equations:

$$M_x(\xi,\eta) = -\frac{D.S_{dw}}{H^2}\left[r^2.f(\xi).f''(\eta) + v.f''(\xi).f(\eta)\right] \quad (37)$$

$$M_z(\xi,\eta) = -\frac{D.S_{dw}}{H^2}\left[v.r^2.f(\xi).f''(\eta) + f''(\xi).f(\eta)\right] \quad (38)$$

$$Q_x(\xi,\eta) = -\frac{D.S_{dw}}{H^3}\left[r^3.f(\xi).f'''(\eta) + r.f''(\xi).f(\eta)\right] \quad (39)$$

$$Q_z(\xi,\eta) = -\frac{D.S_{dw}}{H^3}\left[r^2.f'(\xi).f''(\eta) + f'''(\xi).f(\eta)\right] \quad (40)$$

The successive derivatives of $f(\xi)$,$f(\eta)$ are as shown hereafter:

$f(\xi)$	$\frac{\xi^5}{4} - \frac{\xi^3}{2} + \frac{\xi}{4}$	$f(\eta)$	$16\eta^4 - 32\eta^3 + 16\eta^2$
$f'(\xi)$	$\frac{5\xi^4}{4} - \frac{3\xi^2}{2} + \frac{1}{4}$	$f'(\eta)$	$64\eta^3 - 96\eta^2 + 32\eta$
$f''(\xi)$	$5\xi^3 - 3\xi$	$f''(\eta)$	$192\eta^2 - 192\eta + 32$
$f'''(\xi)$	$15\xi^2 - 3$	$f'''(\eta)$	$384\eta - 192$

3 Example, Results, and Discussions

As an example [5] of the application of the proposed method, we take a small square bunker. The bunker is 4.57 m^2 internally in plan and 3.66 m deep from the top of the hopper slopes. The pyramidal hopper is 2.29 m deep internally.

Data:

Size of columns = 305 mm × 305 mm
Number of columns = 4
Grade of concrete of walls, beams, and columns = M15

Fill material:

Angle of internal friction = 35 deg
Weight density of fill material = 7.85 kN/m^3
Thickness of concrete wall = 127 mm

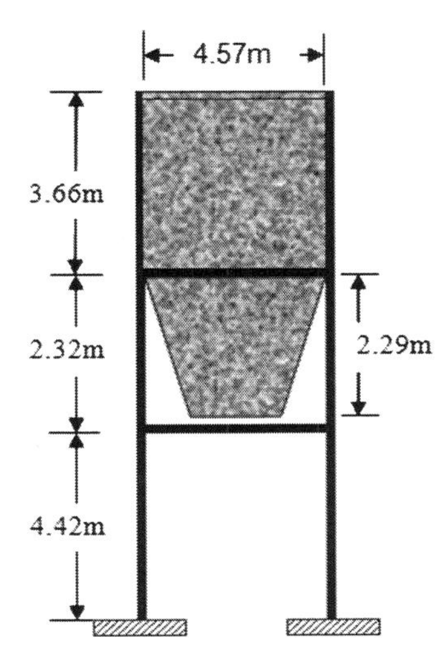

Stiffness of frame:

Stiffness of all columns below tie level $K_1 = 7.76 \times 10^3$ kN/m
Stiffness of all columns above tie level $K_2 = 5.34 \times 10^4$ kN/m

Lumped masses:

$M_1 = 6{,}808$ kg and $M_2 = 114{,}000$ kg

Hence, mass and stiffness matrix of the frame are given as

$$M = \begin{bmatrix} 6.808 \times 10^3 & 0 \\ 0 & 1.14 \times 10^5 \end{bmatrix} \text{kg and } K = \begin{bmatrix} 6.117 \times 10^4 & -5.34 \times 10^4 \\ -5.34 \times 10^4 & 5.34 \times 10^4 \end{bmatrix} \frac{kN}{m}$$

Eigensolution of the above mass and stiffness matrix:

$$T = \begin{Bmatrix} 0.833 \\ 0.065 \end{Bmatrix} \text{s.} \quad \phi = \begin{bmatrix} -0.660 & 0.999 \\ -0.751 & -0.052 \end{bmatrix}$$

Normalized eigenvectors:

$$\phi_N = \begin{bmatrix} -2.544 \times 10^{-3} & 0.012 \\ -2.896 \times 10^{-3} & -6.217 \times 10^{-4} \end{bmatrix}$$

Mode participation factors for two modes of vibrations are given as

$$\kappa_1 = -347.436 \text{ and } \kappa_2 = -249.443$$

Beta factor from IS code:

$$\beta = \frac{ZI}{2R} = 0.06$$

Spectral acceleration coefficients corresponding to two time periods from IS code for 5 % damping ratio:

$$\frac{S_{a1}}{g} = 2.004 \text{ and } \frac{S_{a2}}{g} = 1.972$$

Design acceleration coefficients at the top of frame (node 2) corresponding to two modes of vibration:

$$\frac{D_{a1}}{g} = \kappa_1 \times \beta \times \frac{S_{a1}}{g} \times \phi_{N_{21}} = 0.121 \text{ and } \frac{D_{a2}}{g} = \kappa_2 \times \beta \times \frac{S_{a2}}{g} \times \phi_{N_{22}} = 0.018$$

Amplification factors for two modes of vibration:

$$\Lambda_1 = 1.527 \text{ and } \Lambda_2 = 1.08$$

Amplitude of vibration for first and second modes of vibrations:

$$\Delta_1 = \frac{K_A.\gamma_s.\Lambda_1.H^5}{D(X_1 + 2r^2X_2 + r^4X_3)}X_4 = 5.522 \times 10^{-4} \text{ m}$$

and

$$\Delta_2 = \frac{K_A.\gamma_s.\Lambda_2.H^5}{D(X_1 + 2r^2X_2 + r^4X_3)}X_4 = 3.905 \times 10^{-4} \text{ m}$$

Deflection profile of vertical walls in two modes of vibrations:

$$u_1(\xi,\eta) = 5.522 \times 10^{-4}.f(\xi).f(\eta) \text{ and } u_2(\xi,\eta) = 3.905 \times 10^{-4}.f(\xi).f(\eta)$$

Time periods of walls in two modes of vibration:

$$T_1 = 2.\pi.\sqrt{\frac{u_1(0.447, 0.5)}{g}} = 0.047 \text{ s and } T_2 = 2.\pi.\sqrt{\frac{u_2(0.447, 0.5)}{g}} = 0.040 \text{ s}$$

Spectral acceleration coefficients corresponding to two time periods from IS code for 5% damping ratio:

$$\frac{S_{aw1}}{g} = 1.707 \text{ and } \frac{S_{aw2}}{g} = 1.595$$

Spectral displacements of walls corresponding to two modes of vibration:

$$S_{dw1} = \kappa_1 \times \beta \times \left(\frac{S_{a1}}{g}\right) \times \frac{g}{\omega_1^2} = 1.358 \times 10^{-3} \text{ and}$$
$$S_{dw2} = \kappa_2 \times \beta \times \left(\frac{S_{a2}}{g}\right) \times \frac{g}{\omega_2^2} = 8.968 \times 10^{-4}$$

Hence, spectral displacement profile over the wall is

$$S_{dw}(\xi,\eta) = \sqrt{(S_{dw1} + \Delta_1)^2 + (S_{dw2} + \Delta_2)^2}.f(\xi).f(\eta) = 0.011.f(\xi).f(\eta)$$

Maximum SRSS deflection is $S_{dw}(0.447, 0.5) = 0.794$ mm.
Maximum moments:

$$M_x(0.5,0) = -4.61 \text{ kN - m/m}, M_x(0.5,0.5)$$
$$M_x(0.5,1) = -4.61\text{kN - m/m}.$$

$$M_y(0,0.5) = 0.00\text{kN - m/m}, M_y(0.5,0.5)$$
$$M_y(1,0.5) = -6.39 \text{ kN - m/m}.$$

Maximum shear forces:

$$Q_x(0.5,0) = 6.05 \text{ kN/m}, Q_x(0.5,0.5)$$
$$Q_x(0.5,1) = -6.05 \text{ kN/m}.$$

$$Q_y(0,0.5) = 4.86 \text{ kN/m}, Q_y(0.5,0.5)$$
$$Q_y(1,0.5) = -10.47 \text{ kN/m}.$$

Bending moments and shear force diagram on the wall in horizontal and vertical direction vide Eqs. (37), (38), (39), and (40) are shown below.

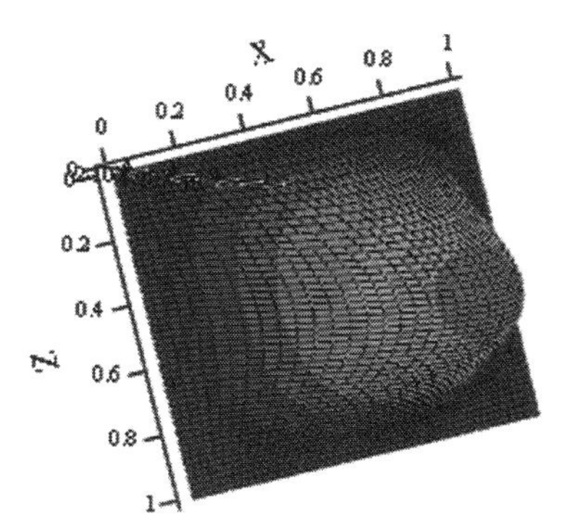

Displacement Profile

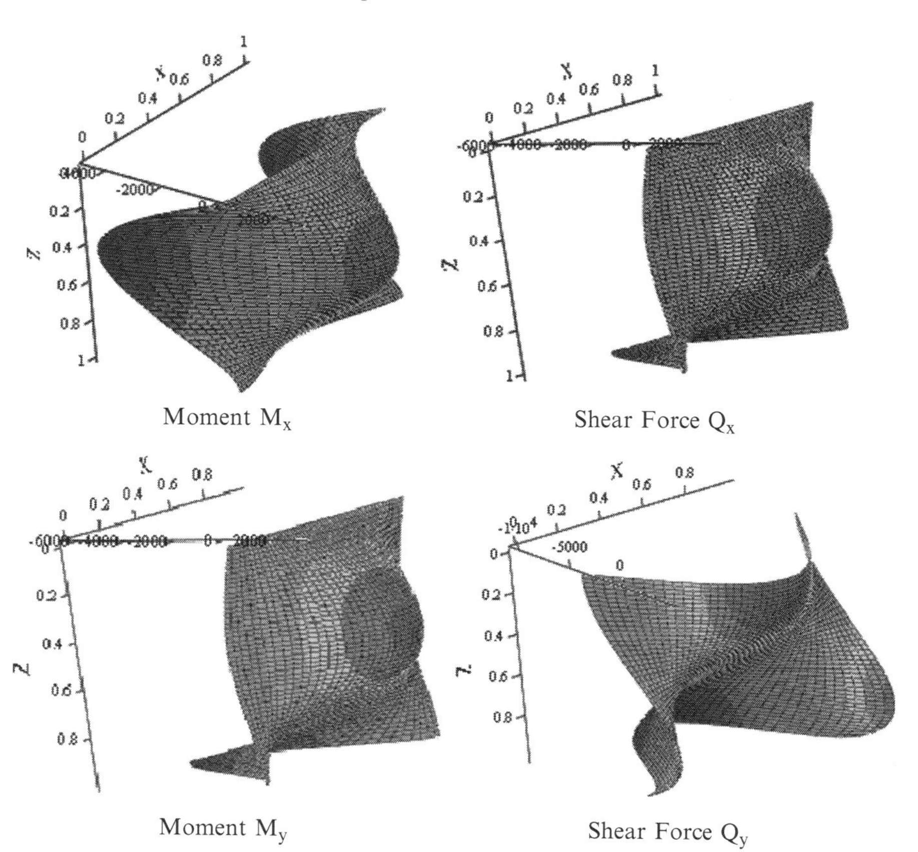

Moment M_x

Shear Force Q_x

Moment M_y

Shear Force Q_y

4 Conclusion

A comprehensive analytical solution based on Galerkin's weighted residual technique is adapted to determine the dynamic response of the bunker walls under seismic force – an important design parameter that is normally ignored in practical design office work till date. Considering the procedure that is analytic in nature does not require a sophisticated software or finite element method to be adapted and can well be carried out by a spreadsheet or using a general utility software like Mathcad with sufficient accuracy.

References

1. Chowdhury I, Dasgupta SP (2008) Dynamics of structures and foundations – a unified approach volume-1. Taylor & Francis Publication, Leiden
2. Clough RW, Penzien J (1985) Structural dynamics. MçGraw-Hill Publication, New York
3. Gray WS, Manning GP (1973) Reinforced concrete water towers bunkers and silos. Concrete Publications limited, London
4. IS 1893 (2002) Indian standard code of practice for earthquake resistant design of structures. Bureau of Indian standards, New Delhi
5. Mallick SK, Rangaswami KS (1980) Design of reinforced concrete structures. Khanna Publishers, New Delhi
6. Timoshenko S, Krieger W (1987) Theory of plates and shells. McGraw-Hill Publication, New York

Probabilistic Risk Analysis in Wind-Integrated Electric Power System Operation

Suman Thapa, Rajesh Karki, and Roy Billinton

Abstract Wind power is regarded as an environment-friendly energy source and the main alternative to the conventional energy resources. Wind power is being rapidly installed in many parts of the world and is considered to be the fastest growing energy source. The uncertain and intermittent nature of wind power, however, creates significant challenges in maintaining the reliability of wind-integrated power systems. The risk in system operation increases as the uncertainty and the amount of wind power connected to the system are increased. While operating a wind-integrated power system, the system operator requires sufficient knowledge of wind power that will be available in the near future to make decisions in committing adequate generation and allocating the reserves in appropriate generating units in order to operate the system economically and within acceptable operating risks. The wind power at a short time in future depends upon the initial condition, and it follows a diurnal pattern. This chapter presents a statistical method using the conditional probability approach to quantify the risks associated with wind power commitment. The impact of rising and falling wind trends in different seasons are considered while evaluating the operating risks.

Keywords Electric power system • Wind power • Reliability • Operating risk • Probabilistic methods

1 Introduction

Wind turbine generators (WTGs) are regarded as one of the most feasible alternatives for reducing the harmful emissions of electricity generation. Wind power is therefore growing rapidly all over the world. Wind power generation

S. Thapa • R. Karki (✉) • R. Billinton
Power System Research Group, University of Saskatchewan, Saskatoon, Canada
e-mail: rajesh.karki@usask.ca

S. Chakraborty and G. Bhattacharya (eds.), *Proceedings of the International Symposium on Engineering under Uncertainty: Safety Assessment and Management (ISEUSAM - 2012)*, DOI 10.1007/978-81-322-0757-3_37,

depends upon the instantaneous wind speed at the wind site which is random and fluctuating. Due to the variable nature of power generation, WTGs are not considered to be dispatchable in the conventional sense. Wind power penetration is defined as the ratio of the total installed capacity of wind power to the total installed capacity of the system. In power systems with small wind penetrations, all the generated wind power can be absorbed by the system as and when available. Wind power penetration is increasing every year and is already contributing appreciable amounts of electricity supply in many electric utilities. The variable and intermittent nature of wind power generation is causing increased reliability concerns during power system operation where wind power penetration is appreciable. A power system operator has the responsibility of continuously satisfying the system demand by committing sufficient generating units. Adequate reserves should also be suitably allocated within the committed units to respond to unit failures and demand fluctuations. The inherent fluctuations in the wind power generation further increases the existing uncertainties during power system operation.

An accurate wind power forecasting is instrumental in determining the appropriate wind power contribution during system operation and to maintain reliability. Wind power forecasting is generally based upon a numeric weather prediction (NWP) [8] and/or statistical model [9]. A method based upon NWP is a very complex process that uses the information from weather stations and satellites and predicts the weather and wind by integrating a large number of equations governing weather. Statistical methods such as autoregressive moving average (ARMA) models can also be used for wind power prediction [4]. A persistence model is a statistical method which assumes that the wind power within the short term in the future will remain the same as at the present time. A persistence model despite being very simple can be very accurate for a short forecasting horizon such as 1–2 h. An ARMA model can be very effective for time horizon up to 4–6 h, while methods based upon the NWP model are more effective for a longer time horizons such as greater than 10 h [10]. A method to develop an ARMA model for a wind site is presented in [2]. An ARMA model is used to simulate the hourly wind speed from the knowledge of the hourly mean and standard deviation of the wind speed at a particular wind site for performing reliability studies of wind-integrated power system from a long-term planning perspective [1, 2, 6]. The ARMA model can also be used to simulate the wind speed data and produce a conditional probability distribution for short-term reliability evaluation in system operation [3]. This chapter utilizes the statistical method based upon the conditional probability approach to quantify the uncertainty of wind power in short future times.

2 Short-Term Wind Risk Model

The hourly wind speed data is simulated using the developed ARMA model and the actual historic mean and standard deviations of the wind speed at any hour using (1).

$$x_t = \mu_t + \sigma_t \times y_t \tag{1}$$

where x_t is the simulated wind speed at hour t, μ_t and σ_t are the mean and standard deviation of the wind speed at time t, respectively, and y_t is the time series value obtained from the ARMA model given by (2).

$$y_t = \emptyset_1 y_{t-1} + \emptyset_2 y_{t-2} + \emptyset_3 y_{t-1} + \cdots\cdots + \emptyset_n y_{t-n} + \alpha_t - \alpha_{t-1}\Theta_1 - \alpha_{t-2}\Theta_2 - \cdots\cdots - \alpha_{t-m}\Theta_m \quad (2)$$

where $\emptyset i (i = 1, \ldots, n)$ and $\Theta j (j = 1, \ldots, m)$ are the autoregressive and the moving average parameters of the model, respectively. $\{\alpha_t\}$ is a normal white noise process with zero mean and a variance of σ^2, i.e. $\alpha_t \in$ NID $(0, \sigma^2)$, where NID denotes normally independently distributed. A wind site located in Toronto, Ontario, Canada, is considered in this study. The ARMA model for the wind site is published in [7] and is presented in Eq. (3).

$$\begin{aligned} y_t =& 0.4709 y_{t-1} + 0.5017 y_{t-2} - 0.0822 y_{t-3} \\ &+ \alpha_t + 0.1876 \alpha_{t-1} - 0.2274 \alpha_{t-2} \\ &\alpha_t \in \mathrm{NID}(0, 0.5508^2) \end{aligned} \quad (3)$$

The knowledge of the initial wind speed is used to produce a conditional wind speed distribution from the hourly simulated data. The wind speed distribution is converted into wind power distribution using the conversion characteristics of a WTG. The speed-power relation of a WTG is nonlinear as given in (4) where P_t is the wind power output in per unit of the rated capacity and V_{ci}, V_r and V_{co} are the cut-in, rated and cut-out wind speeds, respectively. No power is generated when the wind speed is equal to or less than the cut-in speed V_{ci}. A WTG is shut down for safety reasons when the wind speed reaches or crosses the cut-out speed V_{co}. The A, B and C parameters of (4) are presented in [5].

$$\begin{aligned} P_t &= 0 && \text{for } V_{ci} > x_t > V_{co} \\ &= A + Bx_t + Cx_t^2 && \text{for } V_{ci} < x_t < V_r \\ &= 1 && \text{for } V_r < x_t < V_{co} \end{aligned} \quad (4)$$

Wind speed fluctuations generally show a diurnal variation where it starts to rise during the morning until afternoon and falls at night. Figure 1 shows the average hourly wind speed variations on the day of January 1 (day 1) and June 10 (day 161) at the wind site considered. The mean wind speed varies from 19.21 km/h at hour 2 to 26.64 at hour 13, while the hourly standard deviation varies from 8.54 km/h at hour 24 to 12.97 km/h at hour 20 on day 1. The mean wind speed over the day is 22.85 km/h while the mean standard deviation is 11.39 km/h. The hour 8 shows a rising wind trend, while the hour 20 shows a falling wind trend in the next few hours ahead. Day 161 represents a summer day where the mean wind speed varies between 9.18 and 21.89 km/h. The average

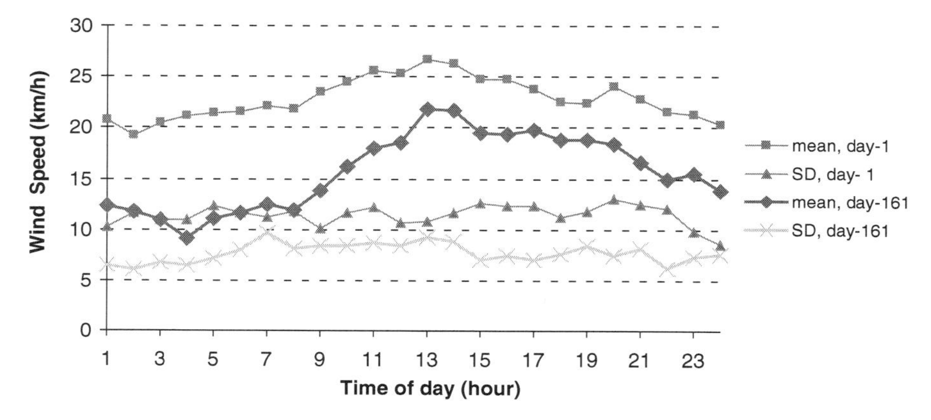

Fig. 1 Average hourly wind speed variations on day 1 and day 161 at Toronto

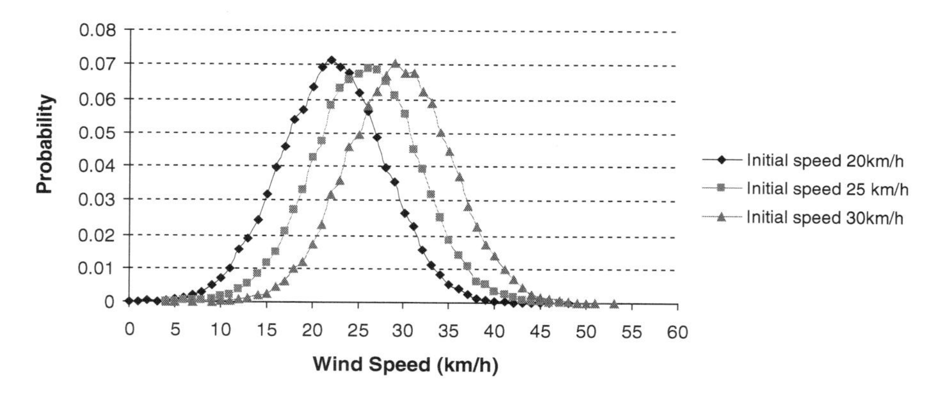

Fig. 2 Wind speed probability distributions at hour 9 conditional to wind speed at hour 8

wind speed over the day is 15.69 km/h and the standard deviation is 7.73 km/h, both of which are relatively lower compared to day 1. The diurnal wind speed variation on day 1 and day 161 is similar.

The ARMA model given in (3) is used to simulate the wind speed data for 800,000 replications to produce the conditional wind speed distributions of the wind speed in the next hour or the next few hours based upon the knowledge of initial wind speed. Figure 2 shows the wind speed probability distributions at hour 9 for three different initial wind speeds of 20, 25 and 30 km/h at hour 8 and while Fig. 3 shows the wind speed probability distribution at hour 21 for the same initial condition at hour 20. It can be seen from Figs. 2 and 3 that the probability distributions are similar to each other and they slide towards the direction of higher or lower wind speed as the initial wind speed is increased or decreased, respectively.

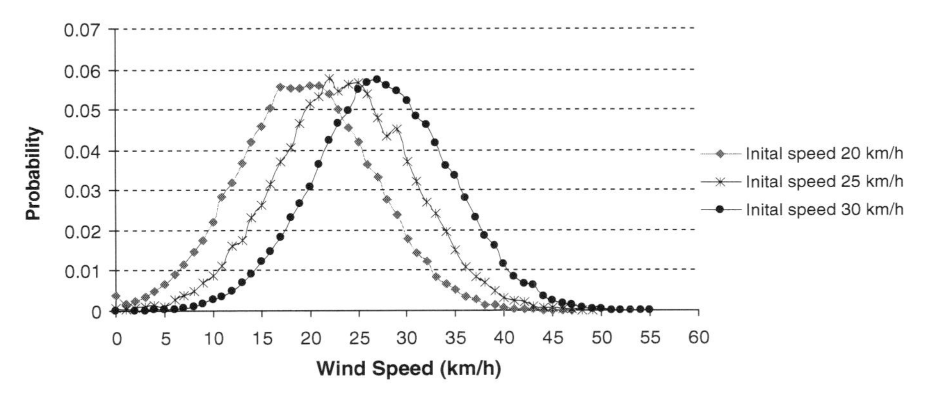

Fig. 3 Wind speed probability distributions at hour 21 conditional on the wind speed at hour 20

Table 1 Statistics of wind speed in the next hour

	Hour 9		Hour 21	
Initial wind speed at hour 8/20 (km/h)	Mean wind speed (km/h)	Standard deviation (km/h)	Mean wind speed (km/h)	Standard deviation (km/h)
20	22.18	5.70	19.51	6.99
25	25.85	5.72	23.51	7.02
30	29.51	5.74	27.53	7.04

Table 1 gives the statistics of the wind speed distributions presented in Figs. 2 and 3. The wind speed expectation in the next hour increases with increase in the initial wind speed which is an indication that the wind speed in the next hour depends upon the initial condition. The distribution moves towards a higher or lower wind speed when the initial wind speed increases or decreases, respectively. It is also evident that the mean wind speed at hour 9 is almost equal to the initial wind speed at hour 8, while the mean wind speed at hour 21 is less than the initial wind speed at hour 20. The diurnal rising and falling wind trend is therefore an important factor in the wind speed or wind power prediction.

3 Evaluation of Wind Power Commitment Risk (WPCR)

Methods based upon persistence models are generally used by the utilities to commit wind power in short future times. Figure 4 presents the wind speed distribution in hour 9 when the initial wind speed in hour 8 is 25 km/h. Figure 4 also shows a wind power curve for a typical WTG having cut-in, rated and cut-out speeds of 15, 50 and 90 km/h, respectively. The wind power cut-off is not shown in the chart as the maximum value on the abscissa is less than the cut-off speed.

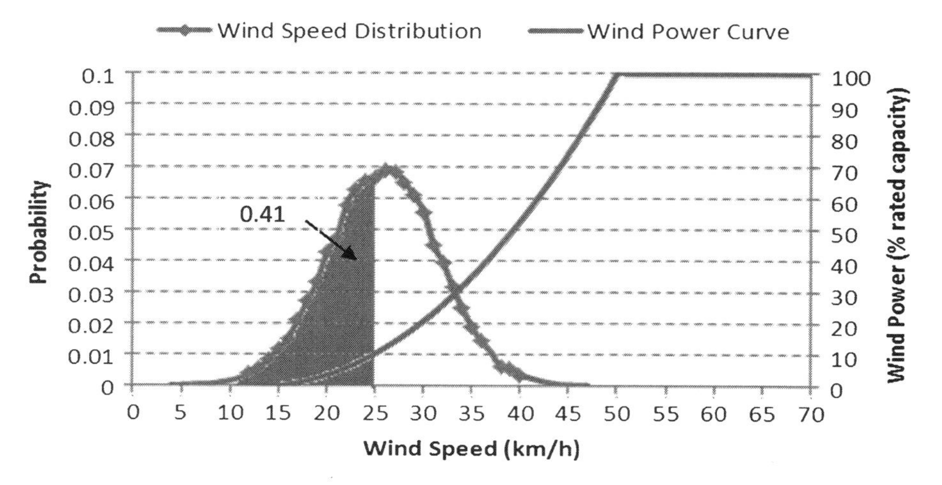

Fig. 4 Wind power commitment risk (WPCR) evaluation

The initial power is 10% of the rated capacity, and this value of wind power is committed in the considered future time according to the persistence model. The shaded area in the distribution gives the probability that the wind power at hour 9 will be less than the committed value. The probability that the actual wind power will be less than the committed value is termed the wind power commitment risk (WPCR) in this chapter and is obtained by cumulating the probabilities of the wind speeds less than the committed value. The WPCR in this case is 0.41.

3.1 Impact of Rising and Falling Wind Trend

Figure 5 presents the cumulative wind speed probability distribution at hours 9, 10 and 11 when the wind speed at hour 8 is 30 km/h. The corresponding wind power output is 20% of the rated capacity. The left ordinate on Fig. 5 gives the WPCR of committing wind power corresponding to the value given by a wind speed in the abscissa. If the wind power commitment is made on the basis of a pure persistence model, the WPCR at hour 9 will be 0.52 as shown by the 100% vertical line in Fig. 5. It can be further observed that the WPCR drops to 0.47 and 0.43 at hour 10 and hour 11, respectively, due to the rising wind trend at these hours. It may be desirable to lower the WPCR by reducing the committed value of wind power.

Table 2 shows the WPCR values by committing 100, 80 and 50% of the wind power available at hour 8 for the lead time considered. The WPCR values at the 100 and 80% commitment level decrease as the lead time is increased. The distributions crossover each other at the wind speed of 26 km/h such that the WPCR of committing wind power below the crossover show an opposite behaviour to that of the ones above it. Table 2 shows that the WPCR of 50% commitment rise as the lead time is increased.

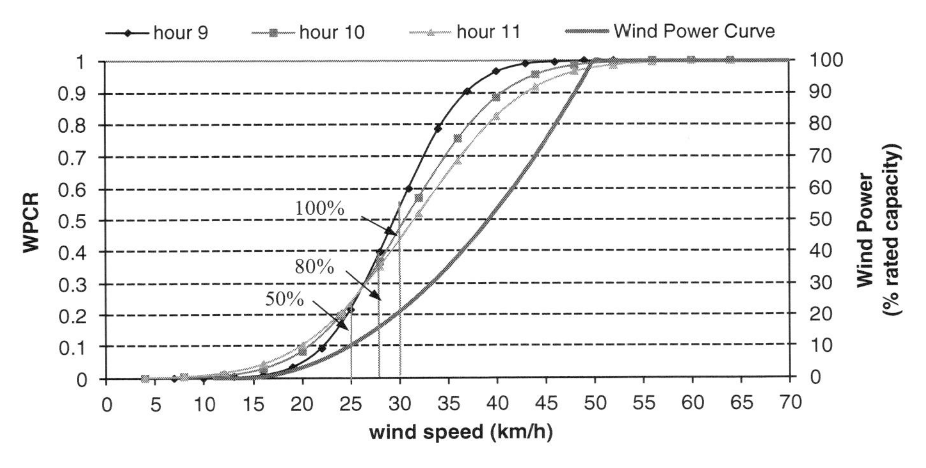

Fig. 5 WPCR analysis for a rising wind trend (initial wind speed = 30 km/h)

Table 2 WPCR for a rising wind trend (initial wind speed = 30 km/h at hour 8)

Hours	WPCR at 100% commitment	80% commitment	50% commitment
9	0.52	0.36	0.19
10	0.47	0.34	0.21
11	0.43	0.33	0.22

Figure 6 shows the cumulative wind speed probability distributions at hours 21, 22 and 23 when the initial wind speed at hour 20 is 30 km/h as in Fig. 5. The distributions in Fig. 6 shift to the left from hour 21 to 23, whereas the distributions in Fig. 5 shift to the right. This indicates that the WPCR of committing a certain value of wind power increases as the lead time is increased. The uncertainty increases as the lead time increases and is further augmented when the wind site is experiencing a falling wind trend. The WPCR values for the three commitments of 100, 80 and 50% of the initial power are shown in Table 3.

The impact of rising and falling wind trends can be observed from the WPCR values given in Tables 2 and 3. The WPCR after any lead time in the future will be higher in situations where the wind site is experiencing a falling wind trend than in the situations when the wind site is experiencing a rising wind trend. A higher value of wind power can be therefore committed during a diurnal rising trend as compared that to during a falling trend.

3.2 *Impact of Seasonality*

Figure 7 presents the WPCR analysis for a lead time of two hours on a winter day represented by day 1 and a summer day represented by day 161. The initial time is

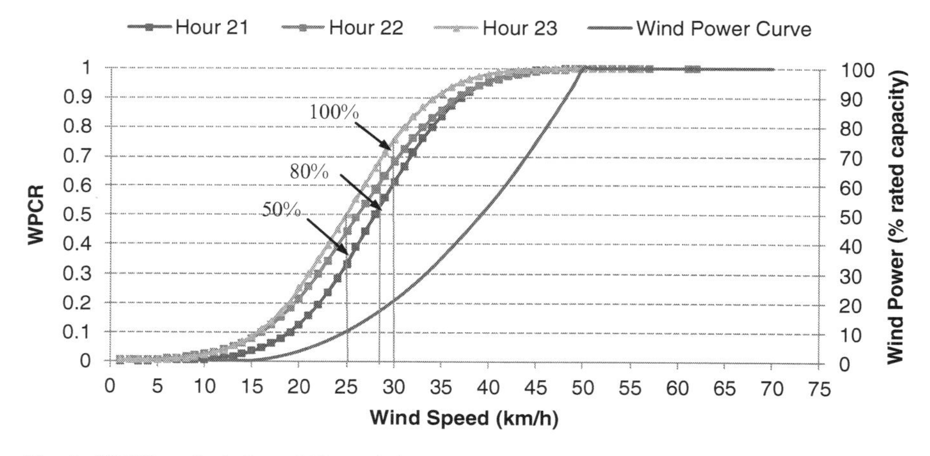

Fig. 6 WPCR analysis for a falling wind trend (initial wind speed = 30 km/h)

Table 3 WPCR for a falling wind trend (initial wind speed = 30 km/h at hour 20)

	WPCR at		
Hours	100% commitment	80% commitment	50% commitment
21	0.61	0.50	0.33
22	0.68	0.59	0.44
23	0.76	0.66	0.50

hour 8, and the initial wind speeds considered are 20 and 25 km/h. The distributions for the two days cross each other at 22 km/h (WPCR = 0.42) and 27 km/h (WPCR = 0.49) for the initial wind speed of 20 and 25 km/h, respectively. It can be seen that day 1 lies to the left of day 161 for wind speeds equal to or less than the initial value. This indicates that the WPCR of committing 100% or less of the initial power will be higher on the winter day than on the summer day. This is mainly because the summer day has lower variability as compared to the winter day, which can be seen from the hourly standard deviations on Fig. 1.

Figure 8 similarly presents the wind speed cumulative probability distribution for hour 22 conditional on the wind speeds at hour 20. The crossovers between the two respective distributions take place at 14 km/h (WPCR = 0.25) and 16 km/h (WPCR = 0.2) for initial wind speeds of 20 and 25 km/h, respectively. The WPCR values for the committed wind power below the crossovers are again lower on day 161 than on day 1.

Table 4 gives the WPCR values when committing 100, 80 and 50% of the initial power on the two days (day 1 and day 161) for the different lead times when the wind site is experiencing a rising wind trend. It can be seen from the table that the

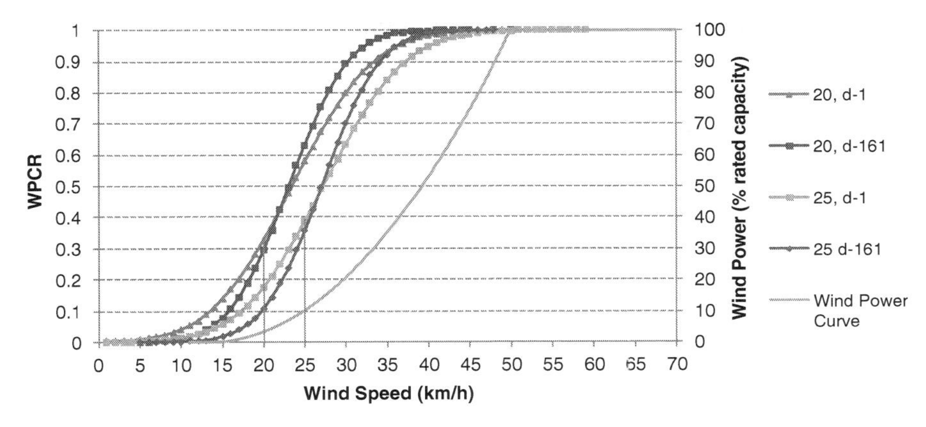

Fig. 7 WPCR analysis for hour 10 on day 1 and day 161

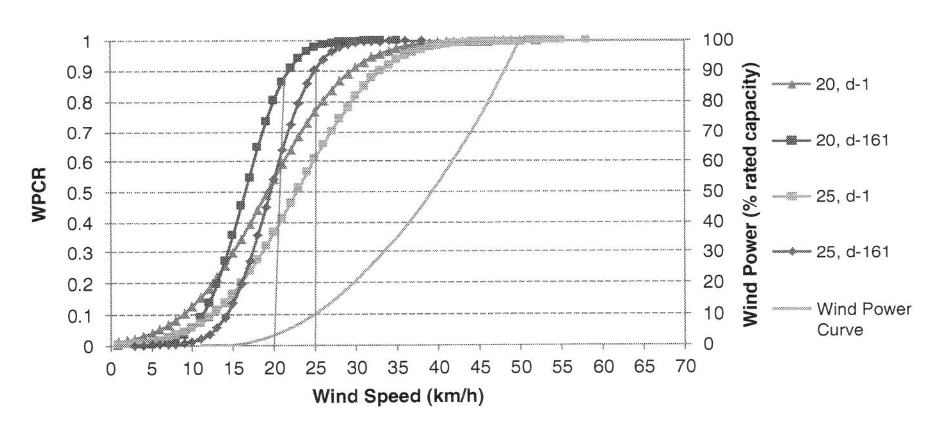

Fig. 8 WPCR analysis for hour 22 on day 1 and day 161

Table 4 WPCR for a rising wind trend (initial wind speed = 25 km/h at hour 8)

	WPCR for day 1 at			WPCR for day 161 at		
Hours	100% commitment	80% commitment	50% commitment	100% commitment	80% commitment	50% commitment
9	0.41	0.34	0.22	0.45	0.36	0.22
10	0.39	0.34	0.25	0.36	0.29	0.19
11	0.36	0.32	0.25	0.31	0.26	0.17

WPCR in hours 10 and 11 are lower in day 161 than in day 1. Table 5 shows the WPCR values on the two days (day 1 and day 161) for the different lead times when the wind site is experiencing a falling wind trend. The WPCR variations in the falling wind trend however are opposite, and the WPCR are greater in day 161 than in day 1. The WPCR during the falling wind trend are relatively high, and it may be desirable to lower the commitment to less than 50% in order to reduce the WPCR.

Table 5 WPCR for a falling wind trend (initial wind speed = 25 km/h at hour 20)

Hours	WPCR for day 1 at			WPCR for day 161 at		
	100% commitment	80% commitment	50% commitment	100% commitment	80% commitment	50% commitment
21	0.56	0.50	0.39	0.67	0.59	0.41
22	0.61	0.56	0.46	0.90	0.86	0.72
23	0.64	0.59	0.48	0.79	0.74	0.61

4 Conclusion

Short-term wind power commitments are dependent on the initial wind power and the lead time considered. A method based upon a conditional probability approach has been employed to quantify the short-term wind speed variations and evaluate the WPCR. Diurnal variations are important factors to be considered when estimating short-term wind power. The variability increases with the increase in lead time. It has been found that rising or falling wind trends can respectively offset or intensify the increase in variability. System operators may therefore need to adjust their commitment from wind farms based upon acceptable WPCR values. The seasonal impact of diurnal variations has been presented using two particular days to represent winter and summer conditions. The results show that the appropriate wind power commitment is highly dependent on the risk criterion deemed acceptable to the system.

References

1. Billinton R, Chowdhury AA (1992) Incorporation of wind energy conversion systems in conventional generating capacity adequacy assessment. IEE Proc Gener Transm Distrib 139 (1):47–56
2. Billinton R, Chen H, Ghajar R (1996) Time-series models for reliability evaluation of power systems including wind energy. Microelectron Reliab 36(9):1253–1261
3. Billinton R, Karki B, Karki R, Ramakrishna G (2009) Unit commitment risk analysis of wind integrated power systems. IEEE Trans Power Syst 24(2):930–939
4. Boone A (2005) Simulation of short-term wind speed forecast errors using a multi-variate ARMA(1,1) Time-series model. Msc thesis, Royal Institute of Technology, Sweden
5. Giorsetto P, Utsurogi KF (1983) Development of a new procedure for reliability modeling of wind turbine generators. IEEE Trans Power Appar Syst PAS-102(1):134–143
6. Karki R, Hu Po (2005) Wind power simulation model for reliability evaluation. In: Proceedings of the Canadian conference on electrical and computer engineering Saskatoon, Saskatchewan, Canada 2005, pp 541–544
7. Karki R, Hu P, Billinton R (2006) A simplified wind power generation model for reliability evaluation. IEEE Trans Energy Convers 21(2):533–540
8. Lange M (2006) Physical approach to short-term wind power prediction. Springer, Berlin
9. Milligan M, Schwartz M, Wan Y (2003) Statistical wind power forecasting models: results for U.S. wind farms. National Renewable Energy Laboratory, Washington, DC
10. Nielsen TS, Joensen A, Madsen H, Landberg L, Giebel G (1998) A new reference for wind power forecasting. Wind Energy 1:29–34. doi: 10.1002/(SICI)1099-1824(199809)1:1<29:: AID-WE10>3.0.CO;2-B

A Frequency Domain Study on the Seismic Response Mitigation of Elevated Water Tanks by Multiple Tuned Liquid Dampers

Soumi Bhattacharyya and Aparna (Dey) Ghosh

Abstract In this paper, an investigation has been carried out on the passive vibration control of elevated water tank structures, subjected to earthquakes, by multiple tuned liquid dampers (MTLDs). An R.C.C. elevated water tank with shaft-type support has been considered. To account for the fluid-structure interaction, the sloshing of water in the water tank container has been modelled by the fundamental convective mode. The remaining water mass has been lumped with that of the container and supporting structure, resulting in a 2-DOF system model for the elevated water tank. The transfer function for the 2-DOF water tank model with MTLDs attached in a parallel configuration has been formulated. The input excitation has been characterized by a white noise power spectral density function (PSDF). The performance of the damper system has been examined on the basis of reduction in the root mean square (rms) value of the structural displacement. The frequencies of the TLDs have been fixed on the basis of tuning to the frequencies corresponding to the peaks of the transfer function curve of the 2-DOF system. This has been studied for three cases, namely, full, half-full and empty conditions of the water tank. The performance of the MTLD system for varying water level in the tank has been examined and has been found to be better in comparison with that for the single TLD case. The geometric parameters of the MTLDs have also been obtained and have been found to be feasible.

Keywords Elevated water tank • Passive damper • Liquid sloshing damper • Multiple tuned liquid damper

S. Bhattacharyya (✉) • A.D. Ghosh
Department of Civil Engineering, Bengal Engineering and Science University, Shibpur, Howrah, India
e-mail: soumibhttchr86@gmail.com; aparrnadeyghosh@gmail.com

S. Chakraborty and G. Bhattacharya (eds.), *Proceedings of the International Symposium on Engineering under Uncertainty: Safety Assessment and Management (ISEUSAM - 2012)*, DOI 10.1007/978-81-322-0757-3_38,

1 Introduction

TLD is a passive energy dissipating system or passive control device, where damping is achieved by the physical properties of the system and no external forces are needed. In TLD, the mass of the damper is provided by liquid (usually water) in a container. Vibration mitigation of a structure is achieved due to the transference of the structural vibrational energy to this liquid when the natural frequency of the liquid motion is nearly equal (i.e. tuned) to the structural frequency. These passive systems impart direct damping to the structure by modifying its frequency response [6]. In 1957 Housner [4] presented an analysis of the hydrodynamic pressures generated in a fluid container when it is subjected to horizontal acceleration. Both the impulsive and convective liquid pressures were taken into consideration. After that, the TLDs have been used in marine structures. The concept of applying TLDs for reduction of vibrations in civil engineering structures began in the mid-1980s. Fujino et al. [2] have conducted an experimental study on the TLD. Since their first applications to ground structures in the 1980s [12, 17], TLDs have become a popular form of inertial damping device [1, 3, 7–9, 14–16]. Currently, both deep and shallow water configurations of TLDs are in application worldwide. The shallow water configurations dissipate energy through viscous action and wave breaking. On the other hand, deep water TLDs require baffles or screens to increase the energy dissipation of the sloshing fluid. Primarily, in case of TLDs, the circular container is suitable for shallow configurations and the rectangular ones for deep water TLDs.

The major limitation of the single damper is that its performance is not robust. In the perfectly tuned condition, the single damper can perform well, but in slightly detuned condition, its efficiency gets reduced. To overcome this situation, instead of using a single damper, *multiple dampers* can be used. In multiple dampers, a range of frequency is selected. The central damper is then tuned to the structural frequency, and other dampers are tuned within the frequency range [10]. Hence, if there is an error in calculating the structural frequency, there is a chance that at least one damper will be tuned optimally in the detuned condition. Hence, in case of detuning, multiple dampers will perform better than the single damper. Some important MTLD applications are Hobart Tower in Tasmania, Atsugi TYG Building, Narita Airport Tower, Yokohama Marine Tower [19], Gold Tower in Kagawa, Shin Yokohama Prince Hotel (SYP) in Yokohama, Nagasaki Airport Tower, Tokyo International Airport Tower at Haneda, Shanghai World Financial Center, etc.

The frequency of an elevated water tank structure will change as the water level in the tank fluctuates. Since the control device under study is a passive system, it is not possible to change the frequency of the TLD once it has been installed. In the present study, the focus is on the application of MTLD (shallow and right circular) for the seismic vibration control of elevated water tanks with shaft-type supports. The structure has been modelled as 2-DOF system in which the sloshing action of the liquid inside the water tank container has been taken into account. The formulation for the transfer function relating the base acceleration and displacement of the MTLD-2-DOF structural system has been presented. A study has been carried out on bimodal control of the structure by MTLD system in frequency domain, and the results have been compared with those obtained for the single TLD cases. The geometric feasibility of the TLD systems has also been investigated.

2 Frequency Domain Study

2.1 *Modelling of Tank with n-TLDs*

The model of an n-TLD-2-DOF structural system has been shown in Fig. 1. The two-mass model suggested by Housner [5] for fixed-base elevated tanks has been considered here [11]. In this model, only the first convective modes of tank water and TLDs have been considered. The convective mass of tank water (m_{cs}) has been considered as an SDOF system. Secondly, the mass (m_{es}) which consists of the remaining water mass in the tank and the mass derived by the weight of container including two-thirds of the supporting structure weight (recommended in ACI 371R) has been considered as another SDOF system. The stiffness and damping of the SDOF system, represented by the mass m_{es}, are k_{es} and c_{es} respectively. The corresponding damping ratio and the natural frequency are denoted by ξ_{es} and ω_{es} respectively. The water mass associated with the fundamental convective or

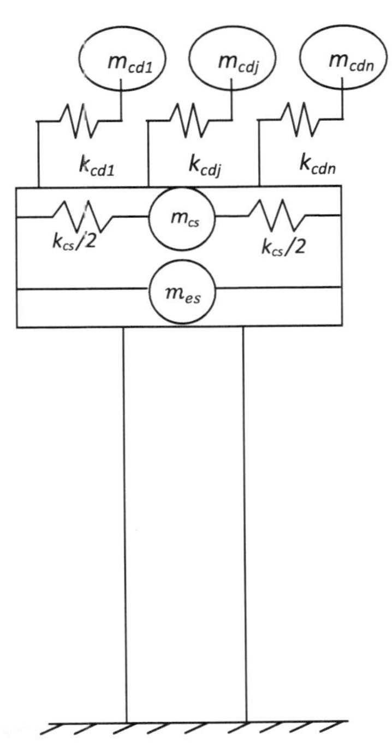

Fig. 1 Mechanical model of n-TLD-2-DOF structural system

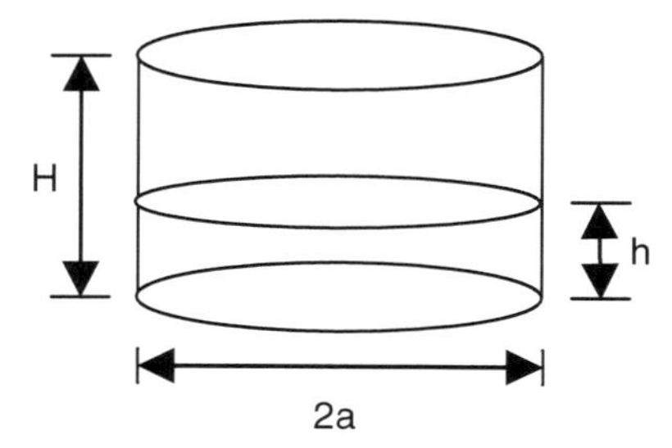

Fig. 2 Definition sketch for circular cylindrical tuned liquid damper (TLD)

sloshing mode of the tank water, m_{cs}, is assumed to be attached to the tank by a linear spring with stiffness k_{cs} and a linear viscous damper with damping coefficient c_{cs}. The damping ratio and the natural frequency of this mode are denoted by ξ_{cs} and ω_{cs} respectively.

n number of TLDs has been used in the study. A right circular cylindrical rigid TLD tank of radius a, shown in Fig. 2, is filled with an incompressible, inviscid liquid of mass density ρ_l up to a height h and has a free upper surface. The convective masses of each (j-th) TLD ($m_{cd}, j = 1$ to n) have been considered as an SDOF system. The liquid mass associated with this convective mode, m_{cdj}, is assumed to be attached to the tank by linear spring with stiffness k_{cdj} and linear viscous damper with damping coefficient c_{cdj}. The damping ratio and the natural frequency of this mode are denoted by ξ_{cdj} and ω_{cdj} respectively. The expressions for m_{cs}, m_{cdj}, ω_{cs} and ω_{cdj} are given by Veletsos and Tang [18].

The remaining water mass of the TLD container is assumed to be lumped with the mass $m_{es,}$ and the total mass is represented by m_{es*} in the frequency domain formulation. Although, in the numerical study, it has been assumed that $m_{es}* = m_{es}$ as the impulsive mass of the TLD liquid is small and the TLD container weight has been neglected. The stiffness and damping of the mass m_{es*} are thus k_{es} and c_{es}, respectively. As per Veletsos and Tang [18], ξ_{cs} and ξ_{cdj} have been taken as 0.01.

2.2 Formulation of Transfer Function of Structure with n-Identical TLDs

The mathematical model of n-TLD-2-DOF structural system has been shown in Fig. 3. Let us consider the n-TLD-2-DOF system subjected to a horizontal base acceleration, $\ddot{z}(t)$. Let $u(t)$ denote the horizontal displacement of m_{es*} relative to ground motion. $x(t)$ denotes the horizontal displacement of the convective mode of the tank liquid relative to the structure. Further, $y_j(t)$ represents the displacement of the convective liquid mass of the j-th TLD relative to the structure, where $j = 1$ to n.

The normalized equation of motion of the sloshing mass of the tank water (m_{cs}) yields the following:

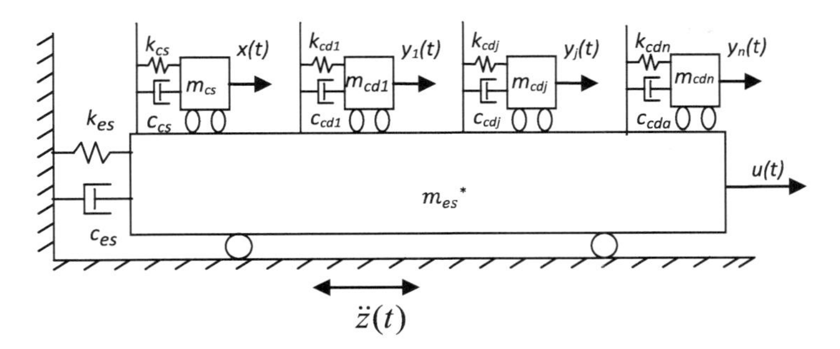

Fig. 3 Mathematical model of n-TLD-2-DOF structural system

$$\ddot{x}(t) + \ddot{u}(t) + 2\xi_{\mathrm{cs}}\omega_{\mathrm{cs}}\dot{x}(t) + \omega_{\mathrm{cs}}{}^2 x(t) = -\ddot{z}(t) \tag{1}$$

where $\frac{c_{\mathrm{cs}}}{m_{\mathrm{cs}}} = 2\xi_{\mathrm{cs}}\omega_{\mathrm{cs}}$ and $\frac{k_{\mathrm{cs}}}{m_{\mathrm{cs}}} = \omega_{\mathrm{cs}}{}^2$.

The normalized equation of motion of sloshing masses of dampers yields the following:

$$\ddot{y}_j(t) + \ddot{u}(t) + 2\xi_{\mathrm{cd}j}\omega_{\mathrm{cd}j}\dot{y}_j(t) + \omega_{\mathrm{cd}j}{}^2 y_j(t) = -\ddot{z}(t) \qquad j = 1, 2, 3 \ldots .n \tag{2}$$

where $\frac{c_{\mathrm{cd}j}}{m_{\mathrm{cd}j}} = 2\xi_{\mathrm{cd}j}\omega_{\mathrm{cd}j}$ and $\frac{k_{\mathrm{cd}j}}{m_{\mathrm{cd}j}} = \omega_{\mathrm{cd}j}{}^2$.

The normalized equation of motion of the n-TLD-2-DOF structural system yields

$$\begin{aligned} &\ddot{u}(t) + 2\xi_{\mathrm{es}}\omega_{\mathrm{es}}\dot{u}(t) + \omega_{\mathrm{es}}{}^2 u(t) - 2\xi_{\mathrm{cs}}\omega_{\mathrm{cs}}\mu_{\mathrm{cses}}\dot{x}(t) - \omega_{\mathrm{cs}}{}^2\mu_{\mathrm{cses}}x(t) \\ &\quad - \sum_{j=1}^{n}\left\{2\xi_{\mathrm{cd}j}\omega_{\mathrm{cd}j}\mu_{\mathrm{cd}j\mathrm{es}}\dot{y}_j(t) + \omega_{\mathrm{cd}j}{}^2\mu_{\mathrm{cd}j\mathrm{es}}y_j(t)\right\} = -\ddot{z}(t) \end{aligned} \tag{3}$$

where $\frac{c_{\mathrm{es}}}{m_{\mathrm{es}*}} = 2\xi_{\mathrm{es}}\omega_{\mathrm{es}}$, $\frac{k_{\mathrm{es}}}{m_{\mathrm{es}*}} = \omega_{\mathrm{es}}{}^2$, $\frac{m_{\mathrm{cs}}}{m_{\mathrm{es}*}} = \mu_{\mathrm{cses}}$ and $\frac{m_{\mathrm{cd}j}}{m_{\mathrm{es}*}} = \mu_{\mathrm{cdjes}}$.

The transfer function relating the displacement of an SDOF system representing mass m_{es} to the base acceleration is given by

$$H_{\mathrm{es}}(\omega) = -\frac{1}{[-\omega^2 + \omega_{\mathrm{es}}{}^2 + 2\xi_{\mathrm{es}}\omega_{\mathrm{es}}i\omega]} \tag{4}$$

The transfer function relating the displacement of an SDOF system representing the mass m_{cs} to the input acceleration is given by

$$H_{\mathrm{cs}}(\omega) = -\frac{1}{[-\omega^2 + \omega_{\mathrm{cs}}{}^2 + 2\xi_{\mathrm{cs}}\omega_{\mathrm{cs}}i\omega]} \tag{5}$$

The transfer function relating the displacement of an SDOF system representing the mass m_{cdj} to the input acceleration is given by

$$H_{cdj}(\omega) = -\frac{1}{\left[-\omega^2 + \omega_{cdj}{}^2 + 2\xi_{cdj}\omega_{cdj}i\omega\right]} \qquad j = 1, 2, 3 \ldots .n \tag{6}$$

On Fourier transforming Eqs. (1), (2), and (3) and by proper substituting, the input-output relation between the base acceleration and the displacement response of the structure equipped with MTLD is obtained:

$$U(\omega) = H_u(\omega)\ddot{Z}(\omega) \tag{7}$$

where

$$H_u(\omega) = \frac{H_{es}(\omega)[\mu_{cses}(\omega^2 H_{cs}(\omega) - 1) + \omega^2 A - B - 1]}{[\{\omega^2 H_{es}(\omega)(\mu_{cses}(\omega^2 H_{cs}(\omega) - 1)) + \omega^2 A - B\} - 1]} \tag{8}$$

is the transfer function relating the displacement of the structure to the input base acceleration.

$$A = \sum_{j=1}^{n} H_{cdj}(\omega)\left(\mu_{cdjes}\right) \tag{9}$$

$$B = \sum_{j=1}^{n} \left(\mu_{cdjes}\right) \tag{10}$$

$U(\omega), X(\omega), Y_j(\omega), \ddot{Z}(\omega)$ are the Fourier transforms of the corresponding time-dependent variables $u(t)$, $x(t)$, $y_j(t)$ and $\ddot{z}(t)$, respectively. If the ground acceleration is characterized by a white noise PSDF of intensity S_0, then the PSDF of the displacement response of the structure, denoted by $S_u(\omega)$, is expressed by Newland (1993)

$$S_u(\omega) = |H_u(\omega)|^2 S_0 \tag{11}$$

The rms value of the displacement response of the structure, σ_u, can be numerically evaluated by computing the square root of the area under the corresponding PSDF curve as given in Eq. (11).

3 Numerical Study

An R.C.C. tank structure with flat top cylindrical container and shaft support has been taken as the example structure to examine the performance of TLD for the passive control of elevated water tank. The relevant data that has been assumed to define the structure are height of shaft support = 30 m, mean diameter of shaft = 3 m, thickness of shaft wall = 0.125 m, inner height of the tank container = 4 m, mean diameter of the tank container = 10 m, thickness of the container wall = 0.2m, thickness of the container bottom = 0.2 m, thickness of the top cover of the container = 0.2 m and grade of concrete = M25. Based on these assumed data and considering the elevated water tank structure as a cantilever, the mass and stiffness of the structure have been calculated for the full, half-full and empty conditions of the tank. The structural damping has been assumed to be 1%. The fundamental natural frequencies and corresponding time period of the two different SDOF systems represented by the masses m_{cs} and m_{es} have been calculated separately for full, half-full and empty conditions of the tank which are presented in Table 1. The transfer functions have been plotted with these frequencies for full (Fig. 4), half-full (Fig. 5) and empty (Fig. 6) conditions of the tank. It is seen that in full and half-full conditions of the tank, the interaction between the structure and the first sloshing mode of water results in two different peaks in the displacement transfer function curve (Table 1). It has been also observed that the first peak is predominant when the tank is full; however, the second peak is predominant when the tank is half full, and the first peak is vanished when the tank is empty. That means, the first peak is representing the fundamental sloshing mode of the tank water, and the second peak is representing the structure along with the impulsive mass of tank liquid.

The fundamental natural frequency of the single TLD (w_{cd}) has been tuned to the frequency corresponding to the predominant mode of each case – full, half-full and empty conditions of the tank. Optimum tuning ratios with respect to minimum rms values of $u(t)$ corresponding to different mass ratios, determined through numerical optimization (Table 2) for full, half-full and empty conditions of the

Table 1 Frequencies of two different SDOF systems for full, half-full and empty conditions of tank before and after the occurrence of fluid-structure interaction

	Natural frequency of the sloshing mass of the tank		Natural frequency of the structure plus impulsive water mass of the tank		Natural frequencies of 2-DOF structural system			
					First peak		Second peak	
Condition of tank	Natural frequency (w_{cs}) (rad/s)	Time period (s)	Natural frequency (w_{es}) (rad/s)	Time period (s)	Natural frequency (rad/s)	Time period (s)	Natural frequency (rad/s)	Time period (s)
Full	1.81	3.47	3.13	2.01	1.65	3.81	3.43	1.83
Half full	1.50	4.18	3.69	1.70	1.44	4.36	3.84	1.64
Empty	–	–	4.03	1.56	–	–	4.03	1.56

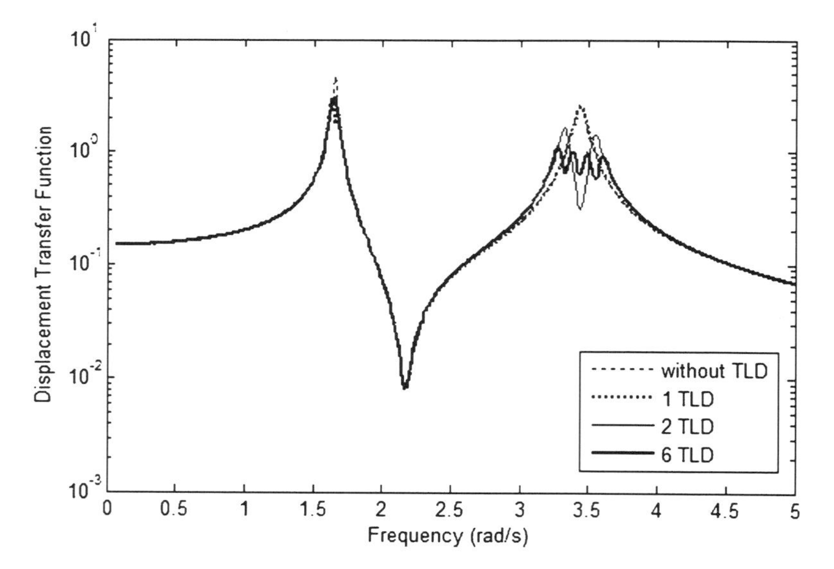

Fig. 4 Displacement transfer function of structure in full condition of the tank

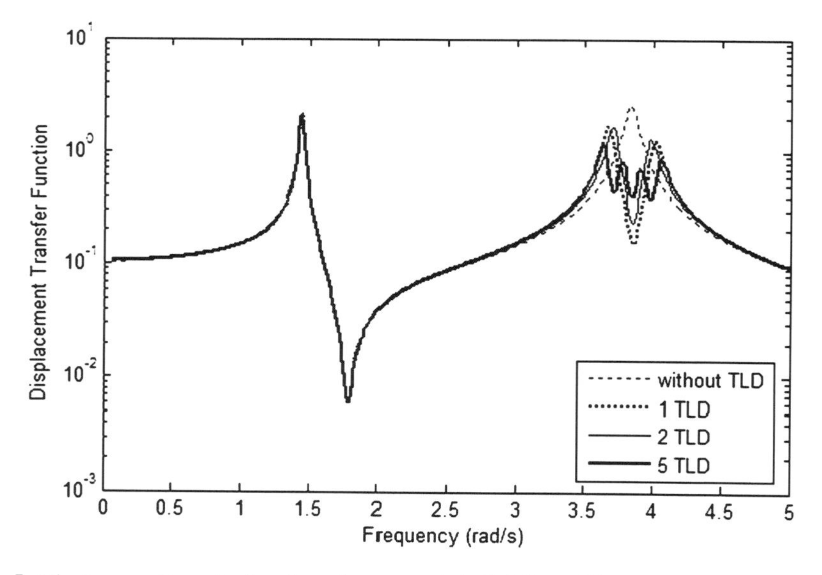

Fig. 5 Displacement transfer function of structure in half-full condition of the tank

tank, are close to unity, but in case of higher mass ratios, the optimum tuning ratio decreases slightly from unity. The influence of different mass ratios (the ratio of total mass of water in TLD to the mass which consists of the mass of the empty container including two-thirds of the supporting structure mass – namely, 1, 2, 3, 4 and 5%) on the percent response reduction of $u(t)$ has been investigated corresponding to optimum tuning ratio (Table 3).

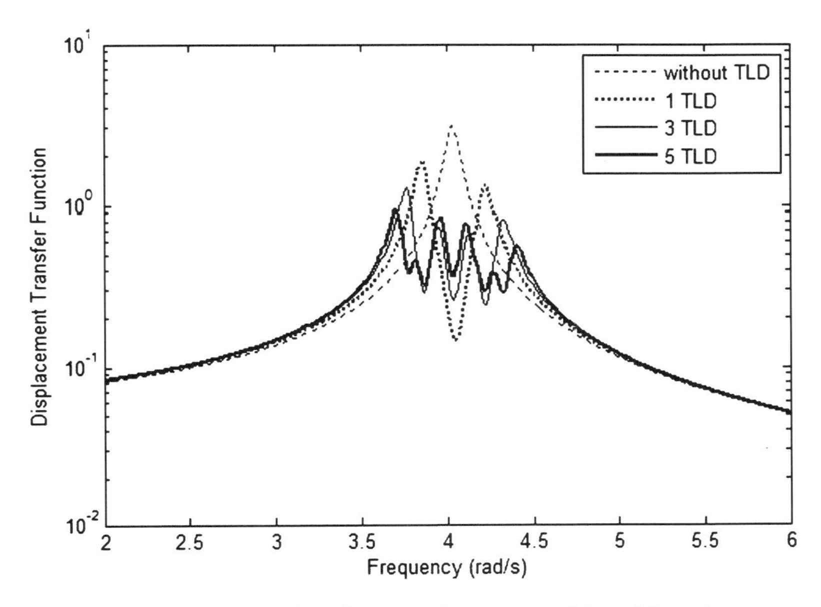

Fig. 6 Displacement transfer function of structure in empty condition of the tank

Table 2 Values of optimum tuning ratio (ν_{opt}) for single TLD for different mass ratios

Mass ratio (%)	Optimum tuning ratio (ν_{opt}) for full condition	Optimum tuning ratio (ν_{opt}) for half-full condition	Optimum tuning ratio (ν_{opt}) for empty condition
1	1.00	0.99	0.99
2	0.99	0.98	0.98
3	0.99	0.98	0.98
4	0.98	0.98	0.98
5	0.98	0.98	0.98

Table 3 Values of percent reduction in rms displacement for single TLD for different mass ratios

Mass ratio (%)	Percent reduction in rms displacement for full condition	Percent reduction in rms displacement for half-full condition	Percent reduction in rms displacement for empty condition
1	6.9058	18.7680	26.7035
2	9.5731	19.3605	27.3095
3	11.0165	19.5576	27.5030
4	11.8011	19.6048	27.5533
5	12.2492	19.6164	27.5644

Representative set of displacement transfer function curves for the cases of STLD tuned to the predominant mode of the full, half-full and empty conditions of the water tank for 2% mass ratio and tuning ratio of unity is given in Figs. 4, 5 and 6, respectively, and has been compared with that of the structure without damper. It has been observed that the magnitude of the peaks of the transfer function curve has been reduced, thereby reducing the displacement reduction of the structure. It has been seen that when the tank is empty, the control performance of the TLD is better due to the absence of sloshing mode whereas the performance of the TLD is worse when the tank is full and the sloshing mode is most effective although the TLD is tuned to the predominant sloshing mode in full condition. It has been also observed that when the damper is tuned to frequencies corresponding to the tank half-full and empty conditions, the control performance of the TLD does not increase significantly with the increase in mass ratio.

To achieve better response reduction, two TLDs of total mass ratio 2% have been tuned to the two modes of the transfer function curve of the 2-DOF structural system for full and half-full conditions. The resulting transfer function curves (Figs. 4 and 5) have four peaks which are reduced in amplitude, and greater response reduction has been achieved (Tables 4 and 5). In the next step, an additional four TLDs of total mass ratio 2% have been tuned corresponding to the frequencies of these four modes which results in larger response reduction (Tables 4 and 5). In case of empty condition, also the same procedure has been adopted, and the results have been shown in Table 6. From Tables 4, 5 and 6, it has been observed that response reductions as high as 21, 29 and 48% for the tank full, half-full and empty conditions may be obtained by the MTLD system as described above, whereas the corresponding response reductions for the STLD were 9.22, 18.7 and 26.5% for the tank full, half-full and empty conditions, respectively. So, it can be concluded that the percent response reductions for displacement have been increased in MTLD cases with respect to the STLD cases for all the three conditions.

Since the water mass in the water tank is a variable quantity, the natural frequency of the SDOF system representing the fundamental sloshing mode of the tank water can also vary between the values corresponding to the full condition and that for the empty condition of the tank, and this will change the values of natural frequencies corresponding to the two peaks of displacement transfer function curves after fluid-structure interaction. Thus, if the TLD system is tuned to a particular frequency of the tank, there will be some amount of detuning as the water content of the tank changes. The variation in the response reduction achieved by the MTLD systems discussed above, in terms of the rms displacement of the structure, over a range of percent tank full, for different values of mass ratios has been observed for three cases (TLD system tuned to the tank full, half-full and empty conditions) and has been compared to that of the single TLD systems (Fig. 7). From Fig. 7, it can be seen that there is a significant improvement in the performance of the damper system by using MTLDs as compared to the STLD over the entire range of percent tank full. The MTLD systems tuned to empty and half-full conditions of the tank provide a very good performance when the tank is 0–50% full, but the performance of this system is very poor when the tank is 100% full. From Fig. 7, it can be observed that the MTLD system tuned to the frequencies corresponding to the full condition of the

Table 4 Frequencies, geometry and percent response reduction for MTLD system in tank full condition

Number of TLDs	Frequencies of TLD (rad/s)	Mass of water in each TLD (m_l) (kg)	Aspect ratio (h/a)	Radius of TLD (a) (m)	Water Height of TLD (m)	Total area required for TLD (%)	Percent reduction in rms value	Figure number
1	1.65	4,149.2	0.17	1.9806	0.3367	15.6906	9.2195	Fig. 4
2	1.65	2,074.6	0.14	1.6771	0.2348	16.1308	17.1251	
	3.43		0.49	1.1046	0.5412			
6	1.64	691.5	0.10	1.3008	0.1301	28.3494	21.1083	
	1.65		0.11	1.2602	0.1386			
	1.67		0.11	1.2602	0.1386			
	3.32		0.33	0.8737	0.2883			
	3.43		0.34	0.8651	0.2941			
	3.55		0.37	0.8410	0.3112			

Table 5 Frequencies, geometry and response reduction for MTLD system in tank half-full condition

Number of TLDs	Frequencies of TLD (rad/s)	Mass of water in each TLD (m_l) (kg)	Aspect ratio (h/a)	Radius of TLD (a) (m)	Water Height of TLD (m)	Total area required for TLD (%)	Percent reduction in rms value	Figure number
1	3.84	4,149.2	0.90	1.1364	1.0227	5.1655	18.7036	Fig. 5
2	1.44	2,074.6	0.11	1.8175	0.1999	17.2971	19.5483	
	3.84		0.64	1.0105	0.6467			
5	1.44	829.8	0.09	1.4318	0.1289	24.7893	28.7327	
	1.45		0.09	1.4318	0.1289			
	3.70		0.42	0.8568	0.3598			
	3.84		0.45	0.8373	0.3768			
	3.98		0.49	0.8139	0.3988			

Table 6 Frequencies, geometry and response reduction for MTLD system in tank empty condition

Number of TLDs	Frequencies of TLD (rad/s)	Mass of water in each TLD (m_l) (kg)	Aspect ratio (h/a)	Radius of TLD (a) (m)	Water Height of TLD (m)	Total area required for TLD (%)	Percent reduction in rms value	Figure number
1	4.03	4,149.2	1.08	1.0694	1.1549	4.5743	26.4926	Fig. 6
3	3.85	1,383.1	0.55	0.9285	0.5107	9.6603	41.5660	
	4.03		0.61	0.8970	0.5472			
	4.22		0.68	0.8651	0.5883			
5	3.76	829.8	0.43	0.8501	0.3655	13.0621	47.7450	
	3.85		0.46	0.8312	0.3823			
	4.03		0.50	0.8084	0.4042			
	4.22		0.55	0.7831	0.4307			
	4.33		0.59	0.7650	0.4514			

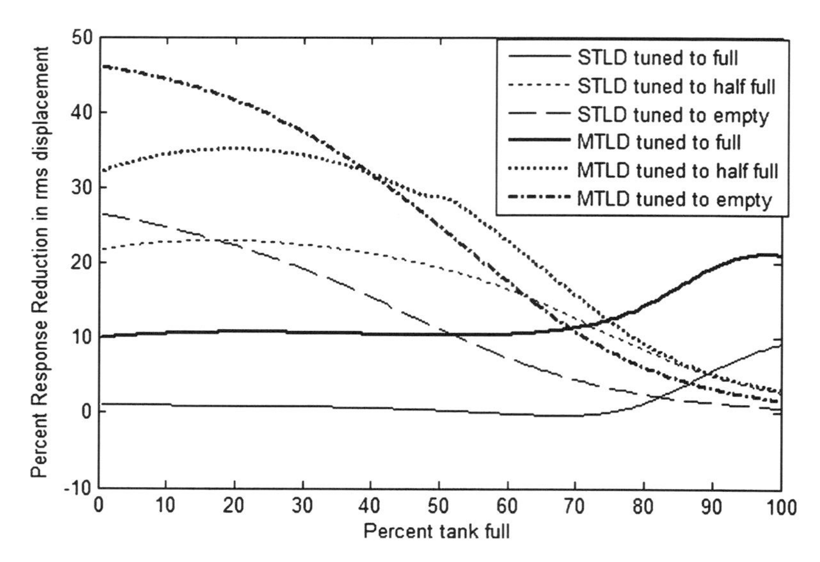

Fig. 7 Comparison of response reduction between STLD and MTLD system over a range of percent tank full for different tuning conditions

tank provides a standard response reduction (10–20%) over the entire frequency range of the structure.

3.1 Geometric Feasibility

It becomes necessary to check the geometric feasibility of the damper system. The parameters of the MTLD system, namely, the aspect ratio of each damper unit to satisfy the shallow criterion, radius and height of each TLD container and the maximum area of the water tank top surface that may be occupied by the MTLD units when tuned to the full (Table 4), half-full (Table 5) and empty (Table 6) conditions of the tank, have been evaluated from the tuning ratio, mass ratio and expression of natural frequency of the convective mass. It has been observed from Tables 4, 5 and 6 that the increase in the value of natural frequency, which is obtained as we move from the full condition to empty condition of the tank, results in an increase in the aspect ratio and decrease in the radius of the TLD. Further, in case of STLD, the aspect ratio is higher which is responsible for lowered performance of the TLD due to reduced sloshing water mass; however, STLD occupies smaller area of the water tank top surface. It has also been seen that if the number of TLD is increased maintaining the total mass ratio same, the response reduction increases although the maximum area of the water tank top surface that may be occupied by the TLD system is within 30%. In case of MTLD system, the size of each TLD unit is smaller than that of STLD unit, which improves their constructability. From these observations, it can be concluded that the geometrical feasibility of the MTLD system is more satisfactory than that of the single TLD system.

4 Conclusion

The effectiveness of an MTLD system for the mitigation of the seismic vibrations of an elevated water tank has been investigated in frequency domain. Fluid-structure interaction has been accounted for by considering the first sloshing mass of the tank water as an SDOF system connected in parallel with the SDOF system representing the water tank structure and remaining water mass. A formulation of the displacement transfer function of a base-excited, viscously damped 2-DOF system equipped with MTLD has been developed in frequency domain. The performance of STLD has been studied first. The drawbacks of the STLD have been overcome by the MTLD systems. In case of MTLD system, bimodal control has been achieved leading to improved response reduction by clustering the TLD frequencies about the frequencies at which the peaks of the transfer function of the structural response occur. The percent response reductions have been obtained as 21% when the tank is full, 29% when the tank is half full and 48% when the tank is empty in the frequency domain, which are significantly greater than the STLD case for the same mass ratio. From the point of view of geometry, the aspect ratios and sizes of the TLD units of MTLD system are smaller than that of STLD which may improve their constructability and maintainability. The provision of more number of TLD units keeping the total mass ratio constant appears to be a more viable option. It has been observed that the MTLD system provides better response reduction over the entire range of structural frequency which varies with the fluctuating water level of the tank from empty to full condition. However, the drop in performance between the two extreme conditions (i.e. full and empty) of the tank remains. Further schemes of designing the MTLD system with the frequency designed in such a way to ensure better robustness are being studied by the authors.

References

1. Fujino Y, Sun LM (1993) Vibration control by multiple tuned liquid dampers. J Struct Eng 119 (12):3482–3502
2. Fujino Y, Pacheco BM, Chaiseri P, Sun LM (1988) Parametric studies on Tuned Liquid Damper (TLD) using circular containers by free-oscillation experiments. Struct Eng/Earthq Eng JSCE 5(2):381–391
3. Fujino Y, Sun LM, Pacheco BM, Chaiseri P (1992) Tuned Liquid Damper (TLD) for suppressing horizontal motion of structures. J Eng Mech 118(10):2017–2030

4. Housner GW (1957) Dynamic pressures on accelerated fluid containers. Seismological Soc Am 47(1):15–35
5. Housner GW (1963) The dynamic behavior of water tanks. BullSeismological Soc Am 53 (2):381–387
6. Kareem A (1983) Mitigation of wind induced motion of tall buildings. J Wind Eng Ind Aerodyn 11:273–284
7. Kareem A (1990) Reduction of wind induced motion utilizing a tuned sloshing damper. J Wind Eng Ind Aerodyn 36(2):725–737
8. Kareem A (1993) Liquid tuned mass dampers: past, present and future. In: Proceedings of the Seventh U.S. National conference on wind engineering, Los Angeles
9. Kareem A, Hsieh CC, Tognarelli MA (1994) Response analysis of offshore systems to nonlinear random waves part I: wave field characteristics. In: Proceeding of the special symposium on stochastic dynamics and reliability of nonlinear ocean systems ASME, New York, pp 39–54
10. Kareem A (1995) Performance of multiple mass dampers under random loading. J Struct Eng 121(2):348
11. Livaoglu R, Dogangun A (2006) Seismic behaviour of cylindrical elevated tanks with a frame supporting system on various subsoil. Indian J Eng Mater Sci 14(2):133–145
12. Modi VJ, Welt F (1987) On the vibration control using nutation dampers. In: King R (ed) Proceedings of the international conference on flow induced vibrations, BHRA, London, pp 369–376
13. Newland DE (1993) An introduction to random vibration, spectral and wavelet analysis. Longman, Scientific and Technical, London
14. Sakai F, Takaeda S (1989) Tuned liquid column damper – new type device for suppression of building vibrations. In: Proceedings international conference on high rise buildings, Nanjing, China, March 25–27
15. Tait MJ, Isyumov N, El Damatty AA (2008) Performance of Tuned Liquid Dampers. J Eng Mech 134(5):417–427
16. Tait MZ, Deng X (2010) The performance of structure-tuned liquid damper systems with different tank geometries. Struct Control Health Monit 17:254–277
17. Tamura Y, Fujii K, Ohtsuki T, Wakahara T, Koshaka R (1995) Effectiveness of tuned liquid dampers under wind excitations. Eng Struct 17(9):609–621
18. Veletsos AS, Tang Y (1990) Soil-structure interaction effects for laterally excited liquid storage tanks. Earthq Eng Struct Dyn 19:473–496
19. Wakahara T, Shimada K, Tamura Y (1994) Practical application of tuned liquid damper for tall buildings. ASCE Structural Congress and IASS International Symposium, Atlanta, USA, pp 851–856

Pavement Performance Modelling Using Markov Chain

S.K. Suman and S. Sinha

Abstract Pavement performance modelling is an essential element of a pavement management system (PMS). The model developed plays a critical role in several aspects of the PMS including financial analysis. In developing countries like India, PMS is the needed approach for the optimum utilisation of the available scarce resources. Pavement management system is concerned with optimal use of materials in time and space, leading to cost optimisation.

This chapter focuses on methodology involved in the prediction of pavement condition using probabilistic techniques. Since traffic loading, pavement materials, construction methods and environmental condition are not deterministic, therefore, probabilistic techniques are used. Markov chains have the property that probabilities involving the process will evolve in the future, depending only on the present state of the process and so are independent of the events in the past. The state of the transition matrix will be defined based on the overall pavement quality indices (OPQI), and element of the transition matrix will be determined by using several methods. Steady-state transition matrix will be obtained from one-step transition matrix.

The probabilistic model requires only a minimal amount of data such as pavement class, pavement condition of two consecutive years and pavement length. OPQI shall be utilised as index, and the present technique in pavement management systems will create good systems which may lead to more savings of the road maintenance funds and enhance the ability of the road network to provide better level of service at network level.

Keywords Pavement management system • Performance • Overall pavement quality index (OPQI) • Markov chain

---

S.K. Suman (✉) • S. Sinha
Department of Civil Engineering, National Institute of Technology Patna, Patna, Bihar, India
e-mail: sksuman77@yahoo.co.in; sanjeev_bangkok@yahoo.com

S. Chakraborty and G. Bhattacharya (eds.), *Proceedings of the International Symposium on Engineering under Uncertainty: Safety Assessment and Management (ISEUSAM - 2012)*,
DOI 10.1007/978-81-322-0757-3_39,

1 Introduction

Pavement performance models simulate the deterioration process of pavement conditions and provide forecasting of pavement condition over time. It plays a vital role in pavement management system. The pavement performance models have several uses in pavement management system: (1) to analyse the conditions and determine the maintenance and repair requirements, (2) to decide the useful service life of the pavement on the basis of reduction in performance value to a certain terminal value, (3) to estimate long-range funding requirements to preserve the pavements, (4) to provide major inputs to perform life cycle cost analysis to compare the economics of various maintenance and repair alternatives and (5) to study the effects of various budget levels on future pavement condition.

In order to take decision of the optimum cost of maintenance and repair works of the pavement surface, there is a need to develop a comprehensive pavement performance index. A reliable pavement condition index can be achieved when all the pavement condition indicators will be incorporated simultaneously. This index includes factors related to all pavement indicators such as surface distresses, riding quality, skid resistance and deflection.

The pavement surface condition is affected by parameters like traffic axle loads, environmental conditions and moisture content that are themselves uncertain in nature; hence, the rate of pavement deterioration is uncertain. Modelling uncertainty requires the use of probabilistic operation research techniques.

In this context, the use of Markov chain in the prediction of model captures the uncertainty behaviour of pavement deterioration. The most important advantage of this technique is that it requires only 2 years or any two-threshold-justified temporal data.

The aim of this study is to highlight the approach of technique used in developing the pavement performance model.

2 Methodology

The first part of this section deals with an overview of overall pavement quality index, and the second part describes the procedures for development of pavement performance model based on Markov chain theory.

2.1 *Overall Pavement Quality Index (OPQI)*

OPQI is the combination of indices such as pavement quality distress-based index, pavement quality roughness-based index, pavement quality structural capacity-based index and pavement quality skid resistance-based index. This combined index is expected to give a good representation of the pavement condition and will be able to predict the performance accurately. It will also enhance the process of optimising the life cycle costs in the selected maintenance and repair option through utilising the acceptance levels of the aforesaid four condition indicator indices. Utilising such an index in pavement management systems will create systems that lead to savings of the road maintenance funds and enhance the ability of the road network to provide better service at network level [1–6].

For more accurate estimation of pavement condition, OPQI is used which describes the pavement structural and functional capacities of the road section, taking into consideration all data collected for the surface condition. Therefore, it is envisaged that this index combines all distress types (severity and density), roughness, effective structural capacity and skid resistance value, according to the relative importance of each condition indicator. As per Shiyab [7], Eq. (1) was arrived based on theoretical concept that comprises all pavement surface conditions:

$$\mathrm{OPQI}_k = 10\sum_{i=1}^{i=n}\left[1-\left(1-\frac{\mathrm{CI}_i}{10}\right)*W_{i,k}\right] \quad (1)$$

where OPQI_k = overall pavement quality index on a scale of 1–10, CI = condition indicator or distress index on a scale of 1–10, K = k-th pavement performance index, i = i-th distress or condition indicator out of the total number of the 'n' condition indicators = total number of distress types or condition indicators included in the performance index and $W_{i,k}$ = the impact or the relative weight of each distress type or condition indicator.CI incorporates index for every type of distress (e.g. RI, rutting index; ACI, alligator cracking index; BI, bleeding index; PCHI, patching index; PTI, potholes index), roughness index (RI), deflection index (ESCI) and skid resistance index (SKRI). Distress index may be obtained by using the method of Paver Systems. Deflection and roughness indices may be obtained by using Eqs. (2) and (3), respectively:

$$\text{Deflection Index} = \frac{\text{Percent (Predicted) deflection}}{\text{Maximum permissible deflection}} \quad (2)$$

$$\text{Roughness Index} = \frac{\text{Percent (Predicted) roughness}}{\text{Maximum permissible roughness}} \quad (3)$$

2.2 Markov Chain

A Markov chain is a special type of discrete-time stochastic process, when the state of a system X_{t+1} at time $t + 1$ depends on the state of the system X_t at some previous time t but does not depend on how the state of the system X_t was obtained. This can be expressed as $P(X_{t+1} = j|X_t = i)$, where P is the probability of the state at time $t + 1$ being j given that the state at time t was i, assuming that the probability is independent of time. This assumption is formally known as the stationary assumption.

This stationary assumption is used because of the limited time period of the data collected. If data were collected over a large period, the quantity and type of materials used might change over time and influence how a typical pavement section would determine. A Markov chain can be summarised through a probability transition matrix and the initial state probabilities.

To model pavement deterioration with time, it is necessary to establish a transition probability matrix (TPM), denoted by P. The general form of P is given below:

$$P = \begin{bmatrix} p_{11} & p_{12} & p_{13} & \cdots & \cdots & \cdots & \cdots & p_{1n} \\ p_{21} & p_{22} & p_{23} & \cdots & \cdots & \cdots & \cdots & p_{2n} \\ \cdots & \cdots & \cdots & \cdots & \cdots & \cdots & \cdots & \cdots \\ \cdots & \cdots & \cdots & \cdots & \cdots & \cdots & \cdots & \cdots \\ p_{n-11} & p_{n-12} & p_{n-23} & \cdots & \cdots & \cdots & \cdots & p_{n-1n} \\ p_{n1} & p_{n2} & p_{n3} & \cdots & \cdots & \cdots & \cdots & p_{nn} \end{bmatrix}$$

The matrix contains all of the information necessary to model the movement of the process among the condition states. The transition probabilities p_{ij} indicate the probability of the portion of the network in condition i moving to condition j in one duty cycle.

Two more conditions apply to the process when it is used to simulate pavement deterioration. First, $p_{ij} = 0$ for $i > j$, signifying the belief that roads cannot improve in condition without first receiving treatment. Second, $p_{nn} = 1$, signifying a holding state whereby roads that have reached their worst condition cannot deteriorate further. Consequently, in pavement deterioration, the general form of the transition matrix P is denoted by P^1:

$$P^1 = \begin{bmatrix} p_{11} & p_{12} & p_{13} & \cdots & \cdots & \cdots & \cdots & p_{1n} \\ 0 & p_{22} & p_{23} & \cdots & \cdots & \cdots & \cdots & p_{2n} \\ 0 & 0 & p_{33} & \cdots & \cdots & \cdots & \cdots & p_{3n} \\ \cdots & \cdots & \cdots & \cdots & \cdots & \cdots & \cdots & \cdots \\ 0 & 0 & 0 & 0 & 0 & 0 & p_{(n-1)(n-1)} & p_{(n-1)n} \\ 0 & 0 & 0 & 0 & 0 & 0 & 0 & 1 \end{bmatrix}$$

A further restriction allowing the condition to deteriorate by no more than one state in one duty cycle is commonly used in pavement deterioration modelling. The transition probability matrix is then denoted by P^2:

$$P^2 = \begin{bmatrix} p_{11} & p_{12} & 0 & 0 & \cdots & \cdots & \cdots & 0 \\ 0 & p_{22} & p_{23} & 0 & \cdots & \cdots & \cdots & 0 \\ 0 & 0 & p_{33} & p_{34} & \cdots & \cdots & \cdots & 0 \\ \cdots & \cdots & \cdots & \cdots & \cdots & \cdots & \cdots & 0 \\ 0 & \cdots & \cdots & \cdots & \cdots & \cdots & p_{(n-1)(n-1)} & p_{(n-1)n} \\ 0 & 0 & 0 & \cdots & \cdots & \cdots & \cdots & 1 \end{bmatrix}$$

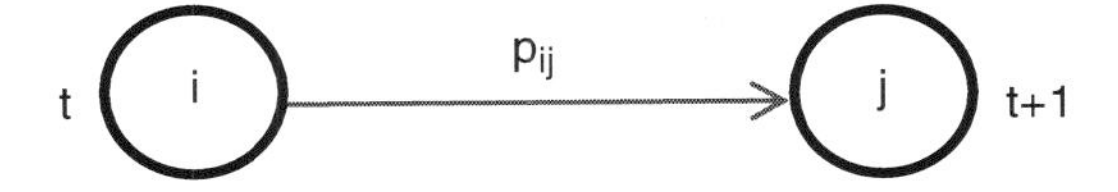

Fig. 1 The transition between two states

The entry of 1 in the last row of the transition matrix corresponding to state 10 (PCI of 0–10) indicates a holding or trapping state. The pavement condition cannot transit from this state unless repair action is performed.

The state vector for any duty cycle t is obtained by multiplying the initial state vector X (0) by the transition matrix P raised to the power of t. Thus,

$$X(1) = X(0) \cdot P$$
$$X(2) = X(1) \cdot P = X(0) \cdot P^2$$
$$\cdots\cdots\cdots\cdots\cdots\cdots$$
$$\cdots\cdots\cdots\cdots\cdots\cdots$$
$$X(t) = X(t-1) \cdot P = X(0) \cdot P^t$$

With this procedure, if the transition matrix probabilities can be estimated, the future state of the road at any duty cycle t can be predicted.

After the several cycle operations, rows of the matrix have identical entries; the reason is that probabilities in any row are the steady-state probabilities for the Markov chain, i.e. the probabilities of the state after enough time has elapsed that the initial state is no longer relevant.

2.2.1 Markov Transition Probability Matrices

These are useful in representing the change in condition of the system from one state to another over time. Change in condition is nothing but the transition from one state to another over time. So, essentially, the key elements of any Markov transition matrices are states and transitions. An interesting example of how a Markov transition matrix may be constructed is given below.

The classic example for a Markov process is a frog sitting in a pond filled with lily pads. In this example, each pad in the pond represents a state of the system. If there is a finite number of a pad in the pond, the system we are describing is a finite state system. If we were to check the pond every 5 minutes to observe the frog's location, each epoch in the model would be equivalent to 5 minutes in real time. The likelihood of the frog making a transition from pad i to j is p_{ij}. Figure 1 shows a simple schematic describing the transition from one state to the next.

There are several methods that can be used to estimate the present transition probabilities ($P_{i,i}$ and $P_{i,i+1}$). These methods are based either on the experience and adjustment of pavement experts or on sound engineering principles. Application of engineering principles requires feedback on pavement performance as obtained from field assessment of pavement distress.

Three methods are presented herewith for computing the transition probability matrix:

1. The first method is to apply the very basic definition of transition probabilities, i.e. if N_0 pavement sections are initially found in state i and N_f sections existed in state i after one transition, the transition probabilities can be estimated using

$$P_{i,i+1} = \frac{N_0 - N_f}{N_0} \tag{4}$$

and

$$P_{i,i} = 1 - P_{i,i+1} = \frac{N_f}{N_0} \tag{5}$$

2. The second method is based on estimating the service periods (D_i) in years that a pavement section is going to stay in state i. Let (t) be the length of time interval in years between successive transitions. Then, one simple equation can estimate the transition probabilities as follows:

$$P_{i,i+1} = \frac{t}{D_t} \leq 1.0 \tag{6}$$

where $\sum D_i = T$, where T is either the service life estimated from actual pavement performance records or the analysis period used in the design of pavement.

3. The standard approach is to observe, from historical data, the way in which a road network deteriorates over time and use this to estimate p_{ij} using the equation below:

$$P_{ij} = \frac{N_{ij}}{N_i} \tag{7}$$

where N_{ij} = number of road sections in the network that moved from condition i to condition j during one duty cycle and N_i = total number of road sections that started the year in condition i. The proportions are likely to vary from year to year, thereby requiring an average to be determined for each p_{ij} to ensure accuracy in the model.

2.3 Illustrative Example

A case study is conducted to demonstrate the procedure of the Markov chain-based pavement performance evaluation methodology. A bituminous-topped pavement of 12.70 km is used to generate the condition matrix by using Eq. (7). Suppose a 4.20-km length of the pavement lies between 8.1 and 10 condition ratings, i.e. in

Table 1 Condition state vector using probability distribution of pavement

Condition state	Corresponding condition rating	Length of the pavement section under consideration (km)	Probability distribution (%)
Very good	8.1–10	4.20	33.07
Good	6.1–8	3.40	26.77
Fair	4.1–6	2.80	22.05
Bad	2.1–4	1.60	12.60
Very bad	<2	0.70	5.51
Total length (km)		12.70	100.00

Table 2 Pavement condition rating distribution

Sl. no.	Condition state transition	Corresponding condition rating	Number of pavement sections
1	Very good → very good	8.1–10 → 8.1–10	665
2	Very good → good	8.1–10 → 6.1–8	29
3	Good → good	6.1–8 → 6.1–8	30
4	Good → fair	6.1–8 → 4.1–6	18
5	Fair → fair	4.1–6 → 4.1–6	110
6	Fair → bad	4.1–6 → 2.1–4	28
7	Bad → bad	2.1–4 → 2.1–4	49
8	Bad → very bad	2.1–4 → <2	15
9	Very bad → very bad	<2 → <2	56
10	Total no. of sections (100 m)		1,000

good condition state, then its probability to remain in this state is 33.07%. The pavement length of 4.20 km that comes under very good condition state is decided on the basis of Eq. (1) (Table 1).

Hence, condition matrix (1×5) based on overall pavement condition is given below:

$$\text{Condition matrix} = \begin{bmatrix} 0.3307 & 0.2677 & 0.2205 & 0.1260 & 0.0551 \end{bmatrix}$$

Pavement transition matrix has been generated by using 1,000 sections of the pavement, whereas the length of the each section considered is 100 m. The process involved in arriving the transition matrix followed Tables 2 and 3. Finally, a transition probability matrix of 5×5 is arrived as shown in Table 4.

Pavement condition can be predicted by using condition probability matrix (1×5) and transition probability matrix (5×5). A horizon year condition matrix of 1×5 will come after multiplication of condition probability matrix (1×5) and transition probability matrix (5×5). Pavement condition performance graph has been plotted over the 12 years as shown in Fig. 2.

Table 3 Pre-transition probability matrix for overall pavement condition

		Future pavement condition in year (t + 1)					
		Very good	Good	Fair	Bad	Very bad	Total no. of sections
Present pavement condition in year (t)	Very good	665	29	0	0	0	694
	Good	0	30	18	0	0	48
	Fair	0	0	110	28	0	138
	Bad	0	0	0	49	15	64
	Very bad	0	0	0	0	56	56
	Total no. of sections						1,000

Table 4 Transition probability matrix for overall pavement condition

		Future pavement condition in year (t + 1)				
		Very good	Good	Fair	Bad	Very bad
Present pavement condition in year (t)	Very good	0.958	0.042	0	0	0
	Good	0	0.625	0.375	0	0
	Fair	0	0	0.797	0.203	0
	Bad	0	0	0	0.766	0.234
	Very bad	0	0	0	0	1

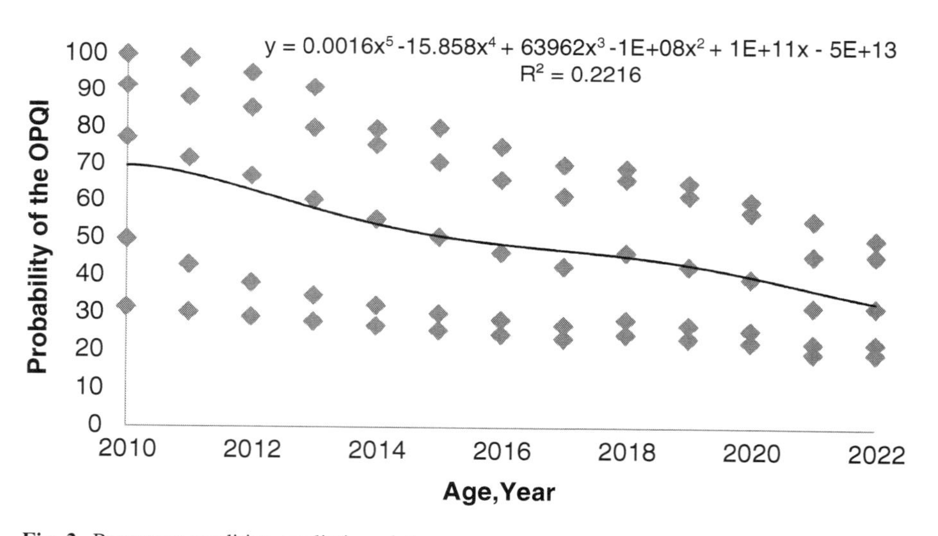

Fig. 2 Pavement condition prediction plot

2.3.1 Validation of Model

The developed model can be validated by mean absolute error (MAE), root mean squared error (RMSE), mean absolute relative error (MARE) or coefficient of regression (R^2). In this chapter, validation is done by comparing the observed condition rating in the year 2011 with the condition rating predicted by the deterioration model developed for the same year. The reason for comparing the condition rating in year 2011 is that this is the only year for which the data is available.

In year 2011, the pavement condition rating obtained by using the model is 6.8 in 1–10 scale, whereas the observed value for the same year is 6.1. This result shows that the developed model is able to predict future condition rating with a reasonable degree of accuracy.

3 Conclusion

This chapter demonstrates the overall process for developing the pavement performance model based on Markov chain. The illustrative example gives the idea how Markov chain can be used for prediction of pavement condition. Base year pavement conditions have been identified using overall pavement quality index. Markov chain process is applied for predicting the horizon year pavement conditions. The developed model shows the satisfactory result with a reasonable degree of accuracy. Therefore, developed model can be used for prediction of pavement performance both at network and project level. Further, this model can be used by the road construction department in planning of maintenance and repair of roads.

References

1. Butt AA, Shahin MY, Feighan KJ, Carpenter SH (1987) Pavement performance prediction model using the Markov Process. Transportation Research Record, TRB, Washington, DC, 1123, pp 12–19
2. Hillear FS, Lieberman GJ (2009) Introduction to operations research, 8th edn. Tata McGraw Hill Education Private Ltd, New Delhi
3. Joseph NT, Chou YJ (2001) Pavement performance analysis applying probabilistic deterioration methods. Transportation Research Record, TRB, Washington, DC, 1769, 01–2962, pp 20–27
4. MORTH (2004) Guidelines for maintenance management of primary, secondary and urban roads. Indian Roads Congress Publications, New Delhi
5. Ortiz-Garcia JJ, Costello SB, Snaith MS (2006) Derivation of transition probability matrices for pavement deterioration modelling. J Transp Eng ASCE 132(2):141–161
6. Shahin MY (1994) Pavement management for airports roads and parking lots. Kluwer Academic Publishers, Boston/London
7. Shiyab AMSH (2007) Optimum use of the flexible pavement condition indicators in pavement management system. Ph.D. thesis, Curtin University of Technology. www.adt.curtin.edu.au/thesis/available/adt

Uncertainty of Code-Torsional Provisions to Mitigate Seismic Hazards of Buildings with Setback

Somen Mahato, Prasanta Chakraborty, and Rana Roy

Abstract Static torsional provisions in most of the seismic codes require that the earthquake-induced lateral force at each story be applied at a distance equal to design eccentricity (e_d) from the center of resistance of the corresponding story. Such code-torsional provisions, albeit not explicitly stated, are generally believed to be applicable to the regularly asymmetric buildings. Examined herein is the applicability of such code-torsional provisions to buildings with setback. A set of low-rise setback systems are analyzed using static torsional provisions and by response-spectrum-based procedure. A comparison of the response in terms of frame shear located at the perimeter and diaphragm displacements suggests the degree of reliability of code-torsional provisions for low-rise setback building.

Keywords Irregular • Torsion • Setback • Code provisions • Seismic • Low-rise building

1 Introduction

Geometry of the structure is often dictated by the architectural and functional requirements, whereas the safety of the same with optimum economy – key design aim – is ensured by structural engineers. For instance, a stepped form (setback systems) of building is often adopted by architect for adequate daylight and ventilation in the lower stories in an urban locality with closely spaced tall

S. Mahato • P. Chakraborty
Department of Civil Engineering, Bengal Engineering and Science University, Shibpur, Howrah 711103, India
e-mail: somen07@gmail.com

R. Roy (✉)
Department of Aerospace Engineering and Applied Mechanics, Bengal Engineering and Science University, Shibpur, Howrah 711103, India
e-mail: rroybec@yahoo.com

S. Chakraborty and G. Bhattacharya (eds.), *Proceedings of the International Symposium on Engineering under Uncertainty: Safety Assessment and Management (ISEUSAM - 2012)*, DOI 10.1007/978-81-322-0757-3_40,

buildings. Such setback structures form an important subclass of irregular structures wherein irregularities are characterized by discontinuities in the distribution of mass, stiffness, and strength over the height of the building.

Seismic codes permit equivalent static procedure, usually for regular buildings, and the dynamic analysis (response-spectrum analysis or response-history analysis) for systems with significant irregularities in plan and elevation. In equivalent static analysis, the design base shear is estimated as a product of seismic weight and codified seismic coefficient associated to fundamental vibration period of the systems. Such seismic coefficient, on the one hand, accounts for the importance and ductility capacity of the structure while, on the other, represents the type of soil, seismic activity of the region, *etc*. Building codes [14] generally, for asymmetric systems, specify that the earthquake-induced lateral force at each story be statically applied with an eccentricity equal to design eccentricity (e_{d}) relative to the center of resistance of the corresponding story. Such design eccentricities are outlined in the form of primary design eccentricity, $e_{\mathrm{d}1j}$, and secondary design eccentricity, $e_{\mathrm{d}2j}$, at any typical story j, as

$$e_{\mathrm{d}1j} = \alpha e_j + \beta D \text{ and } e_{\mathrm{d}2j} = \delta e_j - \beta D \tag{1}$$

where D is the plan dimension of the building normal to the direction of ground motion and e_j is the static eccentricity at j-th story. α and δ are the coefficients aimed to be so calibrated that reasonable agreement in response between the equivalent static analysis specified by code and dynamic analysis is achieved. For each element, the value of e_{d} leading to the larger design force is to be used. The first part is a function of static eccentricity – real distance between center of mass (CM) and center of resistance. Dynamic amplification factor α in $e_{\mathrm{d}1}$ is intended to compensate for the dynamic effect of torsional response through static analysis. Factor δ incorporated in $e_{\mathrm{d}2}$ specifies the portion of the torsion-induced so-called negative shear that can be reduced for the design of stiff-side elements. The second part, referred to as accidental eccentricity, is expressed as a fraction of plan dimension, *i.e.*, βD (normal to the direction of ground motion and is introduced to account for the differences between the actual and perceived eccentricities, torsional ground motion, and other imponderables).

Equivalent static procedure, recommended for regular buildings, using the concept of eccentricity and equivalent static loading “remains a basic approach” to allow for the effect of torsion. However, a lack of unanimously acceptable definition of center of resistance for multistory buildings often appears to be a major crux to implement such static procedure. A search for such resistance center defined elsewhere (e.g., [7, 11, 12, 18, 26, 27, 31, 33]) reveals a number of alternatives. Response computed on the basis of such different resistance centers, although located at variance, is often observed to be in agreement [10].

This observation indicates that the traditional notion of applicability of code-torsional provisions to regularly asymmetric systems (where center of mass and center of resistance are aligned along two vertical lines separated by a constant distance) may be over-restrictive. Thus, the applicability of code-torsional provisions is reviewed herein in the context of buildings with setback where the center of resistance may dramatically vary story-wise. The significance of this undertaking seems obvious as the dynamic analysis, hard to perform and interpret, recommended for such systems may be bypassed subject to the adequacy of the code-static provisions – the focus of the current investigation.

2 Buildings with Setback: Research Progress

Studies (*e.g.*, [2, 13, 21]) up to mid-1980s on seismic response and relevant code provisions of systems with symmetric setback have been reviewed in the literature [34]. Research progress for systems with irregularity in elevation is scarce primarily owing to the relative difficulty to characterize such systems [20]. A simple definition to measure irregularity of such systems has been proposed and used in the recent works [19, 29]. Illustrations therein demonstrate the possibility of higher damage potential in the vicinity at and below the setback. A subsequent study [30] also corroborates such observation. Studies [4, 19] show that the participation of higher modes may be significant, and inter-story drift also magnifies in upper stories of such systems. The relative vulnerability associated to mass, strength, and stiffness irregularities has been examined elsewhere [1]. A simple method for the analysis of torsionally coupled buildings with setback has been developed recently [5, 20]. Barring very few useful attempts [35], relative paucity of experimental works in the relevant field is apparent to date.

A succinct survey [32] reveals that seismic response of structures with setback is kept unheeded in major seismic design codes (such as [3, 9, 17]) and dynamic analysis is recommended. The codes also require the base shear obtained from the dynamic analysis (and thereby, other response quantities) to be scaled up to that corresponds to the code-specified empirical formula.

In recognition of the apparent complexity of dynamic analysis in regular design and the promise of code-static approach [10], adequacy of the latter is explored in the context of low-rise setback systems. Response of such systems is computed with representative values of α and δ employing two convenient height-wise distributions of lateral load. Response so computed is compared to the same obtained due to response-spectrum-based analysis to realize the adequacy of code-static procedure.

2.1 *Details of Structural Systems*

Structures are idealized as rigid diaphragm (rigid in plane and flexure) model with three degrees of freedom at each floor, two translations in two mutually orthogonal directions and one in-plane rotation. Mass is assumed to be lumped at the center of mass (CM) of each floor coinciding with the geometric center of the deck.

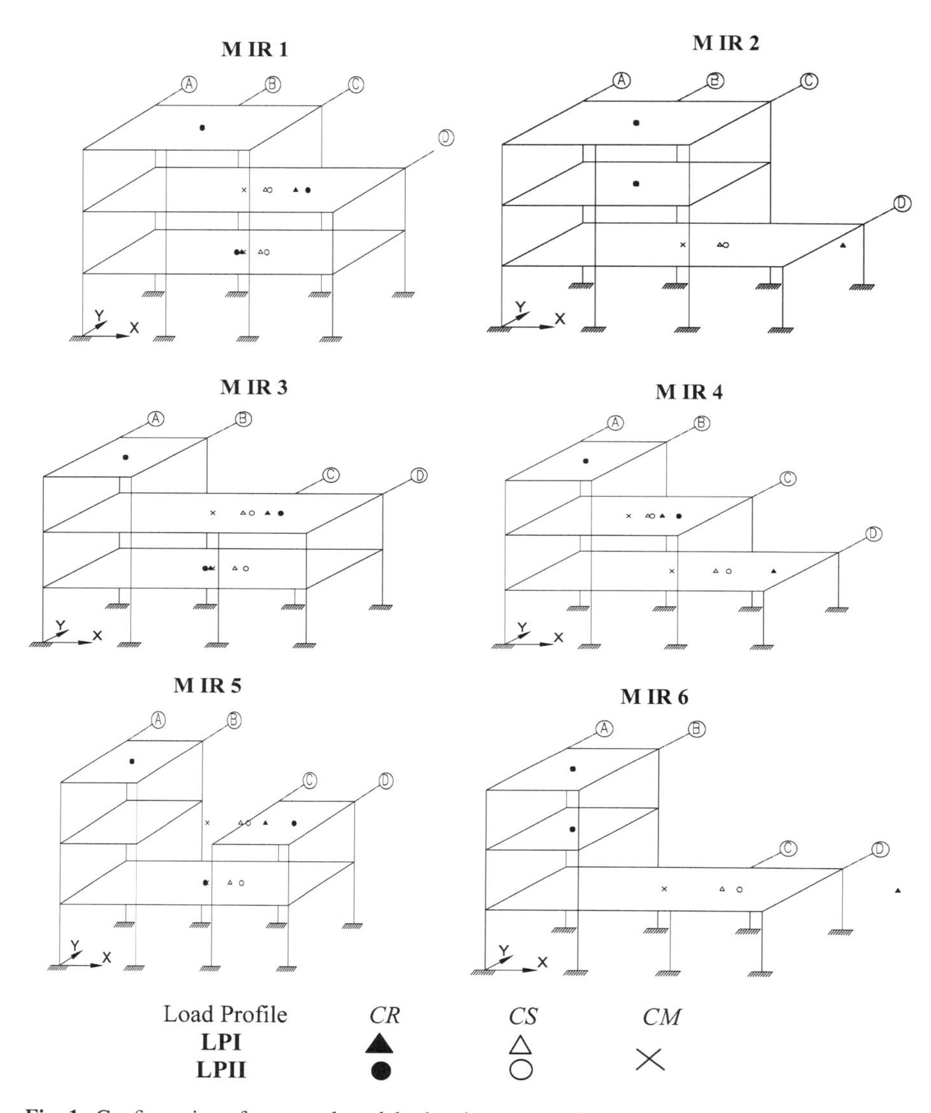

Fig. 1 Configuration of structural models showing center of mass (*CM*), center of rigidity (*CR*), and shear centre (*CS*)

Focusing on low-rise systems, three-story models with different feasible forms of setback are considered. Such systems annotated as M-IR 1 through MIR 6 are schematically presented in Fig. 1.

Irregularity indices (Φ_b, Φ_s) of the systems, proposed and utilized elsewhere [19, 29], are computed as follows and furnished in Table 1 to recognize the nature of elevation irregularity.

Table 1 System characteristics and basic seismic design parameters

			Irregularity index		Dynamic characteristics			Seismic weight (kN)			
Sl. no.	Category	Model identification	Φ_b	Φ_s	Mode 1	Mode 2	Mode 3	Story 1	Story 2	Story 3	Design base shear (kN) [UBC 97]
1	Irregular	M-IR1	1.25	1.25	0.367[a]	0.244	0.135	1,350	1,350	900	620
					0.858[b]	0.057	0.068				
2		M-IR2	2.00	1.25	0.362	0.217	0.140	1,350	900	900	540
					0.828	0.028	0.136				
3		M-IR3	1.25	2.00	0.333	0.234	0.122	1,350	1,350	450	540
					0.814	0.118	0.050				
4		M-IR4	1.75	1.75	0.326	0.183	0.121	1,350	900	450	465
					0.770	0.161	0.055				
5		M-IR5	1.75	1.75	0.303	0.218	0.120	1,350	900	450	465
					0.727	0.207	0.057				
6		M-IR6	2.00	2.00	0.306	0.158	0.134	1,350	450	450	385
					0.701	0.217	0.079				

EI for all column = $8.54 \times 10^7 \text{Nm}^2$

[a]Values in first row represent period of vibration in each mode

[b]Values in second row represent participating mass ratio for excitation in Y-direction (Refer to Fig. 1)

$$\Phi_b = \frac{1}{n_b - 1} \sum_{i=1}^{i=n_b-1} \frac{H_i}{H_{i+1}} \text{ and } \Phi_s = \frac{1}{n_s - 1} \sum_{i=1}^{i=n_s-1} \frac{L_i}{L_{i+1}} \quad (2)$$

where n_s is the number of story, n_b is the number of bay in the first story, L is the length of bay (5m), and H (3.5m) is the height of the respective story. Length of the bay in the direction normal to setback is also kept equal to 5m.

Location of center of resistance varies as per different definitions and is also known to be dependent on the distribution of lateral load. In this context, generalized center of rigidity (denoted hereinafter as CR) and shear center (denoted hereinafter as CS) in each story are located following the procedure outlined in the literature [33]. Each of such resistance centers is located considering the height-wise distribution of lateral load as $w_i H_i^k \Big/ \sum_{i=1}^{n_s} w_i H_i^k$, where w_i and H_i are the weight and height of i story, n_s is the total number of story, and k is an exponent. Values of k are chosen as 1.0 and 2.0 in load profile LP I and LP II, respectively. Resistance centers so identified along with the center of mass (CM) are marked story-wise in respective building models (Fig. 1). Distance between CR and CM is referred to as floor eccentricity (e_R), while story eccentricity (e_S), defined as distance between CS and CM, is summarized in Table 2. In view of the lesser sensitivity of shear center and hence story eccentricity to the lateral load distribution, shear center (CS) is consistently used as reference resistance center and story eccentricity (e_S) as measure of asymmetry. In fact, the same is also reckoned to be a more reasonable measure of asymmetry [33]. Frame A and frame D (as shown in Fig. 1) are designated as flexible and stiff sides, respectively, observing the relative position of CM and CS.

2.1.1 Dynamic Properties

Dynamic properties of the systems are computed corresponding to translational (Y) and torsional degrees of freedom by standard eigenvalue solution. A review of the natural periods, corresponding mode shapes, and "participating mass ratio" (shown in percentage) in each mode (relevant to the excitation in Y-direction only), presented in Table 1, furnishes useful dynamic characteristics of the buildings.

The participating mass ratio for n-th mode is computed as $(f_{yn})^2 / M_y$ where f_{yn} $(f_{yn} = \varphi_n{}^T m_y)$ is the participation factor in which m_y is the load corresponding to unit acceleration; M_y is the total unrestrained mass in Y-direction. The mode shapes (φ) are normalized such that $\varphi_n^T M \varphi_n = 1$ in which M is the global mass matrix [28].

Observing fundamental periods of the buildings, it seems that the same gradually increases with the reduction of irregularity (irregularity is maximum in MIR 6 and minimum in M–IR 1). Fundamental period of such systems, on the other hand, is computed as 0.43s. as per the empirical formula $T_a = 0.0731 h^{3/4}$outlined in UBC

Table 2 Eccentricities of structural systems under two different load distributions

		Load profile: LP I						Load profile: LP II					
		Floor eccentricity [e_R] (m)			Story eccentricity [e_S] (m)			Floor eccentricity [e_R] (m)			Story eccentricity [e_S] (m)		
Sl. no.	Model identification	Floor 1	Floor 2	Floor 3	Story 1	Story 2	Story 3	Floor 1	Floor 2	Floor 3	Story 1	Story 2	Story 3
1	M-IR1	0.00[a]	2.56[b]	−0.13[c]	0.00	1.28	1.00	0.00	3.87	−0.38	0.00	1.54	1.37
2	M-IR2	0.00	0.00	8.39	0.00	0.00	3.16	0.00	0.01	21.77	0.00	0.00	4.09
3	M-IR3	0.01	2.56	−0.13	0.01	1.71	1.25	0.00	3.85	−0.43	0.00	2.20	1.88
4	M-IR4	0.00	1.92	5.90	0.00	1.10	2.5	0.00	2.90	14.40	0.00	1.40	3.30
5	M-IR5	0.01	2.50	−0.05	0.01	2.20	1.50	0.00	3.80	−0.20	0.00	2.70	2.30
6	M-IR6	0.00	0.00	12.6	0.00	0.00	3.20	0.00	0.00	32.80	0.00	0.00	4.10

[a]At CM
[b]To the right of CM
[c]To the left of CM

97 where h is the overall height of the building (taken as 10.5m). This shows that the building period of this class of systems may generally be around 20% shorter than that computed by code-specified empirical formula. Thus, the code-specified empirical formula for building periods requires to be augmented since a higher estimate of period may often result in underestimation of design shear. However, it is intuitive that the effect of nonstructural elements such as infill wall may contribute to lower the period.

Vibration modes are expectedly coupled for all irregular systems. Fundamental mode of vibration appears to be primarily translational, and the second one is torsion dominated (refer to the corresponding values of participating mass ratio). Thus, the class of buildings chosen is torsionally stiff (TS) – also confirmed from the associated mode shapes. It is evident that, as irregularity of the structure decreases, contribution of torsion-dominated second mode is consistently diminishing while the participation of the translation-dominated fundamental mode increases. Thus, the simple irregularity index (Φ_b, Φ_s) seems consistent with the dynamic characteristics of the systems at least qualitatively implying the rationality of the irregularity index from a conceptual standpoint.

3 Method of Analysis

The response of the structures excited in Y-direction is computed utilizing equivalent static lateral load method. The base shear (V_0) is computed by multiplying the spectral ordinate relevant to period of the system, empirically determined, by seismic weight (refer to Table 1). Seismic coefficients C_a and C_v are chosen as 0.24 and 0.32, respectively. Considering zone factor Z equals to 0.2, occupancy importance factor as unity, and response reduction factor for OMRF as 3.5, design base shear is computed as per relevant guideline of UBC 97 (UBC 97). Design base shear so calculated is distributed over the building height as per LP I and LP II. α and δ (in Eq. (1)) recommended in major seismic codes generally take the value of 1.0 or 1.5 and 0.5 or 1.0, respectively. Three sets of values of α and δ are considered following international codes, viz., IS: 1893–2000, MEXICO 90, NZS 1170.5-2004, and NBCC-90. Static lateral load analysis is conducted utilizing two locations of CS and hence considering story eccentricity (e_s) associated to LP I and LP II.

The second approach employs a dynamic response-spectrum analysis (using design spectrum of UBC 97) with complete quadratic combination (CQC) for modal responses. Adequate number of modes is considered so that at least 95% of the total seismic mass is captured. Response obtained from this dynamic analysis is scaled by a factor equal to V_0/V_{dyna}, where V_{dyna} is the base shear from dynamic analysis.

4 Results and Discussions

From static analysis, maximum response in terms of frame shear, diaphragm displacement is computed employing two locations of CS relevant to LP I and LP II. Three combinations of α and δ, *viz.*, $\alpha = 1.0$, $\delta = 0.5$ (NBCC-90); $\alpha = 1.5$, $\delta = 1.0$ (IS 1893–2002, Mexico-87); and $\alpha = 1.0$, $\delta = 1.0$ (NZS-92), are used.

Figure2a presents the height-wise variation of normalized frame shear in flexible side (frame A) and stiff side (frame D) located at the perimeter of the building (as the effect of torsion is maximum in the edge) relevant to the distribution of design base shear as per LP I (static analysis). Response of flexible side that is considering $\alpha = 1.0$ (NZS-92) is observed to consistently underestimate the response. However, it appears that the response of the flexible side at the base of the building can often be well predicted (with an error limit of −7 to +11) by equivalent lateral load method using $\alpha = 1.5$. However, the same may underestimate the response in the upper stories immediately above the setback in particular. This implies a higher concentration of force in the upper-story elements and in the vicinity of the setback, perhaps owing to the participation of higher modes that corroborates earlier works [8, 30]. $\delta = 0.5$ (NBCC-90) may often overestimate (around 65%), while $\delta = 1.0$ may underestimate the stiff-side response by about 20%.

Figure 2b, on the other hand, describing representative results out of comprehensive case studies, for a height-wise distribution of design base shear as per LP II (in static analysis), reflects a substantial conservatism. It may be recalled that the value of the exponent k involved in the definition of load distribution profile has been recommended [15] as unity (as in LP I) for buildings with fundamental period lesser than 0.5s. The observations of the present investigation support such recommendation.

It may be stated that, while frame shear may be used directly in design, diaphragm displacement profile may also be useful to envisage the damage potential due to pounding and may also be used to estimate inter-story drift – indicator of nonstructural damage. Maximum lateral displacements of the frames (frame A to frame D) computed for all possible combinations of codified amplification factors are normalized to the companion quantities obtained through CQC analysis.

Thus, the normalized displacement profile of the diaphragms, as presented in Fig. 3, may be regarded as the diaphragm displacement factor envelope. It is observed that such factor, in the flexible side, may be as high as 1.25 (or even more, *e.g.*, 1.6 in MIR 4) at story 1 and reduces to some extent (though significant increase is noticed) in higher stories (refer to cases related to LP I). However, the same displays a propensity to decrease toward stiff side and may often underestimate in the perimeter frame. Deviation in diaphragm displacement factor envelope, although exhibits similar trend, is generally higher in case of LP II. This is in line with the earlier response scenario in terms of frame shear parameter.

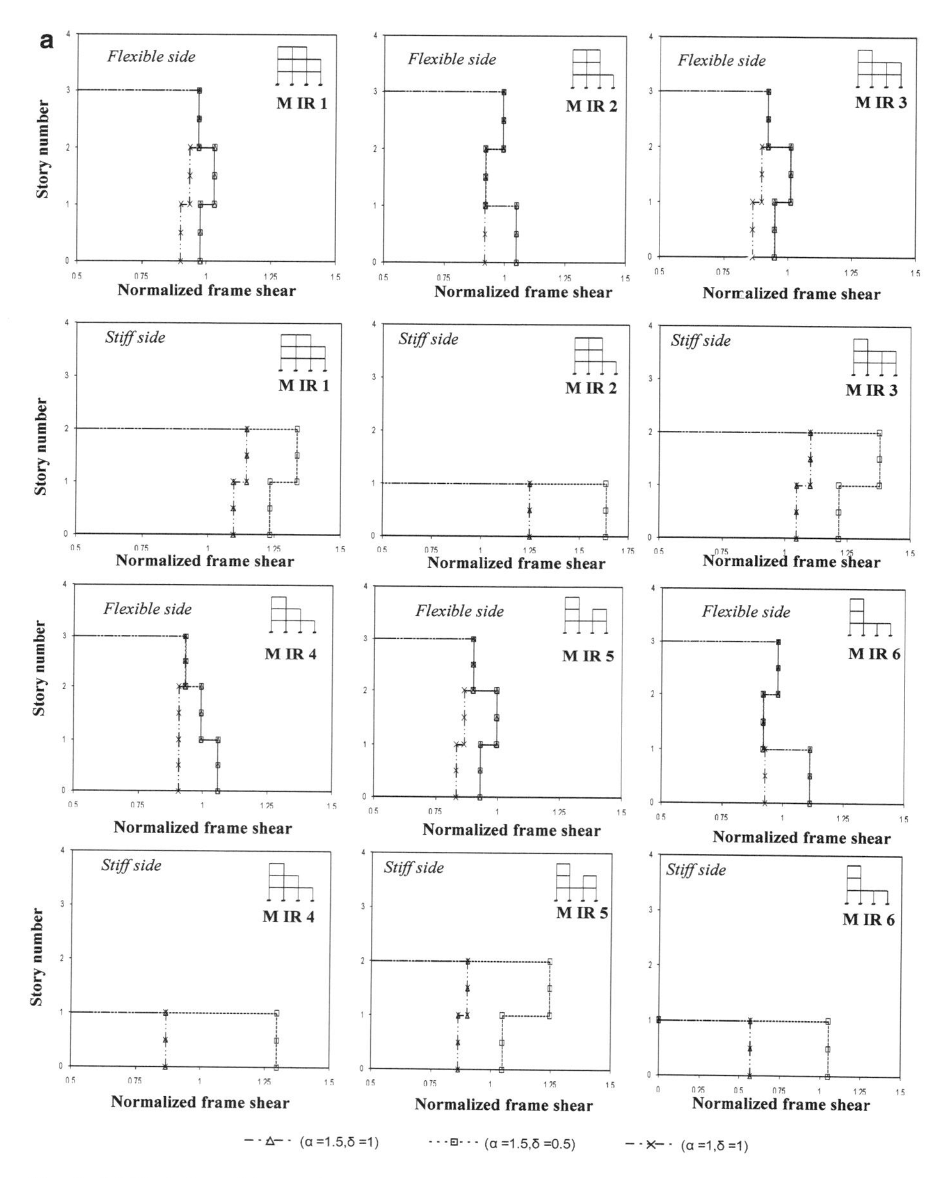

Fig. 2 (**a**) Variation of frame shear in flexible side (frame A) and stiff side (frame D) located at perimeter of code-designed systems (height-wise distribution of loads as per LP I) normalized to CQC response

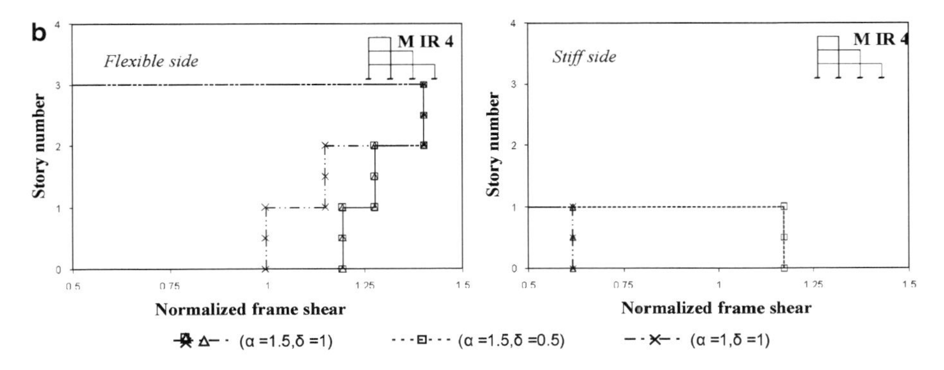

Fig. 2 (continued) (**b**) Variation of frame shear in flexible side (frame A) and stiff side (frame D) located at perimeter of code-designed systems (height-wise distribution of loads as per LP II) normalized to CQC response

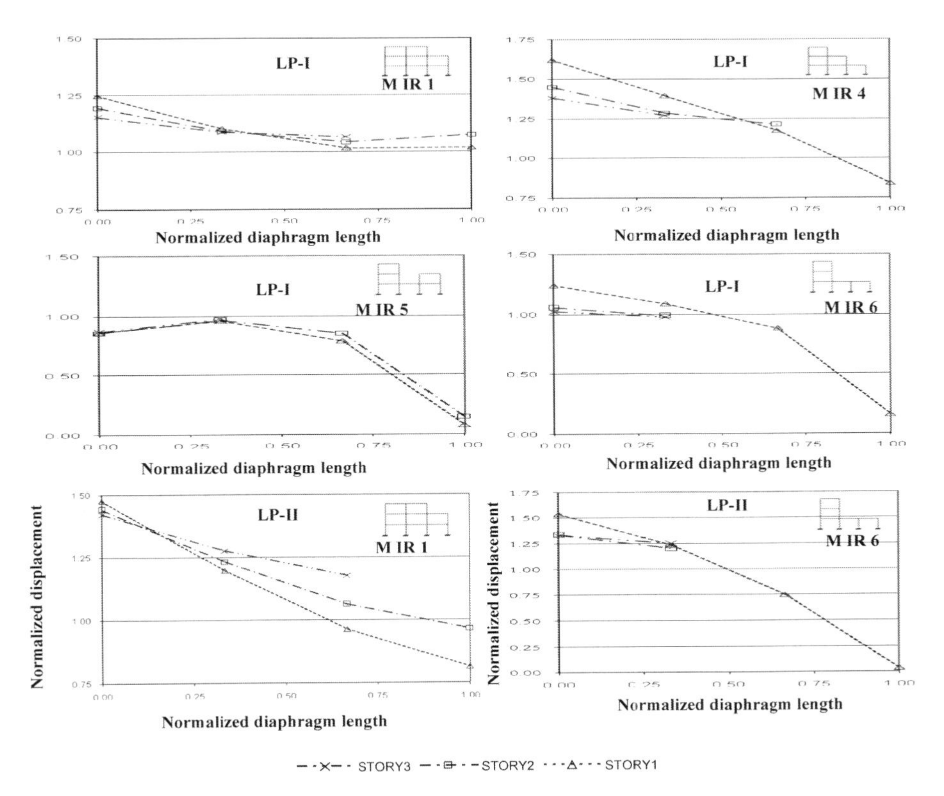

Fig. 3 Variation of diaphragm displacement envelope for code-designed systems (height-wise distribution of load as per *LP I* and *LP II*) normalized to CQC counterpart

5 Summary and Conclusions

In the context of difficulty of performing dynamic analysis in routine seismic design, codified torsional provisions, applied as adjunct to static lateral load analysis, are recognized as important tool. Such provisions, although not explicitly stated, are generally believed to be applicable to buildings with regular asymmetry. The present study examines the reliability of such codified standards for short period systems. To this end, the present investigation systematically examines the response of a number of low-rise systems covering representative forms of elevation irregularity through equivalent static lateral and response-spectrum-based methods. A comparison of the response reveals the following broad conclusions:

Direct design parameters such as frame shear may be reasonably predicted for flexible side of the buildings. Height-wise load distribution complying with LP I and relevant amplification factor $\alpha = 1.5$ should be considered for low-rise systems with setback. However, the code-static procedure does not appear promising for stiff side.

Code-torsional provisions are found to yield a substantially magnified prediction of diaphragm displacement for both the types of lateral load distribution profile adopted. Such magnification is higher toward the base of system and attains its peak at the first story level. However, since, for the purpose of design, the drift of the uppermost story is of major concern, the code-static procedure may be used to envisage the same with proper engineering judgment. Thus, the relevance of code-torsional provisions in the context of inter-story drift or floor lateral displacement – often regarded as important design consideration – is subject to be introspected further.

In view of the above observations, it is perceived that the detailed investigation is deemed essential to consolidate the applicability of code-torsional provisions for medium- to low-rise setback systems to make a final remark. Influence of nonstructural elements should also be accounted in such attempts. Further, codes generally specify to include accidental eccentricity which seems to be inconsistent in view of the relative stability of the location of center of mass in the context of other seismic design parameters [25]. Thus, the present study examines the applicability of code-torsional provisions at the exclusion of accidental eccentricity, and a rigorous study on the influence of the same is currently underway by the authors. Seismic design strategy inherently relies on ductile response, and hence, the performance of such code-designed systems in the post-elastic range of vibration also deserves to be explored in details.

References

1. Al-Ali AAK, Krawinkler H (1998) Effects of vertical irregularities on seismic behavior of building structures. Report no. 130. The John A. Blume Earthquake Engineering Center, Department of Civil and Environmental Engineering, Stanford University, Stanford, USA

2. Aranda GR (1984) Ductility demands for R/C frames irregular in elevation. In: Proceedings of the eighth world conference on earthquake engineering, San Francisco, USA, vol 4, pp 559–566
3. ASCE 7 (2005) Minimum design loads for buildings and other structures. American Society of Civil Engineers, Reston
4. Athanassiadou CJ (2008) Seismic performance of R/C plane frames irregular in elevation. Eng Struct 30:1250–1261
5. Basu D, Gopalakrishnan N (2008) Analysis for preliminary design of a class of torsionally coupled buildings with horizontal setbacks. Eng Struct 30:1272–1291
6. Chandler AM (1986) Building damage in Mexico City earthquake. Nature 320(6062):497–501
7. Cheung VWT, Tso WK (1986) Eccentricity in irregular multistory building. Can J Civil Eng 13(1):46–52
8. Cheung VWT, Tso WK (1987) Lateral load analysis for buildings with setback. J Struct Eng 113(21237):209–227
9. Eurocode 8 (2004) Design of structures for earthquake resistance, part-1: general rules, seismic actions and rules for buildings. European Committee for Standardization (CEN), Brussels
10. Harasimowicz AP, Goel RK (1998) Seismic code analysis of multi-storey asymmetric buildings. Earthq Eng Struct Dyn 27(2):173–185
11. Hejal R, Chopra AK (1987) Earthquake response of torsionally-coupled buildings. Report no. UCB/EERC-87/20. University of California, Berkeley
12. Humar JL (1984) Design for seismic torsional forces. Can J Civil Eng 12(2):150–163
13. Humar JL, Wright EW (1977) Earthquake response of steel-framed multistorey buildings with set-backs. Earthq Eng Struct Dyn 5(1):15–39
14. International Association for Earthquake Engineering (1997) Regulations for seismic design – a world list. IAEE, Tokyo
15. International Code Council (2003) International building code, 2003 edn. ICC, Falls Church
16. International Conference of Building Officials (1997) Uniform building code, 1997 edn. International Conference of Building Officials, Whittier
17. IS 1893–1984 (2002) Indian standard criteria for earthquake resistant design of structures. Bureau of Indian Standards, New Delhi
18. Jiang W, Hutchinson GL, Chandler AM (1993) Definitions of static eccentricity for design of asymmetric shear buildings. Eng Struct 15(3):167–178
19. Karavasilis TL, Bazeos N, Beskos DE (2008) Seismic response of plane steel MRF with setbacks: estimation of inelastic deformation demands. J Constr Steel Struct 64:644–654
20. Kusumastuti D, Reinhorn AM, Rutenberg A (1998) A versatile experimentation model for study of structures near collapse applied to seismic evaluation of irregular structures. Technical report MCEER-05-0002
21. Moehle JP, Alarcon LF (1986) Seismic analysis methods for irregular buildings. J Struct Eng ASCE 112(1):35–52
22. National Research Council of Canada (1990) National building code of Canada. Associate Committee on the National Building Code, Ottawa
23. National University of Mexico (1987) Design manual for earthquake engineering to the construction regulations for the Federal District of Mexico City, Mexico
24. New Zealand standard NZS 4203 (1984) Code of practice for general structural design loadings for buildings. Standards Association of New Zealand, Willington
25. Paulay T (2001) Some design principles relevant to torsional phenomena in ductile buildings. J Earthq Eng 5(3):273–308
26. Poole RA (1977) Analysis for torsion employing provisions of NZRS 4203 1974. Bull N Z Soc Earthq Eng 10(4):219–225
27. Riddell R, Vasquez J (1984) Existence of centres of resistance and torsional uncoupling of earthquake response of buildings. In Proceedings of the 8th world conference on earthquake engineering, vol 4, pp 187–194

28. SAP 2000 (2007) Integrated software for structural analysis and design. Version 11.0. Berkeley, CA. Computers and Structures, Inc.
29. Sarkar P, Prasad MA, Menon D (2010) Vertical geometric irregularity in stepped building frames. Eng Struct 32:2175–2182
30. Shahrooz BM, Moehle JP (1990) Seismic response and design of setback buildings. J Struct Eng ASCE 116(5):1423–1439
31. Smith BS, Vezina S (1985) Evaluation of centres of resistance of multistorey building structures. Proc Inst Civil Eng Part 2 Inst Civil Eng 79(4):623–635
32. Soni DP, Mistri BB (2006) Qualitative review of seismic response vertically irregular building frames. J Earthq Technol 43(4):121–132
33. Tso WK (1990) Static eccentricity concept for torsional moment estimations. J Struct Eng ASCE 116(5):1199–1212
34. Wood SL (1986) Experiments to study the earthquake response of reinforced concrete frames with setbacks. Thesis presented to the University of Illinois, Urbana, 111, in partial fulfillment of the requirements for the degree of Doctor of Philosophy
35. Wood SL (1992) Seismic response of RC frames with irregular profiles. J Struct Eng ASCE 118(2):545–566

Slope Failure Probability Under Earthquake Condition by Monte Carlo Simulation: Methodology and Example for an Infinite Slope

Jui-Pin Wang and Duruo Huang

Abstract A new approach in evaluating the slope failure probability under earthquake condition was proposed in this study. Unlike the use of a deterministic seismic coefficient in a pseudostatic analysis, the uncertainties of earthquake magnitude, location, frequency, and seismic-wave attenuation were taken into account in the new approach. The probability distributions of the earthquake parameters follow those described in probabilistic seismic hazard analysis (PSHA). Along with the considerations of the uncertainties of slope parameters, such as slope angle, slope height, and soil/rock properties, the slope failure probability can be estimated by a probabilistic analysis. In particular, the new approach used Monte Carlo simulation (MCS) in the analysis, in which random parameters were generated with prescribed probability distributions and statistics. With an n-trial MCS being performed, slope failure probability is equal to the ratio between the trial of slope failure and total trials. In this study, the approach was also demonstrated by a benchmark PSHA example integrated with a hypothetical infinite slope. For such a slope under the setup of seismicity, its failure probability increased to 8.3% in a 50-year condition, from 0.16% in a one-year condition. The increase in slope failure probability resulted from a higher earthquake frequency with respect to a longer duration of interest.

Keywords Slope failure probability • Earthquake • Monte Carlo simulation

J.-P. Wang (✉) • D. Huang
Department of Civil & Environmental Engineering, Hong Kong University of Science and Technology, Kowloon, Hong Kong
e-mail: jpwang@ust.hk; drhuang@ust.hk

S. Chakraborty and G. Bhattacharya (eds.), *Proceedings of the International Symposium on Engineering under Uncertainty: Safety Assessment and Management (ISEUSAM - 2012)*,
DOI 10.1007/978-81-322-0757-3_41,

1 Introduction

Earthquake prediction is controversial [3]. A number of research teams have spent countless efforts on earthquake prediction, but the result is not satisfactory; otherwise, the recent earthquake-induced disasters would have been prevented or mitigated. Accordingly, some researchers draw the conclusion about earthquake prediction. That is, the prediction of exact time, location, and size of coming catastrophic earthquakes seems impossible [3]. However, seismic hazard prediction is acceptable by the scientific community [3]. In brief, seismic hazard is a probability-based ground motion. In other words, the seismic hazard prediction is analogue to weather forecast, estimating the probability for a potential event. Accordingly, an earthquake-resistant design of civil engineering can be developed with the desired conservatism for a given type of structures.

A few approaches have been developed for seismic hazard assessments. One of the most popular methods is probabilistic seismic hazard analysis (PSHA), developed in the late 1960s [1]. In the past couple decades, a number of PSHA studies were performed on evaluating earthquake potentials for some regions [8, 12]. Prescribed by a few technical guidelines, PSHA has been used for developing earthquake-resistant designs for critical structures prescribed in industry [5, 15].

Slope stability under earthquake condition is commonly evaluated by the pseudostatic analysis [13], in which earthquake loading owing to excited ground accelerations becomes an extra force. In the pseudostatic analysis, the horizontal seismic coefficient, k_h ($=a_h/g$, where a_h is horizontal earthquake motion), is prescribed by a constant. As an example, in the seismic-stability analysis of a Himalayan rock slope, the horizontal seismic coefficient k_h was selected as 0.31 g based on the maximum credible earthquake around the region [10]. By design, the value ought to be adequately conservative, but the level of conservatism was unknown or unquantified by an exceedance probability or by a recurrence rate.

This study presents the methodology in evaluating the slope stability under earthquake condition, in which the uncertainty of earthquake occurrence is considered, along with those from material properties and slope characteristics. The overviews of PSHA and Monte Carlo simulation (MCS) are described in this chapter. Demonstrated by a benchmark PSHA example integrated with a hypothetical infinite slope, the slope failure probability is evaluated under earthquake condition.

2 Overviews of PSHA

PSHA considers the uncertainties of earthquake size, location, and seismic-wave attenuation and provides the annual rate for a ground motion estimate. With the use of the Poisson distribution, the exceedance probability for such a motion within the duration of interest can be estimated. As an example, the United States

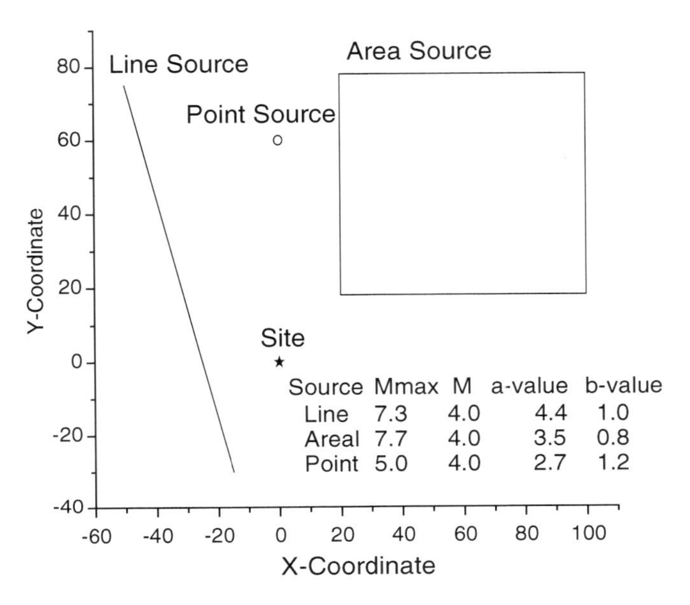

Fig. 1 Setup of the benchmark PSHA example (After Kramer [9])

Geological Survey has used PSHA to develop the national hazard maps with the update in every 6 years [14].

The assumptions and techniques for analyzing earthquake uncertainty in PSHA are described in the following. A uniform distribution is considered for earthquake recurrence in space. In other words, any location within a seismic zone shares an identical probability for the next occurrence. The uncertainty of earthquake magnitude is derived with the use of conditional probability, and accordingly the density function of earthquake magnitude can be developed [4, 11]. Seismic-wave attenuation is also uncertain and can be governed by a so-called ground motion model, which prescribes a predictive, probability-based relationship between earthquakes and ground motions.

Figure 1 shows the setup of a benchmark PSHA example [9]. Three seismic sources are present. Note that the coordinate is not in a longitude-latitude system, so that the distance (d) between two points can be calculated by $d = \sqrt{(x_1 - x_2)^2 + (y_1 - y_2)^2}$. The recurrence parameters, a-value and b-value, and maximum magnitude ($M_{\max}$) and magnitude threshold (M_0) are also summarized in Fig. 1. In addition, the ground motion model used in the example is as follows:

$$\ln Y = 6.74 + 0.859M - 1.8 \times \ln(R + 25) \tag{1}$$

where ln$Y$ denotes the logarithm of PGA in unit of gal; M and R denote magnitude and distance, respectively.

According to the a-value, b-value, M_0, and M_{max}, the annual rates of earthquake (v) can be found as 2.52, 1.99, and 0.008 for the line, area, and point sources, respectively [11]. Note that the number of earthquakes (N) in PSHA is considered a constant, prescribed by annual rate v. However, this consideration is unrealistic, since it is very unlikely that earthquakes would recur periodically in time. Therefore, the number of earthquakes is better regarded as a random variable, and it can be modeled by the Poisson process [6] with a given annual rate. With annual rate equal to 2.52, for instance, the respective probabilities are 8 and 28% for zero earthquake and one earthquake within one year. As a result, the number of events is the additional variable, so that a total of four earthquake variables govern earthquake motion distributions at a site of evaluation.

3 Pseudostatic Analysis for an Infinite Slope in C-Φ Condition

Figure 2 shows the systematic diagram of an infinite slope without considering pore pressure. With the factor of safety (FOS) being defined as the ratio of resistant force to driving force, it was derived as follows:

$$\text{FOS} = \frac{\gamma h \tan\varphi(\cos\beta - k_h \sin\beta) + \frac{c}{\cos\beta}}{\gamma h(\sin\beta + k_h \cos\beta)} \tag{2}$$

where c = cohesion, φ= angle of internal friction, k_h = horizontal seismic coefficient, β = slope angle, h = slope height, and γ = unit weight. Accordingly, the FOS for such a slope is governed by the six parameters. As an example, the FOS is equal to 1.79 in a static condition ($k_h = 0$), with those parameters summarized in Table 1. Figure 3 shows the relationship between critical FOSs and horizontal seismic coefficients. At a critical FOS equal to 1.0, the slope is subject to failure as k_h greater than 0.43.

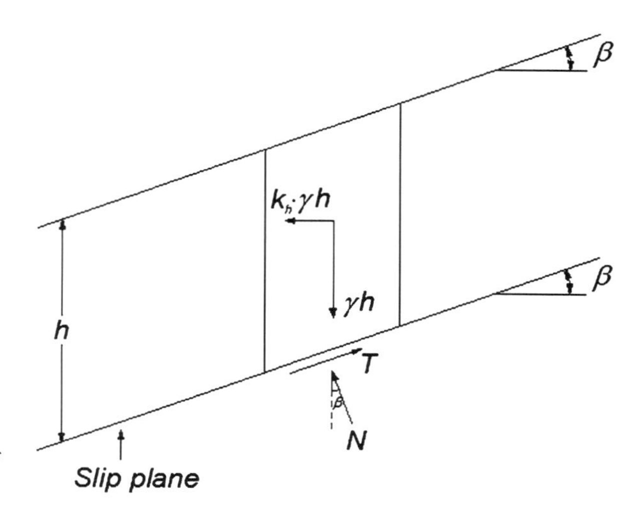

Fig. 2 Systematic diagram of an infinite slope

Table 1 Statistical characteristics of slope parameters

	c [kN/m^2]	φ [°]	h [m]	γ [kN/m^3]	β [°]
Mean value	15	35	5	27	25
COV[a] [%]	10	10	5	5	1
Probability models	Lognormal	Lognormal	Lognormal	Lognormal	Lognormal

[a]COV = coefficient of variation

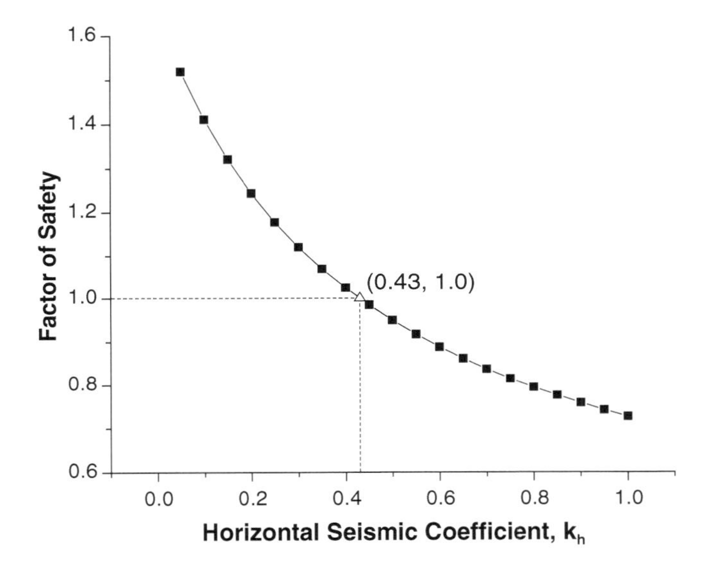

Fig. 3 Relationship between critical FOSs and horizontal seismic coefficients

4 Slope Failure Probability by Monte Carlo Simulation Under Earthquake Condition

4.1 Monte Carlo Simulation

MCS is one of the widely used probabilistic analyses. Its essential is to generate random values following prescribed statistics and probability distributions. With a number of trials being performed, the slope failure probability (P_F) is estimated as follows:

$$P_F = \frac{\sum_{i=1}^{n} F_i \leq F_C}{n} \tag{3}$$

where n is the number of trials (or sample size) in MCS and F_i and F_C denote the FOS in the ith trial and critical factor of safety, respectively. The nominator of the expression dictates the number of trials presenting a scenario of slope failure.

4.2 Randomization of Ground Motions

In this analysis, the randomization of the five slope parameters (c, φ, γ, β, h) is relatively straightforward, with their respective statistics (mean value and COV) and probability distribution being provided. However, this is not the case in the randomization of ground motion, as it is related to the four earthquake parameters as mentioned. Its randomization is described in detail in the following.

For a magnitude-distance ground motion model expressed by $f(M, R)$, such as Eq. 1, given $M = m_0$ and $R = r_0$, the mean value and SD of the logarithm of ground motion ($\ln Y$) are equal to $f(m_0, r_0)$ and σ^*, respectively. Since $\ln Y$ follows a normal distribution [9], the relationship between Z (standard normal variate), M, R, and Y can be derived as

$$\begin{aligned} Z &= \frac{\ln Y - f(M,R)}{\sigma *} \\ &\Rightarrow \ln Y = Z \times \sigma * + f(M,R) \\ &\Rightarrow Y = \exp(Z \times \sigma * + f(M,R)) \end{aligned} \tag{4}$$

Therefore, a random ground motion is governed by the three earthquake variables that are randomly generated. For a random magnitude, it can be generated with respect to the prescribed density function used in PSHA. To ensure a high precision in simulation, the magnitude density functions were developed with the use of the magnitude interval as small as 0.05. Figure 4 shows the functions for the three seismic zones. A random distance can be obtained with a random location being generated in advance. Unlike the two parameters, a random z that controls the variability in ground motion attenuation can be generated in a straightforward manner as randomly generating slope parameters, since Z follows a normal distribution with mean and SD equal to 0 and 1, respectively.

4.3 Randomization of Maximum Ground Motions

In failure assessments under earthquake condition, a deterministic criterion is usually followed. That is, catastrophic failure only occurs when the design motion is exceeded by an earthquake motion. In other words, the number of earthquake motions less than the design motion that a structure has experienced is not a matter to failure, and as long as the structure can survive under a maximum-motion condition, the failure probability is literally zero. Therefore, the distribution of the maximum motion is of interest in earthquake-resistant design.

Considering n earthquakes within a given seismic zone, the maximum ground motion ($Y_{\max}$) is as follows:

$$Y_{\max} = MAX\{Y_1, Y_2, \ldots, Y_n\} \tag{5}$$

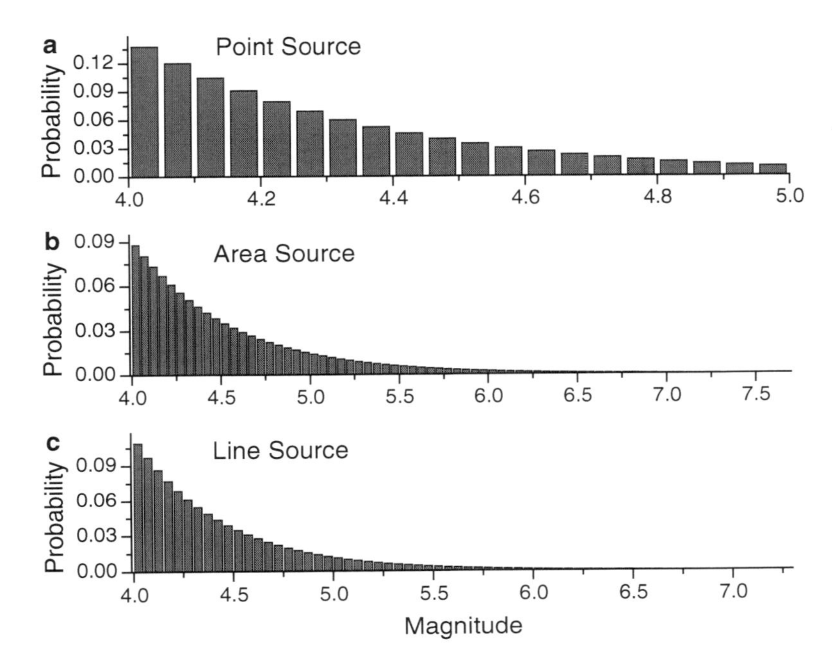

Fig. 4 Magnitude probability density functions for the three seismic sources

where Y can be obtained from Eq. 4. Normally, multiple seismic sources are present in a region as the benchmark PSHA example. Extended from Eq. 5, the ultimate maximum motion ($\tilde{Y}_{\max}$) in an m-source condition can be expressed as follows:

$$\tilde{Y}_{\max} = MAX\{Y^1_{\max}, Y^2_{\max}, \ldots, Y^m_{\max}\} \tag{6}$$

where $Y^i_{\max}$ can be obtained from Eq. 5.

4.4 *Program Tool*

A program was developed in-house for the computer-aided analysis. The program was compiled with the use of the spreadsheets and "macros" (user-defined functions and subroutines) in Excel. The uniform-number generator in Excel that randomly generates values (U) between 0 and 1 was used. For those variables with given probability distributions, such as slope parameters, their respective random values (R) can be converted from U with the use of inverse probability function, $f^{1}()$, as follows:

$$R = f^{-1}(U) \tag{7}$$

On the other hand, for earthquake magnitude with a density function back-calculated from conditional probability, random magnitudes were generated by the mapping rule as follows:

$$M = \begin{cases} m_1; & U \leq \tilde{p}_1 \\ m_{i+1}; & \tilde{p}_i < U \leq \tilde{p}_{i+1} \end{cases} \quad (8)$$

where m_i and $\tilde{p}_i$are the ith earthquake magnitude and cumulative probability, which can be determined from the density function shown in Fig. 4.

5 Results of the Numerical Example

5.1 *Statistical Characteristics of Slope Parameters*

The statistical characteristics, such as COV and probability distribution, of slope parameters (e.g., slope angle) are also summarized in Table 1. A lognormal distribution was used for simulating the distributions of material strength, such as cohesion and angle of internal friction [2, 7]. Without information available regarding probability distributions with respect to slope angle and slope height, both were assumed to follow lognormal distributions as well. The levels of variations for the slope parameters are estimated accordingly. A low variation was assigned for slope angle, since the angle in an infinite slope should keep constant.

5.2 *Results and Discussions*

Figure 5 shows the relationship between failure probability and critical factors of safety in four different durations. The slope failure probability increases with a longer duration being considered. With critical FOS equal to 1.0, the failure probability is increased to 8.3% in 50 years, from 0.16% in one year. A linear relationship in the $\log(P_F) - F_c$ space was found.

Figure 6 shows the relationship between slope failure probabilities and durations of evaluation. Linear relationships with a positive correlation were found between the two variables, especially when the critical factor of safety is close to 1.0. As the critical factor of safety was increased to as high as 1.5, the relationship between failure probability and duration becomes more nonlinear.

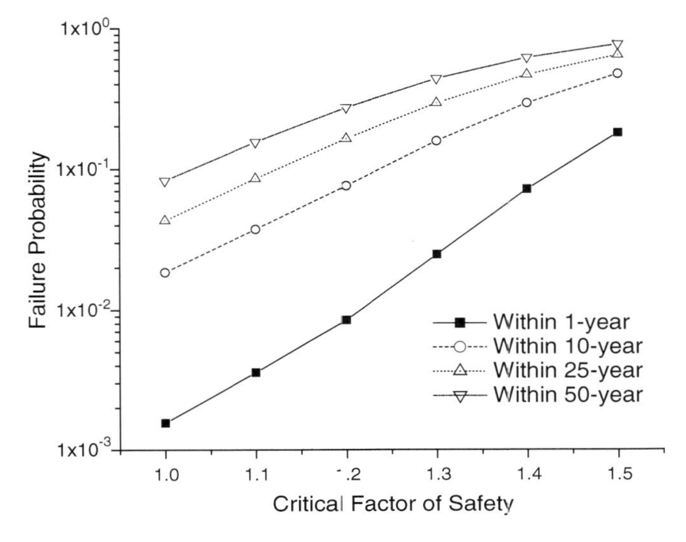

Fig. 5 Relationship between failure probability and critical factors of safety for the slope shown in Fig. 3

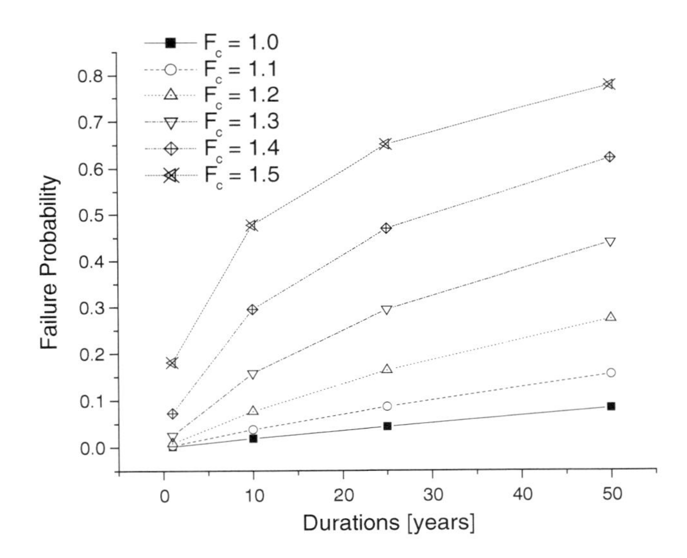

Fig. 6 Relationship between failure probability and duration for the slope shown in Fig. 3

6 Conclusions

A new approach in the assessment of slope failure probability under earthquake condition was described and proposed in this study. The new approach considers the uncertainties of earthquake parameters, such as magnitude, location, frequency, and seismic-wave attenuation, along with the uncertainty of slope parameters. The essential of the probabilistic analysis is with the use of MCS. The slope failure probability is governed by the ratio between the trials with a slope failure scenario and total trials. The approach was demonstrated with the use of a benchmark PSHA example integrated with a hypothetical slope. The slope failure probabilities were successfully evaluated for four different durations, and a positive correlation between failure probability and duration was found.

References

1. Cornell CA (1968) Engineering seismic risk analysis. Bull Seism Soc Am 58(5):1583–1606
2. Duzgun HSB, Yucemen MS, Karpuz C (2003) Probabilistic modeling of plane failure in rock slopes. Appl Stat Probab Civ Eng 1:1255–1262
3. Geller RJ, Jackson DD, Kagan YY, Mulargia F (1997) Earthquakes cannot be predicted. Science 275(5306):1616
4. Gutenberg B, Richter CF (1944) Frequency of earthquakes in California. Bull Seism Soc Am 34(4):1985–1988
5. International Atomic Energy Agency (IAEA), Vienna (2002) Evaluation of seismic hazards for nuclear power plants safety guide, Safety standards series no. NS-G-3.3
6. Jafari MA (2010) Statistical prediction of the next great earthquake around Tehran, Iran. J Geodyn 49:14–18
7. Jimenez R, Sitar N, Chacon J (2006) System reliability approach to rock slope stability. Int J Rock Mech Min Sci 43(6):847–859
8. Kebede F, Van Eck T (1997) Probabilistic seismic hazard assessment for the Horn of Africa based on seismotectonic regionalization. Tectonophysics 270:221–237
9. Kramer AL (1996) Geotechnical earthquake engineering. Prentice-Hall, Inc., Upper Saddle River
10. Latha GM, Garaga A (2010) Seismic stability analysis of a Himalayan rock slope. Rock Mech Rock Eng 43(6):831–843
11. McGuire RK, Arabasz WJ (1990) An introduction to probabilistic seismic hazard analysis. Geotech Environ Geophys Soc Explor Geophys 1:333–353
12. Sokolov VY, Wenzel F, Mohindra R (2009) Probabilistic seismic hazard assessment for Romania and sensitivity analysis: a case of joint consideration of intermediate-depth (Vrancea) and shallow (crustal) seismicity. Soil Dyn Earthq Eng 29(2):364–381
13. Terzaghi K, (1950) Mechanics of landslides. Bull Geol Soc Am Berkeley Volume:83–123
14. United States Geological Survey (2008) United States geological survey national seismic hazard maps. http://earthquake.usgs.gov/hazards/products/conterminous/2008/maps/
15. United States Nuclear Regulatory Commission (USNRC) (2007) A performance-based approach to define the site-specific earthquake ground motion, Regulatory Guide 1.208. Washington, DC

Polynomial Chaos in Bootstrap Filtering for System Identification

P. Rangaraj, Abhijit Chaudhuri, and Sayan Gupta

Abstract The objective of this chapter is to develop a computationally efficient approach for system identification. An algorithm, the bootstrap filter in conjunction with polynomial chaos expansion, is proposed for identification of system parameters. The central idea of the proposed method is the introduction of response through polynomial chaos expansion in the filtering algorithm. Appreciable performance of the proposed algorithm is been provided by considering the problem of the identification of the properties of a single degree of freedom system.

Keywords System identification • Polynomial chaos • Spectral representation • Particle filter • Probabilistic collocation

1 Introduction

Several approaches are discussed in the literature for structural system identification. They can be broadly classified into deterministic and probabilistic approaches. Deterministic approaches to system identification have considerable limitations over probabilistic approaches. Probabilistic identification techniques can be grouped into two categories. The first includes the well-known Kalman filter and its variants [8, 7], and the second includes the Monte Carlo simulation-based algorithms, also referred to as particle filters [3, 6, 10]. The Kalman filter is an efficient method for estimating the state of the system from noisy measurement data using a simple predictor-corrector approach. It consists of a set of mathematical equations that provides a recursive means to estimate the state of the process. The filter can be used to estimate the system state only when the process model

P. Rangaraj (✉) • A. Chaudhuri • S. Gupta
Department of Applied Mechanics, Indian Institute of Technology Madras, Chennai 600036, TN, India
e-mail: rangaraj.p138@gmail.com; abhijit.chaudhuri@iitm.ac.in; sayan@iitm.ac.in

S. Chakraborty and G. Bhattacharya (eds.), *Proceedings of the International Symposium on Engineering under Uncertainty: Safety Assessment and Management (ISEUSAM - 2012)*,
DOI 10.1007/978-81-322-0757-3_42,

and measurement model equations are linear. When the process equations are non-linear, Kalman filters cannot be used directly. Instead, an adaptive strategy that enables the use of Kalman filter has to be adopted. However, in most parameter identification problems, the process model and measurement model equations are nonlinear and hence, Kalman filter-based techniques can be used only in an approximate form.

Particle filtering algorithms work on the same principle of predictor-corrector approach used in the Kalman filters. However, unlike in the case of Kalman filters, here the process of Bayesian updating is carried out using Monte Carlo simulations. Since particle filtering algorithms are based on Monte Carlo simulations, there are no restrictions on the model and/or the measurement equations being linear. As a result, particle filtering algorithms are more widely applicable.

2 System Identification

To analyze and identify system parameters of a dynamic system from measurements, the system equations are written in the form as

$$\mathbf{X}_{k+1} = f(\mathbf{X}_k, \theta_k, \mathbf{w}_k) \tag{1}$$

$$\mathbf{Z}_k = f(\mathbf{X}_k, \theta_k, \mathbf{v}_k). \tag{2}$$

Equation (1) is the process model equation that relates the states of the system at successive time instants in a recursive manner, while Eq. (2) is the measurement model equation that relates the measurements to the state of the dynamical system. Here, $\mathbf{X}_k$ is a n-dimensional vector representing the state space associated with the system at the kth time instant, θ_k denotes the vector of system parameters at time instant k, $\mathbf{Z}_k$ are measurements at time k, and $\mathbf{w}_k$ and $\mathbf{v}_k$ represent the process model noise and measurement model noise, respectively, $f(\cdot)$ denotes the functional form that relates the state vector $\mathbf{X}_k$, and $\mathbf{X}_{k+1}$ and $g(\cdot)$ represent the corresponding functional relationship between $\mathbf{X}_k$ and the measurements $\mathbf{Z}_k$. In the Bayesian approach to dynamic state estimation, one attempts to construct the posterior probability density function (pdf) of the state based on information available from measurements. Since this pdf embodies all available statistical information, it may be said to be the complete solution to the estimation problem. For many problems, an estimate is required every time that a measurement is received. In this case, a recursive filter is a convenient solution. A recursive filtering approach means that received data can be processed sequentially rather than as a batch so that it is not necessary to store the complete data set nor to reprocess existing data if a new measurement becomes available. Such a filter consists of essentially two stages: prediction and update. The prediction stage uses the system model to predict the state pdf forward from one measurement time to the next. The update operation uses the latest measurement to modify the prediction pdf. This is achieved using Bayes theorem, which is the mechanism for updating knowledge about the target state in the light of extra information from new data. Particle filters are based on these approach of prediction-updation. It is a technique for implementing a recursive Bayesian filter by Monte Carlo simulations.

2.1 Bootstrap Particle Filtering

In this study, we use the bootstrap particle filter proposed in Gordon [6]. The objective of the bootstrap filter is to estimate the current state of the system $\mathbf{X}_k$, from the available measurements $\mathbf{D}_k$, where the set of k measurements from initial time is represented as $\mathbf{D}_k = \{\mathbf{Z}_1,\mathbf{Z}_2,\ldots,\mathbf{Z}_k\}$. We aim to explore the application of dynamic state estimation procedures to achieve parameter estimation. Since both $\mathbf{X}_k$ and $\mathbf{Z}_k$ are corrupted by noise $\mathbf{w}_k$ and $\mathbf{v}_k$, complete information of the state is possible if the pdf of $\mathbf{X}_k$ conditioned on the available measurements $\mathbf{D}_k$ and denoted as $p(\mathbf{X}_k|\mathbf{D}_k)$ is available. The key steps in the implementation of the algorithm are as follows [6]:

1. Assume $p(\mathbf{X}_k|\mathbf{D}_{k-1})$ is known. Generate the vector of random samples $\{\mathbf{X}_{k-1}\}_{i=1}^{N}$ from the pdf $\{\mathbf{X}_{k-1}\}_{i=1}^{N}$.
2. Generate samples $\{\mathbf{w}_{k-1}\}_{i=1}^{N}$ from the known distribution of $p(\mathbf{w}_{k-1})$.
3. Each sample is passed through the system model in Eq. (1) to obtain the predictions for the state at time step k. Thus,

$$\mathbf{X}_k^{*(i)} = f_{k-1}(\mathbf{X}_{k-1}^{(i)}, \mathbf{w}_{k-1}^{(i)}). \tag{3}$$

4. Once the measurements $\mathbf{Z}_k$ is available, evaluate the likelihood of each prediction $\{\mathbf{X}_k^{*(i)}\}_{i=1}^{N}$ and obtain a normalized weight for each sample, given by

$$q_i = \frac{p(\mathbf{Z}_k \mid \mathbf{X}_k^{*(i)})}{\sum_{j=1}^{N} p(\mathbf{Z}_k \mid \mathbf{X}_k^{*(j)})}. \tag{4}$$

5. Define a discrete distribution over $\{\mathbf{X}_k^{*(i)}\}_{i=1}^{N}$, with probability mass q_i associated with element i.
6. Resample N times from the discrete distribution to generate samples $\{\mathbf{X}_k^{*(i)}\}_{i=1}^{N}$ so that for any j, $P[\mathbf{X}_k^{j} = \mathbf{X}_k^{*(i)}] = q_i$.

The above steps of prediction and update form a single iteration of the recursive algorithm. The above algorithm was proposed to predict the state of a dynamical system [6]. However, in this study, our focus is on identifying the parameters of a dynamical system. One way to approach the problem is to augment the state vector with the vector of parameters that need to be estimated. The problem with such anapproach is that parameter identification and state estimation of the system result in an increase of the size of the vectors that need to be estimated. This results in wastage of significant computational effort in estimating the state vectors.

Instead, if the system parameters are assumed to be independent parameters and the measurement equation is expressed as a function of these parameters, then significant reduction in computational efforts could be achieved. This has been discussed by Nasrellah and Manohar [10]. Here, the model equation is written as

$$\theta_{k+1} = \theta_k + \mathbf{w}_k, \tag{5}$$

while the measurement equation is given by

$$\mathbf{Z}_k = g(\theta_k) + \mathbf{v}_k, \tag{6}$$

where the variables are as already defined earlier. It is worth commenting here on the process noise $\mathbf{w}_k$. If it is assumed that the system parameters θ_k do not change with time, then the process equation should be $\theta_{k+1} = \theta_k$. Now, if the bootstrap filter algorithm (BFA) is applied, subsequent resampling would lead to the generation of a particular sample which has the greatest weightage q_k among the N initial sampled realizations for θ_k. As the algorithm progresses, all the resampled data would converge to a particular realization and would lead to degeneracy. To overcome this problem, a small noise $\mathbf{w}_k$ is added to jitter the samples at each time step. This would enable a greater variation of resampled data at time t_{k+1} around the realization of the sample with the greatest weight at time t_k. This procedure also has the effect of increasing the least count accuracy of the algorithm in predicting the system parameters. The effect of jittering on the resampled data with artificial noise has been recommended in the literature [10]. A more exhaustive study on this has also been reported in [9].

The computational costs of the bootstrap filtering algorithm depend on

(a) The number of particles N used at each iteration
(b) The number of times the model is updated using the BFA

Here, ideally one should apply the BFA to each measurement data. This would imply that the BFA implementation rate would be equal to the measurement sampling rate. Moreover, it must be noted that application of BFA at each time step requires N calls to the process model. Thus, if there are M measurement points and N particles at each BFA step, the system would need to be solved NM times. This could be computationally very costly especially for complex structures, where the solution of a model equation could involve time in terms of hours. There is therefore a need to investigate the possibility of reducing the computational costs. This chapter focuses on the development of a polynomial chaos-based approach to reduce computational costs.

3 Polynomial Chaos Expansion

Polynomial chaos expansion (PCE) [4, 5] is a spectral uncertainty quantification tool, which is based on the homogeneous chaos theory of Wiener [11]. The stochastic input in PCE is represented spectrally by employing orthogonal polynomial functionals from the Askey scheme as basis in the random space. In its original form, it employs Hermite polynomials as basis from the generalized Askey scheme and Gaussian random variables. It can be shown that an optimum convergence is achieved for Gaussian inputs. As per the Cameron-Martin theorem [1], a random process $X(t, \theta)$ (as function of random event θ) which is second-order stationary can be written as

$$X(t,\theta) = \hat{a}_0\psi_0 + \sum_{i_1=1}^{\infty} \hat{a}_{i_1}\psi_1(\xi_{i_1}(\theta)) + \sum_{i_1=1}^{\infty}\sum_{i_2=1}^{i_1} \hat{a}_{i_1 i_2}\psi_2(\xi_{i_1}(\theta), \xi_{i_2}(\theta)) + \ldots, \quad (7)$$

where $\psi_n(\xi_{i1}, \xi_{i2}, \ldots, \xi_{in})$ denotes the Hermite polynomial of order n in terms of n-dimensional independent standard Gaussian random variables $\xi = (\xi_{i1}, \xi_{i2}, \ldots \xi_{in})$ with zero mean and unit variance. The above equation is the discrete version of the original Wiener polynomial chaos expansion, and the continuous integrals are replaced by summations. For notational convenience, Eq. (7) can be written as

$$X(t,\theta) = \sum_{j=0}^{\infty} a_j(t)\phi_j(\xi(\theta)), \quad (8)$$

where a_j denotes the deterministic coefficients of the random process $X(t, \theta)$ and ϕ_j denotes the Hermite polynomials. ξ's are independent standard Gaussian random variables with zero mean and unit variance. The 1-D Hermite polynomials can be expressed in the recursive form as

$$\phi_n = \xi\phi_{n-1} - (n-1)\phi_{n-2}, \quad (9)$$

where the first few Hermite polynomials are

$$\phi_0 = 1,\ \ \phi_1 = \xi,\ \ \phi_2 = \xi^2 - 1,\ \ \phi_3 = \xi^3 - 3\xi,\ \ \phi_4 = \xi^4 - 6\xi^2 + 3. \quad (10)$$

An approximation of $X(t, \theta)$ can be obtained by truncating the series to p terms as

$$X(t,\theta) = \sum_{j=0}^{p} a_j(t)\phi_j(\xi(\theta)), \quad (11)$$

where

$$p = \frac{(n+n_p)!}{n!n_p!} - 1 \quad (12)$$

for n number of random variable and a polynomial order n_p. Equation (11) is referred to as p-order PCE expansion.

4 Spectral Decomposition of the Response

To illustrate how PCE can be used with BFA, we consider the single degree of freedom system excited by a harmonic excitation and governed by the equation

$$m\ddot{x} + c\dot{x} + kx = f(t) = A\cos(\omega t), \tag{13}$$

where m, c, k are the system mass, damping, and stiffness; $f(t)$ is the forcing function; and A and ω represent the amplitude and frequency of the forcing function, respectively. We assume that the response measurement at discrete time instants is available and that the objective of the study is to estimate damping parameter c. We use BFA to estimate c.

In the absence of any information, we model c as a random variable that is normally distributed with mean c_0 and standard deviation c_1. The spectral representation of c can be expressed as

$$c = c_0 + c_1\xi = \sum_{i=0}^{1} c_i\phi_i. \tag{14}$$

Substituting Eq. (14) in Eq. (13), it is possible to represent the response $x(t, \theta)$ as a PCE. Assuming the response

$$x(t,\theta) = \sum_{j=0}^{p} x_j(t)\phi_j(\xi(\theta)) \tag{15}$$

and substituting in Eq. (13), we get

$$m\left(\sum_{j=0}^{p}\ddot{x}_j(t)\phi_j(\xi)\right) + \left(\sum_{i=0}^{1} c_i\phi_i(\xi)\right)\left(\sum_{j=0}^{p}\dot{x}_j(t)\phi_j(\xi)\right) + k\left(\sum_{j=0}^{p}x_j(t)\phi_j(\xi)\right) = f(t). \tag{16}$$

Simplifying Eq. (16) leads to

$$m\sum_{j=0}^{p}\ddot{x}_j(t)\phi_j(\xi) + \sum_{i=0}^{1}\sum_{j=0}^{p} c_i\dot{x}_j\phi_i(\xi)\phi_j(\xi) + k\sum_{j=0}^{p}x_j\phi_j(\xi) = f(t). \tag{17}$$

Here, the unknowns are the deterministic coefficients, $\{x_j\}$. It must be noted that the system being linear and c being assumed Gaussian, the response $x(t)$ is also Gaussian, and hence, Eq. (15) will contain only two terms. However, in deriving the methodology, we consider a more general case where the system could be nonlinear and the response non-Gaussian. Thus, in developing the formulation, we consider the more general case where the size of the vector $\{x_j\}$ is $p > 2$. To obtain the estimates of x_j, we decouple the Eq. (17). This can be carried out in two approaches. This is discussed in the following sections.

4.1 Classical Galerkin PCE

Using Galerkin projection on the Eq. (17), which involves multiplying by ϕ_k and taking expectations on both sides of Eq. (17), we get

$$m\sum_{j=0}^{p}\ddot{x}_j(t)<\phi_j\phi_k> + \sum_{i=0}^{1}\sum_{j=0}^{p}c_i\dot{x}_j<\phi_i\phi_j\phi_k> + k\sum_{j=0}^{p}x_j<\phi_j\phi_k> = f(t)<\phi_k>, \tag{18}$$

where the expectation operator $< \cdot >$ is defined as

$$<\phi_l, \ldots \phi_k> = \int_{-\infty}^{\infty}\phi_l \ldots \phi_k w(\xi)d\xi. \tag{19}$$

Here, $w(\xi)$ is the weighting function. For Hermite polynomials, the weighting function $w(\xi)$ is the Gaussian probability density function and is of the form

$$w(\xi) = \frac{1}{\sqrt{2\pi}}\exp^{-}\left(\frac{1}{2}\xi^T\xi\right). \tag{20}$$

The Hermite polynomials are orthogonal with respect to this weighting function in the Hilbert space. The polynomial chaos forms a complete orthogonal basis in the L_2 space of real-valued functions depending on the Gaussian random variables; hence, the inner product of two orthogonal polynomial can be replaced by the identity

$$<\phi_l\phi_k> = <\phi_l^2>\delta_{lk}, \tag{21}$$

where δ_{lk} is the Kronecker delta function, given as

$$\delta_{lk} = \begin{cases} 1 & \text{if } l = k \\ 0 & \text{otherwise.} \end{cases}$$

The inner product terms in Eq. (18) can be evaluated analytically prior to computations and substituted in the equation. The resulting system leads to a set of coupled deterministic differential equation in terms of the chaos coefficients $\{x_j\}$. The Galerkin approach is also called the intrusive approach as it modifies the system governing equations in terms of the chaos coefficients.

4.2 Nonintrusive Projection Method

A number of nonintrusive variants of PCE have been developed in the literature. In this study, we consider the stochastic projection method [2]. Here, the chaos expansions are not substituted in the governing equations; instead, samples of the solutions are used to evaluate the coefficients directly using a projection formula. As a result, this approach can utilize the existing deterministic code and hence the name nonintrusive. Here, the response is approximated by a truncated series as shown in Eq. (11). The Hermite polynomials are statistically orthogonal, i.e., they satisfy $< \phi_l \phi_k > = 0$ for $l \neq k$. Thus, the expansion coefficients can be directly evaluated as

$$x_j(t) = \frac{< x(t, \xi(\theta)) \phi_j >}{< \phi_j^2 >}. \tag{22}$$

The main difficulty here lies in evaluating the expectation in the numerator of the above expression. A Gauss-Hermite quadrature will be suitable for evaluating the above as the domain is $(-\infty, \infty)$ and the weight is Gaussian pdf. The quadrature points are the zeros of the Hermite polynomials of chosen order. A number of deterministic runs are performed at the quadrature points. It is to be noted that the number of deterministic runs is still much lower than Monte Carlo simulations that would be necessary if we use BFA without PCE. We refer this step as a pseudo-Monte Carlo simulation approach, and it consists of the following steps:

1. The samples of the random variables in the problem are generated based on the ξ values which correspond to the Gauss-Hermite quadrature points.
2. The realizations of the system response $x(t, \theta)$ are then used to estimate the deterministic coefficients, $x_j(t)$'s, in Eq. (22) using the Gauss-Hermite quadrature rule.
3. The final response is obtained by reconstructing these determined coefficients back into the chaos expansion.

5 Polynomial Chaos Approach to Bootstrap Filtering

Using the PCE approach discussed in the preceding section, it is possible to bypass the necessity for performing computationally costly structural analysis at each step of bootstrap filtering. The basic steps involving coupling PCE with the bootstrap filter algorithm are as follows:

1. The measurements $\mathbf{X}_k$ are simulated for the known parameters. This serves as the reference for the filtering of particles.
2. The random parameter of the system, which is to be identified, is represented in a PCE for a given mean and standard deviation as shown in Eq. (14).

3. Using the available system data and excitation, response is calculated for the whole time step from the methods discussed in the previous section.
4. The normalized weights for prediction at first time step is calculated for all particles as mentioned in the filtering algorithm.
5. From the normalized weights, the resampling is done on ξ rather than the random parameter. This enables the corresponding change in both the response and the parameter.
6. The response is updated using the resampled ξ's and is used for the analysis at next time step.
7. This processing is carried out for the whole time step, enabling the random parameter updation based on ξ's.

The above method is efficient in that it avoids construction of PCE for the response at each time step, which can be computationally very costly. This is made possible by assuming that ξ to be the primary random variable rather than c. Thus, the calculated chaos coefficients for the system response at step 3 remain unchanged throughout the analysis. It is to be noted that, in BFA, the structural analysis is done for all particles at every time step, whereas in PCE approach, the structural analysis is carried out only once throughout the whole analysis. This brings about a significant reduction in the computational effort especially when dealing with large problems, where performing structural analyses many times consumes massive computational time.

If the bootstrap filtering is applied corresponding to each measurement data, then the number of structural analysis required per second of data would be NM if M is the sampling rate for data and N is the number of particles at each filtering step. Studies have shown [9, 10] that instead, if filtering is carried out every K steps of data, there is not much difference in the quality of the predictions of the parameters to be identified. However, this brings about significant reduction in the computations as the number of times the structural analyses is required is now $\frac{MN}{K}$. However, it may be argued that here, one omits some of the measurements, and in effect, one is wasting the available data from the measurements. To address this issue, we applied the BFA using the following four distinct methods and compared the quality of the predictions for the parameters.

Method 1: Bootstrap filter is carried out at all available data points. The number of times the structural analysis is carried out is NM.

Method 2: Bootstrap filter is carried out at every K available data points. The number of times the structural analysis is carried out is $\frac{MN}{K}$. However, here, we are not using the information available from the measurements and are in effect wasting available data.

Method 3: The likelihood is computed at all available data points. However, the resampling is done every K steps from the mean likelihood calculated from the preceding $\frac{M}{K}$ steps. The number of times the structural analysis is carried out is $\frac{MN}{K}$, but here, we use all the information available from the measurement data.

Method 4: Here, the mean of the measurements of preceding $\frac{M}{K}$ steps is used to calculate the likelihood at every K time step and is similar in all aspects to Method 3. Thus, it uses all the available information from the measurements. As there is a change only in calculation of likelihood, the number of structural analysis done remains same as the above method being $\frac{NM}{K}$.

6 Results

The proposed methodology was applied to a single degree of freedom system to estimate the uncertain damping parameter. The system was subjected to a harmonic loading of $A\cos(wt)$. The numerical parameters of the system considered are $m = 10$ kg, $k = 100$ N/m, $c = 20$ Ns/m, $A = 100$, $w = 1\pi$ rad/s. Sampling rate is assumed to be 100 s^{-1}. The measurements are obtained by solving the deterministic equations by adaptive fourth order Runge-Kutta algorithm. Figure 1 shows the time history of the response for 40 s. Here, the full line represents the true response and the dotted line represents the measurement which has been obtained when the response is seeded with Gaussian noise with 1 % variance. The number of measurement points for the assumed sampling rate is 4,001. We now approach the

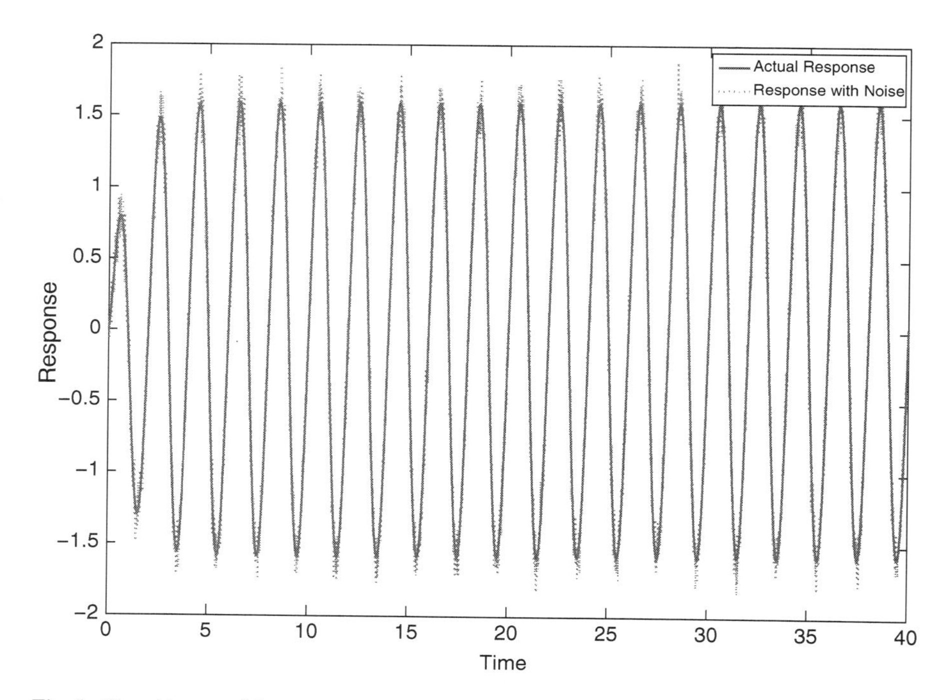

Fig. 1 Time history of the system response

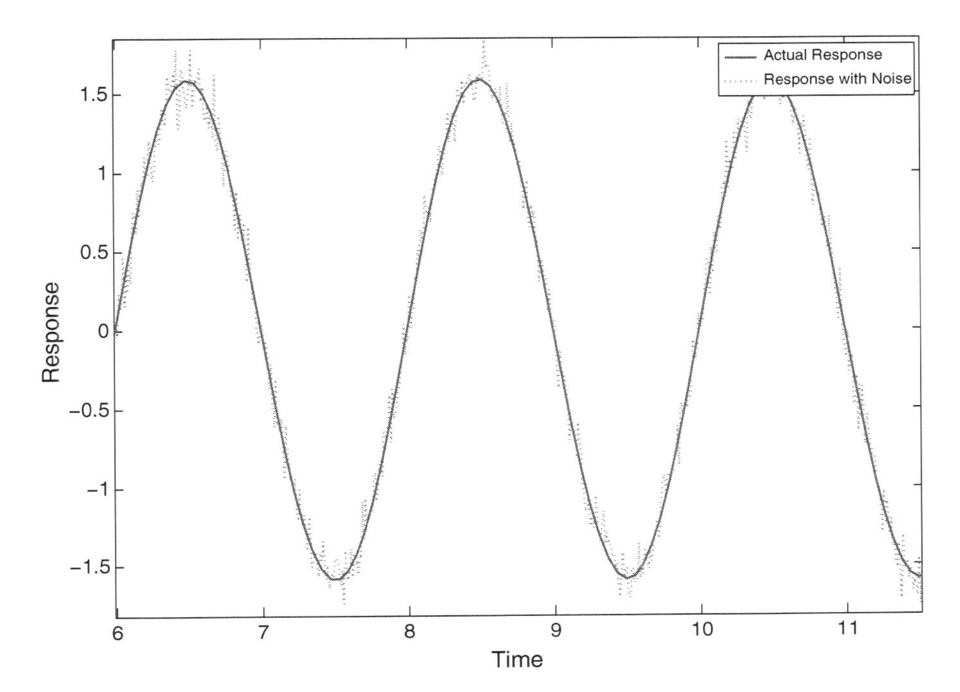

Fig. 2 Zoomed view of the time history of the response

problem assuming only the noisy measurement data is available and the parameter c is to be estimated. Initially, it was assumed c to be Gaussian distributed with mean 25 and standard deviation 5. The number of particles considered in this example is 500. The measurements and the process equations were given as input to the bootstrap filter. The analysis was carried out by using the four methods mentioned earlier (Fig. 2). Figures 3–6 show the estimated value of damping as a function of time by four methods. The horizontal line indicates the actual value of damping, and the curve with circles indicates the estimate of damping through ordinary BFA without PCE. The curve with stars and boxes indicates the analysis through PCE with Galerkin and nonintrusive approaches, respectively. We see that in all figures, initially there are little fluctuations in the estimated values of c, but as more and more measurements are incorporated, the estimates stabilize. The summary of the analysis by all the four methods is been shown in Table 1.

In method 1, filtering is carried out at all time steps; the structural analysis to calculate the system response for ordinary BFA is performed for 2,000,500 times. In the same method using Galerkin approach, the structural analysis is performed only once, whereas in nonintrusive approach, it is carried out thrice for the corresponding quadrature points. In method 1, it is seen that ordinary BFA takes almost five times the time taken when compared to its PCE counterpart. The estimated values obtained through PCE approaches are in fairly good agreement with the actual value. Thus, it is very clear that PCE-based bootstrap filtering is much quicker and efficient than ordinary filtering.

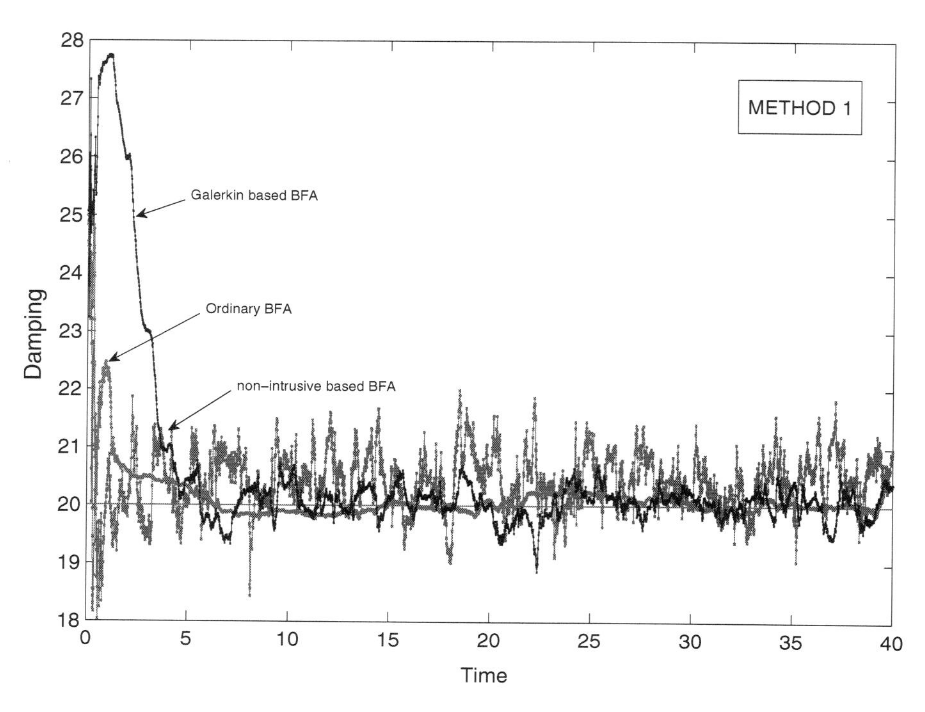

Fig. 3 Estimate of c by method 1

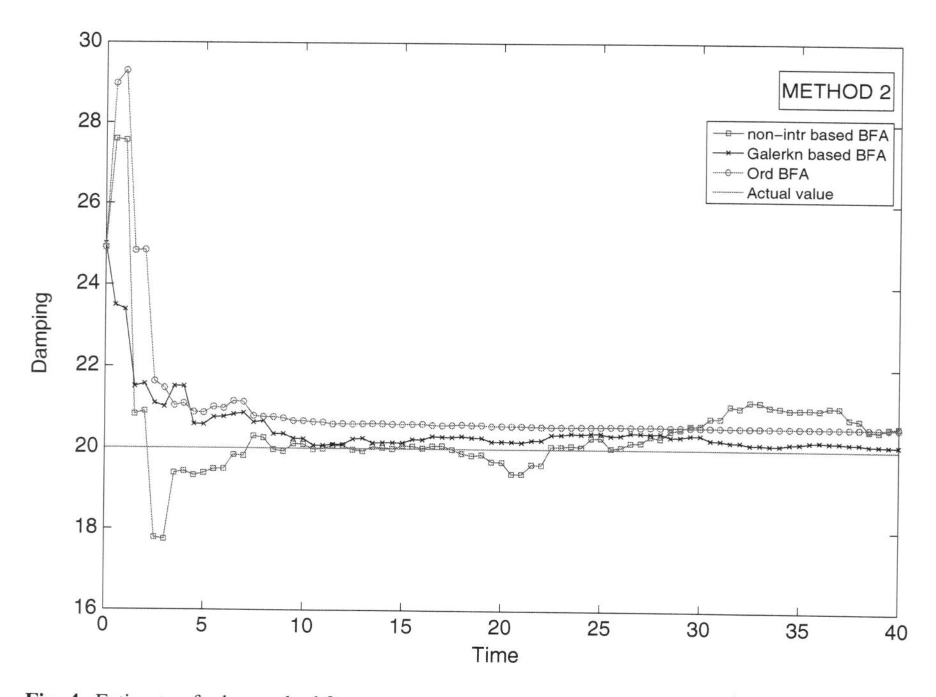

Fig. 4 Estimate of c by method 2

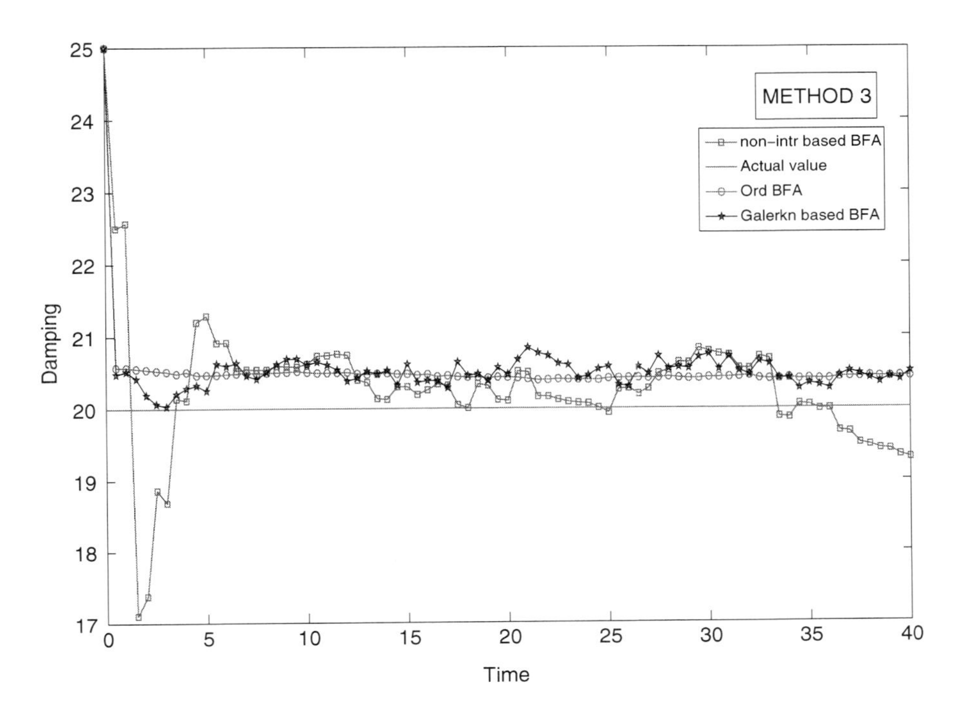

Fig. 5 Estimate of c by method 3

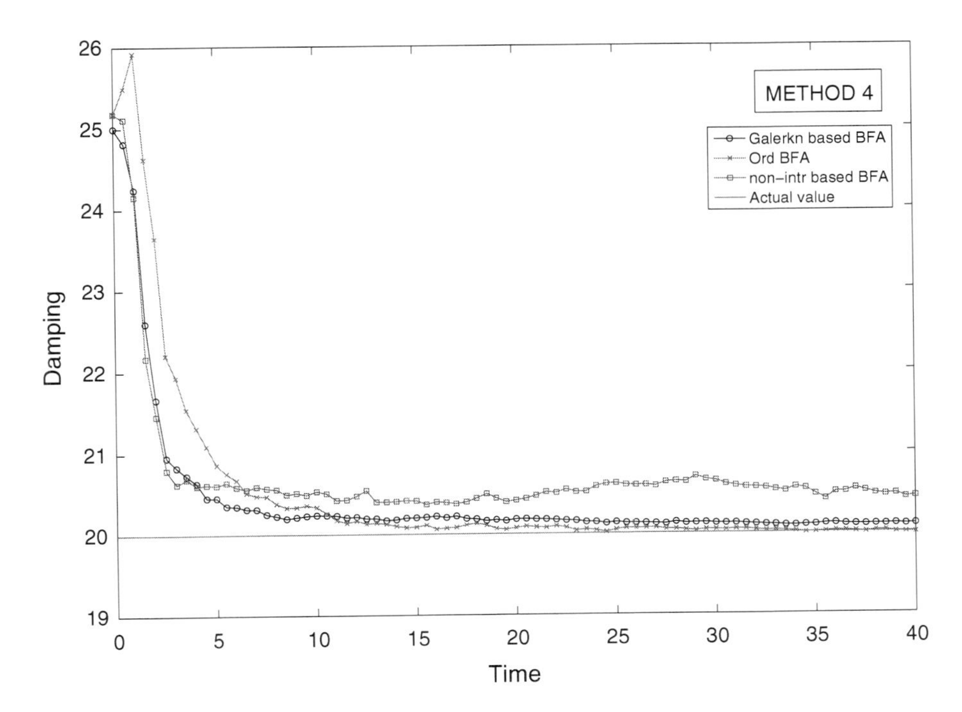

Fig. 6 Estimate of c by method 4

Table 1 Comparison of the performance of methods 1–4

Method	A	B	C	D	E	F
Method 1	Galerkin-based app.	4,001	1	19.97	− 0.15	157.26
	BFA with nonintrusive app.	4,001	3	20.59	2.95	230.3
	Ordinary BFA	4,001	2,000,500	21.94	9.7	906.77
Method 2	Galerkin-based app.	80	1	20.5	2.5	9.07
	BFA with nonintrusive app.	80	3	20.3	1.5	8.32
	Ordinary BFA	80	40,000	20.1	0.5	40.1
Method 3	Galerkin-based app.	80	1	20.5	2.5	17.42
	BFA with nonintrusive app.	80	3	20.24	1.2	14.07
	Ordinary BFA	80	40,000	20.31	1.55	116.12
Method 4	Galerkin-based app.	80	1	20.03	0.15	9.87
	BFA with nonintrusive app.	80	3	20.55	2.75	8.87
	Ordinary BFA	80	40,000	20.12	0.6	39.92

A methodology, *B* number of times BFA is implemented, *C* number of structural analyses, *D*estimated value, *E*% error, *F* CPU time in seconds

In methods (2)–(4), the BFA is carried out at every 50 time steps. As seen from the table below, these methods are quicker when compared to BFA by method 1 without compromising on accuracy. The number of filtering executed in those analyses is 80, which is much lesser than 4,001 to that used in method 1. This enables a considerable decrease in the number of structural runs computed in the algorithm. It is reflected on the time taken to execute the same process with method 1 and the other three methods. For instance, the Galerkin approach in method 2 consumes almost 15 times lesser time when compared to its similar approach using method 1. Analogous pattern is been observed in nonintrusive-based approach, indicating BFA in conjunction with PCE results in faster analysis. On dealing with problems with higher complexity, doing many structural analysis would consume a large computational effort; thus, in such cases, PCE-based BFA provides an easier way to analyze the problem.

The developed algorithm was found to be robust in yielding good results while handling uncertain parameter with larger standard deviation. Also, the algorithm still produced appreciable results on widening the gap between the two successive filtering.

7 Concluding Remarks

A computationally efficient technique for system identification using bootstrap filtering in conjunction with PCE has been successfully implemented. The used numerical example is a simple model problem to understand the mechanism of coupling PCE with bootstrap particle filtering. The PCE-based approach has been shown to be more efficient than Monte Carlo-based approaches for system identification. Various versions of the PCE approach-based bootstrap particle filtering have been investigated to investigate the efficiency and accuracy of the proposed methods. Further work involving more complicated problems is currently in progress.

References

1. Cameron RH, Martin WT (1947) The orthogonal development of nonlinear functionals in series of Fourier–Hermite functionals. Ann Math 48:385–392
2. Desai A, Sarkar S (2010) Analysis of a nonlinear aeroelastic system with parametric uncertainties using polynomial chaos expansion. Math Probl Eng 2010:1–2. Article ID 379472
3. Doucet A, de Freitas JF, Gordon NJ (2001) An introduction to sequential Monte Carlo methods. In: Sequential Monte Carlo methods in practice. Springer, New York
4. Ghanem R, Spanos P (1991) Stochastic finite elements: a spectral approach. Springer, New York
5. Ghanem R, Spanos P (1993) A stochastic Galerkin expansion for nonlinear random vibration analysis. Probab Eng Mech 8:255–264
6. Gordon NJ, Salmond DJ, Smith AFM (1993) Novel approach to nonlinear/non-Gaussian Bayesian state estimation. IEE Proc F 140:107–113
7. Grewal MS, Andrews AP (2001) Kalman filtering: theory and practice using matlab, 2nd edn. Wiley, New York
8. Kalman RE (1960) A new approach to linear filtering and prediction problems. ASME J Basic Eng 82:34–45
9. Morla L (2011) Parameter estimation in vibrating structures using particle filtering algorithm vibration analysis. M.Tech thesis, Indian Institute of Technology, Madras
10. Nasrellah HA, Manohar CS (2010) A particle filtering approach for structural system identification in vehicle structure interaction problems. J Sound Vib 329:1289–1309
11. Wiener N (1938) The homogeneous chaos. Am J Math 60(4):897–936
12. Xiu D, Karniadakis GE (2002) The Weiner-Askey polynomial chaos for stochastic differential equations. SIAM J Sci Comput 24:619–644